Geophysical Monograph Series

Including
IUGG Volumes
Maurice Ewing Volumes
Mineral Physics Volumes

Geophysical Monograph Series

Geophysical Monograph 197

Auroral Phenomenology and Magnetospheric Processes: Earth and Other Planets

Andreas Keiling
Eric Donovan
Fran Bagenal
Tomas Karlsson

Editors

Ⓢ American Geophysical Union
Washington, DC

Library of Congress Cataloging-in-Publication Data

Auroral phenomenology and magnetospheric processes : Earth and other planets / Andreas Keiling, Eric Donovan, Fran Bagenal, and Tomas Karlsson, editors.
    pages cm. – (Geophysical monograph, ISSN 0065-8448 ; 197)
  Includes bibliographical references and index.
  ISBN 978-0-87590-487-0
  1. Auroras. 2. Magnetospheric physics. I. Keiling, Andreas, editor of compilation.
  QC971.A77 2012
  538'.768–dc23
                        2012022748

  ISBN: 978-0-87590-487-0
  ISSN: 0065-8448

Cover Image: Earth's aurora seen from the International Space Station (ISS). Photo credit ISS Expedition 23 crew, Image Science and Analysis Laboratory, NASA. (insets) Aurora at Jupiter and Saturn through the eyes of the Hubble Space Telescope. (left) Photo credit NASA, European Space Agency, and J. Clarke (Boston University). (right) Saturn's aurora shown for 3 days, with each image 2 days apart. Photo credit Z. Levay (Space Telescope Science Institute).

CONTENTS

PREFACE

The aurora is an ever-present phenomenon in the Earth's upper atmosphere. We know that the terrestrial aurora is powered by the interaction of the solar wind–magnetosphere system, and we understand how the light is produced through the collision of precipitating charged particles and upper atmospheric ions and neutrals. There are still, however, huge gaps in our knowledge about the electrons that cause discrete auroras, about the mechanisms responsible for pitch angle scattering of electrons and protons responsible for diffuse aurora, and about what physical processes are responsible for the spatiotemporal structuring of discrete and diffuse aurora. Theories have been developed to explain these auroral phenomenologies, but they often fall short in key aspects. Furthermore, while we recognize some familiar behavior in the auroral emissions of other planets, there are also striking differences that test our basic ideas of auroral processes.

A number of significant advances have been made recently in the fields of instrumentation and observation. At the same time, the current constellation of terrestrial spacecraft and ground-based observatories are providing unprecedented opportunities. Spacecraft missions to other planets as well as ground- and space-based telescopes have also provided a wealth of data on planetary auroras. In addition, there have been tremendous advances in the capabilities to model many of the relevant processes. All of this sets the stage for a fresh look at the aurora and its relationship to magnetospheric processes. This is the subject of this monograph. While in the past terrestrial and planetary auroras have been largely treated in separate books, our intention with this monograph is to take a holistic approach by treating the aurora as a fundamental process and incorporating the phenomenology, physics, and relationship to the magnetosphere for all planets, including the Earth, more equally.

After a tutorial on comparative auroral physics in section 1, the logical flow of the subsequent sections of this book is such that we begin in section 2 with the auroral observations. Shape, form, and dynamic behavior of the aurora are linked to

Auroral Phenomenology and Magnetospheric Processes: Earth and Other Planets
Geophysical Monograph Series 197
10.1029/2012GM001251

the mechanisms responsible for the structure and characteristics of the precipitation that causes the aurora. Not only can the small-scale auroral phenomenology enable valuable insights about the corresponding magnetospheric processes but even on global scales the interhemispheric conjugacy of aurora (or lack thereof) conveys important information about the global magnetospheric topology and dynamics. The section on auroral phenomenology is composed of reviews and new results from both ground-based and spaceborne optical imagers, addressing topics as diverse as pulsating aurora, omega bands, auroral streamers, transpolar arcs, and substorm aurora. It also covers auroras at Jupiter and Saturn through the eyes of the Hubble telescope and auroras at Mars and those created by moons of the giant planets.

While optical imaging data obtained by ground-based and spaceborne cameras have shown many types of spatial distribution of auroral forms, other types of instruments, such as magnetometers and ionospheric radars, have also revealed drastic changes in the ionospheric conditions in association with these auroral forms. It is clear that a full understanding of auroral phenomenology requires an understanding of the underlying ionospheric electrodynamics, which is the topic of section 3. The active role of the ionosphere must also be taken into account when describing processes in the outer magnetosphere since their effects are modulated by magnetosphere-ionosphere coupling and by the ionosphere itself. Ionosphere electrodynamics is equally important at the giant planets where intense auroras signify strong currents that couple magnetospheric plasma to the rotating planet.

Auroral particle acceleration remains a key topic in magnetospheric and auroral physics. Numerous auroral missions have enabled significant progress on the nature of the acceleration processes. For example, it is today a general consensus that both quasi-static and wave electric fields contribute to acceleration of electrons producing discrete aurora. However, despite being one of the most intensely studied regions of the magnetosphere, much is still to be learned about the auroral acceleration region. Recent progress in simulations and multispacecraft observations in the auroral acceleration region are leading the way to a more complete understanding that has broad implications and applications for auroras on other planets as well. Section 4 opens a small window into these recent developments in the form of reviews and research papers.

In the past, investigations of the roles of outer-magnetospheric processes in generating auroral structures (with perhaps the exception of field line resonances) have been more limited, largely because of the lack of overlap between ground and space observatories. For example, there have been few reported outer-magnetospheric signatures that can be linked directly to any particular auroral form. While recent new spacecraft missions, such as Cluster and THEMIS, and extended arrays of ground-based observatories have made it possible to observe this connection more routinely at Earth, it is still a major challenge to relate aurora with magnetospheric data at other planets. Section 5 highlights recent observations, numerical simulations, and theoretical models of this connection.

This sequence completes the chain of interactions to establish the relationship between auroral phenomenology and magnetospheric processes. Understanding this connection will result in a more complete explanation of the aurora itself and will also further the goal of being able to interpret the global auroral distributions as a dynamic map of the magnetosphere. We emphasize that the topics touched upon in each section are not intended to be a comprehensive listing of all currently investigated auroral phenomena since far more space would be needed. In fact, one could easily fill another volume with topics that are not covered in this volume.

This book is the result of the Chapman Conference on the "Relationship Between Auroral Phenomenology and Magnetospheric Processes," held in Fairbanks, Alaska, from 27 February to 4 March 2011, with an attendance of around 140 scientists. This conference provided a forum where auroral scientists debated the implications of recent observational, theoretical, and modeling studies. It also stimulated cross-cutting discussion among scientists specializing in all aspects of terrestrial and planetary auroras.

Thirty years ago, a Chapman conference on a related topic was also held in Fairbanks, albeit during the summer. Corresponding results can be found in the AGU monograph *Physics of Auroral Arc Formation* (S.-I. Akasofu and J. R. Kan, editors). The reader might find it interesting to compare the advances that have been made since then, in both understanding and instrumentation, which can be seen, among other things, in the improved data quality and simulation outputs. Back then, the conference convener was Syun Akasofu, who was the keynote speaker at the recent Chapman conference in Fairbanks.

Conference participants were fortunate not only to have been sharing knowledge and ideas about the aurora but also to have witnessed its majestic appearance first-hand after the science sessions. On the conference opening day, a G2-class geomagnetic storm occurred, triggering spectacular auroras over Alaska during the entire conference week. After Dirk Lummerzheim provided ad hoc aurora forecasts at the end of each day's science sessions, participants watched and took photos, as evidenced on the conference website, late into the night, only to join the meeting again first thing the following morning. More than 20 attendees, young and old, had never seen the aurora in person before attending this conference and were treated to a dramatic first experience.

We are truly grateful to the many people who made this meeting a success. All the attendees brought the energy and enthusiasm that made the meeting fun and rewarding. The other program committee members (Dave Knudsen, Dirk Lummerzheim, Göran Marklund, Kirsti Kauristie, Masafumi Hirahara, Robert Rankin, and Vassilis Angelopoulos) contributed greatly to the conference organization. AGU staff provided outstanding support throughout the organization of and during the conference. The authors of the 37 papers provided the substance for this book, and the many volunteers who reviewed those papers have given the book scientific credibility. The Alaska Geophysical Institute led us on a memorable tour of the rocket range near Poker Flat, and Bill Bristow handled the logistics of the student travel grants. The National Science Foundation and the University of Calgary provided additional funding that made the student support and field trips possible. Thank you to all!

Andreas Keiling
University of California, Berkeley

Eric Donovan
University of Calgary

Fran Bagenal
University of Colorado

Tomas Karlsson
Royal Institute of Technology

Section I
Introduction

Comparative Auroral Physics: Earth and Other Planets

Barry Mauk

The Johns Hopkins University Applied Physics Laboratory, Laurel, Maryland, USA

Fran Bagenal

Laboratory for Space and Atmospheric Sciences, University of Colorado, Boulder, Colorado, USA

Here we review selected similarities and differences between the structures and processes associated with the generation of the aurora of strongly magnetized planets within the solar system. Our ultimate objective is to use a comparative approach to determine which aspects of auroral phenomena represent universal features and which aspects are particular to the special conditions that prevail at any one planet. We begin by providing a high-level review of selected fundamental auroral processes operating at Earth as a precursor to discussing selected similar processes and regions at other planets. We then discuss the broad characteristics of the space environments of different planets (Earth, Jupiter, Saturn, Uranus, and Neptune) with an eye toward determining the factors that dictate similarities and differences between the respective auroral systems. With a focus on discrete auroral processes, we finally discuss comparisons between the different systems on the basis of (1) magnetospheric current systems, (2) mechanisms of current closure within the distant regions of the magnetospheres, (3) particle acceleration, (4) ionospheric feedback, and (5) satellite systems.

1. INTRODUCTION

A central question of planetary space science in general and auroral physics in particular is: What aspects are universal and what aspects are specific to the conditions that prevail at any one planet? Universal aspects are those that one might invoke when addressing any distant astrophysical system. In general, the processes generating the most intense aurora represent the most powerful means by which energy and momentum are transported between a planet's space environment, or magnetosphere, and its upper atmosphere and ionosphere. At Jupiter, for example, such processes cause Jupiter's distant space environment to spin up to substantial

Auroral Phenomenology and Magnetospheric Processes: Earth and Other Planets

Geophysical Monograph Series 197

10.1029/2011GM001192

fractions of the planetary corotational angular rates out to distances as large as 100 Jovian radii, while at the same time causing Jupiter to shed tiny amounts of angular momentum. To the extent that such processes can be invoked over broad parametric states, it is not too great a stretch to conclude that such processes may be involved with the shedding of angular momentum in other distant astrophysical settings, for example, during the periods of planetary formation when magnetic fields still hold sway within the collapsing clouds [e.g., *Mauk et al.*, 2002a].

Findings achieved over the last several decades have revealed that, at least superficially, auroral processes are indeed universal in the sense of being active over a broad spectrum of planetary systems (see Earth, Jupiter, Saturn, and Uranus in Figure 1; see also chapters in this monograph on aurorae on Mars by *Brain and Halekas* [this volume], at Jupiter by *Clarke* [this volume], and at Saturn by *Bunce* [this volume]). Small systems like the Earth that are driven by the solar wind (the wind of ionized gases emanating from the Sun), large

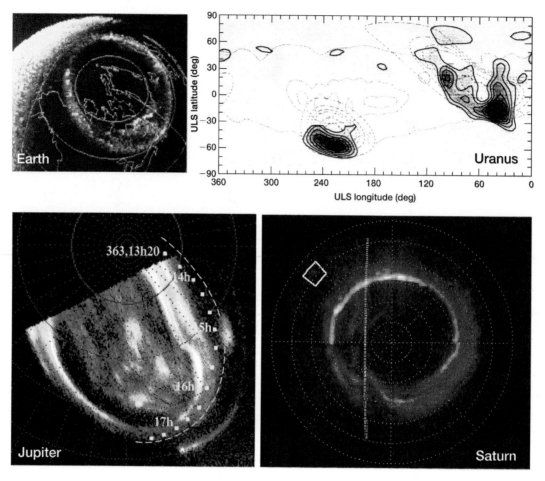

Figure 1. Selected auroral UV images from the Earth (Polar Spacecraft [*Frank and Craven*, 1988], Jupiter (Hubble [from *Mauk et al.*, 2002b] [see *Clarke*, this volume]), Saturn (Cassini Ultraviolet Imaging Spectrograph Subsystem (UVIS) [from *Pryor et al.*, 2011]), and Uranus [*Herbert and Sandel*, 1994; *Herbert*, 2009]). For Jupiter and Saturn, there are planetary latitude lines for both images at 10° latitude intervals. The Uranus image shows a completely different projection and therefore is difficult to compare with the others. However, the symmetry with respect to the magnetic poles can be ascertained by comparisons with the dashed contours, which show the projected positions of the magnetospheric L values, specifically $L = 2, 3, 4, 5, 10, 20,$ and $30\ R_U$. This image comprises a synthesis of Voyager measurements of the UV aurora. Peak emission intensities are less than 500 R.

rotationally driven systems like that of Jupiter, and systems like Saturn with space environments dominated by neutral gas, all have revealed dramatic rings of auroral emissions encircling the magnetic polar axes (Figure 1). While the sizes, power levels, and parametric states of these systems are dramatically different (Table 1) [*Bagenal*, 2009], similarities persist even when the focus is on the details of the planet/space-environment interactions.

In this introduction, we begin by examining some of the fundamental physical processes and regions that have been identified within the Earth's auroral system to set the stage for discussing other planets. The first two sections (sections 1.1 and 1.2) focus on the processes that generate just one

type of aurora, discrete aurora, which represents the most intense and structured aurora and which requires active particle acceleration along the magnetic field lines. Discrete aurora is also where the major fraction of our focus is with the comparisons between different planets. Other types of aurora are discussed and placed into context in section 1.3.

Because the sampling of processes acting at other planets is so sparse, we depend substantially on our understanding of the Earth auroral processes to make judgments about what is happening on these other planets. A phenomenon that has received substantial renewed attention over the last decade, and which garnered controversial discussion at the Chapman Conference from which this volume was initiated, is the

Table 1. Selected Parameters Regarding the Planets of the Solar System [*Bagenal*, 2009]

	Earth	Jupiter	Saturn	Uranus	Neptune
Distance from Sun (AU)	1	5.2	9.5	19	30
Radius (km)	6373	71,400	60,268	25,600	24,765
Spin period (sidereal day)	0.997	0.41	0.44	−0.72	0.67
∠(S, N-Ecliptic) (deg)	23.5	3.1	26.7	97.9	29.6
Surface field (nT)	30,600	430,000	21,400	22,800	14,200
Dipole tilt (deg)	9.92	−9.4	~0.0	−59	−47
Magnetopause location (R_p)	8–12	63–92	22–27	18	23–26
Nominal IMF (nT)	8	1	0.6	0.2	0.1
Nominal solar wind density (1 cm^{-3})	7	0.2	0.07	0.02	0.006
Auroral Emission Power (W)	10^{10}	10^{12}	10^{11}	5×10^9	2–8×10^7
Open magnetic flux (GWb)	0.5–1	250–720 (model)	15–50		

"Alfvénic aurora." This auroral process is thought to be powered by electromagnetic waves, specifically Alfvén waves that propagate with periods of seconds to tens of seconds within the ionized gases or plasmas that connect the distant magnetosphere to the polar ionosphere (it is understood that even quasi-static auroral structures may be mediated by Alfvén waves with much longer periods). Controversies about this dynamic auroral contribution to the Earth's aurora are similar to discussions that have taken place about the relative roles of turbulence and quasi-static sources of auroral energies at Jupiter, as we shall discuss. Because of that connection, and also because of our perception of gaps in the present literature concerning this topic, we spend some time in section 1.2 discussing the possible relative roles of quasi-static and Alfvén wave sources of auroral power transmission at Earth. In section 2, we make direct comparisons between the auroral processes at Earth and other planets, with a focus on discrete auroral processes.

1.1. Strong Auroral Coupling Processes Revealed at Earth

Figure 2 (after *Lundin et al.* [1998]) provides a traditional view of the generation of discrete auroral discharge phenomena consisting of (1) the generation of electrical currents and voltages within the magnetized plasma that comprise the distant magnetosphere, (2) the diversion of those electrical currents along magnetic field lines toward the polar auroral regions, (3) the generation of impedances and parallel electric fields along the magnetic field lines at low altitudes to midaltitudes as a result of the sparsity of charge carriers in the regions just above the ionosphere, (4) the acceleration of charged particles out of the regions of parallel impedance onto the upper atmosphere and out into the distant magnetosphere, (5) the excitation and ionization of atoms and molecules within the upper atmosphere by the accelerated electrons resulting in strong auroral emissions and enhance-

ments in the electrical conductivity of the ionosphere, (6) the closure of the upgoing and downgoing electric current through the partially conducting ionosphere, and (7) the associated heating through ohmic dissipation of the upper atmosphere and the generation of upper atmospheric winds through the collision of current-carrying ions and neutral atmospheric constituents (see *Mauk et al.* [2002a] for a more detailed discussion of Figure 2).

Multiple processes have been invoked for the generation of the midaltitude impedances and parallel electric fields along magnetic fields [e.g., *Borovsky*, 1993; *Lysak*, 1993] (section 4 of this volume), including stationary electrostatic shock-like structures called double layers, larger-scale electric fields supported by magnetic mirror effects that arise because of the converging magnetic field lines, anomalous

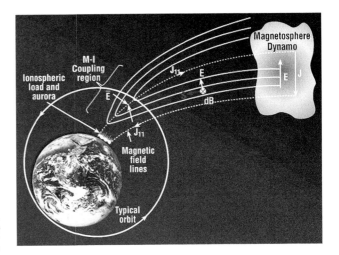

Figure 2. Schematic of the Earth's auroral magnetosphere-ionosphere coupling circuit showing the three key regions and a Freja or FAST spacecraft-like orbit used to sample the midaltitude coupling region. After *Lundin et al.* [1998].

resistivity caused by particle interactions with various wave modes, and parallel electric fields that arise from Alfvén waves propagating at large angles to the magnetic field. Some of these mechanisms are intrinsically time dependent, contrary to the "static" representation given in Figure 2.

1.2. Auroral Energy Flow at Earth

One of the intrinsically time-dependent mechanisms that has received substantial recent attention is the so-called Alfvén wave generator [*Wygant et al.*, 2000; *Keiling et al.*, 2002, 2003; *Watt and Rankin*, this volume]; this process is nicely illustrated in the Figure 1 of *Wygant et al.* [2000]. Reviews on the importance of Alfvén waves generally in auroral and magnetospheric phenomena are provided by *Stasiewicz et al.* [2000] and *Keiling* [2009]. Because the Alfvén wave generator concept has not been reviewed in the context of comparative magnetospheres, and because the argument for supporting the importance of this mechanism is commonly used in the context of planetary magnetospheres, specifically comparing quantitatively the source and dissipation of energy, we spend some time discussing it here.

Alfvénic auroral processes were invoked on the basis of the observation of earthward propagating Alfvén waves at radial distances of 4 to 6 Earth radii (R_E), but at latitudes that map magnetically to the vicinity of the outer boundaries of the population of plasmas that reside within the interior of the antisunward, comet-like magnetic tail of the Earth's magnetosphere, called the plasma sheet. Alfvén wave events are observed with earthward energy fluxes from several to ~100 ergs cm^{-2} s^{-1} when those power density values are mapped (with the funneling amplification associated with the convergence of the magnetic field lines) to auroral altitudes [*Wygant et al.*, 2000; *Keiling et al.*, 2002, 2003]. The energy transport is by means of the Poynting vector, represented in Gaussian units as $\mathbf{S} = (c/4\pi) \cdot \mathbf{dE} \times \mathbf{dB}$, where $\mathbf{dE}$ and $\mathbf{dB}$ are the wave fields of the observed parallel-propagating Alfvén waves (note: we will denote the magnitude of the Poynting vector as simply the Poynting flux and will denote the area-integrated energy transport rate as the Poynting fluence). These power density levels, again levels achieved after amplification by the substantial funneling of the magnetic field lines, are compared with the power densities associated with the electron distributions that are observed to generate discrete auroral emissions. *Keiling et al.* [2003] concluded that a substantial fraction (although not all) of the discrete auroral energy dissipations may be powered by these fluctuating Alfvén waves.

A weakness in this conclusion is that this source of energy has not been properly compared with competitive sources of energy, only with the dissipation of energy at the near-Earth "footprints" of the aurora. For a single striking Alfvén wave event, *Wygant et al.* [2002] performed a direct comparison between the Poynting vector magnitudes associated with the static field-aligned electric currents and those values associated with the propagating Alfvén waves, again in the vicinity of the boundary of the plasma sheet populations. These authors showed that the wave-carried Poynting vector magnitude was 1 to 2 orders of magnitude greater than that associated with the more static currents and fields. This comparison has limited value in deciding between the different auroral power sources, however, because the Poynting fluence traditionally thought to be associated with the static-current generation of discrete aurora likely propagates through a different region of space than that associated with the observed Alfvén waves.

Because a proper "apples to apples" comparison between Alfvén wave energy sources and other sources of energy for the discrete aurora has not been presented, it is instructive to examine the flow of energy associated with static currents and fields traditionally thought to be associated with auroral acceleration. Indeed, that energy is also carried by a Poynting flux vector, but a static version (elaborated by *Kelley et al.* [1991]). What is important to recognize is that outside of the regions of power generation and power dissipation, most of the Poynting fluence is not colocated with the field-aligned currents that propagate from the magnetospheric generator to the auroral ionosphere. That Poynting fluence resides between the two current sheets that carry the upward and downward currents. The nonintuitive nature of this finding is discussed, for example, by *Feynman et al.* [1964], who also points out that the Poynting vector representation of energy flow is not unique. However, it is the representation that has been adopted overwhelmingly by the space science community. Within the context of the Poynting vector representation, the validity of where the energy flow takes place can be demonstrated with the simple thought experiment shown in Figure 3a. With this configuration, we generally "bookkeep" the energy dissipation within the resistors (R) as: $P = I \cdot V = V^2/R$, where P is the power dissipation per meter along the x direction (into the page), I is the current per meter in the x direction, R is the electrical resistance per meter along the x-direction, and V is the voltage. But the energy is actually carried by the Poynting fluence that flows between the two plates. One may simply construct the Poynting vector ($c\mathbf{E} \times \mathbf{B}/4\pi$) using the techniques of elementary electricity and magnetism (Gausses law and Ampere's law) to get $\mathbf{E} = -\mathbf{z}V/d$ and $\mathbf{B} = \mathbf{x}(4\pi/c)I$, where ($\mathbf{x}, \mathbf{y}, \mathbf{z}$) are the unit vectors that form the Cartesian coordinate system. By integrating this Poynting flux across the area between the two plates formed by $A = L \cdot d$, where L is the unit distance of integration along the x direction, one finds that indeed $P = I \cdot V = V^2/R$, just as we found with our bookkeeping formula.

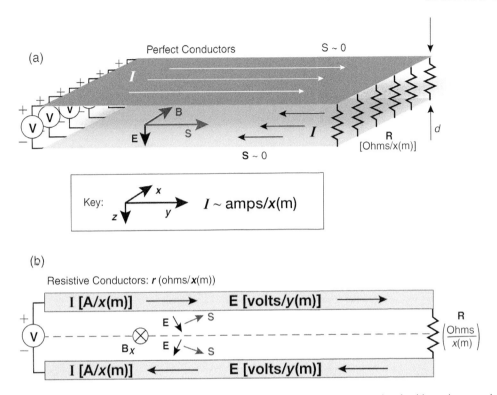

Figure 3. Thought experiments designed to help understand the flow of energy associated with static auroral current systems. See text for details.

The way in which the static current system provides power to the auroral acceleration process is illustrated in Figure 3b, which shows the system examined in Figure 3a, but viewed edge-on. In this case, we also consider current-carrying plates that have some electrical resistance to them. Here one sees that the Poynting flux now no longer flows parallel to the plates but flows across the surface at an angle and into the resistive plates. The plates will heat up in association with the dissipation of electric power, but the flow of energy that provides this heat energy is, within the framework that we have chosen, the Poynting flux that flows through the sides of the plates, not the flow of energy along the current-carrying plates.

So, returning to Figure 2, we see that the Poynting flux that flows predominantly between the two current systems (upward and downward) does not flow along the magnetic field lines but rather along the contours of constant electric potential. Specifically, the Poynting flux can focus in on the region where there are components of the electric field that are parallel to the magnetic field and that provide the principal power source for the auroral acceleration that occurs at those positions.

How large is this static current Poynting flux? With perpendicular electric fields ($\sim$0.5 V m^{-1}) and the perpendicular magnetic fields (200 nT) measured at low altitudes by the FAST mission as reported by *Carlson et al.* [1998], power density values of 100 ergs cm^{-2} appear easy to come by. So it is clear that the Alfvén wave Poynting flux by no means dominates over the Poynting flux for static fields and currents. However, we do not know the relative ranking of these two sources when it comes to efficiency of conversion from electromagnetic energy to particle energy. The Alfvén wave Poynting flux can certainly be an important contributor, consistent with the finding of *Keiling et al.* [2003]. Also, nothing in this discussion specifically demonstrates that the Alfvén wave Poynting flux cannot be one of the drivers, through some conversion process, of the static current and field configurations observed at lower latitudes. But we see that much more is needed than arguments that simply compare the quantity of power available from a possible power source with the quantity of power dissipation. We will return to this topic when we discuss auroral power generation at Jupiter.

1.3. Auroral Regions and Regimes at Earth

Several different auroral regimes are of interest (Figure 4) besides the discrete auroral component that we have been

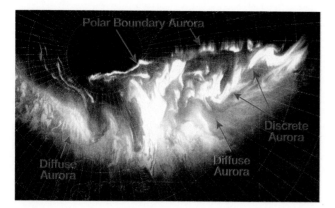

Figure 4. Key regions of the Earth's aurora. The diffuse aurora is identified by the spatial homogeneity of the emission and the smooth and unstructured nature of the spectra of the precipitating electrons. The discrete aurora is identified by the spatially structured character of the emissions and the structured nature of the spectra of the precipitating electron distributions, often showing peaked features indicative of acceleration along field lines. The polar boundary aurora is a discrete auroral feature that resides near the boundary between closed and open field lines. Image from the Defense Meteorological Satellite Program (DMSP).

discussing. While these auroral regimes are expected to have a certain latitudinal ordering, statistical distributions show that there is much overlap (Figure 5) [*Newell et al.*, 2009]. We have not yet mentioned the diffuse aurora that generally resides at the lowest latitudes (Figures 4 and 5a). Diffuse electron aurora, with emissions that are relatively spatially uniform and with unstructured precipitating electron spectra, are thought to result from the scattering of hot electrons that are trapped in the magnetic field of the distant magnetosphere into the magnetic loss cone (comprising those charged particles whose magnetic mirror points reside within the Earth's atmosphere or below). The scattering occurs as a result of strong interactions between the trapped particles and various kinds of plasma waves that reside within the trapped plasma populations. The wave modes thought to be responsible for the scattering are electron cyclotron harmonic waves and/or "chorus" whistler mode waves [*Horne et al.*, 2003; *Ni et al.*, 2008; *Meredith et al.*, 2009; *Su et al.*, 2010]. Interesting

Figure 5. Statistical study of the different kinds of Earth aurora. Shown are binned and averaged particle energy depositions as determined from the particle spectrometers on the low-altitude polar DMSP spacecraft. The different kinds of energy depositions are determined and cataloged according to the characters of the shapes of the particle energy spectra. The cataloging and binning is automated using a neural network algorithm. The "GW" values shown below the color bars are the power in gigawatts of particle energy deposited as integrated over each entire image. From *Newell et al.* [2009].

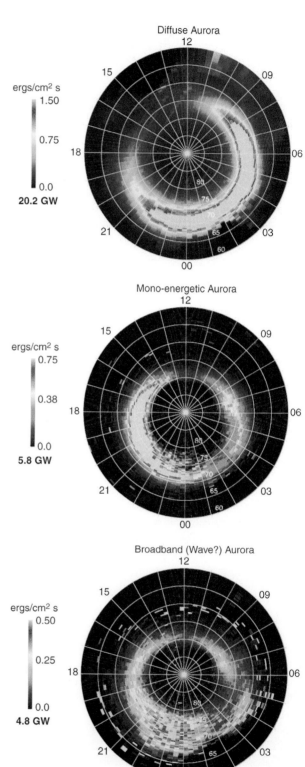

dynamic features of the diffuse electron aurora are discussed by *Lessard* [this volume] and by *Li et al.* [this volume], and proton diffuse auroras are discussed by *Donovan et al.* [this volume]. Note that the overall energy carried into the diffuse electron auroral regions is larger than that provided by any other component (Figure 5), although the intensity is well below that provided by the discrete processes.

At midlatitudes (Figures 4 and 5b) are the so-called discrete auroral emissions that traditionally are thought to be synonymous with the monoenergetic auroral acceleration, which in turn is thought to be the result of the quasi-static current and field configurations discussed above in reference to Figure 2 [e.g., *Carlson et al.*, 1998]. With the quasi-static discrete auroral mechanisms, there are two different regions (Figure 2) that are of substantial interest: (1) the region of upward currents that engender downward accelerated electrons (and upward accelerated ions) and strong discrete auroral emissions, and (2) the region of downward currents that engender powerful upward accelerated electrons that are commonly detected near the equatorial regions of the magnetosphere. The upward accelerated electron distributions constitute a powerful tool for mapping auroral regions to the distant magnetosphere, as we shall see.

The aurora at higher latitudes (Figures 4 and 5c) is where the Alfvén wave processes [*Keiling et al.*, 2002; *Schriver et al.*, 2003; *Chaston et al.*, 2003], discussed in section 1.2, may contribute to the discrete auroral emissions. At the highest latitudes are the "polar boundary auroral emissions" (Figure 4) that may be driven by the Alfvén wave processes described here, but could also be a consequence of the quasi-steady electric currents associated with the open-closed boundary. This boundary is between lower-latitude closed magnetic field lines that have both of their ends connected to the ionosphere and the higher-latitude open magnetic field lines with one end connected to the ionosphere and the other end connected to interplanetary space. This auroral boundary is thought to be connected to distant regions where magnetic energies are converted to plasma and particle heating through "magnetic reconnection" [*Bunce*, this volume]. The reader will note that issues of which physical mechanisms are responsible for specific observed phenomenological features remain rich areas for research.

As a final note, in the history of the study of auroral emissions and features coming from terrestrial and other planetary systems, it has often been assumed that strong aurora occur predominantly near but inside the boundary between open and closed field lines. Figure 4 shows high-latitude auroral emissions (polar boundary aurora) that likely map close to that transition boundary. However, while there is present controversy surrounding the premise that transients within the boundary auroral regions provide a trigger for

features occurring at lower latitudes [*Lyons et al.*, 2010; *Nishimura et al.*, 2010; *Lyons et al.*, this volume], it is clear that the strongest discrete emissions occur well equatorward of that transition region. Strong discrete auroral emissions during such geomagnetic disturbances, called magnetic storms and substorms, are thought to map to the vicinity of 9 to 12 R_E at Earth [e.g., *Akasofu et al.*, 2010, and references therein], while the reconnection sites that may or may not provide the stimulus for strong auroral breakups are thought to occur in the vicinity of 20 R_E and beyond [e.g., *Nagai et al.*, 2005]. The distances between 20 and 9 to 12 R_E certainly cannot be considered "near."

2. COMPARING PLANETARY AURORAL SYSTEMS

2.1. An Approach to Comparing Planetary Magnetospheres

In the discussions that follow, we compare electromagnetic parameters between several of the strongly magnetized planets using an "electrical circuit" approach, and more often than not, we compare the electric currents and electric fields of these respective systems. For the valid reasons mentioned below, it has become unfashionable in recent times to take this circuit approach and, specifically, to speak of electric fields and currents, following the publication of the now famous work by *Vasyliūnas* [2001] and also later discussions [e.g., *Vasyliūnas*, 2011, and references therein]. The values of the circuit approach are (1) it is easy to conceptualize the strong interactions between very different components of a complex system, for example, spanning regions that are controlled by kinetic factors and those dominated by magnetohydromagnetic factors and (2) the historical literature is presently dominated by such approaches, and any review such as this must incorporate them. The disadvantage of this approach is that it is valid only for quasi-static situations, by which we mean that the time scales for changes must be much slower than the Alfvén wave transit times for the region of consideration [*Vasyliūnas*, 2011]. We note that Alfvén transit times are also important for time-stationary configurations for systems that include, for example, the outer portions of Jupiter's huge magnetosphere, where the time for the transit of an Alfvén wave from the inner to the outer reaches of the system is a substantial fraction of Jupiter's rotation period. It is undoubtedly true that future advances in our understanding of planetary auroral phenomena will require such nonsteady approaches as those advocated by *Vasyliūnas* [2011].

So, despite the limitations mentioned above, the crude conceptual framework that we consider in this chapter is provided in Figure 6. Our purpose in showing this too simple figure is not to argue about or defend the particular way that we have connected up the different boxes, but to place

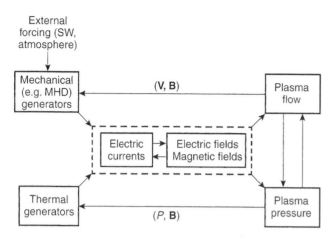

Figure 6. An electrical circuit framework for discussing differences between the electromagnetic environments and auroral systems of the strongly magnetized planetary systems. See text for a discussion of the deficiencies and criticisms of the electrical circuit approach. The purpose of this too-simple diagram is to place thermal (pressure) effects on a more equal footing with dynamical (flow) effects than has been evident in the literature at extraterrestrial magnetospheres.

thermal and dynamical effects (shown with the bottom and top feedback loops in the figure) on a more equal footing than has been evident in much of the literature at extraterrestrial magnetospheres.

2.2. Comparing Planetary Magnetospheres

Given that the auroras at some different planets have strong superficial similarities (Figure 1), it is of interest to understand how the corresponding magnetospheric systems are similar and how they are different. At the highest levels, there are several different conditions that seem to drive important differences between known planetary magnetospheric systems. Two of these conditions are (1) the relative strength of the plasma flows generated within the magnetosphere by the solar wind and by planetary rotation and (2) the presence or absence of a strong internal source of plasma.

With regard to the first of these conditions, the interaction between the fast-flowing solar wind and the magnetosphere, in the form of magnetic reconnection and flows driven inside but in the vicinity of the outer boundary of the magnetosphere, generates electrical currents on the boundary which close in various places within the magnetosphere and ionosphere. Those interior currents and their divergences generate electric fields and plasma motions deep within the interior of the magnetosphere. Empirically, the interior electric field is a fraction of the solar wind electric field ($\mathbf{E_{sw}} = \mathbf{V_{sw}} \times \mathbf{B_{sw}}/c$), with magnitude E_{sw}, and traditionally and heuristi-

cally, researchers have spoken of an electric field $\mathbf{E_p}$ that "penetrates" across magnetospheric boundaries, even while that characterization is highly imprecise. Traditionally, the strength and direction of the externally driven electric field within the interior of the magnetosphere is represented as:

$$\mathbf{E_p} \sim f \cdot \mathbf{E_{sw}} = \overrightarrow{\nabla}(f \cdot \mathbf{E_{sw}} \cdot R \cdot \sin[LT]), \quad (1)$$

where f is the empirically estimated fraction of the external (to the magnetosphere) solar wind electric field that ends up inside the magnetosphere (at Earth $f \sim 0.1$), $\mathbf{V_{sw}}$ is the solar wind velocity (~400 km s^{-1}, assumed to be uniform), $\mathbf{B_{sw}}$ is the magnetic field within the solar wind (~8 nT at Earth, assumed to be uniform), and c is the speed of light. The right-hand portion of equation (1) reformulates the interior electric field in the form of the gradient of a potential. Here Φ_{sw} is the electric potential whose gradient yields a uniform cross-magnetosphere electric field, R is the geocentric radial distance, and LT is the local time expressed in radians. This solar wind–generated electric field is traditionally to be compared with the rotational electric field. When the conducting ionosphere, frictionally dragged by the rotating upper atmosphere, rotates within the planet's magnetic field, a $\mathbf{V} \times \mathbf{B}/c$ electric field is generated within the ionosphere. Under the ideal condition that the magnetic field lines (when populated with plasmas) act as nearly perfect conductors, and when opposing equatorial forces and accelerations are small, the equatorial rotational electric field becomes:

$$\overrightarrow{E}_{rot} = (\overrightarrow{\Omega} \times \overrightarrow{R}) \times \overrightarrow{B}/c = \overrightarrow{\nabla}(\Phi_{rot}) = \overrightarrow{\nabla}\left(\frac{\Omega \cdot B_O}{c \cdot R}\right), \quad (2)$$

where Ω is the planetary rotation vector aligned with the planet's spin axis and Φ_{rot} is the equatorial electric potential that results when the planetary magnetic field B is a dipolar configuration with a normalization strength constant B_o (as in equatorial $B = B_o/R^3$) and with the dipole moment aligned with Ω. Combining rotational and solar wind electric potentials yields (see various approaches and discussions by *Axford and Hines* [1961], *Nishida* [1966], *Brice* [1967], *Kavanagh et al.* [1968], *Chen* [1970], *Brice and Ioannidis* [1970], and *Vasyliūnas* [1975]):

$$\Phi_{tot} = \frac{\Omega \cdot B_O}{c \cdot R} + f \cdot \mathbf{E_{sw}} \cdot R \cdot \sin(LT), \quad (3)$$

which, when plotted for contours of constant Φ_{tot}, evaluated using the parameters in Table 1, yields the patterns like those shown in Figure 7 (T. W. Hill contribution to the review by *Mauk et al.* [2009]) for Earth, Jupiter, and Saturn. These diagrams, representing the patterns of flow for low-energy plasmas and particles (representing the $\mathbf{E} \times \mathbf{B}/c$ drift) [*Parks*, 1991], ignore the deviations near the magnetosphere boundaries and

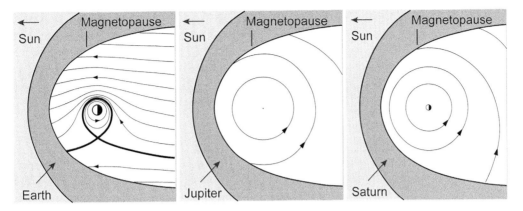

Figure 7. Simple model prediction of equatorial cold plasma flow patterns within the magnetospheres of the Earth, Jupiter, and Saturn. Deviations close to the magnetopause and within the deep magnetotail are not modeled here. Figure 7 provided by T. W. Hill for the review of *Mauk et al.* [2009 , Figure 11.15] of Saturn's magnetospheric processes. Reprinted with kind permission from Springer Science + Business Media.

within the deep tail. In consideration of the criticisms of the unfashionable use of electric field representations in section 2.1, we note that T. W. Hill (again in the review by *Mauk et al.* [2009]) derives these flow patterns from a consideration of the summation of flows rather than with the historical approach of using electric fields. The plots in Figure 7 indicate that the Earth's magnetosphere is powered predominantly by the solar wind and that the magnetospheres of Jupiter and Saturn are powered predominantly by rotation. At Saturn, the role of the solar wind is controversial and may be more important than is indicated by Figure 7 for driving auroral phenomena [*Cowley et al.*, 2004; *Bunce et al.*, 2008; *Bunce*, this volume].

Another factor that seems to be critical in understanding similarities and differences between planetary magnetospheres and their auroral systems is the presence or absence of a strong internal source of plasma, such as the volcanic action of Jupiter's satellite Io (at 5.9 R_J) and the venting activities of Saturn's satellite Enceladus (at ~4.0 R_S). Some of the emitted gases are ionized and energized by being picked up by the rapidly corotating plasma. Because these plasmas are generated near the rapidly rotating planet, and therefore near the peak of a centrifugal potential hill that falls with increasing radial distance, further energization occurs as the plasmas move outward. Some of the energy associated with the internal generation and transport of these new plasmas is tapped to drive various magnetospheric processes, including dramatic auroral displays. The generation, heating, transport, and loss of the gases and plasmas at Jupiter and Saturn remain poorly understood (see review by *Bagenal and Delamere* [2011]).

Table 2 categorizes all of the magnetized planets of the solar system with respect to our two conditions: (1) solar wind influence and (2) the presence or absence of a strong internal

source of plasma. Table 2 was created to provide evidence for the hypothesis that these two conditions are deterministic with regard to the presence or absence of dynamic injection-like phenomena within the respective magnetospheres. Injections are sudden planetward plasma transport events that occur over a limited range of longitudes. At Earth, they are associated with geomagnetic disturbance events called substorms. While Table 2 does seem to order the planets with respect to dynamics (injection-like phenomena occur in magnetospheres that are either powered by the solar wind or by centrifugal energies of strong, internally generated plasma), an outstanding mystery with regard to the occurrence of strong auroral phenomena is Uranus. Uranus was powered by the solar wind because of the Sun-aligned spin axis at the time of the Voyager 2 encounter [*Selesnick and McNutt*, 1987]; this condition is not generally true of Uranus, just true at the time of the Voyager 2 encounter. That magnetospheric phenomena at Uranus were

Table 2. Sorting the Planets According to Solar Wind Influence and Internal Plasma Sources

Planet	Injections?	Solar Wind Dominance?	Strong Internal Source?
Mercury	yes	yes	no
Earth	yes	yes	maybe: atmosphere
Jupiter	yes	no (rotation)	yes (Io)
Saturn	yes	no (rotation with sw triggering?)	yes (Enceladus)
Uranus	yes	yes (peculiar orientation)	no
Neptune	no (none observed)	no	no: Triton is "middle" source

driven by the solar wind during the Voyager 2 encounter is supported by observations of solar wind–driven flow configurations [*Selesnick and McNutt*, 1987], strong dynamic injection phenomena [*Mauk et al.*, 1987; *Belcher et al.*, 1991], whistler/chorus plasma wave emissions that were more intense than Voyager observed at any of the other planets [*Kurth and Gurnett*, 1991], and radiation belt electrons as intense as those observed during supermagnetic storms at Earth [*Mauk and Fox*, 2010]. Yet, auroral emissions with the high powers and ordered (ringed) structures of the sort observed at Earth, Jupiter, and Saturn were not observed at Uranus) [*Herbert and Sandel*, 1994; *Herbert*, 2009] (Figure 1 compare power levels in Table 1). So there are factors that control the occurrence or absence of intense auroral phenomena; factors that have not yet been identified. Possibly, the constantly changing geometry associated with the large magnetic axis tilt (Table 1) and planetary rotation, given an interplanetary magnetic field not aligned with the planet-Sun line, has a role to play.

On the other hand, at Neptune, because the rotational forcing is much larger than the solar wind forcing despite the period modulations, given the large tilt of the magnetic axis [*Selesnick*, 1990], and also because of the absence of a strong internal source of plasma, the aurora is expected to be relatively inactive, and indeed, its auroral emissions are far below those observed at other planets, even lower than those observed at Uranus (Table 1) [*Bishop et al.*, 1995].

A referee to this chapter thoughtfully suggested a third global-controlling parameter for comparing magnetospheres: the amount of solar wind flow energy that impinges on the cross section of the magnetosphere. With this parameter, the referee argues, the relative weakness of Uranus' aurora relative to those of the other active planets is understandable. A puzzle is that other aspects of Uranus' magnetosphere, discussed in the previous paragraph (radiation belt intensities, whistler mode activity), are as energetic as those of the Earth in its most active state.

The auroral emissions that do occur at Uranus and Neptune are thought to be most closely associated with the diffuse aurora at Earth (section 1.3) in that they have been interpreted in the context of scattering of magnetospheric particles onto the atmosphere without the additional energization that accompanies the other auroral processes [*Herbert and Sandel*, 1994; *Bishop et al.*, 1995]. For the rest of this chapter, we focus most of our attentions on the discrete auroral processes at Earth, Jupiter, and Saturn.

2.3. Comparing Auroral Current Systems

Here we describe the differences between auroral current systems driven by the solar wind (Earth), and those driven predominantly by rotation (Jupiter and perhaps Saturn). The relationship between global current systems and magnetospheric regions and dynamics is addressed in section 5 of this volume.

The Earth's aurora current system is driven by strong coupling between the flowing magnetized solar wind and the magnetosphere. Aspects of those current systems are shown in Figure 8 [*Cowley*, 2000; *Stern*, 1984]. On the dayside magnetopause (the boundary between the interplanetary medium and the Earth's magnetosphere), magnetic reconnection (a process that connects interplanetary magnetic field lines together with the Earth-connected field lines and converts magnetic energy to plasma heating and flow) is thought to allow the motional ($\mathbf{V} \times \mathbf{B}/c$) electric field of the solar wind to effectively penetrate inside the magnetosphere. Thus, momentum from the solar wind is coupled to the magnetosphere, drives a two-cell flow pattern within the ionosphere (Figure 8b), and maintains a system of upgoing and downgoing magnetic field-aligned electric currents called region 1 and region 2 (Figures 8a and 8b). How the region 1 system of current sheets, thought to close in the vicinity of the magnetopause on the dayside (Figure 8a), connects across the antisunward, comet-like magnetic tail is uncertain, but one solution is suggested in Figure 8c [*Stern*, 1984]. A dynamic version of the diversion of the cross-tail current into the ionosphere shown with this shunting process is also associated with dynamical events within the magnetosphere giving rise to auroral breakups associated with geomagnetic substorms. The region 2 currents are thought to be closed by the hot ion populations (ring current populations) trapped within the Earth's middle and inner magnetosphere (Figure 8a). So within any one meridional plane, there is a system of upgoing and downgoing electric currents (regions 1 and 2) that mimics the pair of currents sketched in Figure 2. However, during active conditions, the auroral regions are highly structured (Figure 4) [e.g., *Gorney*, 1991], and there are often multiple pairs of upgoing and downgoing currents [*Elphic et al.*, 1998]. How such structuring comes about is a mystery. Note that statistically (Figure 5) the occurrence of strong discrete aurora (and indeed the Alfvénic aurora as well) maximizes in the premidnight region, consistent with the current-flow sense of the region 1 currents (upward currents associated with downward electron acceleration).

Jupiter's auroral current system is driven by rotational energy combined with the production and outward transport of iogenic plasma [*Hill*, 2001; *Cowley and Bunce*, 2001]. These rotationally symmetric currents close through the ionosphere to generate a large-scale meridional current system like that illustrated in Figure 9a [*Hill*, 1979; *Vasyliūnas*, 1983]. A consequence of the current closure is that the rotation of the ionosphere is coupled to the rotation of the equatorial plasmas, and the equatorial plasmas are

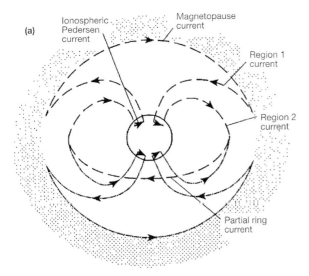

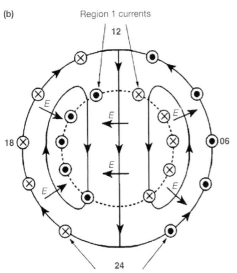

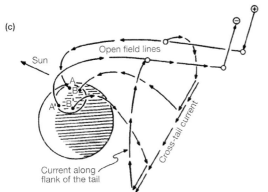

accelerated to a substantial fraction of the rigid rotation speed [*Hill*, 1979]. Rotational speeds as a function of radial distance stay at higher levels than the *Hill* [1979] theory would suggest (taking into account ionized mass outflow from the regions of the moon Io), indicating that modifications engendered by magnetic field-aligned electric fields and auroral precipitation (particle impacts on the ionosphere which increases conductivity) are substantial [e.g., *Ray et al.*, 2010; *Ray and Ergun*, this volume].

Just as we find at Earth, observations at Jupiter of particle acceleration features (section 2.5) indicate that the auroral currents are much more structured than suggested by Figure 9a, with multiple pairs of upward and downward currents occurring [*Mauk and Saur*, 2007]. A notional current profile as a function of magnetospheric *L* at some unspecified, nonequatorial latitude is sketched in Figure 8b. *Saur et al.* [2003] have suggested that the structuring is so pervasive on multiple scales that turbulent processes may be the prime energy conversion mechanism for the generation of Jupiter's aurora. This notion is supported by the power densities and spatial distribution (matching the mapped auroral distribution) of the magnetic turbulent spectrum (see Figure 10). More specifically, *Saur et al.* [2003] argue that there is a sufficient source of energy within the magnetic turbulence to power Jupiter's main aurora. We focus on this suggestion because it is highly reminiscent of the "Alfvénic aurora" discussion in section 1.2 about the Earth's aurora. Just as has been done in the case of the Earth, the argument is supported principally on the basis of energy source (rather than a specific mechanism for energy dissipation) and on the magnetic mapping of structures from the magnetosphere to the auroral dissipation regions. Not only does the region of turbulence at Jupiter map well to the regions of auroral emissions, but the energies available for dissipation from that turbulence are sufficient to provide all of the energy needed to power the aurora. The role of turbulent waves in transporting energy from the magnetosphere to the auroral regions, and in possibly helping to drive the auroral current system, is a ripe area for research on both the Earth and Jupiter and likely on other systems as well.

Figure 8. Schematics of the solar wind–driven auroral current system at Earth. (a) A view toward the Sun with the inner boundary of the shaded region representing the outer boundary of the magnetosphere. (b) A view of the Earth's Northern Hemisphere ionosphere. The crosses and dots represent magnetic field-aligned currents flowing into and out of the ionosphere. Figures 8a and 8b are from *Cowley* [2000]. (c) The antisunward, comet-like magnetic tail of Earth's magnetosphere extends to the right. Figure 8c is from *Stern* [1984].

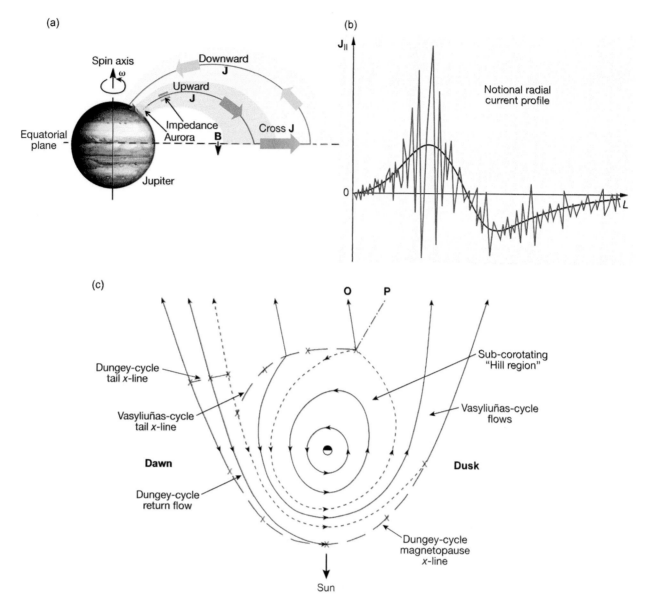

Figure 9. Auroral current systems at Jupiter. (a) Currents within a meridional plane. The structures shown are azimuthally symmetric. (b) A notional radial cut through the currents in Figure 9a at midlatitudes. Figures 9a and 9b are from *Mauk and Saur* [2007]. (c) A theoretical equatorial flow pattern at Jupiter [from *Cowley et al.*, 2003; see *Badman and Cowley*, 2007].

Continuing with Jupiter, Figure 9c [*Badman and Cowley*, 2007] shows theoretical flow patterns both within the inner regions discussed above and also in the more distant regions where solar wind effects may have a role to play, particularly within the magnetic tail. A key feature is the tail reconnection line (labeled Vasyliūnas cycle in Figure 9c) [*Vasyliūnas*, 1983] where field lines populated with dense plasmas from Io disconnect and flow down the tail. The figure shows a second, distinct reconnection line (labeled Dungey cycle in Figure 9c) that accommodates the return flow associated with solar wind–driven motions. It is clear that at Jupiter, a very small portion of the large-scale pattern is driven by solar wind forcing, but the current debate is whether there is a distinct channel (labeled Dungey cycle return flow in the diagram of Figure 9c), whether open flux is closed and returned mixed-in with the Vasyliūnas cycle [*Badman and Cowley*, 2007], or whether the solar wind actions are confined to a viscous boundary layer [*McComas and Bagenal*, 2007; *Delamere and Bagenal*, 2010]. *Delamere* [this volume] addresses Jovian auroral signatures associated with the solar wind interaction at Jupiter.

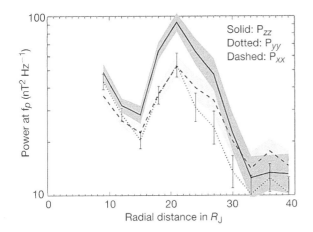

Solid: P_{zz}
Dotted: P_{yy}
Dashed: P_{xx}

Figure 10. Distribution of the total measured magnetic turbulence power within the equatorial regions of Jupiter' magnetosphere both parallel (z, solid line) and perpendicular (x and y, dashed and dotted) to the local magnetic field direction calculated from magnetic fluctuations on the basis of weak turbulence theory involving Alfvén waves. *Saur et al.* [2003] propose that such turbulence may represent a key power source for Jupiter's aurora.

At Saturn, it is argued [*Bunce et al.*, 2008; *Bunce*, this volume] that the solar wind forcing has a more prominent role in the outer magnetosphere than at Jupiter. It is further argued that rotational forcing is insufficient to generate intense aurora at Saturn and that magnetic reconnection within the deep magnetotail is moderated by the solar wind and is the driver of intense auroral emissions and dynamics at Saturn. Various positions on the role of the solar wind in generating Saturn's aurora are discussed broadly in the review by *Kurth et al.* [2009], and we will not summarize them here. We show in the discussions that follow, however, some examples of auroral phenomena that map to the deep interior of Saturn's magnetosphere, contrary to the models referenced above, which model Saturn's aurora as powered by tail reconnection and mapping to positions close to the site of the reconnection. It is now clear that Saturn's auroral configuration and dynamics are more complicated than any one model can accommodate. This finding should be no surprise, since the same thing can be said for the Earth. At Earth, many observers believe that magnetic reconnection driven by the solar wind within the magnetic tail is a prime mover of auroral energetics and dynamics, but it is clear that the most intense auroral phenomena often occur well equatorward of the reconnection site (Figure 4). Unlike at Earth, at Saturn, there is still the open question of what the ultimate source of power is for the most intense aurora. Is it rotational energy that the solar wind helps trigger and moderate, or is it solar wind energy input itself?

2.4. Current Closure

An important aspect of the differences between the global currents of different auroral systems is how the currents close within the distant equatorial magnetosphere. In discussing such current closures, we again point out differences and similarities between systems driven by the solar wind and systems driven by rotation.

Using the guiding center approach in analyzing the motions of particles within a magnetic field, the total current density $\mathbf{J}_\perp$ perpendicular to the magnetic field can be written as [*Parks*, 1991]:

$$\mathbf{J}_\perp = \frac{\mathbf{b}}{B} \times \nabla_\perp(P_\perp) + (P_\parallel - P_\perp)\frac{\mathbf{b} \times (\mathbf{b} \cdot \nabla)\mathbf{b}}{B} + (m \cdot n)\frac{\mathbf{b}}{B} \times \frac{d\mathbf{V}}{dt},$$
(1)

where, $\mathbf{b}$ is the unit magnetic field vector, B is magnetic field strength, P is the particle pressure, the symbols $\perp$ and "$\parallel$" indicate parameters measured perpendicular and parallel to the magnetic field direction, m is the average mass per ion, n is the number density, $m \cdot n$ is the mass density, and $\mathbf{V}$ is the flow velocity. Note that the $d\mathbf{V}/dt$ operation is a total derivative that includes both the explicit time dependence and the time-stationary convective contribution [$d\mathbf{V}/dt = \partial\mathbf{V}/\partial t + (\mathbf{V} \cdot \nabla)\mathbf{V}$]. The first of the three terms of equation (4) is the diamagnetic current driven by gradients in the hot plasma pressure. The second term is what remains of the currents from guiding center drifts that arise from the presence of gradients and curvatures within the magnetic field configuration after partial cancellation from terms associated with magnetization (contributions from $\nabla \times \mathbf{M}$, where $\mathbf{M}$ is the magnetic moment per volume of the plasma medium; the diamagnetic current is one of the magnetization current contributions). The third term represents currents associated with the acceleration of the plasma population. Notice that for an isotropic distribution ($P_\parallel = P_\perp$), the second term is zero, leaving only the diamagnetic and acceleration terms. Equation (4) shows only currents perpendicular to the magnetic field direction, but it is, of course, the divergence of the perpendicular currents ($\nabla \cdot \mathbf{J}_\perp$) that yields the parallel currents that close through the auroral ionosphere.

For the Earth's magnetosphere, region 2 currents are thought to be closed by the diamagnetic (first term of equation (4)) current closure term (Figure 8a) [see *Cowley*, 2000], with field-aligned currents generated by divergences resulting from transport-engendered asymmetries. The region 1 currents on the dayside are thought to be closed by the acceleration term (third term of equation (4)) associated with the sheared solar wind flow in the vicinity of the magnetopause boundary between the Earth's magnetic field and the solar wind on the dayside. However, a great uncertainty is associated with the region 1 currents and the transient substorm currents that cross the magnetic tail regions. In the vicinity of the boundary between open and closed field lines within the magnetotail, a region thought to be regulated by magnetic reconnection, flow gradients engendered by the reconnection process may close the currents associated with the boundary aurora (Figure 4). Planetward and equatorward of that boundary, some models tap into the deceleration of the reconnection-generated earthward flows, combined with the adiabatic heating of compression as the plasmas flow earthward, to drive auroral currents [e.g., *Zhang et al.*, 2007; *Keiling et al.*, 2009; *Pu et al.*, 2010]. The relative roles of the acceleration term and the diamagnetic term in this process are uncertain. Determining the mechanism of current closure at the base of the magnetotail for strong dynamical auroral emission processes is one of the outstanding questions surrounding auroral physics at Earth.

For the nonterrestrial planets like Jupiter and Saturn, it is useful to separate the rotation term from the acceleration term. Specifically, under the assumption that there are no explicit time dependencies, one may jump into a rotational frame of reference using the standard textbook [e.g., *Fowles and Cassidy*, 1993] decomposition of the d**V**/d*t* term to yield:

$$\mathbf{J}_\perp = \frac{\mathbf{b}}{B} \times \nabla_\perp(P_\perp) + (P_\parallel - P_\perp)\frac{\mathbf{b} \times (\mathbf{b} \cdot \nabla)\mathbf{b}}{B} + (m \cdot n)\frac{\mathbf{b}}{B}$$
$$\times [\mathbf{\Omega}_{pl} x(\mathbf{\Omega}_{pl} \times \mathbf{R})] + (m \cdot n)\frac{\mathbf{b}}{B} \times (2 \cdot \mathbf{\Omega}_{pl} \times \mathbf{U}_{\mathrm{rad}}), \quad (5)$$

where $\mathbf{\Omega}_{pl}$ is the rotational rate vector of the plasmas around the planet's spin axis (not necessarily the rotational rate vector of the planet itself), and $\mathbf{U}_{\mathrm{rad}}$ is the radial flow velocity of the plasma within that rotating frame of reference. Note that one may transform into the rotational frame that rotates rigidly with the planet, but for that formulation, there is an additional acceleration term associated with the deviation from rigid corotation. The transformation used here in equation (5) has the disadvantage of being useful only at one particular radial position with a plasma rotation rate of $\mathbf{\Omega}_{pl}$ (see a more complete treatment by *Vasyliūnas* [1983]).

The last two terms of equation (5) make sense if one considers the guiding center response of gyrating charged particles. In the presence of an electric field (**E**), plasmas flow with the well-known drift velocity: $c\,\mathbf{E} \times \mathbf{B}/B^2$. For an externally applied force (**F**) that acts only on mass rather than on charge, the drift velocity is $c \cdot m \cdot \mathbf{F} \times \mathbf{B}/(qB^2)$, where q is charge and m is mass, and where **F** is assumed to have the units force mass^{-1}. While the electric current associated with the $\mathbf{E} \times \mathbf{B}$ drift is zero, the electric current for the mass-dependent $\mathbf{F} \times \mathbf{B}$ drift is $(n \cdot m) \cdot \mathbf{F} \times \mathbf{B}/B^2$. With this understanding, we see that the third term of equation (5) is

the $\mathbf{F} \times \mathbf{B}$ current associated with the centrifugal force (negative of the centripetal acceleration) and the fourth term is the $\mathbf{F} \times \mathbf{B}$ current associated with the Coriolis force due to outward flows of plasma that are continually generated by Io at Jupiter or Enceladus at Saturn.

For the conventional view of Jupiter's middle magnetosphere, which focuses on flow structure and dynamics [*Vasyliūnas*, 1983], it is the third term of equation (5) that provides the azimuthal currents that distort the magnetic field configuration away from the dipolar magnetic configuration toward the extended magnetodisc configuration. However, the diamagnetic currents are known to contribute substantially [*Mauk and Krimigis*, 1987; *Paranicas et al.*, 1991], and beyond 20 R_J, it has been found that the second term of equation (5), the so-called anisotropy term, has perhaps a dominant role [*Mauk and Krimigis*, 1987; *Paranicas et al.*, 1991; *Frank and Paterson*, 2004]. For the closure of the auroral current depicted in Figure 9a, it is the fourth term, the Coriolis term, that provides the radial, near-equatorial closure currents, to the extent that the flow configuration is thought to drive the auroral processes.

Historically, magnetospheric current closure associated with outer planet auroral current systems has been examined from the perspective of flow dynamic mechanisms [*Hill*, 2001; *Cowley and Bunce*, 2001; *Cowley et al.*, 2004; *Bunce et al.*, 2008], both flow dynamics associated with rotation and those associated with magnetic reconnection processes deep in the magnetic tail. It is thought that current closure by pressure-driven diamagnetic currents plays at least a minor role for Jupiter's aurora in providing, for example, the current closure for lower-latitude auroral patches equatorward of the main auroral ring (Figures 11c and 11d) associated with dynamic injection phenomena within the middle to inner magnetosphere [*Mauk et al.*, 2002b]. At Earth, such near-planet hot plasma injections generate magnetic field-aligned discharges, again, presumably associated with pressure-driven currents (Figures 11a and 11b) [*Mauk and Meng*, 1991]. The configuration (Figures 11b and 11d) of upgoing currents coming from one azimuthal boundary of the equatorial plasma cloud, and the downgoing currents coming from the other azimuthal boundary, comes naturally from the perpendicular diamagnetic current's scaling with the term $\nabla P/B$ (equation (5)). Along the contours of constant pressure (P), it is along the azimuthal boundaries of the injected clouds where $\nabla P/B$ diverges because of the variation of B, giving rise to the field-aligned currents. At Saturn, pressure-driven current contributions may be even larger. Specifically, *Mitchell et al.* [2009a] showed that a major auroral breakup-like display (Figure 12) was strongly correlated in time and space with a major middle-magnetosphere ion injection event centered near 13 R_S and revealed by

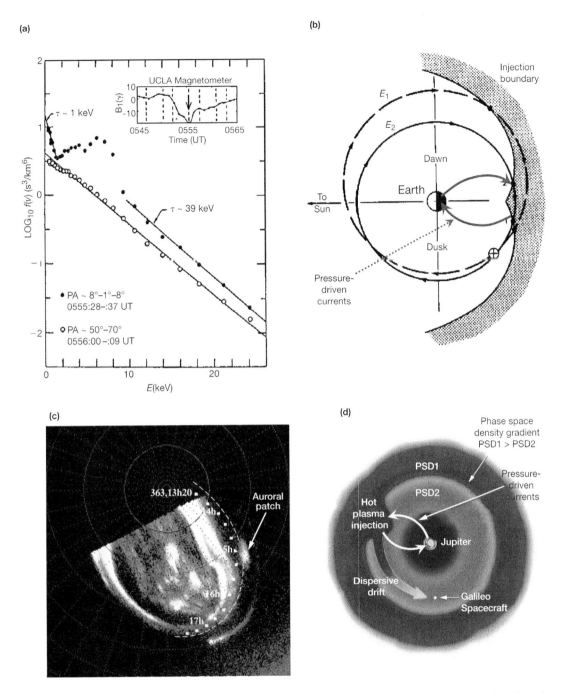

Figure 11. Magnetic field-aligned electrical beaming and magnetic field perturbations (a) associated with a hot plasma injection within the Earth's middle (geosynchronous) magnetosphere [*McIlwain*, 1975], thought to be associated with aurora emissions as diagnosed with auroral X-rays [*Mauk and Meng*, 1991]. These beams are interpreted here (b) as being associated with pressure-gradient-driven closure currents associated with the spatial configuration of the injected distributions. (c) Transient aurora at Jupiter, also associated with hot plasma injections [from *Mauk et al.*, 2002b] may also (d) be associated with hot plasma pressure-gradient current closure.

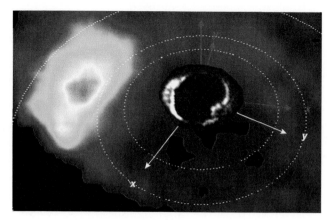

Figure 12. This is one frame of a movie that shows the correlation of the dynamics of a hot ion population (high-pressure region) as imaged with ~50 keV energetic neutral atoms (ENA) at Saturn by the Cassini magnetospheric imaging instrument, and the dynamics of a bright auroral storm occurring in Saturn's polar atmosphere as simultaneously imaged with the Cassini UVIS. The auroral image has been artificially inserted into the middle of the ENA image. The entire movie shows the simultaneous brightening of the ENA and UV emissions, centered about 45° anticlockwise from midnight and then the correlated rotation of both structures around dawn and into the dayside regions. The Sun is along the x axis shown in the figure. The ENA bright region is centered near ~13 R_S (between the dotted circle of the moon Rhea's orbit near 8.7 R_S and the dotted circle of the moon Titan's orbit near 20.3 R_S). Reprinted from *Mitchell et al.* [2009a], copyright 2009, with permission from Elsevier.

energetic neutral atom images, as both the ion injection feature and the auroral breakup feature rotated over several hours from the postmidnight regions into the dayside regions (Figure 12). On the basis of these features, a natural hypothesis is that pressure gradients are responsible for the current closure for the imaged auroral configurations for this event.

The source of the substantial populations of energetic particles is a major issue at Jupiter and Saturn, and the role of pressure-driven currents within the nonterrestrial planet auroral current systems is one of the great unanswered questions. It is significant that thermal energies dominate over the kinetic energy of flow velocities throughout the regions of both Jupiter's and Saturn's magnetospheres that connect to their aurora (Figure 13) [*Bagenal and Delamere*, 2011].

2.5. Particle Acceleration

2.5.1. Electron Acceleration. For the static auroral current systems, there are two regions of interest with regard to particle acceleration processes (Figure 2). The upward current region generates downward accelerated electrons, which excite the intense discrete auroral emissions [e.g., *Carlson et al.,*

1998]. These coherent distributions, often with monoenergetic peaks at ~1 keV to sometimes 30 keV energies at Earth, have not been observed within nonterrestrial planets because space probes have yet to visit regions with sufficiently low altitude and high latitude. Visiting such regions at Jupiter is a principal goal of the Juno mission, with Jupiter orbit insertion in 2016. Importantly, what has been observed on nonterrestrial planets are the upward accelerated electron distributions associated with the downward current regions [*Carlson et al.*, 1998; *Ergun et al.*, 1998]. These distributions have broad energy distributions (without a sharp peak in the energy spectra) and are narrowly confined to the magnetic field direction. Significantly, these upward accelerated electron distributions are observed in the near-equatorial regions and provide a powerful technique for mapping discrete auroral processes. They have been observed at Earth, Jupiter, and Saturn, and in the vicinity

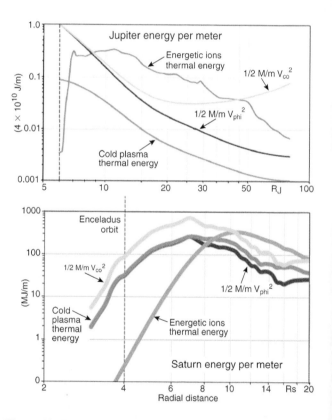

Figure 13. Energy profiles for (top) Jupiter and (bottom) Saturn where the energy density is summed over cylindrical annuli of width 1 m, and M/m is the total mass of plasma per cylindrical meter. The kinetic energy is shown for both rigid corotation and for observed $\mathbf{V}_{phi}$ profiles. The significance of this figure is that it shows that thermal energy densities are either comparable to, or dominate over, flow energy densities within the regions that map magnetically to the most intense auroral emission regions. From *Bagenal and Delamere* [2011].

of several of the satellites of these systems (Figure 14). They are interpreted in each environment as being associated with auroral acceleration [*Klumpar et al.*, 1988; *Carlson et al.*, 1998; *Williams et al.*, 1996; *Mauk et al.*, 2001; *Frank and Paterson*, 2002; *Mauk and Saur*, 2007; *Saur et al.*, 2006]. The mechanism of upward acceleration is thought to be stochastic acceleration through interactions with a multiplicity of small-scale electrostatic structures [*Ergun et al.*, 1998]. It is un-

known whether or not this process is driven in the distant magnetosphere by the Alfvénic auroral generator discussed in section 1.2, the generator of quasistationary auroral currents or some other process.

At Earth (Figure 14a), the equatorial beams were observed by *Klumpar et al.* [1988] at ~9 R_E and were attributed to the consequences of downward accelerated electron beams. *Carlson et al.* [1998] reinterpreted these beams, on the basis

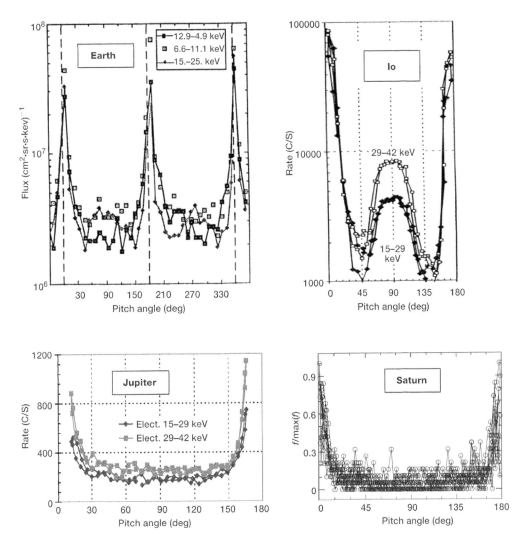

Figure 14. Equatorial magnetic field-aligned electron beams observed at Earth, Jupiter, Saturn, and in the vicinity of Jupiter's satellite Io. All have been associated with upward auroral electron acceleration by *Carlson et al.* [1998] (Earth panel from *Klumpar et al.* [1988]), at Jupiter by *Mauk and Saur* [2007], at Saturn by *Saur et al.* [2006] and at Io by *Williams et al.* [1996] and *Mauk et al.* [2001]. Pitch angle is the angle between the particle velocity vector and the magnetic field vector. The reader should exercise care, since the top two plots have logarithmic *y* axes, whereas the bottom plots have linear *y* axes. Also, the Earth plot shows a complete spacecraft spin, and the angles between 180° and 360° represent a second sampling of true pitch angles between 180° and 0°. Only the distribution observed near Jupiter's moon Io has a trapped population with pitch angle near 90°, presumably resulting from the localized magnetic field minimum detected very close to the moon.

of discoveries made with the FAST mission, as being the equatorial manifestation of the upward accelerated electron beams associated with the downward leg of the auroral electric currents (Figure 2). At Jupiter (Figure 14c), equatorial electron beams have been observed sporadically throughout the broad regions of downward currents in the global auroral current system (Figure 9a), which led to the conclusion that the current systems were highly structured (Figure 9b) [*Mauk and Saur*, 2007]. At Saturn (Figure 14d), equatorial electron beams have been observed as close to the planet as ~10 R_S, which led to the conclusion [*Saur et al.*, 2006] that at least some discrete auroral processes occur in regions much closer to Saturn than would be expected if the driver of auroral processes is primarily the divergence of flow in the vicinity of the boundary between open and closed field lines. Electron beams have been observed within the plasma wakes of both the Jupiter satellites Io (Figure 14b) and Callisto and have been attributed, again, to auroral current systems associated with the interactions between the conducting moons and the rapidly rotating magnetospheric plasmas (section 2.7) [*Williams et al.*, 1996, 1999; *Frank and Paterson*, 1999; *Mauk et al.*, 2001; *Mauk and Saur*, 2007]. More

recently, they have been observed in the vicinity of Saturn's satellite Enceladus [*Pryor et al.*, 2011] (see section 2.7). It would appear that the upward acceleration of electrons over a broad distribution of energies (not shown here) is a universal aspect of intense auroral processes wherever they occur. The differences are in the energies that are achieved. At Earth, energies up to 30 keV are reported, whereas at Saturn and Jupiter, energies >200 keV are common.

2.5.2. Ion Acceleration. At Earth, upgoing ion "conic" distributions are observed on high-latitude, low-altitude regions of the magnetic field lines that carry the upward electric currents and provide the downward accelerated electron distributions that generate intense aurora [*Shelley and Collin*, 1991; *Carlson et al.*, 1998]. Conic-shaped distributions result from low-altitude acceleration perpendicular to the magnetic field combined with the parallel acceleration that follows from the magnetic mirror force that pushes the particles into the distant magnetosphere. Only the Cassini mission has been at the right place with the right instrumentation to view such distributions at a nonterrestrial planet, Saturn (Figure 15) [*Mitchell et al.*, 2009b]. Here not only were very energetic

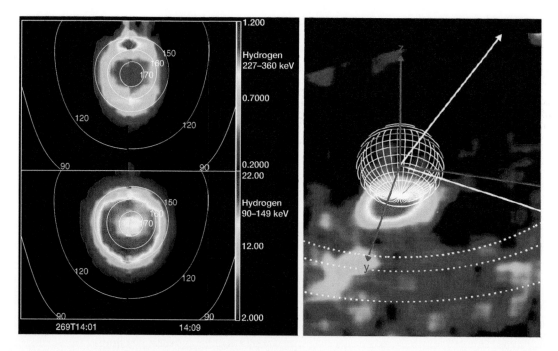

Figure 15. (left) A different representation of pitch angle distributions, this time for ions, where the white contours represent values of the pitch angle in degrees, and the colored intensities represent the intensity of the particle distributions at those pitch angles. Shown are upward propagating ion conic distributions measured at Saturn's high-latitude auroral regions. (right) Energetic neutral atom image of low-altitude auroral ion acceleration at Saturn's southern hemisphere. The bright region just under Saturn's southern pole likely represents the location of auroral acceleration, a conclusion supported by the fact that the ion emissions are protons or proton-related, without such heavy ions as oxygen or nitrogen observed elsewhere. From *Mitchell et al.* [2009b].

(~20 to >220 keV) upgoing ion conic distributions observed (Figure 15a), but the probable ion energization region was simultaneously imaged directly with energetic neutral atom imaging (Figure 15b). A significant difference between the observations at Earth and Saturn for the ions, as with the electrons, is the energies involved, with the Saturn ion conic energies extending up in energy by a factor of 20 to 100 higher than the same acceleration process operating at Earth.

2.6. Ionospheric Feedback

An important element in the auroral current system is the modification of the conductivity caused by the impact of accelerated charged particles onto the upper atmosphere. Such a modification can lead to a feedback process whereby an increase in auroral currents leads to an increase in conductivity, which in turn leads to further increases in auroral currents, etc. [*Watanabe and Sato*, 1988]. The importance of such a feedback process has not been established at Earth because its efficacy depends on the relative impedances of the magnetospheric current sources and the impedance of the ionosphere. These issues are addressed in section 3 and elsewhere in this volume. From a comparative standpoint, the role of the ionospheric response to auroral processes has recently been highlighted. One of the outstanding issues at Jupiter is why the magnetospheric plasmas continue to rotate at a substantial fraction of the planet's rotation rate to distances much larger than anticipated from core theoretical ideas involving plasma outflow, conservation of angular momentum, and uniform ionospheric conductivity [*Hill*, 1979; *Vasyliūnas*, 1983]. Increases to ionospheric conductivity are one way that the coupling between the planet and the distant space environment can be enhanced, thereby enhancing the rotational coupling [*Nichols and Cowley*, 2005]. *Ray et al.* [2010] and *Ray and Ergun* [this volume] describe a model that included both ionospheric conductivity enhancements and magnetic field-aligned electric fields and show that such effects can dramatically enhance rotational coupling.

The beauty of the Jupiter's auroral system compared with the system at Earth is that there is a very simple metric to test one's models: Do the model rotational flows at specified radial distances match the observations? While multiple processes can still influence the answer, leading to uncertainties remaining in the relative importance of those different processes, there exists no such simple metric at Earth. Jupiter provides an important test case.

2.7. Satellite Systems

One of the wonderful aspects of the nonterrestrial magnetospheric systems is the presence of electrically conducting

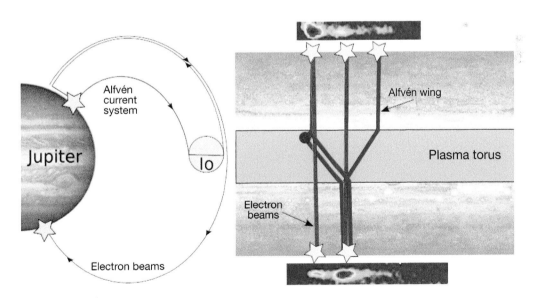

Figure 16. (top and bottom right) Images of northern and southern hemisphere auroral spots. The brightness of the spots as a function of position is interpreted on the basis of the Earth observations, whereby the brightest aurora emissions are generated by the downward acceleration of electrons in the upward (with respect to Jupiter) electric current regions, and the dimmer emissions are generated by the upward acceleration of electrons in the regions of downward electric currents. The upward accelerated electrons stimulate auroral emissions on the hemisphere opposite from where the acceleration occurred. Further details can be found in the source by *Bonfond et al.* [2008].

satellites that provide a whole new set of auroral systems. At Jupiter, auroral emissions are observed at the Jupiter magnetic foot points of the satellites Io (see Figure 1), Europa, and Ganymede [*Clarke et al.*, 2002; *Bonfond*, this volume; *Hess and Delamere*, this volume], and strong magnetic field-aligned electron beams were observed in the wake of Jupiter's satellite Callisto, indicative of the existence of an auroral current system, but with auroral emissions perhaps too weak to observe [*Mauk and Saur*, 2007], particularly occurring among the strong main auroral emissions. The electron beams observed near Callisto are similar to those observed in the plasma wake of Io (Figure 14b).

A highly significant finding (Figure 16) was reported by *Bonfond et al.* [2008], where direct evidence was discovered of the consequences of downward electron acceleration (in what is believed to be the upward current region) generating intense auroral emission associated with the satellite Io and the simultaneous generation of upward acceleration electrons (in the region of downward currents) generating auroral emissions in the opposite hemisphere. The ordering of the auroral phenomena engendered by the rapid rotation of the planet seems to provide a cleaner slate in sorting out the various mechanisms associated with the generation of auroral emissions than do the more chaotic conditions at Earth (Figure 4).

The satellite Enceladus at Saturn also generates a small-scale auroral current system [*Pryor et al.*, 2011] as illustrated in Figure 17 (Figure 17 was generated and provided by A. M. Rymer; the Enceladus spot is also highlighted in Figure 1). The inserted particle distribution (elevated above Enceladus) shows an upward (from Saturn) ion beaming distribution that was anticipated from the Earth aurora and from recent global auroral observations at Saturn (section 2.5), but that has not been reported in association with the other satellite interaction measurements. *Gurnett and Pryor* [this volume] report on other details of the Enceladus interactions.

2.8. Other Processes

With this brief review, we have been able to compare only limited aspects of auroral phenomena among the different magnetized planets. A most glaring omission is our failure to address plasma wave and radio wave emission processes that are directly associated with auroral acceleration. Voyager epoch comparisons of plasma waves measured at Earth and on the nonterrestrial planets were performed by *Kurth and Gurnett* [1991]. Radio and plasma waves specific to auroral processes at Jupiter are discussed by *Clarke et al.* [2004], and those specific to Saturn are discussed by *Kurth et al.* [2009] and *Mauk et al.* [2009], and in all of these discussions, comparisons between the different planets are discussed. We

Figure 17. Auroral emissions measured by the Cassini UVIS instrument remapped onto visible image of Saturn and placed within the context of an artist's conception of the interaction between Saturn and Saturn's moon Enceladus, including an artist's conception of the gas and dust plumes coming out of Enceladus' southern polar regions. Levitated above Enceladus is a measured pitch angle distribution of the ion distributions that have been observed over Enceladus' polar regions (see the caption to Figure 15 for an explanation of the inserted pitch angle distribution). This figure was generated and provided by A. M. Rymer to highlight the discovery of the auroral spot at Saturn generated by the interaction sketched here. This discovery is reported by *Pryor et al.* [2011].

recommend these and other sources to the reader. Our discussions of the upper atmospheric and ionospheric consequences of auroral processes have also been minimal at best. Several articles in the *Geophysical Monograph 130* [*Mendillo et al.*, 2002] provide the reader with reasonable starting points.

3. CONCLUSIONS

From observations taken to date, a preliminary conclusion can be drawn. When magnetospheric processes generate more electric current along magnetic field lines than can be carried by the existing populations, the response of the space environment appears to be at least superficially similar between the very different planetary systems: generation of magnetic field-aligned electrical impedance, parallel electric fields, particle acceleration, auroral emissions, ionospheric modification, and momentum coupling between the upper atmosphere and the space environment. This conclusion, however, must remain preliminary until spacecraft have been able to probe the regions below the auroral acceleration region at a system other than the Earth. That is a prime objective of the late phase of the Cassini mission at Saturn and the Juno mission at Jupiter. The large differences among the various systems appear to be in the mechanisms by which the global system of electric currents is generated. Of course, there remain great uncertainties and controversies about how those current systems are generated, as other chapters in this volume demonstrate.

REFERENCES

Akasofu, S.-I., A. T. Y. Lui, and C.-I. Meng (2010), Importance of auroral features in the search for substorm onset processes, *J. Geophys. Res.*, *115*, A08218, doi:10.1029/2009JA014960.

Axford, W. I., and C. O. Hines (1961), A unifying theory of high-latitude geophysical phenomena and geomagnetic storms, *Can. J. Phys.*, *39*, 1433–1464.

Badman, S. V., and S. W. H. Cowley (2007), Significance of Dungey-cycle flows in Jupiter's and Saturn's magnetospheres, and their identification on closed equatorial field lines, *Ann. Geophys.*, *25*, 941–951.

Bagenal, F. (2009), Comparative planetary environments, in *Heliophysics: Plasma Physics of the Local Cosmos*, edited by C. J. Schrijver and G. L. Siscoe, pp. 360–398, Cambridge Univ. Press, New York.

Bagenal, F., and P. A. Delamere (2011), Flow of mass and energy in the magnetospheres of Jupiter and Saturn, *J. Geophys. Res.*, *116*, A05209, doi:10.1029/2010JA016294.

Belcher, J. W., R. L. McNutt Jr., J. D. Richardson, R. S. Selesnick, E. C. Sittler Jr., and F. Bagenal (1991), The plasma environment of Uranus, in *Uranus*, edited by J. T. Bergstrahl, E. D. Miner, and M. S. Matthews, p. 780, Univ. of Ariz. Press, Tucson.

Bishop, J., S. K. Atreya, P. N. Romani, G. S. Orton, B. R. Sandel, and R. V. Yelle (1995), The middle and upper atmosphere of Neptune, in *Neptune and Triton*, edited by D. P. Cruikshank, pp. 427–487, Univ. of Ariz. Press, Tucson.

Bonfond, B. (2012), When moons create aurora: The satellite footprints on giant planets, in *Auroral Phenomenology and Magnetospheric Processes: Earth and Other Planets, Geophys. Monogr. Ser.*, doi:10.1029/2011GM001169, this volume.

Bonfond, B., D. Grodent, J.-C. Gérard, A. Radioti, J. Saur, and S. Jacobsen (2008), UV Io footprint leading spot: A key feature for understanding the UV Io footprint multiplicity?, *Geophys. Res. Lett.*, *35*, L05107, doi:10.1029/2007GL032418.

Borovsky, J. E. (1993), Auroral arc thicknesses as predicted by various theories, *J. Geophys. Res.*, *98*(A4), 6101–6138.

Brain, D., and J. S. Halekas (2012), Aurora in Martian minimagnetospheres, in *Auroral Phenomenology and Magnetospheric Processes: Earth and Other Planets, Geophys. Monogr. Ser.*, doi:10.1029/2011GM001200, this volume.

Brice, N. M. (1967), Bulk motion of the magnetosphere, *J. Geophys. Res.*, *72*(21), 5193–5211.

Brice, N. M., and G. A. Ioannidis (1970), The magnetospheres of Jupiter and Earth, *Icarus*, *13*(2), 173–183, doi:10.1016/0019-1035(70)90048-5.

Bunce, E. J. (2012), Origins of Saturn's auroral emissions and their relationship to large-scale magnetosphere dynamics, in *Auroral Phenomenology and Magnetospheric Processes: Earth and Other Planets, Geophys. Monogr. Ser.*, doi:10.1029/2011GM001191, this volume.

Bunce, E. J., et al. (2008), Origin of Saturn's aurora: Simultaneous observations by Cassini and the Hubble Space Telescope, *J. Geophys. Res.*, *113*, A09209, doi:10.1029/2008JA013257.

Carlson, C. W., et al. (1998), FAST observations in the downward auroral current region: Energetic upgoing electron beams, parallel potential drops, and ion heating, *Geophys. Res. Lett.*, *25*(12), 2017–2020.

Chaston, C. C., J. W. Bonnell, C. W. Carlson, J. P. McFadden, R. E. Ergun, and R. J. Strangeway (2003), Properties of small-scale Alfvén waves and accelerated electrons from FAST, *J. Geophys. Res.*, *108*(A4), 8003, doi:10.1029/2002JA009420.

Chen, A. J. (1970), Penetration of low-energy protons deep into the magnetosphere, *J. Geophys. Res.*, *75*(13), 2458–2467.

Clarke, J. T. (2012), Auroral processes on Jupiter and Saturn, in *Auroral Phenomenology and Magnetospheric Processes: Earth and Other Planets, Geophys. Monogr. Ser.*, doi:10.1029/2011GM001199, this volume.

Clarke, J. T., et al. (2002), Ultraviolet emissions from the magnetic footprints of Io, Ganymede, and Europa on Jupiter, *Nature*, *415*, 997–1000.

Clarke, J. T., D. Grodent, S. W. H. Cowley, E. J. Bunce, P. Zarka, J. E. P. Connerney, and T. Satoh (2004), Jupiter's aurora, in *Jupiter: The Planet, Satellites and Magnetosphere*, edited by F. Bagenal, T. E. Dowling, and W. B. McKinnon, pp. 639–670, Cambridge Univ. Press, New York.

Cowley, S. W. H. (2000), Magnetosphere-ionosphere interactions: A tutorial review, in *Magnetospheric Current Systems, Geophys. Monogr. Ser.*, vol. 118, edited by S. Ohtani et al., pp. 91–106, AGU, Washington, D. C., doi:10.1029/GM118p0091.

Cowley, S. W. H., and E. J. Bunce (2001), Origin of the main auroral oval in Jupiter's coupled magnetosphere-ionosphere system, *Planet. Space Sci.*, *49*, 1067–1088.

Cowley, S. W. H., E. J. Bunce, T. S. Stallard, and S. Miller (2003), Jupiter's polar ionospheric flows: Theoretical interpretation, *Geophys. Res. Lett.*, *30*(5), 1220, doi:10.1029/2002GL016030.

Cowley, S. W. H., E. J. Bunce, and J. M. O'Rourke (2004), A simple quantitative model of plasma flows and currents in Saturn's polar ionosphere, *J. Geophys. Res.*, *109*, A05212, doi:10.1029/2003JA010375.

Delamere, P. A. (2012), Auroral signatures of solar wind interaction at Jupiter, in *Auroral Phenomenology and Magnetospheric Processes: Earth and Other Planets, Geophys. Monogr. Ser.*, doi:10.1029/2011GM001180, this volume.

Delamere, P. A., and F. Bagenal (2010), Solar wind interaction with Jupiter's magnetosphere, *J. Geophys. Res.*, *115*, A10201, doi:10.1029/2010JA015347.

Donovan, E., E. Spanswick, J. Liang, J. Grant, B. Jackel, and M. Greffen (2012), Magnetospheric dynamics and the proton aurora, in *Auroral Phenomenology and Magnetospheric Processes: Earth and Other Planets, Geophys. Monogr. Ser.*, doi:10.1029/2011GM001241, this volume.

Elphic, R. C., et al. (1998), The auroral current circuit and field-aligned currents observed by FAST, *Geophys. Res. Lett.*, *25*(12), 2033–2036.

Ergun, R. E., et al. (1998), FAST satellite observations of electric field structures in the auroral zone, *Geophys. Res. Lett.*, *25*(12), 2025–2028.

Feynman, R. P., R. B. Leighton, and M. Sands (1964), *The Feynman Lectures on Physics*, vol. 2, Addison-Wesley, Reading, Mass.

Fowles, G. R., and G. L. Cassiday (1993), *Analytical Mechanics*, 5th ed., W. B. Saunders, Philadelphia, Pa.

Frank, L. A., and J. D. Craven (1988), Imaging results from Dynamics Explorer 1, *Rev. Geophys.*, *26*(2), 249–283.

Frank, L. A., and W. R. Paterson (1999), Intense electron beams observed at Io with the Galileo spacecraft, *J. Geophys. Res.*, *104*(A12), 28,657–28,669, doi:10.1029/1999JA900402.

Frank, L. A., and W. R. Paterson (2002), Galileo observations of electron beams and thermal ions in Jupiter's magnetosphere and their relationship to the auroras, *J. Geophys. Res.*, *107*(A12), 1478, doi:10.1029/2001JA009150.

Frank, L. A., and W. R. Paterson (2004), Plasmas observed near local noon in Jupiter's magnetosphere with the Galileo spacecraft, *J. Geophys. Res.*, *109*, A11217, doi:10.1029/2002JA009795.

Gorney, D. J. (1991), An overview of auroral spatial scales, in *Auroral Physics*, edited by C.-I. Meng et al., p. 325, Cambridge Univ. Press, New York.

Gurnett, D. A., and W. R. Pryor (2012), Auroral processes associated with Saturn's moon Enceladus, in *Auroral Phenomenology and Magnetospheric Processes: Earth and Other Planets, Geophys. Monogr. Ser.*, doi:10.1029/2011GM001174, this volume.

Herbert, F. (2009), Aurora and magnetic field of Uranus, *J. Geophys. Res.*, *114*, A11206, doi:10.1029/2009JA014394.

Herbert, F., and B. R. Sandel (1994), The Uranian aurora and its relationship to the magnetosphere, *J. Geophys. Res.*, *99*(A3), 4143–4160.

Hess, S. L. G., and P. A. Delamere (2012), Satellite-induced electron acceleration and related auroras, in *Auroral Phenomenology and Magnetospheric Processes: Earth and Other Planets, Geophys. Monogr. Ser.*, doi:10.1029/2011GM001175, this volume.

Hill, T. W. (1979), Inertial limit on corotation, *J. Geophys. Res.*, *84*(A11), 6554–6558.

Hill, T. W. (2001), The Jovian auroral oval, *J. Geophys. Res.*, *106*(A5), 8101–8107, doi:10.1029/2000JA000302.

Horne, R. B., R. M. Thorne, N. P. Meredith, and R. R. Anderson (2003), Diffuse auroral electron scattering by electron cyclotron harmonic and whistler mode waves during an isolated substorm, *J. Geophys. Res.*, *108*(A7), 1290, doi:10.1029/2002JA009736.

Kavanagh, L. D., Jr., J. W. Freeman Jr., and A. J. Chen (1968), Plasma flow in the magnetosphere, *J. Geophys. Res.*, *73*(17), 5511–5519.

Keiling, A. (2009), Alfvén waves and their roles in the dynamics of the Earth's magnetotail: A review, *Space Sci. Rev.*, *142*(1–4), 73–156.

Keiling, A., J. R. Wygant, C. Cattell, W. Peria, G. Parks, M. Temerin, F. S. Mozer, C. T. Russell, and C. A. Kletzing (2002), Correlation of Alfvén wave Poynting flux in the plasma sheet at 4–7 R_E with ionospheric electron energy flux, *J. Geophys. Res.*, *107*(A7), 1132, doi:10.1029/2001JA900140.

Keiling, A., J. R. Wygant, C. A. Cattell, F. S. Mozer, and C. T. Russell (2003), The global morphology of wave Poynting flux: Powering the aurora, *Science*, *299*, 383–386.

Keiling, A., et al. (2009), Substorm current wedge driven by plasma flow vortices: THEMIS observations, *J. Geophys. Res.*, *114*, A00C22, doi:10.1029/2009JA014114. [Printed 115(A1), 2010].

Kelley, M. C., D. J. Knudsen, and J. F. Vickrey (1991), Poynting flux measurements on a satellite: A diagnostic tool for space research, *J. Geophys. Res.*, *96*(A1), 201–207.

Klumpar, D. M., J. M. Quinn, and E. G. Shelley (1988), Counterstreaming electrons at the geomagnetic equator near 9 R_E, *Geophys. Res. Lett.*, *15*(11), 1295–1298.

Kurth, W. S., and D. A. Gurnett (1991), Plasma waves in planetary magnetospheres, *J. Geophys. Res.*, *96*, 18,977–18,991.

Kurth, W. S., et al. (2009), Auroral processes, in *Saturn from Cassini-Huygens*, edited by M. Dougherty, L. Esposito and S. Krimigis, p. 333, Springer, New York.

Lessard, M. R. (2012), A review of pulsating aurora, in *Auroral Phenomenology and Magnetospheric Processes: Earth and Other Planets, Geophys. Monogr. Ser.*, doi:10.1029/2011GM001187, this volume.

Li, W., J. Bortnik, Y. Nishimura, R. M. Thorne, and V. Angelopoulos (2012), The origin of pulsating aurora: Modulated whistler mode chorus waves, in *Auroral Phenomenology and Magnetospheric Processes: Earth and Other Planets, Geophys. Monogr. Ser.*, doi:10.1029/2011GM001192, this volume.

Lundin, R., G. Haerendel, and S. Grahn (1998), Introduction to special section: The Freja Mission, *J. Geophys. Res.*, *103*(A3), 4119–4123.

Lyons, L. R., Y. Nishimura, Y. Shi, S. Zou, H.-J. Kim, V. Angelopoulos, C. Heinselman, M. J. Nicolls, and K.-H. Fornacon (2010), Substorm triggering by new plasma intrusion: Incoherent-scatter radar observations, *J. Geophys. Res.*, *115*, A07223, doi:10.1029/2009JA015168.

Lyons, L. R., Y. Nishimura, X. Xing, Y. Shi, M. Gkioulidou, C.-P. Wang, H.-J. Kim, S. Zou, V. Angelopoulos, and E. Donovan (2012), Auroral disturbances as a manifestation of interplay between large-scale and mesoscale structure of magnetosphere-ionosphere electrodynamical coupling, in *Auroral Phenomenology and Magnetospheric Processes: Earth and Other Planets, Geophys. Monogr. Ser.*, doi:10.1029/2011GM001152, this volume.

Lysak, R. L. (Ed.) (1993), *Auroral Plasma Dynamics, Geophys. Monogr. Ser.*, vol. 80, 291 pp., AGU, Washington D. C., doi:10.1029/GM080.

Mauk, B. H., and N. J. Fox (2010), Electron radiation belts of the solar system, *J. Geophys. Res., 115*, A12220, doi:10.1029/2010JA015660.

Mauk, B. H., and S. M. Krimigis (1987), Radial force balance within Jupiter's dayside magnetosphere, *J. Geophys. Res., 92*(A9), 9931–9941.

Mauk, B. H., and C.-I. Meng (1991), The aurora and middle magnetospheric processes, in *Auroral Physics*, edited by C.-I. Meng et al., p. 223, Cambridge Univ. Press, Cambridge, U. K.

Mauk, B. H., and J. Saur (2007), Equatorial electron beams and auroral structuring at Jupiter, *J. Geophys. Res., 112*, A10221, doi:10.1029/2007JA012370.

Mauk, B. H., S. M. Krimigis, E. P. Keath, A. F. Cheng, T. P. Armstrong, L. J. Lanzerotti, G. Gloeckler, and D. C. Hamilton (1987), The hot plasma and radiation environment of the Uranian magnetosphere, *J. Geophys. Res., 92*(A13), 15,283–15,308.

Mauk, B. H., D. J. Williams, and A. Eviatar (2001), Understanding Io's space environment interaction: Recent energetic electron measurements from Galileo, *J. Geophys. Res., 106*(A11), 26,195–26,208, doi:10.1029/2000JA002508.

Mauk, B. H., B. J. Anderson, and R. M. Thorne (2002a), Magnetosphere-ionosphere coupling at Earth, Jupiter, and beyond, in *Atmospheres in the Solar System: Comparative Aeronomy, Geophys. Monogr. Ser.*, vol. 130, edited by M. Mendillo, A. Nagy, and J. H. White, pp. 97–114, AGU, Washington, D. C., doi:10.1029/130GM07.

Mauk, B. H., J. T. Clarke, D. Grodent, J. H. Waite Jr., C. P. Paranicas, and D. J. Williams (2002b), Transient aurora on Jupiter from injections of magnetospheric electrons, *Nature, 415*, 1003–1005.

Mauk, B. H., et al. (2009), Fundamental plasma processes in Saturn's magnetosphere, in *Saturn from Cassini-Huygens*, edited by M. Dougherty, L. Esposito and S. Krimigis, p. 281, Springer, New York.

McComas, D. J., and F. Bagenal (2007), Jupiter: A fundamentally different magnetospheric interaction with the solar wind, *Geophys. Res. Lett., 34*, L20106, doi:10.1029/2007GL031078.

McIlwain, C. E. (1975), Equatorial electron beams near the magnetic equator, in *The Physics of Hot Plasma in the Magnetosphere*, edited by B. Hultqvist and L. Stenflo, p. 91, Plenum, New York.

Mendillo, M., A. Nagy, and J. H. White (Eds.) (2002), *Atmospheres in the Solar System: Comparative Aeronomy, Geophys. Monogr. Ser.*, vol. 130, 388 pp., AGU, Washington, D. C., doi:10.1029/GM130.

Meredith, N. P., R. B. Horne, R. M. Thorne, and R. R. Anderson (2009), Survey of upper band chorus and ECH waves: Implica-

tions for the diffuse aurora, *J. Geophys. Res., 114*, A07218, doi:10.1029/2009JA014230.

Mitchell, D. G., et al. (2009a), Recurrent energization of plasma in the midnight-to-dawn quadrant of Saturn's magnetosphere, and its relationship to auroral UV and radio emissions, *Planet. Space Sci., 57*, 1732–1742.

Mitchell, D. G., W. S. Kurth, G. B. Hospodarsky, N. Krupp, J. Saur, B. H. Mauk, J. F. Carbary, S. M. Krimigis, M. K. Dougherty, and D. C. Hamilton (2009b), Ion conics and electron beams associated with auroral processes on Saturn, *J. Geophys. Res., 114*, A02212, doi:10.1029/2008JA013621.

Nagai, T., M. Fujimoto, R. Nakamura, W. Baumjohann, A. Ieda, I. Shinohara, S. Machida, Y. Saito, and T. Mukai (2005), Solar wind control of the radial distance of the magnetic reconnection site in the magnetotail, *J. Geophys. Res., 110*, A09208, doi:10.1029/2005JA011204.

Newell, P. T., T. Sotirelis, and S. Wing (2009), Diffuse, monoenergetic, and broadband aurora: The global precipitation budget, *J. Geophys. Res., 114*, A09207, doi:10.1029/2009JA014326.

Ni, B., R. M. Thorne, Y. Y. Shprits, and J. Bortnik (2008), Resonant scattering of plasma sheet electrons by whistler-mode chorus: Contribution to diffuse auroral precipitation, *Geophys. Res. Lett., 35*, L11106, doi:10.1029/2008GL034032.

Nichols, J. D., and S. W. H. Cowley (2005), Magnetosphere-ionosphere coupling currents in Jupiter's middle magnetosphere: Effect of magnetosphere-ionosphere decoupling by field-aligned auroral voltages, *Ann. Geophys., 23*, 799–808.

Nishida, A. (1966), Formation of plasmapause, or magnetospheric plasma knee, by the combined action of magnetospheric convection and plasma escape from the tail, *J. Geophys. Res., 71*(23), 5669–5679.

Nishimura, Y., et al. (2010), Preonset time sequence of auroral substorms: Coordinated observations by all-sky imagers, satellites, and radars, *J. Geophys. Res., 115*, A00I08, doi:10.1029/2010JA015832. [Printed 116(A5), 2011].

Paranicas, C. P., B. H. Mauk, and S. M. Krimigis (1991), Pressure anisotropy and radial stress balance in the Jovian neutral sheet, *J. Geophys. Res., 96*(A12), 21,135–21,140.

Parks, G. K. (1991), *Physics of Space Plasmas*, Westview, Cambridge, Mass.

Pryor, W. R., et al. (2011), The auroral footprint of Enceladus on Saturn, *Nature, 472*, 331–333, doi:10.1038/nature09928.

Pu, Z. Y., et al. (2010), THEMIS observations of substorms on 26 February 2008 initiated by magnetotail reconnection, *J. Geophys. Res., 115*, A02212, doi:10.1029/2009JA014217.

Ray, L. C., and R. E. Ergun (2012), Auroral signatures of ionosphere-magnetosphere coupling at Jupiter and Saturn, in *Auroral Phenomenology and Magnetospheric Processes: Earth and Other Planets, Geophys. Monogr. Ser.*, doi:10.1029/2011GM001172, this volume.

Ray, L. C., R. E. Ergun, P. A. Delamere, and F. Bagenal (2010), Magnetosphere-ionosphere coupling at Jupiter: Effect of field-aligned potentials on angular momentum transport, *J. Geophys. Res., 115*, A09211, doi:10.1029/2010JA015423.

Saur, J., A. Pouquet, and W. H. Matthaeus (2003), An acceleration mechanism for the generation of the main auroral oval on Jupiter, *Geophys. Res. Lett.*, *30*(5), 1260, doi:10.1029/2002GL015761.

Saur, J., et al. (2006), Anti-planetward auroral electron beams at Saturn, *Nature*, *439*, 699–702, doi:10.1038/nature04401.

Schriver, D., M. Ashour-Abdalla, R. J. Strangeway, R. L. Richard, C. Klezting, Y. Dotan, and J. Wygant (2003), FAST/Polar conjunction study of field-aligned auroral acceleration and corresponding magnetotail drivers, *J. Geophys. Res.*, *108*(A9), 8020, doi:10.1029/2002JA009426.

Selesnick, R. S. (1990), Plasma convection in Neptune's magnetosphere, *Geophys. Res. Lett.*, *17*(10), 1681–1684.

Selesnick, R. S., and R. L. McNutt Jr. (1987), Voyager 2 plasma ion observations in the magnetosphere of Uranus, *J. Geophys. Res.*, *92*(A13), 15,249–15,262.

Shelley, E. G., and H. L. Collin (1991), Auroral ion acceleration and its relationship to ion composition, in *Auroral Physics*, edited by C.-I. Meng, M. J. Rycroft, and L. A. Frank, p. 129, Cambridge Univ. Press, New York.

Stasiewicz, K., et al. (2000), Small scale Alfvénic structure in the aurora, *Space Sci. Rev.*, *92*, 423–533.

Stern, D. P. (1984), Magnetospheric dynamo processes, in *Magnetospheric Currents*, Geophys. Monogr. Ser., vol. 28, edited by T. A. Potemra, pp. 200–207, AGU, Washington, D. C., doi:10.1029/GM028p0200.

Su, Z., H. Zheng, and S. Wang (2010), A parametric study on the diffuse auroral precipitation by resonant interaction with whistler mode chorus, *J. Geophys. Res.*, *115*, A05219, doi:10.1029/2009JA014759.

Vasyliūnas, V. M. (1975), Concepts of magnetospheric convection, in *The Magnetospheres of the Earth and Jupiter*, edited by V. Formisano, p. 179, D. Reidel, Boston, Mass.

Vasyliūnas, V. M. (1983), Plasma distribution and flow, in *Physics of the Jovian Magnetosphere*, edited by A. J. Dessler, pp. 395–453, Cambridge Univ. Press, London, U. K.

Vasyliūnas, V. M. (2001), Electric field and plasma flow: What drives what?, *Geophys. Res. Lett.*, *28*(11), 2177–2180, doi:10.1029/2001GL013014.

Vasyliūnas, V. M. (2011), Physics of magnetospheric variability, *Space Sci. Rev.*, *158*, 91–118, doi:10.1007/s11214-010-9696-1.

Watanabe, K., and T. Sato (1988), Self-excitation of auroral arcs in a three-dimensionally coupled magnetosphere-ionosphere system, *Geophys. Res. Lett.*, *15*(7), 717–720.

Watt, C. E. J., and R. Rankin (2012), Alfvén wave acceleration of auroral electrons in warm magnetospheric plasma, in *Auroral Phenomenology and Magnetospheric Processes: Earth and Other Planets*, Geophys. Monogr. Ser., doi:10.1029/2011GM001171, this volume.

Williams, D. J., B. H. Mauk, R. E. McEntire, E. C. Roelof, T. P. Armstrong, B. Wilken, J. G. Roederer, S. M. Krimigis, T. A. Fritz, and L. J. Lanzerotti (1996), Electron beams and ion composition measured at Io and in its torus, *Science*, *274*, 401–403.

Williams, D. J., R. M. Thorne, and B. Mauk (1999), Energetic electron beams and trapped electrons at Io, *J. Geophys. Res.*, *104*(A7), 14,739–14,753, doi:10.1029/1999JA900115.

Wygant, J. R., et al. (2000), Polar spacecraft based comparisons of intense electric fields and Poynting flux near and within the plasma sheet-tail lobe boundary to UVI images: An energy source for the aurora, *J. Geophys. Res.*, *105*(A8), 18,675–18,692, doi:10.1029/1999JA900500.

Wygant, J. R., et al. (2002), Evidence for kinetic Alfvén waves and parallel electron energization at 4–6 R_E altitudes in the plasma sheet boundary layer, *J. Geophys. Res.*, *107*(A8), 1201, doi:10.1029/2001JA900113.

Zhang, H., et al. (2007), TC-1 observations of flux pileup and dipolarization-associated expansion in the near-Earth magnetotail during substorms, *Geophys. Res. Lett.*, *34*, L03104, doi:10.1029/2006GL028326.

F. Bagenal, Laboratory for Space and Atmospheric Sciences, University of Colorado, Boulder, CO 80309, USA. (bagenal@colorado.edu)

B. Mauk, The Johns Hopkins University Applied Physics Laboratory, Laurel, MD 20723, USA. (Barry.Mauk@jhuapl.edu)

Section II
Auroral Phenomenology

Auroral Morphology: A Historical Account and Major Auroral Features During Auroral Substorms

S.-I. Akasofu

International Arctic Research Center, University of Alaska Fairbanks, Fairbanks, Alaska, USA

In the first part, a brief historical description of the development of auroral morphology leading to the concept of the auroral substorm is described, partly based on the author's humble experience. In the second part, the major features of auroral substorms, such as the onset, poleward expansion, westward traveling surges, the omega bands, torches, and patches, are reviewed, adding many features not described adequately by Akasofu (1964). These features occur in the equatorward half of the oval, while the poleward half responds only passively to the equatorward half. All these features must be telling us fundamental magnetosphere-ionosphere coupling processes associated with substorms. If we can learn physics involved in each of the specific activities, it will become a powerful tool in studying magnetosphere-ionosphere coupling processes, taking advantage of the fact that the aurora is visible (or can be imaged), covering the entire polar region and thus over a large region of the magnetosphere. It is shown that we can now make a good progress by a recent breakthrough observation of substorm onset, which indicates that current reduction of the current sheet at about 8 R_E is an important cause of substorms, not by magnetic reconnection at a distant magnetotail. It is hoped that 10 questions in the summary will be useful in advancing both auroral physics and magnetospheric physics.

1. EARLIEST MORPHOLOGICAL STUDIES OF THE AURORA

The first concrete morphological study of the aurora was made by *Loomis* [1860] who produced a *geographic* map, which shows the belt of high auroral sighting. This was an amazing accomplishment itself. This belt is called the auroral *zone*. Subsequently, *Fritz* [1873] produced a very comprehensive compilation of reports of auroral sightings from all over the world. Auroral observations during the first (1882–1883) and second (1929–1930) Polar Years added much more knowledge on the aurora. In those days, most auroral observations were made visually. Figure 1 shows an all-sky sketch of the aurora by *Gyllenskiold* [1886] at Cape Thorden, Sverbard. *Vestine* [1944] produced a revised map of the auroral zone. The auroral zone was further refined by *Hultqvist* [1959] on the basis of International Geophysical Year (IGY) data. It may be noted that a few books on the aurora were published before 1900, including those by *de Mairan* [1754] and *Angot* [1897].

Stormer [1955], as a pioneer of auroral science, made a number of auroral studies, including the determination of height of the aurora. Perhaps, the first comprehensive study of auroral morphology was made by *Fuller* [1935] of the University of Alaska. He took a large number of photographs (glass plates) and was particularly interested in the occurrence of different types of the auroral displays, such as arcs, glows, drapes, rays, and corona during the course of a night. Stormer and Fuller often corresponded and exchanged information. *Heppner* [1954] succeeded Fuller's study of the morphology.

Auroral Phenomenology and Magnetospheric Processes: Earth and Other Planets
Geophysical Monograph Series 197
10.1029/2011GM001156

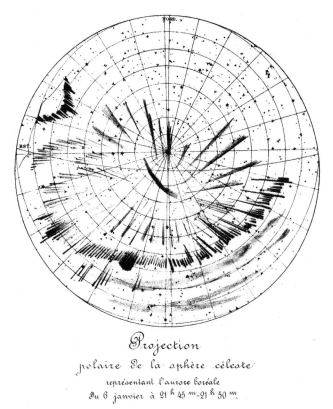

Projection

polaire de la sphère céleste

représentant l'aurore boréale

du 6 janvier à 21ʰ 45ᵐ-21ʰ 50ᵐ.

Figure 1. All-sky sketch of the aurora by *Gyllenskiold* [1886] at Cape Thordsen, Sverbard, during the first Polar Year. He made many series of such a sketch.

2. THE AURORAL OVAL

The modern auroral morphology began when *Feldstein* [1963] determined the geometry of the auroral oval on the basis of the IGY all-sky camera network. He made an extensive, statistical study of the auroral *occurrence* as a function of geomagnetic latitude and magnetic local time (MLT), as well as the *Q* index (a quarter hourly *Kp* index). The belt of the highest occurrence of the aurora appears as an oval shape in the geomagnetic-MLT coordinate. This belt is called the auroral *oval*. (Loomis' auroral *zone* can be considered to be roughly the locus of the midnight part of the oval on a geographic map, as the earth rotates once a day.) The oval expands during active periods and contracts during quiet periods.

Until about the IGY period, Loomis' auroral zone had generally been considered to be the actual belt along which auroral arcs lie. The author, as a graduate student, noticed that in Fairbanks, Alaska, auroral arcs appear first near the northern horizon in the evening sky; they shift gradually southward as the night progresses and recede northward in the morning sky (see Figure 2). At that time, the general opinion on this southward shift was that auroral arcs form first along

the centerline of the auroral zone, geomagnetic latitude 67°, and shift southward after the formation. It so happened at that time that IGY all-sky films arrived from Fort Yukon, Alaska (geomagnetic latitude 67°), located at the centerline of the auroral zone and also from Barrow, located north of the auroral zone. To the author's surprise, the films from both stations showed a similar trend as that in Fairbanks. Thus, the author was greatly puzzled and doubted the validity of the auroral zone as the belt where auroral arcs lie.

Finding the paper by *Feldstein* [1963], the author recognized the solution of what he was puzzled about; roughly speaking, auroral arcs are not shifting, but the Earth and Fairbanks are rotating under the oval once a day as Figure 2 shows. One lesson the author learned at that time was that even a graduate student could doubt a "universally believed or agreed" view. In order to support Felstein's oval, the author conducted various observations, including the operation of the Alaska meridian chain of all-sky cameras (scanning the whole polar sky once a day by using the Earth's rotation: the author's first NSF-funded project) and NASA/U.S. Air Force airborne observations [cf. *Akasofu*, 2007, pp. 43–47]. However, it had been very

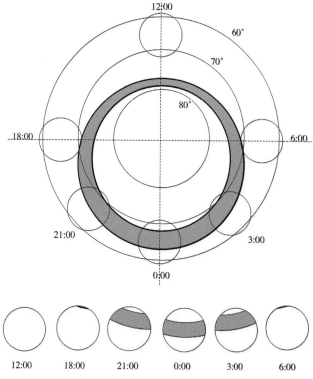

Figure 2. The average auroral oval in the geomagnetic-local magnetic time coordinate. As the Earth and Fairbanks rotate under the oval, auroral arcs appear to shift southward in the evening sky and recede northward in the morning sky. The field of view at Fairbanks is shown by a circle.

difficult for Feldstein and the author to convince auroral researchers, in general, of the validity of the oval, since the auroral zone had been believed to be the belt of the aurora for almost 100 years.

It so happened that two important supports for the oval came unexpectedly from a study by J. A. Van Allen and his group at the University of Iowa, who were determining the intersection line between the outer boundary of the radiation belt and the ionosphere, namely, the projection of the outer boundary of the outer radiation belt on the ionosphere. This line was found to lie close to the auroral oval (see Figure 3). This agreement meant that auroral particles stream into the ionosphere along the outer boundary of the radiation belt not everywhere in the polar region.

Van Allen urged the author to publish the result as a Report of the Department of Physics and Astronomy, University of Iowa. Soon afterward, A. J. Zumuda, of Johns Hopkins University, noticed the author's report and found that the distribution of the field-aligned currents coincides with the shape of the oval. These two physically meaningful pieces of evidence gave important credibility to the shape of the oval. Thus, another lesson the author learned was that auroral observations alone might not be able to convince auroral *specialists* on the aurora, but that some others, which are not directly related to visible auroras, but physically meaningful, could provide unexpected support. Nevertheless, after all, it was the first image of the auroral oval, obtained by C. D. Anger in 1971, using the Canadian satellite ISIS, that convinced everyone of the validity of Feldstein's oval; the long controversy on the oval issues simply faded away.

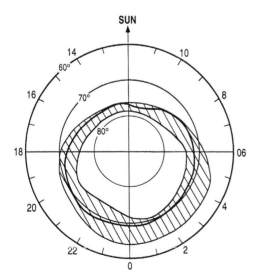

Figure 3. The auroral oval and the intersection line of the outer boundary of the outer radiation belt and the ionosphere.

3. THE BIRTH OF THE CONCEPT OF THE AURORAL SUBSTORM

The study by *Fuller* [1935] became the foundation of the concept of the "fixed pattern." Statistically, the aurora is quiet (mostly in the form of quiet arcs) in the evening sky, active (draperies and corona) in the midnight sky, and patchy (scattered rays) in the morning sky. The Earth and observers located at a fixed point rotate under such a pattern of the aurora, witnessing auroral changes from quiet, active, and patchy forms during the course of a night.

The IGY all-sky camera project, led by C. T. Elvey and S. Chapman at the University of Alaska, allowed us for the first time in history to examine *simultaneously* auroral activity over a large part of the entire polar sky. After much confusion in analyzing the films, it was found that the aurora tends to repeat globally (namely, all along the auroral oval) a particular type of activity every several hours [*Akasofu*, 1964]. The results of this study were described in a draft version with the title "The Auroral Activation."

The concept and model of auroral substorms were formulated by selecting the common features among a great variety of the development of auroral displays. The lesson the author learned in this study was to minimize the number of the common features among a large number of repeating events; in this way, many researchers found those common features in their cases and began to trust the concept.

At the same time, *Akasofu and Chapman* [1962] were analyzing auroral activity during the great geomagnetic storm of 11 February, 1958 (*Dst*, 600 nt). The auroral oval expanded greatly even to the south of the United States-Canada border, and also, the oval expanded and contracted at least four times during the main phase of the storm, just like ocean waves on the seashore. We recognized that such an activity was a very intense case of the auroral "activation." Thus, it was concluded that the auroral *storm* consists of several "activations," which are the *elements* of the auroral storm. For this reason, Chapman proposed to the author that the auroral activation be called "the auroral *sub*storm," saying that "the auroral elementary storm" is too long to pronounce.

However, it was very difficult to convince our colleagues about the concept of auroral substorms in those days because the long-held "fixed pattern concept," which is statistically correct, was firmly held. This was particularly the case, since before the IGY, no one had an opportunity to study simultaneous auroral activities over the entire dark sky; the Earth's rotation does not allow any observer on the earth to stay in the midnight sector for many hours. Thus, the author had to conduct a number of airborne observations to prove the concept, staying in the midnight sector for many hours by flying against the Earth's rotation [see *Akasofu*, 2007].

However, after all, we had to wait for the final proof by the Dynamic Explorer satellite [*Frank et al.*, 1982].

4. MAJOR AURORAL SUBSTORM FEATURES IN EXPLORING MAGNETOSPHERIC PROCESSES

4.1 Initial Brightening and the Subsequent Poleward Expansion

The onset of auroral substorms is signaled by a sudden brightening of an arc. It is difficult to state whether the initially brightening arc (IBA) is a preexisting arc prior to $T = 0$ or the formation of a new arc, but the former is often the case. Typically, this particular arc is located near the equatorward boundary of the auroral oval or at about the boundary between the auroral oval and the diffuse aurora (section 4.3). This boundary is located, in general, between geomagnetic latitudes 60° and 65° during moderately active periods; this rather low-latitude range is an important fact in considering the $T = 0$ process. There are a number of studies near and at substorm onset [see, for example, *Opgenoort et al.*, 1996; *Baumjohan*, 1991; *Henderson*,

2009]. *Akasofu et al.* [2010] found that the equatorward half of the oval tends to contract rapidly just prior to substorm onset, leaving the poleward half behind. This may be an auroral signature of the growth phase (see also section 4.6).

Figure 4 shows the onset and an early epoch of the expansion phase of substorm, an assembly of a typical visual scene (the corona), a series of all-sky photographs to show the time sequence and a typical image taken from the DMSP satellite. Note that they are not simultaneous but are typical examples.

During moderately active periods, bright arcs are present to the north of IBA; many of them are remains of the arcs, which advanced poleward during the expansion phase of previous substorm(s). Those arcs brighten a few minutes after, *not before*, $T = 0$. This feature was described by *Lyons et al.* [2002], who noted, "We also find that arcs poleward of the arc that breaks up appear to be unaffected by substorm onset until expansion-phase auroral activities move poleward to the location of such arcs" [see also *Akasofu*, 1964; *Lui et al.*, 2008; *Donovan et al.*, 2008; *Akasofu et al.*, 2010]. Auroral arcs in the poleward half of the oval respond only

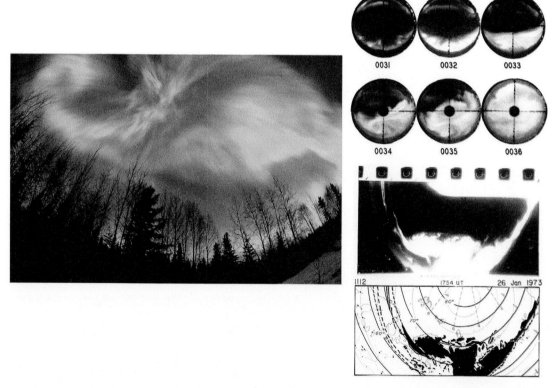

Figure 4. The onset and early epoch of the expansion phase of substorm, an assembly of a typical visual scene (corona), a series of all-sky photographs to show the time sequence, and a typical image taken from the DMSP satellite. Note that they are not simultaneous but are typical examples (photograph courtesy of T. Nakai).

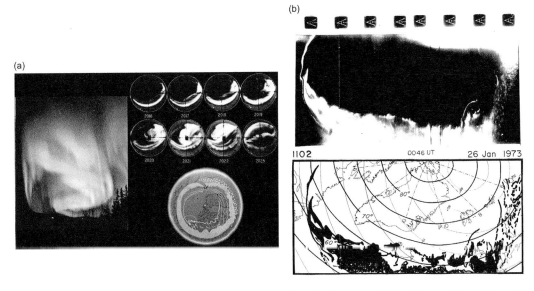

Figure 5. (a) Westward traveling surge (WTS), an assembly of a visual scene, all-sky photographs to show the time sequence, and a sketch of substorm features and WTS (marked by a circle). For an actual DMSP image, see Figure 2b. Note that they are not the simultaneous data but are typical features. (b) The maximum epoch of substorm taken from the DMSP satellite; the geographic location is sketched in the bottom part. The WTS is seen in the evening sector, while patches (section 2.5) are seen in the morning sector.

passively to auroral activities (substorms) in the equatorward half. During an extended quiet period, they fade away, so that they are absent during isolated substorms [*Akasofu*, 1968]. Many satellite studies have confirmed that the initial brightening occurs at about 23 MLT, but all-sky camera data show that the brightening spreads quickly in the east-west direction in a few minutes (Figure 4).

Except for the arc at the front of the expanding bulge, many arc segments behind the front arc drift rapidly equatorward [*Feldstein and Starkov*, 1967; *Snyder and Akasofu*, 1972]. This feature was not particularly mentioned by *Akasofu* [1964]. Thus, physics of the advancing arc (the front of the bulge) and of the equatorward motion of arc segments behind it is crucial in understanding the $T = 0$ processes; the latter may

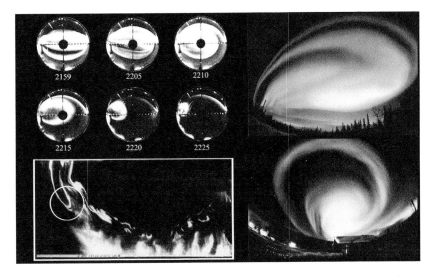

Figure 6. A typical loop structure caused by a WTS, an assembly of visual scenes to show the time sequence, all-sky photographs to show the time sequence, and DMSP photograph (see the circle). Note that they are not simultaneous records but are typical.

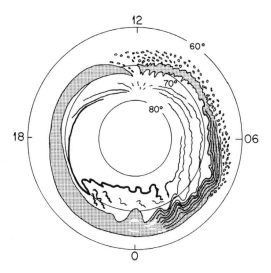

Figure 7. Schematic illustration of auroral features during the maximum epoch of substorm.

be related to the analysis by *Shue et al.* [2008], who found that earthward plasma flows at about 10 R_E are related to substorms, but not those greater than 10 R_E.

4.2. Westward Traveling Surges (WTSs)

The poleward advance of arcs in the midnight sector causes a large-scale wavy structure, which propagates westward (toward the evening sky) along the oval with a speed of about 2 km s^{-1}. Visually, WTSs are a large-scale fold (Figures 5a and 5b).

There are several features that should be noted about WTSs. First, WTSs tend to advance a little poleward of the preexisting arc, "dragging" an arc along with them, forming a beautiful loop-like structure in the sky, which shifts westward (Figure 6). This feature tells us something about the nature of the current sheet (or the electron sheet beam), indicating that the arc-producing structure is imbedded in plasma flow associated with WTSs.

The second feature is that WTSs show a great variety of forms; they have been discussed in terms of the western end of the westward electrojet. The third feature is that they leave behind complicated trailing structures. An example can be seen in the inset of Figure 8. This feature is described in terms of a "streamer" [*Elphinstone et al.*, 1996; *Henderson et al.*, 2002]. Often, they tend to extend equatorward and interact with the diffuse aurora (see also section 4.4).

4.3. The Diffuse Aurora

An important aspect that was not mentioned by *Akasofu* [1964, 1968] is the presence of the diffuse aurora, which surrounds the auroral oval. It is a diffuse glow of width of a few hundred kilometers (Figure 7).

Since its brightness is close to that of the Milky Way and tends to cover a large portion of the sky, it is often difficult to identify and follow it in all-sky camera data or in satellite images (see also section 4.5).

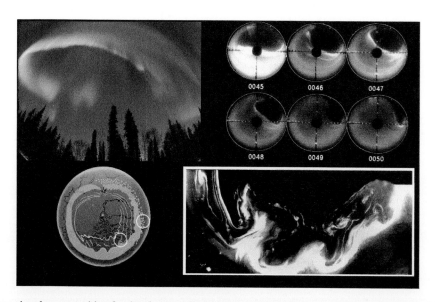

Figure 8. Omega band, an assembly of a visual scene, all-sky photographs to show the time sequence, a sketch, and DMSP image. The two circles indicate an omega band and patches (sections 4.4 and 4.5), respectively. Note that they are not the simultaneous records but are typical.

4.4. Omega Bands and Torches

Contrasting with auroral activities in the evening sector, an arc near the poleward boundary of the oval in the morning sector develops a peculiar fold, which looks like the character "omega," but inverted. This characteristic feature is called the "omega band." The omega band shifts eastward (toward the morning sector) with a speed of a few hundred meters per second (Figure 8). At the same time, the diffuse aurora develops a wavelike structure called a "torch," which tends to shift eastward (but sometimes westward) (Figure 9). In some cases, it is difficult to distinguish between the omega band and a series of torches; the omega band is often located at the poleward boundary of the diffuse aurora. The north-south dimension of torches can be as large as 500 km or even larger, suggesting a very large-scale feature in the internal structure of the magnetosphere. Some of the fea-tures described above have been discussed in terms of interchange instabilities [*Cheng*, 2004].

The torch structure is often activated when the extended streamer from WTSs reaches it during a later stage of substorms. This feature was studied by *Henderson et al.* [2002]. Some of the cases examined by *Nishimura et al.* [2010] may belong to this category.

4.5. Black Auroras and Patches

To the equatorward side of the omega bands, the diffuse aurora begins to split into many thin structures during substorms. The gaps produced by this splitting are initially very narrow and appear as very dark strips in the background of the wide diffuse glow. For this reason, the strips thus produced are called the "black aurora." Eventually, the gaps become wide, and individual arcs are formed (see Figure 7).

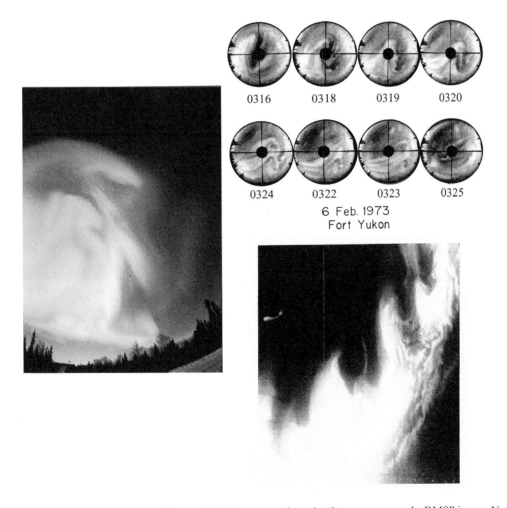

Figure 9. Torches, an assembly of a visual scene, all-sky images to show the time sequence, and a DMSP image. Note that they are not the simultaneous records but are typical.

The arcs thus formed have very complicated folds. When those folds are seen overhead, they appear like a large number of cumulus clouds.

For this reason, they are called "patches"; by careful observation, one can see the ray structure in the patches, indicating that it has a vertical structure (Figure 10). The term "breakup" is often used for substorm onset, but it should be used for this particular display.

Patches drift all the way to the midday sector with a speed of a few hundred meters per second (Figure 7). They pulsate with a period of about 10 s. These features tend to develop after midnight in moderately active periods, but they start to appear in late evening hours during very active periods. The above description of the omega bands, torches, black aurora, and patches requires refinements, and the relationship among them is very complicated (for details, see *Elphistone et al.* [1996]). What is important about the diffuse aurora, torches, and patches is that they tend to occur in rather low latitudes; they are related in the inner part of the magnetosphere.

In ending this section, it is emphasized that auroral activities in the evening sector and the morning sector are entirely different. How this difference is caused by the magnetosphere-ionosphere coupling and manifested in the magnetosphere has not yet received much attention.

4.6. Growth Phase, Plasma/Current Sheet Thinning

It is generally known that the auroral oval expands *as a whole* after the IMF turns southward. As mentioned earlier, the main auroral features during substorms occur mostly in the equatorward half of the oval. After the equatorward expansion of the oval *as a whole*, the equatorward half of the oval tends to contract rapidly before $T = 0$, leaving a dark region between the poleward half and the equatorward half. A typical example is shown in Figure 11a [*Akasofu et al.*, 2010] [see also *Berkey et al.*, 1980]. It is likely that this phenomenon is an auroral manifestation of plasma/current sheet thinning (Figure 11b).

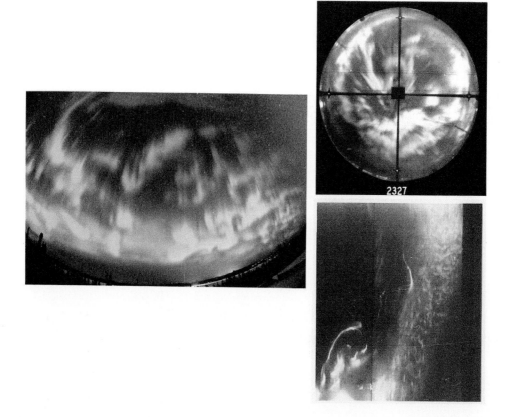

Figure 10. Patches, an assembly of a visual science, all-sky photograph, and DMSP image. Note that they are not the simultaneous records but are typical.

(a)
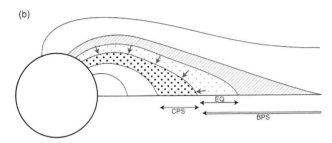

Poker Flat Research Range: 65N, 147W, 19-Sep-2003

(b)

Figure 11. (a) A typical example of meridian scanning photometer data at Poker Flat, Alaska. Three spectral lines, 5577, 4861, and 6300, are combined. The front arc of the expanding bulge at 08:15 UT on 19 September 2003 remained in the poleward sky. After the second activity beginning at 07:10 UT, the equatorward half of the oval contracted, leaving a dark sky. A new substorm activity began at 08:20 UT. (b) A schematic presentation of the contracting equatorward half of that shown in Figure 11a. The plasma sheet consists of two parts, central plasma sheet (CPS) and boundary plasma sheet (BPS). CPS is related to the precipitation in the diffuse aurora, while BPS is related to the auroral oval. A little before $T = 0$, the equatorward half of the oval and earthward half of BPS, dented by equator (EQ), contracts.

5. SUMMARY

It is hoped that the following 10 questions will be useful in initiating a new advancement in auroral physics.

1. Auroral activities across the midnight meridian are very different; what does this indicate?

2. Auroral arcs located poleward of IBA brighten after, not before, $T = 0$; what does this signify?

3. What is the cause of WTSs? They are known to be the western end of the westward auroral electrojet.

4. Behind the poleward expansion, arc segments drift equatorward. What is the cause of the poleward expansion and of the equatorward drift?

5. Auroral activities during substorms occur mainly in the equatorward half of the oval. The poleward half responds to them only passively. What does this mean?

6. Why does the diffuse aurora develop into the black aurora?

7. Why does the streamer activate the torch structure? What is the nature of the streamer?

8. What does the eastward drift of patches tell us? Is it an $E \times B$ drift?

9. What causes the diffuse aurora to develop torches? Is it an interchange instability?

10. The equatorward half of the oval tends to contract a little before $T = 0$. Is it an auroral manifestation of plasma/current sheet thinning?

6. CONCLUDING REMARKS

In establishing a morphological model, several observed facts are chosen and formulated in sequence under a certain criterion among a large number of observed facts. The auroral substorm is such a morphological model; the criterion in this case was the common features among repeated occurrence of auroral activity.

Like theoretical models, morphological models should be challenged with a more reasonable criterion and a better sequencing. Once we can learn physics involved in the major auroral feature during substorms, a physically based criterion may be adopted in establishing a new model in the future.

Recently, it seems that many researchers, theorists, and observers have been bound to prove the only premise that substorm onset is caused as a result of magnetic reconnection at a distance of about 20 R_E in the magnetotail. Because of this trend, the progress of substorm research is somewhat stalled at present and cannot proceed beyond substorm onset issues.

In this respect, it is noted here that *Lui* [2011] reported that the current disruption/dipolarization at about 8 R_E observed by the Time History of Events and Macroscale Interactions during Substorms (THEMIS) satellites, coincided with a substorm onset, which was clearly recorded by the *AU/AL* index and all-sky photographs. This study can be considered to be a breakthrough, indicating that substorms are caused by current reduction, not by magnetic reconnection at a distant tail. It is likely that this feature is reflected in the phenomenon shown in Figures 11a and 11b. We can now advance auroral physics and magnetospheric physics along a new direction. It is hoped that the 10 questions in section 5 will be useful for this purpose.

Acknowledgments. I would like to thank several generations of auroral researchers who have contributed to the development of the study of auroral substorm and magnetospheric substorm.

REFERENCES

Akasofu, S.-I. (1964), The development of auroral substorm, *Planet. Space Sci.*, *12*, 273–282.

Akasofu, S.-I. (1968), *Polar and Magnetospheric Substorms*, 280 pp., D. Reidel, Dordrecht, The Netherlands.

Akasofu, S.-I. (2007), *Exploring the Secrets of the Aurora*, 2nd ed., 288 pp., Springer, New York.

Akasofu, S.-I., and S. Chapman (1962), A large-scale change in the distribution of the auroras during the 11 February 1958 magnetic storm, *J. Atmos. Terr. Phys.*, *24*, 741–742.

Akasofu, S.-I., A. T. Y. Lui, and C.-I. Meng (2010), Importance of auroral features in the search for substorm onset processes, *J. Geophys. Res.*, *115*, A08218, doi:10.1029/2009JA014960.

Angot, A. (1897), *The Aurora Borealis*, 264 pp., D. Appleton, New York.

Baumjohann, W. (1991), Electrodynamics of active auroral forms: Westward traveling surges and omega bands, in *Auroral Physics*, edited by C.-I. Meng, M. Roycroft, and L. A. Frank, pp. 361–367, Cambridge Univ. Press, Cambridge, U. K.

Berkey, F. T., C. D. Anger, S.-I. Akasofu, and E. P. Rieger (1980), The signature of large-scale auroral structure in radio wave absorption, *J. Geophys. Res.*, *85*(A2), 593–606.

Cheng, C. Z. (2004), Physics of substorm growth phase, onset and dipolarization, *Space Sci. Rev.*, *113*, 207–270.

De Mairan M. (1754), *Traite Physique et Historique de L'Aurore Boreale*, 2nd ed., 570 pp., L'Imprimerie Royale, Paris, France.

Donovan, E., et al. (2008), Simultaneous THEMIS in situ and auroral observations of a small substorm, *Geophys. Res. Lett.*, *35*, L17S18, doi:10.1029/2008GL033794.

Elphinstone, R. D., J. S. Murphree, and L. L. Cogger (1996), What is a global auroral substorm?, *Rev. Geophys.*, *34*(2), 169–232.

Feldstein, Y. I. (1963), Some problems concerning the morphology of auroras and magnetic disturbances at high latitudes, *Geomagn. Aeron.*, *3*, 183–192.

Feldstein,Y. I., and G. V. Starkov (1967), Dynamics of auroral belt and polar geomagnetic disturbances, *Planet. Space Sci.*, *15*, 209–229.

Frank, L. A., J. D. Craven, J. L. Burch, and J. D. Winningham (1982), Polar views of the Earth's aurora with Dynamics Explorer, *Geophys. Res. Lett.*, *9*(9), 1001–1004.

Fritz, H. (1873), *Polarlichter*, 255 pp., L. Gerold's Sohn, Vienna.

Fuller, V. R. (1935), A report of work on the aurora borealis for the year 1932–1935, *Terr. Magn. Atmos. Electr.*, *40*, 269–275.

Gyllenskiold, G. (1886), *Observations Failes au Cape Thordsen, Spitzberg*, 409 pp, L'Acad. R. des Sci. des Suede, Stockholm.

Henderson, M. G. (2009), Observational evidence for an inside-out substorm onset scenario, *Ann. Geophys.*, *27*, 2129–2140.

Henderson, M. G., L. Kepko, H. E. Spence, M. Connnors, J. B. Sigwarth, L. A. Frank, H. J. Singer, and Y. Yumoto (2002), The evolution of north-south aligned auroral forms into auroral torch structures: The generation of omega bands and Ps6 pulsations via flow bursts, in *International Conference on Substorms (ICS-6), March 24–29*, edited by R. M. Wiglee, pp. 169–174, Univ. of Washington, Seattle.

Heppner, J. P. (1954), Time sequences and spatial relations auroral activity during magnetic bays at College, Alaska, *J. Geophys. Res.*, *59*(3), 329–338.

Hultqvist, B. (1959), Auroral isochasms, *Nature*, *183*, 1478–1479.

Loomis, E. (1860), On the geographic distribution of auroras in the northern hemisphere, *Am. J. Sci. Arts*, *30*, 89–94.

Lui, A. T. Y. (2011), Reduction of the cross-tail current during near-Earth dipolarization with multisatellite observations, *J. Geophys. Res.*, *116*, A12239, doi:10.1029/2011JA017107.

Lui, A. T. Y., et al. (2008), Determination of the substorm initiation region from a major conjunction interval of THEMIS satellites, *J. Geophys. Res.*, *113*, A00C04, doi:10.1029/2008JA013424. [Printed 115(A1), 2010].

Lyons, L. R., I. O. Voronkov, E. F. Donovan, and E. Zesta (2002), Relation of substorm breakup arc to other growth-phase auroral arcs, *J. Geophys. Res.*, *107*(A11), 1390, doi:10.1029/2002JA009317.

Nishimura, Y., L. Lyons, S. Zou, V. Angelopoulos, and S. Mende (2010), Substorm triggering by new plasma intrusion: THEMIS all-sky imager observations, *J. Geophys. Res.*, *115*, A07222, doi:10.1029/2009JA015166.

Opgenoorth, J., M. A. L. Persson, and A. Olsson (1996), The substorm onset seen with ground-based instrumentation results, problems, future possibilities, in *Proceedings of the Third International Conference on Substorms (ICS), Versailles, 12-17 May*, *ESA Spec Publ., ESA SP-389*, 307.

Shue, J.-H., A. Ieda, A. T. Y. Lui, G. K. Parks, T. Mukai, and S. Ohtani (2008), Two classes of earthward fast flows in the plasma sheet, *J. Geophys. Res.*, *113*, A02205, doi:10.1029/2007JA 012456.

Snyder, A. L., and S.-I. Akasofu (1972), Observations of the auroral oval by the Alaska meridian chain of stations, *J. Geophys. Res.*, *77*(19), 3419–3430.

Stormer, C. (1955), *The Polar Aurora*, 403 pp, Oxford Univ. Press, New York.

Vestine, E. H. (1944), The geographic incidence of aurora and magnetic disturbance, northern hemisphere, *Terr. Magn. Atmos. Electr.*, *49*, 77–102.

S.-I. Akasofu, International Arctic Research Center, University of Alaska Fairbanks, Fairbanks, AK 99775, USA. (sakasofu@iarc.uaf.edu)

Auroral Substorms, Poleward Boundary Activations, Auroral Streamers, Omega Bands, and Onset Precursor Activity

M. G. Henderson

Los Alamos National Laboratory, Los Alamos, New Mexico, USA

We show that poleward boundary intensifications and auroral streamers can evolve into auroral torches and omega bands. We show that quasiperiodic generation of streamers near a given magnetic local time region can produce a quasiperiodic train of auroral torches that successively drift eastward to create an omega band structure. We also show that subsequent streamers can create new torches "on top" of an existing omega band form. These observations explain a number of previously unexplained features of omega bands. In addition, we show that while auroral streamer activity does not typically lead to substorm onsets, streamers can occur prior to auroral onsets, although in such cases, they are usually not colocated with the onset region. Interpreting the auroral streamers as ionospheric manifestations of bursty bulk flows in the tail (due to earthward propagation of low-entropy flux tubes), we surmise that flow braking cannot, by itself, be the cause of the auroral breakup and subsequent expansion.

1. INTRODUCTION

During disturbed magnetospheric conditions, a wide variety of auroral forms and behavior can occur including substorms, poleward boundary intensifications (or PBIs), creation of north-south aligned auroral forms and streamers, and the formation of auroral torches and omega band structures. Although many of these auroral forms have been known about and studied over the past half century, a detailed understanding of how they may (or may not) relate to one another has only emerged over the past few decades or so. The faster progress made in recent years in understanding these auroral forms has come in large part from the ability to image the large-scale auroral distribution from space. Prior to this, ground-based researchers faced the considerable difficulty of trying to follow and understand processes that can operate over distances much greater than the field of view of a single all-sky imager.

Given this difficulty, it is remarkable that *Akasofu* [1964] was able to create a fairly accurate phenomenological picture of how the auroral substorm develops on a global scale that is still relevant today. Indeed, many of the auroral forms that we know about can be seen in his early phenomenological sketches. Notably, he recognized the initial brightening arc at onset, the subsequent poleward, westward, and eastward expansion of the bulge, the development of the westward-traveling surge, the occurrence of bright arcs at the poleward edge of the bulge, and the generation of patches, pulsating auroras, and omega band forms. In addition, although it is not widely recognized, *Akasofu* [1964] and *Akasofu* [1976a] also show the occurrence of auroral forms within the substorm bulge that we now refer to as north-south aligned auroral structures or streamers. This can be seen in Figure 13 of *Akasofu* [1976a]. Note the occurrence of north-south auroral forms in the relatively dim region between the intensified discrete poleward arc system and the brighter structured diffuse aurora near the equatorward edge of the oval in the midnight sector.

Auroral Phenomenology and Magnetospheric Processes: Earth and Other Planets
Geophysical Monograph Series 197
10.1029/2011GM001165

The wavy structured diffuse aurora in Figure 13 of *Akasofu* [1976a] that extends from premidnight to dawn represents the omega band forms that frequently develop during active times. When the wavy poleward protrusions extend more dramatically toward the poleward arc systems, the forms are referred to as "auroral torches" [e.g., *Akasofu*, 1976b; *Oguti et al.*, 1981; *Tagirov*, 1993].

1.1. The Poleward Boundar(ies) and PBIs

It is important to note that when the poleward arc system shown in Figure 13 of *Akasofu* [1976a] represents the poleward edge of the substorm bulge in the early to late phase of the expansion phase, there are still arcs (and hence closed field lines) poleward of it. Thus, during the expansion phase, one must distinguish between the poleward boundary of the substorm bulge and the poleward boundary of the oval itself. Note, however, that late in the life cycle of a substorm, these boundaries will eventually coalesce, and there will be no distinction between them. While Akasofu's earliest schematics of substorm development [e.g., see *Akasofu*, 1965], Figures 1D (p. 15) and 95 (p. 126) clearly show the two boundaries, his later (and most cited) versions do not show the most poleward boundary. The reason for this change is not clear, but its omission may have contributed greatly to the once popular (but incorrect) notion that onsets were situated in the plasma sheet boundary layer.

Since the two boundaries eventually coalesce (and this may or may not happen before the end of the expansion phase), activations and intensifications of the poleward boundary of the bulge are also often the same phenomenon as intensifications of the most poleward boundary. Indeed, when the most poleward boundary is particularly bright and active, one can often trace its origin to a previous substorm expansion. This is one way in which the so-called "double oval" auroral configuration can arise. In terms of a physical model, the coalescence of the boundaries could be thought of as the substorm X line finally taking its place as the new distant X line in the magnetotail (although it may still be situated relatively close to the Earth). In the context of such a model, intensifications at either boundary are likely to involve the same physical mechanisms.

1.2. North-South Aligned Auroral Forms

For the most part, the early pictures developed by Akasofu placed many of the auroral forms that develop during substorms into a correct spatial and temporal context, and this was finally confirmed by the first space-based imagers (ISIS [*Anger et al.*, 1973], Kyokko (Exos A) [*Hirao*, 1978], and DE [*Frank et al.*, 1981]). Nevertheless, details on their dynamics, how they are created, and on how they may be related to one another were still poorly understood. For example, as pointed out by *Belyakova et al.* [1968], even the general equatorward motion of auroral forms within the expanding bulge was not widely appreciated early on. Modifications were also made to Akasofu's original phenomenological substorm model by *Montbriand* [1971] and *Fukunishi* [1973]. In particular, *Montbriand* [1971] added north-south aligned arc segments as an integral part of the expanding bulge. More detailed analyses of the auroral dynamics within the substorm bulge were provided by *Sergeev and Yahnin* [1979] who showed that expansion of the bulge takes place in a stepwise manner in conjunction with the formation of new arcs at the poleward edge of the bulge and the appearance of some north-south aligned structuring of the aurora within the bulge. Later studies by *Kornilova et al.* [1990] and *Sergeev* [1992] focused more on the stepwise episodes of activity that lead to the production of equatorward moving forms, and they interpreted the auroral behavior to be a consequence of periodic bursts of reconnection in the tail.

The first major breakthrough in understanding auroral dynamics in more detail on a truly global scale came with the launch of the Viking spacecraft [*Hultquist*, 1987]. Unlike prior global imagers like ISIS II and DE 1, the Viking imagers [*Anger et al.*, 1987] were capable of taking short-exposure (1 s) snapshot-style images of the entire auroral distribution as fast as one image every 20 s (although 1 min between images was typical). From this rich data set, a detailed picture of how north-south aligned auroral forms and auroral streamers are created following poleward boundary intensifications was revealed [e.g., *Rostoker et al.*, 1987; *Henderson*, 1994; *Henderson et al.*, 1994, 1998].

Figure 1 (from *Henderson* [1994] and *Henderson et al.* [1994]) schematically illustrates how north-south (NS) aligned auroral forms can be ejected from the active poleward edge of the substorm bulge. In this example, the forms are shown emerging from the region near the surge head and extend equatorward to produce torches and structured diffuse aurora. The general motion of the forms is also illustrated with arrows. Often, an eastward propagating intensification of the poleward edge precedes the formation of the streamer structures, and once created, many of the NS forms are observed to drift eastward. However, auroral forms closer to the surge head tend to rotate clockwise as viewed from above (around the Harang discontinuity region) and can drift westward. Although *Akasofu* [1976b] claims that auroral torch structures drift westward, the Viking results show that the vast majority of such forms drift eastward (also see *Oguti et al.* [1981] and *Tagirov* [1993]).

Figure 2 (also from *Henderson* [1994] and *Henderson et al.* [1994]) shows a second example illustrating the

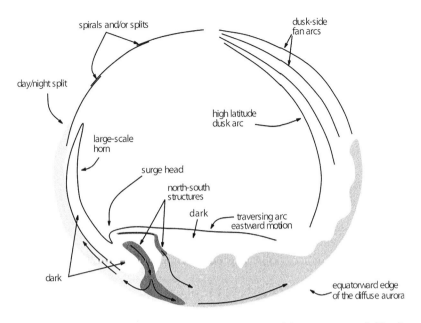

Figure 1. A schematic illustration showing how north-south aligned auroral forms create torch-like forms and structured diffuse aurora [from *Henderson*, 1994, Figure 9.2].

creation of north-south aligned auroral forms and structured diffuse aurora. This case emphasizes details of how the poleward boundary intensification (at the poleward edge of the substorm bulge) occurs and how the streamers are subsequently ejected equatorward. In Figure 2a, the poleward boundary of the bulge (labeled "traversing arc") intensifies (often with an eastward progression) and begins to split or bifurcate. In Figure 2b, the poleward arc system is completely bifurcated with the poleward arc farther poleward than before the PBI occurred. The lower arc system often remains in a more east-west alignment closer to the surge head but degenerates very rapidly into equatorward moving streamers (i.e., NS forms) farther to the east (this behavior is also illustrated for the easternmost streamer shown in Figure 1). As the forms move equatorward in the bulge, they tend to drift as shown in Figure 2c. Forms in the western side of the bulge tend to rotate clockwise (as seen from above) around the Harang discontinuity, while forms farther east tend to drift eastward. Sometimes north-south aligned forms can split as shown in Figure 2c with one part drifting westward and the other drifting eastward [e.g., see *Rostoker et al.*, 1987]. The remnants of the poleward boundary intensification and ejection of auroral streamers are torch-like forms and structured diffuse aurora as shown in Figure 2d.

As illustrated in schematic form in Figure 2b, "indentations" of the poleward arc system are frequently observed to occur above the region where the streamers are ejected equatorward. The bifurcation of the poleward arc, the development of indentations, and the rapid ejection of streamers equatorward were interpreted by *Henderson* [1994] and *Henderson et al.* [1998] to be the ionospheric manifestation of earthward directed bursty bulk flows (BBFs) that had been recently discovered by *Angelopoulos et al.* [1992a] [see also *Angelopoulos et al.*, 1992b, 1994] and are likely created by reconnection at a reconnection site in the tail. The schematic in Figure 2 is closely based on a substorm recorded by Viking near 12:00 UT on 15 October 1986 (an event also studied by *Rostoker et al.* [1987]).

Shortly after the Viking results began to emerge on the auroral dynamics associated with the formation of north-south aligned forms, *Nakamura et al.* [1993] presented similar observations imaged from the ground with an array of eight all-sky TV imagers. Their results on the general dynamics of the poleward boundary and the ejection of north-south aligned auroral forms equatorward into the bulge is shown schematically in Figure 3 and are fully consistent with the Viking results. As with the *Henderson* [1994] and *Henderson et al.* [1994] results, they also interpreted these features to be a result of earthward directed BBFs in the tail. An important aspect of the *Nakamura et al.* [1993] study is that the use of TV cameras allowed them to image the auroral dynamics at 30 frames per second. With such a high temporal resolution, they were able to show that the north-south aligned auroral forms degenerate into diffuse auroral forms, which preceded the occurrence of pulsating auroral patches in the equatorward

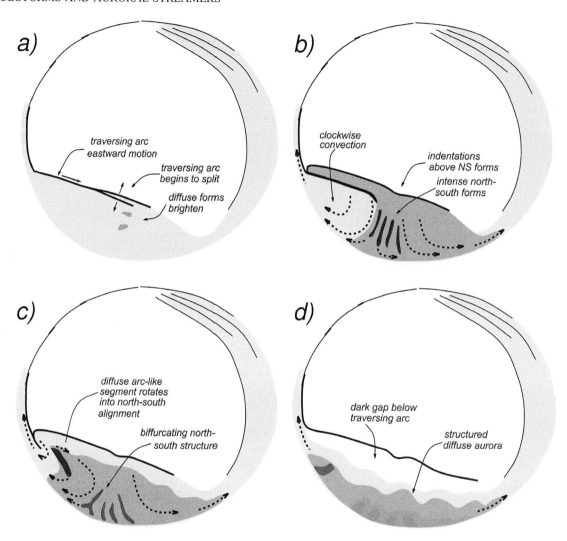

Figure 2. A schematic illustration showing how north-south aligned auroral forms can be generated during the poleward expansion of a substorm [from *Henderson*, 1994, Figure 6.25].

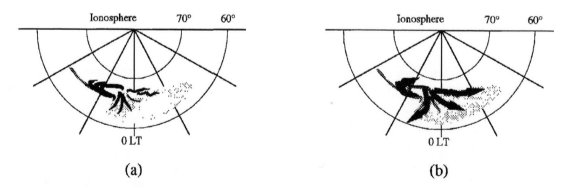

Figure 3. A schematic illustration showing morphology and motion of north-south aligned auroral forms within the substorm bulge [from *Nakamura et al.*, 1993].

regions of the oval and thus concluded that the arrival of north-south auroral forms in the equatorward regions of the bulge were intimately related to the occurrence of pulsating auroral patches.

A number of other studies focusing on north-south aligned auroral forms followed these initial studies. Building on the initial interpretations of *Nakamura et al.* [1993], *Henderson* [1994], and *Henderson et al.* [1994, 1998] that the north-south aligned auroral forms are the ionospheric signatures of BBFs in the tail (and of the interpretations of *Kornilova et al.* [1990] and *Sergeev* [1992] that such activity is related to tail reconnection), several studies were made in an attempt to make a more concrete connection between them [e.g., *Sergeev et al.*, 1999, 2000; *Lyons et al.*, 1999; *Zesta et al.*, 2000, 2006; *Nakamura et al.*, 2001, 2005; *Sergeev*, 2004]. Perhaps the most convincing of these was the *Nakamura et al.* [2001] study in which a strong statistical association was made between the NS forms and the duskward edge of flow bursts in the tail as expected if BBFs are due to depleted flux tube or bubbles [e.g., see *Chen and Wolf*, 1999; *Sergeev et al.*, 1996].

Other studies have focused on further characterization of the north-south aligned auroral forms in the ionosphere and on their connection to other observed quantities [e.g., *Zesta et al.*, 2002; *Sergeev et al.*, 2004; *Partamies et al.*, 2006; *Henderson et al.*, 2002]. Mechanisms other than localized tail reconnection have been proposed as a source of the auroral streamers in the ionosphere. For example, *Liu and Rostoker* [1993] proposed that they could be formed as the results of the entry of depleted flux tube into the plasma sheet from the low-latitude boundary layers due to processes relating to the growth of the Kelvin Helmholtz instability there. This is consistent with the *Chen and Wolf* [1999] depleted flux tube concept for driving BBFs, which can arise from any mechanism that leads to an overall reduction of entropy in a flux tube. Examination of north-south aligned auroral forms have also been used as indicators of magnetic reconnection in order to characterize auroral substorm evolution within the context of a natural or physical mapping scheme [e.g., *Henderson*, 1994; *Sandholt et al.*, 2002; *Henderson*, 2009].

1.3. Association Between Streamers and Omega Bands

Henderson et al. [2002] showed that the equatorward moving streamers that are ejected episodically from the poleward boundary of the oval (or the poleward boundary of the substorm bulge) can evolve directly into torch structures, which contribute to well-defined omega band forms. They concluded that, as a consequence, omega bands can be produced as a direct result of earthward directed BBFs.

These results resolve a number of previously puzzling features of omega bands including their quasiperiodic nature, why they pulsate, and why they occur where and when they do. In addition, *Henderson et al.* [2002, p. 173] proposed a new paradigm in which

rather than producing the poleward/tailward movement of the envelope of activity in the ionosphere/magnetosphere, flow bursts and flow braking instead mediate the coupling between the reconnection site in the tail and the near-Earth magnetosphere that is they couple the poleward component of a double oval configuration with its equatorward component.

They note that

new 'embedded onsets' can be preceded by or perhaps even triggered by flow bursts, but the subsequent expansion of the large-scale envelope of activity is not likely to be a sole consequence of flow braking (or else each and every torch structure that forms from a PBI/NS event would result in an auroral onset and they don't).

1.4. Association Between Streamers and Substorm Onsets

Over the past 40 years or so, a number of studies have shown that auroral breakup can be preceded by the arrival of equatorward moving auroral forms from higher latitudes. Based on detailed analysis of all-sky imager data, *Oguti* [1973] defined the concept of a "contact breakup," which starts when an inclined (or "slanted") arc splits away from a poleward arc system and arrives in the equatorward part of the auroral distribution where the proton aurora resides (such arcs that split away from the poleward boundary were also studied extensively by *Henderson* [1994]) (e.g., see Figure 2a). *Fukunishi* [1973] also showed examples of such contact breakups, but *Oguti* [1973] emphasize that not all breakups are likely to be of this type and that they may represent a distinct subclass of auroral breakups. Much later, *Persson et al.* [1994] presented an event in which auroral arcs propagated equatorward into the onset region prior to onset and continued their equatorward drift during the subsequent poleward expansion of the bulge.

Henderson et al. [2002] presented an event in which a full-blown auroral onset emerged from a torch-like structure that was formed a short while before as the result of the arrival of an auroral streamer in the equatorward region of the auroral distribution. While observations of this sort point to the possibility that onsets can be triggered by prior streamer (i.e., tail flow burst) activity, it is critically important to recognize that the vast majority of streamers definitely do not lead to full-blown auroral substorm breakups. Rather, as discussed above, streamers tend to evolve into torches, structured diffuse pulsating forms, and/or omega bands rather than into substorm onsets. As *Henderson et al.* [2002] point out, this fact, together with the fact that the onset they analyzed emerged from the rapidly growing torch-like undulation with a time delay, strongly argues against the standard flow-braking mechanism for substorm onset and expansion. Instead, *Henderson et al.* [2002] suggest that such activity may be more consistent with the growth of a near-Earth instability like ballooning that becomes unstable due to the arrival of the flow bursts at longitudes that are more susceptible to the instability (e.g., between the western edge of the omega bands and the Harang discontinuity region).

Recently, additional observations of auroral precursor activity have been presented. *Kepko et al.* [2009] presented multispectral all-sky imager observations of an onset event that was preceded by an equatorward-moving patch that emerged from a higher-latitude arc system. As in the *Henderson et al.* [2002] study, the equatorward moving form was attributed to an earthward moving flow burst in the tail, and they conclude that the auroral breakup was likely due to the arrival of the flow burst.

Most recently, based on Time History of Events and Macroscale Interactions during Substorms (THEMIS) observations, *Nishimura et al.* [2010] claim that most auroral breakups are preceded by the arrival of streamers in the equatorward region of the auroral distribution. However, they also suggest that the location of the streamer need not be colocated with the eventual onset location, but rather the streamer can subsequently propagate in the east-west direction to trigger the onset. While these results are interesting, it is still an open question as to whether the events they analyzed are real substorms or not or whether the association between streamers and onsets is truly causal. A cursory analysis of some of their events reveals that they may, in fact, not be real auroral onsets, but rather, they may be the type of auroral activity that develops when PBI and streamers evolve into torches and omega band–type forms. For example, Figure 2 from *Nishimura et al.* [2010] appears to show exactly this type of behavior, which is also very similar to examples shown later in Figures 4 and 5 (e.g., images between 610 and 630 in Figure 4 and between 1151 and 1211 in Figure 5.)

In this paper, examine the evolution of the auroral distribution at high resolution over a 5 h period during a sawtooth-like event in order to demonstrate how many of the auroral features discussed above evolve and relate to one another.

2. OBSERVATIONS

In Figures 4 and 5, we present a sequence of auroral images acquired with the Polar/Visible Imaging System (VIS) low-resolution (LR) camera during an event, which occurred on 11 December 1998. The images show the nightside portion of the northern auroral distribution in 557.7 nm

green line emissions and span a time interval of almost 10.5 h. The time between images is not uniform because we have omitted images taken at other wavelengths. This interval was a sawtooth-like event and contains two embedded auroral onsets. During the intertooth periods, numerous PBIs and streamers were observed, and these provide an excellent opportunity to examine how or if such features are related to the onsets.

At the beginning of the sequence in Figure 4a (04:44:12 UT), the auroral distribution displays a double oval–type configuration and two prominent auroral streamers can be seen extending equatorward from the poleward component of the double oval (blue annotation). By 04:56:25 UT, these structures have evolved into tongues or torches that connect into the lower-latitude portion of the double oval. Similarly, the streamer highlighted in the 04:58:43 frame has evolved into a torch structure by 05:05:01 UT (green annotation), and yet, another such event can be seen in the frames between 05:33:08 and 05:41:45 UT. Starting at about 06:00 UT, the production of streamers intensifies dramatically, and numerous torches and tongues are generated as a result. By 07:00 UT, this activity has largely subsided, and the auroral distribution looks much like it did before 06:00 UT. Note that throughout this intensification, no embedded auroral onset was observed to initiate on the lower component of the double oval configuration.

Figure 4b shows the continuation of Figure 4a and contains the first embedded onset of the sequence, which occurred between 08:11:27 and 08:15:03 UT. Prior to the onset, a number of arcs can be seen moving equatorward to the west of the onset location, and a streamer/torch event can be seen to the east of the onset region. Immediately above the onset location, the auroral emissions are relatively quiescent, and only very faint equatorward moving patches of luminosity can be seen. It is interesting to note that although no prominent streamer activity can be seen feeding directly into the onset region, the initial brightening on the equatorward edge of the oval begins as an intensification on the western edge of the tongue/torch structure that was generated by the streamer activity prior to onset. Note that this streamer activity began at around 07:49:14 UT and was associated with equatorward moving, roughly east-west aligned (or slightly inclined) arcs to the west of the streamer/torch structures.

Following the onset, the substorm-associated auroral emissions expand poleward, westward, and eastward but remain embedded in the closed field line region until the poleward edge of the expanding bulge reaches the poleward edge of the preexisting double oval configuration. Two important features to note are (1) new streamer activity can be seen emerging from the poleward edge of the expanding bulge,

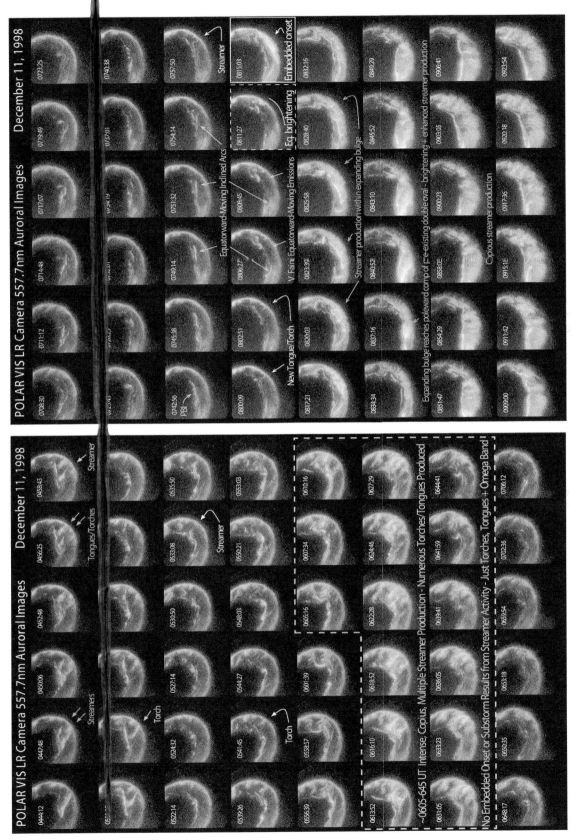

Figure 4. Evolution of the northern auroral distribution on 11 December 1998. Additional examples illustrating how streamers can evolve into torches and omega bands are shown. Despite the intense activity, no auroral onsets occurred during this time period. High-resolution magnetometer data are from the Canadian Carisma array.

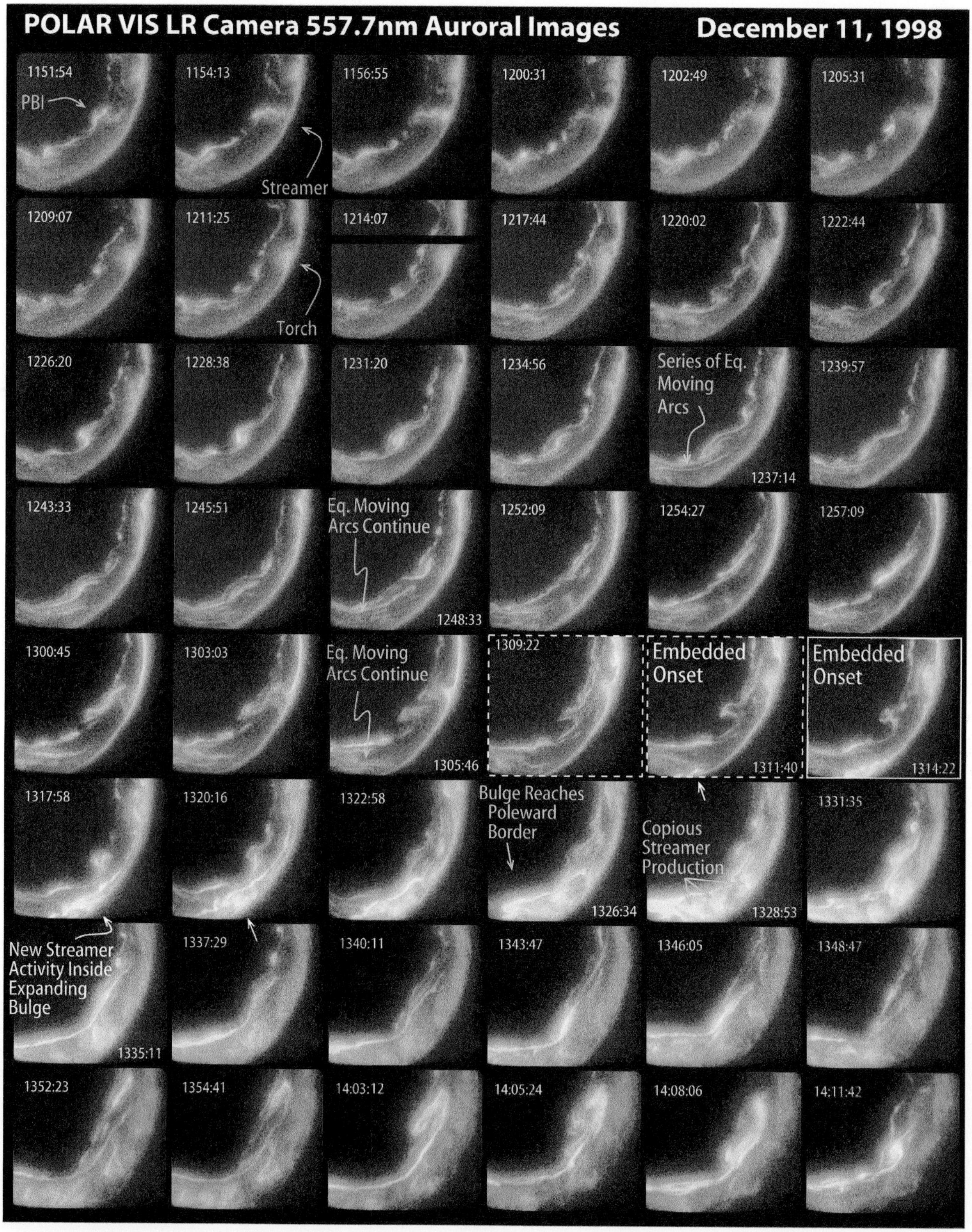

Figure 5. Continuation of the 11 December 1998 sequence. Streamer activity continues, and double oval has thinned dramatically. Another embedded onset occurs between 13:11:40 and 13:14:22 UT. In this case, the onset activity was preceded by a series of equatorward moving arcs. Note that new streamer activity occurs at the still-embedded poleward border of the bulge.

even as it is still embedded in the closed field line region, and (2) a dramatic intensification of streamer production occurs when the bulge finally reaches the poleward limit of preexisting auroral emissions.

Figure 5 shows the continuation of the event starting approximately 2 h following the images in Figure 4. In this final set of images, we can see that the recovery phase of the 08:11 UT substorm and the evolution of the auroral distribution back into a double oval configuration also represent the growth phase for the next embedded onset which, in the field of view (FOV) of the LR camera, occurred between the frames taken at 13:11:40 and 13:14:22 UT. The EC images (not shown) reveal that there was also an onset-like activation that occurred earlier than this much farther to the east. The onset visible in Figure 5 evolved in a manner similar to the previous one in that streamer production occurred within the still-embedded poleward-expanding bulge, and this activity dramatically intensified once the poleward edge of the bulge reached the poleward edge of the preexisting double oval. An interesting feature of this onset is that

inclined equatorward moving arcs can be seen above the onset region prior to onset. This is similar to what was observed prior to the 08:15 UT onset, but in this case, the arcs occur more prominently over the onset region. This type of behavior is also seen in other events [e.g., *Henderson et al.*, 2002], and it is consistent with previous reports of southward drifting arcs observed poleward of an expanding substorm bulge [*Persson et al.*, 1994].

In Figure 6, we present energetic proton and electron differential fluxes at geosynchronous orbit as measured by the Los Alamos National Laboratory (LANL) SOPA particle detectors together with the magnetic field inclination angles at GOES-8 and GOES-10 between 05:00 and 17:00 UT on 11 December 1998. For the electron channels shown in Figure 6a, the energy passbands are (from red to blue) 50–75, 75–105, 105–150, 150–225, 225–315, and 315–500 keV. For the proton channels shown in Figure 6b, the energy passbands are (from orange to cyan) 75–113, 113–170, 170–250, and 250–400 keV. From the proton data shown in Figure 6a, it can be seen that four sawtooth-like injections

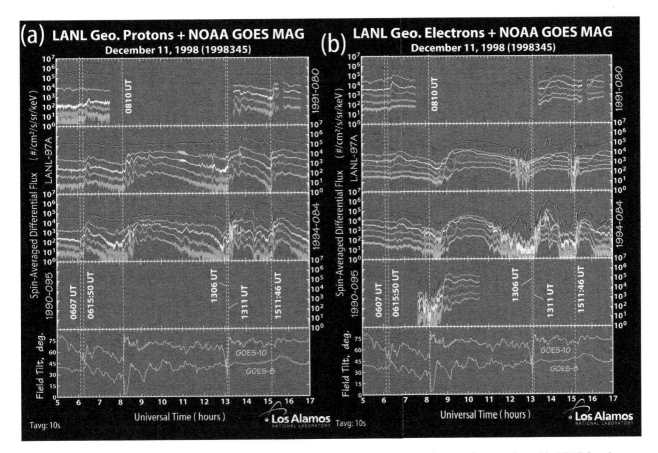

Figure 6. Los Alamos National Laboratory geosynchronous energetic electron and proton data together with GOES 8 and 10 magnetic field inclination angles. Four sawtooth-like injections occurred over the 12 h interval shown.

occurred in the 12 h interval shown. The first three of these correspond to the three activations seen in the Polar/VIS images. Note that the time delay between the second and third injection is somewhat larger than is typical for classical sawtooth events, but aside from this, the event displays all of the other sawtooth characteristics.

The first injection, at 06:07 UT, was clearly sawtooth-like in the proton fluxes. Although the peak was relatively well-dispersed, the start of the flux increases are much less dispersed. Nevertheless, the proton dispersion progressively increases farther to the west and the electron flux increases show energy dispersion that increases toward the east. This is consistent with an injection on the nightside. But what is very interesting here is that this event did not correspond to a substorm onset, but rather a large activation of the poleward boundary combined with copious equatorward ejection of auroral forms (EW inclined, arcs, and streamers). These observations reveal that (1) intense PBI/streamer activity does not necessarily result in an onset and (2) that even a sawtooth-like injection can result despite the lack of an auroral expansion. However, the field inclination angle at GOES-8 and GOES-10 show only a relatively small increase in association with this injection, and this is somewhat atypical for sawtooth injections.

The second sawtooth injection began at roughly 08:10 UT (as determined from 1990-095 electrons), although there were some precursor flux increases seen in the protons as early as ≈07:45 UT. The GOES 10 and GOES 8 field inclination angles responded much more dramatically in association with this injection. At the injection time (08:10 UT), GOES 10 was situated in the premidnight sector and saw a large dipolarization. At this same time, GOES 8, which was situated in the postmidnight sector, saw a sudden stretching (to very small inclination angles) followed several minutes later by a large dipolarization.

The third sawtooth injection was also associated with some precursor activity, and as is typical for sawtooth events, it is somewhat difficult to assign a single time to the flux increases. In Figure 6, lines have been drawn at 13:06 and 13:11 UT, which correspond to the approximate onset times determined from the VIS auroral images. As we can see these times also match the flux increases reasonably well. In addition, at 13:06 UT, the inclination angle at GOES 10 (situated in predawn sector) displayed a large dipolarization. GOES 8 did not see a dipolarization in association with this injection, but it was situated on the dayside near 09:00 MLT during this time period. Thus, the large dipolarization did not extend to the midmorning sector, which is consistent with previous results that show that the strong dipolarization signatures are often approximately confined to the nightside [e.g., *Henderson et al.*, 2006a, 2006b].

3. DISCUSSION

3.1. Torch and Omega Band Formation via Flow Bursts

Analysis of the 11 December 1998 event shows that auroral streamers can evolve directly into auroral torch structures that are connected to the equatorward regions of the oval. The creation of torch structures in this manner leads directly to new "tongues" being added to the eastward drifting omega band forms, i.e., omega bands can be formed as a direct consequence of auroral streamer activity. This behavior is quite common and can be seen in many other events that we have examined. In addition, these findings are completely consistent with the results of a previous study we published several years ago [*Henderson et al.*, 2002].

As discussed by *Henderson et al.* [2002], both well-defined and complex omega band structures can be formed in this way. If streamers are ejected equatorward quasiperiodically from a relatively localized active site on the poleward boundary, the eastward drift within the auroral oval translates into a longitudinally arrayed sequence of torch forms that together form an eastward drifting omega band structure. On the other hand, if the streamers are ejected from different sites along the poleward boundary, or if more than one streamer is ejected at once, very complex omega band structures can be formed. A schematic illustration of this type of activity is shown in Figure 7b. On the top row, a single auroral streamer is shown being ejected equatorward from the poleward boundary, which evolves into a torch structure as it arrives in the equatorward part of the oval. A succession of such occurrences would add tongues to the eastward drifting omega bands like objects dropping onto a conveyor belt. Dynamics of this sort can be seen between 04:45 and 06:00 UT in Figure 4.

A more complex scenario is shown in the lower row of Figure 7b, where a large-scale activation of the poleward boundary is illustrated. This case is very similar to the sequence of images shown between approximately 06:00 and 06:45 UT in Figure 4b. As shown, such activity can result in the formation of very complex omega band structures. It is also interesting to note that the sudden "onset" of such activity could be (and probably has been) misinterpreted as a classical substorm onset if only low-resolution (or low sensitivity) auroral imagery is available.

In a similar manner, streamers can be launched equatorward from the poleward edge of the poleward expanding auroral substorm bulge, even as it is still fully embedded within the closed field line region. Intense and longitudinally arrayed streamer activity can result when the bulge finally reaches the poleward edge of the oval. Such a scenario is illustrated in Figure 7a. Since all of these scenarios can be

(a) Lifecycle of an Embedded Substorm

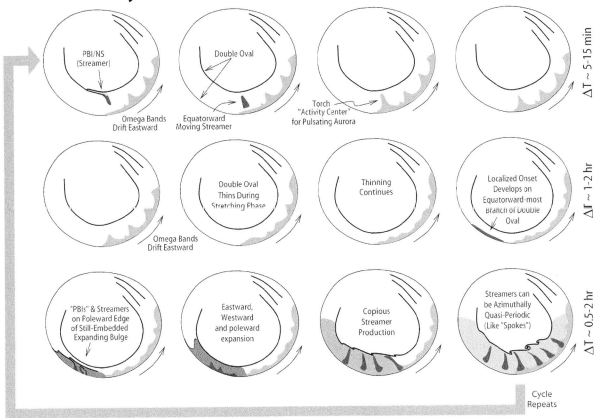

(b) Activation of Poleward Boundary

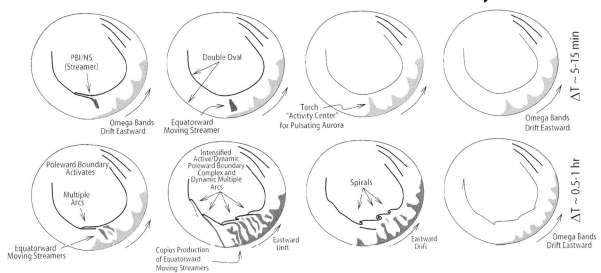

Figure 7. Schematic illustration of how a streamer can evolve into an auroral torch structure. Each such event adds an Ω to the omega band. Both types of disturbances can lead to injection signatures at geosynchronous orbit.

going on at the same time, the torches and omega band structures created can become very complex indeed.

3.2. Relationship of Auroral Streamers to Substorm Onsets

Analysis of this event (and others) illustrates very clearly that auroral streamers generally do not lead directly to substorm onsets in the equatorward portion of the oval. As we have seen, however, they do evolve rapidly into torch/omega band structures that can become quite bright. In addition, the resulting torches can subsequently grow and brighten further (an example of which can be seen in the sequence of images taken between 11:51 and 12:37 UT in Figure 5 (see blue annotation PBI, Streamer, Torch, and subsequent evolution of the equatorward remnants of this activity). This sort of activity could easily be misinterpreted as a streamer-induced onset, but in fact, it was not an onset at all. In general, the vast majority of streamers do not produce, nor are they associated with, auroral substorm onsets. This is not surprising given that (1) substorms typically only occur on a 2–3 h frequency, while streamers recur on a 5–15 min frequency, and there can be multiple streamers launched at once in different regions of the oval, (2) streamers are far more common than substorm onsets, and (3) streamers are also produced inside expanding substorm bulges, which means they could not also produce new onsets at the same time.

Although it is easy to demonstrate that streamers do not generally lead to substorm onsets, this fact does not imply the converse (i.e., that substorm onsets are never caused by or associated with streamer activity). Whether or not streamers *can* be associated with the formation of true onsets is a separate question that we will address shortly. Before we do so, however, it is crucially important to recognize that the fact that streamers are not generally associated with new onsets, by itself, has important implications for current substorm onset models.

In particular, the near-Earth neutral line model of substorm onset initiation associates the brightening and subsequent poleward expansion of the substorm bulge directly to the tailward propagation of magnetic flux pileup via the braking and diversion of fast flows in the near-Earth region caused by a newly created substorm X line. If the same types of flow bursts are also responsible for the auroral streamers, then such a model is not likely to be correct because the vast majority of such events do not result in the hypothesized effect (i.e., streamers do not generally result in onsets). One would have to modify the flow braking model in such a way that flow bursts only *sometimes* lead to onset initiation. On the other hand, there is also the possibility that the flows required in the flow-braking model could be of a type that do not typically lead to auroral streamers.

From Figures 4 and 5, we can see that onsets can indeed sometimes be seen to occur with streamer activity poleward of them, and such observations have been reported before [e.g., *Henderson et al.*, 2002]. During the 08:11 UT onset, the region above the onset location was fairly dim relative to other locations in the oval (this is a fairly common behavior), and there were only relatively faint equatorward moving structures above the onset region. Interestingly, however, a more intense streamer was located just to the west of this region prior to onset. A possible explanation for this sort of behavior is that a much larger-scale (in the cross-tail sense) flow was associated with the dim region and that only the western edge of the flow was visible in the ionosphere as a bright streamer (because streamers are produced as a result of upward field aligned currents at the duskside edge of flows).

The embedded onset at around 13:11 UTC in Figure 5 shows that equatorward moving inclined arcs can appear to lead directly into the onset region. (This behavior is very similar to the onset shown in Figure 2 of *Henderson et al.* [2002].) Although it would seem that such behavior is consistent with the flow-braking scenario of substorm onset, we note the following important caveats: (1) when the onset does begin, it expands in all directions in a manner that is much more explosive than the initial equatorward motion of the inclined arcs, (2) equatorward motion of these arcs was ongoing for quite some time prior to onset (this is similar to the results of *Persson et al.* [1994] who also found southward drifting arcs during the growth phase and into the early expansion phase poleward of the expanding bulge), and (3) a similar occurrence of equatorward moving arcs occurred in the same location between about 11:22 and 12:00 UTC and no onset resulted from that activity. These observations combined with the fact that the vast majority of streamers do not lead to onsets argues strongly in favor of a scenario in which flow bursts (of the sort associated with streamers) can at best cause a destabilization of the inner magnetosphere, which then leads to explosive onset and expansion phase activity. Thus, while it is possible that flow bursts occurring near the time of onset (e.g., after a growth phase period) could be implicated as a possible cause for onset, the ubiquitous nature of streamers means that many such associations could also be purely coincidental.

Another interesting aspect of onsets like those shown in Figure 5 of the present paper and Figure 2 of *Henderson et al.* [2002], is that the inclined auroral forms above the onset region are of the more east-west aligned variety that tend to occur in the western regions of the nightside auroral distribution [*Henderson*, 1994; *Henderson et al.*, 1994]. It is possible that such forms could be associated with different types of flows (e.g., slower, not as bursty, not as structured, or larger scale) or could even be related to flank entry mechanisms, which could bring very different plasma populations into the inner magnetospheric region.

3.3. The Importance of Flow Scale Size

As discussed by *Sergeev et al.* [1996] and *Chen and Wolf* [1999], the cross-tail scale size of depleted flux tubes in the tail likely plays a critical role in the evolution and ultimate fate of BBFs. If the plasma bubbles are too small, differential cross-tail drifts tend to cause them to dissipate before they are able to reach the inner magnetosphere. On the other hand, if the flows are too large scale, they cannot move very fast because they are inefficient at diverting higher-entropy flux tubes ahead (earthward) of them. The bubbles that are most efficient at penetrating deeply toward the Earth have a medium cross-tail scale size of about a R_E.

This has important implications on how the more distant tail couples to the inner magnetosphere during times of enhanced driving. If earthward directed flows correspond to depleted flux tubes of predominantly medium scale sizes, then the coupling is very efficient and fast, and it may be possible for the magnetosphere to forego the need to drastically reconfigure via a substorm (because stresses associated with the so-called "pressure crisis" in the inner magnetosphere are adequately mitigated by the introduction of the depleted flux tubes). This would explain why steady magnetospheric convection (SMC) intervals can arise such that long intervals of enhanced driving can occur that do not lead to substorm onsets. The communication between reconnection in the tail and return flow to the dayside is efficient and responsive enough to occur without the need for a substorm to occur.

On the other hand, if the flows in the tail were associated with much larger-scale depleted flux tubes, they would not be as efficient in terms of earthward penetration. In such cases, the more distant tail would not be able to couple to the inner magnetosphere as efficiently (to alleviate a pressure crisis), and this may lead to the requirement for the magnetosphere to drastically reconfigure in the form of a substorm. A similar situation could arise if the flows were associated with bubbles that were all of very small scale size. In other words, conditions leading to the requirement of an onset could be either a slow monolithic laminar-type flow or a highly turbulent, patchy, and bursty-type of flow.

The foregoing discussion leads to the interesting hypothesis that medium scale-sized bubbles may lead to BBFs that actually tend to *prevent* the need for substorms rather than being their *cause* (e.g., as one might expect from the flow-braking model). This certainly makes sense when one considers that (1) the vast majority of auroral streamers are completely unrelated to substorm onsets and (2) SMC intervals are known to contain high levels of both BBFs and

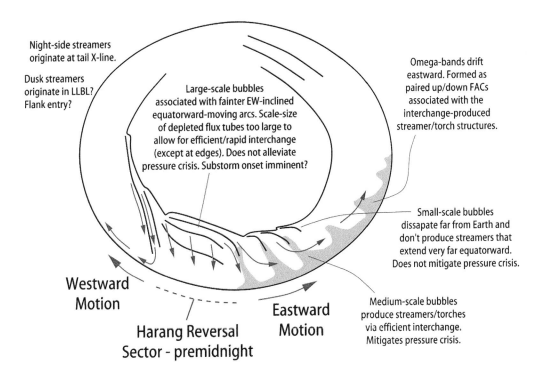

Figure 8. Schematic illustration showing how depleted flux tubes of differing scale-scale may be linked to various auroral forms.

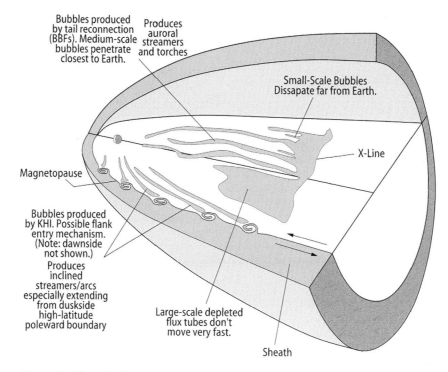

Bubbles produced by tail reconnection (BBFs). Medium-scale bubbles penetrate closest to Earth.

Produces auroral streamers and torches

Small-Scale Bubbles Dissapate far from Earth.

X-Line

Magnetopause

Bubbles produced by KHI. Possible flank entry mechanism. (Note: dawnside not shown.)

Produces inclined streamers/arcs especially extending from duskside high-latitude poleward boundary

Large-scale depleted flux tubes don't move very fast.

Sheath

Figure 9. Schematic illustration showing development of depleted flux tubes in the tail.

auroral streamers. In this scenario, the larger-scale flows are more likely to be related to substorm onsets than the medium-scale flows. Observations of relatively dim emissions poleward of the onset sector coupled with the appearance of streamers on the western edge of the onset region may be consistent with this idea.

These ideas are further illustrated in schematic form in Figures 8 and 9. Figure 8 shows how various ionospheric auroral forms may be related to flows of various scale sizes in the tail as illustrated in Figure 9. As illustrated, medium-scale bubbles tend to create torches and omega bands to the east, while large-scale flows to the west can lead to onset. The type of hypothesized cross-tail asymmetry in flow scale sizes could explain why substorm onsets tend to be a predominantly premidnight phenomenon. On the other hand, it may well be that onsets could also be triggered by any type of earthward directed flow burst if the system is already close to an instability threshold.

4. CONCLUSIONS

A detailed analysis of the 11 December 1998 event leads us to the following conclusions:

1. PBIs and auroral streamers can evolve into auroral torches (or "tongues") and omega bands.

2. The 5–15 min quasiperiodicity of streamer activity combined with eastward drift can add torches to omega band like a conveyor belt. Complex omega band structures can be created in this manner because streamers can move down on top of preexisting omega forms. In addition, more than one torch can be added at once.

3. Auroral streamers do not generally lead to auroral substorm onsets. We have demonstrated that even extremely intense streamer activity need not lead to substorm onsets. This is a crucial constraint for substorm onset initiation theories and appears to be at odds with the widely accepted flow-braking picture of onset initiation.

4. We have demonstrated that auroral streamer and/or equatorward moving arcs can be seen prior to some embedded onsets. However, these auroral forms are not always colocated with the onset region, and similar activity prior to onset can be ongoing for extended intervals before onset occurs. This strongly suggests that earthward directed flows in the tail could be responsible for destabilizing an additional instability that leads to onset. It also strongly argues against the flow-braking picture of onset and expansion phase initiation.

5. Production of auroral streamers occurs on the poleward edge of expanding bulge, even when it is still embedded on a closed field line region. This implies that similar (or the same) physical processes operate there as that which operate at the poleward boundary of the preexisting double oval. This is

consistent with the poleward edge of the expanding bulge being topologically associated with a substorm neutral line.

6. Rapid intensification of streamer production and more rapid expansion of the bulge occurs when its poleward edge reaches the preexisting poleward edge of the double oval. This behavior may explain previous reports of the so-called "poleward leap" as described by *Hones* [1985] and may be caused when the substorm X line begins to process open magnetic field lines (because the reconnection rate would increase dramatically at that point).

7. Dramatic intensification of streamer production can lead to sawtooth-like injections even when no classical equatorward substorm onset occurs. This demonstrates that not all injections can or should be linked with traditional Akasofu-type auroral substorm onsets.

We have proposed a possible scenario in which onsets may be tied to larger-scale flows in the tail that are not as effective as smaller-scale flows in penetrating toward the Earth. The hypothesized inefficacy of these larger-scale flows to reach the inner magnetosphere may lead to enhanced growth of relevant instabilities in the onset sector. This scenario could explain (1) why onsets are predominantly a premidnight phenomenon and (2) why the region above onsets are often relatively dim and void of intense streamer structures (since streamers are generally expected to be located at the western edge of flows, a large-scale flow would be relatively dim in the ionosphere).

Acknowledgments. This research was supported at Los Alamos National Laboratory by NSF GEM grant ATM-0202303.

REFERENCES

Akasofu, S. (1964), The development of the auroral substorm, *Planet. Space. Sci.*, *12*, 273–282.

Akasofu, S. (1965), *Auroral Morphology as Shown by All-Sky Photographs*, Ann. Int. Geophys. Year, vol. 38, 299 pp., Pergamon, Oxford, N. Y.

Akasofu, S. (1976a), Recent progress in studies of DMSP auroral photographs, *Space Sci. Rev.*, *19*, 169–215.

Akasofu, S. (1976b), *Physics of Magnetosphereic Substorms*, D. Reidel, Boston, Mass.

Angelopoulos, V., C. F. Kennel, F. V. Coroniti, R. Pellat, M. G. Kivelson, R. J. Walker, W. Baumjohann, G. Paschmann, and H. Luhr (1992a), Bursty bulk flows in the inner plasma sheet: An effective means of earthward transport in the magnetotail, in *Proceedings of the First International Conference on Substorms (ICS-1)*, ESA Spec. Publ., ESA SP-335, 303.

Angelopoulos, V., W. Baumjohann, C. F. Kennel, F. V. Coroniti, M. G. Kivelson, R. Pellat, R. J. Walker, H. Lühr, and G. Paschmann (1992b), Bursty bulk flows in the inner central plasma sheet, *J. Geophys. Res.*, *97*(A4), 4027–4039.

Angelopoulos, V., C. F. Kennel, F. V. Coroniti, R. Pellat, M. G. Kivelson, R. J. Walker, C. T. Russell, W. Baumjohann, W. C. Feldman, and J. T. Gosling (1994), Statistical characteristics of bursty bulk flow events, *J. Geophys. Res.*, *99*(A11), 21,257–21,280.

Anger, C., T. Fancott, J. McNAlly, and H. S. Kerr (1973), ISIS-II scanning auroral photometer, *Appl. Opt.*, *12*(8), 1753–1766.

Anger, C. D., et al. (1987), An ultraviolet auroral imager for the Viking spacecraft, *Geophys. Res. Lett.*, *14*(4), 387–390.

Belyakova, S. I., S. A. Zaytseva, and M. I. Pudovkin (1968), Development of a polar storm, *Geomagn. Aeron.*, Engl. Transl., *8*, 569–573.

Chen, C. X., and R. A. Wolf (1999), Theory of thin-filament motion in Earth's magnetotail and its application to bursty bulk flows, *J. Geophys. Res.*, *104*(A7), 14,613–14,626.

Frank, L. A., J. D. Craven, K. L. Ackerson, M. R. English, R. H. Eather, and R. L. Carovillano (1981), Global auroral imaging instrumentation for the Dynamics Explorer mission, *Space Sci. Instrum.*, *5*, 369–393.

Fukunishi, H. (1973), Dynamical morphology of proton aurora and electron aurora substorms and phenomenological model of magnetospheric substorms, in *Aeronomy, Jpn. Antarct. Res. Exped. Sci. Rep. Ser. A*, vol. 11, pp. 19–77, Polar Res. Cent., Natl. Sci. Mus., Tokyo.

Henderson, M. G. (1994), Implications of Viking Imager results for substorm models, Ph.D. thesis, Dept. of Phys. and Astron., Univ. of Calgary, Calgary, Alberta, Canada.

Henderson, M. G. (2009), Observational evidence for an inside-out substorm onset scenario, *Ann. Geophys.*, *27*, 2129–2140.

Henderson, M. G., J. S. Murphree, and G. D. Reeves (1994), The activation of the dusk-side and the formation of north-south aligned structures during substorms, in *Proceedings of the Second International Conference on Substorms (ICS-2)*, edited by J. R. Kan, J. D. Craven, and S. Akasofu, p. 37, Geophys. Inst., Univ. of Alaska, Fairbanks.

Henderson, M. G., G. D. Reeves, and J. S. Murphree (1998), Are north-south aligned auroral structures an ionospheric manifestation of bursty bulk flows?, *Geophys. Res. Lett.*, *25*(19), 3737–3740.

Henderson, M. G., L. Kepko, H. E. Spence, M. Connors, J. B. Sigwarth, L. A. Frank, H. J. Singer, and K. Yumoto (2002), The evolution of north-south aligned auroral forms into auroral torch structures: The generation of omega bands and Ps6 pulsations via flow bursts, in *Proceedings of the 6th International Conference on Substorms*, edited by R. M. Winglee, pp. 169–174, Univ. of Washington, Seattle.

Henderson, M. G., G. D. Reeves, R. Skoug, M. F. Thomsen, M. H. Denton, S. B. Mende, T. J. Immel, P. C. Brandt, and H. J. Singer (2006a), Magnetospheric and auroral activity during the 18 April 2002 sawtooth event, *J. Geophys. Res.*, *111*, A01S90, doi:10.1029/2005JA011111.

Henderson, M. G., et al. (2006b), Substorms during the 10–11 August 2000 sawtooth event, *J. Geophys. Res.*, *111*, A06206, doi:10.1029/2005JA011366.

Hirao, K. (1978), Scientific satellite KYOKKO (EXOS-A), *Sol. Terr. Environ. Res. Jpn.*, *2*, 148–152.

Hones, E. W., Jr. (1985), The poleward leap of the auroral electrojet as seen in auroral images, *J. Geophys. Res.*, *90*(A6), 5333–5337.

Hultqvist, B. (1987), The Viking project, *Geophys. Res. Lett.*, *14*(4), 379–382.

Kepko, L., E. Spanswick, V. Angelopoulos, E. Donovan, J. McFadden, K.-H. Glassmeier, J. Raeder, and H. J. Singer (2009), Equatorward moving auroral signatures of a flow burst observed prior to auroral onset, *Geophys. Res. Lett.*, *36*, L24104, doi:10.1029/2009GL041476.

Kornilova, T. A., M. I. Pudovkin, and G. V. Starkov (1990), Fine structure of aurorae near the polar boundary of the auroral prominence in the breakup phase, *Geomagn. Aeron.*, Engl. Transl., *30*, 206–209.

Liu, W. W., and G. Rostoker (1993), On the origin of auroral fingers, *J. Geophys. Res.*, *98*(A10), 17,401–17,407.

Lyons, L. R., T. Nagai, G. T. Blanchard, J. C. Samson, T. Yamamoto, T. Mukai, A. Nishida, and S. Kokubun (1999), Association between Geotail plasma flows and auroral poleward boundary intensifications observed by CANOPUS photometers, *J. Geophys. Res.*, *104*(A3), 4485–4500.

Montbriand, L. E. (1971), The proton aurora and auroral substorm, in *The Radiating Atmosphere*, edited by B. M. McCormac, pp. 366–373, D. Reidel, Hingham, Mass.

Nakamura, R., T. Oguti, T. Yamamoto, and S. Kokubun (1993), Equatorward and poleward expansion of the auroras during auroral substorms, *J. Geophys. Res.*, *98*(A4), 5743–5759.

Nakamura, R., W. Baumjohann, M. Brittnacher, V. A. Sergeev, M. Kubyshkina, T. Mukai, and K. Liou (2001), Flow bursts and auroral activations: Onset timing and foot point location, *J. Geophys. Res.*, *106*, 10,777–10,789, doi:10.1029/2000JA000249.

Nakamura, R., et al. (2005), Localized fast flow disturbance observed in the plasma sheet and in the ionosphere, *Ann. Geophys.*, *23*, 553–566.

Nishimura, Y., L. Lyons, S. Zou, V. Angelopoulos, and S. Mende (2010), Substorm triggering by new plasma intrusion: THEMIS all-sky imager observations, *J. Geophys. Res.*, *115*, A07222, doi:10.1029/2009JA015166.

Oguti, T. (1973), Hydrogen emission and electron aurora at the onset of the auroral breakup, *J. Geophys. Res.*, *78*(31), 7543–7547.

Oguti, T., S. Kokubun, K. Hayashi, K. Tsuruda, S. Machida, T. Kitamura, O. Saka, and T. Watanabe (1981), An auroral torch structure as an activity center of pulsating aurora, *Can. J. Phys.*, *59*, 1056–1062.

Partamies, N., K. Kauristie, E. Donovan, E. Spanswick, and K. Liou (2006), Meso-scale aurora within the expansion phase bulge, *Ann. Geophys.*, *24*, 2209–2218.

Persson, M. A. L., et al. (1994), Near-Earth substorm onset: A coordinated study, *Geophys. Res. Lett.*, *21*(17), 1875–1878.

Rostoker, G., A. T. Y. Lui, C. D. Anger, and J. S. Murphree (1987), North-south structures in the midnight sector auroras as viewed by the Viking imager, *Geophys. Res. Lett.*, *14*(4), 407–410.

Sandholt, P. E., C. J. Farrugia, M. Lester, S. Cowley, S. Milan, W. F. Denig, B. Lybekk, E. Trondsen, and V. Vorobjev (2002), Multistage substorm expansion: Auroral dynamics in relation to plasma sheet particle injection, precipitation, and plasma convection, *J. Geophys. Res.*, *107*(A11), 1342, doi:10.1029/2001JA900116.

Sergeev, V. A. (1992), Tail–aurora direct relationship, in *Proceedings of the First International Conference on Substorms (ICS-1)*, ESA Spec. Publ., ESA SP-335, 277–289.

Sergeev, V. A. (2004), Bursty bulk flows and their ionospheric footprints, in *Multiscale Processes in the Earth's Magnetosphere: From Interball to Cluster*, NATO Sci. Ser., vol. 178, pp. 289–306, Kluwer Acad., Dordrecht, The Netherlands.

Sergeev, V. A., and A. G. Yahnin (1979), The features of auroral bulge expansion, *Planet. Space Sci.*, *27*, 1429–1440.

Sergeev, V. A., V. Angelopoulos, J. T. Gosling, C. A. Cattell, and C. T. Russell (1996), Detection of localized, plasma-depleted flux tubes or bubbles in the midtail plasma sheet, *J. Geophys. Res.*, *101*(A5), 10,817–10,826.

Sergeev, V. A., K. Liou, C.-I. Meng, P. T. Newell, M. Brittnacher, G. Parks, and G. D. Reeves (1999), Development of auroral streamers in association with localized impulsive injections to the inner magnetotail, *Geophys. Res. Lett.*, *26*(3), 417–420.

Sergeev, V. A., et al. (2000), Multiple-spacecraft observation of a narrow transient plasma jet in the Earth's plasma sheet, *Geophys. Res. Lett.*, *27*(6), 851–854.

Sergeev, V. A., K. Liou, P. T. Newell, S.-I. Ohtani, M. R. Hairston, and F. Rich (2004), Auroral streamers: Characteristics of associated precipitation, convection and field-aligned currents, *Ann. Geophys.*, *22*, 537–548.

Tagirov, V. R. (1993), Auroral torch structures: Results of optical observations, *J. Atmos. Terr. Phys.*, *55*, 1775–1787.

Zesta, E., L. R. Lyons, and E. Donovan (2000), The auroral signature of earthward flow bursts observed in the magnetotail, *Geophys. Res. Lett.*, *27*(20), 3241–3244.

Zesta, E., E. Donovan, L. Lyons, G. Enno, J. S. Murphree, and L. Cogger (2002), Two-dimensional structure of auroral poleward boundary intensifications, *J. Geophys. Res.*, *107*(A11), 1350, doi:10.1029/2001JA000260.

Zesta, E., L. Lyons, C.-P. Wang, E. Donovan, H. Frey, and T. Nagai (2006), Auroral poleward boundary intensifications (PBIs): Their two-dimensional structure and associated dynamics in the plasma sheet, *J. Geophys. Res.*, *111*, A05201, doi:10.1029/2004JA010640.

M. G. Henderson, ISR-1, Los Alamos National Laboratory, Los Alamos, NM 87544, USA. (mghenderson@lanl.gov)

A Review of Pulsating Aurora

M. R. Lessard

Space Science Center, University of New Hampshire, Durham, New Hampshire, USA

Pulsating aurora is a class of aurora that is unique in that it is relatively faint yet excited by energetic electrons, of the order of tens of keV. Pulsating patches are typically tens of kilometers in extent, pulsating periods are typically of the order of 8–10 s. While a surge of interest drove a concentrated research effort on pulsating aurora in the 1970s, it has generally not received attention on the same scales as other types of aurora (e.g., aurora associated with substorms, polar cap aurora, cusp aurora, etc.). Recent results, however, show that pulsating aurora is much more widespread than previously thought, implying that it may be more important in terms of magnetosphere-ionosphere coupling than previously thought. In parallel with this work, spacecraft observations have recently led to important conclusions about the generation of pulsating aurora by chorus waves, about the energetics of electron precipitation at geosynchronous orbit, and related topics. While this recent research has answered some of the most fundamental questions about pulsating aurora, it has also led to other, more comprehensive questions. These include, for example, the relationship between pulsating aurora and substorms and, possibly, the radiation belts. Finally, even as these new questions emerge, older questions about fundamental aspects of pulsating aurora (e.g., how are the pulsations controlled, what role does the ionosphere play, what is its relationship to diffuse aurora, etc.) persist.

1. INTRODUCTION

As described by *Störmer* [1955], pulsating aurora was first identified by Sophus Tromholt in late February, 1879, based on observations in Bergen, Norway. *Tromholt* [1880, p. 113] describes the observations as "The arc which had almost disappeared, suddenly reappeared as a rather luminous arc; this lasted some seconds and then it disappeared completely... These pulsations continued for at least a quarter of an hour".

A display of pulsating aurora can be very impressive. Although not as bright or colorful as many auroral arcs and a large westward traveling surge in particular, pulsating aurora often covers the entire sky with intermixed large- and small-scale spatial and temporal variations. The *International Auroral Atlas* [International Union of Geodesy and Geophysics (IUGG), 1963] provides definitions of time-varying auroral displays, where slowly varying forms are broadly classed as *pulsing*, a term that encompasses a few different subclasses. Specifically, *pulsating* aurora involves quasiperiodic modulations of the intensity of extended forms.

In terms of temporal variations alone, pulsating aurora is much different than *flickering* aurora, which tends to develop within discrete, field-aligned arcs and is modulated near 6–8 Hz and, at times, much higher frequencies. *Flaming* aurora is also modulated, but more in the sense that the variation is due to an apparent motion upward. Finally, aurora is classified as *streaming* if the motion appears to travel along an arc or as *fast auroral waves* if the motion appears as a progression of arcs in latitude [*Vallance Jones*, 1974]. Again, these latter definitions imply spatial as well as temporal variation (or perhaps a combination of the two).

Auroral Phenomenology and Magnetospheric Processes: Earth and Other Planets
Geophysical Monograph Series 197
10.1029/2011GM001187

Most often, observers will note the appearance of pulsating aurora in the recovery phase of substorms. During the substorm expansion phase, aurora appears as bright curtains, often with even brighter ("enhanced") lower borders and quite dramatic motion. As a substorm transitions to the breakup phase, the aurora takes the form of dim and diffuse patches (thus the term "breakup") that eventually begin to pulsate [*Akasofu*, 1968; *Duthie and Scourfield*, 1977].

Such routine ground-based observations, however, contribute to two aspects of pulsating aurora that may be misunderstood. One is that pulsating aurora *evolves* directly from diffuse aurora, which does not appear likely. This was demonstrated decades ago by N. Brown (also see *Royrvik and Davis* [1977]), who triangulated the altitudes of these aurora and conclude that pulsating aurora occurs below that of diffuse aurora (as illustrated in Figure 1) from *Brown et al.* [1976]. The difference in altitude of these phenomena is attributed to differences in precipitating electron energies (as noted in Figure 1), which has since been substantiated by various in situ measurements. Still, the strong tendency for these types of aurora to appear in the same region (and sequentially) implies that a connection exists of some sort between them.

A second point of confusion stems from results of statistical studies, showing that pulsating aurora is primarily a dawnside phenomenon. These results appear to contrast with substorm studies that identify pulsating aurora as part of a recovery phase, placing its occurrence region closer to midnight. Below, we return to this topic, where we show that both of these descriptions are accurate, based on large-scale observations of pulsating aurora using the Time History of Events and Macroscale Interactions during Substorms (THEMIS) cameras.

While aurora has been observed scientifically since the 1800s, research efforts dedicated to pulsating aurora have been irregular. While some discoveries can be associated with activities during the International Geophysical Year, research on pulsating aurora was carried out at only a moderate level of effort throughout the 1960s and for most of the 1970s. A peak in pulsating auroral research was reached with a Canadian rocket campaign in 1978 [*McEwen et al.*, 1981], but this was followed by very little work on this topic until late in the 1990s. Recently, research on pulsating aurora has undergone a resurgence, with numerous significant discoveries appearing over the last several years.

Finally, a few noteworthy reviews of pulsating aurora have been published in the past. *Davis* [1978] discusses pulsating aurora and contrasts its characteristics with discrete aurora. *Johnstone* [1978] presents a comprehensive review, focused completely on pulsating aurora. Finally, *Davidson* [1990] provides a later review, one with an emphasis on theoretical aspects.

2. PULSATING PATCH MORPHOLOGY

Pulsating aurora luminosities are typically faint and often subvisual, with a brightness in the range of a few hundred Rayleighs to a few kR at 427.8 nm. Individual patches span tens to hundreds of kilometers and can vary dramatically in shape and size [*Johnstone*, 1978], although a common characteristic is for particular shapes to persist for several minutes, even throughout intervals of varying brightness. Typical pulsating periods range from 2 to 20 s, with an average of 8 s [*Royrvik and Davis*, 1977]. Finally, adjacent patches can perhaps have slightly different periods and may pulsate out of phase with each other.

Unlike discrete auroral arcs, which often extend vertically to several tens of kilometers, pulsating patches tend to be quite thin. *Störmer* [1948] carried out a comprehensive study of pulsating aurora and concludes that its altitude generally is constrained to 90–110 km. A very interesting result was obtained by *Stenbaek-Nielsen and Hallinan* [1979], who triangulated pulsating patches to determine their thicknesses and found them to be as thin as 2 km or less. These authors also note that such thin patches cannot be produced by collisional processes from even a monoenergetic beam of electrons and suggest that ionospheric effects must contribute to the process.

European Incoherent Scatter (EISCAT) radar observations of patch thickness have yielded similar results, with *Kaila et al.* [1989] showing EISCAT observations of enhanced

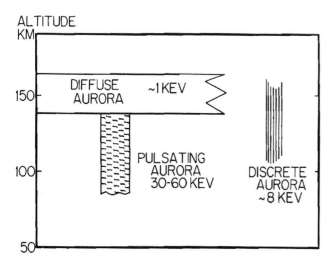

Figure 1. An illustration showing estimated altitudes of auroral forms. Note, in particular, the difference in altitude of diffuse versus pulsating aurora, which implies a significant difference in the electron energies that drive these emissions. From *Brown et al.* [1976].

electron densities that support thicknesses of 8 km and *Wahlund et al.* [1989] showing observations that support thicknesses of 4.5 km or less. More recently, however, *Jones et al.* [2009] used Poker Flat Incoherent Scatter Radar observations (with supporting data from the Rocket Observations of Pulsating Aurora sounding rocket) to determine that in the four cases examined, pulsating patches had a thickness of 15–25 km, significantly thicker than previous observations (see Figure 2). These authors also note a large variability in the altitude of the electron density peaks associated with different events and even during the course of single events.

Spatial structures in pulsating aurora have long been known to exist (see the review by *Davis* [1978]). In extensive surveys of local and global morphology of pulsating aurora, *Royrvik and Davis* [1977] report that pulsating aurorae commonly occur in east-west aligned arcs, in irregular patches, or in quasilinear arrangements, and *Thomas and Rothwell* [1979] show "recurrent propagating forms" in pulsating aurora, wavelike propagation of patches in the meridional direction.

Spatially periodic arc-like forms have also been observed in regions of pulsating aurora, consisting of several narrow bands pulsating with a period of ~10 s and drifting toward the pole at a velocity of similar to 250 m s^{-1} [*Safargaleev et al.*, 2000; *Safargaleev and Osipenko*, 2000]. A similar, perhaps identical, type of structure is described by *Sato et al.* [2002] (see Figure 3), who compare measurements from the FAST satellite to ground-based optical observations of pulsating aurora. The structure they discuss consists of east-west bands of pulsating aurora, embedded in weak inverted-V precipitation.

Pulsating aurora is often observed against a nonpulsating background [*Royrvik and Davis*, 1977; *Stenbaek-Nielsen and Hallinan*, 1979] created by soft electrons [*Smith et al.*, 1980; *McEwen et al.*, 1981; *Sandahl*, 1985]. This background has been identified by *Evans et al.* [1987] as arising from secondary electrons and backscattered electrons produced by the primary high-energy pulsating electrons. Nonetheless, *Evans et al.* [1987, p. 12,305] find that "there appears to be a component of the primary precipitation at energies in the range of 5–20 keV that is not explained by current theories of precipitation pulsations by self-excited generation of VLF waves." This topic is discussed further below.

A tendency for the pulsating period to have a latitudinal dependence, with shorter periods occurring at lower latitudes, has been reported [*Thomas and Stenbaek-Nielsen*, 1981; *Duncan et al.*, 1981; *Creutzberg et al.*, 1981], which supports the idea that the period is somehow related to field-line length, although there are mixed reports about whether the periods actually correspond to bounce periods of the electrons. Other work has shown that variations in pulsating periods may be due to an increase in the time between successive pulsations, i.e., that the "on" time of the pulsations remains relatively constant [*Yamamoto*, 1983, 1988]. This relationship was confirmed by *Viereck and Stenbaek-Nielsen* [1985], who also found the period to correlate with the flux of proton precipitation. This rather surprising observation suggests that protons may play a role in the pulsation mechanism, but none of the processes proposed in the literature include protons.

Important strides have been made more recently on this topic, however. *Li et al.* [2011] show THEMIS spacecraft observations of lower-band chorus and link their modulation both spatially and temporally to pulsating aurora observed from the ground. In addition, these authors are able to correlate the chorus modulations with fluctuations in plasma density, although the cause of these density variations remains unclear.

Finally, we point out that studies of whether or not pulsating auroral patches are magnetically conjugate have produced

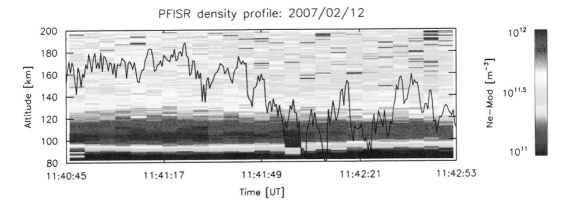

Figure 2. Electron density versus time measured by Poker Flat Incoherent Scatter Radar on 12 February 2007, overplotted with auroral brightness from all-sky camera data. Modulations in brightness of approximately 6 and 20 s periods are both present. From *Jones et al.* [2009].

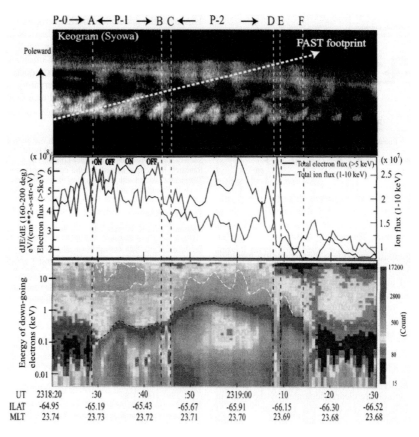

Figure 3. A pulsating auroral event observed by the FAST satellite, as it crossed over Syowa station in Antarctica. (top) A keogram reproduced from the all-sky TV camera. (middle) The total energy flux of downgoing electrons and ions. (bottom) The energy spectrum of the electrons. From *Sato et al.* [2002].

mixed results. Some studies do make such a conclusion [*Belon et al.*, 1969; *Gokhberg et al.*, 1970; *Stenbaek-Nielsen et al.*, 1973; *Davis*, 1978]. However, other studies have shown a lack of conjugacy, with cases of pulsating aurora present in only one hemisphere or with conjugate auroras not pulsating in phase [*Minatoya et al.*, 1995; *Sato et al.*, 1998, 2002, 2004; *Watanabe et al.*, 2007]. The reasons for a lack of conjugacy may be associated with distorted magnetic fields that map individual patches far from the expected interhemispheric foot points, or as *Sato et al.* [2004] suggest, it is also possible that a near-Earth modulation source is present in some cases.

3. ELECTRON PRECIPITATION AND PULSATING AURORA

Pulsating auroral patches are invariably characterized by the precipitation of energetic electrons occurring across regions of limited size (tens of kilometers) and with periodicity the order of a few to several seconds. This precipitation is consistently embedded in a region of larger-scale, lower-energy electrons that excite diffuse aurora as seen from the ground. Sounding

rockets have typically observed pulsations in energetic electron precipitation over energies ranging from a few keV to several tens of keV, with more pronounced modulations occurring at higher energies (typically 30 keV or higher) [*Bryant et al.*, 1969; *Whalen et al.*, 1971; *Bryant et al.*, 1975; *Sandahl et al.*, 1980; *McEwen et al.*, 1981; *Smith et al.*, 1980; *Yau et al.*, 1981; *Saito et al.*, 1992; *Williams*, 2002] (Figure 4). Similar results have been obtained with satellites [*Evans et al.*, 1987; *Sato et al.*, 2002]. Although the pulsation periods and electron energies at which the fluctuations occur vary somewhat from one case to the next, the results correlate very well with optical observations, in general, and also with any specific optical observations made in conjunction with the in situ measurements.

Perhaps the most important conclusion that has been drawn from these experiments is that velocity dispersion in the energetic electrons indicates that they originate from near the equator [*Bryant et al.*, 1971, 1975; *Smith et al.*, 1980; *Yau et al.*, 1981]. One possible exception to this rule is observed in electron data acquired by the recent PARX sounding rocket, which were obtained in conjunction with a pulsation but show no velocity dispersion [*Williams*, 2002].

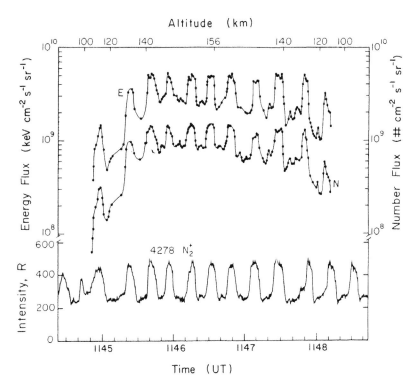

Figure 4. Electron fluxes observed from a sounding rocket (upper two traces) plotted with ground-based optical data (lower trace). Eleven pulsations are clearly shown in all traces. From *McEwen et al.* [1981]. © Canadian Science Publishing or its licensors.

3.1. Diffuse Background

As discussed above, the existence of a nonpulsating, lower-energy population of electrons in pulsating aurora is well established [*Smith et al.*, 1980; *Whalen et al.*, 1971; *Sandahl et al.*, 1980; *Saito et al.*, 1992]. *Evans et al.* [1987] consider whether this population might consist of backscattered electrons that originate from the opposite hemisphere in conjunction with simultaneously occurring pulsations there.

In the model presented in their paper, they partition the nonpulsating electrons into two populations, backscattered primary electrons and secondary electrons, with secondaries having energies below 1 keV, i.e., suggesting that the lower-energy component of the diffuse population consists essentially of secondaries. Since their data, from the NOAA-6 satellite, did not include measurements below 300 eV, they use the empirical relation

$$R_s \approx \frac{0.00785}{E E_0^{3/4}} \, \text{keV}^{-1}$$

to describe this population, where R_s is the number of secondary electrons produced per keV at energy E by an incident electron of energy E_0. Note that detailed observations of the low-energy populations are lacking, in general, with typical measurements in the literature reaching down only to 50 to 500 eV [*McEwen et al.*, 1981; *Sandahl et al.*, 1980; *Smith et al.*, 1980].

Figure 5 shows the prediction of their model during a maximum in pulsation intensity. The spectrum measured by NOAA-6 is shown by the solid curve; the dashed curve represents the backscattered spectrum computed from the downgoing energetic electrons. The dotted curve is the difference between these populations and is thought to be a magnetospheric contribution to the nonpulsating population. Note that the energy range of these particles actually extends up to 20 keV.

The result of their work is that, although a fraction of the nonpulsating electrons consists of backscattered electrons, their modeled intensities are significantly less than the intensities inferred from the NOAA-6 data. They conclude that the remainder (10%–50%) must result from some other mechanism and suggest that "a search for a mechanism to explain the drizzle remains a worthwhile objective."

An answer to this question is provided by *Inan et al.* [1992], who demonstrate that VLF upper-band chorus waves (those with frequencies above one half the local gyrofrequency) are capable of scattering the ~1 keV electrons that drive diffuse aurora. This work was in direct response to that of *Evans et al.* [1987].

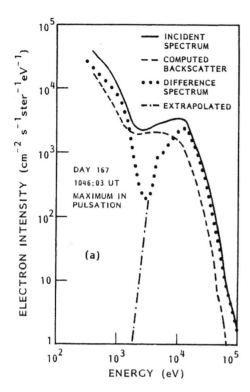

Figure 5. This plot shows the contributions of incident and backscattered electrons during maxima in pulsating aurora (down to 300 eV). The dotted curve is the difference between these populations and is thought to be a magnetospheric contribution. From *Evans et al.* [1987].

A more comprehensive analysis is presented by *Thorne et al.* [2010], who show global observations from the Polar Ionosphere X-ray Imaging Experiment (PIXIE) instrument on the Polar spacecraft, along with CRRES satellite observations of 1.88 keV plasma sheet electrons, electrostatic electron cyclotron harmonic (ECH) and electromagnetic chorus waves. ECH waves are capable of pitch angle scattering electrons (over a narrow range of pitch angles) having energies from a few hundreds of eV up to a few keV. Chorus waves are partitioned between upper-band chorus, which can scatter electrons with comparable energies, but over a much broader range in pitch angles and lower-band chorus (below one half the local gyrofrequency), which also can scatter electrons over a broad range of pitch angles, but with energies of a few keV to ~100 keV. Based on these global statistics and supported by their numerical model (also see *Ni et al.* [2011]), these authors also conclude that chorus waves are responsible for the precipitation of electrons associated with diffuse aurora in the near- and postmidnight region, which coincides with the location of pulsating aurora observations.

3.2. Current System Closure

In spite of the lack of spacecraft observations of currents in pulsating aurora, some progress has been made using ground-based magnetometers. *Arnoldy et al.* [1982] used induction coil data to conclude that the magnetic signature associated with pulsating aurora resulted from ionospheric currents, as opposed to ULF waves. They also determined that these currents must be oriented in the north-south direction. However, noting that the pulsations are excited by energetic precipitation, which must reach down into lower altitudes to regions of enhanced Hall conductivity, and that pulsating aurora tends to occur in the morning sector (where an equatorward ionospheric electric field is typical), they suggest that the horizontal current is evidence of Cowling current, rather than a Pedersen current.

Oguti and Hayashi [1984] used all-sky camera in conjunction with magnetometer data and also concluded that the magnetic signature was a direct result of ionospheric currents. In addition, they carried out a multiple correlation comparison of the two data sets to determine the configuration of currents associated with a pulsating patch. Figure 6 is from their paper and shows the results of their model, which consists of a thin-sheet ionosphere with a vertical magnetic field and uniform enhancements in conductivity resulting from energetic electron precipitation. Figure 6 illustrates how currents associated with a patch can be described as the superposition of two current systems: the field-aligned pair (on the left) and the twin-vortex current (on the right). The twin-vortex current shows the response of the enhanced-conductivity patch to the ambient, large-scale current system exclusive of the current associated with the precipitation. Of course, the field-aligned pair of currents shows the predicted distribution of field-aligned currents associated with the precipitation.

Confirmation of the existence of field-aligned currents was subsequently provided by *Ryoichi et al.* [1985], comparing Magsat and all-sky camera data for two events. These authors show excellent correlations between magnetic perturbations and luminosity on the ground, leading them to conclude that Magsat observed, in each case, a pair of sheet currents associated with the pulsating patch, having a magnitude the order of 20 nT for the event with the better conjunction.

Building on this result, *Hosokawa et al.* [2010] show data from the EISCAT VHF radar in Troms, Norway, acquired during a strong pulsating aurora event. During the event, the authors conclude, an electron Pedersen current was observed in the D region (80–95 km), coincident with pulsating patches (and incident energetic electrons), in addition to the more common ion Pedersen current near 120 km. This

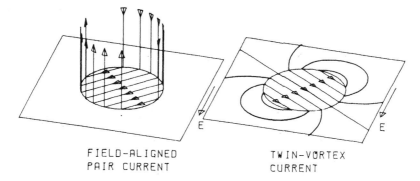

FIELD-ALIGNED
PAIR CURRENT

TWIN-VORTEX
CURRENT

Figure 6. The flow of currents associated with a pulsating auroral patch as determined by *Oguti and Hayashi* [1984], using ground-based magnetometer data.

current, they suggest, could provide the current closure in pulsating patches within the *D* region, consistent with the observations of field-aligned currents as described by *Ryoichi et al.* [1985].

3.3. Ions and Pulsating Aurora

Clearly, in situ observations of pulsating aurora have focused on acquiring properties of electron populations. Typically, positive ions have been measured only at a lower priority, if at all. Interestingly, the observations of ions appear to be quite contradictory.

Although some observations of ions show very low flux levels [*Yau et al.*, 1981; *Evans et al.*, 1987], *Whalen et al.* [1971] successfully measured electrons and ions to conclude that, in between pulsations, the proton flux is softer and more intense than the electrons. Their ion measurement included only those with energies greater than 25 keV, which showed no fluctuations during the pulsating aurora. On the other hand, *Smith et al.* [1980] show ion data from an instrument with a geometric factor larger than that of the electron instrument by a factor of ~50, and note that ion pulsations, most clearly visible in the 26 keV ions, were essentially in phase with the electron pulsations if the ion pulsations were advanced in time by 25 ± 2 s, a lag time derived from a cross correlation of these data. Then, using this delay time and the assumption that the ions and electrons originated from the same source, they estimate the source to be located near the equatorial plane. This result is unique in the literature.

In an apparent contradiction to this work, *Sato et al.* [2002] use FAST data to demonstrate that downgoing 1–10 keV ion and downgoing >5 keV electron fluxes are out of phase and conclude that an oscillating, parallel electric field must be present above (but far from) the FAST satellite. They do not show results of a cross correlation in this paper and did not

shift the ion data several seconds as was done by *Smith et al.* [1980].

From the ground, the variety of results is equally interesting. Early work by *Eather* [1968] compares 4861 Å Hβ to 5577 Å OI emissions and concludes that pulsations do not occur in the proton aurora with periods under a minute. In a similar study, *Creutzberg et al.* [1981] show that proton aurora does not occur coincident with nor equatorward of pulsating aurora and suggest that energetic protons may suppress the process responsible for pulsating electrons. *Viereck and Stenbaek-Nielsen* [1985] carried out a somewhat more extensive study, combining all-sky meridian scanning photometer data. They confirm the *Creutzberg et al.* [1981] result, but also explain that the period of the pulsation appears to be correlated with Hβ intensity, short periods correspond to weak Hβ emissions, while no pulsations were observed in the presence of ≥80 R in Hβ.

4. LARGE-SCALE ASPECTS

A broad perspective on pulsating aurora is possible using the THEMIS cameras in Canada and Alaska. *Jones et al.* [2011] carried out an initial statistical study based on data from two of these cameras, one at Gillam (66.18 magnetic latitude, 332.78 magnetic longitude, magnetic midnight at 06:34 UT) and the other at Fort Smith, (67.38 magnetic latitude, 306.64 magnetic longitude, magnetic midnight at 08:06 UT). These locations are at the same approximate invariant latitude.

The objective of the study was to characterize the temporal and spatial evolution of pulsating aurora on large scales. Using data from September 2007 through March 2008 and noting the relative onset times of pulsating aurora in the cameras, they show that the source region of pulsating aurora drifts or expands eastward, away from magnetic midnight, for premidnight onsets. The spatial evolution is more

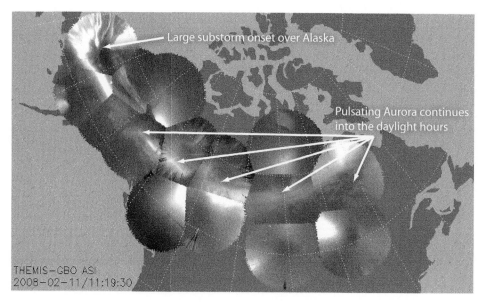

Figure 7. Pulsating aurora observed across North America. This event, as observed with imagers in Canada and Alaska, lasted more than 15 h, spread across 9 h in magnetic local time. From Jones et al. (Pulsating aurora case study of long-lasting, widespread event, submitted to *Journal of Geophysical Research*, 2012).

complicated for postmidnight onsets, but again tends to develop away from magnetic midnight (i.e., westward in this case). The point is that the development of pulsating aurora does not appear to be directly linked to a gradient curvature drift of energetic electrons (which would be eastward) that might have been injected during substorms. This does not mean, however, that there is *no* connection between pulsating aurora and substorms.

Jones et al. [2011] also show that the most probable duration of a pulsating aurora event is roughly 1.5 h, while the occurrence distribution of event durations includes many "extended" events. One example, described in detail, from 11 February 2008, begins with no obvious substorm precursor and expands to eventually cover at least 10 h magnetic local time (MLT). This particular event lasts 15.5 h from its beginning until the camera turnoff for daylight, so the actual duration was even longer than 15.5 h. The spatial extent of this aurora encompasses more than 9 h in MLT. These observations may suggest that pulsating aurora is not simply a substorm recovery phase phenomenon but rather a persistent, long-lived phenomenon that may be temporarily disrupted by auroral substorms.

In fact, out of 84 nights of data in February/March in 2008 and 2009, as well as January 2009, only ~40 contain usable data due to cloud cover and other limiting factors. Still, more than half of the data (~23 nights) present examples of pulsating aurora persisting before, during, and after substorm onsets occurring both close to, and far from, the area of pulsating aurora.

Naturally, substorms occurred during this event, but generally did not disturb pulsating aurora on the dawnside. In fact, pulsating aurora tended to develop with perhaps greater brightness in the regions where substorms occurred, adding to the total intensity of the pulsating aurora event. Figure 7 shows a snapshot of the event, with pulsating aurora persisting over nearly all of Canada, while a substorm occurs over Alaska.

The duration of this event (and many others) is far longer than the typical recovery phase of a substorm, suggesting that pulsating aurora is not simply a recovery phase phenomenon. Still, the data appear to show that pulsating aurora intensifies and/or becomes more widespread when substorms occur, implying that substorms may provide a seed population or that substorms somehow are responsible for the chorus waves that scatter energetic electrons, or perhaps both process occur.

5. SATELLITE OBSERVATIONS

Satellite observations of pulsating aurora have been infrequent, due primarily to instrument limitations. Perhaps the first satellite-based observations of pulsating aurora are described by *Snyder et al.* [1974], who show images from a United States Air Force Defense Meteorological Satellite Program (DMSP) satellite. This satellite flew in a Sun-synchronous orbit at 850 km, aligned in a dawn-dusk orbital plane. Data from its scanning-mirror imager are shown in this paper, spanning 2 days of observations. While these authors do not explicitly identify pulsating aurora in the images, the point of the paper is to validate the "concept of the auroral substorm" and includes the patchy (and presumably pulsating) aurora following a substorm expansion phase.

The images in this paper *are* described as examples of pulsating aurora by *Siren* [1975], however, who show more definitive observations of pulsating aurora in DMSP photographs, where they identified more than 100 cases. Events are identified based on their zebra-stripe appearances, a result of interactions of the pulsating periods with the satellite's 0.56 s line scanning mirror period. In reference to the *Snyder et al.* [1974] paper, *Siren* [1975] extrapolate their own technique to describe their results as evidence of pulsating aurora, noting that the event spans the entire width of the early morning oval.

Data studied by *Siren* [1975] lead these authors to some very important early discoveries, including (1) that the patches must be produced by electron precipitation (based on their sharply defined edges) and (2) that these electrons must have energies much greater than a few hundred eV, since only a small fraction of the emissions are observed at 630 nm. Based on a study of more than 100 events, they also note increased occurrences with greater K_p and a tendency for pulsating aurora to occur more frequently if the oval is displaced equatorward.

More recent studies from low-altitude spacecraft have shed light on the source region of pulsating electrons and calculations that involve velocity dispersion. *Sato et al.* [2002] compare measurements from the FAST satellite to ground-based optical observations of pulsating aurora at Syowa station in Antarctica. The structure they describe consists of east-west bands of pulsating aurora, embedded in weak inverted-V precipitation. A main point of their paper, though, is that a dispersion analysis places the source region between 2 and 6 R_E above FAST, which is far from the equatorial source region that previous rocket missions have shown. They also show that the downgoing, high-energy ion flux modulation was almost out of phase (anticorrelated) with the downgoing high-energy electron flux.

An explanation as to why the dispersion calculation showed such a vastly different result from previous calculations is, perhaps, presented by *Miyoshi et al.* [2010]. These authors use a time-of-flight (TOF) model that considers propagating whistler mode chorus waves distributed along a field line. Since the wave-particle resonant energy depends on magnetic latitude, electrons are potentially scattered over a range of altitudes (and at different times) along this field line. This model is used to estimate the source region of pulsating electrons observed by the REIMEI satellite, which points to a source region near the equator.

Nishiyama et al. [2011] also used data from the REIMEI satellite, including its onboard image data. A total of 29 events were examined using both a standard TOF model and that described by *Miyoshi et al.* [2010]. The standard TOF model indicates the source region to be distributed almost continuously from magnetic latitudes 50° to −20°. The Miyoshi model, on the other hand, is more consistent with previous conclusions that point to a near-equatorial source. In addition, the whistler mode frequencies, derived as part of the Miyoshi model, span a range that encompasses both lower- and upper-band chorus waves, suggesting that both may contribute to pulsating aurora.

From geosynchronous orbit, observations of ELF hiss coincident with the precipitation of energetic (>22 keV) electrons were reported by *Gough and Korth* [1982], using data from the European Space Agency GEOS spacecraft. Since the spacecraft was spin stabilized, the electron spectrometer swept through the 3° loss cone too quickly to actually detect pulsations. Thus, ELF hiss occurrences were compared to those intervals when the loss cone was determined to be partially (or completely) filled as the electron instrument scanned the loss cone. In this case, the observations yield $\kappa^2 = 20.45$, leading the authors to conclude that these effects are, indeed, highly correlated. These measurements were acquired at a time when the satellite was believed to map to a region of pulsating aurora on the ground, although the authors were not able to determine detailed correlations between these data.

Similarly, *Ward et al.* [1982] (also see *Ward* [1983]) identified 10 events where ELF hiss at GEOS-2 was well correlated with all-sky TV observations of pulsating aurora on the ground. They also associated these events with Pi1B pulsations, again observed on the ground.

More comprehensive results have been obtained using data from the THEMIS spacecraft and ground-based cameras. *Nishimura et al.* [2010] use THEMIS satellite data to show a strong correlation between pulsating aurora and lower-band chorus waves by mapping cross-correlated data. This was accomplished by calculating correlation coefficients between wave data and pulsating patches, leading to the conclusion that THEMIS wave observations were well correlated with specific patches seen from the ground. A similar result was obtained by *Liang et al.* [2010], also using THEMIS data and comparing satellite data to ground observations. In addition, however, these authors also note the presence of compressional Pc5 waves and suggest that these waves may play an important role in the process. A number of additional events have been identified by *Nishimura et al.* [2011a] (also see *Nishimura et al.* [2011b]).

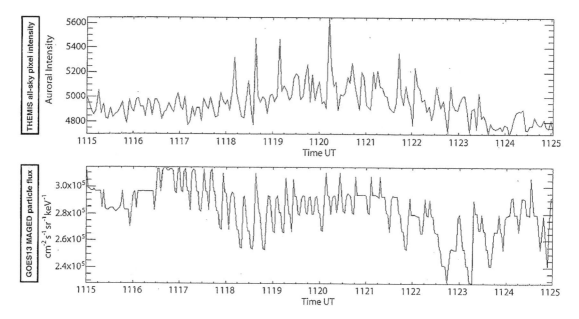

Figure 8. (top) Luminosity fluctuations versus time, extracted from a single pixel in the Time History of Events and Macroscale Interactions during Substorms all-sky camera at The Pas. (bottom) Fluctuations in GOES-13 observations of 30–50 keV electrons. A general correspondence between these fluctuations is clear. From Jaynes et al. (manuscript in preparation, 2012).

Recently, electron observations from GOES-13 on 15 March 2008 reveal clear fluctuations in loss cone precipitation above a region with pulsating aurora that are well correlated (i.e., have correlation coefficients of 0.9 and higher) with luminosity fluctuations observed with THEMIS ground cameras (A. Jaynes et al., Wave-particle interactions at equator drive high-latitude pulsating aurora, manuscript in preparation, 2012, hereinafter referred to as Jaynes et al., manuscript in preparation, 2012). The modulations in precipitation are consistently observed in the telescope that is most aligned to the background magnetic field, as determined by the onboard magnetometer. Of particular interest is the fact that modulations are more prominent in the 30–50 keV channel, but are also observed in the 50–100 keV channel (somewhat more energetic than most sounding rockets have measured). An example is shown in Figure 8, where 30–50 keV electron fluxes are shown, with corresponding fluctuations in luminosity from ground observations. Note, also, the presence of a "double peak" that occurs within the pulsating electron bursts themselves, a feature that is not yet understood.

6. THERMOSPHERE COUPLING

A very different perspective on pulsating aurora is provided by *Wilson et al.* [2005], who used an eight-microphone array

of ground-based infrasonic receivers to correlate atmospheric acoustic pulsations with pulsating aurora (Figure 9). With appropriate propagation delays taken into account, they show good coherence at frequencies from 0.03 to 0.08 Hz and that the waves originate from sources within 35 km of the zenith above the array for a source height of 110 km. The process, they suggest, involves ionospheric heating associated with the electron precipitation that subsequently expands vertically, driving a pressure pulse downward.

This theory is further developed by *de Larquier et al.* [2010], who use an finite-difference time-domain model to estimate the energy flux of the auroral electrons that would provide the requisite heating and conclude that the order of 50 ergs cm^{-2} is needed to explain the infrasonic signatures. Since this energy is not likely provided by precipitating electrons, their conclusion is that Joule heating must also play a role.

Independently, *Oyama et al.* [2010] also considered thermospheric effects, but based on a Fabry-Perot Interferometer (FPI) and an all-sky camera (both at a wavelength of 557.7 nm), as well as the EISCAT UHF radar. FPI observations show fluctuations in both vertical and horizontal thermospheric winds, though perhaps not colocated with the brighter features of pulsating aurora. Ironically, these authors conclude that Joule heating is not adequate in explaining the thermospheric fluctuations, pointing to additional heating possibly associated with observed wave activity.

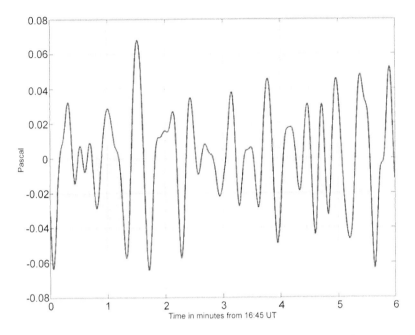

Figure 9. Best beam infrasonic data from 16:45 to 16:51 UT on day 349, 2003. This signal represents the average of the time-lagged waveforms that were band-pass filtered between 0.02 and 0.10 Hz. From *Wilson et al.* [2005].

7. SUMMARY AND OPEN QUESTIONS

On one hand, pulsating aurora has been studied at some level for decades, although perhaps with varying levels of success. On the other hand, pulsating aurora topics have also been overshadowed by auroral topics associated with substorm dynamics, as well as with many other forms of aurora. Fortunately, recent data from spacecraft that are better-equipped to make the needed observations have resulted in important progress on this topic. Of notable importance are the observations in the vicinity of geosynchronous orbit (e.g., THEMIS and GOES-13) and those by the lower-altitude satellite REIMEI, complemented by the broad coverage provided by the THEMIS all-sky cameras. These data sets have provided answers to questions that have lingered since the pioneering sounding rocket observations in the 1970s.

One outcome of these recent results is a realization that pulsating aurora may be far more significant in terms of magnetosphere-ionosphere coupling than previously thought. Early statistical studies of pulsating aurora seemed to show that it is quite common, though these studies were based on limited observations and could not show the extent to which this is the case. Recent studies show that pulsating aurora can occur across Canada and Alaska and that it can persist for hours and hours, which means that the energy transfer from the solar wind to the ionosphere and upper atmosphere is considerable.

Of equal importance is the recent work that helps clarify how the basic process works, including its connection to lower-band chorus and, possibly, upper-band chorus waves. This result, combined with GOES-13 observations of energetic electron precipitation over regions of pulsating aurora confirm that the source is an equatorial one and lay the groundwork for studies that can be expected to fold this knowledge into space weather objectives.

Still, a number of questions exist, some of which remains fundamental to the problem of pulsating aurora. Some of these questions are listed here, with no pretense that it is a complete list. Certainly, as our understanding of pulsating aurora evolves, new questions will emerge.

1. How does the pulsating aurora *process* relate to other large-scale processes, specifically substorm injection and radiation belt dynamics? Do substorms provide the seed populations of energetic electrons as originally suggested by S. Akasofu?

2. What is the total energy involved in pulsating aurora events? How does this compare to substorm expansion phases?

3. What is the spatial extent of occurrences of pulsating aurora? Isolated reports have indicated that it can occur throughout the dayside. Assuming this is correct, under what conditions is this the case?

On smaller scales, a number of questions about the microphysics of pulsating aurora are important.

4. How is pulsating aurora related to diffuse aurora? That is, pulsating aurora appears to invariably be preceded by the onset of diffuse aurora: are they linked via some parent process or by a process internal to the ionosphere?

5. What is the basic process that causes the pulsating?

6. What might the role of the ionosphere be? How do the patches remain so incredibly persistent in terms of their shape and location during an event?

7. Can the inconsistent reports of pulsating protons be understood?

8. Do both lower- and upper-band chorus waves contribute to pulsating aurora as suggested by *Nishiyama et al.* [2011]? This appears to contradict the results of *Nishimura et al.* [2011a].

9. How important is the connection to the thermosphere?

Acknowledgments. We gratefully acknowledge discussions with several members of the auroral science community, including Sarah Jones, Allison Jaynes, Kevin Rychert, Eric Donovan, and Emma Spanswick. Research was supported at the University of New Hampshire by NSF grants ANT-0839938 and ANT-0838910.

REFERENCES

Akasofu, S. I. (1968), *Polar and Magnetospheric Substorms*, 223 pp., D. Reidel, Dordrecht, The Netherlands.

Arnoldy, R. L., K. Dragoon, L. J. Cahill Jr., S. B. Mende, and T. J. Rosenberg (1982), Detailed correlations of magnetic field and riometer observations at $L = 4.2$ with pulsating aurora, *J. Geophys. Res.*, *87*(A12), 10,449–10,456.

Belon, A. E., J. E. Maggs, T. N. Davis, K. B. Mather, N. W. Glass, and G. F. Hughes (1969), Conjugacy of visual auroras during magnetically quiet periods, *J. Geophys. Res.*, *74*(1), 1–28.

Brown, N. B., T. N. Davis, T. J. Hallinan, and H. C. Stenbaek-Nielsen (1976), Altitude of pulsating aurora determined by a new instrumental technique, *Geophys. Res. Lett.*, *3*(7), 403–404.

Bryant, D. A., G. M. Courtier, and A. D. Johnstone (1969), Modulation of auroral electrons at large distances from the Earth, *J. Atmos. Terr. Phys.*, *31*, 579–592.

Bryant, D. A., G. M. Courtier, and G. Bennett (1971), Equatorial modulation of electrons in a pulsating aurora, *J. Atmos. Terr. Phys.*, *33*, 859–867.

Bryant, D. A., M. J. Smith, and G. M. Courtier (1975), Distant modulation of electron intensity during the expansion phase of an auroral substorm, *Planet. Space Sci.*, *23*, 867–878.

Creutzberg, F., R. Gattinger, F. Harris, and A. Vallance Jones (1981), Pulsating auroras in relation to proton and electron auroras, *Can. J. Phys.*, *59*, 1124–1130.

Davidson, G. T. (1990), Pitch-angle diffusion and the origin of temporal and spatial structures in morningside aurorae, *Space Sci. Rev.*, *53*, 45–82.

Davis, T. N. (1978), Observed characteristics of auroral forms, *Space Sci. Rev.*, *22*, 77–113.

de Larquier, S., V. P. Pasko, H. C. Stenbaek-Nielsen, C. R. Wilson, and J. V. Olson (2010), Finite-difference time-domain modeling of infrasound from pulsating auroras and comparison with recent observations, *Geophys. Res. Lett.*, *37*, L06804, doi:10.1029/2009GL042124.

Duncan, C. N., F. Creutzberg, R. L. Gattinger, F. R. Harris, and A. Vallance Jones (1981), Latitudinal and temporal characteristics of pulsating auroras, *Can. J. Phys.*, *59*, 1063–1069.

Duthie, D. D., and M. W. J. Scourfield (1977), Aurorae and closed magnetic field lines, *J. Atmos. Terr. Phys.*, *39*, 1429–1434.

Eather, R. H. (1968), Hydrogen emissions in pulsating auroras, *Ann. Geophys.*, *39*, 5235.

Evans, D. S., G. T. Davidson, H. D. Voss, W. L. Imhof, J. Mobilia, and Y. T. Chiu (1987), Interpretation of electron spectra in morningside pulsating aurorae, *J. Geophys. Res.*, *92*(A11), 12,295–12,306.

Gokhberg, M. B., B. N. Kazak, O. M. Raspopov, V. K. Roldugin, V. A. Troitskaya, and V. I. Fedoseev (1970), On pulsating aurorae in conjugate points, *Geomagn. Aeron.*, *10*, 367–370.

Gough, M. P., and A. Korth (1982), New light on the equatorial source of pulsating aurora, *Nature*, *298*, 253–255.

Hosokawa, K., Y. Ogawa, A. Kadokura, H. Miyaoka, and N. Sato (2010), Modulation of ionospheric conductance and electric field associated with pulsating aurora, *J. Geophys. Res.*, *115*, A03201, doi:10.1029/2009JA014683.

Inan, U. S., Y. T. Chiu, and G. T. Davidson (1992), Whistler-mode chorus and morningside aurorae, *Geophys. Res. Lett.*, *19*(7), 653–656.

Johnstone, A. D. (1978), Pulsating aurora, *Nature*, *274*, 119–126.

Jones, S. L., et al. (2009), PFISR and ROPA observations of pulsating aurora, *J. Atmos. Sol. Terr. Phys.*, *71*, 708–716.

Jones, S. L., M. R. Lessard, K. Rychert, E. Spanswick, and E. Donovan (2011), Large-scale aspects and temporal evolution of pulsating aurora, *J. Geophys. Res.*, *116*, A03214, doi:10.1029/2010JA015840.

Kaila, K., R. Rasinkangas, P. Pollari, R. Kuula, and J. Kangas (1989), High resolution measurements of pulsating aurora by EISCAT, optical instruments and pulsation magnetometers, *Adv. Space Res.*, *9*, 53–56.

Li, W., J. Bortnik, R. M. Thorne, Y. Nishimura, V. Angelopoulos, and L. Chen (2011), Modulation of whistler mode chorus waves: 2. Role of density variations, *J. Geophys. Res.*, *116*, A06206, doi:10.1029/2010JA016313.

Liang, J., V. Uritsky, E. Donovan, B. Ni, E. Spanswick, T. Trondsen, J. Bonnell, A. Roux, U. Auster, and D. Larson (2010), THEMIS observations of electron cyclotron harmonic emissions, ULF waves, and pulsating auroras, *J. Geophys. Res.*, *115*, A10235, doi:10.1029/2009JA015148.

McEwen, D. J., E. Yee, B. A. Whalen, and A. W. Yau (1981), Electron energy measurements in pulsating auroras, *Can. J. Phys.*, *50*, 1106–1115.

Minatoya, H., N. Sato, T. Saemundsson, and T. Yoshino (1995), Absence of correlation between periodic pulsating auroras in geomagnetically conjugate areas, *J. Geomagn. Geoelectr.*, *47*, 583–598.

Miyoshi, Y., K. Sakaguchi, K. Shiokawa, J. Albert, D. Evans, M. Connors, and V. Jordanova (2010), Dual precipitation of relativistic electrons and ring current ions by EMIC waves, paper presented at 38th COSPAR Scientific Assembly, Bremen, Germany, 18–25 July.

Ni, B., R. M. Thorne, R. B. Horne, N. P. Meredith, Y. Y. Shprits, L. Chen, and W. Li (2011), Resonant scattering of plasma sheet electrons leading to diffuse auroral precipitation: 1. Evaluation for electrostatic electron cyclotron harmonic waves, J. Geophys. Res., 116, A04218, doi:10.1029/2010JA016232.

Nishimura, Y., et al. (2010), Identifying the driver of pulsating aurora, Science, 330, 81–84.

Nishimura, Y., et al. (2011a), Multievent study of the correlation between pulsating aurora and whistler mode chorus emissions, J. Geophys. Res., 116, A11221, doi:10.1029/2011JA016876.

Nishimura, Y., et al. (2011b), Estimation of magnetic field mapping accuracy using the pulsating aurora-chorus connection, Geophys. Res. Lett., 38, L14110, doi:10.1029/2011GL048281.

Nishiyama, T., T. Sakanoi, Y. Miyoshi, Y. Katoh, K. Asamura, S. Okano, and M. Hirahara (2011), The source region and its characteristic of pulsating aurora based on the Reimei observations, J. Geophys. Res., 116, A03226, doi:10.1029/2010JA015507.

Oguti, T., and K. Hayashi (1984), Multiple correlation between auroral and magnetic pulsations, 2. Determination of electric currents and electric fields around a pulsating auroral patch, J. Geophys. Res., 89(A9), 7467–7481.

Oyama, S., K. Shiokawa, J. Kurihara, T. T. Tsuda, S. Nozawa, Y. Ogawa, Y. Otsuka, and B. J. Watkins (2010), Lower-thermospheric wind fluctuations measured with an FPI during pulsating aurora at Tromsø, Norway, Ann. Geophys., 28, 1847–1857.

Royrvik, O., and T. N. Davis (1977), Pulsating aurora: Local and global morphology, J. Geophys. Res., 82(29), 4720–4740.

Ryoichi, F., O. Takasi, and Y. Tatsundo (1985), Relationships between pulsating auroras and field-aligned electric currents, Mem. Natl. Inst. Polar Res. Spec. Issue Jpn., 36, 95–103.

Safargaleev, V., and S. V. Osipenko (2000), TV observations of multiple arcs in pulsating and diffuse auroras, Phys. Chem. Earth, Part B, 25, 515–520.

Safargaleev, V., S. V. Osipenko, and A. N. Vasil'ev (2000), Spatially periodic arclike forms in the region of pulsating auroras, Geomagn. Aeron., 40, 716–723.

Saito, Y., S. Machida, M. Hirahara, T. Mukai, and H. Miyaoka (1992), Rocket observation of electron fluxes over a pulsating aurora, Planet. Space Sci., 40, 1043–1047.

Sandahl, I. (1985), Pitch angle scattering and particle precipitation in a pulsating aurora: An experimental study, KGI Rep. 185, Kiruna Geophys. Inst., Kiruna, Sweden.

Sandahl, I., L. Eliasson, and R. Lundin (1980), Rocket observations of precipitating electrons over a pulsating aurora, Geophys. Res. Lett., 7(5), 309–312.

Sato, N., M. Morooka, K. Minatoya, and T. Saemundsson (1998), Nonconjugacy of pulsating auroral patches near L = 6, Geophys. Res. Lett., 25(20), 3755–3758.

Sato, N., D. M. Wright, Y. Ebihara, M. Sato, Y. Murata, H. Doi, T. Saemundsson, S. E. Milan, M. Lester, and C. W. Carlson (2002), Direct comparison of pulsating aurora observed simultaneously by the FAST satellite and from the ground at Syowa, Geophys. Res. Lett., 29(21), 2041, doi:10.1029/2002GL015615.

Sato, N., D. M. Wright, C. W. Carlson, Y. Ebihara, M. Sato, T. Saemundsson, S. E. Milan, and M. Lester (2004), Generation region of pulsating aurora obtained simultaneously by the FAST satellite and a Syowa-Iceland conjugate pair of observatories, J. Geophys. Res., 109, A10201, doi:10.1029/2004JA010419.

Siren, J. C. (1975), Pulsating aurora in high-latitude satellite photographs, Geophys. Res. Lett., 2(12), 557–560.

Smith, M. J., D. A. Bryant, and T. Edwards (1980), Pulsations in auroral electrons and positive ions, J. Atmos. Terr. Phys., 42, 167–178.

Snyder, A. L., S.-I. Akasofu, and T. N. Davis (1974), Auroral substorms observed from above the north polar region by a satellite, J. Geophys. Res., 79(10), 1393–1402.

Stenbaek-Nielsen, H. C., and T. J. Hallinan (1979), Pulsating auroras: Evidence for noncollisional thermalization of precipitating electrons, J. Geophys. Res., 84(A7), 3257–3271.

Stenbaek-Nielsen, H. C., E. M. Wescott, and R. W. Peterson (1973), Pulsating auroras over conjugate areas, Antarct. J., 8, 246–247.

Störmer, C. (1948), Statistics of heights of various auroral forms from southern Norway second communication, J. Geophys. Res., 53, 251–264.

Störmer, C. (1955), The Polar Aurora, Clarendon Press, Oxford, U. K.

Thomas, R. W., and P. Rothwell (1979), A latitude effect in the periodicity of auroral pulsating patches, J. Atmos. Terr. Phys., 43, 1179–1184.

Thomas, R. W., and H. C. Stenbaek-Nielsen (1981), Recurrent propagating auroral forms in pulsating auroras, J. Atmos. Terr. Phys., 43, 243–254.

Thorne, R. M., B. Ni, X. Tao, R. B. Horne, and N. P. Meredith (2010), Scattering by chorus waves as the dominant cause of diffuse auroral precipitation, Nature, 467, 943–946.

Tromholt, S. (1880), Iagttagelser over Nordlys anstillede i Norge, Sverige og Danmark, Christiania Vid. Selsk. Handl., 6.

Vallance Jones, A. (1974), Aurora, Geophys. Astrophys. Monogr., vol. 9, D. Reidel, Dordrecht, The Netherlands.

Viereck, R. A., and H. C. Stenbaek-Nielsen (1985), Pulsating aurora and hydrogen emissions, J. Geophys. Res., 90(A11), 11,035–11,044.

Wahlund, J.-E., H. J. Opgenoorth, and P. Rothwell (1989), Observations of thin auroral ionization layers by EISCAT in connection with pulsating aurora, J. Geophys. Res., 94(A12), 17,223–17,233.

Ward, I. A. (1983), Pulsing hiss and associated phenomena—A morphological study, J. Atmos. Terr. Phys., 45, 289–301.

Ward, I. A., M. Lester, and R. W. Thomas (1982), Pulsing hiss, pulsating aurora and micropulsations, J. Atmos. Terr. Phys., 44, 931–938.

Watanabe, M., A. Kadokura, N. Sato, and T. Saemundsson (2007), Absence of geomagnetic conjugacy in pulsating auroras, *Geophys. Res. Lett.*, *34*, L15107, doi:10.1029/2007GL030469.

Whalen, B. A., J. R. Miller, and I. B. McDiarmid (1971), Energetic particle measurements in a pulsating aurora, *J. Geophys. Res.*, *76*(4), 978–986.

Williams, J. D. (2002), An investigation into pulsating aurora, Ph.D. thesis, Univ. of Wash., Seattle.

Wilson, C. R., J. V. Olson, and H. C. Stenbaek-Nielsen (2005), High trace-velocity infrasound from pulsating auroras at Fairbanks, Alaska, *Geophys. Res. Lett.*, *32*, L14810, doi:10.1029/2005GL 023188.

Yamamoto, T. (1983), On-off characteristics of luminosity fluctuations of pulsating auroras, *Mem. Natl. Inst. Polar Res. Spec. Issue Jpn.*, *26*, 124–134.

Yamamoto, T. (1988), On the temporal fluctuations of pulsating auroral luminosity, *J. Geophys. Res.*, *93*(A2), 897–911.

Yau, A. W., B. A. Whalen, and D. J. McEwen (1981), Rocket-borne measurements of particle pulsation in pulsating aurora, *J. Geophys. Res.*, *86*(A7), 5673–5681.

M. R. Lessard, Space Science Center, University of New Hampshire, Durham, NH 03824, USA. (marc.lessard@unh.edu)

Transpolar Arcs: Summary and Recent Results

Anita Kullen

Space and Plasma Physics, Royal Institute of Technology, Stockholm, Sweden

This review summarizes the current understanding of transpolar arcs. The following topics are covered: (1) transpolar arc (TPA) types, (2) influence of interplanetary magnetic field (IMF) B_y and B_z on shape and motion of TPAs, (3) temporal intensifications of TPAs, (4) substorms and TPAs, (5) solar wind energy coupling, (6) ionospheric convection and source regions of TPAs, (7) interhemispheric differences caused by IMF B_x, (8) interhemispheric differences caused by the Earth dipole tilt, and (9) magnetotail topology during TPAs.

1. INTRODUCTION

The first (ground-based) observations of high-latitude polar arcs have been documented already in the early twentieth century [*Mawson*, 1925]. In the 1970s, it became clear that such small-scale (a few kilometers wide) typically Sun-aligned arcs appear frequently during quiet geomagnetic conditions poleward of the main oval [e.g., *Ismail et al.*, 1977]. Large-scale polar arcs were not discovered until global auroral satellite images became available. In 1982, *Frank et al.* [1982] published an image from the DE 1 satellite, showing an auroral band that stretches over the entire polar cap connecting the nightside oval with its dayside counterpart. As this auroral configuration resembles the Greek letter "theta," *Frank et al.* [1982] suggested the name "theta aurora." Today, the expression "transpolar arc" (TPA) is more commonly used.

In this review, we refer to all polar arcs as TPAs that can be detected on global auroral imagers poleward of the main oval, although not all large-scale polar arcs are (during their entire existence) attached to the noon oval or appear in the center of the polar cap. Many TPAs that seem to be clearly separated from the oval when looking at global imagers are, in fact, attached to one oval side, as can be seen from particle

Auroral Phenomenology and Magnetospheric Processes: Earth and Other Planets
Geophysical Monograph Series 197
10.1029/2011GM001183

detectors [e.g., *Makita et al.*, 1991; *Newell et al.*, 2009]. TPAs can have considerable fine structure. An auroral structure looking like one single TPA on global imagers corresponds often to multiple individual arcs when using ground-based all-sky cameras [*Feldstein et al.*, 1995] or particle detectors [e.g., *Makita et al.*, 1991; *Cumnock et al.*, 2009].

As it is not possible to summarize 30 years of TPA research within a few pages, I refer here to the excellent review paper by *Zhu et al.* [1997] for a comprehensive coverage of the scientific knowledge about polar auroral arcs until 1997. In the present review, the knowledge until the mid-1990s regarding TPA types and their interplanetary magnetic field (IMF) dependence is briefly summarized. The main focus is on research results during the last 15 years.

2. MORPHOLOGY OF TRANSPOLAR ARCS

TPAs separate typically from the dawnside or duskside oval and move poleward. When reaching the noon-midnight meridian, they form a true theta aurora. Some TPAs even move over the entire polar region until they reach the opposite oval side [e.g., *Huang et al.*, 1989; *Cumnock et al.*, 1997]. However, most TPAs do not move considerably poleward after separation from the dawnside or duskside oval. *Murphree and Cogger* [1981] referred to these as oval-aligned arcs. When oval-aligned arcs appear simultaneously on both oval sides, the polar cap area becomes teardrop shaped with the narrow end of the teardrop pointing toward the Sun [*Meng*, 1981]. *Hones et al.* [1989] called such a configuration "horse-collar" aurora. Although TPAs are, in general, Sun-aligned [*Frank et al.*, 1986]

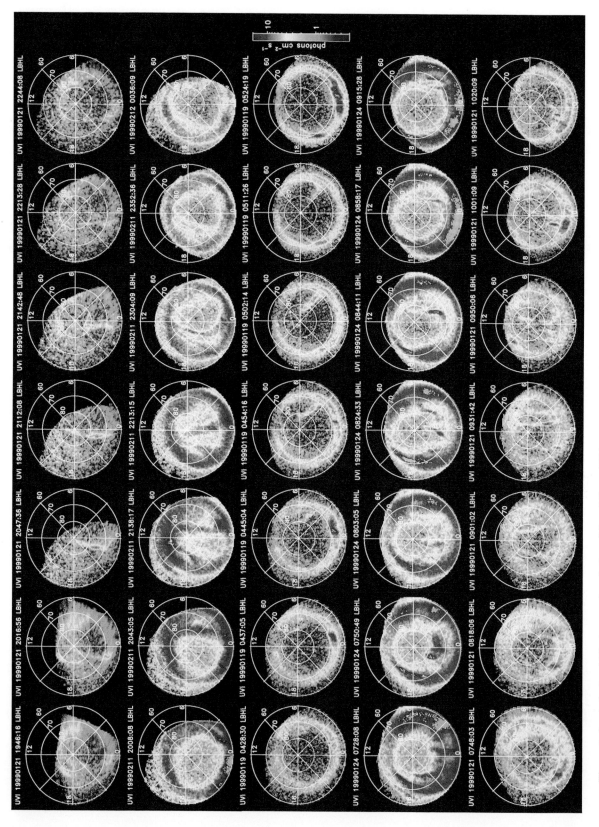

Figure 1. (top to bottom) The temporal evolution of five different TPA types, seen on the Polar UV imager: oval-aligned TPA, moving TPA, bending TPA, nightside-originating TPA, and multiple TPA event. From *Kullen et al.* [2002].

occasionally, they are strongly bent at their nightside oval connection point. Such hook-shaped arcs have been reported by *Ismail and Meng* [1982], *Murphree et al.* [1982], and *Gusev and Troshichev* [1986]. In rare occasions, TPAs evolve from the nightside oval [*Craven et al.*, 1986; *McEwen and Zhang*, 2000; *Kullen et al.*, 2002; *Goudarzi et al.*, 2008]. A few cases have been reported, where several arcs appear simultaneously in the polar cap, which may emerge from both oval sides and/or the nightside oval [*Newell et al.*, 1999; *Kullen et al.*, 2002].

With the launch of the Polar spacecraft in 1996, global UV images of the auroral oval became available with a time resolution of 1–4 min between the images, which allows to study the temporal evolution of TPAs in detail. Based on 200 large-scale polar arcs that were identified in three winter months of Polar UV images, *Kullen et al.* [2002] proposed a new classification of large-scale polar arcs (expanding on an earlier classification by *Gussenhoven*, 1982): oval-aligned TPAs (including those arcs that are clearly separated from the oval, but do not move considerably poleward before they disappear), moving TPAs, bending TPAs, midnight (nightside originating) TPAs, and multiple (three or more arcs) TPA events. Examples of each arc type are given in Figure 1. Bending TPAs (Figure 1, row 3) are extremely faint, hook-shaped arcs where the tailward connection point (often at the dawnside or duskside oval) stays nearly fixed while the dayside part bends more and more into the polar cap, i.e., they do not become transpolar. Due to the weak auroral intensity, these arcs are extremely difficult to discover using global imagers. Bending TPAs should not be confused with the hook-shaped TPAs reported earlier. Often, moving TPAs become, during part of their lifetime, hook-shaped (Figure 1, row 2). During one third of all TPA events, a small oval-aligned arc appears simultaneously on the opposite oval side (see, e.g., the small dusk arc in row 2 of Figure 1).

Newell et al. [2009] suggested a much different classification. They consider there exist only three types of polar arcs: small-scale Sun-aligned arcs, TPAs attached to one oval side, and, in rare cases, true theta aurorae that are clearly detached from both oval sides. This classification has its difficulty as it does not take into account the temporal evolution of TPAs. For example, true theta aurorae start their evolution by separating from one oval side, i.e., during part of their lifetime, they are attached to one oval side as well.

TPAs are rare events. They occur during only 10% of the time. For comparison, small-scale Sun-aligned arcs appear during at least 40% of the time [*Valladares et al.*, 1994]. The average lifetime of TPAs lies at 2 h. However, some oval-aligned and moving TPAs last up to 4–6 h [*Kullen et al.*, 2002; *Cumnock*, 2005].

3. TRANSPOLAR ARC DEPENDENCE ON IMF B_y AND IMF B_z

Different TPA types are not randomly occurring, but they arise in specific situations, as a response of the magnetosphere to specific IMF and solar wind conditions. It is known since a long time that polar arcs are a predominantly northward IMF phenomenon, which includes both, small-scale Sun-aligned arcs [*Lassen and Danielsen*, 1978; *Valladares et al.*, 1994] and TPAs [e.g., *Frank et al.*, 1982, 1986; *Gussenhoven*, 1982; *Ismail and Meng*, 1982]. *Kullen et al.* [2002] showed that, in nearly all cases, IMF B_z is predominantly northward during the last 2 h before (87%) and during TPA events (75%). After a southward turning of the IMF, TPAs fade within 15–65 min [*Rodriguez et al.*, 1997].

The dawn-duskward motion of both small-scale Sun-aligned arcs [*Valladares et al.*, 1994] and TPAs [*Kullen et al.*, 2002] is controlled by the direction of IMF B_y. In the Northern Hemisphere, most TPAs move into the direction of IMF B_y [*Frank et al.*, 1986; *Huang et al.*, 1989]. Only those TPAs that have emerged close to the dusk oval side during duskward IMF (on the dawn oval side during dawnward IMF) remain as oval-aligned TPAs close to the oval [*Elphinstone et al.*, 1990]. This includes even those static TPAs that are separated by quite many degrees latitude from the oval [*Kullen et al.*, 2002]. The few existing interhemispheric studies show that TPAs may exist simultaneously in both hemispheres and that IMF B_y has the opposite effect on location and motion of TPAs in the Southern Hemisphere [*Gorney et al.*, 1986; *Mizera et al.*, 1987; *Obara et al.*, 1988; *Craven et al.*, 1991; *Cumnock et al.*, 2006].

In the mid-1990s, it was discovered that an IMF B_y or B_z rotation may trigger TPAs to move poleward. *Newell and Meng* [1995] proposed that a change from northward to southward IMF is responsible for the occurrence of TPAs. *Cumnock et al.* [1997] assumed that TPAs appear due to a sign change of the IMF B_y component. *Chang et al.* [1998] suggested that either large variations in IMF B_y or an IMF turn to weakly southward IMF may trigger TPAs, a view later also shared by *Newell et al.* [1999]. Most events shown in these early reports about possible IMF triggers appear during varying IMF B_y and B_z conditions, which made it difficult to draw clear conclusions. Finally, *Cumnock* [2005] was able to show with a large statistical study, selecting 55 time periods with at least 3 h northward IMF and one clear IMF B_y sign change that theta aurora almost always occur during such conditions. *Newell et al.* [1997] and *Kullen et al.* [2002] each presented a polar arc event that appears in connection with an IMF B_z southward rotation during constant IMF B_y, which shows that an IMF B_z sign change from northward to southward may trigger a TPA as well. Furthermore, *Kullen et al.*

[2002] discovered that nearly all bending TPAs occur in connection with IMF B_z sign changes.

The rare multiple TPAs events appear commonly during strongly varying IMF conditions [*Kullen et al.*, 2002]. A nice example where several arcs fill the polar cap during the northward phase of a CME storm is presented in *Newell et al.* [1999]. However, it is extremely difficult to discern which IMF change triggers which arc in such events, especially as the reaction delay between solar wind changes and TPA evolution is not exactly known. From observations of solar wind pressure pulses that cause intensifications along a TPA, a time delay of 40 min can be estimated [*Liou et al.*, 2005]. *Kullen et al.* [2002] observed that bending arcs appear soon (1–20 min) after an IMF B_z sign change, while moving TPAs appear on average 60 min after an IMF B_y sign change. This indicates that different creation mechanisms may be responsible for different types of arcs.

4. TEMPORAL INTENSIFICATION OF TRANSPOLAR ARCS

TPAs are never continuously bright. The point of insertion of the TPA into the nightside auroral zone appears often as a bright spot [*Murphree et al.*, 1987]. *Hubert et al.* [2004] observed these in 90% of all studied TPA events. As can be seen in Figure 1, first row, such an auroral intensification may expand along the TPA. Localized brightenings in the nightside oval sector are common during quiet times [*Fillingim et al.*, 2000]. They can be elongated up to several degrees magnetic local time and may exist between 15 and 60 min. About one third of all quiet-time intensifications appear in connection with TPAs [*Kullen and Karlsson*, 2004]. It is, to date, unclear what triggers these [*Hubert et al.*, 2004; *Kullen et al.*, 2010]. *Murphree et al.* [1987] interpreted auroral brightening at the nightside end of a TPA as pseudobreakups, while *Hubert et al.* [2004] assumed these are different from usual pseudobreakups as they are dominated by proton injections.

The magnitude of solar wind density and pressure seems to have no influence on the TPA frequency [*Kullen et al.*, 2002]. High density has only a weak (negative) effect on TPA luminosity [*Kullen et al.*, 2008]. However, a sudden solar wind pressure pulse may change temporarily the brightness of a TPA. *Liou et al.* [2005] shows a pressure pulse event during a preexisting TPA. The pressure pulse leads to a temporal intensification of the auroral oval, which starts at noon within minutes after the pulse hits the magnetopause and spreads along the dawn oval to the nightside sector and further along the TPA. That the TPA brightens as the last part of the oval indicates that TPAs map to far tail regions. It is well known that a pressure pulse may cause a temporal intensification of the auroral oval [*Elphinstone et al.*, 1990;

Zhou and Tsurutani, 1999]. The sudden compression of the magnetosphere by shock impact probably causes more particles to enter the loss cone although the exact mechanism is not known. *Zhou and Tsurutani* [1999] suggested that it is connected to wave particle scattering. *Liou et al.* [2005] proposed a temporal widening of the loss cone due to a magnetic field line reconfiguration as a possible explanation for the auroral enhancement.

5. TRANSPOLAR ARCS AND SUBSTORMS

Although TPAs are a quiet-time phenomenon, they may survive several tens of minutes up to 2 h after substorm onset [*Kullen et al.*, 2002] until they fade gradually in antisunward direction [*Rodriquez et al.*, 1997]. Smaller substorms during predominantly northward IMF have been observed to take place without causing a preexisting TPA to disappear. Instead, the most intense substorm region may spread along the nightside part of the TPA [*Cumnock et al.*, 2000]. These observations fit to the assumption that TPAs are a far tail phenomenon [e.g., *Frank et al.*, 1986; *Elphinstone et al.*, 1990], whose existence is affected only tens of minutes after the onset of large substorms during which the entire tail undergoes large-scale topological changes.

About one third of all TPAs occur at the end of substorm recovery, most of these are oval-aligned TPAs [*Kullen et al.*, 2002]. Typically, IMF has turned northward at onset or during the expansion phase of substorms that are followed by TPA events. Such northward IMF recoveries are very dynamic, and multiple poleward boundary intensifications (PBIs) [*Lyons et al.*, 1999] occur along a strongly deformed poleward oval boundary that has reached highest latitudes. When returning to the ground state, the poleward boundary remains often active, while the region between poleward and equatorward boundary already starts to erode, resulting in a double oval [*Elphinstone et al.*, 1995]. For those recoveries where a double-oval-like structure occurs, the evolving TPA is connected to the poleward oval. PBIs are often observed at the TPA intersection point with the oval and may spread along the arc (note the similarity with quiet-time intensifications appearing on TPAs). On global images, this gives occasionally the impression as if the TPA evolution does not only include a separation from the oval side, but also expands from the nightside toward noon.

In rare cases, a TPA appears to evolve from the nightside oval boundary into the empty polar cap. Such nightside-originating arcs (referred to by *Kullen et al.* [2002] as midnight arcs) appear at the end of a dynamic substorm recovery and develop within minutes from a bright bulge at the oval boundary (maybe a PBI) that extends several degrees latitude into the polar cap (Figure 1, last row). Such TPAs show often a messy, patchy structure. A nice example of such an unusual

TPA is given in the work of *Goudarzi et al.* [2008]. Whether or not such TPAs form at a preexisting (on global imagers subvisual) much poleward expanded oval, remains to be shown. Theoretical studies have shown that the evolution of nightside-originating arcs could be possible on highly curved magnetic field lines where a ballooning instability can occur that could cause plasma sheet filaments to stretch tailward [*Rezhenov*, 1995] or high into the lobes [*Golovchanskaya et al.*, 2006].

6. SOLAR WIND ENERGY COUPLING DURING TRANSPOLAR ARC EVENTS

An important factor for TPAs to occur is that enough energy is available in the solar wind, which is transferred into the magnetosphere. The results of *Cumnock* [2005] demonstrate this nicely. Of the 55 time periods with IMF conditions that are favorable for TPA formation, TPAs occur in all cases except in those two cases where the magnetic solar wind energy flux vB^2 has lowest values [*Kullen et al.*, 2008]. Furthermore, *Kullen et al.* [2008] showed that vB^2 controls the luminosity of (dayside) TPAs. A possible interpretation of these observations is that for low values of vB^2, there is not enough energy available to draw high-latitude field-aligned currents (FAC) that produce aurora, while the magnetosphere topology is such that a TPA could occur.

That high IMF magnitude and fast solar wind speed are favorable for polar arcs to occur has already been reported by *Makita et al.* [1988] and *Gussenhoven* [1982], respectively. *Kullen et al.* [2002] discovered that the combination of high solar wind speed, strong IMF magnitude, and northward IMF gives the highest-occurrence frequency of TPAs. The best correlation with TPA occurrence appears for a solar wind energy coupling parameter, which *Kullen et al.* [2002] referred to as antiepsilon $\sim vB^2 \cos(\theta/2)^4$ with theta defined as the clock angle between IMF B_y and B_z (high for northward, zero for southward IMF). The antiepsilon parameter corresponds to the Akasofu-Perreault epsilon parameter [*Perreault and Akasofu*, 1978] with the sine function replaced by cosine. A similar coupling parameter [epsilon* $\sim (B_y^2 + B_z^2)^{1/2} \cos(\theta/2)$] has already been suggested by *Iijima et al.* [1984] to describe the correlation between solar wind and the intensity of FACs appearing in the dayside polar cap during northward IMF (NBZ currents). This current system is generally believed to be connected to polar arcs [e.g., *Iijima and Shibaji*, 1987].

7. IONOSPHERIC CONVECTION AND SOURCE REGIONS OF TRANSPOLAR ARCS

For northward IMF, reconnection takes place in the high-latitude lobes [e.g., *Reiff and Burch*, 1985; *Crooker*, 1992],

resulting in two or more reverse convection cells at highest latitudes and two viscous cells on lower latitudes that appear completely on closed field lines. The result is sunward flow on the dayside of the polar cap region [*Heppner and Maynard*, 1987]. On the nightside, weak flows with irregular patterns are observed. The global convection pattern during northward IMF is characterized by a variety of configurations with the number of cells increasing with increasing ratio of B_z/B_y. The convection pattern becomes additionally distorted by the occurrence of polar arcs. TPAs are known to be associated with upward FAC, as they nearly always lie on a location with div $E < 0$ [e.g., *Burke et al.*, 1979; *Frank et al.*, 1986; *Nielsen et al.*, 1990]. The FACs associated with TPAs do not always close locally, which has been observed to have an effect on the large-scale convection pattern [*Marklund and Blomberg*, 1991, and references therein]. There exists no general valid model about how the convection pattern changes globally during a TPA event. Based on satellite passes through the dayside polar cap, it has been reported that moving TPAs occur at the convection reversal boundary or inside the sunward part of a distorted two-cell [*Nielsen et al.*, 1990; *Chang et al.*, 1998] or three-cell pattern [*Jankowska et al.*, 1990]. *Cumnock et al.* [1997] observed how a large dominant cell changed to a four-cell pattern, while a TPA moved poleward.

The similar particle characteristics of TPAs and auroral arcs of the main oval and numerous reports of sunward flow on TPAs is the reason why it was commonly assumed that TPAs appear on closed field lines that originate in the plasma sheet or its boundary layer [e.g., *Frank et al.*, 1986; *Peterson and Shelley*, 1984]. As pointed out by *Liou et al.* [2005], most of these TPA studies are based on satellite paths through the dayside polar cap. On the nightside TPA part, a different flow direction has been observed. *Nielsen et al.* [1990] reported a case with sunward flow on the dayside part of the TPA, while the nightside end of the same TPA is colocated with antisunward flow. *Liou et al.* [2005] observed that the plasma flow direction on the dayside TPA was changing with IMF B_y between sunward and dawn-duskward (as expected from convection models), while the strong tailward plasma flow on the nightside TPA was not affected by the IMF orientation (see Figure 2). The existence of TPAs with antisunward flow along the nightside part suggests the low-latitude boundary layer (LLBL) along the magnetotail flanks as a possible source region for these events, an idea already put forward by *Murphree et al.* [1982] for horse-collar aurora. The (during northward IMF thick) LLBL contains plasma sheet-like particles that drift tailward along the outer part of the LLBL. *Liou et al.* [2005] proposed, the entire TPA maps to the LLBL, but the dayside part may be dominated by current states of dayside merging. *Eriksson et*

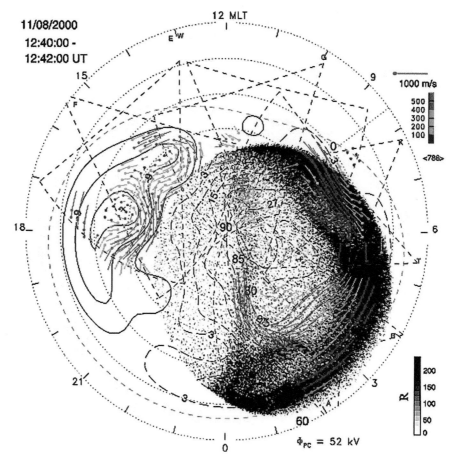

Figure 2. Horizontal ionospheric plasma flows observed with SuperDARN in the Northern Hemisphere during a TPA event (IMF $B_z \sim 7$ nT, IMF $B_y \sim 1$ nT). The colored dots indicate the location of the measurements and the colored "sticks" indicate the magnitude and direction (away from the dot) of the plasma flow. Plotted on the background is a 2 min average UVI image for the same time period. The field of view for each radar is plotted as dashed lines. From *Liou et al.* [2005].

al. [2005] observed separate convection cells at the dayside and nightside parts of a duskside oval-aligned TPA. They suggested two different drivers, with merging poleward of the cusp as the driver of the sunward (dayside) cell. This would mean that dayside and nightside parts of the TPA map to different magnetospheric regions.

8. IMF B_x EFFECTS ON TRANSPOLAR ARCS

IMF B_x has an asymmetric effect on high-latitude lobe reconnection in the different hemispheres. For negative (tailward) IMF B_x, solar wind and magnetospheric field lines become more antiparallel in the Northern Hemisphere than in the Southern Hemisphere. Theoretical studies [e.g., *Crooker*, 1992] indicate that northward IMF coupling with the magnetopause is stronger in one hemisphere when B_x/B_z is large. This has an impact on the polar cap potential. *Taguchi and Hoffman* [1995] observed an IMF B_x effect on the polar

cap potential drop during northward IMF, which they interpreted as a signature for a larger reconnection area in one hemisphere.

The more favorable reconnection in one hemisphere affects the occurrence frequency of high-latitude aurora. Small-scale polar arcs [*Lassen and Danielsen*, 1978], localized high-latitude dayside aurora [*Frey et al.*, 2004; *Frey*, 2007, and references therein], and TPAs [*Kullen et al.*, 2002] are in the Northern Hemisphere more commonly observed during negative than during positive IMF B_x. This has an effect on TPA conjugacy. *Ostgaard et al.* [2003] presented two cases where a bright TPA appears only in the hemisphere with a favorable sign of IMF B_x. The authors suggested that flow shears associated with convection reversal boundaries along TPAs are too weak to draw currents in the hemisphere with nonfavorable reconnection during strong IMF B_x. This is in agreement with observations of a strong IMF B_x dependence of NBZ current intensities [*Iijima et al.*, 1984].

Although not mentioned in their report, in *Ostgaards et al.'s* [2003] second example, polar arcs appear even in the non-favored hemisphere (in their other example, the image quality is too poor to detect similar structures). Instead of one bright TPA, two faint bending-type arcs and one small oval-aligned arc appear at the dawn and dusk oval sides, respectively. This is in agreement with statistical results by *Elphinstone et al.* [1990] and *Kullen et al.* [2002]. *Elphinstone et al.* [1990] found weak polar arcs to occur on both oval sides in the IMF B_x-nonfavored hemisphere and one bright Sun-aligned arc in the B_x-favored hemisphere. *Kullen et al.* [2002] reported that bending arcs are the only TPAs that appear typically in the B_x-nonfavored hemisphere and are commonly accompanied by a small oval-aligned arc on the opposite oval side. This suggests that (in addition to an IMF B_x dependence of FAC strength) interhemispheric differences in the field line topology during nonzero IMF B_x are such that FACs map far into the polar cap only in the B_x-favored hemisphere, but polar arcs may still occur close to the oval in the other hemisphere.

9. DIPOLE TILT EFFECTS ON TRANSPOLAR ARCS

The Earth dipole tilt has a similar effect on the magnetospheric topology as IMF B_x. It is defined as positive when the northern part of the Earth's dipole axis points sunward. A positive dipole tilt not only results in a more favorable magnetic field topology for high-latitude lobe reconnection but also results in a larger solar-illuminated polar cap area in the Northern Hemisphere. The question appears whether seasonal (dipole tilt) effects on TPAs have their cause in the magnetospheric field line topology or the illumination of the ionosphere.

The UV illumination of the ionospheric plasma results in enhanced conductivity. This is known to have a strong influence on the brightness of the oval. While diffuse auroral regions of the oval are brighter in the sunlit hemisphere [*Liou et al.*, 2001; *Shue et al.*, 2001] due to a positive correlation with ionospheric conductivity [*Fujii and Iijima*, 1987], discrete arcs (appearing preferably in the premidnight oval region) are suppressed in sunlight [*Newell et al.*, 1996]. This may be connected to the extremely low conductivity in that region, allowing for particle acceleration via the feedback instability mechanism [*Newell et al.*, 1996].

Kullen et al. [2008] observed that TPAs appearing during completely quiet times are brighter in the sunlit than in the dark hemisphere, i.e., they have the same dipole tilt dependence as diffuse aurora. A clear correlation between TPA luminosity and dipole tilt could only be established in the sunlit hemisphere. Even for large dipole tilts, TPAs do not disappear completely in the dark hemisphere, as can be seen, e.g., from the conjugate TPA example of *Craven et al.* [1991]. No correlation between TPA luminosity and IMF B_x has been found [*Kullen et al.*, 2008]. This suggests that the TPA brightness depends more on the ionospheric conductivity than on a favorable reconnection topology.

There is a general tendency that only the dayside part of the polar cap is directly influenced by the solar wind. TPA luminosity [*Kullen et al.*, 2008], plasma flows on TPAs [*Liou et al.*, 2005], NBZ currents [*Iijima et al.*, 1984], polar rain [*Newell et al.*, 2009], as well as the cross polar cap potential [*Reiff et al.*, 1981] all show clear correlations with solar wind and IMF parameters in the dayside part of the polar cap, but not in the nightside part.

10. MAGNETOTAIL TOPOLOGY DURING TRANSPOLAR ARCS

Meng [1981] proposed already in the early 1980s that polar arcs appearing close to one oval side lie on the boundary of a polewardly expanded closed field line region. *Makita et al.* [1991] confirmed this observationally. As closed field lines map to the plasma sheet or its boundary layer [e.g., *Frank et al.*, 1986], the appearance of a (oval-aligned) TPA would require a highly contracted polar cap and a dawn-dusk asymmetry or a twisting of the tail plasma sheet [*Makita et al.*, 1991]. *Cowley* [1981] proposed that IMF B_y exerts a torque on the magnetosphere about the Earth-Sun line such that the entire plasma sheet becomes increasingly twisted with distance from the Earth. Statistical studies [*Kaymaz et al.*, 1994] and MHD simulations [*Kaymaz et al.*, 1995] confirm such a twist. Later on, it was shown both observationally [*Owen et al.*, 1995] and by simulations [*Walker et al.*, 1999; *Kullen and Janhunen*, 2004] that the tail twist is much larger during northward than during southward IMF, which explains the northward IMF dependence of (oval-aligned) TPAs. The stronger twist occurs due to high-latitude lobe reconnection during northward IMF: B-field lines originating in the Northern Hemisphere are connected to solar wind field lines south of the magnetosphere. When the solar wind drags these field lines in an antisunward direction, the open field lines exert (in case of nonzero B_y) a strong torque on the magnetotail. For positive IMF B_y, the duskside (dawnside) far tail plasma sheet flanks map to the highest latitudes in the Northern (Southern) Hemisphere (see Figure 3a). This explains why most oval-aligned TPAs appear on the duskside (dawnside) of the northern (southern) auroral oval during constant positive IMF B_y.

Frank et al. [1982, 1986] and *Huang et al.* [1989] suggested that TPAs that are clearly separated from the oval (theta aurora) map to a bifurcated tail plasma sheet with a tongue of plasma reaching high into the northern and

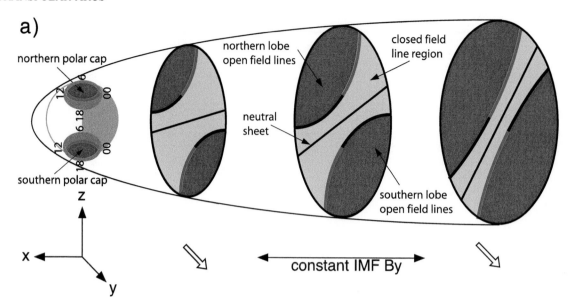

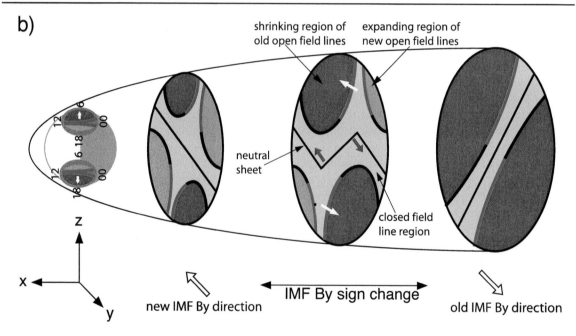

Figure 3. (a) Tail topology for oval-aligned TPAs: An IMF $B_y > 0$ induced (due to northward IMF strong) tail twist causes the high-latitude dusk (dawn) far tail flank to map to highest latitudes in the Northern (Southern) Hemisphere. The blue line at each polar cap boundary indicates the possible location of an oval-aligned TPA. (b) Tail topology for moving TPAs: A (tailward propagating) change in the tail twist, caused by an IMF B_y sign reversal during northward IMF, leads to a bifurcation of the closed field line region in the near-Earth tail that maps to a (polewardly moving) closed-field line strip in the northern and southern polar cap (blue lines) on which a moving TPA may appear.

southern tail lobes. Observation of filamentary plasma sheet extensions into the lobes during a TPA event [*Huang et al.*, 1989] shows that such a plasma sheet topology is possible.

Based on observations that an IMF B_y sign change can trigger a moving TPA [*Cumnock et al.*, 1997], *Kullen* [2000] developed a TPA model, which is based on the idea that after

an IMF B_y rotation, the magnetotail changes its twist not at once, but successively from near-Earth to far-tail regions. Modifying the *Tsyganenko* [1989] magnetosphere model, *Kullen* [2000] showed that opposite twists of near-Earth and far-tail plasma sheet cause a field line topology where closed field lines from the distant tail map to a finger of closed field lines that bifurcates the polar cap (see Figure 3b). The "finger" of closed field lines moves from one oval side to the other, while the new magnetotail twist direction expands tailward. Assuming that TPAs occur on (or at the boundary of) closed field lines, the polewardly moving "finger" represents the possible location of a moving TPA.

MHD simulations with the Winglee model (R. M. Winglee, personal communication, 1998), the Fedder and Lyons model [*Slinker et al.*, 2001], the Gumics-3 MHD code [*Kullen and Janhunen*, 2004], the BATSRUS MHD model [*Naehr and Toffoletto*, 2004], the Tanaka model [*Tanaka*, 2004] and ISM model [*Maynard et al.*, 2003] differ in many details, but all confirm the appearance of closed field line strips in the northern and southern polar cap for an IMF B_y sign reversal during northward IMF (for southward IMF, the tail twist is too weak to cause a polar cap bifurcation). The closed field line strips develop at opposite sides of the oval and move into opposite directions in the different hemispheres, while a rotation of the tail twist into the opposite direction takes place. Thus, the observed conjugacy of TPAs in different hemispheres [e.g., *Craven et al.*, 1991] appears in all MHD models. Possible dipole tilt and IMF B_x effects [*Kullen et al.*, 2008; *Ostgaard et al.*, 2003] on the conjugacy of TPAs have not been studied with these simulations. *Kullen and Janhunen* [2004] examine the reconfiguration of the tail in detail. They find that the twist change does not only propagate tailward (as predicted by *Kullen* [2000]) but also from the flanks to the tail center, which additionally complicates the resulting tail topology (see Figure 3b). Furthermore, the (closed field line region) tail expands during the B_y rotation. As the closed field line strip bifurcating the polar cap maps to the most distant tail regions, an increase of the tail length explains, in part, the long TPA lifetime (however, the MHD model TPAs last only 40 min compared to a 2 h lifetime of real TPAs). The closed field line strips do not reach the dayside oval in any of the different MHD runs. In case this reflects a physical reality, it would mean that the dayside TPA part does not lie on closed field lines. Due to numerical diffusion and low resolution in the tail, it is not possible to reproduce thin FACs or reliable convection maps of the high-latitude ionosphere (the convection pattern deviate extremely between the different MHD codes for similar IMF B_y change runs). Thus, MHD simulations should not be used for detailed studies of the convection and current pattern during TPA events.

As predicted in the conceptual TPA models by *Newell et al.* [1997] and *Chang et al* [1998], a new region of open field lines appears in the simulations at the duskside polar cap close to noon after the reconnection site jumped from dusk to dawn due to the IMF B_y flip. A closed field line strip appears between the old (dawnside) polar cap and the new (duskside) open flux region (see Figure 3b). An expansion of the new open flux region at the expense of the old open flux region is connected to the poleward motion of the closed field line strip in the polar cap.

MHD simulations of an IMF B_z southward turning during nonzero IMF B_y failed to produce any signatures that could explain how a TPA emerges after an IMF B_z south turn [*Maynard et al.*, 2003; A. Kullen, unpublished results with the GUMICS-3 model, 2002]. As suggested by *Newell et al.* [1997] and *Chang et al.* [1998], even in this case, a merging line jump (from northward high latitude to southward dayside reconnection) would lead to a new region of open flux to appear poleward of the cups. How this could be connected to long-lasting TPAs or explain their poleward motion, remains to be shown.

Other recent TPA models involve the concept of a return flow blockage in convection cells due to an IMF B_y sign change, which would lead to a pileup of closed flux that protrudes into the polar cap due to the simultaneous existence of old and new open field line regions [*Tanaka et al.*, 2004] or due to competition between the conjugate hemispheres [*Milan et al.*, 2005].

Despite comprehensive knowledge about TPAs, a global model is still missing that explains consistently how ionospheric convection, TPA-associated currents, particles, and magnetospheric source regions are linked together. Simultaneous imaging of the northern and southern auroral ionospheres with the help of high-resolution global imagers (combined with data from low-altitude orbiting satellites such as DMSP) would help to answer many open questions regarding the ionosphere-magnetosphere coupling during TPA events.

REFERENCES

Burke, W. J., M. C. Kelley, R. C. Sagalyn, M. Smiddy, and S. T. Lai (1979), Polar cap electric field structures with a northward interplanetary magnetic field, *Geophys. Res. Lett.*, *6*(1), 21–24.

Chang, S.-W., et al. (1998), A comparison of a model for the theta aurora with observations from Polar, Wind, and SuperDARN, *J. Geophys. Res.*, *103*(A8), 17,367–17,390.

Cowley, S. W. H. (1981), Magnetospheric asymmetries associated with the y-component of the IMF, *Planet. Space Sci.*, *29*, 79–96.

Craven, J. D., L. A. Frank, C. T. Russell, E. J. Smith, and R. P. Lepping (1986), Global auroral responses to magnetospheric compressions by shocks in the solar wind: Two case studies, in *Solar Wind–Magnetosphere Coupling*, edited by Y. Kamide and J. A. Slavin, p. 367, Terra Sci., Tokyo.

Craven, J. D., J. S. Murphree, L. A. Frank, and L. L. Cogger (1991), Simultaneous optical observations of transpolar arcs in the two polar caps, *Geophys. Res. Lett.*, *18*(12), 2297–2300.

Crooker, N. U. (1992), Reverse convection, *J. Geophys. Res.*, *97*(A12), 19,363–19,372.

Cumnock, J. A. (2005), High-latitude aurora during steady northward interplanetary magnetic field and changing IMF B_y, *J. Geophys. Res.*, *110*, A02304, doi:10.1029/2004JA010867.

Cumnock, J. A., J. R. Sharber, R. A. Heelis, M. R. Hairston, and J. D. Craven (1997), Evolution of the global aurora during positive IMF B_z and varying IMF B_y conditions, *J. Geophys. Res.*, *102*(A8), 17,489–17,497.

Cumnock, J. A., J. F. Spann, G. A. Germany, L. G. Blomberg, W. R. Coley, C. R. Clauer, and M. J. Brittnacher (2000), Polar UVI observations of auroral oval intensifications during a transpolar arc event on December 7, 1996, *Rep. TRITA-ALP-2000-01*, R. Inst. of Technol., Stockholm.

Cumnock, J. A., L. G. Blomberg, I. I. Alexeev, E. S. Belenkaya, S. Y. Bobrovnikov, and V. V. Kalegaev (2006), Simultaneous polar aurorae and modelled convection patterns in both hemispheres, *Adv. Space Res.*, *38*, 1685–1693, doi:10.1016/j.asr.2005.04.105.

Cumnock, J. A., L. G. Blomberg, A. Kullen, T. Karlsson, and K. Sundberg (2009), Small-scale characteristics of extremely high-latitude aurora, *Ann. Geophys.*, *27*, 3335–3347.

Elphinstone, R. D., K. Jankowska, J. S. Murphree, and L. L. Cogger (1990), The configuration of the auroral distribution for interplanetary magnetic field B_z northward, 1. IMF B_x and B_y dependencies as observed by the Viking satellite, *J. Geophys. Res.*, *95*(A5), 5791–5804.

Elphinstone, R. D., et al. (1995), The double oval UV auroral distribution 2. The most poleward arc system and the dynamics of the magnetotail, *J. Geophys. Res.*, *100*(A7), 12,093–12,102.

Eriksson, S., et al. (2005), On the generation of enhanced sunward convection and transpolar aurora in the high-latitude ionosphere by magnetic merging, *J. Geophys. Res.*, *110*, A11218, doi:10.1029/2005JA011149.

Feldstein, Y. I., P. T. Newell, I. Sandahl, J. Woch, S. V. Leontjev, and V. G. Vorobjev (1995), Structure of auroral precipitation during a theta aurora from multisatellite observations, *J. Geophys. Res.*, *100*(A9), 17,429–17,442.

Fillingim, M. O., G. K. Parks, L. J. Chen, M. Brittnacher, G. A. Germany, J. F. Spann, D. Larson, and R. P. Lin (2000), Coincident POLAR/UVI and WIND observations of pseudobreakups, *Geophys. Res. Lett.*, *27*(9), 1379–1382.

Frank, L. A., J. D. Craven, J. L. Burch, and J. D. Winningham (1982), Polar views of the Earth's aurora with Dynamics Explorer, *Geophys. Res. Lett.*, *9*(9), 1001–1004.

Frank, L. A., et al. (1986), The theta aurora, *J. Geophys. Res.*, *91*(A3), 3177–3224.

Frey, H. U. (2007), Localized aurora beyond the auroral oval, *Rev. Geophys.*, *45*, RG1003, doi:10.1029/2005RG000174.

Frey, H. U., N. Østgaard, T. J. Immel, H. Korth, and S. B. Mende (2004), Seasonal dependence of localized, high-latitude dayside aurora (HiLDA), *J. Geophys. Res.*, *109*, A04303, doi:10.1029/2003JA010293.

Fujii, R., and T. Iijima (1987), Control of the ionospheric conductivities on large-scale Birkeland current intensities under geomagnetic quiet conditions, *J. Geophys. Res.*, *92*(A5), 4505–4513.

Golovchanskaya, I. V., A. Kullen, Y. P. Maltsev, and H. Biernat (2006), Ballooning instability at the plasma sheet–lobe interface and its implications for polar arc formation, *J. Geophys. Res.*, *111*, A11216, doi:10.1029/2005JA011092.

Gorney, D. J., D. S. Evans, M. S. Gussenhoven, and P. F. Mizera (1986), A multiple-satellite observation of the high-latitude auroral activity on January 11, 1983, *J. Geophys. Res.*, *91*(A1), 339–346.

Goudarzi, A., M. Lester, S. E. Milan, and H. U. Frey (2008), Multi-instrumentation observations of a transpolar arc in the northern hemisphere, *Ann. Geophys.*, *21*, 201–210.

Gusev, M. G., and O. A. Troshichev (1986), Hook-shaped arcs in dayside polar cap and their relationship to the IMF, *Planet. Space Sci.*, *34*, 489–496.

Gussenhoven, M. S. (1982), Extremely high latitude auroras, *J. Geophys. Res.*, *87*(A4), 2401–2412.

Heppner, J. P., and N. C. Maynard (1987), Empirical high-latitude electric field models, *J. Geophys. Res.*, *92*(A5), 4467–4489.

Hones, E. W., Jr., J. D. Craven, L. A. Frank, D. S. Evans, and P. T. Newell (1989), The horse-collar aurora: A frequent pattern of the aurora in quiet times, *Geophys. Res. Lett.*, *16*(1), 37–40.

Huang, C. Y., J. D. Craven, and L. A. Frank (1989), Simultaneous observations of a theta aurora and associated magnetotail plasmas, *J. Geophys. Res.*, *94*(A8), 10,137–10,143.

Hubert, B., J. C. Gérard, S. A. Fuselier, S. B. Mende, and J. L. Burch (2004), Proton precipitation during transpolar auroral events: Observations with the IMAGE-FUV imagers, *J. Geophys. Res.*, *109*, A06204, doi:10.1029/2003JA010136.

Iijima, T., and T. Shibaji (1987), Global characteristics of northward IMF-associated (NBZ) field-aligned currents, *J. Geophys. Res.*, *92*(A3), 2408–2424.

Iijima, T., T. A. Potemra, L. J. Zanetti, and P. F. Bythrow (1984), Large-scale Birkeland currents in the dayside polar region during strongly northward IMF: A new Birkeland current system, *J. Geophys. Res.*, *89*(A9), 7441–7452.

Ismail, S., and C.-I. Meng (1982), A classification of polar cap auroral arcs, *Planet. Space Sci.*, *30*, 319–330.

Ismail, S., D. D. Wallis, and L. L. Cogger (1977), Characteristics of polar cap sun-aligned arcs, *J. Geophys. Res.*, *82*(29), 4741–4749.

Jankowska, K., R. D. Elphinstone, J. S. Murphree, L. L. Cogger, D. Hearn, and G. Marklund (1990), The configuration of the auroral distribution for interplanetary magnetic field B_z northward, 2. Ionospheric convection consistent with Viking observations, *J. Geophys. Res.*, *95*(A5), 5805–5816.

Kaymaz, Z., G. L. Siscoe, J. G. Luhmann, R. P. Lepping, and C. T. Russell (1994), Interplanetary magnetic field control of magnetotail magnetic field geometry: IMP 8 observations, *J. Geophys. Res.*, *99*(A6), 11,113–11,126.

Kaymaz, Z., G. Siscoe, J. G. Luhmann, J. A. Fedder, and J. G. Lyon (1995), Interplanetary magnetic field control of magnetotail field: IMP 8 data and MHD model compared, *J. Geophys. Res.*, *100*(A9), 17,163–17,172.

Kullen, A. (2000), The connection between transpolar arcs and magnetotail rotation, *Geophys. Res. Lett.*, *27*(1), 73–76.

Kullen, A., and P. Janhunen (2004), Relation of polar auroral arcs to magnetotail twisting and IMF rotation: A systematic MHD simulation study, *Ann. Geophys., 22*, 951–970.

Kullen, A., and T. Karlsson (2004), On the relation between solar wind, pseudobreakups, and substorms, *J. Geophys. Res., 109*, A12218, doi:10.1029/2004JA010488.

Kullen, A., M. Brittnacher, J. A. Cumnock, and L. G. Blomberg (2002), Solar wind dependence of the occurrence and motion of polar auroral arcs: A statistical study, *J. Geophys. Res., 107*(A11), 1362, doi:10.1029/2002JA009245.

Kullen, A., J. A. Cumnock, and T. Karlsson (2008), Seasonal dependence and solar wind control of transpolar arc luminosity, *J. Geophys. Res., 113*, A08316, doi:10.1029/2008JA013086.

Kullen, A., T. Karlsson, J. A. Cumnock, and T. Sundberg (2010), Occurrence and properties of substorms associated with pseudobreakups, *J. Geophys. Res., 115*, A12310, doi:10.1029/2010JA 015866.

Lassen, K., and C. Danielsen (1978), Quiet time pattern of auroral arcs for different directions of the interplanetary magnetic field in the *Y-Z* plane, *J. Geophys. Res., 83*(A11), 5277–5284.

Liou, K., P. T. Newell, and C.-I. Meng (2001), Seasonal effects on auroral particle acceleration and precipitation, *J. Geophys. Res., 106*, 5531–5542.

Liou, K., J. M. Ruohoniemi, P. T. Newell, R. Greenwald, C.-I. Meng, and M. R. Hairston (2005), Observations of ionospheric plasma flows within theta auroras, *J. Geophys. Res., 110*, A03303, doi:10.1029/2004JA010735.

Lyons, L. R., T. Nagai, G. T. Blanchard, J. C. Samson, T. Yamamoto, T. Mukai, A. Nishida, and S. Kokubun (1999), Association between Geotail plasma flows and auroral poleward boundary intensifications observed by CANOPUS photometers, *J. Geophys. Res., 104*(A3), 4485–4500.

Makita, K., C.-I. Meng, and S.-I. Akasofu (1988), Latitudinal electron precipitation patterns during large and small IMF magnitudes for northward IMF conditions, *J. Geophys. Res., 93*(A1), 97–104.

Makita, K., C.-I. Meng, and S.-I. Akasofu (1991), Transpolar auroras, their particle precipitation, and IMF B_y component, *J. Geophys. Res., 96*(A8), 14,085–14,095.

Marklund, G. T., and L. G. Blomberg (1991), On the influence of localized electric fields and field-aligned currents associated with polar arcs on the global potential distribution, *J. Geophys. Res., 96*(A8), 13,977–13,983.

Mawson, D. (1925), Australasian Antarctic expedition 1911–1914, *Sci. Rep., Ser. B*, vol. II, part I, records of the Aurora Polaris, Sydney, Australia.

Maynard, N. C., et al. (2003), Responses of the open–closed field line boundary in the evening sector to IMF changes: A source mechanism for Sun-aligned arcs, *J. Geophys. Res., 108*(A1), 1006, doi:10.1029/2001JA000174.

McEwen, D. J., and Y. Zhang (2000), A continuous view of the dawn-dusk polar cap, *Geophys. Res. Lett., 27*(4), 477–480.

Meng, C.-I. (1981), Polar cap arcs and the plasma sheet, *Geophys. Res. Lett., 8*(3), 273–276.

Milan, S. E., B. Hubert, and A. Grocott (2005), Formation and motion of a transpolar arc in response to dayside and nightside reconnection, *J. Geophys. Res., 110*, A01212, doi:10.1029/ 2004JA010835.

Mizera, P. F., D. J. Gorney, and D. S. Evans (1987), On the conjugacy of the aurora: High and low latitudes, *Geophys. Res. Lett., 14*(3), 190–193.

Murphree, J. S., and L. L. Cogger (1981), Observed connections between apparent polar cap features and the instantaneous diffuse auroral oval, *Planet. Space Sci., 29*, 1143–1149.

Murphree, J. S., C. D. Anger, and L. L. Cogger (1982), The instantaneous relationship between polar-cap and oval auroras at times of northward interplanetary magnetic field, *Can. J. Phys., 60*, 349–356.

Murphree, J. S., L. L. Cogger, C. D. Anger, D. D. Wallis, and G. G. Shepherd (1987), Oval intensifications associated with polar arcs, *Geophys. Res. Lett., 14*(4), 403–406.

Naehr, S. M., and F. R. Toffoletto (2004), Quantitative modeling of the magnetic field configuration associated with the theta aurora, *J. Geophys. Res., 109*, A07202, doi:10.1029/2003JA010191.

Newell, P. T., and C.-I. Meng (1995), Creation of theta-auroras: The isolation of plasma sheet fragments in the polar cap, *Science, 270*, 1338–1341.

Newell, P. T., C.-I. Meng, and K. M. Lyons (1996), Suppression of discrete aurorae by sunlight, *Nature, 381*, 766–767.

Newell, P. T., D. Xu, C.-I. Meng, and M. G. Kivelson (1997), Dynamical polar cap: A unifying approach, *J. Geophys. Res., 102*(A1), 127–139.

Newell, P. T., K. Liou, C.-I. Meng, M. J. Brittnacher, and G. Parks (1999), Dynamics of double-theta aurora: Polar UVI study of January 10–11, 1997, *J. Geophys. Res., 104*(A1), 95–104.

Newell, P. T., K. Liou, and G. R. Wilson (2009), Polar cap particle precipitation and aurora: Review and commentary, *J. Atmos. Sol. Terr. Phys., 71*, 199–215.

Nielsen, E., J. D. Craven, L. A. Frank, and R. A. Heelis (1990), Ionospheric flows associated with a transpolar arc, *J. Geophys. Res., 95*(A12), 21,169–21,178.

Obara, T., M. Kitayama, T. Mukai, N. Kaya, J. S. Murphree, and L. L. Cogger (1988), Simultaneous observations of Sun-aligned polar cap arcs in both hemispheres by EXOS-C and Viking, *Geophys. Res. Lett., 15*(7), 713–716.

Østgaard, N., S. B. Mende, H. U. Frey, L. A. Frank, and J. B. Sigwarth (2003), Observations of non-conjugate theta aurora, *Geophys. Res. Lett., 30*(21), 2125, doi:10.1029/2003GL017914.

Owen, C. J., J. A. Slavin, I. G. Richardson, N. Murphy, and R. J. Hynds (1995), Average motion, structure and orientation of the distant magnetotail determined from remote sensing of the edge of the plasma sheet boundary layer with $E > 35$ keV ions, *J. Geophys. Res., 100*(A1), 185–204.

Perreault, P., and S. I. Akasofu (1978), Study of geomagnetic storms, *Geophys. J. R. Astron. Soc., 54*, 547–573.

Peterson, W. K., and E. G. Shelley (1984), Origin of the plasma in a cross-polar cap auroral feature (theta aurora), *J. Geophys. Res., 89*(A8), 6729–6736.

Reiff, P. H., and J. L. Burch (1985), IMF B_y-dependent plasma flow and Birkeland currents in the dayside magnetosphere, 2. A global model for northward and southward IMF, *J. Geophys. Res.*, *90*(A2), 1595–1609.

Reiff, P. H., R. W. Spiro, and T. W. Hill (1981), Dependence of polar cap potential drop on interplanetary parameters, *J. Geophys. Res.*, *86*(A9), 7639–7648.

Rezhenov, B. V. (1995), A possible mechanism for theta aurora formation, *Ann. Geophys.*, *13*, 698–703.

Rodriguez, J. V., C. E. Valladares, K. Fukui, and H. A. Gallagher Jr. (1997), Antisunward decay of polar cap arcs, *J. Geophys. Res.*, *102*(A12), 27,227–27,247.

Shue, J.-H., P. T. Newell, K. Liou, and C.-I. Meng (2001), Influence of interplanetary magnetic field on global auroral patterns, *J. Geophys. Res.*, *106*, 5913–5926.

Slinker, S. P., J. A. Fedder, D. J. McEwen, Y. Zhang, and J. G. Lyon (2001), Polar cap study during northward interplanetary magnetic field on 19 January 1998, *Phys. Plasmas*, *8*, 1119–1126.

Taguchi, S., and R. A. Hoffman (1995), B_X control of polar cap potential for northward interplanetary magnetic field, *J. Geophys. Res.*, *100*(A10), 19,313–19,320.

Tanaka, T., T. Obara, and M. Kunitake (2004), Formation of the theta aurora by a transient convection during northward interplanetary magnetic field, *J. Geophys. Res.*, *109*, A09201, doi:10.1029/2003JA010271.

Tsyganenko, N. A. (1989), A magnetospheric magnetic field model with a warped tail current sheet, *Planet. Space Sci.*, *37*, 5–20.

Valladares, C. E., H. C. Carlson Jr., and K. Fukui (1994), Interplanetary magnetic field dependency of stable sun-aligned polar cap arcs, *J. Geophys. Res.*, *99*(A4), 6247–6272.

Walker, R. J., R. L. Richard, T. Ogino, and M. Ashour-Abdalla (1999), The response of the magnetotail to changes in the IMF orientation: The magnetotail's long memory, *Phys. Chem. Earth, Part C*, *24*, 221–227.

Zhou, X., and B. T. Tsurutani (1999), Rapid intensification and propagation of the dayside aurora: Large scale interplanetary pressure pulses (fast shocks), *Geophys. Res. Lett.*, *26*(8), 1097–1100.

Zhu, L., R. W. Schunk, and J. J. Sojka (1997), Polar cap arcs: A review, *J. Atmos. Sol. Terr. Phys.*, *59*, 1087–1126.

A. Kullen, Space and Plasma Physics, Royal Institute of Technology, SE-100 44 Stockholm, Sweden. (kullen@kth.se)

Coherence in Auroral Fine Structure

Joshua Semeter

Department of Electrical and Computer Engineering and Center for Space Physics, Boston University, Boston, Massachusetts, USA

Auroras exhibit coherent behavior across a vast hierarchy of scales, from the expansion and contraction of the global auroral ovals to the formation of periodic decameter-scale structure within an auroral breakup. This chapter provides a framework for interpreting motion and periodicity in auroras, with a focus on the finest scales inherent in the phenomenon. Our objective is to connect image plane dynamics to space-time dynamics in the electromagnetic fields of the near-Earth magnetosphere. Our approach appeals to classical notions of phase and group velocity, borrowing some Fourier domain concepts from the field of visual perception. The dynamics of interest can only be observed within a few degrees of the magnetic zenith due to perspective effects, as illustrated through observational examples and numerical simulation.

1. INTRODUCTION

Of all the known auroras in our solar system (Earth, Mars, Saturn, Titan, Triton, Jupiter, Io, Uranus, and Neptune), the terrestrial aurora is the only one that may be studied at the fundamental scales inherent in the phenomenon. All auroras are produced by fluxes of magnetic field-aligned charged particle (electrons and ions). The altitude distribution of optical emissions is governed by the well-established theory of particle penetration into a barometrically increasing gas [*Chamberlain*, 1961]. Thus, when we refer to spatial scales in the aurora, we are invariably referring to scales in the transverse plane. Auroral luminosity provides a measure of the energy imparted on the causative particle flux. Hence, auroral imagery provides a time-dependent map of the parallel electric field responsible for the acceleration. Although much has been learned about large-scale auroral morphology and its relationship to magnetospheric dynamics [e.g., *Johnson et al.*, 1998], our understanding of decameter-scale structure in the aurora remains incomplete. This chapter considers the topic from the perspective of periodicity and motion in the optical forms.

The multiplicity of transverse scales in the aurora is well recognized [*Galperin*, 2002], and spatial periodicity can be found at each scale. The literature on periodic phenomena in auroras is broadly organized according to distortions that develop along an arc (spirals, folds, curls, omega bands) [*Hallinan*, 1976; *Lysak and Song*, 1996; *Trondsen and Cogger*, 1998; *Vogt et al.*, 1999; *Partamies et al.*, 2001] and the formation of parallel arcs [*Atkinson*, 1970; *Sato*, 1978; *Trondsen et al.*, 1997; *Semeter and Blixt*, 2006]. In each category, the periodicity has a self-similar nature across widely disparate scales, suggesting the presence of some unifying principles. One is the connection of auroral vorticity to shear in magnetic and velocity fields [*Lysak and Song*, 1996]. Another is the connection of multiple arc systems to the spreading of power across magnetic lines of force [*Semeter et al.*, 2008].

Even during the most dynamic periods (i.e., substorm breakup), the aurora is never truly turbulent, but rather contains coherent patterns that may only be identified in imagery confined to a narrow range of angles to the magnetic zenith [*Semeter et al.*, 2008]. The prevailing theory holds that fine-scale coherence in the optical phenomenon indicates the presence of coherence in the dispersive Alfvén wavefield modulating, or producing, the suprathermal particle flux [*Stasiewicz et al.*, 2000], but direct observational evidence remains elusive. The purpose of this chapter is to introduce a framework for relating time-dependent image plane morphologies to dynamics in the overlying electromagnetic fields.

Auroral Phenomenology and Magnetospheric Processes: Earth and Other Planets
Geophysical Monograph Series 197
10.1029/2011GM001196

The primary diagnostic tool in this work is high frame rate, narrow-field video. The successful use of video processing in the study of any natural phenomenon requires an understanding of three component problems: (1) spatial aspects (transformation between image and real-world coordinates), (2) spectral aspects (Fourier analysis in space and time), and (3) relating reduced data to physical models. The chapter is organized along these challenges. We begin by addressing perspective effects in mapping image plane morphologies to physical coordinates (section 2). We then introduce a basic mathematical framework for understanding motion and periodicity in auroral video based on the classical concepts of phase and group velocity (section 3). Finally, we connect these results to theoretical models of current sheet and particle flux structuring (section 4).

2. PERSPECTIVE CONSIDERATIONS

When we speak of auroral arc width, we speak of a parameter that is not directly observed. An auroral image represents a map of line-of-sight integrated intensities at a discrete set of angular positions. Assumptions are required to relate this perspective projection to spatial variability in the $B_\perp$ plane. A common analysis step in the processing of auroral imagery is to assume that all emissions arise from a surface of constant altitude (e.g., ~110 km for energetic auroras). This allows one to project auroral morphology onto geophysical coordinates [*Mende et al.*, 2008]. But this map projection is only meaningful for auroras whose horizontal extent is much greater than the vertical extent. In practical terms, this means horizontal scales greater than ~20 km.

Irrespective of this constraint, intuition often leads to a correct interpretation for large-scale auroral features, such as those highlighted in Figure 1. For instance, the space-based images in Figures 1a and 1b represent mostly horizontal morphology, the rayed forms in Figure 1c represent vertical structure, and Figure 1d represents a mixture of horizontal and vertical structure. However, for fine-scale structure, perspective bias becomes extreme, and intuition can fail. The cause of the difficulty lies in the fact that the altitude extent of the aurora is tens of kilometers, independent of horizontal scale. As auroral width decreases, the vertical-to-horizontal aspect ratio increases to the point where the identification of $B_\perp$ scale becomes ambiguous, and spatial periodicity becomes undetectable, for structures lying more than a few degrees from magnetic zenith [*Maggs and Davis*, 1968]. These issues are addressed in more detail below.

2.1. Kinks Versus Rays

Figure 2 illustrates the extreme path length bias in narrow-field imagery of dynamic auroras. The observations were recorded during an auroral breakup over Poker Flat on 23 March 2007 [*Semeter et al.*, 2008]. The left panel shows a sharply kinked auroral form intersecting the magnetic zenith. Careful inspection reveals that this form is composed of at least two parallel arcs of ~100 m width.

The subsequent panels illustrate how perception of this form changes as it propagates to a position ~4° from zenith over the ensuing 2 s. The kinked arc "becomes" a rayed auroral structure, and the parallel forms become undetectable. Indeed, this oblique perspective is far more likely, resulting in misleading terminology such as "rayed aurora" to appear ubiquitously in the literature on auroral breakup. This is a serious and underappreciated consideration, since one might be inclined to attribute an entirely different physical interpretation to the first panel versus the last.

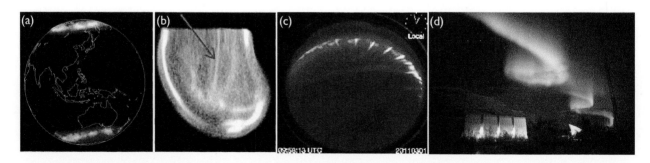

Figure 1. Examples of large-scale coherence in auroras. (a) Northern and southern auroral ovals observed from the Polar spacecraft [*Frank and Sigwarth*, 2003]. (b) Sun-aligned polar auroral filament on Jupiter [*Nichols et al.*, 2009]. (c) Periodic rayed structure forming along an east-west arc observed by all-sky camera at the onset of a substorm that accompanied this conference (1 March 2011). (d) Periodic undulations in the auroral oval captured over the European Incoherent Scatter radar (true color image).

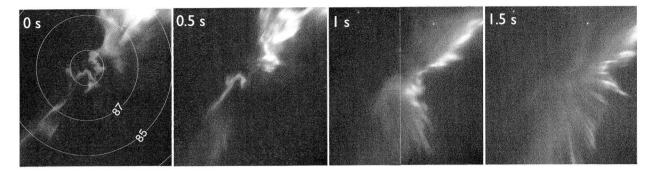

Figure 2. Four frames from a video sequence of an auroral breakup recorded with a narrow-field camera. (left) A sharply kinked but continuous structure in the zenith. Close inspection reveals the presence of two parallel arcs separated by ~100 m. (far right) Recorded 1.5 s later, showing that these same features appear as a series of spatially separated "rays" when viewed at an oblique angle of just ~4°. The parallel structure is also no longer resolvable.

2.2. Observability of $B_\perp$ Structure

Numerical models have been developed to predict auroral volume emission rate as a function of altitude and wavelength [e.g., *Lummerzheim and Lilensten*, 1994; *Strickland et al.*, 1989; *Blelly et al.*, 2005]. Such a model can be used to quantify the constraints on observability of fine-scale variability in the $B_\perp$ plane as a function of transverse scale and instrument parameters, such as wavelength sensitivity and resolution. These issues were discussed by *Maggs and Davis* [1968] in the context of a single arc. Here we pursue a quantitative treatment for the case of a multiple arc structure of the type studied by *Semeter et al.* [2008]. Our approach is to start with a primary auroral electron spectrum, compute the auroral emission response, multiply this by some canonical structure in the $B_\perp$ plan, and finally compute the brightness observed on the focal plan of a camera with a given wavelength sensitivity.

Figure 3 gives an example of such a calculation. Figure 3a shows a representative primary bump-on-tail electron spectrum, in this case, measured by the SIERRA sounding rocket during overflight of an active auroral display [*Lynch et al.*, 2007]. This downward number flux spectrum was used to drive the Transport Carré (TRANSCAR) flux-tube model [*Blelly et al.*, 2005; *Zettergren et al.*, 2007] in order to predict photon volume emission rate as a function of wavelength and altitude, $\varepsilon(\lambda,z)$.

At this point, we introduce a camera model, comprised of an optical filter transmission function $T(\lambda)$ and a detector sensitivity function $S(\lambda)$. This allows us to reduce $\varepsilon(\lambda,z)$ to an effective volume emission profile sensed by the camera,

$$\varepsilon(z) = \int_\lambda T(\lambda')S(\lambda')\varepsilon(\lambda',z)\mathrm{d}\lambda' \quad [\mathrm{m}^{-3}\mathrm{s}]. \qquad (1)$$

Figure 3b plots $\varepsilon(z)$ for the camera configuration used by *Semeter et al.* [2008], which consisted of a BG3 notch filter (transmitting primarily the prompt N_2^+ first negative, N_2 first positive, and N_2 Meinel bands produced by >1 keV primaries) and a silicon-based electron multiplying CCD camera (see Figure 1 of that work for plots of T and S). The emission profile peaks near 105 km.

With this profile in hand, we may investigate image plane projections for any transverse ($B_\perp$) spatial structure and camera angular resolution. Figure 3c shows an idealized representation of the multiarc features studied by *Semeter et al.* [2008] (repeated herein in Figure 4). The arc packet was produced by multiplying $\varepsilon(z)$ by $1 + \cos(2\pi x/0.3)$ for $x \leq 1.2$ km, to produce four arcs with spatial period 300 m. Note that the vertical scale has been compressed by a factor of 30 in this figure. In reality, these features have a vertical-to-horizontal aspect ratio of ~100 to 1.

The gray lines demarcate lines-of-sight (pixel boundaries) through this region at 0.03° intervals. This entire arc packet subtends an angle of <1° as observed from the ground. The image plane projection of this structure is plotted in the black curve of Figure 3d. Although the volume emission rate goes to zero between these synthetic structures, the brightness measured on the detector does not. In order for this to happen, the minimum between the arcs must be directly in the zenith, and one would have to be sure that the field is fully resolved in time (negligent motion blur). The question of whether fine-scale structure is embedded within broader emitting regions is an important one, but this information is difficult to determine definitively using ground-based cameras.

The other curves in Figure 3d show image-pane projections as the edge of the arc packet is displaced horizontally from the observer's position. For a 2.4 km displacement, the spatial periodicity is barely detectable. This distance corresponds to just ~2° from the zenith. Thus, spatial periodicity

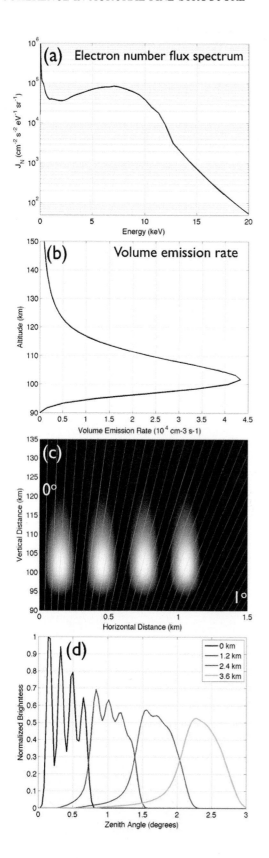

in fine-scale structure can only be studied using observations within a few degrees of the magnetic zenith. The range of observability could potentially be expanded using multistation measurements and tomographic techniques [e.g., *Frey et al.*, 1998; *Semeter et al.*, 1999].

3. TWO TYPES OF AURORAL MOTION

Bearing these observability limitations in mind, we now introduce a mathematical framework for understanding motion and periodicity in the $B_\perp$ plane. Figure 4 shows sample narrow-field snapshots, illustrating the space-time morphologies of interest. The arrows indicate the dominant motions observed in these video sequences. Figure 4a was extracted from the auroral curl video analyzed by *Vogt et al.* [1999]. The red arrow indicates the dissipative rotational motion of the curls, thought to be an optical manifestation of a Kelvin-Helmholtz (K-H) [*Lysak and Song*, 1996] or tearing-mode [*Seyler*, 1990] instability. In either case, the free energy for the instability derives from translational shear in the $E \times B$ flow, indicated by the yellow arrows.

Figures 4b and 4c illustrate a different type of dissipative motion (from *Semeter et al.* [2008]). In Figure 4b, we observe multiple laminar arcs forming, or splitting, within a translating "arc packet." The elemental arcs appear to propagate outward (red arrow) with respect to the aggregate motion of the packet (yellow arrow). Figure 4c, recorded 6 s later as the auroral breakup proceeds, shows the arc packet propagating in the opposite direction. Although parallel structure is still present, the elemental arcs have become sharply kinked, as discussed previously.

When viewing the video sequences associated with Figure 4, one perceives two distinct types of motion. Motions in Figure 4a have a fluid-like character, in the sense that features stretch and deform in a continuous fashion. In the field of optical flow estimation, this notion is formalized as a "data conservation" constraint [*Blixt et al.*, 2006], which simply states that brightness (data) is conserved along a space-time trajectory through an image sequence. This type of motion is illustrated in the top row of Figure 5 (sequence 1), which shows six frames documenting a single turn of an auroral curl.

Figure 3. (a) Representative electron differential number flux spectrum producing visible aurora. (b) Volume emission rate profile corresponding to observations with BG3 filter and a Si CCD, calculated using the TRANSCAR model [*Blelly et al.*, 2005]. (c) An arc packet ($\lambda = 300$ m) constructed by modulating Figure 3b with a horizontal sinusoid. (d) Observed brightness as a function of zenith angle for four different horizontal offsets of the arc packet from the observer location.

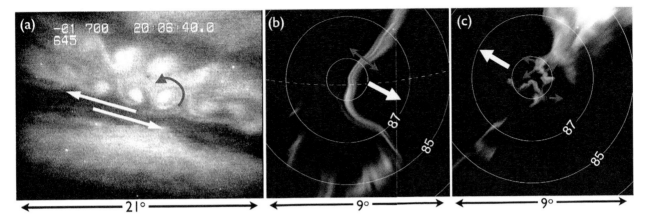

Figure 4. Illustration of two types of auroral motion. (a) Periodically spaced auroral curls forming a vortex street. The red arrow shows the rotation of the curls, the yellow arrow indicates the sheared $E \times B$ motion from which the free energy for the instability is derived [from *Vogt et al.*, 1999]. Parallel auroral forms at substorm (b) onset and (c) breakup. The yellow arrow indicates the general propagation direction of the multiarc structure (group motion). The red arrow indicates the outward propagation of the phase structure in the rest frame of the arc packet (phase motion) [from *Semeter et al.*, 2008].

The motions of the auroras in Figures 4b and 4c are more complex. The elemental arcs in these cases appear and disappear within a defined larger-scale region. The visual impression is one of a translating amplitude envelope. Such motion is also well studied in the field of optical flow estimation where it referred to as the "aperture problem" [*Adelson and Movshon*, 1982]. The lower two rows of Figure 5

show examples of this dynamic. Sequence 2 shows a translating, bifurcating arc packet during the early expansion phase of a substorm. (The duration and field of view are the same as the top row; the improvement in image quality is due to the use of a modern electron multiplying charge-coupled device (EMCCD) sensor.) Sequence 3 was recorded ~6 s later, during the ensuing auroral breakup. The black fiducial

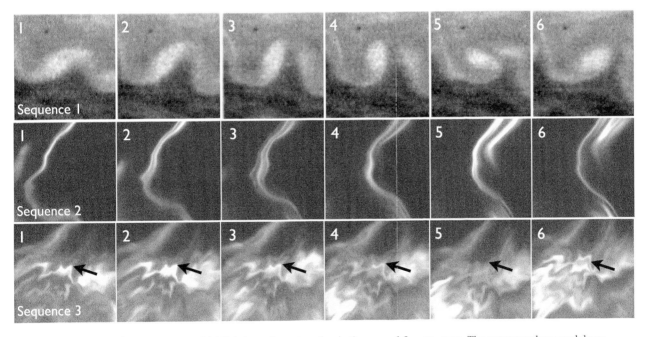

Figure 5. Three image sequences highlighting coherent motion in the auroral fine structure. The sequences have each been cropped to an identical field of view of 5° × 5°. (top) One rotation of an auroral curl exhibiting "Fourier motion" (0.75 s shown). (middle) A translating bifurcating arc packet exhibiting "non-Fourier motion" (0.75 s shown). (bottom) Standing wave dynamic observed during auroral breakup (0.2 s shown).

arrow points to a region directly in the zenith, where temporal modulation of a coherent auroral structure can be seen.

In each of these latter cases, an algorithm to estimate flow velocity would find a violation of the data conservation constraint. *Blixt et al.* [2006] proposed that such violations in model assumptions could provide an automated means of discriminating between auroras generated by different physical mechanisms (e.g., Alfvénic versus inverted V). Our objective here is not classification, but rather establishing a mathematical model for these two classes of auroral motion and relating these models to what we know about acceleration region dynamics. To accomplish this, we appeal to classic notions of phase and group velocity and borrow some terminology from the field of visual perception.

3.1. Auroral Vorticity and Fourier Motion

Fourier analysis provides a valuable framework for assessing motion in video [*Fleet*, 1992]. The two types of motion described above have distinctly different representations in the Fourier domain. Establishing these motion models can provide insight into the overlying magnetospheric wave dynamics driving the phenomena. Consider first motions that satisfy data conservation (Figure 4a). Suppose we have video samples of some field of structures $f(\mathbf{r},t)$, where $\mathbf{r}$ is the position vector, and suppose all features move with a uniform velocity v through the image plane. In the frequency domain, represented by $F(\mathbf{k},\omega)$, all the power lies on a plane intersecting the origin. To better illustrate this, we may consider, without loss of generality, the 1-D case in which the plane becomes a line intersecting the origin described by the equation $k = v_p\omega$, where $k = 2\pi/\lambda$ and ω are the spatial and temporal frequency variables. If the signal is a simple sinusoidal grating,

$$f(x,t) = \sin[k(x - v_p t)], \qquad (2)$$

then $v_p = \omega/k$ describes the motion of points of constant phase and so is referred to as phase velocity. The well-known relationship between spatial frequency, temporal frequency, and phase velocity is illustrated in Figure 6 in both the space-time and k-ω domains.

In a more general case, the features may move at different velocities. In this case, the field is represented by a superposition of Fourier components of the form given in equation (2), the ith component having velocity ω_i/k_i. This motion model also holds if features in an image are distorting (moving at different velocities in different positions) or accelerating (moving at different velocities at different times) if one considers that a video sequence may be segmented in space and time into elemental pieces that follow equation (2). This class of motion is referred to as "Fourier

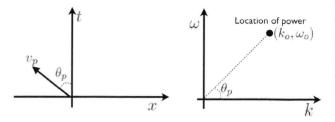

Figure 6. A drifting sinusoid with spatiotemporal frequency (k_o,ω_o) corresponds to an impulse of power in the frequency domain (aperture considerations aside). In the Fourier domain, this function represents a Fourier component that drifts with phase velocity $v_p = k_o/\omega_o$.

motion" in the field of visual perception [*Fleet and Langley*, 1994] and is a special case of the more general model described next.

3.2. Auroral Packets and Non-Fourier Motion

The "Fourier motion" model cannot account for either the bifurcating arc packet or the apparent standing wave structure in Figure 5. In these cases, there are two distinct motions observed, one associated with the fine-scale structure and the other associated with the amplitude envelope. Sequence 2 of Figure 5 shows an unusually well organized example. The auroral form appears as a packet of arcs with an aggregate motion to the right. The elemental phase structure within this packet exhibits a distinctly different dynamic than the amplitude envelop within which they form and fade. An idealized function representing such structure is the modified Gabor function [*Gabor*, 1946] in which a Gaussian envelope $g(x)$ moves with one velocity, while the sinusoidal fine structure $\sin(kx)$ moves with another:

$$f(x,t) = g(x - v_g t)\sin[k(x - v_p t)]. \qquad (3)$$

The amplitude spectrum of equation (3) is illustrated schematically in Figure 7. It consists of two Gaussian distributions centered at $\pm(k,kv_p)$ (only the first quadrant is shown) oriented with slope $v_g t$. The Gaussians are centered at a point corresponding to the dominant mode of the windowed sinusoidal fine structure. The orientation of the power distribution determines the envelope velocity v_g. The slope of the power distribution is given by $v_g = d\omega/dk$, which is referred to as group velocity.

Objects which exhibit non-Fourier motion convey an unusual sensation to the viewer. In the video sequence of Figure 4c, for instance, there is a sensation of rapid motion, even though the group motion is close to 0. Artificial cases of this type can be constructed using the steerable filters of *Freeman et al.* [1991].

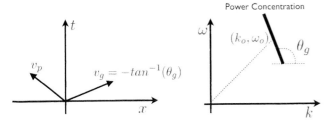

Figure 7. When the drifting sinusoid is modified by a Gaussian amplitude envelope, power lies along a line whose orientation defines the group velocity. Fine-scale structure within the amplitude envelope moves with phase velocity determined by the orientation of the vector intersecting the Gaussian. Phase structure is analogous to a carrier frequency in radio, while motion of the amplitude envelope (group motion) defines the energy flow.

One example is an image of Albert Einstein moving perpetually to the right (group velocity = 0, phase velocity > 0). There are two auroral video sequences (1 and 2) of Figure 5.

4. PHYSICAL INTERPRETATION

The usual goal in studies of dynamic aurora is to relate optical morphology to dynamics in the overlying electromagnetic fields controlling for the phenomenon. Here we review the prevailing theoretical framework for interpreting fine-scale auroral structure in the context of the motion models proposed in section 3.

4.1. Fourier Motion

Theoretical work on fluid-like instabilities observed in the aurora began with *Hallinan and Davis* [1970] and *Hallinan* [1976], who postulated that a U-shaped potential should be unstable to Kelvin-Helmholtz wave growth. Figure 8 is a schematic illustration of the currents, fields, and motions associated with a quasi-static auroral potential structure (after *Lysak and Song* [1996]). The free energy for the instability is derived from the convergence of Poynting flux, which leads to shear in both $E \times B$ velocity v and the induced magnetic field b. The former is thought to govern development of the fine-scale vorticity observed in Figure 4a, with linear growth rate [*Hallinan and Davis*, 1970]

$$\gamma^2 = \frac{(\nabla \times \mathbf{v})_\parallel}{4} \left[e^{-4ka} - (1 - 2ka)^2 \right], \qquad (4)$$

where k is the spatial frequency of the vortices, and a is the width of the shear layer (i.e., the space charge region).

A comprehensive analysis of the Figure 4 data with respect to linear K-H theory was carried out by *Vogt et al.* [1999],

who applied optical flow analysis in order to estimate growth rates, translational shear, and rotational shear. Because of the stratified nature and long life of this particular event, they were able to perform a 1-D space-time analysis of phase motions, similar in concept to Figure 6. Among their findings were (1) the relationship between vorticity and growth rate was inconsistent with equation (4), (2) rotational vorticity was much greater than translational vorticity (velocity shear), and (3) no clear correlation existed between brightness and vorticity, as might be predicted by electrostatic models of auroral acceleration. Although these discrepancies have not been fully accounted for, it is likely that small-scale vorticity is not a pure K-H instability, but rather involves tearing mode and current sheet reconnection [*Seyler*, 1990]. It is clear that development of an accurate Fourier-based motion model provides critical constraints in our efforts to understand this inherently unstable construct (the U-shaped potential) in space plasmas.

4.2. Non-Fourier Motion

For the case of dispersive motion, an analogous schematic may be constructed. The model proposed in Figure 9 was introduced by *Semeter et al.* [2008] in the context of Alfvén resonance cones. But the empirically supported relationship between auroral phase motion (red) and electric field (green) is quite general and may be adapted to a broad range of models. The schematic lies in the plane perpendicular to the dispersive arc packet. The decaying sinusoid represents the

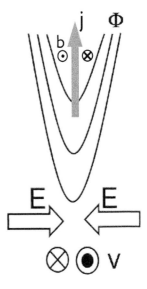

Figure 8. Schematic illustration of electric field, current, and velocity associated with a U-shaped electrostatic potential structure. This construct is inherently unstable to shear instabilities [*Lysak and Song*, 1996].

electric field of an oblique dispersive Alfvénic wave propagating into a medium of increasing conductivity. For illustration, a simple uniform plane wave is shown, and the conversion from shear MHD wave to dispersive Alfvén wave is shown as a discrete point (which may mark the apex of an Alfvén cone). Below this point, wave energy spreads in the transverse direction. The parallel component of E creates, or modifies, the particle flux responsible for the observed auroral dynamic. The amplitude decreases as the wave attenuates in the topside ionosphere. The nominal altitude for the onset of dispersion may be taken as ~1 R_E (or the altitude of the peak Alfvén speed) [*Chaston et al.*, 2003].

The green shading indicates the region where **E** has a component parallel to **B**. Optical aurora occurs where $E_\parallel$ is upward. The space-time variability in the optical aurora is illustrated by the red-shaded features at the bottom. Where the wave amplitude is largest, we expect to find both the largest average energy and the largest total energy flux in the incident electron beam. As such, arcs conjugate to the center of this structure are expected to be brighter, occur at a lower altitude, and have narrower altitude extent [see, e.g., *Zettergren et al.*, 2007]. As the wave attenuates, the emission altitude increases, the emission layer broadens, and the over-

all brightness decreases. This increase in emission altitude with time is a well-known feature of dynamic aurora [*Omholt*, 1971] and is also observed in oblique data such as Figure 2.

It should be noted that Figure 9 represents a snapshot of a dynamic equilibrium. For a propagating wave, the dynamic observed near the magnetic zenith would be one of fine-scale phase structure forming in a periodic fashion, propagating transversely, and fading. In the case of an Alfvén cone, the spreading of wave energy may be symmetric, resulting in symmetric outward motion of auroral structure [*Singh*, 1999] as shown by *Semeter et al.* [2008, Figure 7]. An asymmetric dynamic may result if the dispersion is related to density gradients [*Rankin et al.*, 2005], interaction of plasmas with different properties [*Dahlgren et al.*, 2011], or other cross-scale coupling mechanisms [*Chaston et al.*, 2011].

The possible role of wave reflection in structuring the aurora is not represented in Figure 9. The interference of Alfvén waves within the ionospheric Alfvén resonator [*Belyaev et al.*, 1999] has been proposed as a mechanism for structuring the field-aligned currents at small scales [*Lysak*, 1993] and, hence, the auroral particle flux [*Chaston et al.*, 2002]. Although some difficulties arise in considering the absorption of small-scale wave power in the ionosphere [*Borovsky*, 1993; *Lessard and Knudsen*, 2001], a careful comparison of quantitative predictions with observation, including a specific model for the conversion of wave power to $E_\parallel$ [e.g., *Song and Lysak*, 2006], is needed.

4.3. On the Origin of Fine Structure in $E_\parallel$

As the dominant mode of field-aligned energy transport in the magnetosphere, Alfvén waves likely play an active role for both classes of auroral motion described herein. The association of small-scale auroral structure with dispersive Alfvén waves (DAWs) may be directly inferred from the 1-D dispersion relation, which predicts a ratio of parallel to perpendicular electric field:

$$\frac{E_\parallel}{E_\perp} \propto \frac{k_\parallel}{k_\perp} \frac{(\lambda_e k_\perp)^2}{1 + (\lambda_e k_\perp)^2} \tag{5}$$

[e.g., *Stasiewicz et al.*, 2000], where $k_\parallel$ and $k_\perp$ are the $B_\parallel$ and $B_\perp$ wavenumbers of the disturbance, respectively, and $\lambda_e = 2\pi c/\omega_{pe}$ is the electron inertial length, with ω_{pe} the electron plasma frequency. This ratio is maximum at $k_\perp = 2\pi/\lambda_e$, suggesting that λ_e represents a favored scale for wave-particle coupling within a DAW. This offers some insight into why spatial periodicity in the auroral electron flux emerges so

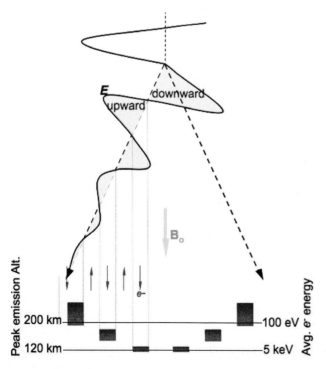

Figure 9. Schematic illustration of electric field, particle flux, and auroral forms associated with a dispersive Alfvén wave. See section 4.2 for a full description.

naturally at λ_e: given some mechanism for the cascade of wave power to smaller scales, equation (5) represents a spatial filter at the electron inertial length.

The cascade to dispersive scales is not surprising for the geospace plasma (or, indeed, for any realistic cosmic plasma). *Seyler* [1990] developed a 3-D reduced MHD model (including inertial effects) demonstrating the development of collisionless tearing, reconnection, and small-scale parallel electric fields as a consequence of interference of incident and reflected waves. The cascade may also occur through simple propagation of MHD waves into regions of inhomogeneous plasma [*Génot et al.*, 2004]. The coupling between fluid and dispersive processes can facilitate this cascade [*Chaston et al.*, 2011]. *Blixt et al.* [2006] developed a robust optical flow technique to automatically determine which of these processes dominates for a given auroral video sequence. The modest quantity of data collected on fine-scale coherence seems to suggest a dominance of either fluid or dispersive motion at any given point in time, but a more comprehensive analysis is needed.

5. FUTURE CHALLENGES

The challenge for the coming decade is to develop instrumentation, infrastructure, and techniques to fill in the many missing details of how fine-scale structure emerges and evolves in the geospace plasma. The emergence of high-speed high-sensitivity imaging sensors, such as EMCCD, scientific-grade complementary metal-oxide-semiconductor, and improved microchannel plate intensifiers, promises to contribute significantly. Careful analysis of multimode observations, in particular optical and radar imaging [*Semeter et al.*, 2005, 2009], is also expected to provide critical new constraints.

Acknowledgments. This work was supported by the National Science Foundation under grants ATM-0538868 and ATM-0547934.

REFERENCES

Adelson, E. H., and J. A. Movshon (1982), Phenomenal coherence of moving visual patterns, *Nature*, *300*, 523–525, doi:10.1038/300523a0.

Atkinson, G. (1970), Auroral arcs: Result of the interaction of a dynamic magnetosphere with the ionosphere, *J. Geophys. Res.*, *75*(25), 4746–4755.

Belyaev, P. P., T. Bösinger, S. V. Isaev, and J. Kangas (1999), First evidence at high latitudes for the ionospheric Alfvén resonator, *J. Geophys. Res.*, *104*(A3), 4305–4317, doi:10.1029/1998JA900062.

Blelly, P.-L., C. Lathuillère, B. Emery, J. Lilensten, J. Fontanari, and D. Alcaydé (2005), An extended TRANSCAR model including ionospheric convection: Simulation of EISCAT observations using inputs from AMIE, *Ann. Geophys.*, *23*, 419–431.

Blixt, E., J. Semeter, and N. Ivchenko (2006), Optical flow analysis of the aurora borealis, *IEEE Trans. Geosci. Remote Sens.*, *3*, 159–163.

Borovsky, J. E. (1993), Auroral arc thicknesses as predicted by various theories, *J. Geophys. Res.*, *98*(A4), 6101–6138.

Chamberlain, J. W. (1961), *Physics of the Aurora and Airglow*, Academic Press, New York.

Chaston, C. C., J. W. Bonnell, C. W. Carlson, M. Berthomier, L. M. Peticolas, I. Roth, J. P. McFadden, R. E. Ergun, and R. J. Strangeway (2002), Electron acceleration in the ionospheric Alfven resonator, *J. Geophys. Res.*, *107*(A11), 1413, doi:10.1029/2002JA009272.

Chaston, C. C., J. W. Bonnell, C. W. Carlson, J. P. McFadden, R. E. Ergun, and R. J. Strangeway (2003), Properties of small-scale Alfvén waves and accelerated electrons from FAST, *J. Geophys. Res.*, *108*(A4), 8003, doi:10.1029/2002JA009420.

Chaston, C. C., K. Seki, T. Sakanoi, K. Asamura, M. Hirahara, and C. W. Carlson (2011), Cross-scale coupling in the auroral acceleration region, *Geophys. Res. Lett.*, *38*, L20101, doi:10.1029/2011GL049185.

Dahlgren, H., B. Gustavsson, B. S. Lanchester, N. Ivchenko, U. Brändström, D. K. Whiter, T. Sergienko, I. Sandahl, and G. Marklund (2011), Energy and flux variations across thin auroral arcs, *Ann. Geophys.*, *29*, 1699–1712, doi:10.5194/angeo-29-1699-2011.

Fleet, D. (1992), *Measurement of Image Velocity*, Kluwer Acad., Norwell, Mass.

Fleet, D., and K. Langley (1994), Computational analysis of non-Fourier motion, *Vision Res.*, *22*, 3057–3079.

Frank, L. A., and J. B. Sigwarth (2003), Simultaneous images of the northern and southern auroras from the Polar spacecraft: An auroral substorm, *J. Geophys. Res.*, *108*(A4), 8015, doi:10.1029/2002JA009356.

Freeman, W. T., E. H. Adelson, and D. J. Heeger (1991), Motion without movement, *Comput. Graphics*, *25*, 27–30.

Frey, H., S. Frey, D. Larson, T. Nygren, and J. Semeter (1998), Tomographic methods for magnetospheric applications, in *Science Closure and Enabling Technologies for Constellation Class Missions*, edited by V. Angelopoulos and P. V. Panetta, pp. 72–77, Univ. of Calif. Press, Berkeley.

Gabor, D. (1946), Theory of communication, *J. IEE*, *95*, 429–457.

Galperin, Y. I. (2002), Multiple scales in auroral plasmas, *J. Atmos. Sol. Terr. Phys.*, *64*, 211–229, doi:10.1016/S1364-6826(01)00085-2.

Génot, V., P. Louarn, and F. Mottez (2004), Alfvén wave interaction with inhomogeneous plasmas: Acceleration and energy cascade towards small-scales, *Ann. Geophys.*, *22*, 2081–2096, doi:10.5194/angeo-22-2081-2004.

Hallinan, T. J. (1976), Auroral spirals, 2. Theory, *J. Geophys. Res.*, *81*(22), 3959–3965.

Hallinan, T. J., and T. N. Davis (1970), Small-scale auroral arc distortions, *Planet. Space Sci.*, *18*, 1735–1736, doi:10.1016/0032-0633(70)90007-3.

Johnson, M. L., J. S. Murphree, G. T. Marklund, and T. Karlsson (1998), Progress on relating optical auroral forms and electric field patterns, *J. Geophys. Res.*, *103*(A3), 4271–4284.

Lessard, M. R., and D. J. Knudsen (2001), Ionospheric reflection of small-scale Alfvén waves, *Geophys. Res. Lett.*, *28*(18), 3573–3576, doi:10.1029/2000GL012529.

Lummerzheim, D., and J. Lilensten (1994), Electron transport and energy degradation in the ionosphere: Evaluation of the numerical solution, comparison with laboratory experiments and auroral observations, *Ann. Geophys.*, *12*, 1039–1051.

Lynch, K. A., J. L. Semeter, M. Zettergren, P. Kintner, R. Arnoldy, E. Klatt, J. LaBelle, R. G. Michell, E. A. MacDonald, and M. Samara (2007), Auroral ion outflow: Low altitude energization, *Ann. Geophys.*, *25*, 1967–1977.

Lysak, R. L. (1993), Generalized model of the ionospheric Alfvén resonator, in *Auroral Plasma Dynamics*, *Geophys. Monogr. Ser.*, vol. 80, edited by R. L. Lysak, pp. 121–128, AGU, Washington, D. C., doi:10.1029/GM080p0121.

Lysak, R. L., and Y. Song (1996), Coupling of Kelvin-Helmholtz and current sheet instabilities to the ionosphere: A dynamic theory of auroral spirals, *J. Geophys. Res.*, *101*(A7), 15,411–15,422.

Maggs, J. E., and T. N. Davis (1968), Measurements of the thicknesses of auroral structures, *Planet. Space Sci.*, *16*, 205–206.

Mende, S. B., S. E. Harris, H. U. Frey, V. Angelopoulos, C. T. Russell, E. Donovan, B. Jackel, M. Greffen, and L. M. Peticolas (2008), The THEMIS array of ground-based observatories for the study of auroral substorms, *Space Sci. Rev.*, *141*, 357–387, doi:10.1007/s11214-008-9380-x.

Nichols, J. D., J. T. Clarke, J. C. Gérard, and D. Grodent (2009), Observations of Jovian polar auroral filaments, *Geophys. Res. Lett.*, *36*, L08101, doi:10.1029/2009GL037578.

Omholt, A. (1971), *The Optical Aurora*, Springer, New York.

Partamies, N., K. Kauristie, T. I. Pulkkinen, and M. Brittnacher (2001), Statistical study of auroral spirals, *J. Geophys. Res.*, *106*(A8), 15,415–15,428, doi:10.1029/2000JA900172.

Rankin, R., R. Marchand, J. Y. Lu, K. Kabin, and V. T. Tikhonchuk (2005), Theory of dispersive shear Alfvén wave focusing in Earth's magnetosphere, *Geophys. Res. Lett.*, *32*, L05102, doi:10.1029/2004GL021831.

Sato, T. (1978), A theory of quiet auroral arcs, *J. Geophys. Res.*, *83*(A3), 1042–1048.

Semeter, J., and E. M. Blixt (2006), Evidence for Alfvén wave dispersion identified in high-resolution auroral imagery, *Geophys. Res. Lett.*, *33*, L13106, doi:10.1029/2006GL026274.

Semeter, J., M. Mendillo, and J. Baumgardner (1999), Multispectral tomographic imaging of the midlatitude aurora, *J. Geophys. Res.*, *104*(A11), 24,565–24,585.

Semeter, J., C. J. Heinselman, G. G. Sivjee, H. U. Frey, and J. W. Bonnell (2005), Ionospheric response to wave-accelerated electrons at the poleward auroral boundary, *J. Geophys. Res.*, *110*, A11310, doi:10.1029/2005JA011226.

Semeter, J., M. Zettergren, M. Diaz, and S. Mende (2008), Wave dispersion and the discrete aurora: New constraints derived from high-speed imagery, *J. Geophys. Res.*, *113*, A12208, doi:10.1029/2008JA013122.

Semeter, J., T. Butler, C. Heinselman, M. Nicolls, J. Kelly, and D. Hampton (2009), Volumetric imaging of the auroral ionosphere: Initial results from PFISR, *J. Atmos. Sol. Terr. Phys.*, *71*, 738–743, doi:10.1016/j.JASTP.2008.08.014.

Seyler, C. E. (1990), A mathematical model of the structure and evolution of small-scale discrete auroral arcs, *J. Geophys. Res.*, *95*(A10), 17,199–17,215.

Singh, N. (1999), Field patterns of Alfven wave resonance cones, *J. Geophys. Res.*, *104*(A4), 6999–7009.

Song, Y., and R. Lysak (2006), Displacement current and the generation of parallel electric fields, *Phys. Rev. Lett.*, *96*, 145002, doi:10.1103/PhysRevLett.96.145002.

Stasiewicz, J., et al. (2000), Small scale Alfvénic structure in the aurora, *Space Sci. Rev.*, *92*, 423–533.

Strickland, D. J., R. R. Meier, J. H. Hecht, and A. B. Christensen (1989), Deducing composition and incident electron spectra from ground-based auroral optical measurements: Theory and model results, *J. Geophys. Res.*, *94*(A10), 13,527–13,539.

Trondsen, T. S., and L. L. Cogger (1998), A survey of small-scale spatially periodic distortions of auroral forms, *J. Geophys. Res.*, *103*(A5), 9405–9415.

Trondsen, T. S., L. L. Cogger, and J. C. Samson (1997), Asymmetric multiple auroral arcs and inertial Alfvén waves, *Geophys. Res. Lett.*, *24*(22), 2945–2948.

Vogt, J., H. U. Frey, G. Haerendel, H. Höfner, and J. L. Semeter (1999), Shear velocity profiles associated with auroral curls, *J. Geophys. Res.*, *104*(A8), 17,277–17,288.

Zettergren, M., J. Semeter, P.-L. Blelly, and M. Diaz (2007), Optical estimation of auroral ion upflow: Theory, *J. Geophys. Res.*, *112*, A12310, doi:10.1029/2007JA012691.

J. Semeter, Department of Electrical and Computer Engineering and Center for Space Physics, Boston University, 8 Saint Mary's St., Boston, MA 02215, USA. (jls@bu.edu)

Ground-Based Aurora Conjugacy and Dynamic Tracing
of Geomagnetic Conjugate Points

Natsuo Sato, Akira Kadokura, and Tetsuo Motoba

National Institute of Polar Research, Tokyo, Japan

Keisuke Hosokawa

Department of Communication Engineering and Informatics, University of Electro-Communications, Tokyo, Japan

Gunnlaugur Bjornsson and Thorsteinn Saemundsson

Science Institute, University of Iceland, Reykjavik, Iceland

We report on highly similar auroras that were simultaneously recorded with all-sky TV cameras situated at two geomagnetically conjugate points at Tjornes in Iceland and at Syowa in Antarctica. During this event, various types of aurora were observed including auroral breakup, curl-type aurora, north-south-directed band-type aurora, and pulsating aurora. We examined their characteristics for their similarity and dissimilarity in terms of shapes, movements, and luminosity variations at both observatories. We were also able to trace the movements and displacement of conjugate auroras in the Northern and Southern Hemispheres with a high spatial-temporal resolution. We discuss their conjugate characteristics with reference to solar wind–magnetosphere-ionosphere coupling processes.

1. INTRODUCTION

The electrons captured in the Earth's magnetosphere are basically constrained to move toward the Northern and Southern Hemispheres along the geomagnetic field lines. Thus, bright nighttime auroras are expected to appear simultaneously in both hemispheres. Observations of interhemispheric conjugate auroras provide a unique opportunity to examine how and where the invisible geomagnetic field lines connect the two hemispheres. Previous studies have demonstrated that auroral features may not always be conjugate [*Belon et al.*, 1969; *Stenbaek-Nielsen et al.*, 1972; *Sato et al.*, 1986, 1998, 2005]. Even when they are conjugate, the auroral features show displacements in latitude and longitude from a global magnetic field model in latitude and longitude [*Stenbaek-Nielsen and Otto*, 1997], the longitudinal displacement being more pronounced [*Sato et al.*, 2005; *Minatoya et al.*, 1995; *Frank and Sigwarth*, 2003; *Østgaard et al.*, 2004; *Stubbs et al.*, 2005; *Motoba et al.*, 2010]. Measurements of such conjugate auroral displacements as well as auroral dynamics should provide useful information for understanding the solar wind–magnetosphere-ionosphere coupling processes and should also help to improve the accuracy of global magnetic field models.

Syowa Station in Antarctica and stations in Iceland form an ideal set of observatories to study geomagnetically conjugate auroras in the auroral zone [*Sato et al.*, 1986, 1998, 2005] (see Figure 1). A campaign of conjugate auroral observations using all-sky TV cameras has been carried out since 1984 during the equinox periods. Simultaneous optical

Auroral Phenomenology and Magnetospheric Processes: Earth and Other Planets
Geophysical Monograph Series 197
10.1029/2011GM001154

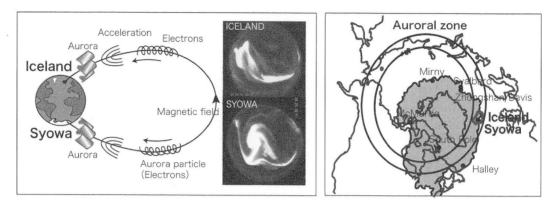

Figure 1. (left) A schematic drawing of conjugate auroral observation at Syowa Station in Antarctica and stations in Iceland (modified from the work of *Sato et al.* [2005]). (right) Mapping of Antarctic Continent into the Northern Hemisphere along geomagnetic field lines (modified from the work of *Lanzerotti* [1987]).

observations from the ground at two conjugate points present many practical problems. Both observatories should be in darkness and both should have fine weather. In spite of these observing limitations, we have found an outstanding pair of similar auroras, the most striking example in the 20 year history of the Syowa-Iceland conjugate campaign. On 26 September 2003, various types of conjugate aurora were observed, including auroral breakup, small-scale curl-type aurora, north-south-directed band-type aurora, and pulsating aurora. We examined these auroras, checking for similarity and dissimilarity in terms of shapes, movements, and lumi-

nosity variations at both observatories. We were also able to trace the temporal movements and displacement of the auroras in the Northern and Southern Hemispheres with a high spatial-temporal resolution. We will discuss their conjugate characteristics with reference to solar wind–magnetosphere and magnetosphere-ionosphere coupling processes.

2. OBSERVATION OF CONJUGATE AURORA

Figure 2 (top) shows snapshot images of aurora obtained by all-sky TV cameras at Syowa and Tjornes on 26 September

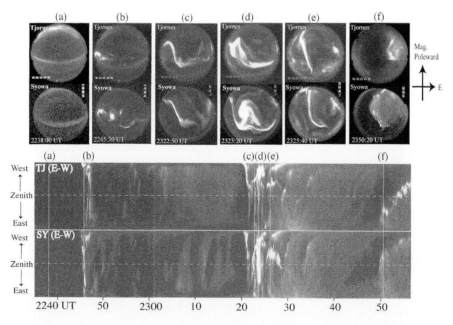

Figure 2. (top) Snapshot conjugate auroral images at different times. (bottom) Auroral Keograms with luminosity as a function of time in the geomagnetic east-west direction, reproduced from the original all-sky TV data. The vertical solid lines marked a–f correspond to the times of auroral snapshot images, which are shown in top plots.

2003. The auroral images are at the times indicated by the vertical solid lines in Figure 2 (bottom). Although the shape and type of aurora changes from time to time, it is clear that the main features remain very similar at both observatories. Figure 2 (bottom) shows auroral Keograms indicating the luminosity variation as a function of time and the distribution along the geomagnetic east-west direction. It is clear that the temporal and spatial variations of the auroras were quite similar at both observatories. The westward drift motion was very similar after ~23:25 UT at both observatories (UT is almost the same as magnetic local time (MLT) at Syowa and Tjornes). In the following, we describe the conjugate characteristics of the temporal variations and spatial structures of different types of aurora.

2.1. Auroral Breakup

Figure 3 shows a time series of snapshot images of an auroral breakup event during the time interval 23:22:30 to 23:23:50 UT. Detailed characteristics are as follows: A bright discrete aurora appeared at the western horizon at about 23:20 UT and expanded eastward as a surge structure (eastward traveling surge). Then, the auroral luminosity became weaker at Syowa in the interval of 23:21:40 to 23:22:30 UT, though the luminosity at Tjornes did not change. A breakup of the aurora started at about 23:23:00 UT at both observatories. We can say that the auroral breakup occurred simultaneously and that the spatial and temporal variations were extremely similar in both hemispheres except for some dissimilarity of shapes and fading phenomena before the main phase of the breakup.

2.2. Small-Scale Curl-Type Aurora

Investigations of conjugate relationships for small-scale auroral distortion phenomena (so-called "curl" auroral phenomena) are interesting. Hallinan and Davis [1970] described the fundamental characteristics of curl auroras as follows: Curls are small-scale spatially periodic distortions of auroral forms and generally have near-circular symmetry, a wavelength of less than 10 km, and a lifetime of less than 2 s. Curls observed in the Northern Hemisphere are counterclockwise as viewed from the ground, while those in the Southern Hemisphere are clockwise. Hallinan and Davis proposed that the cause of the shear flow is the result of $\mathbf{E} \times \mathbf{B}$ drifts in a divergent electric field due to a localized negative charge excess within sheets of precipitating auroral primaries.

Here we present the first-ever conjugate observation of curl-type aurora. Typical curl-type aurora was observed at Syowa when an intense eastward-traveling surge appeared during auroral breakup as described in section 2.1. Figure 4 shows snapshot photos of conjugate all-sky TV images. As seen at Syowa (lower panel), an isolated curl is indicated with an arrow (a), and a periodic series of curls is indicated by arrows (b) at the eastern extremity of the surge. Rotational sense is clockwise as viewed from the ground. These characteristics are consistent with those observed by Hallinan and Davis [1970]. More than 10 curls are seen in this all-sky TV field of view at Syowa. On the other hand, no curl-type aurora was seen at Tjornes. It must be concluded that conjugacy was lacking in this case, as the curl-type aurora was confined to the Southern Hemisphere. If we apply Hallinan and Davis's [1970] model to this event, the electric field caused by excess negative charge must have been much stronger over the Southern Hemisphere than over the Northern Hemisphere.

2.3. North-South-Directed Band-Type Aurora

Figure 5 shows the temporal variation of the north-south (N-S)-directed band-type aurora during the time interval 23:25:20 to 23:25:50 UT. It is found that the start time and the expansion features of the N-S band aurora were almost identical in both hemispheres. On the other hand, the bending shape at the equatorward part of the N-S band aurora only appeared in the Southern Hemisphere. As seen in snapshots at 23:25:40 and 23:25:50 UT, the aurora at Tjornes expanded in a direction straight southward or slightly bending westward. We can conclude that the temporal and spatial variations of the N-S band aurora were highly similar at both observatories. One interesting result of this observation is that the bending feature/direction of the N-S band aurora is not the same in the Northern and Southern Hemispheres [Sato et al., 2005].

2.4. Pulsating Aurora

The geomagnetic conjugacy (or nonconjugacy) of pulsating aurora provides critical information on their generation mechanism [Belon et al., 1969; Davis et al., 1971; Stenbaek-Nielsen et al., 1972; Davis, 1978; Fujii et al., 1987; Sato et al., 1998, 2004; Minatoya et al., 1995].

Watanabe et al. [2007] examined in detail the conjugacy of pulsation aurora using the 26 September 2003 event. During the recovery phase of auroral breakup, a pulsating aurora started to appear at both observatories at about 23:40 UT. In the case of the pulsating aurora event, unlike those discussed in the works of Minatoya et al. [1995] and Sato et al. [1998, 2004], there is no uncertainty in identifying the conjugate regions because the auroras during the breakup showed remarkable conjugacy. Therefore, the nonconjugacy of pulsating

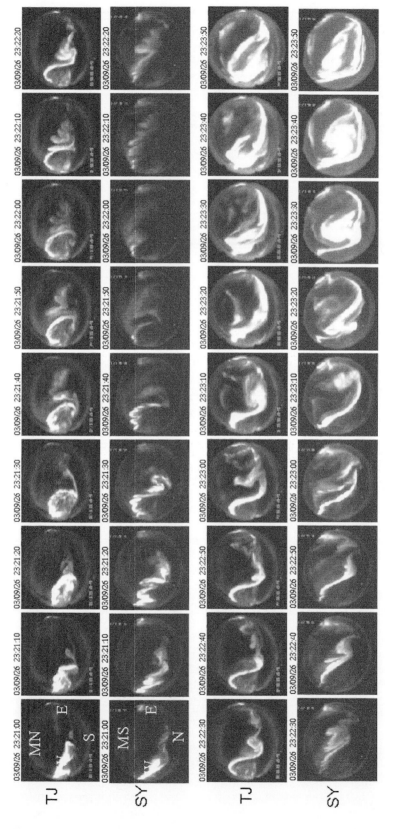

Figure 3. Snapshot conjugate auroral images at 10 s intervals when auroral breakup occurred.

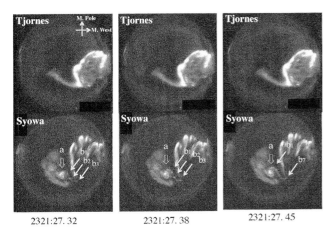

Figure 4. Snapshot conjugate auroral images at intervals of 1/15 s. Arrows indicate small-scale curls observed at Syowa.

aurora can be asserted without doubt. Figure 6 shows summary characteristics of the pulsating aurora. There are two types of nonconjugacy. One is the case in which the pulsating aurora appears in both hemispheres, and the shapes are similar but the pulsation periods are almost always different in the two hemispheres. The other is the case in which the pulsating aurora appears only in one hemisphere ("single-hemisphere pulsating aurora"). These results are basically the same as those of *Sato et al.* [1998]. One additional new point in the present chapter is that single-hemisphere pulsating auroras can occur in the form of an isolated large patch.

3. DYNAMIC TRACING OF CONJUGATE POINTS

Sato et al. [2005] traced the temporal movements and displacement of geomagnetic conjugate points with a high spatial-temporal resolution using the data of 26 September 2003. Their results indicated that the geomagnetic conjugate point of Syowa moved ~200 km in longitude and ~ 50 km in

latitude in 1 h due to changes in topology of the geomagnetic field.

Recently, *Motoba et al.* [2010] examined in detail the relationship between the displacement of a conjugate point and the interplanetary magnetic field (IMF) during the 21 September 2009 substorm event. Figure 7 shows time series of the latitudinal displacement (ΔLat.) (top) and the MLT displacement (ΔMLT) (bottom) of the northern footprint (NF) of Syowa relative to Tjornes, respectively. Both relative displacements (gray solid circles) of the Tjornes-Syowa conjugate points were determined based on a comparison between the conjugate Tjornes-Syowa auroral forms at 14 selected times (see details in section 3.2 of *Motoba et al.* [2010]). Here positive/negative ΔLat. means that the NF of Syowa is poleward/equatorward of Tjornes, while positive/negative ΔMLT means that the NF of Syowa is westward/eastward of Tjornes. The black curve in Figure 7 (bottom) denotes the relative MLT displacement of the conjugate points predicted through *Østgaard et al.'s* [2005] empirical function (referred to as the "Østgaard function"). As *Motoba et al.* [2011] described in section 2.1, the observational result was in reasonable agreement with the result obtained via the Østgaard function when the IMF data at subsolar magnetopause were delayed by 51 min. The gray curves indicate the dynamic relative latitudinal and MLT displacements of the conjugate points calculated via the T96 model [*Tsyganenko and Stern*, 1996], including the 51 min time-shifted IMF parameters. Similar to the Østgaard function, the temporal variation of the relative MLT displacement inferred from the T96 model resembled that obtained from the observations. In contrast, the T96 model significantly underestimated the latitudinal displacement of the conjugate points identified by the conjugate auroral observations, particularly during the initial stage of the substorm, the modeled latitudinal displacement (< 0.5°) was much smaller than the observed one (~3.0°).

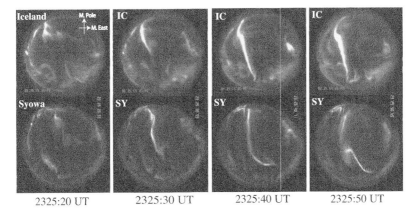

Figure 5. Time series of a N-S-directed band-type aurora obtained every 10 s.

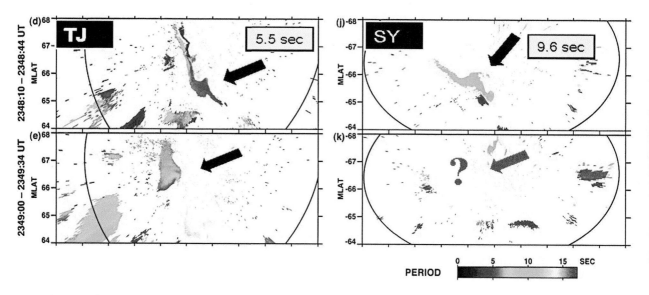

Figure 6. The pulsation period determined from autocorrelation analysis. The black curves indicate a zenith angle of 80°. The arrows indicate pulsating aurora patches referred to in the text.

4. SUMMARY AND DISCUSSION

It is widely understood that a discrete aurora occurs via field-aligned earthward acceleration of electrons at an altitude of ~6000–12,000 km through the magnetosphere-ionosphere coupling processes [e.g., *Borovsky*, 1993]. In general, ionospheric conditions (plasma density and conductivity, etc.) will be different in opposite hemispheres, so there is no guarantee that such acceleration processes are symmetric in both hemispheres [*Newell et al.*, 1996; *Sato et al.*, 1998].

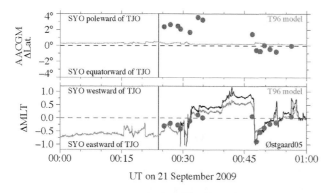

Figure 7. (top) Latitudinal and (bottom) magnetic local time (MLT) displacements of the northern footprint of Syowa relative to Tjornes deduced from conjugate auroral images. Black and gray curves represent relative displacements of the Tjornes-Syowa conjugate points inferred from the Østgaard function [*Østgaard et al.*, 2005] and the T96 model using the time-shifted (51 min) interplanetary magnetic field (IMF) values, respectively.

In this chapter, we considered various types of conjugate auroras.

The observations showed that the spatial and temporal variations of breakup aurora were roughly similar, except for some dissimilarity of shapes and fading before the main phase of auroral breakup. This means that the source region of the breakup must be located near the equatorial region in the magnetosphere and that the acceleration processes caused by the magnetosphere-ionosphere coupling must be symmetric or almost identical in both hemispheres. The probability of having identical physical conditions in the two hemispheres, both in the ionosphere and magnetosphere, must be very small because no comparable symmetric cases have been observed since the Syowa-Iceland conjugate observations began in 1984.

On the other hand, a small-scale curl-type aurora was seen only in the Southern Hemisphere. This means that small-scale field-aligned acceleration was dissimilar (nonconjugacy) in the two hemispheres. A possible explanation may be a strong charge concentration in the Southern Hemisphere if we apply *Hallinan and Davis's* [1970] model. A N-S band aurora showed quite similar spatial and temporal variation in both hemispheres, almost a mirror image. In this case, the magnetosphere-ionosphere coupling must have been symmetric between the hemispheres, that is, the origin and acceleration process and intensity of the auroral electrons were symmetric, as well as the topology of the magnetic field. Such a symmetric mirror image aurora may be regarded as an exceptional case.

The characteristics of the pulsating aurora show that there were two types of nonconjugacy. One is the case in which the pulsating aurora appears in both hemispheres but the pulsation periods are almost always different between the two hemispheres. The other is the case in which the pulsating aurora appears only in one hemisphere ("single-hemisphere pulsating aurora"). These characteristics suggest that the origin of pulsating aurora is not located near the equatorial plane in the magnetosphere, but it would be located near the Earth's ionosphere as suggested by *Sato et al.* [2004]. Although we cannot extend the results in this chapter to all pulsating auroras, the mounting evidence of studies [*Minatoya et al.*, 1995; *Sato et al.*, 1998, 2004] suggests that the conjugacy of pulsating auroras is generally poor. We suspect that early studies were biased for good conjugacy because good conjugacy is much easier to identify than poor conjugacy. As far as we can check, good conjugacy events (both in shape and phase) that explicitly appeared in the literature are only two (the 26 March 1968 event [*Stenbaek-Nielsen et al.*, 1972; *Davis*, 1978] and the 26 September 1984 event [*Fujii et al.*, 1987]). It is worth noting that already in an early study, *Stenbaek-Nielsen et al.* [1973] pointed out the nonconjugacy of pulsating auroras. *Sato et al.* [1998] reported that the overall dynamic variations and activities of pulsating auroras were very similar in the two hemispheres. These facts indicate that there exists a common source of precipitating electrons, presumably near the equatorial plane. At the moment, it is not clear how the equatorial common source and the near-Earth modulation source are related.

The pictures from the two all-sky TV cameras revealed how the field lines joining the Tjornes-Syowa conjugate pair moved in the latitudinal and/or longitudinal directions with a high spatial-temporal resolution. We demonstrated that the relative MLT displacement of the conjugate auroral locations was in a rough agreement with that inferred from the *Østgaard et al.* [2005] empirical relationship, expressed as a linear function of IMF θ_{CA} data by 51 min after the IMF reached the dayside magnetopause. The estimated time delay suggests that the IMF B_y-related reconfiguration (twisting) of the near-Earth tail field connecting Tjornes and Syowa takes about 51 min from the time when IMF B_y encounters the subsolar point. The IMF B_y gives rise to an asymmetry of the field line configuration in the magnetotail, twisting in a clockwise/counterclockwise direction around the x axis (view from tail) for IMF $B_y > 0$/IMF $B_y < 0$ (see Figure 8). Using the near-Earth tail field data from the Cluster spacecraft, *Motoba et al.* [2011] revealed a signature of the IMF B_y-related magnetotail twist with the same reconfiguration time as that deduced from relative MLT displacement of the conjugate auroras. These results suggest that the IMF-induced magnetotail twisting plays an important role in the relative MLT displacement of the conjugate auroral locations. In contrast to MLT displacement, the latitudinal displacement showed a large difference

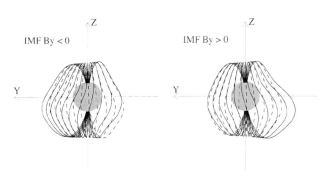

Figure 8. Examples illustrating IMF B_y induced magnetotail field lines in the *YZ* plane estimated from the T96 model (view from tail). (left) Negative IMF B_y (-5 nT). (right) Positive IMF B_y ($+5$ nT). Dashed lines indicate magnetic field lines for the case of no IMF B_y (0 nT).

between the observation and T96 model. It needs further study.

Acknowledgments. We would like to thank the 44th and 50th Japanese Antarctic Research Expedition (JARE) members carrying out the optical operation at Syowa Station. S. Johannesson and A. Egilsson supported the auroral observations in Iceland. THEMIS FGM data were obtained through CDA Web. Solar wind parameters were obtained from the OMNI data site at SPDF, NASA Goddard Space Flight Center. This research was supported by the Grant-in Aid for Scientific Research (B: 13573007, B: 17403009, and B: 21403007) from the Japan Society for the Promotion of Science (JSPS). The preparation of this paper was supported by an NIPR publication subsidy.

REFERENCES

Belon, A. E., J. E. Maggs, T. N. Davis, K. B. Mather, N. W. Glass, and G. F. Hughes (1969), Conjugacy of visual auroras during magnetically quiet periods, *J. Geophys. Res.*, *74*(1), 1–28.

Borovsky, J. E. (1993), Auroral arc thicknesses as predicted by various theories, *J. Geophys. Res.*, *98*(A4), 6101–6138.

Davis, T. N. (1978), Observed characteristics of auroral forms, *Space Sci. Rev.*, *22*, 77–113.

Davis, T. N., T. J. Hallinan, and H. C. Stenbaek-Nielsen (1971), Auroral conjugacy and time-dependent geometry of auroras, in *The Radiating Atmosphere*, edited by B. M. McCormac, pp. 160–169, D. Reidel, Dordrecht, The Netherlands.

Frank, L. A., and J. B. Sigwarth (2003), Simultaneous images of the northern and southern auroras from the Polar spacecraft: An auroral substorm, *J. Geophys. Res.*, *108*(A4), 8015, doi:10.1029/2002JA009356.

Fujii, R., N. Sato, T. Ono, H. Fukunishi, T. Hirasawa, S. Kokubun, T. Araki, and T. Saemundsson (1987), Conjugacies of pulsating auroras by all-sky TV observations, *Geophys. Res. Lett.*, *14*(2), 115–118.

Hallinann, T. J., and T. N. Davis (1970), Small-scale auroral arc distortions, *Planet. Space Sci.*, *18*, 1735–1744.

Lanzerotti, L. J. (1987), Conjugate studies on hydromagnetic waves, *Mem. Natl. Inst. Polar Res. Spec. Issue*, *48*, 121–133.

Minatoya, H., N. Sato, T. Saemundsson, and T. Yoshino (1995), Large displacements of conjugate auroras in the midnight sector, *J. Geomagn. Geoelectr.*, *48*, 967–975.

Motoba, T., K. Hosokawa, N. Sato, A. Kadokura, and G. Bjornsson (2010), Varying interplanetary magnetic field B_y effects on interhemispheric conjugate auroral features during a weak substorm, *J. Geophys. Res.*, *115*, A09210, doi:10.1029/2010JA015369.

Motoba, T., K. Hosokawa, Y. Ogawa, N. Sato, A. Kadokura, S. C. Buchert, and H. Rème (2011), In situ evidence for interplanetary magnetic field induced tail twisting associated with relative displacement of conjugate auroral features, *J. Geophys. Res.*, *116*, A04209, doi:10.1029/2010JA016206.

Østgaard, N., S. B. Mende, H. U. Frey, T. J. Immel, L. A. Frank, J. B. Sigwarth, and T. J. Stubbs (2004), Interplanetary magnetic field control of the location of substorm onset and auroral features in the conjugate hemispheres, *J. Geophys. Res.*, *109*, A07204, doi:10.1029/2003JA010370.

Østgaard, N., N. A. Tsyganenko, S. B. Mende, H. U. Frey, T. J. Immel, M. Fillingim, L. A. Frank, and J. B. Sigwarth (2005), Observations and model predictions of substorm auroral asymmetries in the conjugate hemispheres, *Geophys. Res. Lett.*, *32*, L05111, doi:10.1029/2004GL022166.

Newell, P. T., C.-I. Meng, and K. M. Lyon (1996), Suppression of discrete aurorae by sunlight, *Nature*, *381*, 766–767.

Sato, N., R. Fujii, T. Ono, H. Fukunishi, T. Hirasawa, T. Araki, S. Kokubun, K. Makita, and T. Saemundsson (1986), Conjugacy of proton and electron auroras observed near L=6.1, *Geophys. Res. Lett.*, *13*(13), 1368–1371.

Sato, N., T. Nagaoka, K. Hashimoto, and T. Saemundsson (1998), Conjugacy of isolated auroral arcs and nonconjugate auroral breakups, *J. Geophys. Res.*, *103*(A6), 11,641–11,652.

Sato, N., D. M. Wright, C. W. Carlson, Y. Ebihara, M. Sato, T. Saemundsson, S. E. Milan, and M. Lester (2004), Generation region of pulsating aurora obtained simultaneously by the FAST satellite and a Syowa-Iceland conjugate pair of observatories, *J. Geophys. Res.*, *109*, A10201, doi:10.1029/2004JA010419.

Sato, N., A. Kadokura, Y. Ebihara, H. Deguchi, and T. Saemundsson (2005), Tracing geomagnetic conjugate points using exceptionally similar synchronous auroras, *Geophys. Res. Lett.*, *32*, L17109, doi:10.1029/2005GL023710.

Stenbaek-Nielsen, H. C., and A. Otto (1997), Conjugate auroras and the interplanetary magnetic field, *J. Geophys. Res.*, *102*(A2), 2223–2232.

Stenbaek-Nielsen, H. C., T. N. Davis, and N. W. Glass (1972), Relative motion of auroral conjugate points during substorms, *J. Geophys. Res.*, *77*(10), 1844–1858.

Stenbaek-Nielsen, H. C., E. M. Wescott, and R. W. Peterson (1973), Pulsating auroras over conjugate areas, *Antarct. J. U. S.*, *8*, 246–247.

Stubbs, T. J., R. R. Vondrak, N. Østgaard, J. B. Sigwarth, and L. A. Frank (2005), Simultaneous observations of the auroral ovals in both hemispheres under varying conditions, *Geophys. Res. Lett.*, *32*, L03103, doi:10.1029/2004GL021199.

Tsyganenko, N. A., and D. P. Stern (1996), Modeling the global magnetic field of the large-scale Birkeland current systems, *J. Geophys. Res.*, *101*(A12), 27,187–27,198.

Watanabe, M., A. Kadokura, N. Sato, and T. Saemundsson (2007), Absence of geomagnetic conjugacy in pulsating auroras, *Geophys. Res. Lett.*, *34*, L15107, doi:10.1029/2007GL030469.

G. Bjornsson and T. Saemundsson, Science Institute, University of Iceland, Reykjavik 107, Iceland.

K. Hosokawa, Department of Communication Engineering and Informatics, University of Electro-Communications, Tokyo 182-8585, Japan.

A. Kadokura, T. Motoba, and N. Sato, National Institute of Polar Research, 10-3, Midori-cho, Tachikawa, Tokyo 190-8518, Japan. (nsato@nipr.ac.jp)

Auroral Asymmetries in the Conjugate Hemispheres and Interhemispheric Currents

N. Østgaard

Department of Physics and Technology, University of Bergen, Bergen, Norway

K. M. Laundal

Teknova, Kristiansand, Norway

In this chapter, we first give a short introduction to auroral conjugate studies since the 1960s. We do not intend to give a review of the earlier findings because such papers already exist, but we will describe in some detail new results from the last couple of years based on global imaging in the UV wavelengths by the IMAGE and Polar satellites. We review the results from a paper showing how the interplanetary magnetic field (IMF) clock angle, θ_c, or IMF B_y can be used to organize the substorm locations in the conjugate hemispheres and how they give slightly different information. Then, we report a study where complete asymmetric aurorae were observed. Based on the results from these two papers, we discuss the interhemispheric currents that can be associated with these observations. We present three candidates for producing such currents that are related to IMF orientation, B_y and B_x, and tilt angle. The three candidates are (1) the penetration of IMF θ_c or B_y into the closed magnetosphere, (2) the differences in solar wind dynamo efficiency due to IMF B_x and tilt angle, and (3) the conductivity differences due to tilt angle. We claim that all these candidates are consistent with our recent observations.

1. INTRODUCTION

Our knowledge about how the Earth is coupled to space is, to a large extent, based on measurements from only one polar hemisphere. It is therefore tempting to assume that the aurora borealis (Northern Hemisphere) and aurora australis (Southern Hemisphere) are mirror images of each other because the charged particles producing the aurora follow the magnetic field lines connecting the two hemispheres. There is now much observational evidence that this is not always the case.

It is now well documented that interplanetary magnetic field (IMF) not only controls the dayside aurora, cusp aurora location, and dayside field-aligned current (FAC) systems [e.g., *Sandholt et al.*, 1998; *Sandholt and Farrugia*, 1999; *Zhou et al.*, 2000; *Bobra et al.*, 2004; *Østgaard et al.*, 2005b; *Wing et al.*, 2010] but also affects the location and intensity of the nightside aurora. Although some early studies [e.g., *Hargreaves and Chivers*, 1964; *Belon et al.*, 1969; *Fujii et al.*, 1987] and some more recent studies [*Pulkkinen et al.*, 1995; *Frey et al.*, 1999] focused mostly on similarities between the hemispheres, later studies have reported that both the intensity and location of the aurora can be quite different in the two hemispheres. Displacements have been found both in latitude [*Stenbaek-Nielsen and Otto*, 1997; *Laundal et al.*, 2010a, 2010b] and longitude and ranging from a few hundreds of kilometers [*Sato et al.*, 1986; *Stenbaek-Nielsen*

Auroral Phenomenology and Magnetospheric Processes: Earth and Other Planets
Geophysical Monograph Series 197

and Otto, 1997; *Sato et al.*, 1998; *Frank and Sigwarth*, 2003] up to 1–2 magnetic local time (MLT) sectors [*Burns et al.*, 1990; *Liou et al.*, 2001b; *Liou and Newell*, 2010; *Wang et al.*, 2007; *Østgaard et al.*, 2004, 2005b, 2011a, 2011b]. Latitudinal and longitudinal displacements as well as difference in auroral intensities have been attributed to asymmetric currents [*Stenbaek-Nielsen and Otto*, 1997; *Laundal and Østgaard*, 2009; *Østgaard et al.*, 2011a] or more directly to IMF influence on the magnetospheric field configuration [*Stenbaek-Nielsen and Otto*, 1997; *Vorobjev et al.*, 2001; *Wang et al.*, 2007; *Liou et al.*, 2001b; *Liou and Newell*, 2010; *Østgaard et al.*, 2004, 2005b, 2011a, 2011b]. Nonconjugate auroras, meaning occurrence of auroral features in one hemisphere only or significantly intensity differences, have been attributed to asymmetric FACs resulting from differences in ionospheric conductivity [*Stenbaek-Nielsen et al.*, 1972; *Sato et al.*, 1998]. The most striking example of completely asymmetric aurora was reported by *Laundal and Østgaard* [2009] and will be reviewed in more detail and discussed further in this chapter. Statistical studies also indicate that differences in solar radiation (day/night-winter/summer) cause differences in auroral intensities [*Newell et al.*, 1996; *Liou et al.*, 2001a; *Meng et al.*, 2001; *Newell et al.*, 2010] and that the orientation of the IMF plays an important role in determining different convection patterns in the two hemispheres [*Heppner and Maynard*, 1987]. Statistical studies using global Polar UVI data have revealed significant hemispheric asymmetry in the nightside auroral brightness due to IMF B_y [*Shue et al.*, 2001] and a smaller but still statistical significant intensity asymmetry due to IMF B_x [*Shue et al.*, 2002].

To explore whether the coupling of the two hemispheres to space is symmetric or asymmetric at a given time, we need simultaneous conjugate measurements. In situ conjugate measurements from space are hard to obtain because you do not know if you really are on conjugate field lines. Conjugate optical imaging from the ground is difficult because of the limited view of all-sky images and the fact that you need both clear sky and darkness in both hemispheres [*Sato et al.*, 1998]. To overcome poor observation conditions at certain ground stations, a series of conjugate aircraft flights equipped with all-sky cameras was undertaken [*Stenbaek-Nielsen et al.*, 1972, 1973]. Another approach has been to utilize imaging from space in one hemisphere combined with ground-based optical observations in the other [e.g., *Burns et al.*, 1990; *Vorobjev et al.*, 2001]. The work of *Frank and Sigwarth* [2003] is probably the only report of seeing an auroral onset in both hemispheres by only one camera, the Visible Imaging System (VIS) Earth camera on board the Polar spacecraft. In the mid-1980s, Viking [*Murphree and Cogger*, 1988] and Dynamic Explorer 1 [*Frank et al.*, 1981]

provided the very first simultaneous global imaging from space and discovered that theta aurora could be a conjugate phenomenon [*Craven et al.*, 1991]. The imagers on board DE-1 and Viking also observed the nightside aurora simultaneously in the two hemispheres [e.g. *Pulkkinen et al.*, 1995]. Fifteen years passed before the apsidal precession of the Polar spacecraft orbit and the large field of view of the Polar VIS Earth camera [*Frank et al.*, 1995] and the IMAGE-FUV instruments [*Mende et al.*, 2000] offered a new opportunity to observe the aurora simultaneously in the conjugate hemispheres. With an imaging cadence of 2 min and 1 min for IMAGE-FUV and Polar VIS Earth camera, respectively, both slowly varying phenomena as theta aurora [*Østgaard et al.*, 2003] as well as dynamic features such as cusp precipitation [*Østgaard et al.*, 2005a] and auroral features during substorms [*Østgaard et al.*, 2004, 2005b, 2011a; *Laundal and Østgaard*, 2009; *Laundal et al.*, 2010a] were reported. Our earlier findings were summarized by *Østgaard et al.* [2007]. In this chapter, we will review some of our more recent results comparing nightside auroral features in the conjugate hemispheres and discuss, in particular, the implications of these results regarding interhemispheric currents.

2. ASYMMETRIC SUBSTORM ONSET LOCATION

As mentioned in section 1, early studies indicated that the substorm onset azimuthal locations had a statistical IMF B_y dependence [*Elphinstone et al.*, 1990; *Liou et al.*, 2001b]. Using simultaneous UV imaging data from both hemispheres provided by IMAGE and Polar, *Østgaard et al.* [2004, 2005b] showed that a slightly better correlation was found when IMF clock angle, θ_c, was used instead of IMF B_y. A statistical study [*Østgaard et al.*, 2007] using the substorms identified from several years of IMAGE data [*Frey et al.*, 2004] confirmed the θ_c control. *Wang et al.* [2007] analyzed an extended version of the list of substorms from IMAGE [*Frey and Mende*, 2006] including the year 2005, giving a total of 4192 substorms (instead of 3738 used by *Østgaard et al.* [2007]). Although they mainly focused on the IMF B_y and the solar zenith effect, they also reported that no IMF θ_c control of substorm onset location could be found in the data. Unfortunately, no results were shown, and they did not explain how they performed the unsuccessful search for an IMF θ_c control. *Liou and Newell* [2010] analyzed a new set of substorms identified from Polar UVI data [*Liou*, 2010], which are complementary to substorms identified in the IMAGE data and cover the years from 1996 to 2000 and 2007. Their main result is that substorm onset locations have a dependence on the pairing of IMF B_y/tilt angle, which is a confirmation of what was suggested by *Østgaard et al.*

[2005b], see their Figure 1, where $\theta_c = 90°$ (positive B_y), and positive tilt angle (northern summer) gives the largest positive ΔMLT, which implies the earliest substorms in the Northern Hemisphere. *Liou and Newell* [2010] found a similar dependence on the pairing of IMF B_y/solar zenith angle, consistent with the results from the Northern Hemisphere, but not the Southern Hemisphere by *Wang et al.* [2007]. They also found a weak IMF θ_c control of substorm onset location, but argued that this is not a strong controlling parameter for substorm onset location.

Motivated by the findings (or the lack thereof) in these two studies [*Wang et al.*, 2007; *Liou and Newell*, 2010] and the release of a new and complementary substorm list based on Polar UVI data [*Liou*, 2010], we decided to revisit this problem analyzing the combined data set, which covers 6601 substorms from the years 1996–2005 plus 2007 [*Østgaard et al.*, 2011b]. The main results from that paper are found in Figure 1 (same as Figure 2 of *Østgaard et al.* [2011b]). The solar wind data used to produce these plots are mainly from ACE, but Wind is used when ACE is not available. The data, which can be downloaded from measure. igpp.ucla.edu, were time shifted to $X = 17\ R_E$ using the propagation method described by *Weimer* [2004]. The data were further time shifted using planar propagation from $X = 17\ R_E$

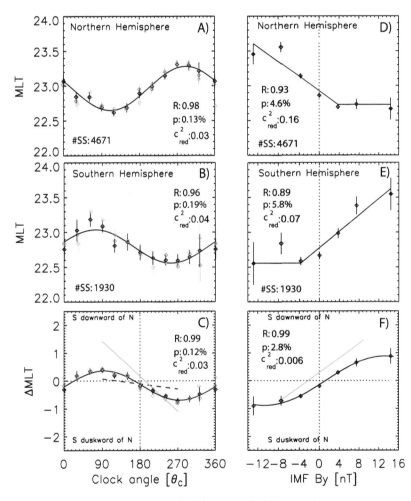

Figure 1. (a) Average substorm magnetic local time (MLT) locations for 30° bins of interplanetary magnetic field (IMF) θ_c in the Northern Hemisphere based on the combined IMAGE and Polar data sets. (b) Same as Figure 1a for Southern Hemisphere. Total number of substorms is shown in lower left corner. (c) ΔMLT between the southern and northern average substorm locations. Black, red, and blue diamonds are for time shifts to 10 R_E/±5 min, 20 R_E/±5 min, and −10 R_E/40 min. (d, e, and f) Same as Figures 1a–1c, but the substorm onset MLTs have been organized in 4 nT bins of IMF B_y (only time shifts to −10 R_E/±5 min). The blue straight lines in Figures 1c and 1f are the linear fits from *Østgaard et al.* [2005b]. The error bars in all plots are standard deviation of the mean values. Figure 1 is identical to Figure 2 of *Østgaard et al.* [2011b].

to $X = -10 \, R_E$ (average of ±5 min and 40 min prior to onset time) and $X = -20 \, R_E$ (average of ±5 min) to investigate different possible impact times from the solar wind to the magnetosphere.

In Figures 1a, 1b, and 1c, it can be seen that the IMF θ_c control the average substorm onset locations in both hemispheres. Due to the larger number of substorms, 30° bins of IMF θ_c could be used. The number of substorms in each bin ranges from 89 to 620 and 33 to 240 in the Northern and Southern Hemispheres, respectively. The different time shifts and averaging (black, red, and blue diamonds) gave almost similar results. The sine functional fits show how the average substorm locations in the two hemispheres are in antiphase. The best correlations were found for $-10 \, R_E$ and ±5 min, with the Spearman correlation coefficients, R_S, of 0.98 (north), 0.96 (south), and 0.99 (relative) and probability in all three panels, p, was <0.2%, and the χ^2_{red} was <0.05. Spearman correlation coefficient was used because it is valid for nonlinear relations. The correlation was found to be significantly poorer for 40 min averaging with R_S down to 0.75, p as high as 1.8%, and χ^2_{red} of 0.3. Time shifts to $-10 \, R_E$ gave slightly better correlation than for $-20 \, R_E$. These high R_S values and low p and χ^2_{red} leave little doubt that the average onset locations and interhemispheric asymmetry are organized as a sine function of IMF θ_c.

These results were all based on average values and cannot be used to predict the location of one single substorm. However, the relative displacement is probably a more robust result that can be used when the location in one hemisphere is known [e.g., *Motoba et al.*, 2010] and can be expressed as

$$\Delta MLT \, (MLT_s - MLT_n) = 0.53 \times \sin(\theta_c - 4.8°) - 0.17. \quad (1)$$

Interestingly, they also found that if only values with total IMF larger than 5 nT were included, the amplitudes increased to 0.41 (north), 0.35 (south), and 0.73 (relative), while the phase shifts and constants were unchanged.

Østgaard et al. [2011b] also examined how the average substorm locations were related to IMF B_y using the same combined data set. This can be seen in Figures 1d, 1e, and 1f. Despite the larger number of substorms, the sampling for large $|B_y|$ values were poor, and therefore, 4 nT bin resolution and integral <-10 nT and >10 nT were used, giving the number of substorms in each bin that ranges from 61 to 1427 in the Northern Hemisphere and 24 to 753 in the Southern Hemisphere. Neither the Northern or the Southern Hemisphere showed a linear relation, as previously reported [*Østgaard et al.*, 2005b; *Wang et al.*, 2007; *Liou and Newell*, 2010], but reveals a saturation effect for positive (negative) B_y values in the North (South). The asymmetry between hemispheres is very well fitted with a sine function ($R_S = 0.99$):

$$\Delta MLT \, (MLT_s - MLT_n) = 0.88 \times \sin\left(\frac{B_y}{12 \, \text{nT}} \times 9.0° - 9.3°\right). \quad (2)$$

The saturation effect could be explained by considering how magnetic flux is added nonuniformly to the magnetotail for negative and positive B_y values [*Khurana et al.*, 1996] and assuming that this nonuniform penetration of IMF B_y extends into the closed magnetosphere. We have illustrated this in Figure 2a by sketching the cross section of the magnetotail and how magnetic flux is added (+) in the northern dawn and the southern dusk for IMF $B_y > 0$. This means that a positive B_y will penetrate only in the southern dusk, and a negative B_y will only penetrate into the northern dusk. Considering also that substorms, on average, are located in the premidnight region, the closed magnetic field lines (arrows) will be affected by $B_y > 0$ in the Southern Hemisphere and not in the Northern Hemisphere, where we observe the saturation effect for $B_y > 0$. For $B_y < 0$ (Figure 2b), it is opposite. The flux is not added to the southern dusk, consistent with the saturation we observe in the Southern Hemisphere for $B_y < 0$. This idea is supported by results from the semiempirical Tsyganenko 96 model (Figures 2c and 2d).

One feature that only could be extracted from the θ_c control is the indirect evidence of nightside reconnection even for strongly northward IMF. Our first studies [*Østgaard et al.*, 2004, 2005b] were limited to only southward IMF, and we explained the IMF θ_c control by the magnetic stress imposed by the IMF on the Earth's magnetic field from the moment the field lines are opened on the dayside, draped down the tail, and until they eventually close through reconnection in the midtail before substorm onset, as depicted by the sketch in Figure 3. Due to stress imposed on the open field lines, only field lines with asymmetric foot point can reconnect in the midtail. In these latter results [*Østgaard et al.*, 2011b], the IMF θ_c control is also apparent for substorms during northward IMF, under which conditions no magnetic flux is opened by lobe reconnection. However, open field lines can still be closed by tail reconnection [*Cowley and Lockwood*, 1992; *Grocott et al.*, 2005]. Thus, due to the tension force imposed by IMF before tail reconnection, the resulting closed field lines will, also in this case, have asymmetric foot points. The sine behavior with maxima at $\theta_c = 90°/270°$ (Figure 1c) is exactly what one would expect from considering dayside and lobe reconnection geometry and tension force on open field lines.

Asymmetric foot points for closed field lines imposed through nightside reconnection by the IMF means that the

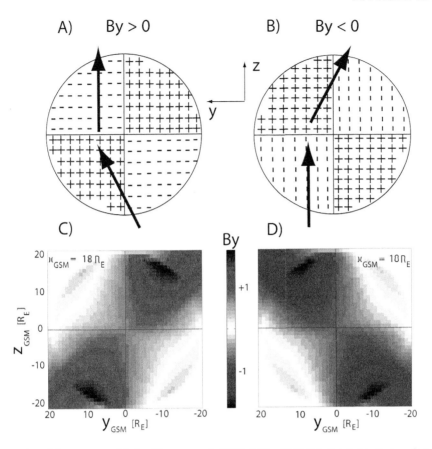

Figure 2. Cross section of the magnetotail [after *Khurana et al.*, 1996]. (a) For IMF $B_y > 0$, $+(-)$ indicates where magnetic flux is added (not added) for IMF $B_y > 0$. Arrows indicate how closed magnetic fields can be affected in the dusk sector. (b) Same as Figure 2a but for IMF $B_y < 0$. (c and d) Results from Tsyganenko 96 model showing how IMF B_y nonuniformly penetrates the magnetotail when (c) IMF $B_y = +3$ nT and (d) IMF $B_y = -3$ nT. Figure 2 is identical to Figure 3 of *Østgaard et al.* [2011b].

closed field lines are twisted and will induce an interhemispheric current. This will be further discussed in section 4.

3. COMPLETELY ASYMMETRIC AURORA

From the years 2002–2003, when VIS had good coverage of the Southern Hemisphere and IMAGE of the Northern Hemisphere, we have identified quite a few examples of auroral features that only appear in one hemisphere. Some of these are small features at the poleward boundary, and some are larger arc structures covering several local time sectors. These are subject to ongoing studies by our group. However, the most striking example of completely asymmetric aurora has been published and was reported by *Laundal and Østgaard* [2009].

The results of this chapter can be seen in Figure 4. In the upper panel, one can see that the aurora in the northern summer hemisphere (IMAGE WIC) has a bright spot at dawn, which is not present in the southern winter hemisphere

(Polar VIS Earth). The opposite is observed in the dusk, the southern aurora is more intense than in the north. The spot at dawn is a transient (10 min), seen in several subsequent images and coincides with a rapid poleward expansion of ~8° in less than 10 min, marked with blue dashed and solid lines in Figure 4a, signifying a local increase in tail reconnection. The dusk aurora in the Southern Hemisphere was significantly brighter than its northern counterpart for more than an hour.

As these observations were made by two different cameras, *Laundal and Østgaard* [2009] paid special attention to the differences in sensitivity and passband, which would give different counts and kilo-Rayleigh (kR) in the two hemispheres. To make sure that the differences were real, they used the information and the modeled instrumental response to electron precipitation given by *Frank and Sigwarth* [2003] and *Frey et al.* [2003]. The modeled response is given for different energies of electrons and assumes a certain atmospheric model. However, the response of the VIS camera is

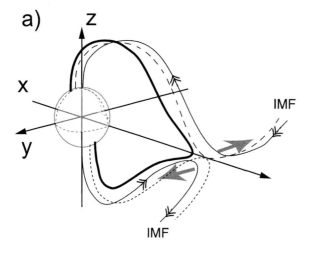

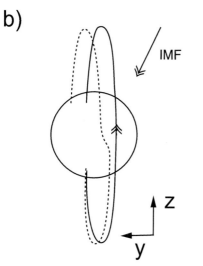

Figure 3. (a) Snapshot of open (solid, dashed, dotted) field lines and one reconnected closed (thick) field line under the influence of the magnetic tension force (thick gray arrows) from IMF $B_y > 0$ and $B_z < 0$ acting toward dawn in the Northern Hemisphere and toward dusk in the Southern Hemisphere. (b) Superposition of the IMF and the symmetric field. Figure 3 is similar to Figure 5 of *Østgaard et al.* [2004], except that we have emphasized that the B_y penetration is largest in the plasma sheet where the total field strength is weakest, as indicated by the dashed field line in Figure 3b.

dominated by emissions from atomic oxygen (OI) at 130.4 nm, while the WIC camera is sensitive to molecular nitrogen (N2) emissions in the Lyman-Birge-Hopfield (LBH) band and a few emission lines of atomic nitrogen. Because emissions in the LBH band is reduced more by molecular oxygen (O2) absorption than the OI emission, and heating of the atmosphere will affect the scale height of N2 more than of O2, a relatively higher intensity would be seen by IMAGE-WIC in the sunlit Northern Hemisphere. This means that the observed

differences in the dusk sector could be underestimated, and the difference at dawn may be exaggerated. Due to the modeled response [*Frey et al.*, 2003], the observed emissions decrease with increasing mean electron energy, and this effect is stronger for the OI line (VIS). This means that high-electron energy could contribute to the asymmetry at dusk. However, the large relative difference observed at 6 MLT, combined with the high absolute intensity in the north, >5.2 kR (with a peak intensity of 10 kR), lead the authors to conclude that the observed asymmetry could not be an instrumental effect.

Laundal and Østgaard [2009] explored various candidates to explain the complete asymmetric aurora. First, differences in magnetic field strength were examined, but the differences where the spots are observed were too small (<10%) to explain the large asymmetries. Then, the effect of difference in solar radiation was explored. As suggested by *Newell et al.* [1996], the dark hemisphere where the conductivity is lower will have brighter aurora. The idea is that the magnetospheric current generator will require a larger potential drop in the hemisphere where the conductivity is lower. From Figure 4, one can see that this would apply to the southern winter (dark) hemisphere. This is consistent with the southern dusk spot, but is not true for the dawn spot. In a recent paper, *Newell et al.* [2010] have found that the local winter/summer intensity asymmetry is most pronounced for the diffuse aurora produced by energetic electron precipitation. However, this would not apply to the dusk spot in the southern winter hemisphere, where there is a statistical minimum of diffuse aurora [*McDiarmid et al.*, 1975; *Newell et al.*, 2010]. Diffuse aurora has its maximum at midnight and morning sectors [*McDiarmid et al.*, 1975; *Østgaard et al.*, 1999]. The morning spot we observe is probably not related to diffuse aurora for several reasons: (1) it is a transient (10 min), while the diffuse morning spot has a duration of 30–60 min [*Akasofu*, 1968; *Østgaard et al.*, 1999], (2) the spot is seen close to the open-closed boundary, while the diffuse aurora caused by drifting energetic electrons is usually observed slightly equatorward of the main auroral oval in the morning sector [*Newell et al.*, 2010], and (3) it should have been most intense in the southern dark hemisphere [*Newell et al.*, 2010]. The next two candidates of explanations that also were the conclusion of the paper, both involve interhemispheric currents. As these will be further explained and discussed in section 4, we will just briefly mention them here. The first candidate is the more effective solar wind dynamo in the Southern Hemisphere due to the combination of positive tilt angle and positive IMF B_x, as suggested by *Cowley* [1981b]. This is consistent with the persistent spot in the southern dusk. The second candidate is the conductivity difference and gradient across the terminator that according to *Benkevich et al.* [2000] will produce an upward current in

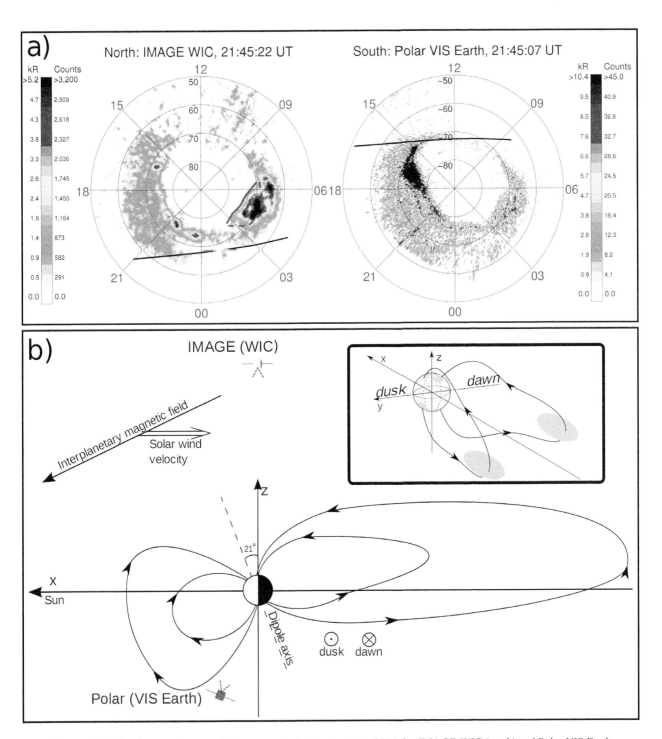

Figure 4. (a) Simultaneous images of the aurora at 21:45 UT 12 May 2001, by IMAGE-WIC (north) and Polar-VIS Earth camera in apex coordinates. The color bars show intensity in both counts and kilo-Rayleigh. The UT times are at the middle of the exposure; the cadences for WIC and VIS are 2 and 1 min. Black curves indicate the terminator. The blue lines show the location of the poleward boundary of the aurora at 21:37 (dashed) and 21:45 (solid) UT. (b) The seasonal conditions, the geometry of the magnetosphere, and the orientation of the IMF, measured by ACE ($|B_y|$ was less than 1 nT). The tilt angle was 21° at 21:45 UT. The coordinate system is Geocentric Solar Magnetospheric. Figure 4 is identical to Figure 1 of *Østgaard et al.* [2009].

the northern dawn and southern dusk. The spot at dawn coincides with a significant increase of tail reconnection in this region, which can explain its transient character. This also suggests that both Alfven waves and plasma flows and shears might be involved in creating the bright aurora.

4. THREE SOURCES OF INTERHEMISPHERIC CURRENTS

In this section, we will discuss the implication of these observations regarding the existence of interhemispheric currents. Such currents have been predicted [*Richmond and Roble*, 1987; *Benkevich et al.*, 2000], but to our knowledge were only observed directly once [*Vallat et al.*, 2005]. They have been suggested to be associated with observed asymmetries [*Cowley*, 1981a; *Stenbaek-Nielsen and Otto*, 1997; *Shue et al.*, 2001; *Østgaard et al.*, 2011a], and the most significant indirect observation of such currents were presented by *Laundal and Østgaard* [2009]. In the following, we will discuss three sources of interhemispheric currents that are consistent with our observations of asymmetric aurora in the conjugate hemispheres.

4.1. Penetration of IMF in the Closed Magnetosphere

That the penetration of IMF into the closed magnetosphere would lead to asymmetric foot points was suggested by *Cowley* [1981a] and followed up by others [e.g., *Khurana et al.*, 1996]. As discussed and shown in section 2, the observational evidence for this is quite overwhelming and convincing. As indicated in Figure 3b, asymmetric foot points means that the field lines are twisted. If we now apply Ampere's law to the twist and stretching of the field lines as seen in the plasma sheet, where we have approximately a Cartesian geometry with Z pointing along the magnetic field, the parallel component can be expressed as

$$\left(\frac{\partial B_y}{\partial x} - \frac{\partial B_x}{\partial y}\right)_{\parallel} \sim \mu_0 J_{\parallel}. \tag{3}$$

While the magnetic stress associated with tail stretching $\left(\frac{\partial B_x}{\partial y}\right)$ implies pairs of upward/downward FAC in both hemispheres, a twist in B_y implies an interhemispheric FAC. In the following, we only consider this (first) term in equation (3). In Figure 5, we show two conceptual sketches of the currents that will result from the penetration of a $B_y < 0$. First, we consider a geometry where the twist is produced in the tail by reconnection and decrease steadily toward the Earth. As the negative B_y becomes smaller in the positive x direction, the first term in equation (3) is positive, indicating an interhemispheric current from south to north. If this current is compa-

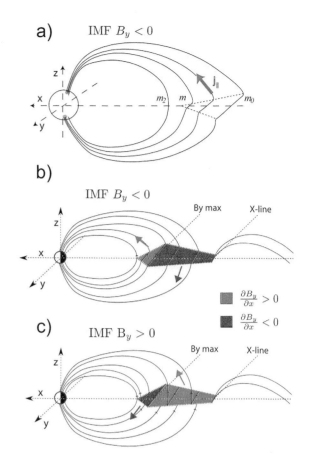

Figure 5. Sketches on how an IMF B_y will penetrate the magnetosphere and extend to close field lines. (a) How the magnetic fields are twisted from the tail and inward with a nontwisted field line earthward of the substorm onset region, similar to Figure 3b of *Østgaard et al.* [2011a] for IMF $B_y < 0$, but with the current vector (blue) pointing from south to north. (b) Similar idea, but here the penetration has a maximum in the region of discrete aurora and decreases toward the tail and at the inner edge of the plasma sheet, giving a south-to-north current at the inner edge of the plasma sheet and a north-to-south current in the auroral region for IMF $B_y < 0$. Figure 5 is similar to Figure 4 of *Stenbaek-Nielsen and Otto* [1997]. (c) Same geometry applied for IMF $B_y > 0$.

rable to (or even just a fraction of) the FACs associated with the auroral potential drops, we would expect to see a slightly brighter aurora in the Southern Hemisphere. This was discussed by *Østgaard et al.* [2011a], but unfortunately, they had the sign of the current wrong. Figure 5a is similar to Figure 3b of *Østgaard et al.* [2011a], but with the direction of the current corrected to flowing from the south to the north. The idea is similar to what was proposed earlier by *Stenbaek-Nielsen and Otto* [1997], but they considered a slightly different penetration geometry. *Stenbaek-Nielsen and Otto* [1997] also assumed the penetration of B_y into the closed magnetosphere to be facilitated by midtail reconnection. However, by assuming that the earthward convection will lead to a pileup of magnetic flux toward the inner edge of the plasma sheet, there will be a region with maximum penetration decreasing slowly toward the tail and a sharper decrease toward the Earth as the magnetic field quickly becomes stiffer. This geometry is supported by in situ measurements by *Wing et al.* [1995] and *Lui* [1984] who found that fractions of IMF B_y of 0.8 and 0.5 penetrate the plasma sheet at geosynchronous and between -10 and -30 R_E, respectively. From this geometry and IMF $B_y < 0$, *Stenbaek-Nilesen and Otto* [1997] derived a north-to-south interhemispheric current in the region of auroral arcs and a south-to-north current at the inner edge of the plasma sheet that maps to the region of region 2 currents, as shown in Figure 5b. Interestingly, if *Østgaard et al.* [2011a] had used the penetration geometry shown in Figure 5b instead of the one in Figure 5a, the rest of the argument, based on integrating the Faraday's law along a loop of two closed field lines, would still be valid. This would also explain why *Frank and Sigwarth* [2003] observed a brighter substorm onset in the south than in the north. They observed the southern onset to be dawnward of the northern consistent with a IMF $B_y > 0$ penetration as shown in Figure 5c inducing an interhemispheric current from south to north, and a brighter aurora should be seen in the south, which is what they did. Furthermore, in the statistical study by *Shue et al.* [2001], using two winter and summer seasons of Polar UVI data (1997 and 1998), they found a significant auroral intensity difference due to IMF B_y consistent with the interhemispheric current described in this section. For IMF $B_y < 0$ implying a current from north to south (Figure 5b), they found the brightest aurora in the north. Direct measurement of an interhemispheric current in the ring current region has been reported by *Vallat et al.* [2005]. In their Figure 14, there is fairly clear evidence of a current flowing from north to south at the equator. The IMF B_y was positive many hours prior to this event (not shown by *Vallat et al.* [2005]), and according to the geometry in Figure 5c, $dB_y/dx < 0$ in this region and would induce a current from north to south in agreement with the CLUSTER observations. The geometry of the IMF B_y penetration that leads to these interhemispheric currents still needs to be confirmed. Such a validation could probably be addressed by using large statistical data sets of ground-based magnetic measurements.

4.2. Hemispherical Differences in Solar Wind Dynamo Efficiency

The Earth's magnetosphere creates an obstacle for the solar wind and its "frozen-in" interplanetary magnetic field. As suggested by the open magnetospheric model [*Dungey*,

1961], reconnection on the dayside leads to field lines with one foot point on the Sun and the other on the Earth. As the solar wind flows past the Earth, magnetic flux is added to the nightside lobes, which eventually lead to the reconnection on the nightside. Although there is no friction in a collissionless plasma, the tension force on magnetic field lines tends to slow down the solar wind, and as first noticed by *Cowley* [1981b], the orientation of the IMF in the *XZ* plane would lead to different strengths of the tension force in the two hemispheres. This is shown in Figure 6, which is similar to Figure 2 of *Cowley* [1981b]. The black arrows indicate that the tension force that slows down the solar wind have different strengths in the two hemispheres. For IMF $B_x > 0$, as shown here, the force is larger in the Southern Hemisphere and slows down the solar wind more rapidly in this hemisphere. This is consistent with the shorter distance, due to the slight downward tilt of the magnetotail, the open flux tubes in the Southern Hemisphere will travel before reconnection in the tail. In the reference frame of the Earth for IMF $B_z < 0$, there will be an electric field of equal strength in the north and south pointing out of the plane as shown by the gray vectors. The slowing down of the solar wind will create a current into plane (black vectors) given by

$$\mathbf{J}_\perp = \frac{\mathbf{B} \times \rho \frac{\partial \mathbf{V}}{\partial t}}{B^2}. \tag{4}$$

With $\mathbf{J} \cdot (-\mathbf{v} \times \mathbf{B}) < \mathbf{0}$, this creates the solar wind dynamo, which is more effective in the Southern Hemisphere for $B_x > 0$. If one now considers that parts of these currents are

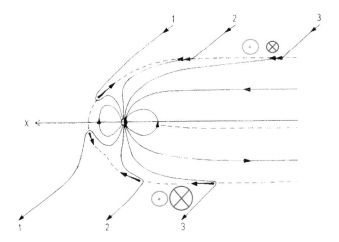

Figure 6. The reconnection geometry for IMF $B_z < 0$ and $B_x > 0$. The black arrows show the magnetic tension force accelerating/slowing down the solar wind sunward/tailward of the cusp. Figure 6 is identical to Figure 2 of *Cowley* [1981b], except for the inserted electric field vectors (gray and out of the plane) and current vectors (black and into the plane). Reprinted with permission from Elsevier.

closed in the ionosphere, as region 1 currents, they will flow out of the dusk and into the dawn regions in both hemispheres. The persistent dusk aurora in the Southern Hemisphere that *Laundal and Østgaard* [2009] observed is consistent with this idea proposed by *Cowley* [1981b]. Based on the same two winter and summer seasons of Polar UVI data (1997 and 1998) that was used to reveal the auroral brightness asymmetry due to IMF B_y, *Shue et al.* [2002] found a smaller but still statistically significant auroral intensity asymmetry due to IMF B_x, when IMF also had a $B_z < 0$ component, and similar B_y intervals were compared. As predicted by the difference in solar wind dynamo efficiency described here, the aurora should be brighter in the dusk-premidnight sector in the Northern Hemisphere for IMF $B_x < 0$. *Shue et al.* [2002] found an overall brighter aurora in the Northern Hemisphere for IMF $B_x < 0$.

4.3. Seasonal and Conductivity Differences

The third candidate for producing interhemispheric currents is the differences due to differences in solar radiation. The mechanism for these currents are redistribution of the 3-D current system due to different ionospheric conductivities in the dark and sunlit conjugate hemispheres and has been modeled by *Richmond and Roble* [1987] and *Benkevich et al.* [2000]. While *Richmond and Roble* [1987] modeled the generation of interhemispheric currents due to the dynamo effect of the thermospheric winds, which are most significant at low latitudes, *Benkevich et al.* [2000] modeled the interhemispheric currents at high latitudes resulting from the large

conductivity gradient near the terminator in the winter hemisphere, as shown in Figure 7. The idea is that due to the low conductivity in the dark hemisphere, the high-latitude currents (region 1) cannot close, but as the two hemispheres are connected by highly conductive magnetic field lines, currents can flow out of the sunlit hemisphere into the region of the large gradient in the dark hemisphere near the terminator and close through the sunlit part of that hemisphere. The strength of these currents between the hemispheres maximizes for maximum tilt angle during winter and summer, but is predicted to be present even at equinox due to diurnal tilt angle variations. The observed dawn and dusk asymmetry observed by *Laundal and Østgaard* [2009] is consistent with this current system. Compared to Figure 7, the seasons were opposite, with the Northern Hemisphere sunlit, but with the view from the tail instead of from the sun, and the interhemispheric currents would have the same orientations as in Figure 7. The dawn spot would be in the Northern Hemisphere, as *Laundal and Østgaard* [2009] observed.

5. SUMMARY

In this chapter, we have reviewed some new results obtained in the last couple of years based on global imaging in the UV wavelengths by the IMAGE and Polar satellites. These results can be summarized as follows.

1. The IMF θ_c controls the average substorm locations in both hemispheres and explains the asymmetry of onset locations between the hemispheres. This is a manifestation of dayside and lobe reconnection geometry and magnetic

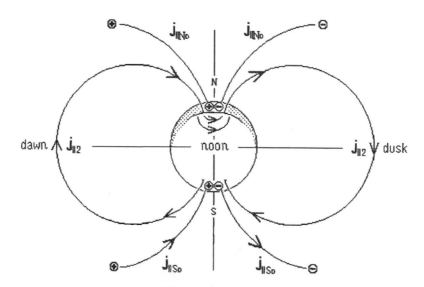

Figure 7. A schematic presentation of the interhemispheric currents that will be produced when the Northern Hemisphere is in darkness. The interhemispheric currents are denoted by $J_{\|2}$, while the high-latitude region 1 currents are denoted by $J_{\|No}$ and $J_{\|So}$, for Northern and Southern Hemispheres, respectively. Figure 7 is identical to Figure 1 of *Benkevich et al.* [2000].

tension on open field lines before tail reconnection resulting in closed field lines with asymmetric foot points. The IMF control is valid for all θ_c angles and is an indirect evidence of tail reconnection even for strongly northward IMF.

2. The relation between IMF B_y and the average substorm locations is also statistically significant. It is not linear, but reveals a saturation effect due to the nonuniform penetration of IMF B_y into the closed magnetosphere.

3. The auroral display in the two hemispheres can be completely asymmetric and is a strong indirect indication of the existence of predicted interhemispheric currents.

Based on these results and supported by some earlier statistical studies and some event studies, we present three candidates for producing such currents, which are related to IMF orientation, B_y and B_x, and tilt angle. The three candidates are (1) the penetration of IMF θ_c/B_y into the closed magnetosphere, (2) the differences in solar wind dynamo efficiency due to IMF B_x and tilt angle, and (3) the conductivity differences due to tilt angle. We claim that all these candidates are consistent with our recent observations.

Acknowledgments. We thank S. B. Mende for the use of IMAGE FUV WIC data, J. B. Sigwarth for the use of Polar VIS Earth Camera data. We thank R. Lepping for the WIND magnetic field data, A. Lazarus for the WIND solar wind data, C. Smith for the ACE magnetic field data and D. McComas for the ACE solar wind data. Thanks to Jone Reistad for making Figures 5b and 5c. This study was supported by the Norwegian Research Council, through the IPY-ICESTAR project 176045/S30.

REFERENCES

Akasofu, S.-I. (1968), *Polar and Magnetospheric Substorms*, 298 pp., D. Reidel, Norwell, Mass.

Belon, A. E., J. E. Maggs, T. N. Davis, K. B. Mather, N. W. Glass, and G. F. Hughes (1969), Conjugacy of visual auroras during magnetically quiet periods, *J. Geophys. Res.*, *74*(1), 1–28.

Benkevich, L., W. Lyatsky, and L. L. Cogger (2000), Field-aligned currents between conjugate hemispheres, *J. Geophys. Res.*, *105*(A12), 27,727–27,737.

Bobra, M. G., S. M. Petrinec, S. A. Fuselier, E. S. Claflin, and H. E. Spence (2004), On the solar wind control of cusp aurora during northward IMF, *Geophys. Res. Lett.*, *31*, L04805, doi:10.1029/ 2003GL018417.

Burns, G. B., D. J. McEwen, R. A. Eather, F. T. Berkey, and J. S. Murphree (1990), Optical auroral conjugacy: Viking UV imager - South Pole station ground data, *J. Geophys. Res.*, *95*(A5), 5781–5790.

Cowley, S. W. H. (1981a), Magnetospheric asymmetries associated with the y-component of the IMF, *Planet. Space Sci.*, *29*, 79–96.

Cowley, S. W. H. (1981b), Asymmetry effects associated with the x-component of the IMF in a magnetically open magnetosphere, *Planet. Space Sci.*, *29*, 809–818.

Cowley, S. W. H., and M. Lockwood (1992), Excitation and decay of solar wind-driven flows in the magnetosphere-ionosphere system, *Ann. Geophys.*, *10*, 103–115.

Craven, J. D., J. S. Murphree, L. A. Frank, and L. L. Cogger (1991), Simultaneous optical observations of transpolar arcs in the two polar caps, *Geophys. Res. Lett.*, *18*(12), 2297–2300.

Dungey, J. W. (1961), Interplanetary magnetic field and the auroral zones, *Phys. Rev. Lett.*, *6*, 47–48.

Elphinstone, R. D., K. Jankowska, J. Murphree, and L. Cogger (1990), The configuration of the auroral distribution for interplanetary magnetic field B_z northward, 1. IMF B_x and B_y dependencies as observed by the Viking satellite, *J. Geophys. Res.*, *95*(A5), 5791–5804.

Frank, L. A., and J. B. Sigwarth (2003), Simultaneous images of the northern and southern auroras from the Polar spacecraft: An auroral substorm, *J. Geophys. Res.*, *108*(A4), 8015, doi:10. 1029/2002JA009356.

Frank, L. A., J. D. Craven, K. L. Ackerson, M. R. English, R. H. Eather, and R. L. Carovillano (1981), Global auroral imaging instrument from the Dynamic Explorer mission, *Space. Sci. Instrum.*, *5*, 369–393.

Frank, L. A., J. B. Sigwarth, J. D. Craven, J. P. Cravens, J. S. Dolan, M. R. Dvorsky, P. K. Hardebeck, J. D. Harvey, and D. W. Muller (1995), The visible imaging system (VIS) for the Polar spacecraft, *Space Sci. Rev.*, *71*, 297–328.

Frey, H. U., and S. B. Mende (2006), Substorm onset as observed by IMAGE-FUV, in *Proceedings of the 8th International Conference on Substorms*, edited by M. Syrjäsuo and E. Donovan, pp. 71–76, Univ. of Calgary, Alberta, Canada.

Frey, H. U., S. B. Mende, H. B. Vo, M. Brittnacher, and G. K. Parks (1999), Conjugate observation of optical aurora with Polar satellite and ground-based cameras, *Adv. Space Res.*, *23*, 1647–1652.

Frey, H. U., T. J. Immel, G. Lu, J. Bonnell, S. A. Fuselier, S. B. Mende, B. Hubert, N. Østgaard, and G. Le (2003), Properties of localized, high latitude, dayside aurora, *J. Geophys. Res.*, *108* (A4), 8008, doi:10.1029/2002JA009332.

Frey, H. U., S. B. Mende, V. Angelopoulos, and E. F. Donovan (2004), Substorm onset observations by IMAGE-FUV, *J. Geophys. Res.*, *109*, A10304, doi:10.1029/2004JA010607.

Fujii, R., N. Sato, T. Ono, H. Fukunishi, T. Hirasawa, S. Kokubun, T. Araki, and T. Saemundsson (1987), Conjugacies of pulsating auroras by all-sky TV observations, *Geophys. Res. Lett.*, *14*(2), 115–118.

Grocott, A., T. K. Yeoman, S. E. Milan, and S. W. H. Cowley (2005), Interhemispheric observations of the ionospheric signature of tail reconnection during IMF-northward non-substorm intervals, *Ann. Geophys.*, *23*(5), 1763–1770.

Hargreaves, J. K., and H. J. A. Chivers (1964), Fluctuations in ionospheric absorption events at conjugate stations, *Nature*, *203*, 963–964.

Heppner, J. P., and N. C. Maynard (1987), Empirical high-latitude electric field models, *J. Geophys. Res.*, *92*(A5), 4467–4489.

Khurana, K. K., R. J. Walker, and T. Ogino (1996), Magnetospheric convection in the presence of interplanetary magnetic field B_y: A conceptual model and simulations, *J. Geophys. Res.*, *101*(A3), 4907–4916.

Laundal, K. M., and N. Østgaard (2009), Asymmetric auroral intensities in the Earth's Northern and Southern hemispheres, *Nature*, *460*, 491–493, doi:10.1038/nature08154.

Laundal, K. M., N. Østgaard, K. Snekvik, and H. U. Frey (2010a), Interhemispheric observations of emerging polar cap asymmetries, *J. Geophys. Res.*, *115*, A07230, doi:10.1029/2009JA 015160.

Laundal, K. M., N. Østgaard, H. U. Frey, and J. M. Weygand (2010b), Seasonal and interplanetary magnetic field-dependent polar cap contraction during substorm expansion phase, *J. Geophys. Res.*, *115*, A11224, doi:10.1029/2010JA015910.

Liou, K. (2010), Polar Ultraviolet Imager observation of auroral breakup, *J. Geophys. Res.*, *115*, A12219, doi:10.1029/2010JA 015578.

Liou, K., and P. T. Newell (2010), On the azimuthal location of auroral breakup: Hemispheric asymmetry, *Geophys. Res. Lett.*, *37*, L23103, doi:10.1029/2010GL045537.

Liou, K., P. T. Newell, and C.-I. Meng (2001a), Seasonal effects on auroral particle acceleration and precipitation, *J. Geophys. Res.*, *106*, 5531–5542.

Liou, K., P. T. Newell, D. G. Sibeck, C.-I. Meng, M. Brittnacher, and G. Parks (2001b), Observation of IMF and seasonal effects in the location of auroral substorm onset, *J. Geophys. Res.*, *106*(A4), 5799–5810.

Lui, A. T. Y. (1984), Characteristics of the cross-tail current in the Earth's magnetotail, in *Magnetospheric Currents*, *Geophys. Monogr. Ser.*, vol. 28, edited by T. A. Potemra, pp. 158–170, AGU, Washington, D. C.

McDiarmid, I. B., J. R. Burrows, and E. E. Budzinski (1975), Average characteristics of magnetospheric electrons (140 eV to 200 keV) at 1400 km, *J. Geophys. Res.*, *80*(1), 73–79.

Mende, S. B., et al. (2000), Far ultraviolet imaging from the IMAGE spacecraft. 3. Spectral imaging of Lyman-α and OI 135.6 nm, *Space Sci. Rev.*, *91*, 287–318.

Meng, C. I., K. Liou, and P. T. Newell (2001), Asymmetric sunlight effect on dayside/nightside auroral precipitation, *Phys. Chem. Earth, Part A*, *26*, 43–47, doi:10.1016/S1464-1917(00)00088-X.

Motoba, T., K. Hosokawa, N. Sato, A. Kadokura, and G. Bjornsson (2010), Varying interplanetary magnetic field B_y effects on interhemispheric conjugate auroral features during a weak substorm, *J. Geophys. Res.*, *115*, A09210, doi:10.1029/2010JA015369.

Murphree, J. S., and L. L. Cogger (1988), The application of CCD detectors to UV imaging from a spinning satellite, *Proc. SPIE Int. Soc. Opt. Eng.*, *932*, 42–49.

Newell, P. T., K. M. Lyons, and C.-I. Meng (1996), A large survey of electron acceleration events, *J. Geophys. Res.*, *101*(A2), 2599–2614.

Newell, P. T., T. Sotirelis, and S. Wing (2010), Seasonal variations in diffuse, monoenergetic, and broadband aurora, *J. Geophys. Res.*, *115*, A03216, doi:10.1029/2009JA014805.

Østgaard, N., J. Stadsnes, J. Bjordal, R. R. Vondrak, S. A. Cummer, D. L. Chenette, G. K. Parks, M. J. Brittnacher, and D. L. McKen-

zie (1999), Global-scale electron precipitation features seen in UV and X rays during substorms, *J. Geophys. Res.*, *104*(A5), 10,191–10,204.

Østgaard, N., S. B. Mende, H. U. Frey, L. A. Frank, and J. B. Sigwarth (2003), Observations of non-conjugate theta aurora, *Geophys. Res. Lett.*, *30*(21), 2125, doi:10.1029/2003GL017914.

Østgaard, N., S. B. Mende, H. U. Frey, T. J. Immel, L. A. Frank, J. B. Sigwarth, and T. J. Stubbs (2004), Interplanetary magnetic field control of the location of substorm onset and auroral features in the conjugate hemispheres, *J. Geophys. Res.*, *109*, A07204, doi:10.1029/2003JA010370.

Østgaard, N., S. B. Mende, H. U. Frey, and J. B. Sigwarth (2005a), Simultaneous imaging of the reconnection spot in the opposite hemispheres during northward IMF, *Geophys. Res. Lett.*, *32*, L21104, doi:10.1029/2005GL024491.

Østgaard, N., N. A. Tsyganenko, S. B. Mende, H. U. Frey, T. J. Immel, M. Fillingim, L. A. Frank, and J. B. Sigwarth (2005b), Observations and model predictions of substorm auroral asymmetries in the conjugate hemispheres, *Geophys. Res. Lett.*, *32*, L05111, doi:10.1029/2004GL022166.

Østgaard, N., S. B. Mende, H. U. Frey, J. B. Sigwarth, A. Aasnes, and J. M. Weygand (2007), Auroral conjugacy studies based on global imaging, *J. Atmos. Terr. Phys.*, *69*, 249–255, doi:10.1016/j.jastp.2006.05.026.

Østgaard, N., B. K. Humberset, and K. M. Laundal (2011a), Evolution of auroral asymmetries in the conjugate hemispheres during two substorms, *Geophys. Res. Lett.*, *38*, L03101, doi:10.1029/2010GL046057.

Østgaard, N., K. M. Laundal, L. Juusola, A. Åsnes, S. E. Håland, and J. M. Weygand (2011b), Interhemispherical asymmetry of substorm onset locations and the interplanetary magnetic field, *Geophys. Res. Lett.*, *38*, L08104, doi:10.1029/2011GL046767.

Pulkkinen, T. I., D. N. Baker, R. J. Pellinen, J. S. Murphree, and L. A. Frank (1995), Mapping of the auroral oval and individual arcs during substorms, *J. Geophys. Res.*, *100*(A11), 21,987–21,994.

Richmond, A. D., and R. G. Roble (1987), Electrodynamic effects of thermospheric winds from the NCAR Thermospheric General Circulation Model, *J. Geophys. Res.*, *92*(A11), 12,365–12,376.

Sandholt, P. E., and C. J. Farrugia (1999), On the dynamic cusp aurora and IMF B_y, *J. Geophys. Res.*, *104*(A6), 12,461–12,472.

Sandholt, P. E., C. J. Farrugia, J. Moen, Ø. Noraberg, B. Lybekk, T. Sten, and T. Hansen (1998), A classification of dayside auroral forms and activities as a function of interplanetary magnetic field orientation, *J. Geophys. Res.*, *103*(A10), 23,325–23,345.

Sato, N., R. Fujii, T. Ono, H. Fukunishi, T. Hirasawa, T. Araki, S. Kokubun, K. Makita, and T. Saemundsson (1986), Conjugacy of proton and electron auroras observed near L = 6.1, *Geophys. Res. Lett.*, *13*(13), 1368–1371.

Sato, N., T. Nagaoka, K. Hashimoto, and T. Saemundsson (1998), Conjugacy of isolated auroral arcs and nonconjugate auroral breakups, *J. Geophys. Res.*, *103*(A6), 11,641–11,652.

Shue, J.-H., P. T. Newell, K. Liou, and C.-I. Meng (2001), Influence of interplanetary magnetic field on global auroral patterns, *J. Geophys. Res.*, *106*(A4), 5913–5926.

Shue, J.-H., P. T. Newell, K. Liou, C.-I. Meng, and S. W. H. Cowley (2002), Interplanetary magnetic field B_x asymmetry effect on auroral brightness, *J. Geophys. Res.*, *107*(A8), 1197, doi:10.1029/2001JA000229.

Stenbaek-Nielsen, H. C., and A. Otto (1997), Conjugate auroras and the interplanetary magnetic field, *J. Geophys. Res.*, *102*(A2), 2223–2232.

Stenbaek-Nielsen, H. C., T. N. Davis, and N. W. Glass (1972), Relative motion of auroral conjugate points during substorms, *J. Geophys. Res.*, *77*(10), 1844–1858.

Stenbaek-Nielsen, H. C., E. M. Wescott, T. N. Davis, and R. W. Peterson (1973), Differences in auroral intensity at conjugate points, *J. Geophys. Res.*, *78*(4), 659–671.

Vallat, C., I. Dandouras, M. Dunlop, A. Balogh, E. Lucek, G. K. Parks, M. Wilber, E. C. Roelof, G. Chanteur, and H. Rème (2005), First current density measurements in the ring current region using simultaneous multi-spacecraft CLUSTER-FGM data, *Ann. Geophys.*, *23*, 1849–1865.

Vorobjev, V. G., O. I. Yagodkina, D. Sibeck, K. Liou, and C.-I. Meng (2001), Aurora conjugacy during substorms: Coordinated Antarctic ground and Polar Ultraviolet observations, *J. Geophys. Res.*, *106*(A11), 24,579–24,591.

Wang, H., H. Lühr, S. Y. Ma, and H. U. Frey (2007), Inter-hemispheric comparison of average substorm onset locations: Evidence for deviation from conjugacy, *Ann. Geophys.*, *25*, 989–999.

Weimer, D. R. (2004), Correction to "Predicting interplanetary magnetic field (IMF) propagation delay times using the minimum variance technique", *J. Geophys. Res.*, *109*, A12104, doi:10.1029/2004JA010691.

Wing, S., P. T. Newell, D. G. Sibeck, and K. B. Baker (1995), A large statistical study of the entry of interplanetary magnetic field Y-component into the magnetosphere, *Geophys. Res. Lett.*, *22*(16), 2083–2086.

Wing, S., S. Ohtani, P. T. Newell, T. Higuchi, G. Ueno, and J. M. Weygand (2010), Dayside field-aligned current source regions, *J. Geophys. Res.*, *115*, A12215, doi:10.1029/2010JA015837.

Zhou, X.-W., C. T. Russell, and G. Le (2000), Local time and interplanetary magnetic field B_y dependence of field-aligned currents at high altitudes, *J. Geophys. Res.*, *105*(A2), 2533–2539.

K. M. Laundal, Teknova, Kristiansand N-4630, Norway.

N. Østgaard, Department of Physics and Technology, University of Bergen, Bergen N-5007, Norway. (nfyno@ift.uib.no)

Auroral Processes on Jupiter and Saturn

John T. Clarke

Center for Space Physics, Boston University, Boston, Massachusetts, USA

This chapter will give an overview of the auroral emissions observed at Jupiter and Saturn and describe the general level of understanding of the physics behind the emissions and controlling electrodynamic processes. Topics include the observed distribution and variations in auroral emissions, the different auroral emission regions on Jupiter, our understanding of the locations within the magnetospheres controlling the auroral emissions, and our understanding of the auroral physics for the different regions and emissions. Relevance to the large-scale structure of the magnetospheres and interaction with the solar wind will also be discussed.

1. INTRODUCTION

Decades of space- and ground-based measurements of the Earth's aurora and magnetosphere have shown that auroral activity is largely controlled by the interaction with the solar wind. The nature of this interaction has been discovered by concentrated measurements from earth-orbiting spacecraft, in parallel with auroral and ionospheric observations from the ground. It is instructive to compare the conditions in the Earth's magnetosphere with those at the giant planets, as listed in Table 1. The size scales of the magnetospheres of the Earth, Jupiter, and Saturn are vastly different, and this has important implications for how their magnetospheres must operate. In particular, the flow of the solar wind past each magnetosphere happens over very different time scales. While the solar wind moves from the magnetopause to the Earth in a few minutes, the corresponding times are an hour for Saturn to a few hours for Jupiter. The time scale for solar wind flow down Jupiter's magnetotail is days to weeks, a period over which the solar wind conditions can change dramatically. The corresponding times to fill the magnetotails with open flux range from a few hours on the Earth to

Auroral Phenomenology and Magnetospheric Processes: Earth and Other Planets
Geophysical Monograph Series 197
10.1029/2011GM001199

many days on Jupiter and Saturn. This suggests that the interaction of the solar wind with the Jovian and Saturnian systems will proceed along much longer time scales, and with different processes, than on the Earth.

2. JUPITER'S AND SATURN'S MAGNETOSPHERES

In contrast to the Earth's magnetosphere, with the strong influence of the solar wind, Jupiter's magnetosphere is dominated by corotating plasma originating from Io [*Dessler*, 1983]. Io's volcanoes and surface sublimation produce a collisionally thick atmosphere, from which ions and electrons are picked up and brought to corotation speed with Jupiter's magnetic field. This hot plasma forms the Io plasma torus, and the resulting impact of plasma on the upstream side of Io releases more plasma [*Bagenal et al.*, 1980]. The plasma energy is mainly perpendicular to the field as a result of the pickup direction, and it is thus closely confined to the centrifugal equator. Plasma slowly drifting outward from Io's orbit fills the magnetosphere with high-β plasma, and neutral sheet currents strongly flatten the overall shape of the magnetosphere. Field-aligned currents needed to enforce corotation of the plasma lead to strong field-aligned potentials in Jupiter's upper ionosphere, and the accelerated particles produce the main auroral oval [*Cowley and Bunce*, 2001; *Ray and Ergun*, this volume]. This maps to regions in the middle magnetosphere close to the distance of breakdown of corotation, and the main aurora is thus not directly connected with the solar wind boundary [*Hill*, 2001].

Table 1. Comparison of Magnetospheres and Aurora

Planet/Rotation Period (h)	Magnetic Moment/ Equatorial Field Strength (G)	Magnetosphere Cross-Section Radii/Diameter (km)	Solar Wind Travel Time, Bow to Planet at 500 kms^{-1} (min)	Average Auroral Brightness/Input Power (W)
Earth/24	1 (normalized)/0.3	17 R_E 2 × 10^5	2	1–100 kR (1–100 × 10^9)
Jupiter/10	20,000/4	140 R_J 200 × 10^5	200	10–1000 kR (few × 10^{13})
Saturn/11	580/0.2	20 R_S 40 × 10^5	40	1–100 kR (1–10 × 10^{11})

Saturn's magnetosphere is intermediate between the cases of the Earth and Jupiter. The corotating plasma from the rings and icy satellites gives a β ~ 1, so that the influence of the solar wind should be comparable to that of the internal plasma [*Kurth et al.*, 2009]. Saturn's magnetic field is also coaligned with the rotation axis. Estimates of the auroral brightness that would be produced by the Jupiter-like process of field-aligned currents from corotating plasma suggest that this process would not lead to the observed bright main auroral oval, suggesting that Saturn's aurora could originate closer to the boundary with the solar wind like the Earth's [*Cowley and Bunce*, 2001].

By contrast with the detailed measurements on the Earth, observations of the aurora on Jupiter and Saturn have only occurred a few times a year, with in situ spacecraft measurements even less often. As a first step to determine the physical cause and effect of planetary aurora, a large Hubble Space Telescope (HST) observing program was initiated, with simultaneous measurements by Cassini on Saturn, the New Horizons flyby of Jupiter, and extrapolation of solar wind conditions from 1 AU to Jupiter and Saturn. These results will be summarized after a brief introduction to the history of the observations and our present understanding of the two planets.

3. OBSERVATIONS OF THE AURORA ON JUPITER AND SATURN

Jupiter's intense nonthermal radio emissions were well known in the 1950s and implied a strong magnetic field and likely auroral processes. Jupiter's auroral emission was first clearly detected at FUV wavelengths in 1979 by the Voyager 1 ultraviolet spectrometer instrument [*Broadfoot et al.*, 1979]. Subsequent observations with the IUE satellite in Earth orbit for 16 years established the basic properties of the distribution of the UV auroral ovals, the spectrum of the emissions, and the relatively constant nature of Jovian auroral activity [*Clarke et al.*, 1980; *Skinner and Moos*, 1984; *Livengood et al.*, 1992]. IR thermal emissions of CH$_4$ and

C$_2$H$_2$ from the lower auroral atmosphere were discovered in the 1980s [*Caldwell et al.*, 1988; *Kostiuk et al.*, 1993], and bright near-IR emissions from the auroral regions were identified and associated with previously unknown spectral features of H$_3^+$ in 1989 [*Trafton et al.*, 1989; *Drossart et al.*, 1989; *Stallard et al.*, this volume]. The Earth-orbiting Einstein X-ray observatory discovered soft X-ray emissions from Jupiter [*Metzger et al.*, 1983], and further observations with Röntgen satellite [*Waite et al.*, 1994] and Chandra [*Gladstone et al.*, 2002] established the locations and spectra of the X-ray emissions [*Cravens and Ozak*, this volume]. The Galileo spacecraft imager has been used to capture high-resolution visible wavelength images of nightside auroral emissions [*Ingersoll et al.*, 1998]. Much progress has been made since the early 1990s through increasingly sensitive observations with the HST (Figure 1). These observations have established the morphology of Jupiter's multiple auroral processes with high spatial and time resolution. Reviews of the many observations are given by *Bhardwaj and Gladstone* [2000] and *Clarke et al.* [2004]. Jupiter's auroral processes are far more energetic than on the Earth or even

Figure 1. Hubble Space Telescope (HST) UV image of Jupiter's northern auroral zone, showing the three different emission regions. From *Clarke et al.* [2004]. Copyright © 2004 Cambridge University Press. Figures 1 and 3 are both log stretches in intensity to emphasize the fainter emissions.

on Saturn (Table 1), while typical energies of the incoming particles are several tens of keV [*Grodent et al.*, 2001; *Gérard et al.*, 2002].

UV observations of Saturn by the Pioneer 10 and IUE spacecraft gave indications of UV auroral emissions from the polar regions [*Judge et al.*, 1980; *Clarke et al.*, 1981], and Voyager 1 again provided the first clear spectral and spatial evidence for auroral emissions [*Broadfoot et al.*, 1981]. Continued IUE observations indicated that bright aurora on Saturn are infrequent, while spectra from Voyager showed that the auroral emissions were relatively unabsorbed leaving the atmosphere, limiting the energy of the incoming particles in the range of 1–10 keV. Saturn's auroral emissions are much fainter than Jupiter's, and IR auroral emissions have proven difficult to detect from ground-based observations. Faint H_3^+ emissions are now observed on a regular basis [*Stallard et al.*, 1999], while thermal hydrocarbon emissions have not been observed, and only faint X-rays have been detected from Saturn. More recently, the Cassini mission has led to many new insights into the workings of the Saturnian magnetosphere from both UV observations and in situ particle and field data [*Gombosi et al.*, 2009].

The excitation of the auroral emissions from the hydrogen atmosphere of Jupiter has been modeled [*Grodent et al.*, 2001; *Gustin et al.*, 2009], and information about the energy of the incoming particles can be derived from the degree of atmospheric absorption of the outgoing UV emissions in addition to the relative intensity of H_2 band emissions. The stronger the absorption, the deeper the particles penetrate, and the larger their initial energies, although most of the emission is produced by secondary electrons. Typical energies of the incident particles are a few tens of keV for Jupiter and 1–10 keV for Saturn, comparable to or more energetic than on the Earth. Values for the total auroral energy are derived by summing the auroral emissions across the polar regions and assuming an ~10% efficiency of production of UV emission from the total incident particle energy, as implied by detailed modeling of the degradation of energy of the incoming particles. The energy is distributed between collisional ionization, dissociation, heating, and the production of emissions at all wavelengths.

4. AURORAL PROCESSES AT JUPITER

The earliest IUE spectral observations showed that Jupiter's main auroral emissions rotated with the planet, since the observed auroral intensity varied as the highly nondipolar and tilted magnetic field carried the auroral zones in and out of the line of sight from the Earth with the planet's rotation. With the advent of UV images of Jupiter's aurora taken with HST, it became clear that there are three independent regions of auroral emissions, in the sense that the emissions vary independently of each other and map to clearly different regions in the magnetosphere [*Delamere*, this volume; *Vogt and Kivelson*, this volume]. The degree of correlation of emissions from different latitude ranges has been presented by *Nichols et al.* [2009]. Auroral emissions have been detected from the magnetic footprints of Io, Europa, and Ganymede [*Connerney et al.*, 1993; *Clarke et al.*, 2004; *Bonfond*, this volume; *Hess and Delamere*, this volume]. These emission features remain at the magnetic footprints of the satellites, rather than rotating with the planet (as the main oval does), thus their location reveals the geometry of Jupiter's local magnetic field. They also exhibit variations in intensity as each satellite moves in and out of the corotating torus plasma [*Gerard et al.*, 2006; *Serio and Clarke*, 2008; *Wannawichian et al.*, 2010]. There is a clear "main oval" in the auroral emissions, although in contrast with the Earth's main oval, this oval maps to the middle magnetosphere, *not* to the vicinity of the boundary with the solar wind. These emissions are fairly steady on time scales of minutes to hours and clearly rotate with the planet rather than being roughly fixed in local time, as on the Earth. In contrast, the emissions that appear poleward of the main oval are rapidly variable, with several discrete features identified including the polar flares described below.

4.1. Jupiter's Main Oval

Since the Pioneer and Voyager spacecraft flybys of Jupiter, the basic structure of the magnetosphere has been known [*Dessler*, 1983]. The magnetosphere is enormous and loaded with plasma originally from Io. The corotating plasma flattens the shape of the magnetosphere and dominates the energy of the system [*Vasyliunas*, 1983]. It was proposed early on that strong field-aligned currents were needed to maintain the corotation of the plasma and that these currents with associated field-aligned potentials could accelerate particles to produce bright auroral emissions [*Cowley and Bunce*, 2001; *Hill*, 2001; *Southwood and Kivelson*, 2001; *Ray and Ergun*, this volume] (see Figure 2). Observational confirmation that the incident particles producing the main oval originate in the middle magnetosphere came from observations of auroral emissions from the magnetic footprints of three satellites. Jupiter's magnetic field has not been mapped close to the planet, thus there are substantial uncertainties in connecting latitudes with distances from the planet. The satellite footprints permit one to accurately trace the distance of the main auroral oval in comparison with the latitudes of these footprints. The mapping of the main auroral oval to distances of 20–30 R_J in the middle magnetosphere is consistent with the relatively constant auroral emissions,

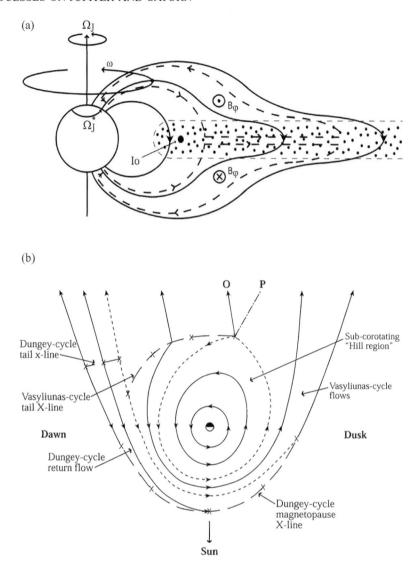

Figure 2. (a) Cross section of the configuration of the Jovian magnetosphere. Reprinted from *Cowley and Bunce* [2001], with permission from Elsevier. Currents are indicated by dashed lines, magnetic field by the solid lines, and induced magnetic field from plasma corotation is shown in and out of the plane. (b) Polar view of the mapping of different regions in Jupiter's north auroral zone into the magnetosphere. From *Clarke et al.* [2004]. Copyright © 2004 Cambridge University Press.

since the plasma is constantly produced and drifts slowly outward [*Grodent et al.*, 2003a]. Even if the solar wind was to diminish to near zero, this auroral process would continue. This is in marked contrast to the Earth's auroral processes, which are closely related to the interaction with the solar wind.

4.2. Satellite Magnetic Footprints

The locations of the auroral emissions from the satellite magnetic footprints can be used to help "map" the geometry of the magnetic field near the planet [*Connerney et al.*, 1998]. In particular, one region in the north appears to indicate a localized region of low field, likely due to a magnetic anomaly (the "Dessler anomaly") [*Grodent et al.*, 2008]. Studies of the location and brightness of Io's auroral footprint have shown a persistent pattern of brightness increases when Io is center in the plasma torus [*Gérard et al.*, 2006; *Serio and Clarke*, 2008; *Wannawichian et al.*, 2010]. However, the implied vertical scale height of the plasma does not appear to be consistent with other measurements; thus, while the trend has been consistent over a decade's time, the data are not fit well by simple models. The footprint auroral emissions show an interesting structure, especially in the case of Io, which can be related to the details of the electro-dynamic interaction near the satellite [*Bonfond et al.*, 2009].

4.3. Auroral Storms and Polar Emissions

Extremely bright auroral emissions have frequently been observed along Jupiter's main oval near local dawn. The emissions rise to several MRayleighs in brightness over a period of ~ 1 h and persist for a few hours. These "dawn storm" emissions remain centered along the main oval near magnetic dawn, while other main oval emissions rotate past. If the main oval maps to the middle magnetosphere (20–25 R_J), then how do the conditions in the solar wind control the location of these storms? One suggestion has been a Kelvin-Helmholtz instability at the solar wind boundary [*Prangé et al.*, 1998], but this should map to higher latitudes than where the dawn storms are observed.

Another persistent feature of the Jovian polar aurora observed in high time resolution data are the bright "polar flares" observed in the polar regions both north and south [*Waite et al.*, 2001]. These emissions rise from near background to tens of MRayleighs on a time scale of tens of seconds, then fade away on a somewhat longer time scale. The emissions often appear near local noon, but there is an observational bias toward detecting features on the dayside in the oblique observing geometry from the Earth. These flares have a duty cycle on the order of 15%, they can be accompanied by strong X-ray emissions, and it has been proposed that they map to the dayside magnetopause boundary [*Waite et al.*, 2001]. The polar auroral emissions are generally more rapidly variable than either the main oval or satellite footprints, coming and going on time scales of tens of seconds. These emissions may be related more closely to the open/closed field line boundary or to regions in the magnetotail that are relatively devoid of plasma [*Grodent et al.*, 2003b].

4.4. Dependence of Jovian Aurora on Solar Wind Conditions

Jupiter's overall auroral power is driven mainly by the brightness of the main oval, with increases seen from specific regions at different times. Based on two historical events and six events from 2 month long HST auroral observing campaigns in 2007, the arrival of a solar wind shock appears to be consistent with a brightening of the main oval, subject to some uncertainty in the arrival times of solar wind features at the planet [*Gurnett et al.*, 2002; *Pryor et al.*, 2005; *Clarke et al.*, 2009; *Nichols et al.*, 2009]. By contrast, solar wind

velocity increases with a pressure *decrease* have *not* been seen to correlate with auroral brightening (based on three events from the campaign). Dawn storms can occur at times of quiet solar wind conditions, based on two events from the campaign, raising serious questions about the nature of the processes (see below). It appears that Jupiter's auroras are relatively more constant than those on the Earth and on Saturn, yet they still show some apparent response to solar wind conditions.

5. SATURNIAN AURORAL PROCESSES

On Saturn, the auroral morphology is dominated by a main oval, which is often not symmetric but does appear more or less centered about the combined rotational and magnetic poles. Bright emission regions observed along the ovals come and go on a few hour time scales and appear to move at a fraction of Saturn's rotation rate [*Gérard et al.*, 2004]. The main oval appears nominally near 75° latitude, but drifts a great deal from day to day. The overall auroral oval has been found to be offset by ~2° toward midnight and ~1° toward dawn, while no significant offset from the rotation axis in a frame of reference fixed to Saturn has been discovered [*Nichols et al.*, 2008]. The oval has been further seen to "wobble" about the average center location, and fits to this motion have resulted in periods consistent with the Saturn kilometric radiation (SKR) rotation period and changing with time, also similar to the SKR period. With regard to the satellites, auroral emissions have been detected only from the magnetic footprint of Enceladus, and those are faint and variable [*Pryor et al.*, 2011; *Gurnett and Pryor*, this volume]. Without the footprint evidence of the mapping distance into the magnetosphere, one cannot accurately determine from the auroral images the distance to which the main ovals map. In one case, Cassini measurements of field-aligned currents were obtained with simultaneous auroral images, indicating auroral processes near the solar wind boundary [*Bunce et al.*, 2008; *Cowley et al.*, 2008]. Overall though, it is not well established from the auroral observations alone where the open/closed field line boundary falls in the auroral regions on either Jupiter or Saturn. This is in clear contrast to that of the Earth's auroral regions, where the auroral oval appears slightly equatorward of the open/closed field line boundary.

In the case of Saturn, theoretical estimates of the strength of currents in and out of the ionosphere required to enforce plasma corotation led to the conclusion that the associated field-aligned potentials would not be sufficient to produce the observed auroral brightness [*Cowley et al.*, 2004; *Bunce*, this volume]. It was concluded that the main auroral oval was likely to map to the open/closed field line boundary and be produced by the interaction with the solar wind. The higher variability of Saturn's aurora compared with the case on Jupiter is generally consistent with this hypothesis. These authors further suggested associations between various regions in the auroral zone and mapped regions in the magnetosphere. Observations of energetic neutral atom events in the middle magnetosphere have suggested that at least, at times, auroral processes closer to the planet are also important [*Mitchell et al.*, 2009].

An interesting line of research involves the discovery of clear correlations between the arrival of shock fronts in the solar wind and the intensification of both Saturn's UV auroral emissions and the nonthermal SKR (Figure 3). Coordinated HST observations of the aurora with Cassini measurements of the solar wind as it approached Saturn in late 2003 and early 2004, plus Cassini measurements of the SKR, have shown a clear correlation between the UV aurora and the SKR emissions [*Kurth et al.*, 2005]. Increases in both emissions also were recorded at the arrival times of two solar wind shocks at Saturn, suggesting a causal connection [*Clarke et al.*, 2005; *Crary et al.*, 2005]. During the 2 month long HST observing campaigns in 2007 and 2008, there was

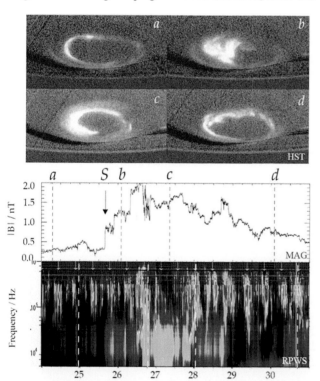

Figure 3. (top, a–d) HST UV images of Saturn in January 2004 showing the progression of an auroral storm. The storm in Figure 3b was shortly after the arrival of a solar wind shock, (middle) marked "S" in the measurement of interplanetary magnetic field. (bottom) The event also corresponded with an intensification of the SKR emission. From *Kurth et al.* [2009].

a one to one correspondence between auroral activity, SKR emission, and size of the oval [*Clarke et al.*, 2009]. From combined observations now covering almost 3 months duration, UV emission and SKR radiation are correlated with solar wind pressure, and auroral brightenings begin at times of solar wind shocks (six events have been observed). A strong connection to solar wind pressure is consistent with a source region in the outer magnetosphere, unlike the case on Jupiter, yet there is the added evidence for the importance of processes in the middle magnetosphere. There is clearly more to be learned about Saturn's auroral dynamics.

6. CHALLENGES FOR PHYSICAL UNDERSTANDING

There are many areas in which our understanding of auroral processes on the giant planets falls short. A few outstanding questions are highlighted below:

Why would both Jupiter and Saturn aurora brighten in response to solar wind pressure increases? Jupiter's main oval maps to the middle magnetosphere, while Saturn's maps in theory to the outer magnetosphere, yet both respond with brightenings when solar wind shock fronts arrive on each planet. During a period of enhanced auroral activity, Jupiter's main oval gets brighter and at times more extended in latitude, with patchy clumps of emission along the oval. The overall brightening sometimes starts with a dawn storm, which may begin with a brightening observed poleward of the main oval. The observed time scale for auroral events on Jupiter is 1–2 days. During an auroral intensification on Saturn, the main oval gets brighter, and the brightest emission is shifted poleward of the quiet oval latitude. The dawnside initially fills in with bright emissions, with an observed time scale for auroral events of 2–4 days. A weaker event may accompany a smaller increase in solar wind pressure. The extent to which these characteristics relate to the interaction with the solar wind is unclear.

What is the principal driving physics of the Jovian main ovals? Jupiter is the prototype of a magnetosphere controlled by internal plasma and planetary rotation, yet the aurora appears to be at least somewhat affected by the solar wind. How does the solar wind exert any effect on the region producing the main oval, deep within the magnetosphere? If the main oval is driven by corotation breakdown currents, it has been predicted that the outward drifting plasma content would decrease with increased solar wind pressure, giving weaker currents and fainter auroral emissions [*Southwood and Kivelson*, 2001]. Yet the opposite is observed, with shock intensifications in the solar wind and compression of the Jovian magnetosphere corresponding to increases in total auroral power. For Saturn, it has been proposed that the main oval maps out close to boundary with solar wind, but this has yet to be established by measurement. The distance to which the main oval maps was determined from satellite footprint emissions on Jupiter, but emission has only been detected from the footprint of Enceladus on Saturn, too close to Saturn to be helpful.

How do we explain the other types of auroral storms observed mainly on Jupiter? Jupiter's dawn storms start near local dawn close to the latitude of the main oval, which maps into the middle magnetosphere far from the boundary with the solar wind. In addition, these storms have been seen more than once to evolve into an overall brightening of the main oval. There is evidence from modeling that Kelvin-Helmholtz instabilities form easily on the dusk flank of the magnetosphere. Could a disturbance on the flank propagate inward and disturb conditions in the middle magnetosphere? The current sheet could build up over time until such a disturbance scattered energetic particles into the loss cone, leading to bright aurora. These storms are detected roughly once every few weeks and last a few hours. Jupiter also displays active region flares, with low-level flares in the polar region that occur all the time, and high-intensity events with a ~10% duty cycle. These events appear at a consistent location in the polar region, implying a mapping to a consistent location within the magnetosphere. Whether this location is fixed in local time has not yet been well determined.

What controls Saturn's auroral storms? Is there more than one kind of auroral storm on Saturn? How does the solar wind influence the auroral activity?

To what distance in the magnetosphere does the auroral oval map? On Jupiter, we used satellite footprints to determine the mapping. Is there anything on Saturn that will serve this purpose, perhaps in measurements from the Cassini spacecraft?

Understanding these different auroral storms on Jupiter and Saturn is critical to help us understand auroral processes both on the Earth and on exoplanets. Please note that all the HST imaging data from the 2007/2008 campaigns can be viewed and downloaded from the Planetary Atmospheres and Space Sciences Group website (http://www.bu.edu/csp/PASS/main.html).

Acknowledgments. This work is based in part on observations with the NASA/ESA Hubble Space Telescope, obtained at the Space Telescope Science Institute, which is operated by AURA for NASA. The work was supported by grants HST-GO-10862.01-A and HST-GO-12176.01-A from the Space Telescope Science Institute to Boston University.

REFERENCES

Bagenal, F., J. D. Sullivan, and G. L. Siscoe (1980), Spatial distribution of plasma in the Io torus, *Geophys. Res. Lett.*, *7*(1), 41–44.

Bhardwaj, A., and G. R. Gladstone (2000), Auroral emissions of the giant planets, *Rev. Geophys.*, *38*(3), 295–353, doi:10.1029/1998RG000046.

Bonfond, B. (2012), When moons create aurora: The satellite footprints on giant planets, in *Auroral Phenomenology and Magnetospheric Processes: Earth and Other Planets, Geophys. Monogr. Ser.*, doi:10.1029/2011GM001169, this volume.

Bonfond, B., D. Grodent, J.-C. Gérard, A. Radioti, V. Dols, P. A. Delamere, and J. T. Clarke (2009), The Io UV footprint: Location, inter-spot distances and tail vertical extent, *J. Geophys. Res.*, *114*, A07224, doi:10.1029/2009JA014312.

Broadfoot, A. L., et al. (1979), Extreme ultraviolet observations from Voyager 1 encounter with Jupiter, *Science*, *204*, 979–982.

Broadfoot, A. L., et al. (1981), Extreme ultraviolet observations from Voyager 1 encounter with Saturn, *Science*, *212*, 206–211.

Bunce, E. J. (2012), Origins of Saturn's auroral emissions and their relationship to large-scale magnetosphere dynamics, in *Auroral Phenomenology and Magnetospheric Processes: Earth and Other Planets, Geophys. Monogr. Ser.*, doi:10.1029/2011GM001191, this volume.

Bunce, E. J., et al. (2008), Origin of Saturn's aurora: Simultaneous observations by Cassini and the Hubble Space Telescope, *J. Geophys. Res.*, *113*, A09209, doi:10.1029/2008JA013257.

Caldwell, J., H. Halthore, G. Orton, and J. Bergstralh (1988), Infrared polar brightenings on Jupiter IV. Spatial properties of methane emission, *Icarus*, *74*, 331–339.

Clarke, J. T., H. W. Moos, S. K. Atreya, and A. L. Lane (1980), Observations from Earth orbit and variability of the polar aurora on Jupiter, *Astrophys. J.*, *241*, L179–L182.

Clarke, J. T., H. W. Moos, S. K. Atreya, and A. L. Lane (1981), IUE detection of bursts of H Ly α emission from Saturn, *Nature*, *290*, 226–227.

Clarke, J. T., D. Grodent, S. W. H. Cowley, E. J. Bunce, P. Zarka, J. E. P. Connerney, and T. Satoh (2004), Jupiter's aurora, in *Jupiter: The Planet, Satellites and Magnetosphere*, edited by F. Bagenal, T. E. Dowling, and W. B. McKinnon, pp. 639–670, Cambridge Univ. Press, Cambridge, U. K.

Clarke, J. T., et al. (2005), Morphological differences between Saturn's ultraviolet aurorae and those of Earth and Jupiter, *Nature*, *433*, 717–719.

Clarke, J. T., et al. (2009), Response of Jupiter's and Saturn's auroral activity to the solar wind, *J. Geophys. Res.*, *114*, A05210, doi:10.1029/2008JA013694.

Connerney, J. E. P., R. L. Baron, T. Satoh, and T. Owen (1993), Images of excited H_3^+ at the foot of the Io flux tube in Jupiter's atmosphere, *Science*, *262*, 1035–1038.

Connerney, J. E. P., M. H. Acuña, N. F. Ness, and T. Satoh (1998), New models of Jupiter's magnetic field constrained by the Io flux tube footprint, *J. Geophys. Res.*, *103*(A6), 11,929–11,939.

Cowley, S. W. H., and E. J. Bunce (2001), Origin of the main auroral oval in Jupiter's coupled magnetosphere-ionosphere system, *Planet Space Sci.*, *49*, 1067–1088.

Cowley, S. W. H., E. J. Bunce, and J. M. O'Rourke (2004), A simple quantitative model of plasma flows and currents in

Saturn's polar ionosphere, *J. Geophys. Res., 109*, A05212, doi:10.1029/2003JA010375.

Cowley, S. W. H., C. Arridge, E. Bunce, J. Clarke, A. Coates, M. Dougherty, J.-C. Gérard, D. Grodent, J. Nichols, and D. Talboys (2008), Auroral current systems in Saturn's magnetosphere: Comparison of theoretical models with Cassini and HST observations, *Ann Geophys., 26*, 2613–2630.

Crary, F. J., et al. (2005), Solar wind dynamic pressure and electric field as the main factors controlling Saturn's aurorae, *Nature, 433*, 720–722.

Cravens, T. E., and N. Ozak (2012), Auroral ion precipitation and acceleration at the outer planets, in *Auroral Phenomenology and Magnetospheric Processes: Earth and Other Planets, Geophys. Monogr. Ser.*, doi:10.1029/2011GM001159, this volume.

Delamere, P. A. (2012), Auroral signatures of solar wind interaction at Jupiter, in *Auroral Phenomenology and Magnetospheric Processes: Earth and Other Planets, Geophys. Monogr. Ser.*, doi:10.1029/2011GM001180, this volume.

Dessler, A. J. (Ed.) (1983), *Physics of the Jovian Magnetosphere*, Cambridge Univ. Press, Cambridge, U. K.

Drossart, P., et al. (1989), Detection of H_3^+ on Jupiter, *Nature, 340*, 539–541.

Gérard, J.-C., J. Gustin, D. Grodent, P. Delamere, and J. T. Clarke (2002), Excitation of the FUV Io tail on Jupiter: Characterization of the electron precipitation, *J. Geophys. Res., 107*(A11), 1394, doi:10.1029/2002JA009410.

Gérard, J.-C., D. Grodent, J. Gustin, A. Saglam, J. T. Clarke, and J. T. Trauger (2004), Characteristics of Saturn's FUV aurora observed with the Space Telescope Imaging Spectrograph, *J. Geophys. Res., 109*, A09207, doi:10.1029/2004JA010513.

Gérard, J.-C., A. Saglam, D. Grodent, and J. T. Clarke (2006), Morphology of the ultraviolet Io footprint emission and its control by Io's location, *J. Geophys. Res., 111*, A04202, doi:10.1029/2005JA011327.

Gladstone, G. R., et al. (2002), A pulsating auroral X-ray hot spot on Jupiter, *Nature, 415*, 1000–1003.

Gombosi, T. I., et al. (2009), Saturn's magnetospheric configuration, in *Saturn from Cassini-Huygens*, edited by M. K. Dougherty, L. W. Esposito, and S. M. Krimigis, pp 203–256, Springer, Dordrecht, Netherlands, doi:10.1007/978-1-4020-9217-6_9.

Grodent, D., J. H. Waite Jr., and J.-C. Gérard (2001), A self-consistent model of the Jovian auroral thermal structure, *J. Geophys. Res., 106*(A7), 12,933–12,952, doi:10.1029/2000JA900129.

Grodent, D., J. T. Clarke, J. Kim, J. H. Waite Jr., and S. W. H. Cowley (2003a), Jupiter's main auroral oval observed with HST-STIS, *J. Geophys. Res., 108*(A11), 1389, doi:10.1029/2003JA009921.

Grodent, D., J. T. Clarke, J. H. Waite Jr., S. W. H. Cowley, J.-C. Gérard, and J. Kim (2003b), Jupiter's polar auroral emissions, *J. Geophys. Res., 108*(A10), 1366, doi:10.1029/2003JA010017.

Grodent, D., B. Bonfond, J.-C. Gérard, A. Radioti, J. Gustin, J. T. Clarke, J. Nichols, and J. E. P. Connerney (2008), Auroral evidence of a localized magnetic anomaly in Jupiter's northern hemisphere, *J. Geophys. Res., 113*, A09201, doi:10.1029/2008JA013185.

Gurnett, D. A., and W. R. Pryor (2012), Auroral processes associated with Saturn's moon Enceladus, in *Auroral Phenomenology and Magnetospheric Processes: Earth and Other Planets, Geophys. Monogr. Ser.*, doi:10.1029/2011GM001174, this volume.

Gurnett, D. A., et al. (2002), Control of Jupiter's radio emission and aurorae by the solar wind, *Nature, 415*, 985–987, doi:10.1038/415985a.

Gustin, J., J.-C. Gérard, W. Pryor, P. D. Feldman, and G. Holsclaw (2009), Characteristics of Saturn's polar atmosphere and auroral electrons derived from HST/STIS, FUSE and Cassini/UVIS spectra, *Icarus, 200*, 176–187, doi:10.1016/j.icarus.2008.11.013.

Hess, S. L. G., and P. A. Delamere (2012), Satellite-induced electron acceleration and related auroras, in *Auroral Phenomenology and Magnetospheric Processes: Earth and Other Planets, Geophys. Monogr. Ser.*, doi:10.1029/2011GM001175, this volume.

Hill, T. W. (2001), The Jovian auroral oval, *J. Geophys. Res., 106*(A5), 8101–8107, doi:10.1029/2000JA000302.

Ingersoll, A. P., A. R. Vasavada, B. Little, C. D. Anger, S. J. Bolton, C. Alexander, K. P. Klaasen, W. K. Tobiska, and the Galileo SSI Team (1998), Imaging Jupiter's aurora at visible wavelengths, *Icarus, 135*, 251–264.

Judge, D. L., F. Wu, and R. Carlson (1980), Ultraviolet photometer observations of the Saturnian system, *Science, 207*, 431–434.

Kostiuk, T., P. Romani, F. Espenak, T. A. Livengood, and J. J. Goldstein (1993), Temperature and abundances in the Jovian auroral stratosphere 2. Ethylene as a probe of the microbar region, *J. Geophys. Res., 98*(E10), 18,823–18,830.

Kurth, W. S., et al. (2005), An Earth-like correspondence between Saturn's auroral features and radio emission, *Nature, 433*, 722–725.

Kurth, W. S., et al. (2009), Auroral processes at Saturn,, in *Saturn From Cassini-Huygens*, edited by M. K. Dougherty, L. W. Esposito, and S. M. Krimigis, pp. 333–374, Springer, Dordrecht, Netherlands, doi:10.1007/978-1-4020-9217-6_12.

Livengood, T. A., H. W. Moos, G. E. Ballester, and R. M. Prangé (1992), Jovian ultraviolet auroral activity, *Icarus, 97*, 26–45.

Metzger, A. E., D. A. Gilman, J. L. Luthey, K. C. Hurley, H. W. Schnopper, F. D. Seward, and J. D. Sullivan (1983), The detection of X rays from Jupiter, *J. Geophys. Res., 88*(A10), 7731–7741.

Mitchell, D. G., et al. (2009), Recurrent energization of plasma in the midnight-to-dawn quadrant of Saturn's magnetosphere, and its relationship to auroral UV and radio emissions, *Planet. Space Sci., 57*, 1732–1742, doi:10.1016/j.pss.2009.04.002.

Nichols, J. D., J. T. Clarke, S. W. H. Cowley, J. Duval, A. J. Farmer, J.-C. Gérard, D. Grodent, and S. Wannawichian (2008), Oscillation of Saturn's southern auroral oval, *J. Geophys. Res., 113*, A11205, doi:10.1029/2008JA013444.

Nichols, J. D., J. T. Clarke, J. C. Gérard, D. Grodent, and K. C. Hansen (2009), Variation of different components of Jupiter's auroral emission, *J. Geophys. Res., 114*, A06210, doi:10.1029/2009JA014051.

Prangé, R., D. Rego, L. Palliser, J. E. P. Connerney, P. Zarka, and J. Queinnec (1998), Detailed study of FUV Jovian auroral features with the post-COSTAR HST faint object camera, *J. Geophys. Res.*, *103*(E9), 20,195–20,215.

Pryor, W. R., et al. (2005), Cassini UVIS observations of Jupiter's auroral variability, *Icarus*, *178*, 312–326, doi:10.1016/j.icarus.2005.05.021.

Pryor, W. R., et al. (2011), Discovery of the Enceladus auroral footprint at Saturn, *Nature*, *472*, 331–333, doi:10.1038/nature09928.

Ray, L. C., and R. E. Ergun (2012), Auroral signatures of ionosphere-magnetosphere coupling at Jupiter and Saturn, in *Auroral Phenomenology and Magnetospheric Processes: Earth and Other Planets, Geophys. Monogr. Ser.*, doi:10.1029/2011GM001172, this volume.

Serio, A. W., and J. T. Clarke (2008), The variation of Io's auroral footprint brightness with the location of Io in the plasma torus, *Icarus*, *197*, 368–374.

Skinner, T. E., and H. W. Moos (1984), Comparison of the Jovian north and south pole aurorae using the IUE observatory, *Geophys. Res. Lett.*, *11*(11), 1107–1110.

Southwood, D. J., and M. G. Kivelson (2001), A new perspective concerning the influence of the solar wind on the Jovian magnetosphere, *J. Geophys. Res.*, *106*(A4), 6123–6130, doi:10.1029/2000JA000236.

Stallard, T., S. Miller, G. E. Ballester, D. Rego, R. D. Joseph, and L. M. Trafton (1999), The H_3^+ latitudinal profile of Saturn, *Astrophys. J.*, *521*, L149–L152.

Stallard, T., S. Miller, and H. Melin (2012), Clues on ionospheric electrodynamics from IR aurora at Jupiter and Saturn, in *Auroral Phenomenology and Magnetospheric Processes: Earth and Other Planets, Geophys. Monogr. Ser.*, doi:10.1029/2011GM001168, this volume.

Trafton, T., D. F. Lester, and K. L. Thompson (1989), Unidentified emission lines in Jupiter's northern and southern 2 micron aurorae, *Astrophys. J.*, *343*, L73–L76.

Vasyliunas, V. M. (1983), Plasma distribution and flow, in *Physics of the Jovian Magnetosphere*, edited by A. J. Dessler, pp. 395–453, Cambridge Univ. Press, Cambridge, U. K.

Vogt, M. F., and M. G. Kivelson (2012), Relating Jupiter's auroral features to magnetospheric sources, in *Auroral Phenomenology and Magnetospheric Processes: Earth and Other Planets, Geophys. Monogr. Ser.*, doi:10.1029/2011GM001181, this volume.

Waite, J. H., Jr., F. Bagenal, F. Seward, C. Na, G. R. Gladstone, T. E. Cravens, K. C. Hurley, J. T. Clarke, R. Elsner, and S. A. Stern (1994), ROSAT observations of the Jupiter aurora, *J. Geophys. Res.*, *99*(A8), 14,799–14,809.

Waite, J. H., Jr., et al. (2001), An auroral flare at Jupiter, *Nature*, *410*, 787–789.

Wannawichian, S., J. T. Clarke, and J. D. Nichols (2010), Ten years of Hubble Space Telescope observations of the variation of the Jovian satellites' auroral footprint brightness, *J. Geophys. Res.*, *115*, A02206, doi:10.1029/2009JA014456.

J. T. Clarke, Center for Space Physics, Boston University, Boston, MA 02215, USA. (jclarke@bu.edu)

Aurora in Martian Mini Magnetospheres

David Brain

Laboratory for Atmospheric and Space Physics, University of Colorado, Boulder, Colorado, USA

Jasper S. Halekas

Space Sciences Laboratory, University of California, Berkeley, California, USA

Auroral processes are active at Mars, which lacks a global dynamo field. Observations of relatively faint UV emission in crustal magnetic fields demonstrate that upper atmospheric species are excited in cusp regions, presumably by incident particles directed along cusp flux tubes. Measurement of particles and fields above the crustal fields, including counterstreaming energized ion and electron populations, suggest that electrons are accelerated downward into the atmosphere, at least partly by a quasi-static field-aligned potential. A variety of additional mechanisms may be contributing to aurora, including day-night transport of ionospheric particles, waves and reconnection, and particle acceleration in the magnetotail current sheet. Future observations and modeling should help to distinguish between the different mechanisms, constrain the variation in auroral acceleration and brightness at different wavelengths, correlate auroral activity with external conditions, and determine the importance of auroral processes for upper atmospheric electrodynamics and atmospheric escape.

1. INTRODUCTION

Mars is one of the earliest solar system bodies to have been visited by spacecraft, and one of the most frequently visited. So it may seem surprising that aurora were not discovered there until the twenty-first century, after the discovery of aurora on seven (Jupiter, Saturn, Uranus, Neptune, Io, Europa, and Ganymede) other solar system bodies, including three smaller than Mars and with more tenuous atmospheres [*Mauk and Bagenal*, this volume].

Aurora at Mars was not reported until 2005 for two reasons. First, no spacecraft successfully visited Mars with the right combination of instrumentation and observing plan to detect faint, small-scale, UV emission until Mars Express (MEX) arrived in late 2003. The early Mariner spacecraft, 6 and 7 (flybys) and 9 (orbiter), carried UV spectrometers with the appropriate wavelength range and spectral resolution, but did not prioritize nightside observations. Subsequent orbiting spacecraft, such as Viking 1 and 2, Phobos 2, Mars Global Surveyor (MGS), and Mars Odyssey lacked any UV instruments at all. Only MGS passed sufficiently close to Mars to easily detect the particle and field signatures of auroral acceleration, and these were not recognized until after the first report of UV auroral emission [*Bertaux et al.*, 2005].

Second, until the late 1990s, there was little reason to suspect that auroral processes were active at Mars. The magnetic field observations made by early spacecraft to visit Mars suggested that it was a Venus-like unmagnetized planet with a magnetosphere induced via interaction between the flowing solar wind and the conducting planetary ionosphere [*Luhmann et al.*, 1992]. Though there had been reports [*Fox and Stewart*, 1991] of UV auroral-like emission at Venus, which lacks a planetary magnetic field, most studies assume that aurora result from the acceleration of charged particles

Auroral Phenomenology and Magnetospheric Processes: Earth and Other Planets
Geophysical Monograph Series 197
10.1029/2011GM001201

along *planetary* magnetic field lines into an atmosphere. Thus, while planetward acceleration of electrons into the Martian atmosphere was considered as a process for forming a thin nightside ionosphere [*Haider et al.*, 1992; *Fox et al.*, 1993], Mars was believed to lack the planetary magnetic field to focus precipitating particles. Hence, there was little reason to search for faint Martian aurora using in situ spacecraft or terrestrial telescopes.

Magnetic field measurements from MGS confirmed in 1997 that Mars lacked a significant global magnetic field and also revealed the presence of localized regions of intense crustal magnetization [*Acuña et al.*, 1998]. Likely to have been created in the presence of an ancient global dynamo, crustal fields are strongest and most concentrated beneath the oldest portions of the Martian crust. Closed loops of crustal magnetic field extend well above the main peak of the Martian ionosphere (near 120 km) and the Martian exobase region near 200 km [*Mitchell et al.*, 2001]. Though nonglobal, the discovery of crustal fields revealed the possibility that aurora could occur at Mars.

Given the historical and scientific context outlined above, we review studies of Martian aurora that have progressed over the last 6–7 years. In sharp contrast to terrestrial aurora, it is currently possible to become familiar with most or all of the relevant literature on Martian aurora rather quickly. Here we describe the observations of auroral emission and particle acceleration at Mars (section 2), the mechanisms that may explain the observations (section 3), and the consequences that these observations have for ongoing and future investi-

gation of the Martian upper atmosphere, plasma environment, and climate (section 4).

2. OBSERVATIONS

2.1. UV Emission

The first observation of aurora at Mars was fortuitous. In August 2004, the Spectroscopy for the Investigation of Characteristics of the Atmosphere of Mars (SPICAM) UV spectrometer on MEX conducted limb scans of the nightside upper atmosphere in a successful effort to detect nightglow [*Bertaux*, 2005]. During one of the scans (Figure 1a), a large but spatially confined increase in UV brightness was recorded in all five groups of pixels read from the instrument during the observation [*Bertaux et al.*, 2005]. Further analysis showed that the emission came along a line of sight that passed through a region of nearly vertical crustal magnetic field, analogous to the terrestrial polar cusps. The duration of the increase in brightness was used to infer the width of the emitting region (~30 km), and the time offset between the brightness increase in the five groups of pixels was used to infer the distance between the spacecraft and the emitting region (~450 km). The distance corresponded to the location of vertical crustal magnetic field at an altitude of ~130 km, near the typical altitude of the main ionospheric peak.

The average SPICAM spectrum recorded shortly before and after the unusual observation was subtracted from the observation to obtain a spectrum of the emission coming

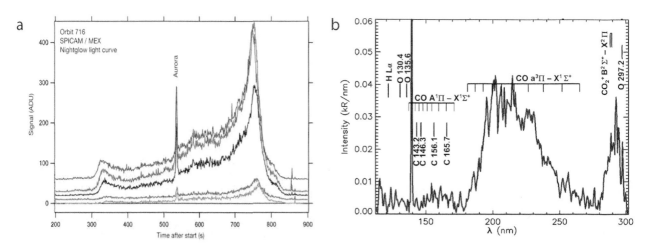

Figure 1. (a) SPICAM count rate in each of five spatial pixels as a function of time for Mars Express (MEX) on 11 August 2004. From *Bertaux et al.* [2005]. Reprinted by permission from Macmillan Publishers Ltd: *Nature*, copyright 2005. The observation took place on the nightside, and brightness increased as MEX progressed toward the dayside. A spike in brightness associated with auroral emission is identified ~525 s after the start of the observation. (b) The spectrum of the aurora is obtained by subtracting average spectral information from the measurements immediately before and after the brightness spike [from *Leblanc et al.*, 2006].

from the cusp [*Bertaux et al.*, 2005; *Leblanc et al.*, 2006]. Emission was identified from species common in the Martian upper atmosphere (Figure 1b). The CO Cameron Bands ($a^3\Pi - X^1\Sigma^+$; 180–240 nm) were brightest, at ~2 kR, and CO_2^+ doublet emission ($\tilde{B}^2\Sigma_u^+ - \tilde{X}^2\Pi_g$; 289 nm) was also clearly identified with brightness ~200 R. A marginal detection of emission from atomic oxygen ($^1S-^3P$; 297.2 nm) with brightness ~90 R was also reported. From this observation, it was inferred that Martian upper atmospheric species emitted in a narrow crustal field cusp region after excitation by a flux of incident particles. Mars, for the first time, was known to have aurora.

The discovery of aurora prompted MEX and the SPICAM team to conduct a nightside observing campaign dedicated to identifying auroral emission. Nine additional auroral events from six orbits have been reported, and all occurred near the strong crustal magnetic fields in the southern hemisphere of Mars [*Leblanc et al.*, 2008]. Correlation of the emission regions with maps of the average Martian magnetic field topology produced in the work of *Brain et al.* [2007] showed that aurora occur near the boundary between open and closed magnetic field lines, similar to terrestrial aurora. All but one of the events was recorded using a nadir viewing geometry. All events were substantially fainter in the Cameron bands (~105–825 R) than the first event, although the first event was recorded when SPICAM was limb pointing, which may have allowed for a longer path length through the emission. Emission by CO_2^+ at 289 nm was detected for all but one event (10–160 R), and emission from atomic oxygen at 297.2 nm was not reported. It was not possible to obtain a reliable estimate of how often auroral emission can be detected due to the varying orbit geometry of MEX; though 66 orbits were examined (most or all near crustal fields), more than half had very unfavorable SPICAM viewing geometry.

2.2. Particles and Fields

Reports of auroral emission prompted examination of in situ particle and field measurements for evidence of auroral plasma processes near the Martian crustal fields. MGS carried both a vector magnetometer and an electrostatic analyzer designed to measure suprathermal electrons and from 1999 through 2006 was in a nearly circular orbit at ~400 km altitude with fixed local time near 02:00 local time on the nightside. Examination of these observations revealed hundreds of events in which an energized electron population (peak energy: ~200 eV to 4 keV) was evident, analogous to energized electrons measured in terrestrial auroral regions [*Brain et al.*, 2006]. One such event near a strong crustal field region is shown in Figure 2. An energized electron population is evident in regions where the horizontal components of

magnetic field are small compared to the vertical component (i.e., in a cusp region). The electron pitch angle distributions reveal one-sided loss cones in these regions (with fewer electrons moving upward from below the spacecraft), indicative of an open magnetic field topology. The energized electrons are observed to be near regions with two-sided loss cones, indicative of a closed field topology. Nearby perturbations in one of the horizontal magnetic field components are consistent with vertical (field-aligned) currents with current density (~1 $\mu A\ m^{-2}$) comparable to terrestrial field-aligned currents. Overall, the event reveals an auroral-like energized incident electron population in a crustal field cusp region, near a field-aligned current region and the boundary between open and closed fields. It is possible that such a population could impact the atmosphere, causing UV emission in cusps.

Hundreds of "auroral" events have been identified in MGS observations, with ~13,000 individual auroral-like energized electron energy spectra. Given the repeatable orbit of MGS during the period in which these events were recorded, the energized electron distributions could be correlated with geographic location and external conditions without orbital bias. Energized electron distributions are measured predominantly in the southern hemisphere of Mars, near strong and moderate crustal magnetic fields (Figure 3). Both their existence in observations and their characteristics are observed to vary with location, Martian season, solar wind pressure, and the clock angle of the upstream interplanetary magnetic field [*Brain et al.*, 2006]. Further, the most energetic events are more likely to be observed during periods of disturbed solar wind conditions (typically the result of a passing coronal mass ejection).

Caution must be taken when interpreting the particle and field results described above. First, not all of the "auroral events" are similar. For example, *Halekas et al.* [2008] noted that each of the MGS events could be classified as one of three types: localized events predominantly near strong crustal fields (similar to Figure 2 above), extended (longer duration) events located mostly near moderate or weaker crustal fields, and current sheet events where energized electrons were identified in current sheets on the Martian nightside. The typical properties of the electron distributions and local magnetic topology differ for each of the three types of events, suggesting the possibility that the physical mechanisms responsible for the observations differ as well. Next, it is not certain that all of the MGS observations result in the deposition of energized electrons into the Martian upper atmosphere. Since the magnetic field below the spacecraft should increase when it is above crustal fields, it is likely that some (or even most) of the incident flux is adiabatically reflected before it can reach the collisional atmosphere. Finally, though the number of "auroral events" in each location

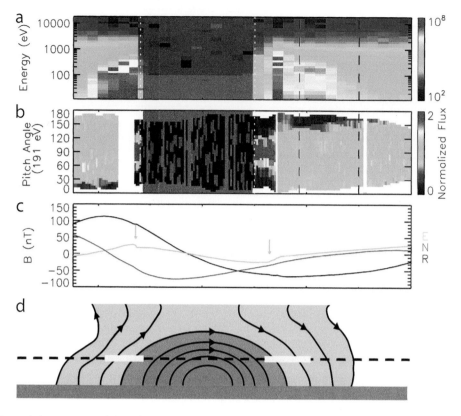

Figure 2. Auroral electron event observed over 6.5 min at 400 km altitude near strong crustal fields by *Brain et al.* [2006]. (a) Electron energy flux versus time. (b) The 191 eV electron pitch angle distribution versus time (with each distribution normalized separately). (c) Radial (black), eastward (green), and northward (red) vector magnetic field components. (d) A cartoon of the spacecraft trajectory through open and closed magnetic field lines, highlighting where auroral-like electron distributions are observed.

varies with external conditions, it is not clear whether changes in external conditions prevent energized electron distributions from forming, or simply move flux tubes containing energized electrons to locations outside of the 02:00 local time orbit of MGS.

While MGS measured suprathermal electrons and magnetic field from a circular orbit, MEX measures suprathermal electrons and ions from an elliptical orbit. Using the Analyzer of Space Plasma and Energetic Atoms (ASPERA-3) ion mass analyzer and electrostatic analyzer, *Lundin et al.* [2006b, 2006a] have identified a number of "inverted-V" events near local midnight. The events are characterized by downward traveling electrons and upward traveling planetary ions, each with a peak in the energy distribution ranging from tens to hundreds of eV for electrons and hundreds to few keV for ions (Figure 4a). The events occur when the spacecraft is above moderate or strong crustal fields, though it can be difficult to unambiguously associate the events with specific crustal field cusps due to the usually higher altitude

of the spacecraft (up to ~8000 km) compared to MGS. Events are clustered near midnight, with a preference for the "premidnight" sector.

Since the ASPERA measurements are made at a variety of altitudes, it is also possible to explore the altitude variation in peak energy of the accelerated particle distributions. Within individual events, the energy of the electron beams takes up an increasingly larger fraction of the total (ion + electron) beam energy as altitude decreases [*Lundin et al.*, 2006a]. The same is true generally, as shown in Figure 4b. The ion beam energy relative to the total beam energy increases with altitude, consistent with a region extending up to ~2000 km altitudes that accelerates ions upward and electrons downward. One mechanism that could achieve this is a quasi-static field-aligned potential, as has been proposed for terrestrial aurora.

Further, ASPERA measures ion mass in addition to ion energy. Therefore, the observations reveal additional clues about the possible particle acceleration mechanisms. Analysis

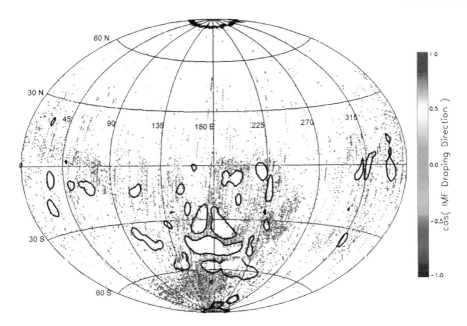

Figure 3. Geographic locations of ~13,000 peaked electron distributions recorded by Mars Global Surveyor from 400 km altitude at 02:00 local time. The location of each distribution is colored according to the cosine of a proxy for the clock angle of the upstream interplanetary magnetic field (essentially the orientation of the external magnetic field). Locations of crustal field regions that are typically closed to the solar wind at 400 km are outlined in black.

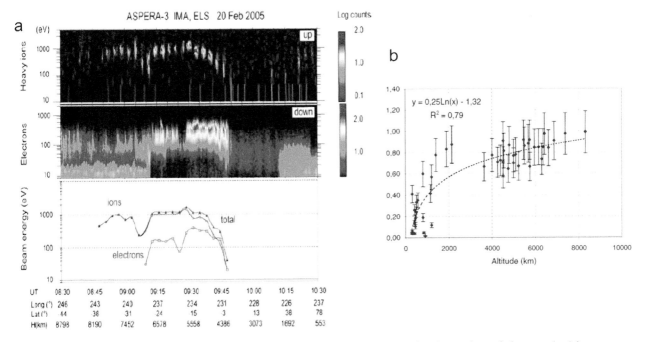

Figure 4. Inverted-V particle events observed by MEX ASPERA. (a) Time series observations of planetary (top) heavy ions and (middle) electrons, along with the (bottom) peak energy of the particle beams. (b) Fraction of the total beam energy (ion + electron) taken up by ions. From *Lundin et al.* [2006a].

of the ion beams for some ASPERA events shows that lower-mass species (e.g., O^+) have lower energy than high-mass species (e.g., CO_2^+) [*Lundin et al.*, 2006a]. If the ions were all accelerated from the same initial energy, then the observations imply that a mass-dependent acceleration mechanism must operate on these flux tubes; that is, a field-aligned electrostatic potential is not capable of completely explaining the observations.

One main challenge in observational studies of Martian auroral processes is demonstrating that the three different classes of observations (UV emission, MGS electrons and magnetic field, MEX ions and electrons) are related. Fortunately, MEX carries both SPICAM and ASPERA. It has been possible to correlate the UV emission with the particle measurements in a few instances [*Leblanc et al.*, 2008]. During some auroral emission events, ASPERA measured narrow electron beams at a time when SPICAM was nadir oriented, observing enhanced UV emission coming from a magnetic cusp region directly below the spacecraft. Further, the Mars Advanced Radar for Subsurface and Ionosphere Sounding instrument on MEX simultaneously recorded an enhancement in the total electron content of the upper atmosphere below the spacecraft. Accurate correlations between in situ particle and UV measurements are complicated by curvature of magnetic field lines below the spacecraft. Regardless, these observations lend credence to the idea that accelerated downgoing electrons measured at spacecraft altitude encounter the atmosphere and lead to both enhanced ionization and emission.

3. MECHANISMS

Auroral emission occurs at Mars in crustal magnetic field cusp regions, as do auroral-like particle acceleration processes. The emission and particle acceleration signatures are related. But the observations, so far, suggest that Martian aurora is relatively weak (faint emission, small acceleration) by terrestrial standards. So how do the mechanisms responsible for aurora in the small-scale crustal fields compare to mechanisms discussed for terrestrial and other planetary aurora? The cartoon in Figure 5 provides an overview of the various mechanisms that are discussed at present for Mars.

The observed brightness ratio between different auroral emission lines can yield insight into the energy of the particle population responsible for the emission. Analysis of the ratio of CO Cameron band emission to CO_2^+ doublet emission for the auroral emission reported by *Bertaux et al.* [2005] suggested a relatively low-energy (tens of eV) incident electron population excited the upper atmosphere [*Leblanc et al.*, 2006]. It was noted that dayside ionospheric photoelectrons have nearly the correct energy distribution to explain the emission and that the auroral emission could be airglow due to photoelectrons transported from day to night (Figure 5a). Events observed subsequently have a variety of brightness ratios, which are inversely correlated with the peak energy of the electron beam observed above the atmosphere [*Leblanc et al.*, 2008]. Peak energies range from 40 to 350 eV for events where emission and electron beams are observed simultaneously, suggesting that at least some of the events

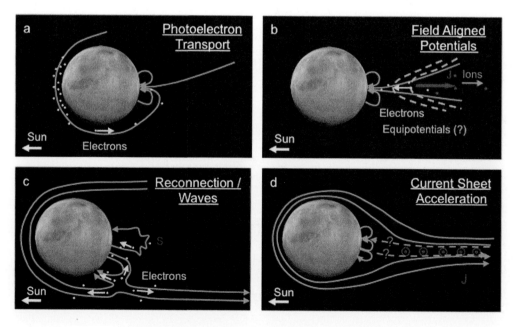

Figure 5. Possible mechanisms for Martian aurora.

cannot be explained by photoelectron transport. Further, *Liemohn et al.* [2007] mapped the magnetic field topology for the *Bertaux et al.* [2005] event using a global plasma model and found that the modeled magnetic field lines connected to the emission site do not cross the Martian terminator. Instead, auroral particles for this event should come from the Martian tail or nightside magnetosheath. Still, low-energy electrons appear to be responsible for a subset of the observations, and transport of electrons from day to night should occur [*Haider et al.*, 2002; *Liemohn et al.*, 2007; *Uluşen and Linscott*, 2008].

Many of the observations presented in section 2 are consistent with the existence of quasi-static field-aligned potentials above cusp regions on the Martian nightside (Figure 5b), analogous to terrestrial auroral particle acceleration [*Marklund*, 2009]. Peaked "inverted-V" particle energy distributions are reminiscent of charged particles observed in field-aligned current regions at Earth [*Brain et al.*, 2006; *Lundin et al.*, 2006b]. Counterstreaming ions and electrons suggest that the acceleration region has been sampled at Mars and extends upward to ~2000 km altitudes or higher [*Lundin et al.*, 2006a], but the typical lower boundary of this region has not been determined. Furthermore, magnetic field perturbations observed near cusp boundaries are consistent with the presence of field-aligned currents above crustal fields [*Brain et al.*, 2006].

Despite the similarities between the Martian and terrestrial measurements, there are still a number of open questions about the ability of this picture to explain any or all of the observations. For example, how and where do the field-aligned currents close? Are auroral acceleration regions long-lived, or do they constantly form and dissipate? Martian crustal field strengths in the ionosphere are weaker than the Earth's magnetic field, with the consequence that the Pedersen conductivity is much higher [*Dubinin et al.*, 2008]. This work concludes that a region of strong parallel electric field is required to sustain field-aligned potential drops above the ionosphere. Alternatively, magnetic field stresses at high altitudes may be periodically dissipated through connection with the conducting ionosphere. Auroral particle signatures have been observed repeatedly in the same locations on several repeating MEX orbits, over periods of weeks, implying that the mechanism is likely to be stable [*Dubinin et al.*, 2009].

Field-aligned potentials are only one possible mechanism for accelerating particles (Figure 5c). For example, based on the mass dependence of the planetary ion energy in inverted-V structures, *Lundin et al.* [2006a] proposed that waves may play a significant role in accelerating ions. Angular ion distributions in the acceleration region reveal evidence for heating or acceleration transverse to the magnetic field [*Dubinin et al.*, 2009]. Magnetic reconnection is also known to occur on the nightside of Mars [*Eastwood et al.*, 2008]. Reconnection has been at least indirectly demonstrated to occur in crustal field cusp regions as well [*Brain*, 2006]. Reconnection near crustal fields is a leading candidate mechanism for the creation of electron conic pitch angle distributions observed on the Martian nightside [*Uluşen et al.*, 2011] and should provide a source of energy for electrons passing near the reconnection diffusion region. The extent to which plasma waves and reconnection contribute to particle acceleration and aurora at Mars is not well determined at present, however.

A final intriguing mechanism is illustrated in Figure 5d. Acceleration of electrons in current sheets on the Martian nightside may result in their propagation into the nightside atmosphere. Many of the most energetic peaked electron distributions recorded by MGS occur in current sheets [*Halekas et al.*, 2008], and current sheets are often observed at low altitudes on the nightside [*Halekas et al.*, 2006]. It is not certain whether the observed current sheets are the main magnetotail current sheet induced at Mars or whether they are independently associated with crustal fields. Their observed geographic distribution is not strongly correlated with crustal field location, suggesting the main magnetotail current sheet extends to very low altitudes on the nightside. However, it is not clear whether or how the observed current sheets provide access for accelerated electrons to the nightside upper atmosphere.

All of the four mechanisms illustrated in Figure 5 may occur at Mars and likely contribute to the observations of auroral processes. Continued analysis and future observations should help to reveal which acceleration mechanisms dominate, how they operate, and where the accelerated particles originate.

4. CONSEQUENCES AND FRONTIERS

4.1. Consequences

Three main types of observation related to Martian aurora have been reported to date, and each suggests further consequences of auroral processes.

Reports of UV auroral emission in cusps naturally lead to questions of whether visible aurora may occur at Mars (Figure 6a), as they do at Earth. Early estimates are divided on this point. Certainly, there are species in the Martian upper atmosphere that emit at visible wavelengths. The question is whether the relevant transitions are sufficiently excited to create visible emission that is bright enough for spacecraft instruments to distinguish. Based on the energy flux of inverted-V electron distributions and an assumption of an

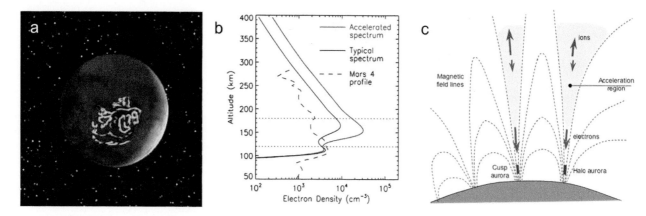

Figure 6. Consequences of Martian aurora. (a) Visible aurora might be visible at midlatitudes (image courtesy M. Holmstrom/ European Space Agency). (b) Localized "patchy" ionospheres exist in cusps [*Fillingim et al.*, 2007]. (c) Atmospheric ions escape along auroral flux tubes [*Lundin et al.*, 2006b].

earthlike upper atmospheric composition, *Lundin et al.* [2006b] estimated that the oxygen green line at 557.7 nm should produce as much as 2–80 kR of emission, easily observable by instruments or even the naked eye. This estimate assumes that there are inverted-V events at Mars that are more energetic than those that have been observed so far. A more cautious estimate (still assuming a terrestrial atmospheric composition) of ~220 R is obtained by using the observed 557.7 nm/297.2 nm emission ratio at Earth [*Slanger et al.*, 2006] and applying it to the brightest UV observation at Mars. Finally, models of the brightest (and first reported) emission event assuming a Martian upper atmospheric composition yield a green line brightness of only 30 R [*Bertaux et al.*, 2005]. This issue is most likely to be resolved by direct observation of cusp regions, either from cameras on the surface or from orbiting spacecraft.

Measurements of downward traveling electrons suggest that they influence the upper atmosphere in localized regions on the nightside (Figure 6b). Auroral emission is only one consequence of this interaction between incident plasma and atmospheric species. Another consequence is enhanced ionization in cusp regions, as both modeled [*Fillingim et al.*, 2007] and measured [*Safaeinili et al.*, 2007; *Leblanc et al.*, 2008]. The localized "patchy" ionosphere may have densities >10% of the subsolar ionospheric density. Further, their presence will create gradients in ionospheric density that leads to significant horizontal plasma transport and electric fields [*Fillingim et al.*, 2007]. This transport, in turn, should lead to Joule heating of the upper atmosphere as charged particles attempt to flow in the neutral background.

Measurements of upward traveling ions in crustal field cusps suggest that auroral flux tubes may lead to enhanced escape rates from the Martian ionosphere (Figure 6c). Atmo-

spheric escape is a major topic in studies of Martian climate evolution [*Jakosky and Phillips*, 2001]. Present estimates of the "auroral contribution'" to ion escape vary. A back of the envelope calculation based on terrestrial analogy yields an upper bound to the O^+ escape rate on auroral flux tubes of ~5 × 10^{25} s^{-1} [*Ergun et al.*, 2006]. However, an estimate based on ASPERA observations of energized electron and ion distributions leads to an upper bound escape rate 2 orders of magnitude lower, at ~5 × 10^{23} s^{-1} [*Dubinin et al.*, 2009]. Auroral flux tubes in the present epoch occupy a small fraction of the upper atmosphere. Auroral ion escape rates are likely to have been larger long ago, when Mars had a global dynamo magnetic field and shortly afterward when crustal fields may have occupied a larger fraction of the Martian crust.

4.2. Frontiers

There are several frontiers for auroral research at Mars in the coming years. Chief among these is continued analysis of the existing observations of UV emission, accelerated ions and electrons, and magnetic field. MEX continues to make measurements from Mars orbit and is likely to contribute new information about auroral variability under different external conditions. Perhaps most exciting among the potential new observations is a determination of the influence of solar activity on auroral brightness.

There is also a need for improved modeling of auroral processes at Mars. At present, global plasma models lack either the spatial resolution or the necessary physical assumptions to adequately simulate auroral acceleration processes. Localized electron transport models have been developed and applied to open crustal field flux tubes [*Seth et al.*, 2002; *Liemohn et al.*, 2006; *Lillis et al.*, 2009], but currently lack a complete picture of the physics operating on those flux tubes (e.g., ion motion and waves). Models of auroral emission are reasonably sophisticated, however, and should be applied more vigorously to the existing measurements to provide constraints on how often visible aurora are measurable at Mars or whether aurora might be observed on the dayside of the planet.

One future set of measurements has the potential to better constrain auroral mechanisms and brightness. The MAVEN spacecraft mission, scheduled to arrive in the fall of 2014, will carry a full complement of particle and field instruments to measure in situ accelerated ion and electron populations, magnetic fields, and the background upper atmosphere. It will also carry a UV spectrometer capable of measuring aurora with greater sensitivity and spectral resolution than SPICAM. Combined with MAVEN measurements of the Sun and solar wind, the mission has the potential to distinguish between the various auroral mechanisms described in section 3.

In summary, given the presence of crustal "mini magnetospheres" near Mars, it is not surprising that auroral processes appear to be active there. The aurora is faint and the particle acceleration is relatively weak compared to typical terrestrial aurora. But this does not mean that Martian aurora should be dismissed as a curiosity. On the contrary, the study of auroral processes in an "end-member" situation such as Mars provides has the potential to better constrain our understanding of the limits of auroral processes everywhere.

Acknowledgments. D. Brain acknowledges a useful discussion with A.I.F. Stewart on the Mariner UV observations. This effort was supported by NASA grant NNX08AK95G.

REFERENCES

Acuña, M., et al. (1998), Magnetic field and plasma observations at Mars: Initial results of the Mars global surveyor mission, *Science*, *279*(5357), 1676–1680.

Bertaux, J. L. (2005), Nightglow in the upper atmosphere of Mars and implications for atmospheric transport, *Science*, *307*(5709), 566–569, doi:10.1126/science.1106957.

Bertaux, J.-L., F. Leblanc, O. Witasse, E. Quemerais, J. Lilensten, S. A. Stern, B. Sandel, and O. Korablev (2005), Discovery of an aurora on Mars, *Nature*, *435*(7), 790–794, doi:10.1038/nature03603.

Brain, D. A. (2006), Mars Global Surveyor measurements of the Martian solar wind interaction, *Space Sci Rev*, *126*(1), 77–112, doi:10.1007/s11214-006-9122-x.

Brain, D. A., J. S. Halekas, L. M. Peticolas, R. P. Lin, J. G. Luhmann, D. L. Mitchell, G. T. Delory, S. W. Bougher, M. H. Acuña, and H. Rème (2006), On the origin of aurorae on Mars, *Geophys. Res. Lett.*, *33*, L01201, doi:10.1029/2005GL024782.

Brain, D. A., R. J. Lillis, D. L. Mitchell, J. S. Halekas, and R. P. Lin (2007), Electron pitch angle distributions as indicators of magnetic field topology near Mars, *J. Geophys. Res.*, *112*, A09201, doi:10.1029/2007JA012435.

Dubinin, E., G. Chanteur, M. Fraenz, and J. Woch (2008), Field-aligned currents and parallel electric field potential drops at Mars. Scaling from the Earth' aurora, *Planet. Space Sci.*, *56*(6), 868–872, doi:10.1016/j.pss.2007.01.019.

Dubinin, E., M. Fraenz, J. Woch, S. Barabash, and R. Lundin (2009), Long-lived auroral structures and atmospheric losses through auroral flux tubes on Mars, *Geophys. Res. Lett.*, *36*, L08108, doi:10.1029/2009GL038209.

Eastwood, J. P., D. A. Brain, J. S. Halekas, J. F. Drake, T. D. Phan, M. Øieroset, D. L. Mitchell, R. P. Lin, and M. Acuña (2008), Evidence for collisionless magnetic reconnection at Mars, *Geophys. Res. Lett.*, *35*, L02106, doi:10.1029/2007GL032289.

Ergun, R. E., L. Andersson, W. K. Peterson, D. Brain, G. T. Delory, D. L. Mitchell, R. P. Lin, and A. W. Yau (2006), Role of plasma waves in Mars' atmospheric loss, *Geophys. Res. Lett.*, *33*, L14103, doi:10.1029/2006GL025785.

Fillingim, M. O., L. M. Peticolas, R. J. Lillis, D. A. Brain, J. S. Halekas, D. L. Mitchell, R. P. Lin, D. Lummerzheim, S. W. Bougher, and D. L. Kirchner (2007), Model calculations of electron precipitation induced ionization patches on the nightside of Mars, *Geophys. Res. Lett.*, *34*, L12101, doi:10.1029/2007GL029986.

Fox, J. L., and A. I. F. Stewart (1991), The Venus ultraviolet aurora: A soft electron source, *J. Geophys. Res.*, *96*(A6), 9821–9828.

Fox, J. L., J. F. Brannon, and H. S. Porter (1993), Upper limits to the nightside ionosphere of Mars, *Geophys. Res. Lett.*, *20*(13), 1339–1342.

Haider, S. A., J. Kim, A. F. Nagy, C. N. Keller, M. I. Verigin, K. I. Gringauz, N. M. Shutte, K. Szego, and P. Kiraly (1992), Calculated ionization rates, ion densities, and airglow emission rates due to precipitating electrons in the nightside ionosphere of Mars, *J. Geophys. Res.*, *97*(A7), 10,637–10,641.

Haider, S. A., S. P. Seth, E. Kallio, and K. I. Oyama (2002), Solar EUV and electron-proton-hydrogen atom-produced ionosphere on Mars: Comparative studies of particle fluxes and ion production rates due to different processes, *Icarus*, *159*(1), 18–30, doi:10.1006/icar.2002.6919.

Halekas, J. S., D. A. Brain, R. J. Lillis, M. O. Fillingim, D. L. Mitchell, and R. P. Lin (2006), Current sheets at low altitudes in the Martian magnetotail, *Geophys. Res. Lett.*, *33*, L13101, doi:10.1029/2006GL026229.

Halekas, J. S., D. A. Brain, R. P. Lin, J. G. Luhmann, and D. L. Mitchell (2008), Distribution and variability of accelerated electrons at Mars, *Adv. Space Res.*, *41*(9), 1347–1352, doi:10.1016/j.asr.2007.01.034.

Jakosky, B. M., and R. J. Phillips (2001), Mars' volatile and climate history, *Nature*, *412*(6), 237–244.

Leblanc, F., O. Witasse, J. Winningham, D. Brain, J. Lilensten, P.-L. Blelly, R. A. Frahm, J. S. Halekas, and J. L. Bertaux (2006), Origins of the Martian aurora observed by Spectroscopy for Investigation of Characteristics of the Atmosphere of Mars (SPICAM) on board Mars Express, *J. Geophys. Res.*, *111*, A09313, doi:10.1029/2006JA011763.

Leblanc, F., et al. (2008), Observations of aurorae by SPICAM ultraviolet spectrograph on board Mars Express: Simultaneous ASPERA-3 and MARSIS measurements, *J. Geophys. Res.*, *113*, A08311, doi:10.1029/2008JA013033.

Liemohn, M. W., et al. (2006), Numerical interpretation of high-altitude photoelectron observations, *Icarus*, *182*(2), 383–395, doi:10.1016/j.icarus.2005.10.036.

Liemohn, M. W., Y. Ma, A. F. Nagy, J. U. Kozyra, J. D. Winningham, R. A. Frahm, J. R. Sharber, S. Barabash, and R. Lundin (2007), Numerical modeling of the magnetic topology near Mars auroral observations, *Geophys. Res. Lett.*, *34*, L24202, doi:10.1029/2007GL031806.

Lillis, R. J., M. O. Fillingim, L. M. Peticolas, D. A. Brain, R. P. Lin, and S. W. Bougher (2009), Nightside ionosphere of Mars: Modeling the effects of crustal magnetic fields and electron pitch angle distributions on electron impact ionization, *J. Geophys. Res.*, *114*, E11009, doi:10.1029/2009JE003379.

Luhmann, J. G., C. T. Russell, L. H. Brace, and O. L. Vaisberg (1992), The intrinsic magnetic field and solar-wind interaction of Mars, in *Mars*, edited by H. H. Kieffer et al., *Rep. A93-27852 09-91*, pp. 1090–1134, Univ. of Ariz. Press, Tucson.

Lundin, R., et al. (2006a), Auroral plasma acceleration above Martian magnetic anomalies, *Space Sci. Rev.*, *126*(1), 333–354, doi:10.1007/s11214-006-9086-x.

Lundin, R., et al. (2006b), Plasma acceleration above Martian magnetic anomalies, *Science*, *311*(5), 980–983, doi:10.1126/science.1122071.

Marklund, G. T. (2009), Electric fields and plasma processes in the auroral downward current region, below, within, and above the acceleration region, *Space Sci. Rev.*, *142*(1–4), 1–21, doi:10.1007/s11214-008-9373-9.

Mauk, B., and F. Bagenal (2012), Comparative auroral physics: Earth and other planets, in *Auroral Phenomenology and Magnetospheric Processes: Earth and Other Planets*, *Geophys. Monogr. Ser.*, doi:10.1029/2011GM001192, this volume.

Mitchell, D. L., R. P. Lin, C. Mazelle, H. Rème, P. A. Cloutier, J. E. P. Connerney, M. H. Acuna, and N. F. Ness (2001), Probing Mars' crustal magnetic field and ionosphere with the MGS Electron Reflectometer, *J. Geophys. Res.*, *106*(E10), 23,419–23,427, doi:10.1029/2000JE001435.

Safaeinili, A., W. Kofman, J. Mouginot, Y. Gim, A. Herique, A. B. Ivanov, J. J. Plaut, and G. Picardi (2007), Estimation of the total electron content of the Martian ionosphere using radar sounder surface echoes, *Geophys. Res. Lett.*, *34*, L23204, doi:10.1029/2007GL032154.

Seth, S. P., S. A. Haider, and K. I. Oyama (2002), Photoelectron flux and nightglow emissions of 5577 and 6300 Å due to solar wind electron precipitation in Martian atmosphere, *J. Geophys. Res.*, *107*(A10), 1324, doi:10.1029/2001JA000261.

Slanger, T. G., P. C. Cosby, B. D. Sharpee, K. R. Minschwaner, and D. E. Siskind (2006), $O(^1S \rightarrow {}^1D, {}^3P)$ branching ratio as measured in the terrestrial nightglow, *J. Geophys. Res.*, *111*, A12318, doi:10.1029/2006JA011972.

Ulusen, D., and I. Linscott (2008), Low-energy electron current in the Martian tail due to reconnection of draped interplanetary magnetic field and crustal magnetic fields, *J. Geophys. Res.*, *113*, E06001, doi:10.1029/2007JE002916.

Ulusen, D., D. A. Brain, and D. L. Mitchell (2011), Observation of conical electron distributions over Martian crustal magnetic fields, *J. Geophys. Res.*, *116*, A07214, doi:10.1029/2010JA016217.

D. Brain, Laboratory for Atmospheric and Space Physics, University of Colorado, Boulder, CO 80309-0392, USA. (David.Brain@lasp.colorado.edu)

J. S. Halekas, Space Sciences Laboratory, University of California, Berkeley, CA 94720, USA.

When Moons Create Aurora: The Satellite Footprints on Giant Planets

B. Bonfond

Laboratoire de Physique Atmosphérique et Planétaire, Université de Liège, Liège, Belgium

Satellite footprints are localized auroral emissions in the upper atmosphere of Jupiter (Saturn) near the magnetic field lines linking to satellites, Io, Europa, and Ganymede (Enceladus). They are the auroral signatures of the strong electromagnetic interactions taking place between these moons and the intensely magnetized, rapidly rotating planets they orbit. The Io and Europa spots have been shown to be followed by an extended tail. This might also be the case for the Ganymede and Enceladus emissions, although not yet unambiguously observed. Moreover, the main Io spot is accompanied by secondary spots attributed either to reflections of the plasma waves generated at Io on the Io plasma torus boundary or to electrons accelerated in one hemisphere but precipitating in the opposite one. While the horizontal extent of the spots gives a hint of the size of the interaction region in the equatorial plane, the vertical profile of the footprints provides clues to the energy distribution of the precipitating electrons. Moreover, the location of the footprints can be used as constraints for magnetic field models. Finally, the brightness of the footprints is a valuable diagnostic of the interaction mechanism and has been observed to vary at different time scales, each one tentatively associated with a different process.

1. INTRODUCTION

Jupiter is a source of intense radio emissions, the observation (from Earth) of which was found to be highly correlated with the orbital position of the satellite Io [*Bigg*, 1964]. This led *Goldreich and Lynden-Bell* [1969] to propose an electromagnetic interaction between Jupiter and Io that ultimately gives rise to radio emissions from the foot of the Io flux tube in Jupiter's upper atmosphere. Io's volcanism supplies a neutral torus that ultimately feeds a dense plasma that migrates outward, forming an equatorial plasma sheet. Jupiter's magnetic dipole is tilted some 9.6° from its rotation axis. Cold iogenic plasma corotating with the magnetic field is confined by centrifugal forces to the centrifugal equator, a plane about two thirds of the way between the rotation and

magnetic equators [*Gledhill*, 1967]. Consequently, as the plasma sheet rotates with Jupiter, it passes over orbiting satellites twice per rotation where the centrifugal and rotation planes intersect. Viewed in a frame of reference at rest with the plasma sheet, the satellites move up and down in harmonic motion about the centrifugal equator, as a function of their System III longitude.

At Saturn, Enceladus's cryovolcanism is also a major plasma source in the Kronian magnetosphere [*Pontius and Hill*, 2006]. Enceladus is also embedded in a plasma torus, but it always remains constantly 0.04 Saturn radius below its center because Saturn's rotation axis and its magnetic dipole are nearly coaligned, but the magnetic equator is offset relative to the rotational one.

In brief, the basic scenario to generate auroral footprints is the following one. The satellites constitute obstacles to the corotating plasma flow, which overtakes their much slower prograde orbital motion. This perturbation propagates along the magnetic field lines as Alfvén waves, and the locus of the perturbed points is the Alfvén wing. On their way to the planet, the waves cause the acceleration of

Auroral Phenomenology and Magnetospheric Processes: Earth and Other Planets
Geophysical Monograph Series 197
10.1029/2011GM001169

electrons, which finally precipitate into the planetary atmosphere and trigger aurora [*Hess and Delamere*, this volume]. The auroral spots may appear downstream of the foot of an undisturbed magnetic field line traced from the satellite to the top of Jupiter's atmosphere, due to complexities of the interaction. For example, in the ideal Alfvén wing scenario [*Neubauer*, 1980; *Goertz*, 1980], the wing is inclined with respect to the background magnetic field because of the combination of the flux tube motion relative to the satellite and the finite propagation speed of the Alfvénic disturbance. The downstream shift of the most prominent ("main") spot along the footpath, is called the "lead angle." However, its measurement is delicate because of the limited accuracy of the magnetic field models [*Bonfond et al.*, 2009]. This terse summary does not do justice to the complexity and to the diversity of the phenomena taking place close to the satellites (see reviews by *Kivelson et al.* [2004], *Saur et al.* [2004], and *Jia et al.* [2009]).

The first satellite footprint to be detected was the Io footprint, as its H_3^+ emissions were identified in the infrared domain by *Connerney et al.* [1993]. This finding was confirmed in the FUV (120–170 nm) by Hubble Space Telescope (HST) observations [*Prangé et al.*, 1996; *Clarke et al.*, 1996]. No other footprint has been observed in the IR domain since. Moreover, the Io footprint is also the only one that has been detected in visible wavelengths (by the Galileo and New Horizons probes) [*Vasavada et al.*, 1999; *Gladstone et al.*, 2007]. The Europa and Ganymede footprints were simultaneously discovered in 2002 with the HST in the UV range [*Clarke et al.*, 2002]. The Enceladus footprint was discovered in 2011 with the EUV and FUV channels of the Ultraviolet Imaging Spectrograph (UVIS) instrument aboard the Cassini probe [*Pryor et al.*, 2011; *Gurnett and Pryor*, this volume] (Figure 1).

2. MORPHOLOGY OF THE SATELLITE FOOTPRINTS

Up to now, the Europa, Ganymede, and Enceladus footprints have been seen as single spots in each hemisphere. However, based on images from NASA's Infrared Telescope Facility (IRTF), *Connerney and Satoh* [2000] reported an Io footprint made of five equally spaced spots. Moreover, on UV images, up to three Io footprint spots can be distinguished along with an extended downstream tail (Figure 2) [*Clarke et al.*, 2002; *Gérard et al.*, 2006; *Bonfond et al.*, 2008]. For each satellite, the locus of these features forms a closed contour in a System III reference frame, which is called the satellite footpath. The Europa spot is also sometimes followed by a short tail [*Grodent et al.*, 2006]. Furthermore, these Io spots appear to move with respect to each other as Io moves up and down inside the plasma torus. The multiplicity of these spots first suggested that they were due to reflections of the Alfvén waves on the density gradient at the plasma torus boundary as illustrated by the blue trajectories in Figure 2 [*Neubauer*, 1980; *Gurnett and Goertz*, 1981]. The main Alfvén wing (MAW) spot would be located at the feet of a direct Alfvén wing, while subsequent ones would be related to reflected Alfvén wings (RAW). Another explanation could be that the multiple spots are not the direct counterpart of RAW, but an interference pattern due to the multiple reflections [*Jacobsen et al.*, 2007]. The finding of a faint spot emerging upstream of the brightest one in one hemisphere while only downstream spots are seen in the opposite hemisphere challenged these hypotheses. *Bonfond et al.* [2008] suggested that, additionally to the MAW spot and to the RAW spot, one of the spots, called the transhemispheric electron beam (TEB) spot, is caused by electrons accelerated away from the planet in one hemisphere and precipitating in the opposite hemisphere (Figure 2). This scenario does not only account both qualitatively

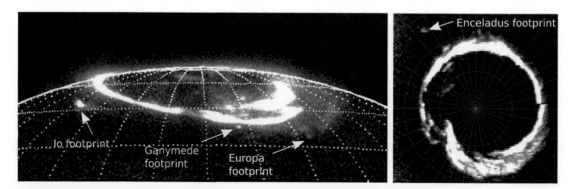

Figure 1. (left) Image of the northern FUV aurora of Jupiter acquired with the Hubble Space Telescope's ACS camera on 7 February 2006. The Io, Europa, and Ganymede footprints are simultaneously visible. (right) Polar projection of Saturn's FUV aurora as observed by Cassini's UVIS instrument on 26 August 2008 (D. Grodent and the Cassini/UVIS team). The Enceladus footprint is visible more equatorward than any other auroral emission.

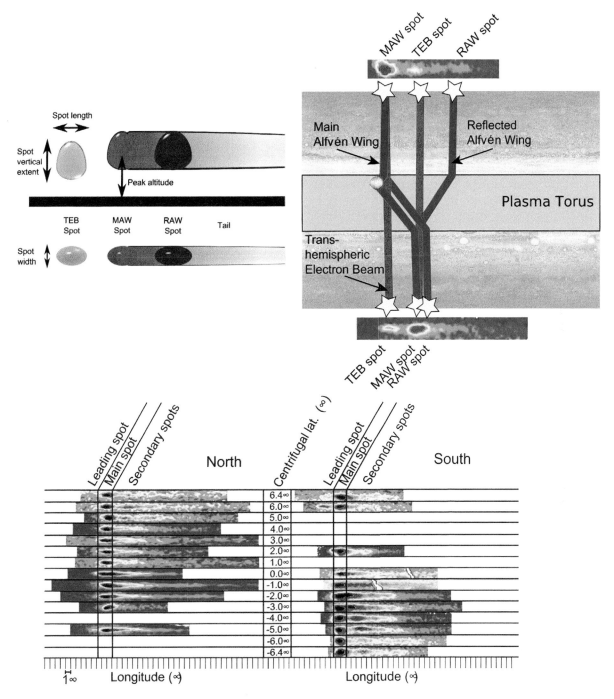

Figure 2. (top left) A sketch of the Io footprint as seen from the side (top diagram) and from above (bottom diagram). The Io footprint is considered to be formed by three spots and an extended downstream tail. (top right) Geometry of the Alfvén wave propagation and their reflection against the inner boundary of the torus. The direct Alfvén wing generates the main Alfvén wing spots, while the reflected wings create the reflected Alfvén wing spots. In contrast to the Alfvén waves, the electron beams are not affected by the high torus density, which enables them to propagate rapidly from one hemisphere to the other, generating the transhemispheric electron beam spots. (bottom) Io footprint morphology as a function of the centrifugal latitude of Io in the torus. The color table of each stripe is scaled individually for a clearer illustration of the morphology. The longitudes are not measured on the planet but mapped to the equatorial plane along the magnetic field lines according to the VIP4 model for an easier comparison of both hemispheres. Adapted from *Bonfond et al.* [2008] and *Bonfond* [2010].

and quantitatively for the evolution of the interspot distances, but it also explains the occurrence of bidirectional electron beams close to Io [*Williams and Thorne*, 2003]. Indeed, *Jacobsen et al.* [2010] computed the magnetic field line bending due to the Io-magnetosphere interaction with a nonlinear 3-D MHD model and showed that the scenario of the TEBs would account for the location of the electron beams as observed when Galileo flew by Io. *Saur et al.* [2002] and *Dols et al.* [2008] discovered that these energetic electron beams were an additional ionization source for Io's stagnant plasma wake. Moreover, *Wilkinson* [1998] argued that the periodicity of Io-related decametric radio emissions could arise from a similar mechanism. Several models [*Hill and Vasyliūnas*, 2002; *Delamere et al.*, 2003; *Ergun et al.*, 2009] describe the Io footprint tail as caused by a steady state process due to the progressive reacceleration of the plasma downstream of Io. On the other hand, MHD simulations indicate that it might actually be the result of multiple reflections of the Alfvén waves [*Jacobsen et al.*, 2007].

The presence of electron beams and their probable relationship with the auroral footprint has also been highlighted at Enceladus [*Pryor et al.*, 2011]. Unlike Jupiter, the Kronian magnetic field dipole axis is perpendicular to the satellites' orbital plane within measurement accuracy, and Enceladus remains ~10 Enceladus radii southward from the torus center. Consequently, the distance between the MAW spot and the TEB spot should remain small and constant.

3. LOCATION OF THE SATELLITE FOOTPRINTS

Mapping a position in the magnetosphere to its ionospheric foot can be challenging if the field is not well known along any portion of its path, which routinely occurs in the outer magnetosphere of the planets, linking to polar magnetic latitudes. However, when moons generate auroral footprints, the direct connections between the two phenomena provide indisputable landmarks in the magnetosphere. For example, *Radioti et al.* [2009] used tabulated Ganymede footprint positions to discuss the mapping of the so-called "equatorward diffuse emissions" in the equatorial plane.

Connerney et al. [1998] used the location of the Io footprint as observed by the IRTF IR telescope and by the HST to build the VIP4 magnetic field model. This model is made of a fourth-order multipole, plus an external contribution from an axisymmetric ring current. In order to constrain the weight of the different spherical harmonic coefficients, they used Pioneer 10 and 11 and Voyager 1 and 2 magnetic field measurements and tuned the coefficients such that the mapping of the Io orbit would fall as close as possible to the Io footpath (i.e., the locus of the Io footprint locations in a System III fixed reference frame). While this model prescribed that any point

on the Io orbit would map to the Io footpath, the exact footprint location along this footpath was not constrained by observations. Based on a much larger number of high-resolution HST images, *Hess et al.* [2011b] built a fifth-order multipole model, called VIPAL, constrained not only by the Io footpath, but also by the locations of the MAW spot along it.

The comparison of the shape of the footpaths of the different moons can also provide some useful information. *Grodent et al.* [2008] noticed that, in the northern hemisphere, the Io footpath on one side and the Europa and the Ganymede footpaths on the other side appear to diverge around 100° in System III longitude. These authors suggested that this behavior could be related to a localized magnetic anomaly. They showed that they could fit well the three observed northern footpaths by adding a weak dipole located ~17,000 km under the surface to a fourth-order spherical harmonics magnetic field model.

While the footprints can be used as constraints for modeling the internal magnetic field, they can also be useful for mapping features from the outer magnetosphere. *Vogt et al.* [2011] used the Ganymede footpath as a starting point to build a magnetic mapping model of Jupiter's auroral features to magnetospheric sources based on magnetic flux equivalence (see also *Vogt and Kivelson* [this volume]).

For a given System III longitude, the location of the Io footprint does not appear to change significantly from one epoch to another, but the same conclusion cannot be drawn for the Ganymede footprint. *Grodent et al.* [2008] analyzed two images acquired 5 years apart, and they noted a significant equatorward shift of 2° for the Ganymede footprint and of 3° for the main auroral emission, while the Io footprint location remains the same. They suggested that these motions stemmed from increased plasma sheet ring current, which caused an increased stretching of the magnetic field lines. The most probable cause for an increasing ring current is an enhancement of Io's volcanic plasma supply, which likely increased the density of the plasma sheet as well as the mass outflow rate.

4. SIZE OF THE SATELLITE FOOTPRINTS

On high-sensitivity HST FUV images, the spots usually appear roughly circular because the observation line of sight is generally perpendicular to the footpath and because the vertical scale height is similar to the length along the footpath. However, based on a subset of observations where the Io footprint was coaligned with the line of sight, *Bonfond* [2010] found that the Io footprint was a less than 200 km wide curtain, which corresponds to twice the size of Io at most. Based on another set of HST images on which the Io footprint is approximately perpendicular to the line of sight, *Bonfond* [2010] measured the full width at half maximum of the MAW spot and of the TEB spot to be ~850 km long. This value corresponds to three to four times the projected size of Io along magnetic field lines. This indicates that the interaction region probably encompasses both Io and part of its wake, as suggested by *Clarke et al.* [2002]. It should, however, be noted that the ~3000 km long spots discussed in the latter paper are the result of the juxtaposition of the three spots and not to a particularly extended unique one. On visible images acquired with Galileo's solid-state imaging (SSI) camera and with New Horizons' Long Range Reconnaissance Imager (LORRI) camera, the footprint looks like a ~450 km wide circular patch [*Vasavada et al.*, 1999] or like a 400 km wide and 1000 km high spot [*Gladstone et al.*, 2007], respectively. *Bonfond* [2010] suggested that the smaller size of the visible spot could be caused by the limited sensitivity of the instruments. A similar effect has been observed when comparing the size of the Io footprint main spot obtained from the low-sensitivity faint object camera (FOC) aboard HST, also ~400 km [*Prangé et al.*, 1996], on one hand, and from the more sensitive STIS and ACS cameras aboard the same spacecraft on the other. As far as the Io footprint tail is concerned, its brightness progressively decreases downstream with an e-folding distance of ~21,000 km [*Bonfond et al.*, 2009].

Some estimates of size of the other footprints' spots have also been published. It should, however, be noted that these studies did not discuss the discrimination between the vertical extent of the emissions and their horizontal width or the blurring caused by the motion of the footprint during the exposure with the same level of detail as those concerning the Io footprint. As far as Europa's footprint is concerned, *Grodent et al.* [2006] measured the spot's full width at half maximum to be ~1100 km long. This distance is much longer than the projected diameter of the moon (~75 km), indicating that the interaction region is probably much larger than the satellite itself. Moreover, these authors also observed a faint ~7500 km long tail following the spot when Europa is close to the torus center. This result suggests that this auroral feature is the signature of an extended plasma plume downstream of Europa [*Kivelson et al.*, 1999]. *Grodent et al.* [2009] estimated the surface of the Ganymede footprint to cover ~5 × 10^5 km^2. Mapped back in the equatorial plane, this surface would correspond to an 8 to 20 Ganymede radii wide region. They concluded that the interaction region is not restricted to Ganymede but also includes its mini-magnetosphere owing to its internal magnetic field. Finally, the Enceladus footprint observed by *Pryor et al.* [2011] also appears to map to a region extending up to 20 Enceladus radii (R_E) downstream of the moon. Depending whether the

extent of the spot corresponded to the vertical extent of the spot or to its latitudinal width, the authors noted that the radial extent of the interaction region would range from 0 to 20 R_E. The large footprint extent also suggests that the interaction region covers the extended neutral plume exhausted by the geysers rather than the satellite alone.

5. VERTICAL EXTENT OF THE IO FOOTPRINT

The isolation of the Io footprint from the other auroral emissions at Jupiter allows direct observations of its vertical extent above the planetary limb (associated with the 1 bar level). The more energetic precipitating electrons are, the more deeply into the atmosphere they penetrate, and the lower the auroral emissions will be. As far as the tail emissions are concerned, the altitude of the brightness peak is ~900 km above limb, and it remains constant whatever the distance from the spots, indicating that the energy of the precipitating electrons does not change with the distance from the spots. The scale height of the tail emissions is on the order of ~430 km. The peak altitude and the scale height of the MAW spots are approximately similar [*Bonfond*, 2010]. *Bonfond et al.* [2009] concluded that the electron distribution that best matches the data was a kappa distribution with a mean energy of 1 keV and a spectral index of 2.3. This estimate of the mean energy contrasts with the 55 keV deduced from measurements of the attenuation of FUV emissions below 135 nm following methane absorption [*Gérard et al.*, 2002]. This discrepancy most probably arises from our poor knowledge of the Jovian auroral atmosphere composition profile. *Hess et al.* [2010] modeled the energy distribution of electrons accelerated by inertial Alfvén waves, and resulting power law energy distributions are consistent with the broadness of the MAW spot vertical profile. *Delamere et al.* [2003] proposed that, contrary to the spots, the tail emissions are generated by a steady state electrodynamic process related to the acceleration of the initially stagnant plasma at Io back to corotation with the planet. In this steady state framework, *Ergun et al.* [2009] concluded that the tail emissions were generated by electrons accelerated by a quasi-static electric field on the order of 1 kV, in accordance with the mean energy deduced from the tail altitude. Nevertheless, *Bonfond et al.* [2009] argued that such a quasi-static electric field would lead to quasi-monoenergetic electrons, which would then create a narrower vertical extent than observed. While the altitudes of the main spot and of the tail are similar, the TEB spot vertical profile appears to peak 200 km lower in the atmosphere. As a consequence, the impinging electron energy is thus expected to be approximately four times higher than for the other two features. This result confirms that the different spots have different origins, and the model of *Hess et al.* [2010] indeed predicted different energy distributions for the electrons directly precipitating into the atmosphere and those accelerated toward the opposite hemisphere.

6. BRIGHTNESS OF THE SATELLITE FOOTPRINTS

Using increasingly large image data sets, recent studies of the Io footprint main spot brightness indicated that the brightness of the main spot appeared to peak when Io is close to the torus center [*Gérard et al.*, 2006; *Serio and Clarke*, 2008; *Wannawichian et al.*, 2010]. *Wannawichian et al.* [2010] estimated that the brightness peaked at 107° and at 287° ± 8° because the denser plasma would generate a stronger interaction. When Io approached the torus northern or southern boundaries, the brightness appeared to be down to ~10 times lower. An alternative reason for the enhanced brightening of the main spot when Io is close to the torus center is the fact that it is merged with the TEB spot at that time.

Bonfond et al. [2007] observed variations of the Io footprint on time scales of a minute and found brightness variations up to 50%. Moreover, *Hess et al.* [2009] identified fluctuations with a similar time scale when analyzing Io-related S-burst radio emissions. Indeed, their results suggest the presence of vertically drifting double layer structures with a regeneration time of 200 s. The electric field of these double layers does not seem to be the main cause for the electron precipitation [*Hess and Delamere*, this volume]. However, a possible scenario is that these variable acceleration structures could provide the electrons with some additional energy and thus trigger the fast brightness variations.

Wannawichian et al. [2010] also studied the Europa footprint brightness, but they did not find evidence for variations of the footprint brightness with the centrifugal latitude.

As far as the Ganymede footprint is concerned, *Grodent et al.* [2009] identified three different time scales for brightness variations. As in the Io case, the longest time scale (~ 10 h) is related to the location of Ganymede in the plasma sheet, the footprint being approximately twice brighter when Ganymede approaches the center. The second time scale ranges between 10 and 40 min and has been tentatively associated with interactions between the Ganymede mini-magnetosphere and localized magnetospheric inhomogeneities, such as plasma injections. The shortest time scale is on the order of 1–2 min, and two scenarii have been proposed. Either these variations are triggered by bursty reconnections at the Ganymede magnetopause [*Jia et al.*, 2010], or they are related to double layer regeneration as suggested for the Io footprint.

The Enceladus footprint is the most elusive of all footprints. It has only been detected in a few percent of the UVIS observations, which implies that its brightness only occasionally reaches the instrument detection threshold [*Pryor et al.*, 2011]. Even when it is detected, its brightness appears to vary by a factor of 3 within 5 h. This behavior was attributed to the variability of Enceladus's cryovolcanism.

Another quantity to extract from footprint images is the integrated emitted power, which is directly related to the total precipitated power. *Hess et al.* [2010] compared this quantity to the available power in the interaction region for Io, and *Hess et al.* [2011a] extended this study to Europa and Enceladus. Computing the amount of energy escaping the torus and the efficiency of the power transmission to the precipitating particles, they concluded that the large-scale Alfvén waves need to filament into smaller structures to be able to generate the observed auroral brightness.

7. CONCLUSIONS

The strength of their magnetic field, their rapid rotation, and the presence of generous internal plasma sources make the Jovian and Kronian magnetospheres very different from the Earth's magnetosphere. One of the consequences of these three elements is the occurrence of localized aurora close to the feet of the field lines passing through the satellites Io, Europa, Ganymede at Jupiter, and Enceladus at Saturn.

The direct relationship between the satellites and the auroral footprints provide unique landmarks in the planetary magnetosphere, which have been used to improve the Jovian magnetic field models and investigate the variability of the ring current.

A major recent advance in the field is the finding that the Alfvén waves generated by the strong moon-magnetosphere interactions accelerate electrons in both directions along the field lines. This mechanism simultaneously explains the multiplicity of the Io footprint spots and the detection of electron beams affecting the ionization processes near the satellite. It also accounts for the vertical extent of the different features of the Io footprint.

Whatever the satellite under consideration is, the size of its footprint spots appears to map to a region much wider than the moon. This is a clear indication that the satellite-magnetosphere interactions are not restricted to the satellites themselves, but more likely include either parts of the neutral cloud that surrounds and follows them in the case of Io, Europa, and Enceladus or its mini-magnetosphere for Ganymede. It is, however, surprising that Io appears to be the satellite with the smallest interaction region, and further analyses of the other footprints' sizes should be carried out to reduce measurement uncertainties.

Finally, studies of the Io and Ganymede footprint brightness indicate that the position of the satellites in the Jovian magnetic field does strongly control the footprint brightness. However, other variation time scales have been identified, suggesting that many other processes are simultaneously at play. The huge emitted power of these footprints, up to a few GW for the Io footprint, a few 100 MW for the Europa footprint, and a few MW for the Enceladus footprint [*Hess et al.*, 2011a], suggests filamentation of the Alfvén waves as they travel from the satellites to the planetary ionospheres.

Acknowledgments. The author thanks J.-C. Gérard and D. Grodent for their help and constructive comments during the preparation of this manuscript. This study was supported by the PRODEX program managed by ESA in collaboration with the Belgian Federal Science Policy Office.

REFERENCES

Bigg, E. K. (1964), Influence of the satellite Io on Jupiter's decametric emission, *Nature*, *203*, 1008–1010.

Bonfond, B. (2010), The 3-D extent of the Io UV footprint on Jupiter, *J. Geophys. Res.*, *115*, A09217, doi:10.1029/2010JA015475.

Bonfond, B., J.-C. Gérard, D. Grodent, and J. Saur (2007), Ultraviolet Io footprint short timescale dynamics, *Geophys. Res. Lett.*, *34*, L06201, doi:10.1029/2006GL028765.

Bonfond, B., D. Grodent, J.-C. Gérard, A. Radioti, J. Saur, and S. Jacobsen (2008), UV Io footprint leading spot: A key feature for understanding the UV Io footprint multiplicity?, *Geophys. Res. Lett.*, *35*, L05107, doi:10.1029/2007GL032418.

Bonfond, B., D. Grodent, J.-C. Gérard, A. Radioti, V. Dols, P. A. Delamere, and J. T. Clarke (2009), The Io UV footprint: Location, inter-spot distances and tail vertical extent, *J. Geophys. Res.*, *114*, A07224, doi:10.1029/2009JA014312.

Clarke, J. T., et al. (1996), Far-ultraviolet imaging of Jupiter's aurora and the Io "footprint," *Science*, *274*, 404–409.

Clarke, J. T., et al. (2002), Ultraviolet emissions from the magnetic footprints of Io, Ganymede and Europa on Jupiter, *Nature*, *415*, 997–1000.

Connerney, J. E. P., and T. Satoh (2000), The H_3^+ ion: A remote diagnostic of the Jovian magnetosphere, *Philos. Trans. R. Soc. London, Ser. A*, *358*(1774), 2471–2483.

Connerney, J. E. P., R. Baron, T. Satoh, and T. Owen (1993), Images of excited H_3^+ at the foot of the Io flux tube in Jupiter's atmosphere, *Science*, *262*, 1035–1038.

Connerney, J. E. P., M. H. Acuña, N. F. Ness, and T. Satoh (1998), New models of Jupiter's magnetic field constrained by the Io flux tube footprint, *J. Geophys. Res.*, *103*(A6), 11,929–11,939.

Delamere, P. A., F. Bagenal, R. Ergun, and Y.-J. Su (2003), Momentum transfer between the Io plasma wake and Jupiter's ionosphere, *J. Geophys. Res.*, *108*(A6), 1241, doi:10.1029/2002JA009530.

Dols, V., P. A. Delamere, and F. Bagenal (2008), A multispecies chemistry model of Io's local interaction with the Plasma Torus, *J. Geophys. Res.*, *113*, A09208, doi:10.1029/2007JA012805.

Ergun, R. E., L. Ray, P. A. Delamere, F. Bagenal, V. Dols, and Y.-J. Su (2009), Generation of parallel electric fields in the Jupiter–Io torus wake region, *J. Geophys. Res.*, *114*, A05201, doi:10.1029/2008JA013968.

Gérard, J.-C., J. Gustin, D. Grodent, P. Delamere, and J. T. Clarke (2002), Excitation of the FUV Io tail on Jupiter: Characterization of the electron precipitation, *J. Geophys. Res.*, *107*(A11), 1394, doi:10.1029/2002JA009410.

Gérard, J.-C., A. Saglam, D. Grodent, and J. T. Clarke (2006), Morphology of the ultraviolet Io footprint emission and its control by Io's location, *J. Geophys. Res.*, *111*, A04202, doi:10.1029/2005JA011327.

Gladstone, G. R., et al. (2007), Jupiter's nightside airglow and aurora, *Science*, *318*, 229–231, doi:10.1126/science.1147613.

Gledhill, J. A. (1967), Magnetosphere of Jupiter, *Nature*, *214*, 155–156, doi:10.1038/214155a0.

Goertz, C. K. (1980), Io's interaction with the plasma torus, *J. Geophys. Res.*, *85*(A6), 2949–2956.

Goldreich, P., and D. Lynden-Bell (1969), Io, a Jovian unipolar inductor, *Astrophys. J.*, *156*, 59–78.

Grodent, D., J.-C. Gérard, J. Gustin, B. H. Mauk, J. E. P. Connerney, and J. T. Clarke (2006), Europa's FUV auroral tail on Jupiter, *Geophys. Res. Lett.*, *33*, L06201, doi:10.1029/2005GL025487.

Grodent, D., B. Bonfond, J.-C. Gérard, A. Radioti, J. Gustin, J. T. Clarke, J. Nichols, and J. E. P. Connerney (2008), Auroral evidence of a localized magnetic anomaly in Jupiter's northern hemisphere, *J. Geophys. Res.*, *113*, A09201, doi:10.1029/2008JA013185.

Grodent, D., B. Bonfond, A. Radioti, J.-C. Gérard, X. Jia, J. D. Nichols, and J. T. Clarke (2009), Auroral footprint of Ganymede, *J. Geophys. Res.*, *114*, A07212, doi:10.1029/2009JA014289.

Gurnett, D. A., and C. K. Goertz (1981), Multiple Alfven wave reflections excited by Io: Origin of the Jovian decametric arcs, *J. Geophys. Res.*, *86*(A2), 717–722.

Gurnett, D. A., and W. R. Pryor (2012), Auroral processes associated with Saturn's moon Enceladus, in *Auroral Phenomenology and Magnetospheric Processes: Earth and Other Planets*, Geophys. Monogr. Ser., doi:10.1029/2011GM001174, this volume.

Hess, S., and P. A. Delamere (2012), Satellite-induced electron acceleration and related auroras, in *Auroral Phenomenology and Magnetospheric Processes: Earth and Other Planets*, Geophys. Monogr. Ser., doi:10.1029/2011GM001175, this volume.

Hess, S., P. Zarka, F. Mottez, and V. B. Ryabov (2009), Electric potential jumps in the Io-Jupiter flux tube, *Planet. Space Sci.*, *57*, 23–33, doi:10.1016/j.pss.2008.10.006.

Hess, S. L. G., P. Delamere, V. Dols, B. Bonfond, and D. Swift (2010), Power transmission and particle acceleration along the Io flux tube, *J. Geophys. Res.*, *115*, A06205, doi:10.1029/2009JA014928.

Hess, S. L. G., P. A. Delamere, V. Dols, and L. C. Ray (2011a), Comparative study of the power transferred from satellite-magnetosphere interactions to auroral emissions, *J. Geophys. Res.*, *116*, A01202, doi:10.1029/2010JA015807.

Hess, S. L. G., B. Bonfond, P. Zarka, and D. Grodent (2011b), Model of the Jovian magnetic field topology constrained by the

Io auroral emissions, *J. Geophys. Res.*, *116*, A05217, doi:10.1029/2010JA016262.

Hill, T. W., and V. M. Vasyliunas (2002), Jovian auroral signature of Io's corotational wake, *J. Geophys. Res.*, *107*(A12), 1464, doi:10.1029/2002JA009514.

Jacobsen, S., F. M. Neubauer, J. Saur, and N. Schilling (2007), Io's nonlinear MHD-wave field in the heterogeneous Jovian magnetosphere, *Geophys. Res. Lett.*, *34*, L10202, doi:10.1029/2006GL029187.

Jacobsen, S., J. Saur, F. M. Neubauer, B. Bonfond, J.-C. Gérard, and D. Grodent (2010), Location and spatial shape of electron beams in Io's wake, *J. Geophys. Res.*, *115*, A04205, doi:10.1029/2009JA014753.

Jia, X., M. G. Kivelson, K. K. Khurana, and R. J. Walker (2009), Magnetic fields of the satellites of Jupiter and Saturn, *Space Sci. Rev.*, *152*, 271–305, doi:10.1007/s11214-009-9507-8.

Jia, X., R. J. Walker, M. G. Kivelson, K. K. Khurana, and J. A. Linker (2010), Dynamics of Ganymede's magnetopause: Intermittent reconnection under steady external conditions, *J. Geophys. Res.*, *115*, A12202, doi:10.1029/2010JA015771.

Kivelson, M. G., K. K. Khurana, D. J. Stevenson, L. Bennett, S. Joy, C. T. Russell, R. J. Walker, C. Zimmer, and C. Polanskey (1999), Europa and Callisto: Induced or intrinsic fields in a periodically varying plasma environment, *J. Geophys. Res.*, *104*(A3), 4609–4625.

Kivelson, M. G., F. Bagenal, W. S. Kurth, F. M. Neubauer, C. Paranicas, and J. Saur (2004), Magnetospheric interactions with satellites, in *Jupiter: The Planet, Satellites and Magnetosphere*, edited by F. Bagenal, T. E. Dowling, and W. B. McKinnon, pp. 513–536, Cambridge Univ. Press, Cambridge, U. K.

Neubauer, F. M. (1980), Nonlinear standing Alfvén wave current system at Io: Theory, *J. Geophys. Res.*, *85*(A3), 1171–1178.

Pontius, D. H., Jr., and T. W. Hill (2006), Enceladus: A significant plasma source for Saturn's magnetosphere, *J. Geophys. Res.*, *111*, A09214, doi:10.1029/2006JA011674.

Prangé, R., D. Rego, D. Southwood, P. Zarka, S. Miller, and W. Ip (1996), Rapid energy dissipation and variability of the Io-Jupiter electrodynamic circuit, *Nature*, *379*, 323–325, doi:10.1038/379323a0.

Pryor, W. R., et al. (2011), The auroral footprint of Enceladus on Saturn, *Nature*, *472*, 331–333, doi:10.1038/nature09928.

Radioti, A., A. T. Tomás, D. Grodent, J.-C. Gérard, J. Gustin, B. Bonfond, N. Krupp, J. Woch, and J. D. Menietti (2009), Equatorward diffuse auroral emissions at Jupiter: Simultaneous HST and Galileo observations, *Geophys. Res. Lett.*, *36*, L07101, doi:10.1029/2009GL037857.

Saur, J., F. M. Neubauer, D. F. Strobel, and M. E. Summers (2002), Interpretation of Galileo's Io plasma and field observations: I0, I24, and I27 flybys and close polar passes, *J. Geophys. Res.*, *107*(A12), 1422, doi:10.1029/2001JA005067.

Saur, J., F. M. Neubauer, J. E. P. Connerney, P. Zarka, and M. G. Kivelson (2004), Plasma interaction of Io with its plasma torus, in *Jupiter: The Planet, Satellites and Magnetosphere*, edited by F. Bagenal, T. E. Dowling, and W. B. McKinnon, pp. 537–560, Cambridge Univ. Press, Cambridge, U. K.

Serio, A. W., and J. T. Clarke (2008), The variation of Io's auroral footprint brightness with the location of Io in the plasma torus, *Icarus*, *197*, 368–374, doi:10.1016/j.icarus.2008.03.026.

Vasavada, A. R., A. H. Bouchez, A. P. Ingersoll, B. Little, C. D. Anger, and Galileo SSI Team (1999), Jupiter's visible aurora and Io footprint, *J. Geophys. Res.*, *104*(E11), 27,133–27,142.

Vogt, M. F., and M. G. Kivelson (2012), Relating Jupiter's auroral features to magnetospheric sources, in *Auroral Phenomenology and Magnetospheric Processes: Earth and Other Planets*, Geophys. Monogr. Ser., doi:10.1029/2011GM001181, this volume.

Vogt, M. F., M. G. Kivelson, K. K. Khurana, R. J. Walker, B. Bonfond, D. Grodent, and A. Radioti (2011), Improved mapping of Jupiter's auroral features to magnetospheric sources, *J. Geophys. Res.*, *116*, A03220, doi:10.1029/2010JA016148.

Wannawichian, S., J. T. Clarke, and J. D. Nichols (2010), Ten years of Hubble Space Telescope observations of the variation of the Jovian satellites' auroral footprint brightness, *J. Geophys. Res.*, *115*, A02206, doi:10.1029/2009JA014456.

Wilkinson, M. H. (1998), Evidence for periodic modulation of Jupiter's decametric radio emission, *J. Geophys. Res.*, *103*(E9), 19,985–19,991.

Williams, D. J., and R. M. Thorne (2003), Energetic particles over Io's polar caps, *J. Geophys. Res.*, *108*(A11), 1397, doi:10.1029/2003JA009980.

B. Bonfond, Laboratoire de Physique Atmosphérique et Planétaire, Université de Liège, B-4000 Liège, Belgium. (b.bonfond@ulg.ac.be)

Section III
Aurora and Ionospheric Electrodynamics

Auroral Arc Electrodynamics: Review and Outlook

Octav Marghitu

Institute for Space Sciences, Bucharest, Romania

A necessary step toward unveiling the relationship between auroral phenomenology and magnetospheric processes is understanding the simplest and perhaps most widespread form of aurora, the discrete auroral arc. Current continuity and anisotropic Ohm's law in the ionosphere provide the basic tools to investigate the arc. After a general introduction into high-latitude and arc electrodynamics, three models are explored in more detail, corresponding to specific terms in the current closure equation: the 1-D thin uniform arc (in altitude and longitude, respectively), the 2-D thick uniform arc, and the 2-D thin nonuniform arc. The examination of the 1-D model is focused on the contributions made by polarization and field-aligned current to the ionospheric current closure. The relative importance of the two mechanisms depends on the complete auroral current circuit, which is briefly addressed as well. The 2-D thick uniform model enables a closer exploration of the Cowling effect, while the 2-D thin nonuniform model concentrates on the conductance gradients along the arc. The 2-D features are likely to be more prominent in the Harang region, where the formation of Cowling channels is related to substorms, while the termination of the large-scale electrojets breaks the 1-D symmetry. The various arc features are assembled together in a tentative 3-D arc model, whose evolution is qualitatively described during the substorm cycle. Quantitative progress in the definition of the 3-D arc is expected from newly developed ground-based techniques and upcoming spacecraft missions.

1. INTRODUCTION

Aurora is the most spectacular effect of the complex interaction between the collisionless, hot, and tenuous magnetospheric plasma and the collisional, cold, and dense ionospheric plasma. The typical setup of the auroral current circuit associated with the bright, discrete aurora, includes a magnetospheric generator, providing the required energy, the ionospheric-thermospheric load, where this energy is dissipated, and an auroral acceleration region (AAR), where a fraction of the electromagnetic field energy is converted into particle

energy, precipitating further in the ionosphere. In the polar regions of the Earth, all these major constituents are connected together in a coupled magnetosphere-ionosphere-thermosphere (M-I-T) system by magnetic field lines and by the field-aligned currents (FAC) flowing along the field lines.

The I-T provides the passive load where the energy is dissipated, but it also feeds back to the magnetosphere, contributing actively to the M-I-T dynamics. The active I-T role in the formation of aurora started to be recognized already 40 years ago by *Atkinson* [1970], who was the first to describe the mechanism of the feedback instability, explored later by *Sato* [1978], *Lysak* [1986], and others. In this particular (and illustrative) case, the FAC modifies the ionospheric conductance, this changes the ionospheric current, and the divergence of the ionospheric current feeds back to the FAC. As pointed out by *Mauk and Bagenal* [this volume], such basic plasma processes have a universal character, and auroral electrodynamics

Auroral Phenomenology and Magnetospheric Processes: Earth and Other Planets
Geophysical Monograph Series 197
10.1029/2011GM001189

Figure 1. Quiet arc over Alaska. Photograph credit J. Curtis.

as investigated on Earth is relevant also for other planetary systems [e.g., *Ray and Ergun*, this volume; *Stallard*, this volume].

The epitome of the aurora is the auroral arc (Figure 1), perhaps the most widespread and "simplest" auroral form. The present review is intended to check how simple the "simple" arc is from an ionospheric perspective. Although some of the features to be discussed are rather general, the arc prototype to be explored is located in the nightside auroral oval, is rather wide, perhaps a few km to a few 10 km, and is reasonably steady, on a time scale of ~10–100 s. A comprehensive view over the extended range of spatial and temporal scales covered by aurora is provided by other papers in this volume, from dynamic, thin filaments and arc systems [e.g., *Lanchester and Gustavsson*, this volume; *Kaeppler et al.*, this volume], to more slowly varying meso-scale and large-scale structures [e.g., *Lyons et al.*, this volume; *Zou et al.*, this volume].

Auroral arcs are typically described in terms of 1-D, infinite stripes of increased ionospheric conductance. An upward FAC sheet above the arc is connected to a downward FAC sheet near the arc, while a divergence-free electrojet (EJ) flows along the arc (Figure 2a). In this case, both the FAC closure and the EJ are driven by an electric field normal to the arc, as Pedersen and Hall current, respectively. The 1-D configuration, including sometimes a (fairly small) tangential electric field, is often realized in the evening and morning sectors of the auroral oval and was studied extensively in the past, based on radar, rocket, and satellite data [e.g., *de la Beaujardiere et al.*, 1977; *Evans et al.*, 1977; *Marklund*, 1984]. As demonstrated by *Marklund* [1984], the 1-D arc model is suitable for a detailed examination of the relative contributions of the polarization and FAC to the ionospheric current closure. The 1-D arc current system is reproduced to a certain extent on oval scale, where downward and upward (thick) FAC sheets [*Iijima and Potemra*, 1976] are connected by meridional Pedersen currents [*Sugiura*, 1984], while large-scale eastward and westward electrojets (EEJ and WEJ) flow along the oval in the evening and morning sector, respectively [*Baumjohann*, 1983].

In contrast to the 1-D model, real arcs can exhibit also 2-D features: the FAC can close not only normal to the arc but also along the arc, via both Pedersen and Hall currents, while the electric field can have a significant component along the arc (Figure 2b). One possibility to extend the 1-D model toward two dimensions is to relax the assumption that the EJ is divergence-free [*Marghitu et al.*, 2004, 2009, 2011]. Appropriate conditions for diverging EJs, with impact also on arc scale, are realized in the (late) evening to midnight sector, in the so-called Harang region (HR) [*Harang*, 1946; *Heppner*, 1972; *Koskinen and Puklkkinen*, 1995], where both the EEJ and WEJ terminate. In the HR, the evening and morning sectors overlap, with the electric field and current

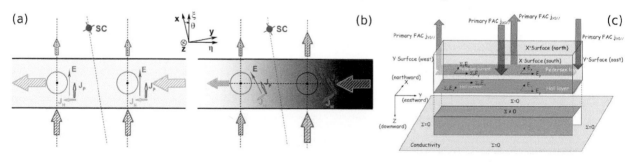

Figure 2. (a) Thin uniform, (b) thin nonuniform, and (c) thick uniform arc models. In Figures 2a and 2b, the conductance, field-aligned currents (FAC), ionospheric electric field, and ionospheric current are indicated by the gray shade, circles, solid arrows, and hatched arrows, respectively. Red and green show the Pedersen and Hall components of the current. For simplicity, the conductance variation in normal direction is not shown in Figures 2a and 2b (but is taken into account in the text). Note that in Figures 2a and 2b (ξ, η) is associated with the arc and (x, y) with the satellite overpass, while in Figure 2c, (x, y) is associated with the arc. Figures 2a and 2b are after *Marghitu et al.* [2011]. Figure 2c is after *Fujii et al.* [2011].

pattern changing from evening-like to morning-like in west-east (zonal) and south-north (meridional) direction. The termination of the large-scale EJs implies coupling with both meridional currents [*Kamide*, 1978], in particular during quiet conditions, and FAC [*Baumjohann*, 1983; *Fujii et al.*, 1994], prevailing during disturbed conditions.

The 1-D and 2-D arc models introduced so far rely on a "thin" ionosphere, whose thickness is considered negligible compared to the length of the magnetic field line in the magnetosphere. It turns out, however, that in order to properly consider the ionospheric current closure, one should also take into account the ionospheric thickness [*Amm et al.*, 2008]. One step in this direction [*Amm et al.*, 2011; *Fujii et al.*, 2011] is to assume that the Hall and Pedersen current flow in thin layers at different altitudes (Figure 2c), consistent with the respective profiles of the Hall and Pedersen conductance (Figure 3). While originally Figure 2c was meant to illustrate the electrodynamic configuration of a Cowling channel, it also appears appropriate to illustrate the "thick" arc. Note that despite the 3-D view, the model in Figure 2c is uniform along the arc (even if, in this case, the arc is finite). The 2-D arc models, illustrated by Figures 2b and 2c, appear to be complementary to each other and may help to build up an as yet undeveloped 3-D model of the arc.

After a short introduction into high-latitude ionospheric electrodynamics in section 2, we shall explore in some detail

the 1-D thin uniform, 2-D thick uniform, and 2-D thin nonuniform arc, in sections 3, 4, and 5, respectively. A qualitative discussion of a 3-D thick nonuniform arc model and its tentative evolution during the substorm cycle is provided in section 6. The paper concludes by a concise summary in section 7.

2. IONOSPHERIC ELECTRODYNAMICS IN THE AURORAL REGION

This section provides a brief summary of concepts and formulas to be used later in the paper. Comprehensive reviews of high-latitude ionospheric electrodynamics are available, for example, in the work of *Kelley* [1989] or *Paschmann et al.* [2003].

2.1. General Considerations

The current conduction in the ionosphere, contributed by both electrons and ions, is quantified by the anisotropic Ohm's law:

$$\mathbf{j} = \sigma_\| \mathbf{E}'_\| + \sigma_P \mathbf{E}'_\perp + \sigma_H \mathbf{e}_B \times \mathbf{E}'_\perp, \quad \mathbf{e}_B = \mathbf{B}/B, \quad (1)$$

where **B** is the magnetic field, $\sigma_\|$, σ_P, σ_H are, respectively, the parallel, Pedersen, and Hall conductivity, $\mathbf{j}_P = \sigma_P \mathbf{E}'_\perp$ and $\mathbf{j}_H = \sigma_H(\mathbf{e}_B \times \mathbf{E}'_\perp)$ are the Pedersen and Hall currents. **E**' is the electric field in the reference system of the neutral atmosphere, $\mathbf{E}' = \mathbf{E} + \mathbf{u} \times \mathbf{B}$, with **u** the neutral wind velocity. The symbols "∥" and "⊥" are understood with respect to the magnetic field.

By taking into account the Lorentz force and the collisions with neutral atoms in the equations of motion for electrons and ions, $\sigma_\|$, σ_P, σ_H are found to have the following expressions [*Chapman*, 1956; *Brekke and Moen*, 1993]:

$$\sigma_\| = \frac{ne}{B}\left(\frac{1}{\nu_{en}/\omega_{ge}} + \frac{1}{\nu_{in}/\omega_{gi}}\right)$$
$$\sigma_P = \frac{ne}{B}\left(\frac{\nu_{en}/\omega_{ge}}{1+(\nu_{en}/\omega_{ge})^2} + \frac{\nu_{in}/\omega_{gi}}{1+(\nu_{in}/\omega_{gi})^2}\right), \quad (2)$$
$$\sigma_H = \frac{ne}{B}\left(\frac{1}{1+(\nu_{en}/\omega_{ge})^2} - \frac{1}{1+(\nu_{in}/\omega_{gi})^2}\right),$$

where n is the plasma density, ν_{in} and ν_{en} are the ion-neutral and electron-neutral collision frequencies, ω_{gi} and ω_{ge} are the ion and electron gyrofrequencies ($\omega_{ge,gi} = eB/m_{e,i}$). Because at E layer altitudes, where the current flows, the mass difference between the main ion constituents, O_2^+ and NO^+, is small, the ionosphere can be represented by one equivalent ion species, with density $n_i = n_e = n$.

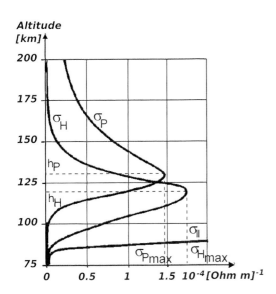

Figure 3. Altitudinal profiles of $\sigma_\|$, σ_P, and σ_H. While the numerical values are orientative, and the details of the profiles can vary, the maximum Hall conductivity is always reached below the maximum Pedersen conductivity. After *Kertz* [1971, Figure 89], with kind permission from Springer Science+Business Media B.V.

The variation of the conductivity with altitude, sketched in Figure 3, reflects both the changing plasma density, n, and the varying ratios of collision frequency to gyrofrequency, ν_{en}/ω_{ge} and ν_{in}/ω_{gi}. The combined effect of these factors leads to the concentration of the ionospheric current in the E layer, where both plasma density and collision frequencies are high enough. At lower altitudes, in the D layer, plasma density is rather low, while at higher altitudes, in the F layer, the collision frequencies decrease too much. The parallel conductance takes very high values, so that for many applications, the field-aligned potential drop in the ionosphere can be disregarded. Based on the respective contribution of the ion and electron terms to σ_P and σ_H, one notes also that Pedersen current is dominated by ion transport at higher altitudes, while the Hall current relies on electron flow at lower altitudes (as indicated by the maxima of σ_P and σ_H in Figure 3).

Since the I-T end of the M-I-T system is comparatively quite thin (a few 100 km compared to magnetic field lines of a few 10,000 km to several 100,000 km), the I-T is often regarded as a conductive thin layer. In addition, at auroral latitudes, the magnetic field is almost normal to the ionosphere, within 10°–20°, and this small difference is, in general, neglected. Since the magnetic field lines can be considered as equipotentials, Ohm's law can be integrated over the height of the ionosphere, resulting in

$$\mathbf{J}_\perp = \Sigma_P \mathbf{E}'_\perp + \Sigma_H \mathbf{e}_B \times \mathbf{E}'_\perp \qquad (3)$$

with

$$\Sigma_P = \int \sigma_P \, dz, \quad \Sigma_H = \int \sigma_H \, dz$$

the height-integrated Pedersen and Hall conductivities or conductances. In the following, we shall disregard the neutral winds, assuming that $\mathbf{E}' \equiv \mathbf{E}$, and omit the "$\perp$" symbol. Thus, by $\mathbf{E}$ and $\mathbf{J}$, we shall understand the 2-D ionospheric electric field and current. Although neutral winds in the E layer [e.g., Brekke et al., 1994; Nozawa and Brekke, 1995] are associated with typically small electric fields, their influence can become significant when the auroral electric fields are also small. Since neutral wind information is often missing in auroral studies, this feature was not explored systematically, but it may receive more attention in the future (see section 6).

The most dynamic factor in causing variations of the conductance is the plasma density. The gyrofrequencies, $\omega_{e,i}$, are essentially constant, while the collision frequencies, $\nu_{e,i}$, depend mainly on the neutral atmosphere [e.g., Kelley, 1989, Appendix B.1], which varies on longer time scales. Thus, the variation of the conductance scales with the variation in n, whose evolution is governed by the continuity equation

$$\partial n/\partial t + \nabla \cdot (n\mathbf{v}) = q - \alpha(n^2 - n_0^2), \qquad (4)$$

where $\mathbf{v}$ is the velocity of the plasma flow, α is the recombination coefficient, and n_0 is the background ionization. The source term q depends mainly on solar irradiation in the sunlit ionosphere and on the energy flux of the precipitating particles in the dark ionosphere. If the divergence of the ionization flux and the background ionization can be neglected, one obtains the steady state solution

$$n(z) = \sqrt{\frac{q(z)}{\alpha(z)}} \qquad (5)$$

valid on time scales longer than the recombination time, $\tau_{rec} = 1/\alpha n \simeq 1 - 100$ s. While equation (5) ignores the respective contributions of the electrons and ions to the evolution of ionization, a careful analysis is required for a detailed understanding of this process [Yoshikawa et al., 2011].

Proxies for the conductance induced by particle precipitation were provided, e.g., by Robinson et al. [1987] (for electrons, Σ^e) and Galand and Richmond [2001] (for protons, Σ^p):

$$\Sigma_P^e = \frac{40\bar{E}}{16 + \bar{E}^2} \Phi_E^{1/2}$$
$$\frac{\Sigma_H^e}{\Sigma_P^e} = 0.45\bar{E}^{0.85} \qquad (6)$$

$$\Sigma_P^p = 5.7\Phi_E^{1/2}$$
$$\frac{\Sigma_H^p}{\Sigma_P^p} = 0.45\bar{E}^{0.3}, \qquad (7)$$

where Φ_E is the energy flux in mW m^{-2}, and $\bar{E}$ is the average energy in keV, $\bar{E} = \Phi_E/\Phi_N$, with Φ_N the number flux. The reader is warned against possible confusion between $\bar{E}$, the particle energy, and $\mathbf{E}$, the electric field. The total conductance scales with the total energy flux and can be approximated by

$$\Sigma_{P,H} = \sqrt{\Sigma_{P,H}^{e^2} + \Sigma_{P,H}^{p^2}}. \qquad (8)$$

Besides Ohm's law and a conductance estimate, current closure equation is also required, in order to fully resolve the electrodynamics of the thin ionosphere. The secondary current driven by the electric field associated with the buildup of polarization charge follows this buildup almost instantaneously (on a time scale of the order $\varepsilon_0/\sigma_P < 10^{-4}$ s), therefore, charge conservation turns into current continuity:

$$\nabla \cdot \mathbf{j} = 0. \qquad (9)$$

By integrating equation (9) over the height of the ionosphere, and assuming that no current can flow in the neutral atmosphere below, ionospheric current closure writes

$$j_{\parallel} = \nabla \cdot \mathbf{J} = \nabla \cdot \mathbf{J}_P + \nabla \cdot \mathbf{J}_H, \qquad (10)$$

which can be further processed to

$$j_\parallel = \Sigma_P \nabla \cdot \mathbf{E} + \nabla \Sigma_P \cdot \mathbf{E} - (\nabla \Sigma_H \times \mathbf{E})_\parallel - \Sigma_H (\nabla \times \mathbf{E})_\parallel. \quad (11)$$

The operator ∇ is understood in the plane of the 2-D ionosphere. The ionospheric electric field is typically assumed to be electrostatic, and therefore, the fourth term in equation (11) is neglected. However, as demonstrated by *Yoshikawa* [2002a, 2002b], the inductive electric field is instrumental for feeding magnetic energy to the rotational part of the ionospheric current system (defined by $\nabla \cdot \mathbf{J}^{rot} = 0$ and dominated typically by the Hall current). As estimated by *Yoshikawa* [2002b], this process can take between a few seconds and a few minutes, depending on the spatial scale. For arc scales, the time is likely to be in the seconds range; therefore, we shall disregard this effect in the following.

2.2. Arc-Related Considerations

One can cast equation (11) into a form better suited for arc studies, by splitting the conductance gradients and the electric field into normal and tangential components, ξ and η (see Figure 2):

$$\begin{aligned} j_\parallel &= \Sigma_P \nabla \cdot \mathbf{E} + \partial \Sigma_P / \partial \xi E_\xi \\ &\quad + \partial \Sigma_P / \partial \eta \, E_\eta - \partial \Sigma_H / \partial \xi E_\eta + \partial \Sigma_H / \partial \eta E_\xi. \end{aligned} \quad (12)$$

By disregarding the variation in η and the divergence of the Hall current, one is left, as a first approximation, with the simplest 1-D arc model:

$$j_\parallel = dJ_{P_\xi}/d\xi = \Sigma_P dE_\xi/d\xi + d\Sigma_P/d\xi E_\xi. \quad (13)$$

Equation (13) indicates that when the conductance has significant variations, as is always the case for auroral arcs, current continuity is achieved either by FAC ($d\Sigma_P/d\xi E_\xi = j_\parallel$) or by polarization ($d\Sigma_P/d\xi E_\xi = -\Sigma_P dE_\xi/d\xi$). The relationship between polarization and FAC in providing the ionospheric current closure for the 1-D arc is investigated closer in section 3.

As shown by equation (11), when the gradient of the Hall conductance, $\nabla \Sigma_H$, is not parallel to the electric field, there is also a Hall term contributing to the current closure. The contribution related to the gradient across the arc, fourth term on the right-hand side (r.h.s.) of equation (12), underlies the Cowling effect and can result also both in FAC and in accumulation of charges that generate a polarization electric field. While the Cowling effect can be incorporated in the 1-D arc model, a proper treatment requires a thick ionosphere, taking into account the different altitudinal profiles of the Pedersen and Hall currents. A simple configuration to achieve this goal, consisting of two thin layers at the altitudes where $\mathbf{J}_P$ and $\mathbf{J}_H$ reach their respective maxima [*Amm et al.*, 2011; *Fujii et al.*, 2011], will be addressed in section 4.

Alternatively, one can add the second dimension in the plane of the thin ionosphere by considering also the variation of the conductance along the arc, third, and fifth r.h.s. terms in equation (12). Since variations along the arc are still presumed to be significantly smaller than across the arc, the contribution of $\partial E_\eta / \partial \eta$ to the first r.h.s. term of equation (12) will be neglected. As discussed in section 5, conductance variation along η can be incorporated by relaxing the assumption that the EJ is divergence-free. A technique to analyze the data and an event study will be presented to illustrate this case.

Before turning to the discussion of the arc models, it is appropriate to mention briefly the investigations of 2-D aurora not relying on arc symmetry. While the elongated arc shape makes possible fairly complete approaches based on 1-D data collected across the arc, e.g., by satellites, rockets, or radar scans, the exploration of 2-D aurora requires 2-D coverage of the observed data. For the time being, 2-D coverage can be provided only by ground observations, most often on medium and large scale (see, for example, the reviews by *Untiedt and Baumjohann* [1993] and *Vanhamäki and Amm* [2011]. Since rather recently, 2-D dynamic small-scale structures can be explored as well from the ground, by advanced optical and radar techniques [e.g., *Lanchester et al.*, this volume; *Semeter*, this volume].

3. THIN UNIFORM 1-D ARC

The 1-D arc model implies a thin ionosphere and variations of the physical parameters just across the arc. The consequences of the 1-D current closure are explored first, and then, some M-I coupling implications are briefly summarized.

3.1. Ionospheric Current Closure

With the FAC density expressed by Ampére's law, $j_\parallel = dH_\eta/d\xi$, equation (10) becomes

$$\frac{d}{d\xi}(H_\eta - J_\xi) = 0, \quad (14)$$

which integrates to

$$H_\eta(\xi) - J_\xi(\xi) = c_0 = H_\eta(\xi_0) - J_\xi(\xi_0) \equiv H_{\eta_0} - J_{\xi_0}. \quad (15)$$

The reference point ξ_0 is located well outside of the arc, where the parameters take background values, while the point ξ is arbitrary. By using Ohm's law to replace J_ξ and J_{ξ_0}, one obtains the electric field at ξ as

$$E_\xi = \frac{\Sigma_{P_0}}{\Sigma_P} E_{\xi_0} + \frac{\Sigma_H - \Sigma_{H_0}}{\Sigma_P} E_{\eta_0} + \frac{H_\eta - H_{\eta_0}}{\Sigma_P}. \quad (16)$$

$E_\eta = E_{\eta_0}$ is constant because $\partial E_\eta/\partial \xi = \partial E_\xi/\partial \eta = 0$ and $\partial E_\eta/\partial \eta = 0$.

Equation (16) indicates that the ionospheric electric field associated with the arc is determined by two factors: the ionospheric polarization induced by the variation in conductance (first and second r.h.s. terms) and the field-aligned current (third r.h.s. term). The close relationship between FAC and polarization, as means to provide current continuity in the auroral ionosphere, was addressed theoretically already by *Boström* [1964] and *Coroniti and Kennel* [1972], together with possible M-I coupling implications (see section 3.2). Later on, a systematic investigation of several arcs was performed by *Marklund* [1984]. He classified the observations according to the dominant current closure mechanism and was able to organize thus the rich variety of electric field signatures in the vicinity of auroral arcs. The numerical examination of equation (16), performed in Section 6.1 of the work of *Paschmann et al.* [2003], provides a concise summary of these signatures, reproduced here in Figures 4 and 5. Depending on the orientation of the background electric field with respect to the arc and on the magnitude of the FAC, the electric field and conductance patterns associated with the arc can be correlated, anticorrelated, or non-correlated, with various degrees of asymmetry.

For the polarization arcs (Figure 4), the electric field in equation (16) is dominated by the first (Pedersen) or second (Hall) term. In this case, sharp conductance gradients, e.g., at the edges of the arc, are associated with buildup of polarization charge because part of the intense current inside the arc cannot be transported to the low conductance ionosphere outside of the arc. The secondary electric field induced by the polarization charge serves to restore current continuity. Note that for a "thin" ionosphere, only the current carriers make the difference between the Pedersen and Hall terms: ions moving along the normal electric field and electrons moving normal to the tangential electric field, respectively.

For the Birkeland current arcs (Figure 5), the variation of the electric field is dominated by the third term in equation (16). The reader may feel confused by the fact that the FAC plays a key role in shaping the arc-associated electric field only for some of the arcs, while the increase in conductance inside the arc is always generated by FAC. However, the sheet current carried by the FAC is typically of the order of one to a few 0.1 A m^{-1}, while the ionospheric current can easily reach values of 1 A m^{-1} or larger (e.g., by an electric field of 50 mV m^{-1} and a conductance of 20 mho, which are certainly not extreme). Thus, even if the FAC drives the variation in conductance, the excess ionospheric current driving the buildup of the polarization electric field can still be significantly larger than the ionospheric current closing the FAC. The third term in equation (16) dominates only when the ambient electric field is weak, for example, near the convection reversal (CR) boundary. However, since the most structured and dynamic auroras are observed to occur near the CR, close to the polar cap boundary in the evening sector and in the HR (see section 5), in the premidnight and midnight sectors, the fraction of Birkeland current arcs is certainly significant.

FAST observations [*Elphic et al.*, 1998] showed that broad upward currents associated with auroral arcs are often connected to narrow intense return currents, flowing at the side(s) of the arc. A synthetic scheme that emerged based on these observations is shown in Figure 6. In the return current region, the plasma density is decreased (because both the electrons and the ions move away, as carriers of the downward FAC and Pedersen current, respectively) [e.g., *Karlsson and Marklund*, 1998]; therefore, the electric field overshoots the background level, to keep the current continuous [e.g., *Aikio et al.*, 1993]. This effect is reflected also by the first r.h.s. term of equation (16), if Σ_P is smaller than Σ_{P_0}. Note that the electric field overshoot visible in Figure 5 is required to carry the enhanced ionospheric current fed by the downward FAC; the intensification due to plasma depletion comes on top.

3.2. Auroral Current Circuit

While polarization and FAC provide the two basic mechanisms to ensure current continuity in the ionosphere, their relative importance in a specific event depends not only on ionospheric processes but also on M-I coupling details. The *Type 1* configuration of the auroral current circuit suggested by *Boström* [1964] (Figure 7) emphasizes the polarization, required to compensate the FAC blockage above the arc, while for *Type 2*, current continuity relies on FAC sheets closing across the arc. Although some of the features associated with the two configurations were in the meanwhile revised (like the missing field-aligned potential drop for Type 2), the points made by *Boström* [1964] proved to be essentially correct, at a time when the existence of field-aligned currents was still doubted!

An outstanding illustration of the Type 1 configuration is provided by the substorm current wedge [*McPherron et al.*, 1973], while the large-scale "region 1" and "region 2" FAC [*Iijima and Potemra*, 1976] or the smaller-scale arc current system illustrate the Type 2. The correlation between E_ξ and H_η, sometimes very high [e.g., *Sugiura*, 1984], is an effect of the Type 2 configuration and an immediate consequence of equation (16), provided that the third term prevails, and Σ_P is constant (this assumption is, of course, questionable in darkness, but some correlation may still survive on oval scale).

The fundamental question of quantifying the relationship between polarization and FAC was addressed from an M-I coupling perspective by *Lysak* [1986], who provided a unifying framework including both the time-varying ionosphere,

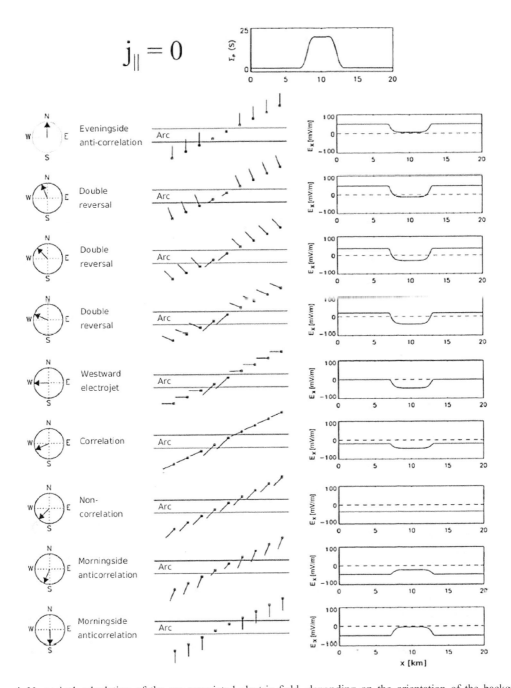

Figure 4. Numerical calculation of the arc-associated electric field, depending on the orientation of the background electric field, when $j_\parallel = 0$. (top left) The conductance profile used for the calculations. The other panels present electric field results. Each line shows the direction of the electric field, a name for the respective pattern (according to Table 1 of *Marklund* [1984]), a vector plot of the electric field across the arc, and the poleward component of the electric field, E_x, as provided by equation (16). Note that the direction normal to the arc is labeled x instead of ξ. The background electric field is equal to 50 mV m^{-1}, and its direction is varied from poleward to equatorward via westward in steps of 22.5°. After *Paschmann et al.* [2003, Figure 6.2], with kind permission from Springer Science+Business Media B.V.

the dynamic and nonuniform flux tube, and the magnetospheric generator. *Lysak* [1986] integrated former results on the ionospheric feedback and its effect upon the development of the auroral arc [*Sato*, 1978, and references therein] and upon the westward traveling surge [*Rothwell et al.*, 1984], explaining, for example, the fast motion of the aurora during substorms. *Lysak* [1986] was also able to show that for short time scales, shorter than ~1 min (equal to the Alfvén travel time from the

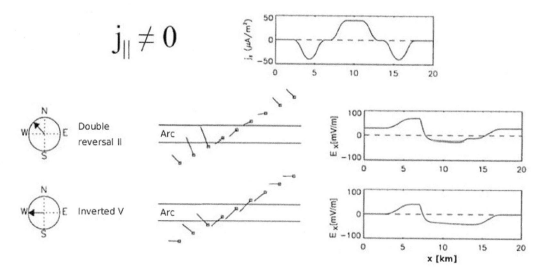

Figure 5. Numerical calculation of the arc-associated electric field for two cases of $j_\parallel \neq 0$. (top) The FAC profile. The conductance and the format of the electric field results are the same as in Figure 4. After *Paschmann et al.* [2003, Figure 6.5], with kind permission from Springer Science+Business Media B.V.

ionosphere to the magnetospheric generator and back), the feedback instability is governed by the ionosphere interaction with the nonuniform flux tube. Considering the ionospheric reflection coefficient of the Alfvén wave, $R = (\Sigma_A -$

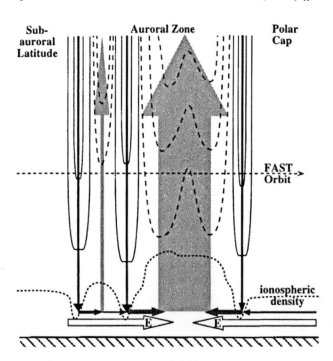

Figure 6. The low-altitude end of the auroral current circuit, including broad upward FAC and narrow downward FAC regions, ionospheric FAC closure by Pedersen currents, density depletions associated with the downward FAC, and auroral acceleration region (AAR) equipotential contours in both FAC regions. After *Elphic et al.* [1998].

$\Sigma_P)/(\Sigma_A + \Sigma_P) \simeq -1$ (because $\Sigma_A \simeq 1$ mho, the Alfvén wave admittance is, in general, much smaller than $\Sigma_P \simeq 10$ mho), the auroral arc can be regarded as almost completely polarized on such short time scales. A variation in the electric field carried by an incoming Alfvén wave is canceled by the induced polarization associated with the reflected wave. Such rapid variations, however, are beyond the scope of the present review.

On longer time scales, as soon as the equilibrium of the M-I system is achieved, the relationship between polarization and FAC depends on the specific features of the magnetospheric generator. A possible way to model the generator is by attributing to it a Pedersen-type conductance Σ_G [*Lysak*, 1985], whose variation between very small ($\Sigma_G \ll \Sigma_P$) and very large ($\Sigma_G \gg \Sigma_P$) values is equivalent to a smooth change from a purely current to a purely voltage generator. In the first case, demonstrated to fit better with smaller scales, the ionospheric electric field has to adjust, by polarization, to the variation of the ionospheric conductance, in order to match the FAC imposed from the magnetosphere. In the second case, better suited to larger scales, the electric field is fixed, and the variation of ionospheric conductance results in the modification of the ionospheric current, feeding or being fed by FAC. Of course, in specific events, neither the characteristics of the generator, nor the actual relationship between polarization and FAC, are known beforehand.

4. THICK UNIFORM 2-D ARC

When a fraction of the polarization buildup is driven by the tangential electric field, the fourth r.h.s. term in equation

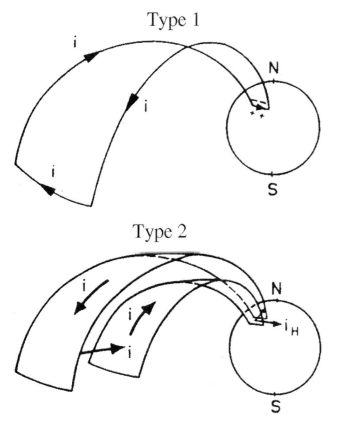

Type 1

Type 2

Figure 7. The two configurations of the auroral current circuit predicted by *Boström* [1964].

(12), the EJ along the arc (or, in general, along the high conductivity channel) can become much larger than the current driven by the primary electric field. Assuming a configuration with the primary electric field in tangential direction, E_η^p, this will drive a primary Pedersen current along the arc, $J_{\eta_P}^p = \Sigma_P E_\eta^p$, and a primary Hall current across the arc, $J_{\xi_H}^p = \Sigma_H E_\eta^p$. If there is no discharge of the polarization by FAC (that is, if the efficiency of the Cowling mechanism is $\alpha_C = 1$, see below), the secondary electric field associated with the charge buildup, E_ξ^s, drives a secondary Pedersen current that balances the primary Hall current, $J_{\xi_P}^s = \Sigma_P E_\xi^s = \Sigma_H E_\eta^p$. At the same time, E_ξ^s drives a secondary Hall current along the arc, $J_{\eta_H}^s = \Sigma_H E_\xi^s = (\Sigma_H^2/\Sigma_P)E_\eta^p$, that adds to the primary Pedersen current, such that the total current becomes

$$J_\eta = J_{\eta_P}^p + J_{\eta_H}^s = \Sigma_P(1 + (\Sigma_H/\Sigma_P)^2)E_\eta^p = \Sigma_C E_\eta \quad (17)$$

with Σ_C the Cowling conductivity. If the ratio Σ_H/Σ_P is large enough, the Cowling effect can enhance the effective conductance by an order of magnitude (e.g., $\Sigma_C/\Sigma_P = 10$ for $\Sigma_H/\Sigma_P = 3$).

While this textbook explanation of the Cowling effect assumes that the polarization discharge by FAC is completely

blocked, in reality, both the polarization and FAC contribute (as before) to the current closure. The efficiency of the Cowling mechanism is defined, in general terms, by [e.g., *Amm et al.*, 2011]

$$\nabla \cdot \mathbf{J}_P^s = -\alpha_C \nabla \cdot \mathbf{J}_H^p, \quad (18)$$

where $\alpha_C = 1$ corresponds to complete FAC blockage and $\alpha_C = 0$ implies that the primary Hall current is fully coupled to the FAC. The situations in between, $0 < \alpha_C < 1$, correspond to a partial Cowling channel.

In order to understand the details of the current closure associated with a partial Cowling channel, one has to take into account that the Pedersen and Hall currents have different altitudinal profiles (Figure 3). In the model proposed by *Fujii et al.* [2011], the two currents flow in two thin sheets separated in altitude (Figure 2c), and a fraction α_C of the diverging primary Hall current feeds an ionospheric loop closed by the secondary Pedersen current. For $\alpha_C = 1$, the divergence of the Hall current is fully closed in the ionospheric loop, while for $\alpha_C = 0$, there is no ionospheric loop, and the diverging Hall current is fully coupled to the FAC. Note that an observed $\alpha_C = 0$ may also mean that the Hall current is actually divergence-free; therefore, the Cowling mechanism is absent (see below).

The FAC coupling to the Hall current associated with a partial Cowling channel raises also the problem of the energy flow. While the Poynting flux carried by the FAC cannot be dissipated by the Hall current, the model advanced by *Fujii et al.* [2011] explains how the energy is eventually dissipated by the secondary Pedersen current. An apparent difficulty of this process is that in a complete Cowling channel, $\alpha_C = 1$, there is no FAC energy supply to feed the dissipation. However, in a time-dependent framework, energy buildup and energy consumption are not necessarily simultaneous. As a matter of fact, *Fujii et al.* [2011] assume electrostatic conditions, which implies that the energy build up process has already occurred [*Yoshikawa*, 2002a, 2002b]. The complete Cowling channel can still dissipate the energy accumulated during the buildup phase.

An outstanding example of Cowling channel is the equatorial EJ [*Chapman*, 1956]. In this case, polarization charge builds up because the vertical Hall current is prevented to flow outside of the ionospheric layer. However, in the auroral ionosphere, the configuration is different, with more elusive limitations of the Hall current flow at the "edges" of the arc or EJ. Similar to the relationship between FAC closure by Pedersen currents and polarization, discussed in the previous section, the efficiency of the Cowling mechanism depends not only on ionospheric processes but also on the magnetospheric end of the current circuit. While a comprehensive theory is not available yet, experimental evidence obtained by *Amm et al.* [2011] indicates that the Cowling efficiency is likely to be correlated with the activity level. Since the Cowling mechanism relies on the diverging component of the Hall current, this study confirms also that the Hall current is essentially divergence-free during quiet times.

Even if *Amm et al.* [2011] address mesoscale features, based on radar and ground magnetic field data, it is reasonable to think that the Cowling mechanism operates also on arc scales and shows a similar correlation with arc-related "activity." Since mesoscale activity, quantified by the magnetic disturbance, scales with the energy dumped in the ionosphere, the efficiency of the Cowling mechanism on arc scale may well be related to the energy flux of the precipitating electrons.

5. THIN NONUNIFORM 2-D ARC

In this section, the exploration of equation (12) continues with the terms that depend on variations in η because of nonuniformity along the arc. We concentrate on the variations in conductance and neglect the variation in the electric field, $\partial E_\eta / \partial \eta$, which contributes to polarization. While this contribution can be important near the ends of the arc, its effect is presumably small in rest, where the polarization associated with $\partial E_\xi / \partial \xi$ dominates. Following *Marghitu et al.* [2004, 2009, 2011], we introduce a simple 2-D arc model, whose key difference with respect to the 1-D model is the diverging EJ. The model is implemented by the **auroral arc electrodynamics** (ALADYN) technique and illustrated with an arc event in the HR. A number of features of the ionospheric current closure are then explored by a semiquantitative approach.

5.1. Model, Technique, Event Study

5.1.1. Model and technique. If the source-free EJ, $\partial J_\eta / \partial \eta = 0$, is replaced with a divergent EJ, $\partial J_\eta / \partial \eta = c_1 \neq 0$, where c_1 is assumed constant in normal direction, the current continuity equation (10) becomes

$$j_\parallel - \frac{\partial J_\xi}{\partial \xi} = \frac{\partial J_\eta}{\partial \eta} = c_1, \qquad (19)$$

which yields, by integration along ξ,

$$H_\eta - J_\xi = c_0 + c_1 \xi. \qquad (20)$$

Both c_0 and c_1 are assumed independent of η, assumption motivated in particular for satellites on polar orbits, crossing the arc (and oval) typically close to normal direction. In this case, the satellite displacement along the arc is small compared to the length scale of the EJ, and c_0, c_1 can indeed be considered as constant.

While equation (20) is very similar to the 1-D equation (15) and the formal electric field solution can be easily obtained, the term $c_1 \xi$ introduces one more degree of freedom, which makes obtaining actual values for the electric field more difficult. In addition, the tangential electric field, E_η, is not always known (which may raise difficulties also for the 1-D arc model, equation (16)).

Observed satellite data can be processed by the ALADYN technique, introduced by *Marghitu et al.* [2004] and updated by *Marghitu et al.* [2011]. As before, Ohm's law is used to replace J_ξ, and E_η is constant, $E_\eta = b_0$. In addition, E_x is expressed as a series expansion:

$$E_x = E_{0_x} + \sum_{i=1}^{n_x} a_i G_i, \qquad (21)$$

where G_i is the Legendre polynomial of order i, and E_{0_x} is the average ionospheric electric field, $E_{0_x} = \int E_x \, dx / L$ (with E_x the measured electric field and L the mapped length of the satellite path). Compared to section 3, the electric field at some initial point is replaced by the more robust average electric field, whose error is smaller (because the error of the average is smaller than the error of individual data points). The order n_x of the series expansion depends on the conductance profile, with larger values needed for higher variability of the conductance (within the constraint of the data resolution).

With simple algebra, equation (20) is cast into

$$\frac{\Sigma_P}{\cos \theta} \sum_{i=1}^{n_x} a_i G_i - (\Sigma_H - \Sigma_P \tan \theta) b_0 + c_0 + c_1 x \cos \theta =$$
$$H_y \cos \theta - H_x \sin \theta - \frac{\Sigma_P E_{0_x}}{\cos \theta}, \qquad (22)$$

where $\Sigma_P, \Sigma_H, E_{0_x}, H_x, H_y$ can be inferred from the measured data and the parameters (a_i, b_0, c_0, c_1) can be, in principle, derived by fit. Note that when b_0 and c_1 are fixed, equation (22) provides an approximation (whose accuracy depends on n_x) of the unique E_x solution and unique c_0 constant, in the same way as equation (16) provides the unique E_x solution when $E_\eta \equiv b_0$ is fixed.

In practice, the fit equation (22) is solved by assuming a divergence-free EJ, $c_1 = 0$, for a set of different b_0 values, over a sliding window moved at a certain time step. For each b_0, one obtains thus a sliding series of fitted electric field solutions, E_x^f, whose root mean square difference against the (mapped) measured electric field, E_x^m, is computed as δE_x. At the same time, one obtains also a sliding series of c_0 values, which can be used to check the divergence of the EJ, $c_1 = \Delta c_0 / \Delta x \cos \theta$, depending as well on b_0. A positive/negative variation of c_0 indicates a positive/negative divergence of the EJ. As demonstrated below, the two sets of profiles, δE_x and c_0, provide also an indication on the most likely range for b_0.

While the ALADYN technique enables the examination of 2-D arc (and oval) features, the assumptions made still require an elongated geometry and reasonable steady state conditions, at least on time scales comparable to the width of the sliding window, $\gtrsim 10$ s. The geometry and dynamics of the aurora can be best judged from conjugate optical data, but ground and in situ magnetic field data may also provide this information (however, with less accuracy). ALADYN results can also suffer from errors in conductance (in particular when the conductance is low), significant neutral winds, errors in the arc inclination θ, and arc curvature significantly larger than the curvature of the latitude circle [*Marghitu et al.*, 2011].

5.1.2. Event study: FAST orbit 1859.

Figures 8 and 9 summarize optical and FAST data, together with ALADYN results, for a relatively quiet evening event, observed during the growth phase of a small substorm [*Marghitu et al.*, 2009, 2011]. An outstanding feature of this event is the close proximity of the CR and FAC reversal boundaries, both of which are encountered near 8:22 (Figures 9d and 9e). This configuration prevents the standard Type 2 current closure and requires FAC coupling to the EJs, a qualitative result, which is substantiated by the ALADYN analysis. At the same time, an (weaker) eastward and a (stronger) westward EJ equatorward and poleward of the CR, respectively, indicate an event observed in the HR.

Figures 9f and 9g show δE_x and c_0 depending on b_0, as obtained over a sliding window of 15 s moved in steps of 1 s (similar results were obtained for windows of 10 and 20 s). In each plot, b_0 is varied from -40 to 20 mV m^{-1}, in steps of 5 mV m^{-1}, including presumably the actual value of the tangential electric field (typically small and negative). A priori,

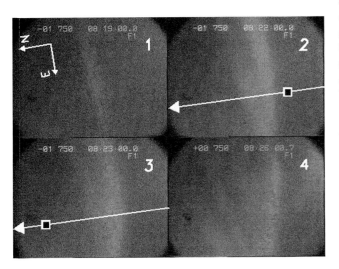

Figure 8. Arc event analyzed by ALADYN. The four selected frames show a stationary arc during the ~1 min FAST overpass, but a gradual development on a ~10 min time scale. After *Marghitu et al.* [2011].

when varying b_0 from -40 mV m^{-1} (westward) to 20 mV m^{-1} (eastward), a minimum in δE_x is expected to correspond to the actual b_0. When the conductance is large, and the results more accurate, this is indeed the case, with the minimum δE_x reached for the cyan-green lines (b_0 from -20 to 0 mV m^{-1}). However, a certain amount of variability is always associated with c_0, in the range of 0.1–0.2 A m^{-1}. This is comparable to the FAC sheet current and, therefore, significant. According to equation (15), one can conclude that $H_\eta - J_\xi$ varies along the satellite footpoint, which implies a diverging EJ.

The dependence of the overall fit quality and c_0 variability on b_0 can be examined closer in Figures 9h and 9i, which indicate that both parameters reach their minima for b_0 between -20 and -10 mV m^{-1}. Figure 9j shows the associated three profiles of c_0. The average EJ divergence over the time intervals corresponding to the downward and upward FAC, roughly 8:19–8:22 and 8:22–8:23, are equal to about 0.2 A m^{-1}/480 km $\simeq$ 0.4 μA m^{-2} and -0.2 A m^{-1}/160 km $\simeq -1.3$ μA m^{-2}. These values are in good agreement with the respective FAC densities, suggesting that, on average, the downward FAC feeds the eastward EJ, while the upward FAC is fed by the westward EJ. Figure 9j shows also that a small change in b_0 can result in a significant variation of the small-scale current structure where the conductance is high. However, the effect on the large-scale trend is rather limited.

At the peak of Σ_H, from 8:22:10 to 8:22:20, c_0 has an abrupt decrease of ~0.1 A m^{-1}. This value is comparable to the decrease of the FAC H_p in Figure 9e and corresponds to a current density of ~3.3 μA m^{-2}. At this time, $\Sigma_P \simeq 15$ mho (Figure 9c); therefore, the Pedersen tangential current is $J_{\eta P} \simeq 0.15$ A m^{-1}, assuming $E_\eta \simeq -10$ mV m^{-1}. If the EJ divergence were achieved only at the expense of the Pedersen current, its variation length scale should be equal to about 0.15 A m^{-1}/3.3 μA m^{-2}, that is, some 50 km. This length appears to be quite short, suggesting that the tangential Hall current may contribute as well to the FAC closure, a point to be explored further below.

5.2. Ionospheric Current Closure

If the polarization charge associated with $\partial E_\eta / \partial \eta$ is negligible, as assumed also by ALADYN, the divergence of the EJ is caused by longitudinal gradients in conductance. If one can assume in addition that the FAC (and the number flux, Φ_N) is uniform along η, then according to equations (6) and (7), the gradients in conductance depend solely on the variation with η of the average particle energy, $\bar{E}(\eta)$. This second assumption is often supported by the magnetic field signature, consistent with a current sheet geometry (e.g., for FAST orbit 1859 explored above). Since structured aurora is more commonly associated with electron precipitation, we concentrate

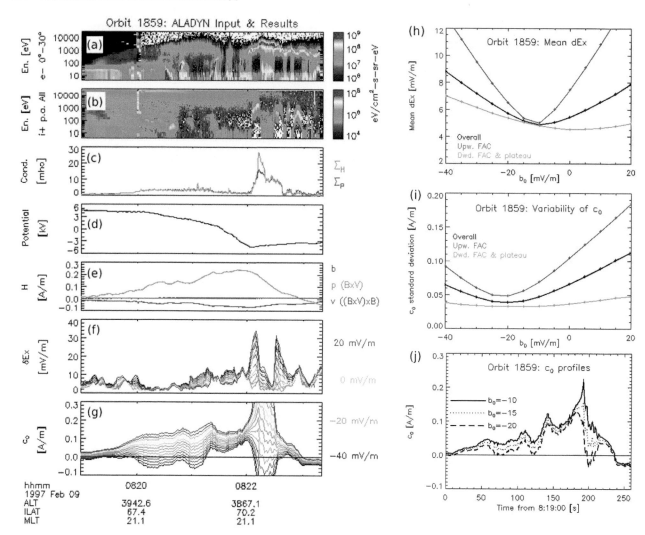

Figure 9. FAST data conjugate with the arc in Figure 8 together with ALADYN results. FAST data: (a) electron and (b) ion time-energy spectrograms, (c) Pedersen and Hall conductances, (d) ionospheric potential along the FAST footpoint, and (e) magnetic field perturbation. ALADYN results: (f, g) δE_x and c_0 for b_0 between -40 and 20 mV m^{-1}, varied in steps of 5 mV m^{-1}, (h, i) mean δE_x and the standard deviation of c_0, computed for each line in Figures 9f and 9g, respectively, over the whole FAC region (black), upward FAC (red), and downward FAC (green), and (j) the c_0 profile for $b_0 = -20$, -15, and -10 mV m^{-1}. After *Marghitu et al.* [2011, Figures 3 and 4].

here on the electron-induced conductance, equations (6), but a similar approach is possible also for proton precipitation.

With $\Phi_E = \bar{E}\Phi_N$, equations (6) write

$$\Sigma_P^e = \frac{40\bar{E}^{1.5}}{16 + \bar{E}^2}\Phi_N^{1/2}$$
$$\Sigma_H^e = \frac{18\bar{E}^{2.35}}{16 + \bar{E}^2}\Phi_N^{1/2}. \tag{23}$$

For a given orientation of the electric field, E_η/E_ξ, the Pedersen and Hall components of the EJ become (disregarding the sign)

$$|J_{\eta P}| = \Sigma_P|E_\xi|\ |E_\eta/E_\xi|$$
$$|J_{\eta H}| = \Sigma_H|E_\xi|. \tag{24}$$

After replacing Σ_P and Σ_H by equations (23) and dropping the common factors, $\Phi_N^{1/2}$ and E_ξ, the relative variation of $J_{\eta H}$ and $J_{\eta P}$ with $\bar{E}$ can be read in Figure 10a. The key information in this figure is that for electron energies above some 4–5 keV, $|dJ_{\eta H}/d\bar{E}| \gg |dJ_{\eta P}/d\bar{E}|$; therefore, the divergence of the EJ at high and moderate electron energies relies essentially on the Hall current. If the experimental evidence suggests FAC-EJ coupling, as it is the case for our FAST event, the FAC closure along the arc can only be achieved by Hall current. The

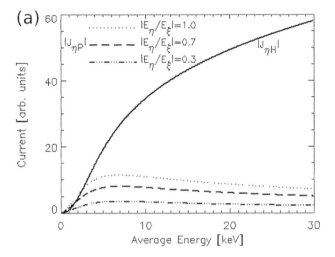

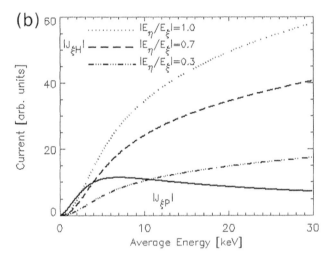

Figure 10. The relative contributions of the Pedersen and Hall components (absolute values) to the (a) longitudinal and (b) meridional currents, depending on electron energy. After *Marghitu et al.* [2011], with energy range extended to 30 kV. It is assumed that the FAC (and number flux) is homogeneous in both directions; therefore, the relative intensity of the currents $|J_{\xi P}|$, $|J_{\xi H}|$, $|J_{\eta P}|$, $|J_{\eta H}|$ depends only on the electron energy and on the ratio $|E_\eta/E_\xi|$. For $|E_\eta/E_\xi| = 1$, $|J_{\eta P}| = |J_{\xi P}|$ and $|J_{\eta H}| = |J_{\xi H}|$.

association of this feature with higher electron energies is consistent with increased auroral activity and nonsteady current flow.

To complete the exploration of the 2-D current closure near the arc, one can compute the Pedersen and Hall components of J_ξ by the same procedure as for J_η. In this case,

$$|J_{\xi P}| = \Sigma_P |E_\xi|$$
$$|J_{\xi H}| = \Sigma_H |E_\xi| \quad |E_\eta/E_\xi|, \qquad (25)$$

and the results are presented in Figure 10b. This time, the FAC is assumed to be uniform across the arc, assumption motivated when the FAC profile is smoother than the conductance profile (compare, e.g., the variations of $j_\parallel$ in Figure 9e, as indicated by the changes in the slope of H_p, with the variations of the conductance in Figure 9c). The polarization, $\partial E_\xi/\partial \xi$, is again neglected. However, this condition is much stronger now, considering the sharper gradients across the arc. Although the polarization is less likely to be indeed negligible, Figure 10b still provides useful information.

The Pedersen current is seen to vary at low energies, as before, and to be relatively flat at higher energies. This does not mean that the Pedersen current cannot close the FAC at higher electron energies, but points out the importance of the neglected polarization in this energy range, corresponding to more disturbed conditions. Conversely, for low energies, it appears that polarization is less critical for the Pedersen current closure. On the other hand, the information provided on the Hall current closure is rather limited. The variation of $J_{\xi H}$ in Figure 10b cannot be compared with the variation of $J_{\eta H}$ in Figure 10a because the variation length scales of $\bar{E}$ along ξ and η, $\lambda_\xi^{\bar{E}}$ and $\lambda_\eta^{\bar{E}}$, are not known.

However, a qualitative discussion of the Hall current divergence is still possible. If the electric field is electrostatic, a divergence-free Hall current, as expected for quiet aurora, implies that the gradient of the Hall conductance is parallel to the electric field (see equation (11)). By ignoring the FAC nonuniformity and assuming that Σ_H depends solely on the electron energy, $\bar{E}$, this condition can be written as

$$\frac{\lambda_\xi^{\bar{E}}}{\lambda_\eta^{\bar{E}}} \simeq \left| \frac{E_\eta}{E_\xi} \right|, \qquad (26)$$

where the partial derivatives were replaced by the variation length scales, $\partial/\partial(\xi, \eta) \simeq 1/\lambda_{\xi,\eta}^{\bar{E}}$.

If the tangential electric field is much smaller than the normal electric field, $E_\eta \ll E_\xi$, as expected in the evening and morning sectors, equation (26) indicates that quiet arcs are very elongated, consistent with observations. When the tangential electric field is significant, as expected in the HR, equation (26) suggests that steady state arcs should be less elongated. An alternative interpretation is that a mismatch is more likely; therefore, the Hall current is more likely to diverge and perhaps to couple with the FAC, in good agreement with the dynamic character of the HR. The auroral activity in the HR is thought to be closely related to the substorm onset [e.g., *Nielsen and Greenwald*, 1979; *Zou et al.*, 2009], even if the details of this relationship are not fully understood [e.g., *Weygand et al.*, 2008]. Small-scale perturbations in the magnetosphere that drive locally the M-I system out of the steady state can result in violations of equation (26). Such perturbations are related, e.g., to bursty bulk flows, and recent observations that the substorm onset can be triggered by streamers that reach equatorward arcs in the HR [*Nishimura et al.*, 2010; *Lyons et al.*, 2010] are consistent with the alternative interpretation of equation (26).

6. OUTLOOK: TENTATIVE MODEL OF THE 3-D ARC DURING SUBSTORM CYCLE

Based on the arc models in the previous sections, a tentative scenario for the evolution of the 3-D arc during the substorm cycle can be formulated as follows (see also *Marghitu et al.*, 2011]:

1. The quiet arc is likely to be associated with a divergence-free Hall current, little polarization, and FAC closure by Pedersen current. In the evening and morning sectors, where the electric field is mostly normal to the arc, the configuration of the auroral current circuit is of Type 2 (Figure 7), and the arc can be described by the 1-D model. In the HR, where the tangential electric field can be significant, the configuration can be mixed, Type 1/Type 2, with FAC sheets (Type 2) connected to the Pedersen component of the EJ (Type 1). In this case, the 2-D thin nonuniform model is required.

2. As soon as the substorm starts to grow, small amounts of polarization and Hall current divergence may begin to develop. At the same time, the divergence-free Hall current can also grow slowly, associated with inductive buildup of magnetic energy. Since growth phase arcs are typically located in the premidnight sector, in or near the HR, the electric field is likely to have a tangential component. Polarization and FAC can couple to both Pedersen and Hall currents. At this stage, it may already be that some arc features can only be captured by a 3-D model.

3. At onset and shortly afterward, most of the Hall current divergence may couple to the FAC, and the inductive buildup of magnetic energy is strongly enhanced. Polarization may build up as well, in parallel with strong Alfvénic activity. Only a 3-D model can fully describe the onset arc.

4. Later on, during the expansion phase, polarization plays a major role, and a complete Cowling channel may develop along the arc. The dynamics of such a process is still to be unveiled and represents another challenge for the 3-D arc model.

5. During the recovery phase, the M-I system returns to steady state, and the energy stored inductively is dissipated. The Cowling channel may survive for a while, but with less and less Hall current divergence to feed the polarization. Eventually, the arcs can again be described by the simple 1-D or 2-D models.

As mentioned already in section 2.2, advanced techniques based on ground observations have opened new possibilities

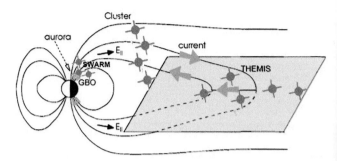

Figure 11. Sketch of the Swarm, Cluster, and THEMIS spacecraft probing, respectively, the low-altitude I-T, the topside AAR (indicated by $E_\parallel$), and the inner plasma sheet.

to study aurora and are expected to contribute also to the understanding of the 3-D arc. Significant advances in this direction are expected as well from the upcoming three-satellite Swarm mission, scheduled for launch at mid-2012. With two satellites on parallel orbits at ~490 km and a third satellite at ~530 km, Swarm will make possible a close examination of the ionospheric current closure associated with aurora. The two side-by-side Swarm spacecraft will provide a platform for systematic investigations of the gradients along the arc, while all three satellites, when flying in close formation, will enable studies of 2-D auroral electrodynamics, including effects on the M-I-T coupling. One effect less studied, so far, is the influence of the neutral winds, in particular in sensitive regions with small electric fields, like the vicinity of the CR boundary. The low-altitude data provided by Swarm will complement the observations made by Cluster near the topside AAR and by Time History of Events and Macroscale Interactions during Substorms (THEMIS) in the inner plasma sheet. Thus, multipoint data will be available from all the key regions of the auroral current circuit (Figure 11): the generator region (probed by THEMIS), the AAR (probed by Cluster), and the I-T load (probed by Swarm and ground-based observatories).

7. SUMMARY

On time scales long enough to ignore inductive effects, auroral arc electrodynamics is controlled by current closure and Ohm's law. The details of the current closure, achieved by polarization and FAC, depend both on ionospheric and magnetospheric processes. While the ionospheric processes can be described with a certain accuracy, the magnetospheric end of the auroral current circuit is still poorly understood. Arc models in one or two dimensions are able to capture some of the observed ionospheric features, but a complete 3-D description is still to be developed. New ground-based techniques and satellite missions will help in reaching this goal.

Acknowledgments. The author is pleased to acknowledge fruitful discussions at the International Space Science Institute (ISSI), Bern, in the framework of the POLARIS team project. The work at ISS Bucharest was supported by the project ECSTRA, ESA Contract 4200098048, carried on under the PECS programme, and by the M-ICAR grant of the Romanian National Authority for Scientific Research, CNCS-UE-FICDI, project number PN-II-ID-PCE-2011-3-1013.

REFERENCES

Aikio, A., H. Opgenoorth, M. Persson, and K. Kaila (1993), Ground-based measurements of an arc-associated electric field, *J. Atmos. Terr. Phys.*, *55*, 797–808.

Amm, O., A. Aruliah, S. C. Buchert, R. Fujii, J. W. Gjerloev, A. Ieda, T. Matsuo, C. Stolle, H. Vanhamäki, and A. Yoshikawa (2008), Towards understanding the electrodynamics of the 3-dimensional high-latitude ionosphere: Present and future, *Ann. Geophys.*, *26*, 3913–3932.

Amm, O., R. Fujii, K. Kauristie, A. Aikio, A. Yoshikawa, A. Ieda, and H. Vanhamäki (2011), A statistical investigation of the Cowling channel efficiency in the auroral zone, *J. Geophys. Res.*, *116*, A02304, doi:10.1029/2010JA015988.

Atkinson, G. (1970), Auroral arcs: Result of the interaction of a dynamic magnetosphere with the ionosphere, *J. Geophys. Res.*, *75*(25), 4746–4755.

Baumjohann, W. (1983), Ionospheric and field-aligned current systems in the auroral zone: A concise review, *Adv. Space Res.*, *2*, 55–62.

Boström, R. (1964), A model of the auroral electrojets, *J. Geophys. Res.*, *69*(23), 4983–4999.

Brekke, A., and J. Moen (1993), Observations of high latitude ionospheric conductances, *J. Atmos. Terr. Phys.*, *55*, 1493–1512.

Brekke, A., S. Nozawa, and T. Sparr (1994), Studies of the *E* region neutral wind in the quiet auroral ionosphere, *J. Geophys. Res.*, *99*(A5), 8801–8825.

Chapman, S. (1956), The electrical conductivity of the ionosphere: A review, *Nuovo Cimento*, *4*, suppl. 4, 1385–1412, doi:10.1007/BF02746310.

Coroniti, F., and C. Kennel (1972), Polarization of the auroral electrojet, *J. Geophys. Res.*, *77*(16), 2835–2850.

de la Beaujardiere, O., R. Vondrak, and M. Baron (1977), Radar observations of electric fields and currents associated with auroral arcs, *J. Geophys. Res.*, *82*(32), 5051–5062.

Elphic, R. C., et al. (1998), The auroral current circuit and field-aligned currents observed by FAST, *Geophys. Res. Lett.*, *25*(12), 2033–2036.

Evans, D. S., N. C. Maynard, J. Trøim, T. Jacobsen, and A. Egeland (1977), Auroral vector electric field and particle comparisons, 2, Electrodynamics of an arc, *J. Geophys. Res.*, *82*(16), 2235–2249.

Fujii, R., R. A. Hoffman, P. C. Anderson, J. D. Craven, M. Sugiura, L. A. Frank, and N. C. Maynard (1994), Electrodynamic parameters in the nighttime sector during auroral substorms, *J. Geophys. Res.*, *99*(A4), 6093–6112.

Fujii, R., O. Amm, A. Yoshikawa, A. Ieda, and H. Vanhamäki (2011), Reformulation and energy flow of the Cowling channel, *J. Geophys. Res.*, *116*, A02305, doi:10.1029/2010JA015989.

Galand, M., and A. D. Richmond (2001), Ionospheric electrical conductances produced by auroral proton precipitation, *J. Geophys. Res.*, *106*, 117–125.

Harang, L. (1946), The mean field of disturbance of polar geomagnetic storms, *Terr. Magn. Atmos. Electr.*, *51*, 353–380.

Heppner, J. (1972), The Harang discontinuity in auroral belt ionospheric currents, *Geofys. Publ.*, *29*, 105–120.

Iijima, T., and T. A. Potemra (1976), The amplitude distribution of field-aligned currents at northern high latitudes observed by Triad, *J. Geophys. Res.*, *81*(13), 2165–2174.

Kaeppler, S. R., et al. (2012), Current closure in the auroral ionosphere: Results from the Auroral Current and Electrodynamics Structure rocket mission, in *Auroral Phenomenology and Magnetospheric Processes: Earth and Other Planets, Geophys. Monogr. Ser.*, doi:10.1029/2011GM001177, this volume.

Kamide, Y. (1978), On current continuity at the Harang discontinuity, *Planet. Space Sci.*, *26*, 237–244.

Karlsson, T., and G. Marklund (1998), Simulations of effects of small-scale auroral current closure in the return current region, *Phys. Space Plasmas*, *15*, 401–406.

Kelley, M. (1989), *The Earth's Ionosphere, Int. Geophys. Ser.*, vol. 43, Academic Press, San Diego, Calif.

Kertz, W. (1971), *Einführung in die Geophysik. II. Obere Atmosphäre und Magnetosphäre, Hochschultaschenbücher*, vol. 535, Bibliographisches Institut, Mannheim, Germany.

Koskinen, H. E. J., and T. I. Pulkkinen (1995), Midnight velocity shear zone and the concept of Harang discontinuity, *J. Geophys. Res.*, *100*(A6), 9539–9547.

Lanchester, B., and B. Gustavsson (2012), Imaging of aurora to estimate the energy and flux of electron precipitation, in *Auroral Phenomenology and Magnetospheric Processes: Earth and Other Planets, Geophys. Monogr. Ser.*, doi:10.1029/2011GM001161, this volume.

Lyons, L. R., Y. Nishimura, Y. Shi, S. Zou, H.-J. Kim, V. Angelopoulos, C. Heinselman, M. J. Nicolls, and K.-H. Fornacon (2010), Substorm triggering by new plasma intrusion: Incoherent-scatter radar observations, *J. Geophys. Res.*, *115*, A07223, doi:10.1029/2009JA015168.

Lyons, L. R., Y. Nishimura, X. Xing, Y. Shi, M. Gkioulidou, C.-P. Wang, H.-J. Kim, S. Zou, V. Angelopoulos, and E. Donovan (2012), Auroral disturbances as a manifestation of interplay between large-scale and mesoscale structure of magnetosphere-ionosphere electrodynamical coupling, in *Auroral Phenomenology and Magnetospheric Processes: Earth and Other Planets, Geophys. Monogr. Ser.*, doi:10.1029/2011GM001152, this volume.

Lysak, R. L. (1985), Auroral electrodynamics with current and voltage generators, *J. Geophys. Res.*, *90*(A5), 4178–4190.

Lysak, R. L. (1986), Coupling of the dynamic ionosphere to auroral flux tubes, *J. Geophys. Res.*, *91*(A6), 7047–7056.

Marghitu, O., B. Klecker, G. Haerendel, and J. McFadden (2004), ALADYN: A method to investigate auroral arc electrodynamics from satellite data, *J. Geophys. Res.*, *109*, A11305, doi:10.1029/2004JA010474.

Marghitu, O., T. Karlsson, B. Klecker, G. Haerendel, and J. McFadden (2009), Auroral arc and oval electrodynamics in the Harang region, *J. Geophys. Res.*, *114*, A03214, doi:10.1029/2008JA013630.

Marghitu, O., C. Bunescu, T. Karlsson, B. Klecker, and H. C. Stenbaek-Nielsen (2011), On the divergence of the auroral electrojets, *J. Geophys. Res.*, *116*, A00K17, doi:10.1029/2011JA016789.

Marklund, G. (1984), Auroral arc classification scheme based on the observed arc-associated electric field pattern, *Planet. Space Sci.*, *32*, 193–211.

Mauk, B., and F. Bagenal (2012), Comparative auroral physics: Earth and other planets, in *Auroral Phenomenology and Magnetospheric Processes: Earth and Other Planets*, Geophys. Monogr. Ser., doi:10.1029/2011GM001192, this volume.

McPherron, R. L., C. T. Russell, and M. P. Aubry (1973), Satellite studies of magnetospheric substorms on August 15, 1968: 9. Phenomenological model for substorms, *J. Geophys. Res.*, *78*(16), 3131–3149.

Nielsen, E., and R. A. Greenwald (1979), Electron flow and visual aurora at the Harang discontinuity, *J. Geophys. Res.*, *84*(A8), 4189–4200.

Nishimura, Y., L. Lyons, S. Zou, V. Angelopoulos, and S. Mende (2010), Substorm triggering by new plasma intrusion: THEMIS all-sky imager observations, *J. Geophys. Res.*, *115*, A07222, doi:10.1029/2009JA015166.

Nozawa, S., and A. Brekke (1995), Studies of the *E* region neutral wind in the disturbed auroral ionosphere, *J. Geophys. Res.*, *100*(A8), 14,717–14,734.

Paschmann, G., S. Haaland, and R. Treumann (Eds.) (2003), *Auroral Plasma Physics*, Space Sci. Ser. of ISSI, vol. 15, Kluwer, Dordrecht, The Netherlands.

Ray, L. C., and R. E. Ergun (2012), Auroral signatures of ionosphere-magnetosphere coupling at Jupiter and Saturn, in *Auroral Phenomenology and Magnetospheric Processes: Earth and Other Planets*, Geophys. Monogr. Ser., doi:10.1029/2011GM001172, this volume.

Robinson, R. M., R. R. Vondrak, K. Miller, T. Dabbs, and D. Hardy (1987), On calculating ionospheric conductances from the flux and energy of precipitating electrons, *J. Geophys. Res.*, *92*(A3), 2565–2569.

Rothwell, P. L., M. B. Silevitch, and L. P. Block (1984), A model for the propagation of the westward traveling surge, *J. Geophys. Res.*, *89*(A10), 8941–8948.

Sato, T. (1978), A theory of quiet auroral arcs, *J. Geophys. Res.*, *83*(A3), 1042–1048.

Semeter, J. (2012), Coherence in auroral fine structure, in *Auroral Phenomenology and Magnetospheric Processes: Earth and Other Planets*, Geophys. Monogr. Ser., doi:10.1029/2011GM001196, this volume.

Stallard, T., S. Miller, and H. Melin (2012), Clues on ionospheric electrodynamics from IR aurora at Jupiter and Saturn, in *Auroral Phenomenology and Magnetospheric Processes: Earth and Other Planets*, Geophys. Monogr. Ser., doi:10.1029/2011GM001168, this volume.

Sugiura, M. (1984), A fundamental magnetosphere-ionosphere coupling mode involving field-aligned currents as deduced from DE-2 observations, *Geophys. Res. Lett.*, *11*(9), 877–880.

Untiedt, J., and W. Baumjohann (1993), Studies of polar current systems using the IMS Scandinavian magnetometer array, *Space Sci. Rev.*, *63*, 245–390.

Vanhamäki, H., and O. Amm (2011), Analysis of ionospheric electrodynamic parameters on mesoscale—A review of selected techniques using data from ground-based observation networks and satellites, *Ann. Geophys.*, *29*, 467–491.

Weygand, J. M., R. L. McPherron, H. U. Frey, O. Amm, K. Kauristie, A. Viljanen, and A. Koistinen (2008), Relation of substorm onset to Harang discontinuity, *J. Geophys. Res.*, *113*, A04213, doi:10.1029/2007JA012537.

Yoshikawa, A. (2002a), How does the ionospheric rotational Hall current absorb the increasing energy from the field-aligned current system?, *Geophys. Res. Lett.*, *29*(7), 1133, doi:10.1029/2001GL014125.

Yoshikawa, A. (2002b), Excitation of a Hall-current generator by field-aligned current closure, via an ionospheric, divergent Hall-current, during the transient phase of magnetosphere–ionosphere coupling, *J. Geophys. Res.*, *107*(A12), 1445, doi:10.1029/2001JA009170.

Yoshikawa, A., A. Nakamizo, O. Amm, H. Vanhamäki, R. Fujii, Y.-M. Tanaka, T. Uozumi, K. Yumoto, and S. Ohtani (2011), Self-consistent formulation for the evolution of ionospheric conductances at the ionospheric *E* region within the M-I coupling scheme, *J. Geophys. Res.*, *116*, A09223, doi:10.1029/2011JA016449.

Zou, S., L. R. Lyons, C.-P. Wang, A. Boudouridis, J. M. Ruohoniemi, P. C. Anderson, P. L. Dyson, and J. C. Devlin (2009), On the coupling between the Harang reversal evolution and substorm dynamics: A synthesis of SuperDARN, DMSP, and IMAGE observations, *J. Geophys. Res.*, *114*, A01205, doi:10.1029/2008JA013449.

Zou, S., L. R. Lyons, and Y. Nishimura (2012), Mutual evolution of aurora and ionospheric electrodynamic features near the Harang reversal during substorms, in *Auroral Phenomenology and Magnetospheric Processes: Earth and Other Planets*, Geophys. Monogr. Ser., doi:10.1029/2011GM001163, this volume.

O. Marghitu, Institute for Space Sciences, P.O. Box MG-23, RO-77125 Bucharest-Măgurele, Romania (marghitu@gpsm.spacescience.ro)

Mutual Evolution of Aurora and Ionospheric Electrodynamic Features Near the Harang Reversal During Substorms

Shasha Zou

Department of Atmospheric, Oceanic and Space Sciences, University of Michigan, Ann Arbor, Michigan, USA

Larry R. Lyons and Yukitoshi Nishimura

Department of Atmospheric and Oceanic Sciences, University of California, Los Angeles, California, USA

Substorms are one of the fundamental elements of geomagnetic activity, involving complex magnetosphere-ionosphere coupling processes and auroral activity. We briefly review recent progress on the relationship between the aurora and ionospheric electrodynamic features in the vicinity of the Harang reversal during substorms with a focus on observations from ground-based radars. A combination of ground-based radars and optical measurements allows us to study the mutual evolution of the aurora and ionospheric electrodynamics, such as convection flows and current systems. Observations from the Super Dual Auroral Radar Network (SuperDARN) and Poker Flat Incoherent Scatter Radar (PFISR) show that nightside convection flows exhibit repeatable and distinct variations at different locations relative to the substorm-related auroral activity near the Harang reversal region. The Harang reversal represents a key feature of magnetospheric and ionospheric convection. Taking advantage of the simultaneous flow and electron density measurements from PFISR, we found that the substorm field-aligned current (FAC) systems are at least partially closed by the region 2 FACs via meridionally directed Pedersen currents. By synthesizing these observations, a 2-D comprehensive view of the nightside ionospheric electrodynamical features, including electrical equipotentials, flows and FACs, and their evolution associated with substorms has been constructed. We also present recent progress on the ionospheric flows associated with possible auroral precursors. The equatorward component of the flows just poleward of the onset arc and within the auroral zone is found to increase about 4–5 min before the onset. This time scale is consistent with statistical study on the basis of auroral observations.

1. INTRODUCTION

Theories about magnetic reconnection [*Dungey*, 1961] and viscous interaction [*Axford and Hines*, 1961] together provide a basis for understanding the generation of the large-scale convection in both the magnetosphere and the ionosphere. For both northward and southward interplanetary magnetic field (IMF), convection has a more coherent pattern on the dayside than on the nightside. For example, unlike the

Auroral Phenomenology and Magnetospheric Processes: Earth and Other Planets

Geophysical Monograph Series 197

10.1029/2011GM001163

dayside convection, the nightside convection pattern is not a mirror image for opposite IMF B_y directions [*de la Beaujardière et al.*, 1985, 1986]. In particular, there is a frequently observed and distinct feature of the nightside ionospheric convection pattern called the Harang reversal.

Harang [1946] discovered a boundary separating positive and negative ΔH perturbations on the nightside using ground-based magnetometers. This boundary represents the transition region where the influence of the eastward electrojet is taken over by that of the westward electrojet in the same meridian plane. *Davis* [1962] found a related boundary existing in auroral motions on the nightside. Later, this boundary was named "Harang discontinuity" by *Heppner* [1972] in recognition of *Harang*'s finding. Since the auroral electrojets are mainly Hall currents, the electric fields are expected to be poleward in the eastward electrojet region and equatorward in the westward electrojet region, assuming uniform conductance. Therefore, across the Harang discontinuity latitudinally, electric fields and thus $\vec{E} \times \vec{B}$ convection flows should reverse their directions. In terms of convection flows and electric fields, this is not a strict physical discontinuity, but rather a clockwise rotation of these vectors from higher to lower latitudes, as shown by direct electric field measurements from a rocket experiment [*Wescott et al.*, 1969] and by convection flow measurements from the Scandinavian Twin Auroral Radar Experiment (STARE) coherent scatter radars [*Nielsen and Greenwald*, 1979]. Therefore, in terms of electric field and convection flows, the term "Harang reversal" is more appropriate than the "Harang discontinuity."

The Harang reversal has been studied by using various instruments, such as ground magnetometers [e.g., *Baumjohann et al.*, 1981; *Amm*, 1998; *Weygand et al.*, 2008], low-altitude polar-orbiting satellites [e.g., *Maynard*, 1974; *Heppner and Maynard*, 1987; *Gjerloev et al.*, 2003], ground-based coherent scatter radars [e.g., *Nielsen and Greenwald*, 1979; *Baumjohann et al.*, 1981; *Kunkel et al.*, 1986; *Koskinen and Pulkkinen*, 1995; *Bristow et al.*, 2001, 2003; *Bristow and Jenson*, 2007; *Zou et al.*, 2009a; *Grocott et al.*, 2010], and incoherent scatter radars (ISR) [e.g., *Banks et al.*, 1973; *Kamide*, 1978; *Kamide and Vickrey*, 1983; *Robinson et al.*, 1985; *Zou et al.*, 2009b]. These observations have revealed that the reversal can be very dynamic in nature. It can cover a few hours of magnetic local time (MLT) in the dusk to premidnight sector, sometimes even intruding into the postmidnight sector, and its latitudinal location is associated with the strength of geomagnetic activity, moving equatorward as the level of activity increases [e.g., *Maynard*, 1974].

Although the Harang reversal is a feature in the high-latitude ionospheric convection, its generation has been attributed to the dawn-dusk pressure asymmetry in the near-Earth plasma sheet according to theoretical and simulation studies [e.g., *Erickson et al.*, 1991; *Gkioulidou et al.*, 2009, 2011].

Erickson et al. [1991] used the Rice Convection Model (RCM) with simplified MLT-dependent boundary conditions and a *Hilmer* magnetic field model to study the formation of the Harang reversal. As convection increases, energetic particles are transported from the distant tail into the near-Earth plasma sheet and energized. Energetic ions in the plasma sheet are the dominant contributor to plasma pressure, and they drift westward due to gradient and curvature drifts. Additionally, the magnetotail has a finite width, and the dawnside low-latitude boundary layer is not as good a supplier of energetic ions to the plasma sheet as is the distant tail [*Spence and Kivelson*, 1990, 1993]. As a result, a dawn-dusk plasma pressure gradient with higher pressure at dusk is formed. This azimuthal pressure gradient plus the gradient of flux tube volume, which is roughly radially outward, generates an upward field-aligned current (FAC) [*Vasyliunas*, 1970]. This upward FAC requires converging Pederson current and thus converging electric fields in the ionosphere. These converging electric fields add to the convection electric field and give the north-south-directed electric fields that form the Harang reversal. The argument for forming the dawn-dusk pressure asymmetry is essentially that used for explaining the generation of the region 2 FAC system, although *Erickson et al.* [1991] emphasized the dawnside depletion effect.

Gkioulidou et al. [2009] conducted RCM simulations with a nonforce-balanced Tsyganenko 96 (T96) magnetic field model and realistic plasma sheet particle boundary conditions on the basis of 11 years Geotail observations [*Wang et al.*, 2007] to investigate the formation of the Harang reversal and its relationship with the region 2 FAC system. They found that an overlap in MLT of the region 2 upward and downward FACs is necessary for its formation. More recently, *Gkioulidou et al.* [2011] replaced the T96 magnetic field model with a modified Dungey force-balanced magnetic field solver and compared the results with those given by *Gkioulidou et al.* [2009]. They found that with self-consistent magnetic field, the azimuthal pressure gradient and the region 2 FACs are weaker, and characteristics of the Harang reversal, including its location and latitudinal range, are more consistent with observations.

2. RELATIONSHIP BETWEEN THE HARANG REVERSAL AND SUBSTORMS

Substorms are one of the fundamental geomagnetic disturbances, involving global-scale reconfigurations of the magnetosphere and auroral activity in the ionosphere. Despite the fact that they have been studied for over four decades, the substorm onset process remains a controversial topic.

2.1. Location of Substorm Onset Relative to the Center of the Harang Reversal

The substorm onset controversy is due in part to the limitation of observational capability and mapping uncertainty between the magnetosphere and the ionosphere. Given the limited number of satellites in the magnetosphere, it is difficult to determine exactly where and when the substorm onset initiates and to obtain a 2-D distribution of important physical quantities. Remote sensing of the ionospheric manifestations of substorm onset, such as the breakup arc, offers an opportunity to locate the onset relative to 2-D large-scale auroral morphology and to unique particle precipitation and convection features. This enables morphological mapping of the onset from the ionosphere to the magnetosphere. The Harang reversal is one of the unique convection structures that can be used for morphological mapping of the substorm onset location.

Observations that associate the Harang reversal with the substorm onset auroral activity can be traced back to the 1970s. Benefited by the development of the STARE radars, *Nielsen and Greenwald* [1979] were able to study the relationship between 2-D plasma flow in the *E* region and auroral arcs near the Harang reversal for the first time. They found that the substorm breakup arc can either be an existing arc in the center of the Harang reversal or a new arc suddenly appearing equatorward of it. *Baumjohann et al.* [1981] later used a combination of STARE and the Scandinavian Magnetometer Array to study the current system associated with a multiple onset substorm. They noticed that the breakup arcs occurred slightly equatorward of the Harang reversal. *Gjerloev et al.* [2003] showed that an optical substorm onset occurred in the region where overlap of the two electrojets was observed by the FAST spacecraft in the premidnight sector during the proceeding substorm growth phase and concluded that, based on this event, substorm onset should occur in the Harang reversal region. *Bristow et al.* [2003] conducted a case study of an isolated substorm using a comprehensive set of instruments and found that, during the growth phase, the breakup arcs were equatorward of the Harang reversal inferred from the ground magnetometers and from the radar. *Weygand et al.* [2008] used equivalent ionospheric currents inferred from ground-based magnetometer observations to identify the location of the Harang reversal and compared it with that of the optical onset list based on the IMAGE observations developed by *Frey et al.* [2004]. They found that approximately two thirds of substorm onsets do not occur within or near the Harang reversal

identified in the growth phase. Combining auroral images obtained by the IMAGE Wideband Imaging Camera (WIC) and convection flows derived from the Super Dual Auroral Radar Network (SuperDARN) radars [*Greenwald et al.*, 1995; *Chisham et al.*, 2007], *Zou et al.* [2009a] found that the substorm onsets are located near the center of the Harang reversal. Later, *Zou et al.* [2009b, 2010] supported the results using Poker Flat Incoherent Scatter Radar (PFISR) observations. The superposed epoch analysis of more than a thousand substorm onsets from the work of *Grocott et al.* [2010] showed that statistically, the substorm onset tends to occur near the eastern edge and either in or slightly equatorward of the center of the Harang reversal, where equatorward flows divert to azimuthal flows. On the other hand, *Bristow* [2009] claim that the onset occurs ~4° equatorward of the convection reversal boundary.

Based on the above observations, substorm onset seems to occur near the Harang reversal region, but a consensus has not been reached on the exact location of substorm onset relative to the center of the Harang reversal. There is a need to compare the substorms that occur within the center of the Harang reversal with those that do not. In addition to the possible natural variability of their relative locations, there are also other possibilities that may be responsible for the discrepancy that one should bear in mind when conducting the comparison. For example, the inference of the location of the Harang reversal can be different depending on the type of measurements used. Using simultaneous measurements of the electric fields from the Chatanika ISR and magnetic perturbations from the Alaska meridian chain, *Kamide and Vickrey* [1983] found that the electric Harang reversal measured by the Chatanika ISR was displaced 1°–2° poleward of the magnetic Harang reversal determined using ground magnetometer measurements. A case study conducted by *Kunkel et al.* [1986] further confirmed this finding. This displacement may be a result of the imbalance between the strength of the eastward and westward electrojets [*Kamide and Vickrey*, 1983] and/or may be because of the influence of strong FACs that flow upward out of the electric Harang reversal [*Koskinen and Pulkkinen*, 1995]. Because the magnetic Harang reversal is affected by conductivity, electric field, as well as the shape of the reversal, *Kamide and Vickrey* [1983] suggested that the electric Harang reversal offers a better way to effectively describe the electrodynamics near this region and to study the relevant magnetosphere-ionosphere coupling process. Additionally, the breakup arc can be as thin as tens of kilometers, and this imposes a rigid requirement for the spatial resolution of the instrument used to determine the flow and electric field in the surrounding region. A novel technique to derive the convection flows at 30 km spatial resolution over a 100 km × 100 km field was developed by

Semeter et al. [2010] using 26 beams of PFISR. However, this method to study the electrodynamics in the vicinity of the breakup arc is limited by temporal resolution of ~2 min, which is not enough for resolving the dynamics around substorm onset time. In addition, the field of view (FOV) of PFISR is quite limited, and thus, it is difficult to capture a substorm with onset located right within the FOV.

2.2. Evolution of Ionospheric Electrodynamic Features Near the Harang Reversal During Substorms

Since the Harang reversal represents a critical region in the magnetospheric and ionospheric convections, monitoring the evolution of ionospheric electrodynamics near the Harang reversal during substorms can shed light on the dynamics of the corresponding region in the magnetosphere, in particular, the preconditioning of that region that facilitates the onset of substorms.

Weimer [1999] statistically demonstrated that the nightside flow reversal is more prominent during substorms than that during nonsubstorm periods using satellite electric field data. *Bristow et al.* [2001] and *Bristow and Jensen* [2007] used the SuperDARN radars in the Northern Hemisphere to study the nightside convection pattern during substorms. They found that the meridional flows rotate to become more zonally aligned during the growth phase and enhance a couple of tens of minutes before the onset, while the flows decrease and return back to being more meridional at onset. These features led the authors to conclude that there must be a relation between the flow enhancements and the substorm onset.

As pointed out by *Gjerloev and Hoffman* [2001], the global auroral images can be treated as a natural coordinate system, which can allow us to organize events with respect to key auroral features. *Zou et al.* [2009a, 2009b] investigated the evolution of the convection flows and FAC closure at three meridian planes, i.e., west of, at, and east of the optical onset determined by IMAGE and Time History of Events and Macroscale Interactions during Substorms (THEMIS) [*Angelopoulos*, 2008] all-sky imager (ASI) array [*Mende et al.*, 2008]. In Figure 1, observations from SuperDARN, PFISR, and ground magnetometers at these three local times are gathered to show an overall picture of the typical electrodynamic features near the Harang reversal during substorms. An auroral image taken by IMAGE is shown to represent a generic substorm onset in the nightside auroral zone (Figure 1a). The three chosen meridians are marked by purple, yellow, and orange lines. Figures 1b–1d show time series of merged convection flow vectors derived from the SuperDARN line-of-sight velocity measurements at these meridians. Figures 1e–1g display PFISR and

ground magnetometer observations at the three meridians. These observations are from different substorm events, but each of them shows typical features for its location. In each plot, the magenta vertical line indicates the substorm onset time.

Figures 1b–1d show that during the substorm growth phase, the Harang reversal gradually forms in all three locations, indicating wide local time coverage. Evolution of convection flows after onset depends strongly on their location relative to the onset, suggesting strong coupling to the local auroral activity. In Figure 1b, there is evidence of a clockwise vortex traveling through the selected MLT. A 2-D snapshot of the convection flows confirms this and suggests its association with the leading edge of the westward-traveling surge [*Zou et al.*, 2009a]. In Figure 1c, the westward convection flows equatorward of the center of the Harang reversal (i.e., sub-auroral polarization stream (SAPS)) [*Foster and Burke*, 2002; *Foster and Vo*, 2002] increased significantly after onset. This increase is associated with enhanced poleward electric field and Pedersen currents, which are required to support the suddenly increased upward FACs in the onset region. Poleward of the onset and within the bulge, flows became more meridionally aligned. In Figure 1d, after onset, southeastward flows became nearly azimuthally aligned, while the flows equatorward of the shear center decreased.

The SuperDARN radars, data from which are shown in Figures 1b–1d, have a wide spatial coverage, but these coherent scatter radars do not provide information about ionospheric parameters other than flows, such as electron density. Sudden increase or decrease of electron density in the ionosphere can contribute to the identification of the polarity of FACs [e.g., *Karlsson and Marklund*, 1998; *Marklund et al.*, 2001]. *Zou et al.* [2009b, 2010] used PFISR, THEMIS ASI array, and ground magnetometers to comprehensively analyze the ionospheric electrodynamics in those regions. In Figures 1e–1g, from top to bottom, flow magnitude, flow vectors, and electron density from two PFISR beams are shown together with ground magnetograms. Detailed explanation of the format can be found in the figure caption. It is revealing to compare the observations obtained to the west (Figure 1e) and east of the onset (Figure 1g). Because of the limited FOV of PFISR, either pure westward flows equatorward of the center of the Harang reversal or pure eastward flows poleward of it are observed. In Figure 1e, the large equatorward gradient of the flow magnitude and the poleward gradient of the E region electron density together indicate the existence of a pair of FACs with upward FAC at higher latitude and downward FAC at lower latitude, and they are closed through enhanced poleward Pederson currents. Exactly opposite flows and electron density gradients can be seen in Figure 1g. Similarly, this suggests another pair of FACs with downward FAC at higher latitude and upward FAC at lower latitude, and they are closed through equatorward Pederson currents. The decrease in the electron density and the large eastward flows, which developed soon after onset and are associated with downward FAC, are terminated by the intrusion of the eastward-moving auroral bulge [*Zou et al.*, 2009b].

The observations obtained at the onset local time fall between these two situations. In Figure 1f, convection flow changed its direction from mainly westward to eastward as the radar rotated with the Earth. The Harang reversal moved into the radar FOV from higher latitude a few minutes before the onset. The onset (green circle) later initiated near the center of the flow shear [*Zou et al.*, 2010].

Figure 2 shows a schematic plot of electrodynamical features and their evolution associated with the expansion phase of substorms. Features are based on the observations of *Zou et al.* [2009a, 2009b] and the RCM simulations under enhanced convection shown in the work of *Gkioulidou et al.* [2009], and are referenced to the auroral onset. From top to bottom, each diagram represents ionospheric equipotentials, convection flows, and FACs before, at, and after onset in geomagnetic coordinates, respectively. The *x* axis represents the east-west span of the Harang reversal extending approximately a few hours of MLT, and the *y* axis covers roughly 10° in latitude. Note that Figure 2 is a schematic plot, and the latitudinal and MLT coverage for individual events can be highly variable. In each diagram, the solid arrows (both magenta and blue) denote the convection flows, the dashed black arrows denote the Pedersen currents, and circles with dots (crosses) represent upward (downward) FAC.

In Figure 2a, a well-defined Harang reversal can be seen, and the dashed line highlights the center of the shear. Yellow (blue) shading roughly represents the region of proton (electron) precipitation. They overlap each other except for a narrow wedge-shaped region at lower latitude. Based on the work of *Gkioulidou et al.* [2009], upward region 2 FACs are approximately distributed at and poleward of the Harang shear, while downward region 2 FACs are roughly equatorward of it. The boundary separating the upward and downward FACs is slightly equatorward of the center of the reversal.

In Figure 2b, the breakup arc, represented by the orange-shaded region, is placed near the center of the Harang reversal. SAPS flows (large magenta arrow) are observed equatorward of the onset region. An enhanced eastward flow channel, denoted by the larger blue arrow, forms northeast of the breakup arc at onset. The flow-ionization relation shown in Figure 1 suggests that FACs of the substorm current wedge formed at onset are at least partially closed by the

region 2 FACs via meridionally directed Pedersen currents (black arrows with dashed line). As shown in Figure 2c, during the expansion phase, the breakup arc expands to form an auroral bulge. Shear flows become more meridionally aligned in regions where the auroral expansion reaches, and ionization is enhanced. The substorm and region 2 FAC closure is also observed farther away from the onset location. The pink- and cyan-shaded regions, respectively, indicate the regions where positive and negative magnetic H perturbations are usually observed. In Figures 1e–1g, the locations of ground magnetometers used are also labeled in geomagnetic coordinates, and they are at approximately the same meridian as PFISR. Therefore, the location of either positive or negative magnetic perturbations can be located easily by comparing their latitude with convection flows shown in the second panel in each plot.

2.3. Preliminary Results From THEMIS ASI and SuperDARN

Recently, *Nishimura et al.* [2010] analyzed the THEMIS ASI data for hundreds of substorm events and found that substorm onsets are frequently preceded by equatorward extension of a north-south auroral structure (auroral streamer), which develops from poleward boundary intensifications (PBIs) at the polar cap boundary, toward the onset latitude. Because of the linkage of fast magnetotail flows to PBIs and streamers [e.g., *Lyons et al.*, 1999; *Zesta et al.*, 2006], the authors suggest that substorm onset may be preceded by enhanced earthward plasma flows associated with enhanced reconnection in the magnetotail. Supportive evidence observed by the ISRs and THEMIS satellite are reported in companion papers by *Lyons et al.* [2010a, 2010b] and *Xing et al.* [2010]. Observations from the Northern Hemisphere SuperDARN and THEMIS ASI array are used to study the mutual evolution of the convection flows and aurora for those substorm events provided by *Nishimura et al.* [2010]. In the following paragraphs, we present two examples that were selected mainly because of the good availability of the SuperDARN echoes.

Figures 3a–3d show four selected auroral images of the THEMIS ASI observations of a substorm onset at ~08:02 UT on 3 February 2008, with line-of-sight (LOS) convection flow velocity superimposed as dots. Blue colors represent flow toward the radar and red/yellow colors represent flow away from the radar. Figures 3i and 3j show time series of the LOS velocity from beams 0 and 5, respectively, of the Saskatoon (Sas) radar with the magenta line indicating onset time. Locations of these two beams are shown in Figure 3a as two semitransparent lines. In Figures 3a–3c, a hook-like aurora, named "Harang aurora" by *Nishimura et al.* [2010],

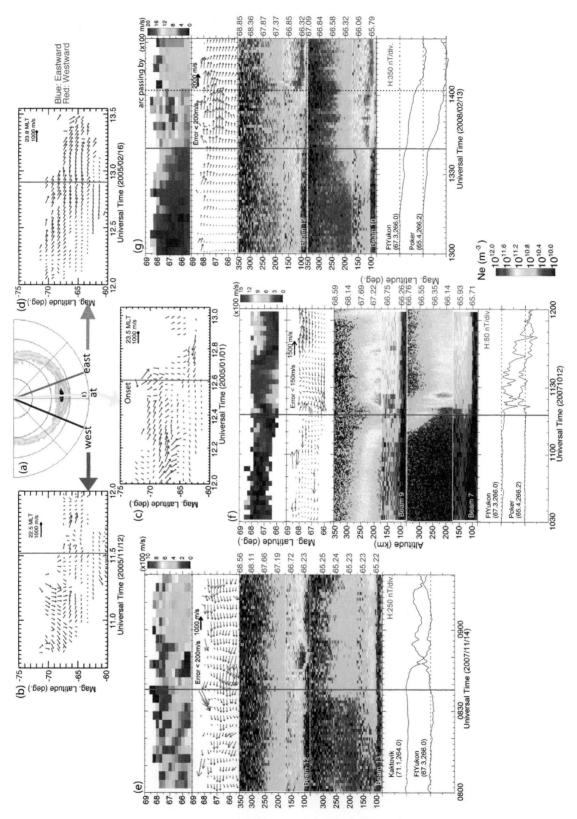

Figure 1

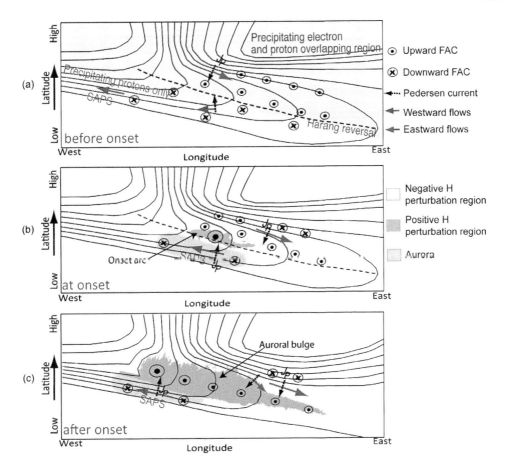

Figure 2. Schematic plots of equipotentials, convection flows, and currents near the Harang reversal region (a) before, (b) at, and (c) after substorm onset (based on the work of *Zou et al.* [2009b]). In Figure 2a, color-shaded areas represent regions characterized with different precipitating particle species. In Figures 2b and 2c, pink-/cyan-shaded areas represent regions characterized with positive/negative magnetic *H* perturbation, and orange represents aurora.

because it resembles the shape of the Harang flow reversal, can be clearly seen in the northwest part of the image mosaic, while diffuse aurora is apparent in the lower-latitude regions. As shown in Figures 3b and 3c, the LOS flows observed east of the Harang aurora changed from mainly toward the radar at higher latitudes to away from it at lower latitudes, implying a flow reversal from largely equatorward to azimuthal (illustrated by two green arrows). Comparing the location of the Harang aurora in Figures 3a–3c, one can see that it

moved slightly equatorward. This equatorward motion is probably associated with the enhanced equatorward flows, which initiated ~5 min before onset (Figures 3i–3j). As shown in Figure 3c, the onset arc initiated at ~08:02 UT near the center of the flow shear (~67.3° magnetic latitude) and just poleward of the preexisting diffuse aurora band, and it further expanded after onset (Figure 3d).

Figures 3e–3h show THEMIS ASI and SuperDARN observations in the same format as Figures 3a–3d but for a

Figure 1. (opposite) SuperDARN and PFISR observations west of, at, and east of substorm onset for six different events. (a) Center image is selected to represent a generic substorm auroral onset. (b–d) Time series of SuperDARN merged convection vectors at a fixed magnetic local time [from *Zou et al.*, 2009a]. (e–g) (top to bottom) PFISR observations of convection flow magnitude, flow vector, electron density from two radar beams, and magnetic *H* component perturbations from two ground magnetometers (based on the works of *Zou et al.* [2009b, 2010]). In the electron density plots, altitude (magnetic latitude) is indicated on the left (right) *y* axis. These two electron density plots approximately show electron density distribution from higher to lower latitude. In all plots, red (blue) arrows represent vectors with eastward (westward) component, and the magenta vertical line indicates substorm onset time.

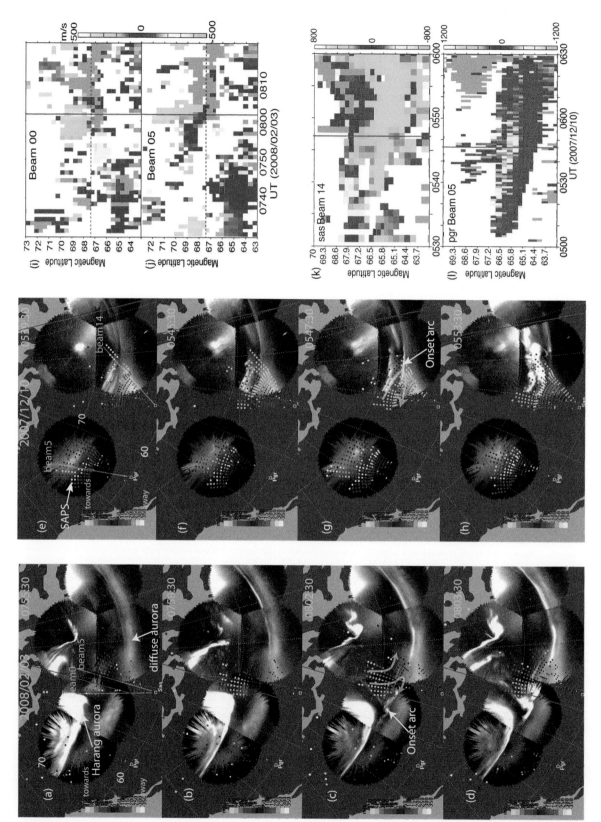

Figure 3

substorm onset at ~05:46 UT on 10 December 2007. Figures 3k and 3l show time series of the LOS velocity from beam 14 of the Sas radar and beam 5 of the Prince George radar, respectively, with the magenta lines indicating the onset time. In Figures 3f–3h, the Harang reversal can be seen roughly at about the same latitude as a thin growth phase arc, which later broke up and initiated substorm onset. Figure 3k shows that the center of the shear was located at ~67° magnetic latitude, and strongly enhanced equatorward flows toward the onset arc can be seen ~4 min before the onset, similar to the previous case. This enhancement is not observed by other beams farther to the west, which may suggest the narrowness of the corresponding flow in the magnetotail. In Figure 3l, a clear narrow channel of SAPS formed before the onset and lasted to the expansion phase.

These high-resolution THEMIS and radar observations are consistent with the previous studies [*Zou et al.*, 2009a, 2009b] and reveal important new features, such as the ~4–5 min equatorward flows within the auroral zone toward the onset arc. This time scale is consistent with the statistical study of auroral images given by *Nishimura et al.* [2010]. Based on the scenario suggested by the authors, this equatorward flow might be the ionospheric counterpart of a low-entropy plasma channel in the equatorial plane. Some detailed descriptions of such low-entropy plasma channels are given by *Wolf et al.* [2009, and references therein], and more about the proposed scenario is given by *Lyons et al.* [this volume]. In order to fully explore the possibility, further study is needed, and simultaneous electron density or precipitating particle measurements probably from a fortunate LEO satellite pass or incoherent scatter radars are required.

3. DISCUSSIONS AND SUMMARY

The combination of ground-based radars and auroral imagers has been used to study the ionospheric electrodynamics associated with auroral arcs for many years. The availability of satellite-based imaging and continental-scale ground-based ASI array enables us to put local measurements into global context and to understand the evolution of these electrodynamic features at different stages of geomagnetic disturbances.

In this chapter, we have reviewed recent progress on the relationship between the aurora and ionospheric electrodynamic features near the Harang reversal during substorms with a focus on observations from ground-based radars and summarized the observations in a schematic sketch. This sketch is not meant to encompass all features for all cases. One aspect of the ionospheric electrodynamic features that was treated casually in the present study and requires much further study is the direct measurement of FAC distribution in the vicinity of the Harang reversal during the late growth phase of the substorm and the location of onset arc relative to the FACs. This is also the most difficult part mainly because of instrumental limitation, i.e., one often only gets infrequent measurements of FACs from LEO satellites or rockets. 2-D FAC data from the Active Magnetosphere and Planetary Electrodynamics Response Experiment (AMPERE) project [e.g., *Waters et al.*, 2001] may be useful in this aspect, but whether the instrument sensitivity is high enough for studying substorm growth phase needs further test.

We have also qualitatively discussed the current closure relationship between the region 2 FACs and the FACs of the substorm current wedge. Quantitative analysis regarding the extent to which the region 2 FACs contribute to the closure of the substorm FAC would be interesting, which will require utilization of comprehensive model calculations. This current closure relationship observed in the ionosphere suggests that in the magnetotail, the region 2 FACs should also partially close the substorm FACs through equatorial currents with the radial component. Such a current circuit has also been suggested by *Ohtani et al.* [1990] on the basis of spacecraft observations.

Acknowledgments. The research at the University of Michigan was supported by NSF AGS-1111476. The research at UCLA was supported by NSF ATM-0646233 and AGS-1042255, NASA grant NNX09AI06G. We thank S. Mende, H. Frey, and E. Donovan for the use of IMAGE WIC data and the THEMIS ASI and the CSA for logistical support in fielding and data retrieval from the GBO stations. We thank J. M. Ruohoniemi, N. Nishitani, and G. Sofko for access and support of the SuperDARN data. We thank M. Nicolls and C. Heinselman for access and support of the PFISR data. We also thank the Geophysical Institute of the University of Alaska for the use of the ground magnetometer data.

Figure 3. (opposite) (a–d) Selected auroral images taken by the THEMIS ASI array for a substorm onset at ~08:02 UT on 3 February 2008. SuperDARN observations from the Saskatoon (Sas) and Prince George (Pgr) radars are superimposed on top of the auroral image. Blue colors represent flow toward the radar, and red/yellow colors represent flow away from the radar. The flow velocities in each plot are taken ±30 s of the auroral image shown. (e–h) Same as Figures 3a–3d but for a substorm onset at ~05:46 UT on 10 December 2007. (i–j) Time series of line-of-sight (LOS) velocity measured by beams 0 and 5 of the Sas radar on 3 February 2008. The two beams are highlighted by semitransparent yellow lines in Figure 3a. (k–l) Time series of LOS velocity measured by beam 14 of the Sas radar and beam 5 of the Pgr radar on 10 December 2007. The two beams are highlighted by semitransparent yellow lines in Figure 3e.

REFERENCES

Amm, O. (1998), Method of characteristics in spherical geometry applied to a Harang discontinuity situation, *Ann. Geophys.*, *16*, 413–424, doi:10.1007/s00585-998-0413-2.

Angelopoulos, V. (2008), The THEMIS mission, *Space Sci. Rev.*, *141*(1–4), 5–34, doi:10.1007/s11214-008-9336-1.

Axford, W. I., and C. O. Hines (1961), A unifying theory of high-latitude geophysical phenomena and geomagnetic storms, *Can. J. Phys.*, *39*, 1433–1464.

Banks, P. M., J. R. Doupnik, and S.-I. Akasofu (1973), Electric field observations by incoherent scatter radar in the auroral zone, *J. Geophys. Res.*, *78*(28), 6607–6622.

Baumjohann, W., R. J. Pellinen, H. J. Opgenoorth, and E. Nielsen (1981), Joint two-dimensional observations of ground magnetic and ionospheric electric fields associated with auroral zone currents: Current systems associated with local auroral breakup, *Planet. Space Sci.*, *29*, 431–447.

Bristow, W. A. (2009), Relationship between substorm onset locations and nightside convection pattern features, *J. Geophys. Res.*, *114*, A12202, doi:10.1029/2009JA014576.

Bristow, W. A., and P. Jensen (2007), A superposed epoch study of SuperDARN convection observations during substorms, *J. Geophys. Res.*, *112*, A06232, doi:10.1029/2006JA012049.

Bristow, W. A., A. Otto, and D. Lummerzheim (2001), Substorm convection patterns observed by the Super Dual Auroral Radar Network, *J. Geophys. Res.*, *106*, 24,593–24,609.

Bristow, W. A., G. J. Sofko, H. C. Stenbaek-Nielsen, S. Wei, D. Lummerzheim, and A. Otto (2003), Detailed analysis of substorm observations using SuperDARN, UVI, ground-based magnetometers, and all-sky imagers, *J. Geophys. Res.*, *108*(A3), 1124, doi:10.1029/2002JA009242.

Chisham, G., et al. (2007), A decade of the Super Dual Auroral Radar Network (SuperDARN): Scientific achievements, new techniques and future directions, *Surv. Geophys.*, *28*, 33–109, doi:10.1007/s10712-007-9017-8.

Davis, T. N. (1962), The morphology of the auroral displays of 1957–1958. 1. Statistical analyses of Alaska data, *J. Geophys. Res.*, *67*(1), 59–74.

de la Beaujardière, O., V. B. Wickwar, J. D. Kelly, and J. H. King (1985), Effect of the interplanetary magnetic field Y component on the high-latitude nightside convection, *Geophys. Res. Lett.*, *12*(7), 461–464.

de la Beaujardière, O., V. B. Wickwar, and J. H. King (1986), Sondrestrom radar observations of the effects of the IMF By component on polar cap convection, in *Solar Wind-Magnetosphere Coupling*, edited by Y. Kamide and J. A. Slavin, pp. 495–505, Terra Sci., Tokyo, Japan.

Dungey, J. W. (1961), Interplanetary magnetic fields and the auroral zones, *Phys. Rev. Lett.*, *6*, 47–48.

Erickson, G. M., R. W. Spiro, and R. A. Wolf (1991), The physics of the Harang discontinuity, *J. Geophys. Res.*, *96*(A2), 1633–1645.

Foster, J. C., and W. J. Burke (2002), SAPS: A new categorization for sub-auroral electric fields, *Eos Trans. AGU*, *83*(36), 393.

Foster, J. C., and H. B. Vo (2002), Average characteristics and activity dependence of the subauroral polarization stream, *J. Geophys. Res.*, *107*(A12), 1475, doi:10.1029/2002JA009409.

Frey, H. U., S. B. Mende, V. Angelopoulos, and E. F. Donovan (2004), Substorm onset observations by IMAGE-FUV, *J. Geophys. Res.*, *109*, A10304, doi:10.1029/2004JA010607.

Gjerloev, J. W., and R. A. Hoffman (2001), The convection electric field in auroral substorms, *J. Geophys. Res.*, *106*, 12,919–12,931, doi:10.1029/1999JA000240.

Gjerloev, J. W., R. A. Hoffman, E. Tanskanen, M. Friel, L. A. Frank, and J. B. Sigwarth (2003), Auroral electrojet configuration during substorm growth phase, *Geophys. Res. Lett.*, *30*(18), 1927, doi:10.1029/2003GL017851.

Gkioulidou, M., C.-P. Wang, L. R. Lyons, and R. A. Wolf (2009), Formation of the Harang reversal and its dependence on plasma sheet conditions: Rice convection model simulations, *J. Geophys. Res.*, *114*, A07204, doi:10.1029/2008JA013955.

Gkioulidou, M., C.-P. Wang, and L. R. Lyons (2011), Effect of self-consistent magnetic field on plasma sheet penetration to the inner magnetosphere: Rice convection model simulations combined with modified Dungey force-balanced magnetic field solver, *J. Geophys. Res.*, *116*, A12213, doi:10.1029/2011JA016810.

Greenwald, R. A., et al. (1995), DARN/SuperDARN: A global view of the dynamics of high-latitude convection, *Space Sci. Rev.*, *71*, 761–796, doi:10.1007/BF00751350.

Grocott, A., S. E. Milan, T. K. Yeoman, N. Sato, A. S. Yukimatu, and J. A. Wild (2010), Superposed epoch analysis of the ionospheric convection evolution during substorms: IMF B_Y dependence, *J. Geophys. Res.*, *115*, A00I06, doi:10.1029/2010JA015728.

Harang, L. (1946), The mean field of disturbance of polar geomagnetic storms, *Terr. Magn. Atmos. Electr.*, *51*(3), 353–380.

Heppner, J. P. (1972), The Harang discontinuity in auroral belt ionospheric currents, *Geophys. Publ.*, *29*, 105–120.

Heppner, J., and N. Maynard (1987), Empirical high-latitude electric field models, *J. Geophys. Res.*, *92*(A5), 4467–4489.

Kamide, Y. (1978), On current continuity at the Harang discontinuity, *Planet. Space Sci.*, *26*(3), 237–244, doi:10.1016/0032-0633(78)90089-2.

Kamide, Y., and J. F. Vickrey (1983), Variability of the Harang discontinuity as observed by the Chatanika radar and the IMS Alaska magnetometer chain, *Geophys. Res. Lett.*, *10*(2), 159–162.

Karlsson, T., and G. Marklund (1998), Simulations of effects of small-scale auroral current closure in the return current region, *Phys. Space Plasmas*, *15*, 401–406.

Koskinen, H. E. J., and T. I. Pulkkinen (1995), Midnight velocity shear zone and the concept of Harang discontinuity, *J. Geophys. Res.*, *100*(A6), 9539–9547.

Kunkel, T., W. Baumjohann, J. Untiedt, and R. A. Greenwaldt (1986), Electric fields and currents at the Harang discontinuity: A case study, *J. Geophys.*, *59*, 73–86.

Lyons, L. R., T. Nagai, G. T. Blanchard, J. C. Samson, T. Yamamoto, T. Mukai, A. Nishida, and S. Kokubun (1999), Association between Geotail plasma flows and auroral poleward boundary intensifications observed by CANOPUS photometers, *J. Geophys. Res.*, *104*(A3), 4485–4500.

Lyons, L. R., Y. Nishimura, Y. Shi, S. Zou, H.-J. Kim, V. Angelopoulos, C. Heinselman, M. J. Nicolls, and K.-H. Fornacon (2010a), Substorm triggering by new plasma intrusion: Incoherent-scatter radar observations, *J. Geophys. Res.*, *115*, A07223, doi:10.1029/2009JA015168.

Lyons, L. R., Y. Nishimura, X. Xing, V. Angelopoulos, S. Zou, D. Larson, J. McFadden, A. Runov, S. Mende, and K.-H. Fornacon (2010b), Enhanced transport across entire length of plasma sheet boundary field lines leading to substorm onset, *J. Geophys. Res.*, *115*, A00I07, doi:10.1029/2010JA015831.

Lyons, L. R., Y. Nishimura, X. Xing, Y. Shi, M. Gkioulidou, C.-P. Wang, H.-J. Kim, S. Zou, V. Angelopoulos, and E. Donovan (2012), Auroral disturbances as a manifestation of interplay between large-scale and mesoscale structure of magnetosphere-ionosphere electrodynamical coupling, in *Auroral Phenomenology and Magnetospheric Processes: Earth and Other Planets, Geophys. Monogr. Ser.*, doi:10.1029/2011GM001152, this volume.

Marklund, G., et al. (2001), Temporal evolution of the electric field accelerating electrons away from the auroral ionosphere, *Nature*, *414*, 724–727.

Maynard, N. C. (1974), Electric field measurements across the Harang discontinuity, *J. Geophys. Res.*, *79*(31), 4620–4631.

Mende, S. B., S. E. Harris, H. U. Frey, V. Angelopoulos, C. T. Russell, E. Donovan, B. Jackel, M. Greffen, and L. M. Peticolas (2008), The THEMIS array of ground-based observatories for the study of auroral substorms, *Space Sci. Rev.*, *141*, 357–387, doi:10.1007/s11214-008-9380-x.

Nielsen, E., and R. A. Greenwald (1979), Electron flow and visual aurora at the Harang discontinuity, *J. Geophys. Res.*, *84*(A8), 4189–4200.

Nishimura, Y., L. Lyons, S. Zou, V. Angelopoulos, and S. Mende (2010), Substorm triggering by new plasma intrusion: THEMIS all-sky imager observations, *J. Geophys. Res.*, *115*, A07222, doi:10.1029/2009JA015166.

Ohtani, S., S. Kokubun, R. Nakamura, R. C. Elphic, C. T. Russell, and D. N. Baker (1990), Field-aligned current signatures in the near-tail region, 2. Coupling between the region 1 and the region 2 systems, *J. Geophys. Res.*, *95*(A11), 18,913–18,927.

Robinson, R. M., F. Rich, and R. R. Vondrak (1985), Chatanika radar and S3-2 measurements of auroral zone electrodynamics in the midnight sector, *J. Geophys. Res.*, *90*(A9), 8487–8499.

Semeter, J., T. W. Butler, M. Zettergren, C. J. Heinselman, and M. J. Nicolls (2010), Composite imaging of auroral forms and convective flows during a substorm cycle, *J. Geophys. Res.*, *115*, A08308, doi:10.1029/2009JA014931.

Spence, H. E., and M. G. Kivelson (1990), The variation of the plasma sheet polytropic index along the midnight meridian in a finite width magnetotail, *Geophys. Res. Lett.*, *17*(5), 591–594.

Spence, H. E., and M. G. Kivelson (1993), Contributions of the low-latitude boundary layer to the finite width magnetotail convection model, *J. Geophys. Res.*, *98*(A9), 15,487–15,496.

Vasyliunas, V. M. (1970), Mathematical models of magnetospheric convections and its coupling to the ionosphere, in *Particles and Fields in the Magnetosphere*, edited by B. M. McCormac, pp. 60–71, D. Reidel, Hingham, Mass.

Wang, C.-P., L. R. Lyons, T. Nagai, J. M. Weygand, and R. W. McEntire (2007), Sources, transport, and distributions of plasma sheet ions and electrons and dependences on interplanetary parameters under northward interplanetary magnetic field, *J. Geophys. Res.*, *112*, A10224, doi:10.1029/2007JA012522.

Waters, C. L., B. J. Anderson, and K. Liou (2001), Estimation of global field aligned currents using the iridium® System magnetometer data, *Geophys. Res. Lett.*, *28*(11), 2165–2168, doi:10.1029/2000GL012725.

Weimer, D. R. (1999), Substorm influence on the ionospheric electric potentials and currents, *J. Geophys. Res.*, *104*(A1), 185–197.

Wescott, E. M., J. D. Stolarik, and J. P. Heppner (1969), Electric fields in the vicinity of auroral forms from motions of barium vapor releases, *J. Geophys. Res.*, *74*(14), 3469–3487.

Weygand, J. M., R. L. McPherron, H. U. Frey, O. Amm, K. Kauristie, A. Viljanen, and A. Koistinen (2008), Relation of substorm onset to Harang discontinuity, *J. Geophys. Res.*, *113*, A04213, doi:10.1029/2007JA012537.

Wolf, R. A., Y. Wan, X. Xing, J.-C. Zhang, and S. Sazykin (2009), Entropy and plasma sheet transport, *J. Geophys. Res.*, *114*, A00D05, doi:10.1029/2009JA014044.

Xing, X., L. Lyons, Y. Nishimura, V. Angelopoulos, D. Larson, C. Carlson, J. Bonnell, and U. Auster (2010), Substorm onset by new plasma intrusion: THEMIS spacecraft observations, *J. Geophys. Res.*, *115*, A10246, doi:10.1029/2010JA015528.

Zesta, E., L. Lyons, C.-P. Wang, E. Donovan, H. Frey, and T. Nagai (2006), Auroral poleward boundary intensifications (PBIs): Their two-dimensional structure and associated dynamics in the plasma sheet, *J. Geophys. Res.*, *111*, A05201, doi:10.1029/2004JA010640.

Zou, S., L. R. Lyons, C.-P. Wang, A. Boudouridis, J. M. Ruohoniemi, P. C. Anderson, P. L. Dyson, and J. C. Devlin (2009a), On the coupling between the Harang reversal evolution and substorm dynamics: A synthesis of SuperDARN, DMSP, and IMAGE observations, *J. Geophys. Res.*, *114*, A01205, doi:10.1029/2008JA013449.

Zou, S., L. R. Lyons, M. J. Nicolls, C. J. Heinselman, and S. B. Mende (2009b), Nightside ionospheric electrodynamics associated with substorms: PFISR and THEMIS ASI observations, *J. Geophys. Res.*, *114*, A12301, doi:10.1029/2009JA014259.

Zou, S., et al. (2010), Identification of substorm onset location and preonset sequence using Reimei, THEMIS GBO, PFISR, and Geotail, *J. Geophys. Res.*, *115*, A12309, doi:10.1029/2010JA015520.

L. R. Lyons and Y. Nishimura, Department of Atmospheric and Oceanic Sciences, University of California, Los Angeles, CA 90095-1565, USA.

S. Zou, Department of Atmospheric, Oceanic and Space Sciences, University of Michigan, Ann Arbor, MI 48109, USA. (shashaz@umich.edu)

Imaging of Aurora to Estimate the Energy and Flux of Electron Precipitation

Betty Lanchester and Björn Gustavsson

Department of Physics and Astronomy, University of Southampton, Southampton, UK

The aurora is fascinatingly rich in structure (down to scale sizes of tens of meters across the magnetic field), with fast motions and change of color. Methods developed over the past several decades have used spectral information, with modeling of the auroral upper atmosphere, to study the changes in energy input of the precipitating electrons and the dependence on composition, in particular atomic oxygen variations. Important insight about the physics of auroral phenomena is obtained with these methods not only in the 1-D region of the field-aligned beam but in a larger 3-D region around the magnetic zenith. This chapter reviews and describes the most important details of (forward) modeling of the aurora and demonstrates the power and robustness of the latest methods. These methods combine emission rate ratios with modeling to estimate the energy of the primary precipitating particle distributions. Important results using such methods with measurements from the multispectral imager Auroral Structure and Kinetics are described as examples of the methods outlined.

1. INTRODUCTION

In his book *The Polar Aurora*, the great Norwegian physicist and mathematician Carl Størmer states: "For future researches it is important to stress the necessity of *simultaneous photographs of the auroral spectrum and of photographic height measurements*, by which it can be determined from which level of the atmosphere the spectrum is originating. This has only been done occasionally" [*Størmer*, 1955, p. 162]. The difficulty of obtaining such measurements of the dynamically changing colors of the aurora remains the reason that measurement and analysis techniques are still being refined today. Essential to the task of determining the energy distribution of the aurora are measurements of selected emissions, isolated from others in wavelength.

Early work using scanning photometers to obtain the primary electron energy spectrum was performed by *Belon et al.*

[1966]. Since then, emission ratios from photometers, both ground-based and on board rockets and aircraft, have been used by many workers to study energy characteristics and composition [e.g., *Rees and Lummerzheim*, 1989; *Rasinkangas et al.*, 1989; *Vallance Jones et al.*, 1991; *Gattinger et al.*, 1991; *Dashkevich et al.*, 2006]. *Semeter* [2003] used all-sky images from three wavelength regions to study the distribution and separation of emissions along tall rays. *Gustavsson et al.* [2008] used spectral ratios of emissions from a camera employing a filter wheel to show that so-called "black" aurora is caused by a reduction of high-energy electron precipitation consistent with reduced pitch-angle scattering.

Why are energy and flux of the aurora important? The changing colors and dynamics of the aurora are indeed inspiring to observe, but the processes that lead to such displays can only be understood if the rapid changes in energy (related to the colors), and flux (related to the brightness), are measured. The temporal and spatial variations of these two properties are the direct response to processes that organize the source particles in the magnetosphere and accelerate them into the upper atmosphere, where changes in temperature, electron and ion densities, and conductivity affect the flow of currents, both horizontal and field-aligned, and lead to feedback effects, which modify and control the

Auroral Phenomenology and Magnetospheric Processes: Earth and Other Planets
Geophysical Monograph Series 197
10.1029/2011GM001161

acceleration processes in the magnetosphere [*Karlsson*, this volume; *Haerendel*, this volume]. Small-scale aurora exists within larger structures of precipitation, and large-scale structures are filled with small-scale structures, so the dynamic spatial changes at all scales must be measured. There is a huge variety of forms and phenomena, all of which may have different generation mechanisms, for example, arcs with shear and counter-streaming features such as curls, folds and ruffs, flickering aurora, filamenting and antifilamenting arcs, flaming rays, boundary arcs, pulsating patches, and regions of "black" or reduced intensity in diffuse aurora [*Sandahl et al.*, 2008]. All of these auroral types are the result of magnetospheric processes with ionospheric feedback [*Marghitu*, this volume; *Lysak and Song*, 2002; *Streltsov and Lotko*, 2008; *Chaston and Seki*, 2010; *Yoshikawa et al.*, 2010].

Størmer [1955, p. 163] also states that "the great importance of the study of the auroral spectrum lies not only in the information about the kind of gases in the upper atmosphere but also about the excitation mechanism of the different spectral lines." Both of these inputs are needed for successful modeling of the auroral energy input. The optical spectrum from the aurora is rich in lines and bands of emission in visible and near-visible wavelengths [*Vallance Jones*, 1974; *Chamberlain*, 1995]. The characteristics of each line or band depend on the temperature and composition of the neutral atmosphere, the resulting ion chemistry interactions, and the cross sections for ionization, excitation, and emission. Ionospheric modeling in conjunction with optical observations has been used by *Rees and Luckey* [1974]. They derived the relations between the intensities of the emissions in 557.7 nm (O ^{1}S), 427.8 nm (N$_2^+$ 1N), and 630.0 nm (O ^{1}D) and the total fluxes and characteristic energies of the precipitating electrons. Major improvements in modeling methods were made by *Strickland et al.* [1976] and *Stamnes* [1981]. *Strickland et al.* [1989] applied these methods to the problem of variability in oxygen density. They used ratios of the emissions at 427.8, 630.0, and 777.4 nm (O 3p ^{5}P) to derive the electron characteristics and to determine a correction factor for atomic oxygen density. This method was then extended to use 844.6 nm (O 3p ^{3}P) emission instead of 777.4 nm by *Hecht et al.* [1985].

In dynamic aurora, ratios of selected emissions can be used to give the energy and flux within the smallest scales. *Lanchester et al.* [1997] estimated energy fluxes as large as 500 mW m^{-2} in filaments of 100 m width, using photometers, radar, and modeling. However, the very small structures were found to be embedded within much wider features with a broader energy distribution, showing that such narrow features cannot be treated in isolation. The aim of this chapter is to show how the methods developed over the past several decades are applied to the latest high-resolution measurements, not only in the 1-D region of the field-aligned beam, but in a larger 3-D region around the magnetic zenith. Section 2 reviews the most important details of (forward) modeling of the aurora. The power and robustness of the latest methods is demonstrated in section 3, methods which combine emission rate ratios with modeling to estimate the energy of the primary precipitating particle distributions. Important results using such methods with measurements from the multispectral imager Auroral Structure and Kinetics (ASK) are described in section 4.

2. FORWARD MODELING OF EMISSIONS

The main source of enhanced optical emissions in aurora is electron precipitation. To model the spatial distribution and temporal variation of these emissions, it is necessary to include all sources of excitation, both from electron impact and from chemical interactions, and to take account of the height and energy variations of the energetic electron flux.

To calculate the direct electron excitation, it is necessary to model the electron transport and production of secondary electrons, to give a coupled set of linear differential equations, which must be solved simultaneously. Precipitating electrons ionize the atmosphere along their trajectories, creating secondary electrons with lower energies. This process increases the electron and ion concentrations and, moreover, leads to heating of the ions and ambient electrons through dissipation of energy. Ion-neutral collisions heat the neutral atmosphere, influencing the chemistry and thus the recombination rates of electrons and ions. This localized energy input affects the chemistry and dynamics of the upper atmosphere; to model that response, solutions of the continuity equations of the minor neutral species O (^{1}D), O (^{1}S), N (^{4}S), N (^{2}P), N (^{2}D), N (^{2}P), NO, and the ions O$_2^+$, NO$^+$, N$_2^+$, and O$^+$ are needed in order to obtain the time- and height-dependent variations in the ionosphere/thermosphere system. Further, it is necessary to solve the neutral, ion, and electron energy equations to find the variations in temperatures.

Many electron transport models exist in the literature and are used in optical studies of the aurora. *Roble and Rees* [1977] provide a thorough overview of the factors that influence the auroral ionosphere and thermosphere. *Basu et al.* [1993] and *Strickland et al.* [1993] include the effects of proton precipitation in a coupled electron-hydrogen-proton transport model. The electron transport problem can also be solved with Monte Carlo methods [e.g., *Sergienko and Ivanov*, 1993; *Solomon*, 1993; *Gattinger et al.*, 1996]. The following description of the key elements of electron transport is based on the work of *Lummerzheim and Lilensten* [1994], who provide a detailed presentation of an electron transport code and a comprehensive review of other work, which solves the transport and energy equations.

The volume emission rate, η_λ, for an emission produced by direct electron impact is

$$\eta_\lambda(s) = \frac{A_\lambda}{\sum A'} n_k(s) \int_0^\infty \sigma'(E) I(E,s) dE, \qquad (1)$$

where s is the distance along the magnetic field direction, σ' is the cross section for production of the state emitting at wavelength λ, n_k is the density of neutral species k, for which we use the Mass Spectrometer Incoherent Scatter (MSIS) model atmosphere [Hedin, 1991], $I(E, s)$ is the electron flux at energy E, A_λ is the Einstein coefficient for emissions of photons with wavelength λ, and $\sum A'$ is the sum of all Einstein coefficients from the excited level.

The column emission rate (or brightness) is then obtained by integrating the volume emission rate along the line of sight:

$$I_\lambda = \int_0^\infty \eta_\lambda(s) \, ds. \qquad (2)$$

The electron fluxes propagated downward from the top of the ionosphere can be calculated with the coupled electron transport equations

$$\mu \frac{\partial I(E,s,\mu)}{\partial s} = -B \, I(E,s,\mu)$$
$$+ n_e(s) \frac{\partial}{\partial E}[L(E) \, I(E,s,\mu)] \qquad (3)$$
$$+ C(E,s,\mu,I)$$
$$+ Q(E,s,\mu,I),$$

where the electron flux, $I(E,s,\mu)$ has units of electrons $m^{-2} s^{-1} eV^{-1}$. The first term on the right represents losses of electrons from energy, E, and the pitch angle cosine, μ, due to elastic scattering changing the pitch angle and by inelastic scattering changing the energy, with

$$B = \sum_k n_k(s) \sigma_k^{tot}(E), \qquad (4)$$

where $\sigma_k^{tot}(E)$ is the total cross section for collisions of electrons with energy E with the k-th species. The second term on the right of equation (3) represents the losses due to energy transfer to ambient electrons of density $n_e(s)$, where $L(E)$ is a function of the thermal electron energy [Swartz et al., 1971]. The third term represents elastic scattering of electrons from other pitch angles, μ', into pitch angle μ:

$$C(E,s,\mu,I) = \sum_k n_k(s) \sigma_k^{el}(E) \cdot \int_{-1}^1 p(E, \mu' \to \mu) I(E,s,\mu') d\mu',$$
$$(5)$$

where $p(E, \mu' \to \mu)$ is the probability that an electron at energy E with pitch angle μ' will scatter to pitch angle μ. The fourth term, $Q(E, s, \mu, I)$, combines all internal sources of electrons, i.e., photoelectrons, secondary electrons, and electrons cascading from higher energies, ε, due to inelastic collisions:

$$Q(E,s,\mu,I) = Q_{local}(E,s,\mu)$$
$$+ \sum_k n_k(s) \sum_j \sigma_j^k(\varepsilon \to E)$$
$$\cdot \int_{-1}^1 p_j^k(\varepsilon, \mu' \to \mu) I(\varepsilon,s,\mu) d\mu' \qquad (6)$$
$$+ \sum_k n_k(s) \int_{E+E^*}^\infty \sigma_{ion}^k(\varepsilon \to E)$$
$$\cdot \int_{-1}^1 p_{ion}^k(\varepsilon, \mu' \to \mu) I(\varepsilon,s,\mu) d\mu' d\varepsilon,$$

where the summations are over all the excited states j of the k-th species. Electrons exciting an atom or a molecule lose a quanta of energy corresponding to the excitation energy, ΔE. Therefore, electrons cascading from energy $\varepsilon = E + \Delta E$ to E give contributions to the second term of Q from a series of discrete energies corresponding to the excitation thresholds of the neutral species. In ionizing collisions, the primary electron energy loss is the sum of the ionization energy and the energy of the secondary electrons. The third term of Q is an integral from $E + E^*$ where E^* is the ionization energy, taking into account production of secondary electrons with increasing energies.

3. THE METHOD OF EMISSION RATIOS

Estimating the energy variation of electron precipitation during an auroral event is a very robust method, which follows from a short chain of well-understood physics, outlined below. Primary electrons penetrate further into the atmosphere with increasing energy, leading to decreasing height of peak energy deposition with corresponding peaks of ionization and production of secondary electrons.

In order to produce auroral emissions, excitation of atoms and molecules is needed. Figure 1 (top) shows the total inelastic cross sections for electron impact with atomic oxygen, σ_{ie}, with the contributions from ionization, σ_{ion}, and excitation, σ_{exc}, as a function of energy (from the work of Itikawa et al. [1986], and references therein). The fraction of the total inelastic cross sections from ionization increases from approximately 0.2 at 20 eV to around 0.9 at 100 eV and above. The same fraction applies for the inelastic cross sections of electrons with O_2 and N_2. In other words, the cross sections for excitation of O, O_2, and N_2 typically peak at energies between 5 and 30 eV. One consequence is that the primary electrons, typically with energies in the keV range, do not directly produce any excited atoms or molecules but mainly produce ions (some in excited states) and secondary

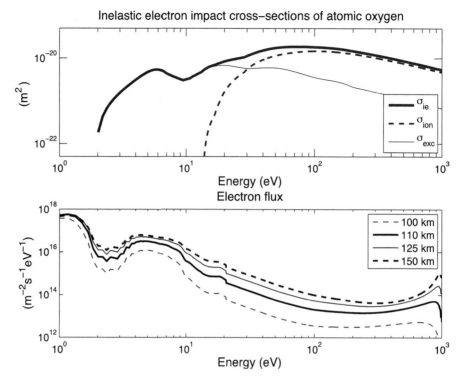

Figure 1. (top) Inelastic cross sections for electron impact with O, showing the relative contributions to the total from ionization and excitation. (bottom) Electron flux at four heights, showing the scaling with height at low energies.

electrons, typically with energies below 50–100 eV. It is the secondary electrons that then excite atoms and molecules.

Crucially, below about 100 eV, the shape of the electron flux does not vary much with height. Figure 1 (bottom) shows the omnidirectional electron flux at four selected heights using a sample monoenergetic electron beam with energy of 1 keV as input to equation (3). More realistic precipitation would show more variation in the shape above 100 eV. Because the flux in the 2–100 eV energy range scales with height, the electron flux, $I(E, s)$, in equation (1) can be separated into two factors, one depending on height and one on energy, with the former being moved outside the integral, which can then be solved. The result is that ratios of emission brightness from equation (2) will depend on the height where the primary electrons produce most of the secondary electrons and on the neutral composition.

Neutral densities increase with decreasing height, as does the proportion of both O_2 and N_2, while that of O decreases. Figure 2 shows emissions from O and O_2^+ resulting from different Maxwellian input energy spectra (top graph), which have peak energies of 1 and 3 keV. The bottom left-hand graph gives the modeled height variation of the two emissions for all energies, and the right-hand graph is the volume emission rate for both emissions. The key fact is that the difference in the height of the peaks for each emission varies

with energy. The emissions chosen here represent two prompt emissions that are used in the ASK instrument (see section 4), 777.4 nm from O(3p ^{5}P) and 562.0 nm from $O_2^+(b^4\Sigma_g^-)$. Their respective column emission rates (brightness) are subsequently referred to as I_{7774} and I_{5620}.

The "robustness" of estimating energies from emissions from different species is that the energy variation in emission ratios ultimately depends on the height variation of the atmospheric composition. Figure 3 (top) shows I_{7774} and I_{5620} for input monoenergetic precipitation of fixed energy flux and varying peak energy. The dotted and dashed lines show the same calculations with different fractions of oxygen density scaled to the MSIS values. Figure 3 (bottom) shows the ratio of the two emissions as a function of energy, with the same oxygen scaling factors. Depending on the relative oxygen density, the ratio changes by a factor of more than 30 between energies of 100 eV and 10 keV.

The effect of variations in atomic oxygen density is most pronounced in ratios of prompt molecular emissions with 630.0 nm emission; the latter has a radiative lifetime of 134 s and is increasingly quenched at lower altitudes where the collision frequency increases. The 630.0 nm emission is therefore most responsive to low-energy precipitation that has its peak energy deposition at 200–250 km. The long lifetime of the O(^{1}D) state constrains the use of 630.0 nm emission to aurora where the

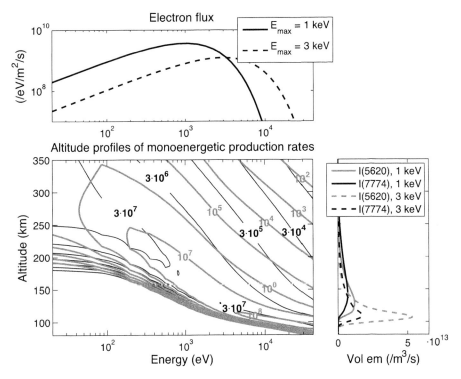

Figure 2. (top) Two primary spectra used as input. (bottom, left) Energy-height variation of monoenergetic production profiles for excited O (black) and O_2^+ (gray) and (bottom right) the resulting height profiles for 777.4 nm (black) and 562.0 nm (gray) emissions.

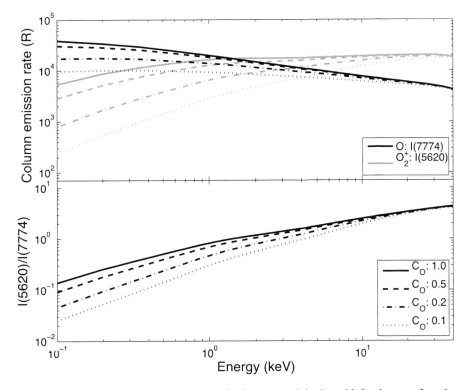

Figure 3. (top) Variation in I_{7774} and I_{5620} for monoenergetic electron precipitation with fixed energy flux. (bottom) Energy variation of I_{7774}/I_{5620}. Oxygen density profiles are scaled with 1 (solid), 0.5 (dashed), 0.2 (dot-dashed), and 0.1 (dotted).

temporal variations are slow compared with the effective O(^{1}D) lifetime, typically 10–30 s. However, during individual auroral events that last for a few seconds up to tens of seconds, the neutral densities will not change noticeably, and changes in optical intensity ratios will accurately reflect variations in energy and flux of the particle precipitation.

From the above, and with some assumption about the spectral shape, for example, Gaussian or Maxwellian, with or without low-energy tails, it is possible to estimate the characteristics of the electron precipitation, both total number flux and average energy. However, the limitation of using pairs of column emission rates is that photons emitted along the line of sight must be emitted from atoms, molecules, and ions excited by electrons resulting from primary electrons with the same spectral shape. This condition is met for observations made in the magnetic zenith, i.e., along the trajectory of the precipitating electrons. For observations of dynamic discrete aurora with fine-scale structures, made away from the magnetic zenith, light from different ranges along the line of sight is produced by precipitation along adjacent magnetic field lines, as depicted in Figure 4. The larger hatched region represents a

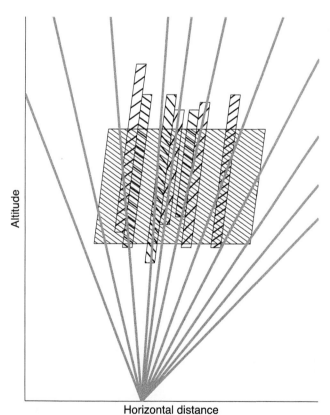

Figure 4. Geometry in the meridian plane showing optical fields of view in gray and regions characterized by different precipitations in hatched regions (not to scale).

diffuse background spectrum surrounding several filaments of different spectral characteristics. Horizontal scales may be as small as tens of meters.

It is possible to overcome the problem of interpreting observations off the magnetic zenith by the following method. The electron precipitation can be estimated along the "top" of the meridian plane. The height variations of, for example, the 777.4 and 562.0 nm emissions can be modeled for a given primary precipitation and compared to brightness variations with elevation angle in the observed images. By extending the modeling of the optical emissions to the full meridional plane inside the field-of-view and constraining the horizontal variation of electron precipitation to be a combination of Gaussian-shaped filaments [e.g., *Semeter et al.*, 2008], a model of the 2-D variation of volume emission rates is obtained. (Here constraining the spatial variations is analogous to constraining the electron spectra to Gaussian or Maxwellian in energy.) By adjusting the parameters of widths, intensities, and electron spectra of the arcs, a fit can be made of the modeled image intensities to the observed. The above method is illustrated in Figure 5, which shows how the volume emission rate profiles of 777.4 and 562.0 nm for two sample Maxwellian spectra of 1 keV (solid) and 3 keV (dashed) from Figure 2 are combined by matrix multiplication with two horizontal shapes in number flux. Figure 5 (middle) shows the resulting volume distributions in both emissions, where the horizontal distance corresponds to the distance south from the location of the field-aligned camera at 100 km height. Figure 5 (bottom) is the same result converted to brightness versus pixels in the observed image. An example of this method applied to a filamenting arc is given in section 4.4.

4. APPLICATIONS AND RESULTS

4.1. The ASK Instrument

One example of what is achievable with state-of-the-art technology is the ASK instrument, consisting of three electron multiplying charge-coupled device (EMCCD) cameras with 3° by 3° field-of-view optics with narrow passband interference filters and two photometers [*Lanchester et al.*, 2009]. With ASK, it is possible to make spectral observations at up to 32 frames per second in the magnetic zenith of selected emissions (e.g., 562.0, 673.0, 732.0, and 777.4 nm), which are not the brightest auroral emissions, but which still provide measurements with good signal-to-noise ratios.

4.2. Comparison of Radar Methods and Optical Ratios

Field-aligned radar measurements of electron density can be used to give the energy and energy flux in aurora. *Vallance*

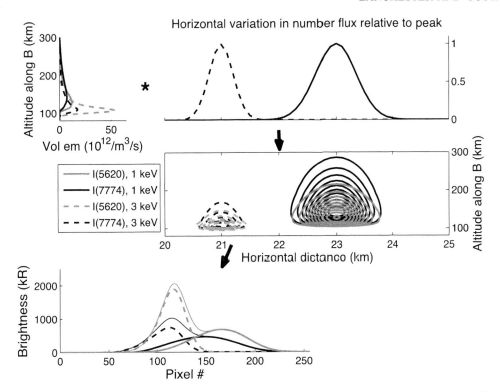

Figure 5. (top) Height profiles of volume emission rates from Figure 2, combined with Gaussian horizontal profiles to give (middle) the resulting volume distributions. Horizontal axis is distance to south of camera position at 100 km height. Vertical axis is magnetic field aligned. (bottom) Resulting image intensities of the modeled aurora in the meridional plane.

Jones et al. [1987] and *Jokiaho et al.* [2008], for example, also derived neutral temperatures from optical and radar data using the height of peak energy deposition. *Lanchester et al.* [2009] used forward modeling of electron densities from the European Incoherent Scatter (EISCAT) mainland radar to validate the transport and ion chemistry model by comparing measured and modeled emissions from the ASK instrument. This model was used by *Dahlgren et al.* [2011] to derive the energy and flux in a narrow arc during the same winter season. A new time-dependent inversion model was applied to the radar electron density profiles, to give independent estimates of the energy and energy flux in the region of the arc. The results of this study are shown in Figure 6, which covers an interval of 10 s with the arc entering the radar field of view at 19:32:03 UT. In Figure 6 (top), the average energy (black line) is estimated from the ratio of ASK emissions, I_{5620}/I_{7774}, integrated over the region of the radar beam at 110 km height. The radar inversion results are shown as gray bars. The energy increases from about 2.5 keV before the arc enters the beam to about 5 keV inside the arc. Figure 6 (bottom) shows the energy flux, derived from the O_2^+ emissions and by integrating over the incoherent scatter radar–estimated electron spectra. Modeling confirms that for a given input energy flux, I_{5620} is nearly constant for energies

above 1 keV. The brightness can therefore be scaled to give the energy flux. The agreement in both graphs gives confidence to the method of using optical ratios, which can be applied in a wider region than that of the radar beam, and at greatly increased temporal and spatial resolution, and with the advantage of being easy to measure automatically during times of clear skies at a comparatively low cost.

4.3. Flickering Aurora

The importance of making temporal observations to match the variation of auroral brightness can be illustrated with recent results on flickering aurora by *Whiter et al.* [2010]. Flickering aurora, varying in intensity typically at 5–15 Hz, is sometimes observed in bright aurora and during auroral breakup events. The flickering is characterized by spots with horizontal size of a few kilometers, which often can appear both to drift and rotate with high apparent velocity during their 1–2 s lifetime.

There is no definitive answer to what causes the modulation of the electron precipitation resulting in flickering. One theory is that interfering electromagnetic ion cyclotron (EMIC) waves cause the flickering by trapping ambient electrons at about 1 R_E where the wave phase velocity is low or zero. As the wave and electrons propagate down the magnetic field line, the

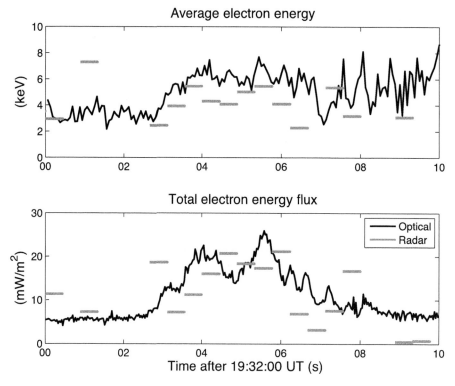

Figure 6. Comparison between (top) average electron energy and (bottom) total electron energy flux estimated from brightness emission ratio $I_{5620}/I_{777.4}$ (black line) and estimates of electron spectra from incoherent scatter radar electron density measurements (gray bars).

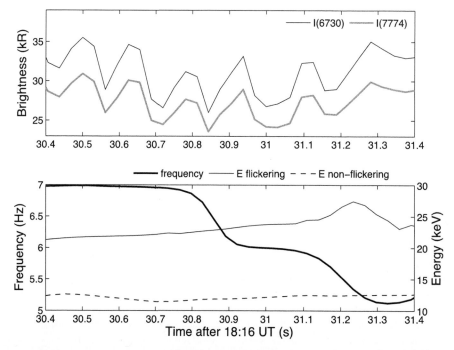

Figure 7. (top) Variation in brightness of two emissions in flickering aurora. (bottom) Variation of flickering frequency over the same interval (black), the corresponding increase in energy of the flickering component (thin line), and the constant nonflickering energy (dashed).

phase velocity increases, and the electrons gain energy. From this model, *Sakanoi et al.* [2005] showed that electrons should be accelerated to higher energies by EMIC waves with lower frequencies, since the peak field-aligned phase velocity of EMIC waves decreases with increasing frequency. Hence, it is straightforward to predict that there should be an increase in the flickering component of I_{5620}/I_{7774} with decreasing frequency, which is exactly what *Whiter et al.* [2010] confirmed, from observations of a number of flickering "chirps." One example, in which the molecular emission is from N_2 1P at 673.0 nm, is shown in Figure 7. Figure 7 (top) shows the two emissions varying over an interval of 1 s. Figure 7 (bottom) shows the decrease in frequency with corresponding increase in energy derived from the emissions, from about 20 to 27 keV.

The other significant result from this analysis is that the peak energy of the flickering component of the precipitation was found to be higher than that of the underlying nonflickering precipitation by 8–15 keV. The near-constant energy of the nonflickering component here is about 12 keV.

4.4. 3-D Distributions From 2-D Images

Occasionally, auroral arcs multiply into filaments of successively smaller scale, typically on time scales of one to a few seconds. This splitting is thought to be caused by dispersive Alfvén waves (i.e., *Semeter et al.* [2008], and references therein). To understand the process, it is necessary to determine the energy and flux of precipitation and how it

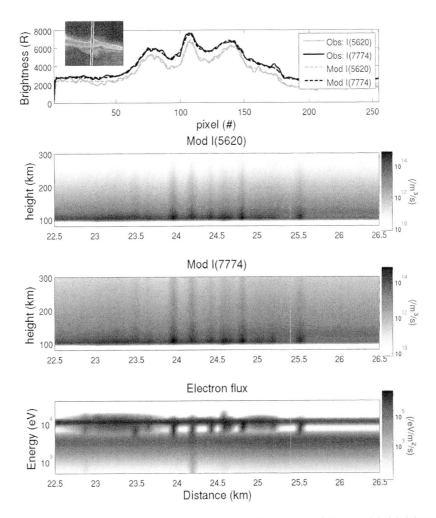

Figure 8. (top) Observed $I_{562.0}$ (solid gray) and $I_{777.4}$ (solid black), and best fitting modeled brightnesses (dashed). Observations are taken as a cut along the meridional plane shown as the white line in the inset image. (middle) Estimated 562.0 and 777.4 nm volume emission rates in the meridional plane. (bottom) Estimated primary electron spectra. The horizontal axis is the distance south from the location of the Auroral Structure and Kinetics instrument in the meridional plane at 106 km; the vertical axis is along the magnetic field.

varies with time, both in the filaments and in the background aurora over the entire horizontal scale of the filamenting arc system. When observations are made around the magnetic zenith in multiple emissions from atomic oxygen and molecular oxygen or nitrogen (either neutral molecules or ions), the energy and flux within the filaments can be estimated, using the method outlined in section 3. An example snapshot of ASK observations of a filamenting arc is shown in Figure 8. Figure 8 (top) shows the observed brightnesses in the magnetic meridian plane as drawn in the image (solid lines) and modeled brightnesses (dashed). Small-scale structures that appear in both I_{7774} and I_{5620} are slightly shifted relative to one another due to parallax and the different height distributions. In Figure 8 (middle two graphs), the estimated volume emission rates show several small-scale filaments with widths <200 m. In Figure 8 (bottom), the estimated energy flux reveals that the filaments are caused by electron precipitation with a large spread in energy, mainly extending toward energies below the peak energy of the spatially wide precipitation of about 10 keV, which fills the field of view. The filaments, as seen in Figure 8 (top, inset), cover a region of about 2–3 km and represent the optically bright arc, which is embedded within a wider region of precipitation, as depicted in Figure 4 (although not to scale).

5. SUMMARY

Returning to the quote of Størmer at the start of this chapter, "the necessity of *simultaneous photographs of the auroral spectrum and of photographic height measurements*" has been demonstrated. It has been shown that multimonochromatic observations provide a powerful method for finding the temporal and spatial evolution of the primary energy spectrum of the precipitation. Using such measurements of auroral structure, combined with ionospheric modeling, and with particular attention to the problems of parallax and the height distribution of the emissions, it is now possible to estimate the energy and flux of the primary electrons within the region surrounding the magnetic zenith. Combining such measurements with ground-based radar increases their value by providing height profiles of electron density in the zenith, which can be used for modeling the input spectrum, but at lower time and space resolution. Although satellite particle measurements give excellent resolution for such spectra above the atmosphere, they will rarely be coincident with the small field of view of the cameras and radar beam. However, as shown by *Sakanoi et al.* [this volume] and *Chaston et al.* [2011], the combination of in situ particle data and optical multimonochromatic measurements from above is a great advance, albeit with lower spatial resolution for the optical data than in the present work.

The methods outlined here require careful evaluation of the ambiguities and uncertainties inherent in the observations and in the assumptions required within the modeling. The cross sections for emissions have uncertainties of about 10%. The atmospheric composition, in particular the density of atomic oxygen, can vary greatly from day to day, maybe up to 50%. Assuming the shape of the precipitating spectra in the model has been validated by a separate process, which fits radar height profiles to combinations of model spectra. A solution has been found to the problems of parallax, which means that 2-D images can be interpreted as emissions in three dimensions close to the zenith. Hence, these observations and methods provide the best available information about highly dynamic and fine-scale processes in ionosphere-magnetosphere interactions, which are relevant to aurora, with an unprecedented combination of temporal and spatial resolution.

Acknowledgments. The ASK instrument was funded by the PPARC of the UK. B.G. is supported by the NERC of the UK. For the data used in the research, we thank the EISCAT and ASK campaign teams for running the instruments during several winter campaigns. We thank the referees for their helpful and encouraging comments and suggestions for this chapter.

REFERENCES

Basu, B., J. R. Jasperse, D. J. Strickland, and R. E. Daniell Jr. (1993), Transport-theoretic model for the electron-proton-hydrogen atom aurora, 1. Theory, *J. Geophys. Res.*, *98*(A12), 21,517–21,532.

Belon, A. E., G. J. Romick, and M. H. Rees (1966), The energy spectrum of primary auroral electrons determined from auroral luminosity profiles, *Planet. Space Sci.*, *14*(7), 597–615.

Chamberlain, J. W. (1995), *Physics of the Aurora and Airglow*, 704 pp., AGU, Washington, D. C., doi:10.1029/SP041.

Chaston, C. C., and K. Seki (2010), Small-scale auroral current sheet structuring, *J. Geophys. Res.*, *115*, A11221, doi:10.1029/2010JA015536.

Chaston, C. C., K. Seki, T. Sakanoi, K. Asamura, M. Hirahara, and C. W. Carlson (2011), Cross-scale coupling in the auroral acceleration region, *Geophys. Res. Lett.*, *38*, L20101, doi:10.1029/2011GL049185.

Dahlgren, H., B. Gustavsson, B. S. Lanchester, N. Ivchenko, U. Brändström, D. K. Whiter, T. Sergienko, I. Sandahl, and G. Marklund (2011), Energy and flux variations across thin auroral arcs, *Ann. Geophys.*, *29*, 1699–1712, doi:10.5194/angeo-29-1699-2011.

Dashkevich, Z. V., V. L. Zverev, and V. E. Ivanov (2006), Ratios of the I 630.0/I 427.8 and I 557.7/I 427.8 emission intensities in auroras, *Geomagn. Aeron.*, *46*(3), 366–370.

Gattinger, R. L., A. Vallance Jones, J. H. Hecht, D. J. Strickland, and J. Kelly (1991), Comparison of ground-based optical observations of N_2 second positive to N_2^+ first negative emission ratios with electron precipitation energies inferred from the Sondre Stromfjord radar, *J. Geophys. Res.*, *96*(A7), 11,341–11,351.

Gattinger, R. L., E. J. Llewellyn, and A. Vallance Jones (1996), On I (5577 Å) and I (7620 Å) auroral emissions and atomic oxygen densities, *Ann. Geophys.*, *14*(7), 687–698, doi:10.1007/s00585-996-0687-1.

Gustavsson, B., M. J. Kosch, A. Senior, A. J. Kavanagh, B. U. E. Brändström, and E. M. Blixt (2008), Combined EISCAT radar and optical multispectral and tomographic observations of black aurora, *J. Geophys. Res.*, *113*, A06308, doi:10.1029/2007JA012999.

Haerendel, G. (2012), Auroral generators: A survey, in *Auroral Phenomenology and Magnetospheric Processes: Earth and Other Planets*, Geophys. Monogr. Ser., doi:10.1029/2011GM001162, this volume.

Hecht, J. H., A. B. Christensen, and J. B. Pranke (1985), High-resolution auroral observations of the OI(7774) and OI(8446) multiplets, *Geophys. Res. Lett.*, *12*(9), 605–608.

Hedin, A. (1991), Extension of the MSIS thermospheric model into the middle and lower atmosphere, *J. Geophys. Res.*, *96*(A2), 1159–1172.

Itikawa, Y., M. Hayashi, A. Ichimura, K. Onda, K. Sakimoto, K. Takayanagi, M. Nakamura, H. Nishimura, and T. Takayanagi (1986), Cross sections for collisions of electrons and photons with nitrogen molecules, *J. Phys. Chem. Ref. Data*, *15*(3), 985–1010.

Jokiaho, O., B. S. Lanchester, N. Ivchenko, G. J. Daniell, L. C. H. Miller, and D. Lummerzheim (2008), Rotational temperature of N_2^+ (0,2) ions from spectrographic measurements used to infer the energy of precipitation in different auroral forms and compared with radar measurements, *Ann. Geophys.*, *26*, 853–866, doi:10.5194/angeo-26-853-2008.

Karlsson, T. (2012), The acceleration region of stable auroral arcs, in *Auroral Phenomenology and Magnetospheric Processes: Earth and Other Planets*, Geophys. Monogr. Ser., doi:10.1029/2011GM001179, this volume.

Lanchester, B. S., M. H. Rees, D. Lummerzheim, A. Otto, H. U. Frey, and K. U. Kaila (1997), Large fluxes of auroral electrons in filaments of 100 m width, *J. Geophys. Res.*, *102*(A5), 9741–9748.

Lanchester, B. S., M. Ashrafi, and N. Ivchenko (2009), Simultaneous imaging of aurora on small scale in OI (777.4 nm) and $N_2$1P to estimate energy and flux of precipitation, *Ann. Geophys.*, *27*, 2881–2891, doi:10.5194/angeo-27-2881-2009.

Lummerzheim, D., and J. Lilensten (1994), Electron transport and energy degradation in the ionosphere: Evaluation of the numerical solution, comparison with laboratory experiments and auroral observations, *Ann. Geophys.*, *12*, 1039–1051.

Tysak, R. L., and Y. Song (2002), Energetics of the ionospheric feedback interaction, *J. Geophys. Res.*, *107*(A8), 1160, doi:10.1029/2001JA000308.

Marghitu, O. (2012), Auroral arc electrodynamics: Review and outlook, in *Auroral Phenomenology and Magnetospheric Pro-cesses: Earth and Other Planets*, Geophys. Monogr. Ser., doi:10.1029/2011GM001189, this volume.

Rasinkangas, R. A., K. U. Kaila, and T. Turunen (1989), Comparison of the lower border of aurorae determined by two optical emission ratio models, *Planet. Space Sci.*, *37*(9), 1117–1126.

Rees, M. H., and D. Luckey (1974), Auroral electron energy derived from ratio of spectroscopic emissions 1. Model computations, *J. Geophys. Res.*, *79*(34), 5181–5186.

Rees, M. H., and D. Lummerzheim (1989), Characteristics of auroral electron precipitation derived from optical spectroscopy, *J. Geophys. Res.*, *94*(A6), 6799–6815.

Roble, R. G., and M. H. Rees (1977), Time-dependent studies of the aurora: Effects of particle precipitation on the dynamic morphology of ionospheric and atmospheric properties, *Planet. Space Sci.*, *25*, 991–1010.

Sakanoi, K., H. Fukunishi, and Y. Kasahara (2005), A possible generation mechanism of temporal and spatial structures of flickering aurora, *J. Geophys. Res.*, *110*, A03206, doi:10.1029/2004JA010549.

Sakanoi, T., Y. Obuchi, Y. Ebihara, Y. Miyoshi, K. Asamura, A. Yamazaki, Y. Kasaba, M. Hirahara, T. Nishiyama, and S. Okano (2012), Fine-scale characteristics of black aurora and its generation process, in *Auroral Phenomenology and Magnetospheric Processes: Earth and Other Planets*, Geophys. Monogr. Ser., doi:10.1029/2011GM001178, this volume.

Sandahl, I., T. Sergienko, and U. Brändström (2008), Fine structure of optical aurora, *J. Atmos. Sol. Terr. Phys.*, *70*, 2275–2292, doi:10.1016/j.jastp.2008.08.016.

Semeter, J. (2003), Critical comparison of OII(732–733 nm), OI (630 nm), and N_2(1PG) emissions in auroral rays, *Geophys. Res. Lett.*, *30*(5), 1225, doi:10.1029/2002GL015828.

Semeter, J., M. Zettergren, M. Diaz, and S. Mende (2008), Wave dispersion and the discrete aurora: New constraints derived from high-speed imagery, *J. Geophys. Res.*, *113*, A12208, doi:10.1029/2008JA013122.

Sergienko, T. I., and V. E. Ivanov (1993), A new approach to calculate the excitation of atmospheric gases by auroral electron impact, *Ann. Geophys.*, *11*, 717–727.

Solomon, S. (1993), Auroral electron transport using the Monte Carlo method, *Geophys. Res. Lett.*, *20*(3), 185–188.

Stamnes, K. (1981), On the two-stream approach to electron transport and thermalization, *J. Geophys. Res.*, *86*(A4), 2405–2410.

Størmer, C. (1955), *The Polar Aurora*, 403 pp., Clarendon, Oxford, U. K.

Streltsov, A. V., and W. Lotko (2008), Coupling between density structures, electromagnetic waves and ionospheric feedback in the auroral zone, *J. Geophys. Res.*, *113*, A05212, doi:10.1029/2007JA012594.

Strickland, D. J., D. L. Book, T. P. Coffey, and J. A. Fedder (1976), Transport equation techniques for the deposition of auroral electrons, *J. Geophys. Res.*, *81*(16), 2755–2764.

Strickland, D. J., R. R. Meier, J. H. Hecht, and A. B. Christensen (1989), Deducing composition and incident electron spectra from ground-based auroral optical measurements: Theory and model results, *J. Geophys. Res.*, *94*(A10), 13,527–13,539.

Strickland, D. J., R. E. Daniell Jr., J. R. Jasperse, and B. Basu (1993), Transport-theoretic model for the electron-proton-hydrogen atom aurora, 2. Model results, *J. Geophys. Res.*, *98*(A12), 21,533–21,548.

Swartz, W. E., J. S. Nisbet, and A. E. S. Green (1971), Analytic expression for the energy-transfer rate from photoelectrons to thermal-electrons, *J. Geophys. Res.*, *76*(34), 8425–8426.

Vallance Jones, A. (1974), *Aurora, Geophys. Astrophys. Monogr.*, vol. 9, D. Reidel, Dordrecht, Netherlands.

Vallance Jones, A., R. L. Gattinger, P. Shih, J. W. Meriwether, V. B. Wickwar, and J. Kelly (1987), Optical and radar characterization of a short-lived auroral event at high latitude, *J. Geophys. Res.*, *92*(A5), 4575–4589.

Vallance Jones, A., et al. (1991), The Aries auroral modelling campaign: Characterization and modelling of an evening auroral arc observed from a rocket and a ground-based line of meridian scanners, *Planet. Space Sci.*, *39*, 1677–1705, doi:10.1016/0032-0633(91)90029-A.

Whiter, D. K., B. S. Lanchester, B. Gustavsson, N. Ivchenko, and H. Dahlgren (2010), Using multispectral optical observations to identify the acceleration mechanism responsible for flickering aurora, *J. Geophys. Res.*, *115*, A12315, doi:10.1029/2010JA 015805.

Yoshikawa, A., H. Nakata, A. Nakamizo, T. Uozumi, M. Itonaga, S. Fujita, K. Yumoto, and T. Tanaka (2010), Alfvenic-coupling algorithm for global and dynamical magnetosphere-ionosphere coupled system, *J. Geophys. Res.*, *115*, A04211, doi:10.1029/ 2009JA014924.

B. Gustavsson and B. Lanchester, Department of Physics and Astronomy, University of Southampton, Southampton SO17 1BJ, UK. (b.s.lanchester@soton.ac.uk)

Current Closure in the Auroral Ionosphere: Results From the Auroral Current and Electrodynamics Structure Rocket Mission

S. R. Kaeppler,[1] C. A. Kletzing,[1] S. R. Bounds,[1] J. W. Gjerloev,[2] B. J. Anderson,[3] H. Korth,[3] J. W. LaBelle,[4] M. P. Dombrowski,[4] M. Lessard,[5] R. F. Pfaff,[6] D. E. Rowland,[6] S. Jones,[6] and C. J. Heinselman[7]

The Auroral Current and Electrodynamics Structure mission consisted of two sounding rockets launched nearly simultaneously from Poker Flat Research Range, Alaska, on 29 January 2009 into a dynamic multiple-arc aurora. The two well-instrumented payloads were flown along very similar magnetic field footprints, at different altitudes, with small temporal separation to measure electrodynamic and plasma parameters above and within the ionospheric current closure region. The higher-altitude payload (360 km apogee) acquired in situ measurements of electro-dynamic and plasma parameters above the current closure region to determine the magnetospheric input signature. The low-altitude payload (130 km apogee) made conjugate observations within the current closure region. Results are presented comparing observations of the electric fields, magnetic fields, and the electron differential energy flux at magnetic foot points common to both payloads. In situ data is compared to ground-based all-sky imager data, which recorded the evolution of the auroral event as the payloads traversed through magnetically conjugate regions. Current measurements derived from the magnetometers on the high-altitude payload observed upward and downward field-aligned currents. The effect of collisions with the neutral atmosphere is investigated to determine if it is a significant mechanism to explain differences in the low-energy electron flux. A calculation of ionospheric conductivity is performed to explain attenuation in electric field observations between the two payloads. The electric fields and magnetic fields of the first auroral crossing are examined in detail and found to have results consistent with the model of an auroral arc.

1. INTRODUCTION

Field-aligned currents associated with particle precipitation that produce the aurora are a key mechanism of the coupling that exists between the magnetosphere and the ionosphere. Upward directed field-aligned current sheets are associated with precipitating auroral electrons [*Arnoldy*, 1974; *Elphic et al.*, 1998] that form inverted-V signatures in the electron differential energy flux [*Frank and Ackerson*, 1971; *Ackerson and Frank*, 1972]. Black aurora [*Marklund*

---

[1]Department of Physics and Astronomy, University of Iowa, Iowa City, Iowa, USA.

[2]Department of Physics and Technology, University of Bergen, Bergen, Norway.

Auroral Phenomenology and Magnetospheric Processes: Earth and Other Planets
Geophysical Monograph Series 197
10.1029/2011GM001177

[3]Applied Physics Laboratory, Johns Hopkins University, Baltimore, Maryland, USA.

[4]Department of Physics and Astronomy, Dartmouth College, Hanover, New Hampshire, USA.

[5]Space Sciences Center, University of New Hampshire, Durham, New Hampshire, USA.

[6]NASA Goddard Space Flight Center, Greenbelt, Maryland, USA.

[7]Center for Geospace Studies, SRI International, Menlo Park, California, USA.

et al., 1994], a lack of visible auroral emission due to upgoing electrons, are typically associated with downgoing field-aligned currents [*Marklund et al.*, 1997; *Elphic et al.*, 1998]. Within the lower ionosphere, currents flow perpendicular to the mean magnetic field that close the magnetospheric-ionospheric circuit by connecting the upward and downward field-aligned currents. Energy is transmitted from the magnetosphere to the ionosphere and is dissipated through Joule heating. Energy may also be dissipated in the ionosphere by friction between atmospheric neutrals and ions, which can be enhanced by auroral particle precipitation.

The perpendicular closure current resides at altitudes where ion and electron collisions with the neutral atmosphere become significant. The departure from $\mathbf{E} \times \mathbf{B}$ drift motion as a result of collisions causes a cross-field velocity difference between ions and electrons, which establishes a perpendicular current. These closure currents flow along channels of enhanced conductivity and are oriented relative to the perpendicular electric field, $\mathbf{E}_\perp$: Pedersen currents ($\mathbf{j}_P$) flow parallel to $\mathbf{E}_\perp$, and Hall currents ($\mathbf{j}_H$) flow perpendicular to $\mathbf{E}_\perp$. Ohm's law empirically relates the perpendicular electric field to the perpendicular current density, as represented by *Richmond and Thayer* [2000]:

$$\mathbf{j}_\perp(z) = \mathbf{j}_P(z) + \mathbf{j}_H(z) = \sigma_P(z)\mathbf{E}_\perp + \sigma_H(z)(\mathbf{B} \times \mathbf{E}_\perp), \quad (1)$$

where $\sigma_P(z)$ and $\sigma_H(z)$ are the height-dependent Pedersen and Hall conductivities (see *Richmond and Thayer* [2000], equations (13) and (14)), and $\hat{\mathbf{B}}$ is the magnetic field unit vector. The Pedersen conductivity peaks at altitudes of approximately 130–150 km, and the Hall conductivity peaks at altitudes below 120 km [*Richmond and Thayer*, 2000]. A commonly invoked assumption is that the Hall and Pedersen conductivities can be integrated over an altitude range [*Fejer*, 1953], which yields the height-integrated Hall and Pedersen conductivities, Σ_H and Σ_P, respectively.

The low altitude of the auroral ionosphere makes this region well suited for sounding rocket observations. Many sounding rocket missions have examined electrodynamics associated with the aurora [*Arnoldy*, 1974, 1977; *Evans et al.*, 1977; *Marklund et al.*, 1982; *Mallinckrodt and Carlson*, 1985; *Kletzing et al.*, 1996; *Sangalli et al.*, 2009]. In an early sounding rocket investigation, *Evans et al.* [1977] presented a very complete study of auroral electrodynamics. *Evans et al.* [1977] calculated Hall and Pedersen conductivities from observed electron densities and found that both conductivities significantly increased as the payload traversed a stable auroral arc. A strong correlation was observed between a reduction in the perpendicular electric field and a similar reduction in the reciprocal of the height-integrated Pedersen conductivity over the auroral arc. This observation suggested that an anticorrela-

tion exists between conductivities and the electric field; that is, an increase in conductivities results in a reduction in the electric field magnitude within the auroral arc. Further sounding rocket missions [*Mallinckrodt and Carlson*, 1985; *Marklund et al.*, 1982; *Kletzing et al.*, 1996] found similar anticorrelations between the electric field and the height-integrated Hall and Pedersen conductivities consistent with the observations of *Evans et al.* [1977]. More recently, *Sangalli et al.* [2009] presented results from the JOULE II rocket mission that examined the connection between the ionosphere and the neutral atmosphere, but at altitudes lower than the previously mentioned studies. JOULE II made in situ observations of ion flow velocities, the $\mathbf{E} \times \mathbf{B}$ drift velocities, the electron density, and in situ calculations of the height-integrated Hall and Pedersen conductivities. When integrated over the altitude range of 110–128 km, *Sangalli et al.* [2009] found that the height-integrated Pedersen conductivity decreased when neutral wind effects were included, but the height-integrated Hall conductivity increased when the neutral winds were included.

A model was developed by *Mallinckrodt* [1985] that examined the current density structure within the auroral ionosphere (80–250 km) to explain anticorrelation observations made by *Mallinckrodt and Carlson* [1985]. Electric fields and field-aligned currents were prescribed at the upper boundary of the model. Two different auroral electrodynamic configurations, as classified by *Marklund* [1984], were used: polarization arcs, in which space charge separation creates a polarization electric field with negligible field-aligned currents, and field-aligned current arcs, in which significant field-aligned currents were present. In the case of polarization arcs, the model suggested that a vortex in the current would develop in the auroral current closure region. In the case of field-aligned current arcs, the closure geometry was U-shaped. The results of *Mallinckrodt* [1985] suggested that current closure geometries depend strongly upon the input auroral configuration.

To understand plasma and electrodynamic observations within the current closure region, for example, phenomena such as current closure geometry and Joule heating, observations of the magnetospheric inputs are required as well as observations within the closure region. Electric fields are generally assumed to map down from the magnetosphere into the ionosphere; however, at current closure altitudes, this assumption becomes less certain. The assumption that the Hall and Pedersen currents are constant with respect to altitude can be applied to large-scale models, but its validity is questionable for detailed models within the current closure region. These observational issues motivated a twin-payload sounding rocket mission that would cross similar magnetic field foot points, at differing altitudes, to obtain in situ observations in the current closure region while also observing the magnetospheric input signatures.

2. MISSION AND INSTRUMENTATION

The Auroral Currents and Electrodynamics Structure (ACES) rocket mission utilized two nearly identical, well-instrumented payloads that flew along similar magnetic field foot points with small longitudinal separation to measure key electrodynamic fields and plasma parameters. Both payloads were launched nearly simultaneously to constrain the temporal-spatial ambiguity inherent of in situ observations and a stable auroral arc was the desired launch condition so that the steady state assumption would be most valid. The first payload (hereafter referred to as "ACES High") was designed to fly at higher altitudes to measure electrodynamic and plasma inputs into the current closure region. The second lower-altitude payload (hereafter referred to as "ACES Low") was designed to obtain in situ plasma and electrodynamic observations within the current closure region. In addition to the rocket-borne payloads, all-sky imagers, and the Poker Flat Incoherent Scatter Radar (PFISR) provided photometric and radar observations of the auroral configuration, respectively.

ACES High and Low successfully launched from Poker Flat Research Range, Alaska, on 29 January 2009 at 09:49:40.0 UT and 09:51:10.0 UT, respectively. The overall geomagnetic conditions preceding the launch were very quiet as a result of the launch window being near the 2008–2009 solar minimum. The payloads were launched into a dynamic multiple-arc aurora located north of Fort Yukon, the approximate apogee location of both payloads. The ground-based magnetometers indicated a 100 nT deflection in the H component observed at Fort Yukon, which suggested that the event was likely a substorm with an electrojet and further indicated that a large-scale current system was present. The launch time of both payloads was approximately 23:00 magnetic local time (MLT), which was in the evening-midnight MLT sector. The Altitude Adjusted Corrected Geomagnetic Model was used to produce the magnetic mapping presented in Figure 1 that shows the times when the payloads crossed similar magnetic field lines that map down to ionospheric foot points at an altitude of 110 km. ACES High reached an apogee of 360 km, and ACES Low attained an apogee of 130 km. The maximum longitudinal separation that mapped down to foot points at 110 km, between both payloads, was 23 km near the end of their respective trajectories.

The ACES payloads included a variety of instruments to measure auroral plasma electron populations. The Electrostatic Electron Pitch Angle Analyzer (EEPAA) was used to obtain full pitch angle differential energy flux distributions of precipitating auroral electrons. The EEPAA is a "top-hat" style electrostatic particle analyzer consisting of 24 anode pads simultaneously measuring 15° pitch angle bins, while the symmetry axis of the detector remains nominally aligned to the

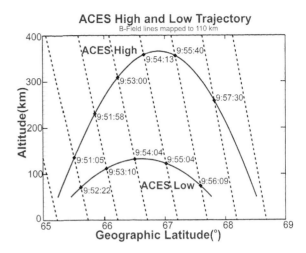

Figure 1. Auroral Currents and Electrodynamics Structure (ACES) High and Low trajectories plotted as functions of altitude and geographic latitude. Dashed lines are magnetic field lines mapped to foot points at 110 km. Times also indicate when both payloads crossed the same magnetic field foot point, respectively.

mean magnetic field. A full energy sweep of 0.1–16 keV was completed in 48 ms to create full electron distributions with high temporal resolution. An electron retarding potential analyzer (ERPA) was aligned parallel to the spin-axis of the payload to measure the electron differential energy flux of the background (0–3 eV) electron populations and the electron temperature. For a payload that is well aligned to the mean magnetic field, the ERPA can make observations of the low-energy field-aligned electrons that carry current. Spherical swept and fixed biased Langmuir probes suspended on booms 1 m from the payload and an impedance probe were used to obtain in situ electron density measurements. Data from the Langmuir probe, impedance probe, and PFISR radar provide the capability for cross calibration between instruments to arrive at a precise absolute electron density.

The ACES payloads also included a suite of instruments to measure electric and magnetic fields. A high-resolution flux-gate magnetometer, aligned with the spin-axis of the payload, measured the DC magnetic field vector. The fluxgate magnetometer could observe magnetic gradients, which are indicative of currents, with resolution better than 50 nT over the background 50,000 nT magnetic field, which was limited by payload coning and spin motion. The double-probe technique was used to obtain the LF (0–5 kHz) and DC electric fields perpendicular to the spin axis of the payload. The spherical double-probe sensors were attached to booms with tip-to-tip lengths of 5.3 and 3.0 m on ACES High and Low, respectively. AC double probes were aligned parallel to the spin axis of the payload to measure HF electric fields with a bandwidth of 0–5 MHz. Cutoffs in the plasma frequency measured by the HF double probes can also be used as an electron density measurement.

The mission was successful, although a few issues arose that affected the data gathered on each payload. During the flight, both payloads spun with a nominal frequency of 0.65 Hz, and the spin axis had a coning precession relative to the mean magnetic field. The ACES High payload coned with a half angle of 3°, but ACES Low coned at a much larger half angle of 13°, which was due to a gas valve leak in the attitude control system. The EEPAA on ACES High had a failure in the energy sweep electronics, that truncated the full sweep range. Therefore, the peak energy observed on ACES High was 500 eV, as opposed to the full 16 keV.

3. RESULTS

Figure 2 presents the electrodynamic and electron plasma data for both ACES High and Low. The top row of Figure 2 presents the spectrograms of the electron differential energy flux. The two middle rows in Figure 2 are the zonal (east-west) and meridional (north-south) DC electric fields after being transformed into a geomagnetic field-aligned coordinate system. The final three rows in Figure 2 present the residual DC magnetic field components in geomagnetic field-aligned coordinates after a third-order polynomial fit was applied to remove the Earth's mean magnetic field. The entire ACES Low data interval, on the right side of Figure 2, is represented by the black line in the margin of the ACES High data, and this represents the conjugate magnetic field lines that were eventually traversed by ACES Low. Figure 3 presents a montage of six all-sky images, from the Fort Yukon all-sky imager, depicting the auroral evolution during the flights. These images show the magnetic foot points of ACES High and ACES Low, especially during the time frame of conjugate data coverage between the payloads.

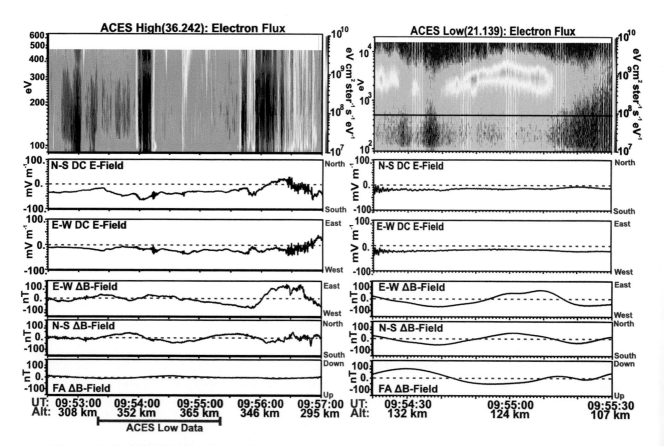

Figure 2. (left) ACES High data; (right) ACES Low data. row 1, electron differential energy flux; rows 2–3, DC electric field data; rows 4–6, residual magnetic field components. The electric field and residual magnetic field contain dashed lines at zero and the positive direction represented eastward, northward, and field-aligned, respectively. Gray bands are regions where data are not available. The black line in the ACES Low electron differential energy flux represents the maximum energy observed by ACES High of 500 eV.

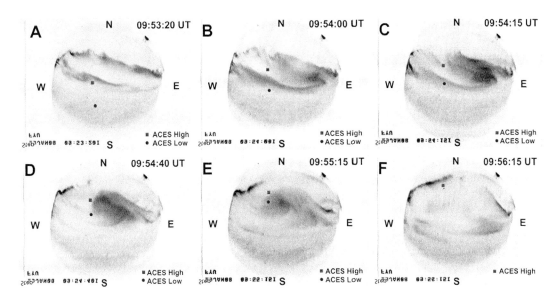

Figure 3. The images from the Fort Yukon all-sky imager showing the evolution of the auroral event on 29 January 2009. The foot points for ACES High and Low after being mapped to 110 km are represented by a red square and blue dot, respectively.

3.1. Electron Flux

The spectrograms of the electron differential energy flux shown on the top panel of Figure 2 for both ACES High and Low cover the pitch-angle range of 15°–30°. This pitch angle range was selected because it was nearly field aligned while remaining relatively immune to pitch-angle coverage gaps caused by payload coning motion. Although the full energy range was not observed on ACES High, the electron flux is ideal for an analysis focused on low-energy precipitating auroral electrons. The data from the all-sky imager have been compared with the electron flux to gain insight into the auroral events ACES High traversed.

ACES High had three auroral crossings over the course of the flight. As shown in Figure 3a, the payload entered a relatively stable auroral arc at 09:53:20 UT and exited the visible region at approximately 09:54:00 UT (hereafter, this event will be referred as the "quasi-static arc"). There was a modest increase in the electron flux at approximately 09:53:20 UT, which correlates well with the entry of the payload into the quasi-static arc. At 09:54:00 UT, there was a depletion of low-energy electron flux that correlates well with the time at which the payload entered the dark region adjacent to the poleward edge of the quasi-static arc, as shown in Figure 3b. At approximately 09:54:15 UT, ACES High passed over a faint auroral arc in the all-sky imager, as shown in Figure 3c, and a time-dispersed arrival of electrons was observed in the electron flux.

The all-sky imager data in Figure 3d indicated that ACES High skirted the westward edge of a large region of dynamic westward moving aurora (hereafter, referred to as the "westward moving" aurora). The electron data indicated that there were moderate levels of low-energy precipitation in this region along with embedded regions of more intense electron flux and isolated time-dispersed electron precipitation. By 09:55:15 UT, ACES High had passed by the bulk of the westward moving aurora and had moved into a region that was devoid of visible aurora and had reduced levels of precipitating electrons, as shown in Figure 3c. At 09:56:15 UT, the all-sky imager in Figure 3f showed that ACES High encountered the final auroral crossing, into a very active poleward arc. The electron flux data indicated that after passing from the region of reduced flux, the payload entered into a region of intense time-dispersed electron precipitation that correlate well with the active poleward arc.

The electron flux for ACES Low indicated that the payload traversed two inverted Vs that were located on magnetic field flux tubes that were very close to those previously crossed by ACES High. As shown in Figure 3b, ACES Low entered the quasi-static arc at 09:54:00 UT, 40 s after ACES High had previously passed through the same region. Figure 3c shows that the visible arc remained spatially stable; however, it began to fade in intensity as the ACES Low payload completed its passage through the arc, exiting at 09:54:25 UT. The correlation between the electron data on ACES Low and the all-sky images suggest that the quasi-static arc produced an inverted V with a peak energy of 4 keV. At 09:54:40 UT, ACES Low began to enter the western edge of the westward moving auroral form that ACES High had previously skirted, as shown by Figure 3d. ACES Low moved more centrally through the dynamic westward moving region, as illustrated by Figure 3e. An inverted V was observed in the electron flux with a similar peak energy of 4 keV that corresponded well with the westward moving region.

3.2. DC Electric Fields

The DC electric fields observed on ACES High and Low were generally southwesterly. For the midnight MLT sector, into which the payloads were launched, the electric fields observed on ACES High and Low are consistent with electric fields formed as the result of plasma convection around the Earth [*Baumjohann*, 1982]. Until 09:55:45 UT, both of the DC electric field components showed low levels of variation, but the meridional field exhibited more variability than the zonal component. After 09:55:45 UT, as the payload entered the region with multiple time-dispersed electron events, the zonal and meridional electric field components reversed direction to more eastward and northward, respectively. Electric and magnetic field perturbations correlate with the time-dispersed arrival of electrons throughout the flight, especially at 09:54:15 UT and after 09:55:45 UT. The time-dispersed electrons are consistent with simulations by *Kletzing and Hu* [2001], and combined with the perturbations in the fields, they suggest that ACES High passed through regions where Alfvén waves accelerated low-energy electrons. However, a more detailed analysis of these observations is outside the scope of this paper.

The DC electric fields on the ACES Low payload showed low levels of variation over the duration of the flight. The westward directed electric field exhibited a maximum variation of 15 mV m^{-1}, whereas, the southward directed field had a maximum variation of 20 mV m^{-1}. The oscillations at the beginning of the data, in both components of the electric field, were an artifact of the payload attitude solution. The total perpendicular electric field magnitude observed on ACES Low was typically half the magnitude of the total perpendicular electric field observed on ACES High.

3.3. Magnetic Fields

The bottom three panels in Figure 2 present the residual magnetic field components for both ACES High and Low. On ACES High before 09:56:00 UT, variations in the zonal and meridional residual magnetic field components were suggestive that field-aligned sheet currents were present. The most notable signature of the presence of field-aligned cur-

rents occurred at 09:56:00 UT, in which there was a 75 nT reversal from north to south in the meridional component, while simultaneously, a reversal of 100 nT from west to east was observed in the zonal component. Reductions in the precipitating electron flux were observed at the location of this magnetic reversal, which suggest a possible downward current.

As a result of payload coning, the ACES Low magnetometer required more filtering to reduce the effects of spinning and coning in the residual magnetic field components. Gradients of the zonal and meridional residual magnetic field components correlate with a reduction in the electron flux at approximately 09:54:30 UT. This magnetic field gradient, combined with a lack of visible aurora in Figure 3d, and a reduction in the electron flux could suggest that downward field-aligned current was present. Shortly thereafter, at 09:54:45 UT, gradients in the zonal and meridional residual magnetic field component correlate with precipitating electrons associated with the large westward moving auroral region. Even at lower altitudes, some component of field-aligned current may be penetrating down to the altitudes of ACES Low.

One of the other notable differences between ACES High and Low resides in the magnitude of the residual field-aligned magnetic component. From Ampère's law, gradients in the residual zonal and meridional magnetic field components would only contribute to field-aligned currents. However, if cross field currents flow, large gradients in the residual field-aligned component would be present. The maximum magnetic variation observed in the residual field-aligned component on ACES High was 25 nT; however, on ACES Low the maximum variation is approximately 145 nT. Moreover, the gradients observed in the residual field-aligned magnetic component on ACES Low aligned well with the payload entering and exiting regions of precipitating auroral electrons. At 09:54:20 UT, as ACES Low exited the first arc, the residual field-aligned component has a large gradient from negative to positive. Shortly thereafter, the residual field-aligned component reverses from large positive to negative values as the payload enters into the westward moving auroral region. The larger gradients observed in the residual field-aligned component on ACES Low suggest the presence of closure current.

3.4. Currents

A calculation of the field-aligned currents, using a sheet approximation, was performed using gradients in the residual magnetic field components observed on ACES High. Figure 4 illustrates the result of this calculation, in which the electron flux and the DC electric fields are plotted above the

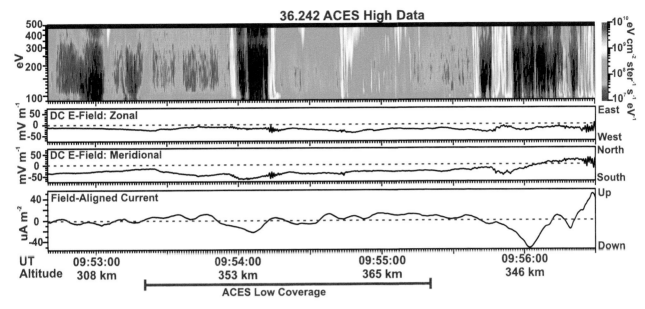

Figure 4. Results from the calculation of field-aligned current and the electron differential energy flux. Upward and downward field-aligned currents are positive and negative, respectively.

field-aligned current for comparison. The current data presented have been smoothed to elucidate the large-scale features. Observations on ACES High agree with results from *Arnoldy* [1977], in which it was found that current sheets tend to be observed toward the edges of auroral arc structures. In addition, modest regions of upward field-aligned current, indicative of precipitating electrons, were observed over regions mapped to inverted-V aurora as deduced from ACES Low data and the regions of visible aurora on the all-sky imagers. The most notable region is the quasi-static arc, but a lower magnitude of upward current is also associated with the westward moving aurora, which ACES High skirted. A lack of precipitating electron flux correlated well with regions of downward field-aligned current, which is consistent with upward moving electrons [*Marklund et al.*, 1994, 1997; *Elphic et al.*, 1998]. The region poleward of the quasi-static arc and the region poleward of the westward moving aurora both have downward currents along with a reduction of electron flux.

4. DISCUSSION

The lack of structure in the low-energy electron flux was one of the most notable differences between observations made on ACES High versus ACES Low. Two explanations can account for this discrepancy. First, the auroral configuration evolved during the time interval when ACES High first traversed a given region to the time when ACES Low passed through the same region. While the first auroral arc

was quasi-static over the traversal time of both payloads, the larger westward moving auroral region was dynamically evolving over short time scales.

The second explanation is that precipitating electrons colliding with atmospheric neutrals were significant at ACES Low altitudes. To determine the effect of ionospheric collisionality, the stopping altitude was determined, which is the altitude at which precipitating electrons have experienced scattering interactions becoming indistinguishable from the background. A procedure was used similar to that described by *Kivelson and Russell* [1995], section 7.2.2, and a model ionosphere mass density from *Kelley* [1989] was used. Figure 5 presents the incident energy of precipitating electrons versus altitude; it was found from this calculation that electrons at 500 eV were typically scattered and became indistinguishable from the background plasma at approximately 160 km. This is 30 km above the apogee of ACES Low, which suggests that collisions are significant enough to diminish structure in the precipitating low-energy electron flux.

The low level of electric field variation observed on ACES Low is likely due to the inverse relationship between the electric field and ionospheric conductivity. As the conductivity in the ionosphere increases, it becomes significant enough to effectively "short" the electric fields in the lower ionosphere. The consequences of ionospheric conductivity are that the magnitude of the electric field is reduced, and it effectively eliminates electric field structure at lower altitudes. An estimate was performed to understand the electric

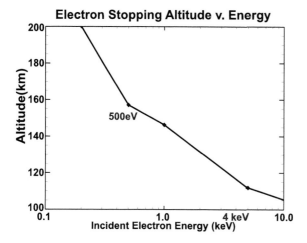

Figure 5. The results from the stopping altitude calculation, which show that precipitating electrons at 500 eV are scattered and become indistinguishable at about 160 km from background ionospheric electrons.

field observations from ACES Low by considering the ionosphere as a uniformly conductive slab ranging in altitude from 120–160 km. From classical electrodynamics, the skin depth of a conducting slab is $\delta = \sqrt{2/(\omega\mu_0\sigma_p)}$ [*Jackson*, 1998], where σ_p is the Pedersen conductivity, and the frequency was chosen to be 1 Hz. Using the magnitude difference between the observed electric fields from both payloads over both arc crossings, along with the altitude of ACES Low relative to the top side of the slab, the value for the Pedersen conductivity (Σ_p) was determined. The height-integrated Pedersen conductivities were further calculated and found to be 2.9 mho for the quasi-static arc and 6.0 mho

for the westward moving aurora. These values are in good agreement with 10 mho from *Kelley* [1989] and 5 mho using empirical relations from *Robinson et al.* [1987].

The orientation of the observed fields within an auroral arc is important for making inferences regarding arc electrodynamics. In a stationary 2-D model of a long arc with large east-west extent, the residual magnetic fields will have maximum variation in the tangential component aligned along the arc, and the electric fields will have maximum variation in the normal component, perpendicular to the length of the arc [*Boström*, 1964]. A minimum variance technique was employed to compare the direction of an arc-oriented coordinate system relative to the fields. The orientation of the normal and tangential directions of the auroral arc at the location of the payload were determined from the all sky images for comparison to the fields.

The minimum variance procedure was carried out on the fields of the quasi-static arc on ACES High, and the results are shown in Figure 6. The gray region represents the arc location and orientation on a geomagnetic longitude and latitude grid. As seen in Figure 6, the residual magnetic fields have a significant tangential component that is well aligned to the arc, which is in good agreement with the 2-D model. This component points in the eastward direction on the equatorward side of the arc and westward on the poleward side of the arc, which is consistent with an upward directed field-aligned current that was observed in this region.

The results for the residual electric field, in which the bulk convective flow has been removed to elucidate variation, are also shown in Figure 6. The electric fields are not as well aligned with the arc, but the fields do have a converging sense pointing toward the center of the arc. This electric field

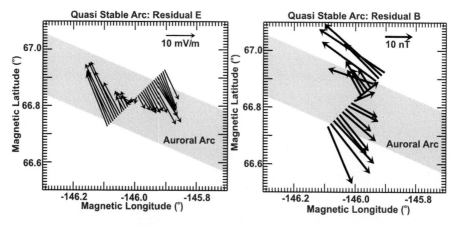

Figure 6. (left) The electric field vectors, after having the convective flow removed, as ACES High traversed the quasi-static arc. The vectors are consistent with what would be expected for a U-shaped potential drop mapped down from high altitudes. (right) The residual magnetic field as ACES High traversed the quasi-static region. The field pattern is consistent with what would be expected for an upward field-aligned current.

signature is consistent with electric fields associated with inverted-V-type aurora [*Marklund*, 1984]. It is worth noting that the all-sky imager observed some vorticity in the arc that was occurring coincidently to the passage of ACES High through the quasi-static arc. In addition, the all-sky imager observed some expansion of the arc width, which could be the result of $\mathbf{E} \times \mathbf{B}$ plasma motion. These two mechanisms could have affected the electric fields that mapped from the magnetosphere and explain the lack of alignment with the arc model of *Boström* [1964].

ACES Low acquired in situ magnetic field measurements that are consistent with the closure current; however, from these observations, it is difficult to deduce the direction and the geometry of the closure current directly from the measurements. First, an ambiguity exists in which both meridional or zonal currents could cause gradients in the residual field-aligned component. However, the crucial parameter is the direction of the currents relative to the DC electric field vector, which defines the direction of the Hall and Pedersen currents. One of the most significant outstanding questions regarding current closure is what percentage of the closure current is due to Hall or Pedersen current. This is especially true with respect to Joule heating, in which energy will be dissipated through Pedersen currents, but not dissipated by Hall currents. Therefore, a prerequisite for any study of current closure must be knowledge of the electric field vector relative to the direction of the cross-field current. Second, there is an ambiguity of the payload location within the closure current itself, which is unknown based on these observations. Even for a simple geometry, payload crossings at different locations in the closure current region could produce similar magnetic field signatures.

To understand the observations from ACES Low, a model could be developed that would address the issues regarding the ambiguity in currents deduced from magnetic field perturbations and the ambiguity of a current closure geometry. The model by *Mallinckrodt* [1985] would be an appropriate starting point. Much of the relevant physics regarding collisional ionization and neutral winds has been included in this model. As has been noted from JOULE II observations [*Sangalli et al.*, 2009], the neutral winds do play a role in the Hall and Pedersen conductivities. Observations of electric fields and field-aligned currents from ACES High, which are assumed to have mapped down from the magnetosphere, could be used as model inputs. The model can then be run with different input parameters, with Hall and Pedersen currents being deduced from the inputs, and changes in the electric field due to the increased ionospheric conductivity. A "test payload" can be flown through the modeled current closure region and observations from ACES Low can be compared to model result. As illustrated in Figures 3a–3c, the quasi-static arc remained spatially stable over nearly a minute, while both payloads crossed it; combined with measurements of field-aligned currents, it appears to be a good candidate for such a model. However, development and results from such a model are outside the scope of this paper.

5. CONCLUSIONS

Two payloads were successfully launched from Poker Flat Research Range, Alaska, on 29 January 2009 into a dynamic multiple-arc aurora. As has been shown, there is approximately a minute of conjugate data coverage between both payloads. Both payloads traversed a quasi-static arc early in their respective flights and later traversed a westward moving auroral form. Toward the end of the flight, only ACES High crossed a very dynamic poleward arc.

Data presented for both payloads included the electron differential energy flux, electric fields, and residual magnetic fields. Field-aligned currents were also derived from residual magnetic field components for the ACES High payload. ACES High observed time-dispersed arrivals of low-energy precipitating electrons and simultaneous observations of perturbations in the electric and magnetic fields, which suggest that ACES High traversed regions where Alfvén waves were present. Upward and downward field-aligned sheet currents were observed and compared to both the electron differential energy flux and all-sky images. It was found that regions with a lack of precipitating electrons and lack of visible aurora in the all-sky imagers were observed at the same time as regions of downward directed field-aligned current. Conversely, regions with precipitating electron flux and visible aurora correlated well with upward field-aligned currents.

ACES Low traversed two inverted Vs with peak energies of 4 keV. One of the most significant features was the lack of structure in the low-energy precipitating electrons. To get a measure on how significant collisional effects might be on ACES Low, a calculation of the stopping altitude suggested that most electrons below 1 keV would appear as indistinguishable from the ionospheric background. Ionospheric conductivity is most likely the dominant mechanism responsible for the low variation in the magnitude of both the electric field components. Using a slab geometry for the ionosphere, a calculation of the skin depth that inferred the height-integrated Pedersen conductivity was in good agreement with typical values associated with the ionosphere. Gradients in the residual magnetic field-aligned component are strongly suggestive that closure currents were flowing at ACES Low altitudes providing ideal inputs for modeling to address the observations.

REFERENCES

Ackerson, K. L., and L. A. Frank (1972), Correlated satellite measurements of low-energy electron precipitation and ground-based observations of a visible auroral arc, *J. Geophys. Res.*, *77*(7), 1128–1136.

Arnoldy, R. L. (1974), Auroral particle precipitation and Birkeland Currents, *Rev. Geophys.*, *12*(2), 217–231.

Arnoldy, R. L. (1977), The relationship between field-aligned current, carried by suprathermal electrons, and the auroral arc, *Geophys. Res. Lett.*, *4*(10), 407–410.

Baumjohann, W. (1982), Ionospheric and field-aligned current systems in the auroral zone—A concise review, *Adv. Space Res.*, *2*, 55–62, doi:10.1016/0273-1177(82)90363-5.

Boström, R. (1964), A model of the auroral electrojets, *J. Geophys. Res.*, *69*(23), 4983–4999.

Elphic, R. C., et al. (1998), The auroral current circuit and field-aligned currents observed by FAST, *Geophys. Res. Lett.*, *25*(12), 2033–2036.

Evans, D. S., N. C. Maynard, J. Trøim, T. Jacobsen, and A. Egeland (1977), Auroral vector electric field and particle comparisons, 2. Electrodynamics of an arc, *J. Geophys. Res.*, *82*(16), 2235–2249.

Fejer, J. A. (1953), Semidiurnal currents and electron drifts in the ionosphere, *J. Atmos. Terr. Phys.*, *4*(4–5), 184–203, doi:10.1016/0021-9169(53)90054-3.

Frank, L. A., and K. L. Ackerson (1971), Observations of charged particle precipitation into the auroral zone, *J. Geophys. Res.*, *76*(16), 3612–3643.

Jackson, J. D. (1998), *Classical Electrodynamics*, 3rd ed., Wiley, New York.

Kelley, M. C. (1989), *The Earth's Ionosphere: Plasma Physics and Electrodynamics*, 487 pp., Academic Press, San Diego, Calif.

Kivelson, M. G., and C. T. Russell (1995), *Introduction to Space Physics*, Cambridge Univ. Press, Cambridge, U. K.

Kletzing, C. A., and S. Hu (2001), Alfvén wave generated electron time dispersion, *Geophys. Res. Lett.*, *28*(4), 693–696, doi:10.1029/2000GL012179.

Kletzing, C. A., G. Berg, M. C. Kelley, F. Primdahl, and R. B. Torbert (1996), The electrical and precipitation characteristics of morning sector Sun-aligned auroral arcs, *J. Geophys. Res.*, *101*(A8), 17,175–17,189.

Mallinckrodt, A. J. (1985), A numerical simulation of auroral ionospheric electrodynamics, *J. Geophys. Res.*, *90*(A1), 409–417.

Mallinckrodt, A. J., and C. W. Carlson (1985), On the anticorrelation of the electric field and peak electron energy within an auroral arc, *J. Geophys. Res.*, *90*(A1), 399–408.

Marklund, G. (1984), Auroral arc classification scheme based on the observed arc-associated electric field pattern, *Planet. Space Sci.*, *32*, 193–211, doi:10.1016/0032-0633(84)90154-5.

Marklund, G., I. Sandahl, and H. Opgenoorth (1982), A study of the dynamics of a discrete auroral arc, *Planet. Space Sci.*, *30*, 179–197, doi:10.1016/0032-0633(82)90088-5.

Marklund, G., L. Blomberg, C. Fälthammar, and P. Lindqvist (1994), On intense diverging electric fields associated with black aurora, *Geophys. Res. Lett.*, *21*(17), 1859–1862.

Marklund, G., T. Karlsson, and J. Clemmons (1997), On low-altitude particle acceleration and intense electric fields and their relationship to black aurora, *J. Geophys. Res.*, *102*(A8), 17,509–17,522.

Richmond, A. D., and J. P. Thayer (2000), Ionospheric electrodynamics: A tutorial, in *Magnetospheric Current Systems*, Geophys. Monogr. Ser., vol. 118, edited by S. Ohtani et al., pp. 131–146, AGU, Washington, D. C., doi:10.1029/GM118p0131.

Robinson, R. M., R. R. Vondrak, K. Miller, T. Dabbs, and D. Hardy (1987), On calculating ionospheric conductances from the flux and energy of precipitating electrons, *J. Geophys. Res.*, *92*(A3), 2565–2569.

Sangalli, L., D. J. Knudsen, M. F. Larsen, T. Zhan, R. F. Pfaff, and D. Rowland (2009), Rocket-based measurements of ion velocity, neutral wind, and electric field in the collisional transition region of the auroral ionosphere, *J. Geophys. Res.*, *114*, A04306, doi:10.1029/2008JA013757.

B. J. Anderson and H. Korth, Applied Physics Laboratory, Johns Hopkins University, Baltimore, MD, USA.

S. R. Bounds, S. R. Kaeppler and C. A. Kletzing, Department of Physics and Astronomy, University of Iowa, Iowa City, IA, USA. (stephen-kaeppler@uiowa.edu)

M. P. Dombrowski and J. W. LaBelle, Department of Physics and Astronomy, Dartmouth College, Hanover, NH, USA.

J. W. Gjerloev, Department of Physics and Technology, University of Bergen, Allegt. 55, Bergen N-5007, Norway.

C. J. Heinselman, Center for Geospace Studies, SRI International, Menlo Park, CA, USA.

S. Jones, R. F. Pfaff and D. E. Rowland, NASA Goddard Space Flight Center, Greenbelt, MD, USA.

M. Lessard, Space Sciences Center, University of New Hampshire, Dover, NH, USA.

Auroral Disturbances as a Manifestation of Interplay Between Large-Scale and Mesoscale Structure of Magnetosphere-Ionosphere Electrodynamical Coupling

L. R. Lyons,[1] Y. Nishimura,[1] X. Xing,[1] Y. Shi,[1] M. Gkioulidou,[1] C.-P. Wang,[1] H.-J. Kim,[1] S. Zou,[2] V. Angelopoulos,[3] and E. Donovan[4]

Both slowly varying large-scale structure and a variety of disturbances occur within the coupled magnetosphere-ionosphere system along nightside plasma sheet magnetic field lines, and both are reflected in the aurora. Recently, there have been expansion and additions to radar systems, enhancements in ground auroral imaging, strategically placed multiple spacecraft of the Time History of Events and Macroscale Interactions during Substorms program, and new model development. Studies using these new capabilities are suggesting that much about the structure and disturbances along plasma sheet field lines may be understood based on a slowly changing large-scale structure describable by region 1 and 2 physics and its interplay with much more dynamic mesoscale structures associated with flow channels emanating from near the polar cap boundary. These new results also suggest a possible unifying view that many auroral disturbances (poleward boundary intensifications, streamers, substorms, and perhaps others) may be related to the mesoscale flow channels. If so, understanding the generation and propagation of the flow channels and their coupling with the large-scale background is critical. Evidence is now fairly strong that the flow channels consist of plasma with lower flux tube–integrated entropy than the surrounding plasma. Also, initial studies have led to the hypothesis that enhanced mesoscale flows from deep within the polar cap and impinging on the nightside polar cap boundary may be important for triggering the flow channels within the plasma sheet, the resulting highly structured flow of plasma into the plasma sheet, perhaps giving a highly structured source of plasma sheet plasma with highly variable properties, including entropy.

[1]Department of Atmospheric and Oceanic Sciences, University of California, Los Angeles, California, USA.

[2]Department of Atmospheric, Oceanic and Space Sciences, University of Michigan, Ann Arbor, Michigan, USA.

[3]Department of Earth and Space Sciences, University of California, Los Angeles, California, USA.

[4]Department of Physics and Astronomy, University of Calgary, Calgary, Alberta, Canada.

Auroral Phenomenology and Magnetospheric Processes: Earth and Other Planets
Geophysical Monograph Series 197
10.1029/2011GM001152

1. INTRODUCTION

Many space weather disturbances are associated with temporal changes and spatial structuring of electric fields and associated currents along plasma sheet magnetic field lines. They can be visually identified via the aurora, which is a manifestation of the electrodynamical coupling between the magnetosphere and ionosphere. Under the slow-flow approximation (neglect of the inertial term in the momentum equation) [*Wolf*, 1983], the electrodynamical coupling arises because magnetospheric particle magnetic drift can drive currents across closed magnetic field lines that are not divergence free when there are pressure gradients [*Heinemann*, 1999; *Lyons et al.*, 2009a]. For current continuity to be maintained within the magnetosphere, this divergence leads to magnetic field-aligned currents (FACs) flowing to and from the ionosphere. FACs can also be driven when there are large spatial and/or temporal changes in particle velocity so that the inertial term cannot be neglected. Ionospheric electric fields adjust to give a divergence of the horizontal ionospheric currents that balances the FAC, giving current continuity in the ionosphere as well as the magnetosphere. Differences between the mapping along magnetic field lines of ionospheric electric fields to the magnetosphere and the magnetospheric electric field are mitigated by Alfven waves [*Southwood and Kivelson*, 1991; *Vogt et al.*, 1999; *Yoshikawa et al.*, 2011]. This leads to electrodynamic feedback between the magnetosphere and ionosphere that leads, under the slow variation assumption [*Wolf*, 1983], to identically mapped electric fields within both regions (except in localized regions of substantial field-aligned electric potential drops). Thus, while coupling with the solar wind is important for determining the overall strength of magnetospheric convection, it does not impose a distribution of electric fields within the ionosphere or magnetosphere. Instead, electrodynamical coupling controls this distribution.

Significant advances in understanding the coupled magnetosphere-ionosphere electrodynamic system are now starting to occur as a result of major improvements in our observational and modeling capabilities. Most important to the discussion here are ground-based radars, most of which have been deployed by the National Science Foundation, the continental-scale, high-resolution all-sky auroral imager (ASI) array [*Mende et al.*, 2008] and the multiple spacecraft constellation of NASA's Time History of Events and Macroscale Interactions during Substorms (THEMIS) program [*Angelopoulos*, 2008], and the Rice Convection Model (RCM) [*Harel et al.*, 1981; *Wolf et al.*, 2007]. Initial studies from these facilities and the RCM have suggested that much about the plasma sheet processes leading to space weather disturbances can be viewed as interplay *between a slowly changing large-scale structure and much more dynamic mesoscale structures* associated with flow channels. In addition, recent analyses have revealed the unexpected possibility that flow structure formed within the polar cap regions of open field lines (which, within the ionosphere, lies poleward of the auroral oval) may be important for driving the plasma sheet mesoscale structures.

In this chapter, we describe the pieces that are starting to emerge of the framework for understanding the dynamics and disturbances along plasma sheet field lines, including the large-scale structure and the mesoscale flow structures and disturbances associated with the flow channels. The possible connection with polar cap convection structure is also addressed.

2. LARGE-SCALE STRUCTURE

The properties of a slowly evolving (~tens of minutes to hours) large-scale structure, extending throughout the nightside plasma sheet and auroral oval, have been studied extensively with the RCM [*Erickson et al.*, 1991; *Jaggi and Wolf*, 1973; *Toffoletto et al.*, 2003; *Wang et al.*, 2004; *Wolf et al.*, 2007; *Gkioulidou et al.*, 2009]. The RCM evaluates the drift transport physics of plasma sheet particles and self-consistently includes the electrodynamic coupling with the ionosphere under a time-dependent magnetic field and variable rate of driving by the solar wind.

An example of the response to enhanced driving by the solar wind as simulated with the RCM [*Gkioulidou et al.*, 2009] is shown within the ionosphere in Figure 1. The two rows show FACs as mapped to the ionosphere and height-integrated Pedersen conductivity in the ionosphere as a function of magnetic latitude (Λ) and magnetic local time (MLT) within 3 h of midnight. Equipotentials are overlaid in each panel. An enhancement of solar wind driving was simulated by increasing the cross polar cap potential drop ($\Delta\Phi_{PCP}$) from 30 to 100 kV just after $T = 0$. At $T = 0$, immediately prior to the increase in $\Delta\Phi_{PCP}$, the model shows that the inner plasma sheet pressure [see *Gkioulidou et al.*, 2009] leads to weak FACs, referred to as region 2 (R2) currents, which are upward at midnight to dawn MLTs and downward at dusk to midnight MLTs and which map to the ionosphere at Λ ~66°– 67°. Electrodynamic coupling between the plasma sheet and the ionosphere associated with these currents leads to strong shielding of electric fields from the region equatorward of the plasma sheet precipitation. Shielding is seen as a bending of equipotential contours away from midnight. This gives enhanced poleward-directed electric fields on the premidnight side (close together equipotentials) equatorward of the inner edge of the electron plasma sheet (identified from the equatorward edge of the height-integrated ionospheric Pedersen conductivity). These fields give strong plasma flows and are referred to as subauroral polarization (SAPS) electric fields. These lie in the region of downward R2 currents and are driven by gradients of the strongest inner magnetosphere ion pressures and low conductance in the subauroral ionosphere.

The SAPS electric field can be seen to move equatorward within the ionosphere along with the electron plasma sheet after the increase in $\Delta\Phi_{PCP}$. The $\Delta\Phi_{PCP}$ increase also leads to a gradual increase in plasma sheet pressures, pressure gradients, and R2 currents. Initially, enhanced electric fields penetrate equatorward of the plasma sheet, but shielding gradually reestablishes. The plasma sheet pressures and equipotentials within the equatorial plane well after the increase in $\Delta\Phi_{PCP}$ are illustrated in Figure 2. Plasma and magnetic field properties of this large-scale system have been verified

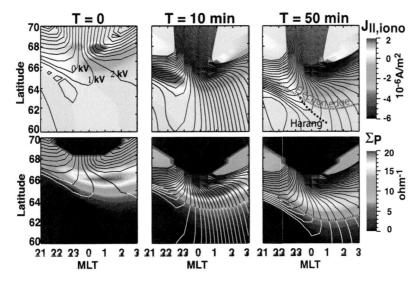

Figure 1. Rice Convection Model (RCM) results within the ionosphere using the Tsyganenko 1996 magnetic field model. Plasma boundary conditions were taken from a statistical analysis of 11 years of Geotail observations for weakly northward interplanetary magnetic field conditions [*Wang et al.*, 2007]. Results are shown just prior to the enhancement in convection ($T = 0$) and for indicated times after the enhancement. Field-aligned currents (FACs) are positive downward. Height-integrated Pederson conductivity in the bottom row reflects the energy flux of plasma sheet electron precipitation. Equipotentials are overlaid in each panel. Dashed lines in the panels for $T = 50$ min identify the Harang reversal [*Gkioulidou et al.*, 2009].

with observations from in situ spacecraft [e.g., *Wang et al.*, 2011, and references therein] and from radar observations from the ground [e.g., *Bristow and Jensen*, 2007; *Bristow et al.*, 2001; *Hughes and Bristow*, 2003; *Lyons et al.*, 2009b; *Zou et al.*, 2009a], demonstrating the importance of the magnetosphere-ionosphere electrical coupling in determining the large-scale particle, current, and electric field structure along plasma sheet field lines.

3. INTERPLAY BETWEEN LARGE-SCALE AND MESOSCALE STRUCTURE

Superposed on the large-scale ionospheric and plasma sheet flow structure are mesoscale perturbations [*Angelopoulos et al.*, 1992] that often have a substantially larger magnitude than the background. These bursts of flow within the plasma sheet are typically localized in the y direction [*Angelopoulos et al.*,

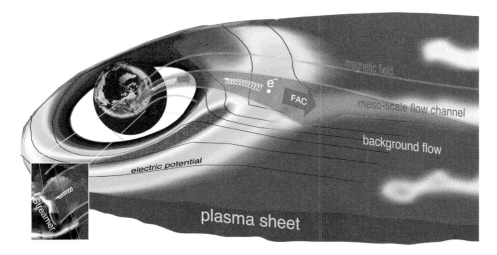

Figure 2. Coupling along magnetic field lines between plasma sheet particles near the equatorial plane and flows and the auroral ionosphere. Plasma pressures and electric potentials of the large-scale background obtained from the RCM well after the increase in $\Delta\Phi_{PCP}$ are shown in the equatorial plane, and the connection between mesoscale flow channels and auroral upward FAC and streamers is illustrated.

1996], so that they can be viewed as channels of earthward flow within the plasma sheet that are superposed on the large-scale pattern as sketched in Figure 2, and they are associated with transport of plasma perpendicular to the magnetic field direction.

The flow channels cause auroral activity that is also localized in longitude. Specifically, poleward boundary intensifications (PBIs) have been related to flow channels that can carry plasma across the nightside separatrix into the plasma sheet [*de la Beaujardière et al.*, 1994; *Lyons et al.*, 1999]. This can be seen in Figure 3, which shows a global auroral image from the Wideband Imaging Camera on the IMAGE spacecraft, with flow vector observations from the global SuperDARN radar array overlaid on the auroral image. This example is shown because of the two bright PBIs (that reach the yellow on the color scale and can be seen at geomagnetic latitude $\Lambda \sim 73°$ on the nightside) and the quite good coverage of radar echoes over the polar cap and auroral oval. Following the format of *Zou et al.* [2009b], two types of ionospheric flow vectors are shown: the solid squares with thicker lines give flow vectors merged from multiple line-of-sight (LOS) velocity observations when they are available and the dots

with thinner lines give the LOS velocities when only one such measurement is available at a given location. Note the two longitudinally localized regions of enhanced flow, identified in Figure 3 by solid white arrows, which appear to move from the polar cap into the plasma sheet adjacent to the two PBIs. The PBIs are to the right of the flow direction, as required for the converging Pedersen currents on the edges of the flow regions to give the upward FAC that are associated with discrete aurora.

Some PBIs develop into equatorward-moving auroral arcs that are roughly north-south oriented and are referred to as streamers. Streamers have been related to channels of enhanced earthward flows [*Rostoker et al.*, 1987; *Sergeev et al.*, 1999, 2000; *Zesta et al.*, 2000; *Henderson et al.*, 2002; *Pitkänen et al.*, 2011], which can extend from deep within the plasma sheet to near its earthward edge. An example of a streamer observed from the THEMIS all-sky camera network of North America is shown in Figure 4, and LOS flow speeds from the SuperDARN radars are overlaid on the aurora image. This example shows a streamer that extends in a southeastward direction from the poleward portion of the auroral oval and then turns westward within the equatorward portion of the oval.

The magenta arrow in Figure 4 illustrates the plasma sheet flow channel that would be expected to be associated with such a streamer. Based on the shape of the streamer, the flow channel appears to be guided by the large-scale background flow, first flowing around the Harang reversal, and then flowing westward within the SAPS region. The radar LOS flows are consistent with this picture, the flows in the eastern Saskatoon (SAS) beams being toward the radar (positive speeds) at the higher latitudes with echoes ($\Lambda \sim 66°–68°$). At $\Lambda \sim 64°–66°$, the LOS flows from the SAS radar switch from being toward the radar in the eastern beams to away from the radar in the western beams, which is as expected from approximately westward flow across the radar field of view (FOV). Based on the LOS flows observed by the Prince George radar, this westward flow appears to have extended well westward approximately parallel to the aurora. Images and flows before the time shown in Figure 4 show that the aurora shape and the LOS flows evolved with time together. This suggests that the flow channels affect the large-scale pattern in addition to being partially guided by the large-scale pattern, thus indicating that the full flow pattern may be viewable as an interplay between the large-scale and mesoscale structures.

PBIs and streamers are space weather disturbances that frequently occur, and the streamer association with mesoscale flow channels and FACs that couple between the plasma sheet and ionosphere is illustrated in Figure 2. The obvious question arises as to why the flow channels move rapidly

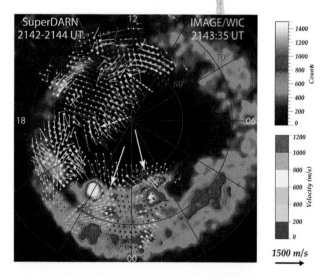

Figure 3. Global auroral image from the Wideband Imaging Camera on the IMAGE spacecraft, with flow vector observations from the global SuperDARN radar array overlaid on the auroral image. Two bright poleward boundary intensifications (PBIs) can be seen at geomagnetic latitude $\Lambda \sim 73°$ on the nightside. Solid squares with thicker lines give flow vectors merged from line-of-sight (LOS) velocity observations when they are available, and the dots with thinner lines give the LOS velocities when only one such measurement is available at a given location. Solid white arrows identify two longitudinally localized regions of enhanced, polar cap flow associated with the PBIs, and dashed white arrows identify two additional localized polar cap flow enhancements.

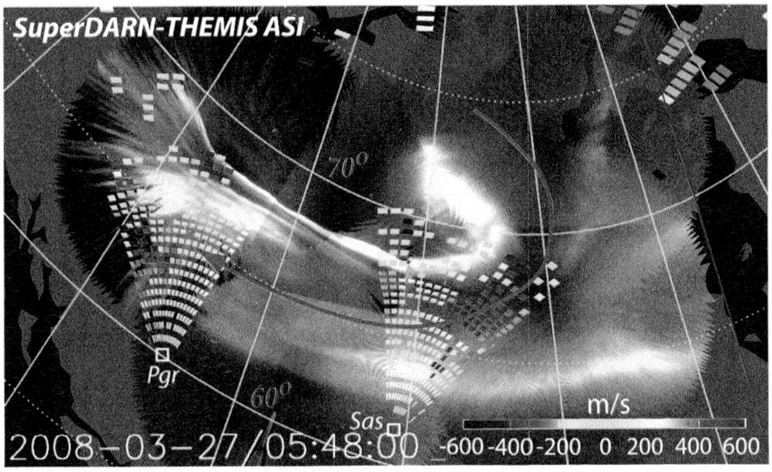

Figure 4. A streamer observed from the Time History of Events and Macroscale Interactions during Substorms (THEMIS) all-sky cameras, with LOS flow speeds from the SuperDARN radars overlaid on the aurora image as color-coded dots. The magenta arrow illustrates the plasma sheet flow channel that would be expected to be associated with the streamer seen in the auroral image.

earthward through the more slowly flowing background plasma sheet. It has been proposed that this motion occurs when there is a longitudinally localized region of plasma within flux tubes that has substantially reduced $PV^{5/3}$ than the surrounding plasma, causing the reduced $PV^{5/3}$ plasma to move earthward via interchange motion [*Chen and Wolf*, 1993; *Pontius and Wolf*, 1990; *Yang et al.*, 2008; *Zhang et al.*, 2009]. Here P is plasma pressure and V is flux tube volume. This proposal for the flow channels has been supported quite well by observations. In particular, the field lines within the flow channels are expected to be more dipolar than the surrounding field, and statistical analyses of observations have shown that is a common feature of flow channels [*Angelopoulos et al.*, 1994; *Dubyagin et al.*, 2010; *Ohtani et al.*, 2004]. Furthermore, *Dubyagin et al.* [2010] have shown that the dipolarization within flow channels occurs together with a reduction in $PV^{5/3}$.

An example showing the connection between flow channels and streamers on 27 February 2009 is shown in Figure 5. Two combined images from four THEMIS ASI stations over central Canada are shown at the top, the blue line showing the magnetic midnight meridian. Two auroral streamers are identified. The first, labeled streamer 1 in the two panels, became discernible in the images ~1 min earlier than the time (07:53:00 UT) of the first panel. The second, labeled streamer 2 in the 07:54:36 panel, formed ~1.5 min after and ~1 h in MLT to the east of streamer 1. Mappings to the auroral oval of two of the THEMIS spacecraft, C and D, are overlaid on the images (as obtained with the Tsyganenko 1996 magnetic field model) [*Tsyganenko*, 1995]. The mappings show C being just to the east of streamer 1 and D being near the

longitude of streamer 2. While azimuthal errors in the mapping can be as much as ~0.5 h in MLT [*Xing et al.*, 2010], C and D, located in the near-Earth plasma sheet at $X_{GSM} = -17$ and -11 R_E, respectively, appear to have been in good positions to measure the plasma associated with streamers 1 and 2, respectively.

The remainder of Figure 5 shows observations from C and D for a 17 min interval that includes the time of the two streamers, as described in the figure caption. Flow channels can clearly be seen at C at a time appropriate for the development of streamer 1 and at D for the development of streamer 2. C also saw two subsequent flow channels at times that are appropriate for subsequent streamers that were seen to have formed near the longitude of C in images subsequent to those shown in Figure 5. All of these flow channels show the characteristic features expected for earthward-moving channels of plasma having lower global entropy $PV^{5/3}$ than was seen previously by the spacecraft. There are abrupt and substantial increases in B_z, which implies a corresponding decrease in V, and simultaneous decreases in P_{tot}, which reflects an equatorial plasma pressure decrease. There are also simultaneous bursts of earthward flow, which is seen in both V_{perp} and V_{ExB}. Consistency between V_{perp} and V_{ExB} demonstrates the accuracy of the calculated velocities and that the earthward flow is due to the electric drift, as is expected from interchange motion. The abrupt B_z and plasma parameter changes seen for flow channels, such as associated with streamers 1 and 2 in Figure 5, have led them to be referred to as "dipolarization fronts" [*Nakamura et al.*, 2001; *Runov et al.*, 2009]. *Runov et al.* [2009] noted the important feature that they appear to move earthward as coherent

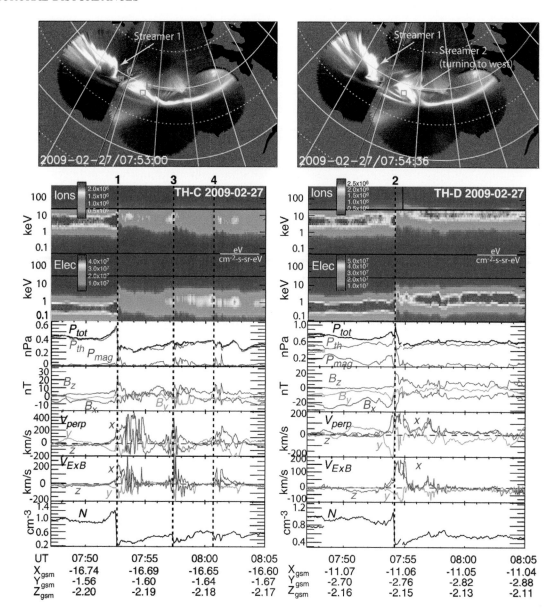

Figure 5. (top) Two combined images from four THEMIS all-sky auroral imager (ASI) stations over central Canada from 27 February 2009. The blue line gives the magnetic midnight meridian, and small squares give the mappings of THEMIS C and D to the auroral oval. (bottom) Observations from THEMIS C and D versus UT for 17 min including the times of the ASI images above. The top two panels show energy flux spectrograms for ions and electrons from the electrostatic analyzer [*McFadden et al.*, 2008] and solid-state detector. The next panels show from top to bottom the pressure (total P_{tot} as defined by *Xing et al.* [2009], thermal P_{th}, and magnetic P_{mag} from the THEMIS magnetometer [*Auster et al.*, 2008]), the GSM components of the magnetic field (B_x, B_y, and B_z), the components of the perpendicular velocity moment $\mathbf{V}_{perp}$, the components of the electric drift velocity $\mathbf{V}_{ExB}$ obtained from the measured electric [*Bonnell et al.*, 2008] and magnetic fields, and the plasma density N.

structures within the $x \sim -10$ to $-20\ R_E$ region of the plasma sheet, and the basic features of the large, abrupt dipolarization front events identified by *Runov et al.* [2009] appear to be the same for other flow channel events, such as those in the earlier studies referred to above.

4. INTERPLAY ASSOCIATED WITH SUBSTORM PREONSET SEQUENCE

The above discussion shows evidence that PBIs and streamers are related to mesoscale channels of earthward-flowing, reduced entropy plasma and suggests that interplay with the large-scale field and plasma system is critical in determining the dynamics on both scales. This interplay becomes particularly interesting for substorms. While most PBIs and streamers are not associated with substorms, they are commonly seen as localized structures during the much larger-scale auroral displays of a substorm expansion phase [*Henderson et al.*, 1998]. Also, they have recently been found to often be a crucial part of the sequence of events that leads to substorm onset [*Nishimura et al.*, 2010a, 2010b], and this finding has been supported by the observations of preonset flow channels within the ionosphere [*Lyons et al.*, 2010a] and the plasma sheet [*Angelopoulos et al.*, 2008; *Lyons et al.*, 2010b; *Xing et al.*, 2010].

The auroral and flow pattern inferred to lead to substorm onset is illustrated in Figure 6. *Nishimura et al.* [2010b] observed that the auroral forms leading to substorm onset frequently approach the onset location from the east or northeast, consistent with the motion of the streamer and adjacent flow channels approximately following the dusk-side convection cell, leading to premidnight onsets as illustrated in Figure 6. Some streamers were observed to move in a pattern that followed the dawn convection cell, such motion generally leading to onsets located from near midnight to postmidnight as is also illustrated in Figure 6. These observations indicate the importance of the large-scale flow pattern in guiding the trajectory of the flow channels and their reduced entropy plasma. The importance of the large-scale pattern was supported by a high probability of Harang aurora, an auroral form having features that move with a pattern that mimics the convection flow around the Harang reversal.

These observations indicate that the large-scale convection pattern has a strong effect on the motion of the preonset auroral forms, and they suggest that the preonset flow channels bring reduced entropy plasma into the near-Earth plasma sheet, leading to onset via an instability that develops as a result of modification of plasma and magnetic field conditions caused by the intruding new plasma of the flow channel. It seems clear that flow channels must penetrate to the near-Earth plasma sheet to lead to a substorm onset, yet most streamers reaching the near-Earth plasma sheet do not lead to an onset. We thus have also investigated differences between auroral streamers reaching the near-Earth plasma sheet that do and do not lead to substorm expansion onset [*Nishimura et al.*, 2011]. While the two types of streamers were found to have similar characteristics, streamers that do not lead to onset were seen to lead to small intensification of a thin growth phase arc, and when an onset-related streamer was observed to reach the equatorward portion of the auroral oval, the preexisting thin growth phase arc was much brighter than at the times of nononset-related streamers. The onset arc is typically near the poleward boundary of a diffuse growth-phase auroral band [*Samson et al.*, 1992], which is likely SAPS region proton aurora. These observations suggest that substorm onset instability is possible only when the preexisting inner plasma sheet pressure and pressure gradient is sufficiently large, which reflects the large-scale plasma distribution, but can be set up by a preceding mesoscale flow channel.

5. POSSIBLE RELATION TO MESOSCALE, POLAR CAP STRUCTURE

As discussed above, the mesoscale flow channels appear to play a crucial role in the dynamics and disturbances of the coupled magnetosphere-ionosphere system. If this is the case, it becomes important to understand the source of these flow channels and why the plasma within the flow channels has reduced $PV^{5/3}$. A potentially critical clue is the observation that flow enhancements leading to PBIs can cross into the

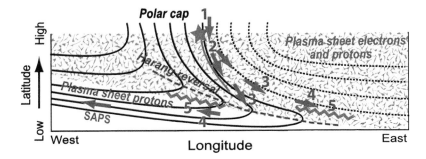

Figure 6. Schematic illustration of motion of preonset auroral forms and their relation to nightside ionospheric convection. The pink star, equatorward extending pink line, and azimuthally extended wavy lines indicate a PBI, an auroral streamer, and onset arcs, respectively. Blue arrows illustrate the plasma flow pattern inferred from preonset auroral motion. Numbers 1–5 show the time evolution of preonset aurora, and yellow and gray areas correspond to proton and electron precipitations. From the work of *Nishimura et al.* [2010b].

auroral oval from the high-latitude region of open polar cap magnetic fields [*de la Beaujardière et al.*, 1994; *Lyons et al.*, 2010b]. Flow crossing into the auroral oval from the polar cap is included in the illustration of Figure 6. It can also be seen in Figure 3, where the strong south-eastward-directed flows heading toward the western edge of the PBI near 01 MLT can be seen extending poleward of the poleward boundary of the aurora to Λ >80°. Further supporting this possibility, *Nishimura et al.* [2010a] found several narrow flow bursts moving from far within the nightside polar cap and leading to PBIs that are followed by streamers, including one that led to a near-Earth substorm onset. This association leads to the hypothesis that enhanced mesoscale flows formed along polar cap field lines may contribute to the triggering of the mesoscale earthward flows along plasma sheet field lines that lead to PBIs, streamers, and substorm onset. Polar cap convection is often viewed as having a slowly varying large-scale structure. However, recent radar observations from within the polar cap are showing that the polar cap flows include significant mesoscale flow structures as does the plasma sheet [*Nishi-*

mura et al., 2010a; *Lyons et al.*, 2011]. This can also be seen in Figure 3, where two localized regions of enhanced flow are identified by dashed arrows in addition to the two localized flow enhancements associated with the nightside PBIs.

The connection between mesoscale flows formed within the polar cap and the triggering of the mesoscale earthward flows that lead to PBIs and streamers is seen clearly in the observations in Figure 7 (based on the work of *Lyons et al.* [2011]). Figure 7 shows LOS flow vectors measured within the polar cap from all beams of the PolarDARN radar (polar cap radar of the SuperDARN network) at Rankin Inlet and the new incoherent scatter radar at Resolute Bay (RISR). The flows are overlaid over a merger of the auroral images from three THEMIS ASIs located equatorward of the flow measurements. A time sequence of selected overlays is shown for the period leading up to a substorm onset at 06:26:30 UT on 21 September 2009. For this event, a flow enhancement is first seen deep within the polar cap in the westward- and poleward-looking beams of RISR, as indicated by the orange arrow in the 06:12:00 UT panel. These flows are directed southeastward,

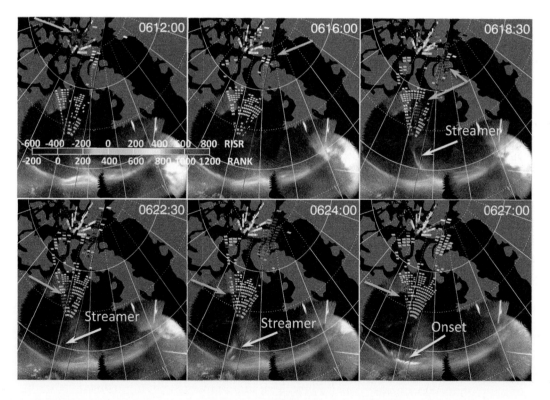

Figure 7. LOS flow vectors measured by the Rankin Inlet PolarDARN radar and Resolute Bay incoherent scatter radar (RISR) overlaid over a merger of the auroral images from three THEMIS ASIs located equatorward of the flow measurements. A time sequence of selected overlays is shown for the period leading up to a substorm onset 06:26:30 UT on 21 September 2009. The RISR and PolarDARN flows, available with 3 and 1 min resolution, respectively, that are overlaid in each panel are from the measurement interval with center time nearest the image time. Yellow arrows identify aurora features discussed in the text, and orange arrows identify equatorward-directed LOS flow channels. The blue line identifies magnetic midnight.

based on RISR flow vectors inferred from the los flows. Then, as indicated in the 06:16 UT panel, an equatorward-directed LOS flow enhancement is seen in the highest latitude echoes of the eastern PolarDARN beams. A longitudinally limited band of equatorward flows are then soon seen (06:18:30 UT panel) centered slightly to the east of the center of the PolarDARN FOV from Λ ~74° to 79°, and this channel of enhanced flow appears to be directly connected to an auroral streamer that is seen extending from Λ ~72° toward the diffuse auroral band lying near the equatorward boundary of the auroral oval. The enhanced flow channel then moved westward within the PolarDARN FOV, and the streamer moved westward along with the flow channel. This westward motion can be seen by comparing the 06:18:30 and 06:22:30 UT panels. The auroral substorm onset is then seen in the 06:27:00 UT panel very near the longitude of the flow enhancement and extending from the streamer longitude (identified in the 06:24:00 UT panel) to about 1 h in MLT to the east.

The flow channel that appears to lead to the streamer and to the substorm onset is narrow in longitudinal extent, much narrower than expected for large-scale convection, and is seen persistently for the 10 min period prior to onset from Λ ~78° to the most equatorward PolarDARN range gate at Λ = 74°. It is tempting to connect this flow channel to the flow enhancements seen earlier at higher latitudes by PolarDARN and RISR-N, suggesting that the enhanced flow propagated from well within the polar cap. While it is not possible to reliably determine if there was a connection to the flows seen at Λ >80° using the available LOS flow measurements, the connection of the flow channel seen at Λ <80° to the streamer and the ensuing onset seems quite clear.

6. SUMMARY

We have summarized evidence that has led us to suggest that the plasma sheet can be viewed as having a large-scale background state, which can be understood using models such as the RCM that include the energy-dependent magnetic drift and electrodynamic coupling with the ionosphere, and mesoscale flow channels, which can lead to disturbances, such as PBIs, streamers, and substorms. Interplay between the large-scale background and the flow channels partially guides the trajectory of the flow channels and plays an important role in controlling the occurrence and features of ensuing disturbances. Also, the interplay with the flow channels affects the large-scale pattern.

We illustrate this framework using the combination of THEMIS ASI images of the aurora in Figure 8 from an unusually

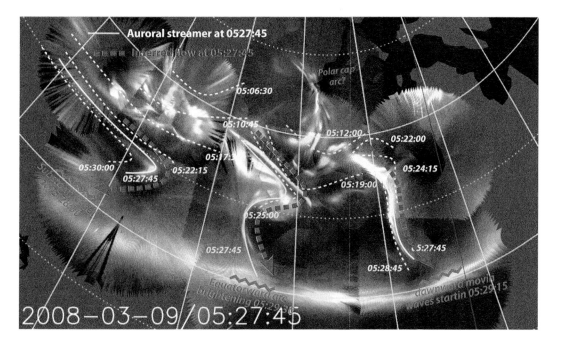

Figure 8. Auroral image at 05:27:45 UT on 9 March 2009 from the THEMIS ASIs, with north to the top. Magnetic latitude circles are drawn as solid lines at 60° and 70°, and longitude lines are drawn at 1 h magnetic local time intervals. The blue line identifies magnetic midnight. Three auroral streamers are identified, the solid yellow curves identifying the streamers at the time of the image, and the dashed yellow curves giving the streamer locations in preceding and subsequent images with the times identified. Red dashed arrows illustrate the flow channels that would be expected to be associated with the streamers at 05:27:45 UT.

clear, disturbed night over North America. The considerable activity seen within the poleward portion of the auroral oval is, we propose, associated with reduced total entropy flow channels. Several auroral streamers are seen simultaneously in the image. Based on our proposed framework, the streamers are like weather fronts, separating plasma masses of different properties. Here they separate higher entropy plasma from lower entropy plasma that is being brought closer to the Earth within the flow channels from the outer plasma sheet boundary. These "fronts" lie along the edge of flow regions (the edge to the right of the flow direction) where upward FAC are associated with discrete aurora, and they slowly move through the background plasma as weather fronts do through the atmosphere. Three streamers (i.e., fronts) are identified in Figure 8, where the solid curves identified the streamers at the time of the image (05:27:45 UT), and the dashed curves show the motion of the fronts as indicated by the streamer locations in preceding and subsequent images. Red dashed arrow illustrates the plasma sheet flow channels that we estimate to have been associated with the streamers at 05:27:45 UT.

Our suggested partial guiding of the inferred flow channels by the large-scale flow can be seen by the streamers at and after the time of the image. The westernmost channel follows the shape of the evening side flow contours, moving first equatorward and dawnward, and then around the Harang electric field reversal to the duskside broad region of diffuse aurora that characterizes the SAPS region [*Zou et al.*, 2009b]. On the other hand, the easternmost flow channel reaches the SAPS region near magnetic midnight and is deflected dawnward as is the dawnside large-scale flow pattern. Each of these flow channels leads to a different type of auroral disturbance when it reaches near the equatorward auroral boundary, as indicated. The results suggest a unifying view that substorms, PBIs, streamers, and other disturbances may be all related to mesoscale flow channels, which come from the distant plasma sheet and couple with the more slowly varying large-scale background. If so, understanding the generation and propagation of these mesoscale flow channels and the coupling with the large-scale background is critical for understanding space weather disturbances.

We have also discussed measurements that have led to the hypothesis that enhanced mesoscale flows from deep within the open polar cap region and impinging on and crossing the nightside polar cap boundary may be important for the triggering of the mesoscale earthward flows along plasma sheet field lines that lead to PBIs, streamers, and substorm onset. The resulting highly structured flow of plasma into the plasma sheet should give a highly structured source of plasma for the plasma sheet, so that the properties of the entering plasma should be quite variable. This could give plasma with reduced global entropy ($PV^{5/3}$) along some of the entering flow channels,

allowing these channels to propagate earthward within the plasma sheet (equatorward with the ionosphere) toward the equatorward portion of the plasma sheet. We have not addressed how mesoscale flow structures may form within the polar cap, but it appears plausible that they should connect to the dayside so that spatial and temporal variations of dayside reconnection could be a possible source of the polar cap structure.

Acknowledgments. This review brings together aspects from several projects at UCLA, which were supported by a number of different grants. These include National Science Foundation grants ATM-0646233 and AGS-1042255 (Lyons, Nishimura), NASA grants NNX09AI06G (Xing), NNX09AQ41H (Gkioulidou), NNX07AF66G (Wang), NNX06AB89G (Kim), and NNX10AL30G (Shi), and NASA contract NAS5-02099 (Angelopoulos). The research at the University of Michigan was supported by NSF AGS1111476 (Zou). We thank S. Mende and H. Frey for the use of the THEMIS ASI and IMAGE WIC data and the CSA for logistical support in fielding and data retrieval from the GBO stations. We thank J. M. Ruohoniemi, N. Nishitani, and G. Sofko for access and support for the SuperDARN and PolarDARN data. We thank M. Nicolls and C. Heinselman for access and support of the RISR data.

REFERENCES

Angelopoulos, V. (2008), The THEMIS mission, *Space Sci. Rev.*, *141*(1–4), 5–34, doi:10.1007/s11214-008-9336-1.

Angelopoulos, V., W. Baumjohann, C. F. Kennel, F. V. Coroniti, M. G. Kivelson, R. Pellat, R. J. Walker, H. Lühr, and G. Paschmann (1992), Bursty bulk flows in the inner central plasma sheet, *J. Geophys. Res.*, *97*(A4), 4027–4039.

Angelopoulos, V., C. F. Kennel, F. V. Coroniti, R. Pellat, M. G. Kivelson, R. J. Walker, C. T. Russell, W. Baumjohann, W. C. Feldman, and J. T. Gosling (1994), Statistical characteristics of bursty bulk flow events, *J. Geophys. Res.*, *99*(A11), 21,257–21,280.

Angelopoulos, V., et al. (1996), Multipoint analysis of a bursty bulk flow event on April 11, 1985, *J. Geophys. Res.*, *101*(A3), 4967–4989.

Angelopoulos, V., et al. (2008), Tail reconnection triggering substorm onset, *Science*, *321*(5891), 931–935, doi:10.1126/science.1160495.

Auster, H. U., et al. (2008), The THEMIS fluxgate magnetometer, *Space Sci. Rev.*, *141*(1–4), 235–264, doi:10.1007/s11214-008-9365-9.

Bonnell, J. W., F. S. Mozer, G. T. Delory, A. J. Hull, R. E. Ergun, C. M. Cully, V. Angelopoulos, and P. R. Harvey (2008), The electric field instrument (EFI) for THEMIS, *Space Sci. Rev.*, *141*(1–4), 303–341, doi:10.1007/s11214-008-9469-2.

Bristow, W. A., and P. Jensen (2007), A superposed epoch study of SuperDARN convection observations during substorms, *J. Geophys. Res.*, *112*, A06232, doi:10.1029/2006JA012049.

Bristow, W. A., A. Otto, and D. Lummerzheim (2001), Substorm convection patterns observed by the Super Dual Auroral Radar Network, *J. Geophys. Res.*, *106*, 24,593–24,609.

Chen, C. X., and R. A. Wolf (1993), Interpretation of high-speed flows in the plasma sheet, *J. Geophys. Res.*, *98*(A12), 21,409–21,419.

de la Beaujardière, O., L. R. Lyons, J. M. Ruohoniemi, E. Friis-Christensen, C. Danielsen, F. J. Rich, and P. T. Newell (1994), Quiet-time intensifications along the poleward auroral boundary near midnight, *J. Geophys. Res.*, *99*(A1), 287–298.

Dubyagin, S., V. Sergeev, S. Apatenkov, V. Angelopoulos, R. Nakamura, J. McFadden, D. Larson, and J. Bonnell (2010), Pressure and entropy changes in the flow-braking region during magnetic field dipolarization, *J. Geophys. Res.*, *115*, A10225, doi:10.1029/2010JA015625.

Erickson, G. M., R. W. Spiro, and R. A. Wolf (1991), The physics of the Harang discontinuity, *J. Geophys. Res.*, *96*(A2), 1633–1645.

Gkioulidou, M., C.-P. Wang, L. R. Lyons, and R. A. Wolf (2009), Formation of the Harang reversal and its dependence on plasma sheet conditions: Rice Convection Model simulations, *J. Geophys. Res.*, *114*, A07204, doi:10.1029/2008JA013955.

Harel, M., R. A. Wolf, P. H. Reiff, R. W. Spiro, W. J. Burke, F. J. Rich, and M. Smiddy (1981), Quantitative simulation of a magnetospheric substorm, 1. Model logic and overview, *J. Geophys. Res.*, *86*(A4), 2217–2241.

Heinemann, M. (1999), Role of collisionless heat flux in magnetospheric convection, *J. Geophys. Res.*, *104*(A12), 28,397–28,410.

Henderson, M. G., G. D. Reeves, and J. S. Murphree (1998), Are north-south aligned auroral structures an ionospheric manifestation of bursty bulk flows?, *Geophys. Res. Lett.*, *25*(19), 3737–3740.

Henderson, M. G., L. Kepko, H. E. Spence, M. Connors, J. B. Sigwarth, L. A. Frank, and H. J. Singer (2002), The evolution of north-south aligned auroral forms into auroral torch structures: The generation of omega bands and ps6 pulsations via flow bursts, in *Sixth International Conference on Substorms*, edited by R. M. Winglee, pp. 169–174, The Univ. of Washington, Seattle.

Hughes, J. M., and W. A. Bristow (2003), SuperDARN observations of the Harang discontinuity during steady magnetospheric convection, *J. Geophys. Res.*, *108*(A5), 1185, doi:10.1029/2002JA009681.

Jaggi, R. K., and R. A. Wolf (1973), Self-consistent calculation of the motion of a sheet of ions in the magnetosphere, *J. Geophys. Res.*, *78*(16), 2852–2866.

Lyons, L. R., T. Nagai, G. T. Blanchard, J. C. Samson, T. Yamamoto, T. Mukai, A. Nishida, and S. Kokubun (1999), Association between Geotail plasma flows and auroral poleward boundary intensifications observed by CANOPUS photometers, *J. Geophys. Res.*, *104*(A3), 4485–4500.

Lyons, L. R., C. Wang, M. Gkioulidou, and S. Zou (2009a), Connections between plasma sheet transport, Region 2 currents, and entropy changes associated with convection, steady magnetospheric convection periods, and substorms, *J. Geophys. Res.*, *114*, A00D01, doi:10.1029/2008JA013743.

Lyons, L. R., S. Zou, C. Heinselman, M. Nicolls, and P. Anderson (2009b), Poker flat radar observations of the magnetosphere-ionosphere coupling electrodynamics of the earthward penetrating plasma sheet following convection enhancements, *J. Atmos. Sol. Terr. Phys.*, *71*(6–7), 717–728, doi:10.1016/j.jastp.2008.09.025.

Lyons, L. R., Y. Nishimura, Y. Shi, S. Zou, H.-J. Kim, V. Angelopoulos, C. Heinselman, M. J. Nicolls, and K.-H. Fornacon (2010a), Substorm triggering by new plasma intrusion: Incoherent-scatter radar observations, *J. Geophys. Res.*, *115*, A07223, doi:10.1029/2009JA015168.

Lyons, L. R., Y. Nishimura, X. Xing, V. Angelopoulos, S. Zou, D. Larson, J. McFadden, A. Runov, S. Mende, and K.-H. Fornacon (2010b), Enhanced transport across entire length of plasma sheet boundary field lines leading to substorm onset, *J. Geophys. Res.*, *115*, A00I07, doi:10.1029/2010JA015831.

Lyons, L. R., Y. Nishimura, H.-J. Kim, E. Donovan, V. Angelopoulos, G. Sofko, M. Nicolls, C. Heinselman, J. M. Ruohoniemi, and N. Nishitani (2011), Possible connection of polar cap flows to pre- and post-substorm onset PBIs and streamers, *J. Geophys. Res.*, *116*, A12225, doi:10.1029/2011JA016850.

McFadden, J. P., C. W. Carlson, D. Larson, M. Ludlam, R. Abiad, B. Elliott, P. Turin, M. Marckwordt, and V. Angelopoulos (2008), The THEMIS ESA plasma instrument and in-flight calibration, *Space Sci. Rev.*, *141*(1–4), 277–302, doi:10.1007/s11214-008-9440-2.

Mende, S. B., S. E. Harris, H. U. Frey, V. Angelopoulos, C. T. Russell, E. Donovan, B. Jackel, M. Greffen, and L. M. Peticolas (2008), The THEMIS array of ground-based observatories for the study of auroral substorms, *Space Sci. Rev.*, *141*(1–4), 357–387, doi:10.1007/s11214-008-9380-x.

Nakamura, R., W. Baumjohann, R. Schödel, M. Brittnacher, V. A. Sergeev, M. Kubyshkina, T. Mukai, and K. Liou (2001), Earthward flow bursts, auroral streamers, and small expansions, *J. Geophys. Res.*, *106*, 10,791–10,802.

Nishimura, Y., et al. (2010a), Preonset time sequence of auroral substorms: Coordinated observations by all-sky imagers, satellites, and radars, *J. Geophys. Res.*, *115*, A00I08, doi:10.1029/2010JA015832.

Nishimura, Y., L. Lyons, S. Zou, V. Angelopoulos, and S. Mende (2010b), Substorm triggering by new plasma intrusion: THEMIS all-sky imager observations, *J. Geophys. Res.*, *115*, A07222, doi:10.1029/2009JA015166.

Nishimura, Y., L. R. Lyons, V. Angelopoulos, T. Kikuchi, S. Zou, and S. B. Mende (2011), Relations between multiple auroral streamers, pre-onset thin arc formation, and substorm auroral onset, *J. Geophys. Res.*, *116*, A09214, doi:10.1029/2011JA016768.

Ohtani, S., M. A. Shay, and T. Mukai (2004), Temporal structure of the fast convective flow in the plasma sheet: Comparison between observations and two-fluid simulations, *J. Geophys. Res.*, *109*, A03210, doi:10.1029/2003JA010002.

Pitkänen, T., A. T. Aikio, O. Amm, K. Kauristie, H. Nilsson, and K. U. Kaila (2011), EISCAT-Cluster observations of quiet-time near-Earth magnetotail fast flows and their signatures in the ionosphere, *Ann. Geophys.*, *29*(2), 299–319, doi:10.5194/angeo-29-299-2011.

Pontius, D. H., Jr., and R. A. Wolf (1990), Transient flux tubes in the terrestrial magnetosphere, *Geophys. Res. Lett.*, *17*(1), 49–52.

Rostoker, G., A. T. Y. Lui, C. D. Anger, and J. S. Murphree (1987), North-south structures in the midnight sector auroras as viewed by the Viking imager, *Geophys. Res. Lett.*, *14*(4), 407–410.

Runov, A., V. Angelopoulos, M. I. Sitnov, V. A. Sergeev, J. Bonnell, J. P. McFadden, D. Larson, K.-H. Glassmeier, and U. Auster (2009), THEMIS observations of an earthward-propagating dipolarization front, *Geophys. Res. Lett.*, *36*, L14106, doi:10.1029/2009GL038980.

Samson, J. C., L. R. Lyons, P. T. Newell, F. Creutzberg, and B. Xu (1992), Proton aurora and substorm intensifications, *Geophys. Res. Lett.*, *19*(21), 2167–2170.

Sergeev, V. A., K. Liou, C.-I. Meng, P. T. Newell, M. Brittnacher, G. Parks, and G. D. Reeves (1999), Development of auroral streamers in association with localized impulsive injections to the inner magnetotail, *Geophys. Res. Lett.*, *26*(3), 417–420.

Sergeev, V. A., et al. (2000), Multiple-spacecraft observation of a narrow transient plasma jet in the Earth's plasma sheet, *Geophys. Res. Lett.*, *27*(6), 851–854.

Southwood, D. J., and M. G. Kivelson (1991), An approximate description of field-aligned currents in a planetary magnetic field, *J. Geophys. Res.*, *96*(A1), 67–75.

Toffoletto, F., S. Sazykin, R. Spiro, and R. Wolf (2003), Inner magnetospheric modeling with the Rice Convection Model, *Space Sci. Rev.*, *107*(1), 175–196.

Tsyganenko, N. A. (1995), Modeling the Earth's magnetospheric magnetic field confined within a realistic magnetopause, *J. Geophys. Res.*, *100*(A4), 5599–5612.

Vogt, J., G. Haerendel, and K. H. Glassmeier (1999), A model for the reflection of Alfvén waves at the source region of the Birkeland current system: The tau generator, *J. Geophys. Res.*, *104*(A1), 269–278.

Wang, C.-P., L. R. Lyons, M. W. Chen, and F. R. Toffoletto (2004), Modeling the transition of the inner plasma sheet from weak to enhanced convection, *J. Geophys. Res.*, *109*, A12202, doi:10.1029/2004JA010591.

Wang, C.-P., L. R. Lyons, T. Nagai, J. M. Weygand, and R. W. McEntire (2007), Sources, transport, and distributions of plasma sheet ions and electrons and dependences on interplanetary parameters under northward interplanetary magnetic field, *J. Geophys. Res.*, *112*, A10224, doi:10.1029/2007JA012522.

Wang, C.-P., M. Gkioulidou, L. R. Lyons, R. A. Wolf, V. Angelopoulos, T. Nagai, J. M. Weygand, and A. T. Y. Lui (2011), Spatial distributions of ions and electrons from the plasma sheet to the inner magnetosphere: Comparisons between THEMIS-Geotail statistical results and the Rice Convection Model, *J. Geophys. Res.*, *116*, A11216, doi:10.1029/2011JA016809.

Wolf, R. A. (1983), The quasi-static (slow-flow) region of the magnetosphere, in *Solar Terrestrial Physics*, edited by L. Carovillano and J. M. Forbes, pp. 303–368, D. Reidel, Dordrecht, Netherlands.

Wolf, R. A., R. W. Spiro, S. Sazykin, and F. R. Toffoletto (2007), How the Earth's inner magnetosphere works: An evolving picture, *J. Atmos. Sol. Terr. Phys.*, *69*(3), 288–302, doi:10.1016/j.jastp.2006.07.026.

Xing, X., L. R. Lyons, V. Angelopoulos, D. Larson, J. McFadden, C. Carlson, A. Runov, and U. Auster (2009), Azimuthal plasma pressure gradient in quiet time plasma sheet, *Geophys. Res. Lett.*, *36*, L14105, doi:10.1029/2009GL038881.

Xing, X., L. Lyons, Y. Nishimura, V. Angelopoulos, D. Larson, C. Carlson, J. Bonnell, and U. Auster (2010), Substorm onset by new plasma intrusion: THEMIS spacecraft observations, *J. Geophys. Res.*, *115*, A10246, doi:10.1029/2010JA015528.

Yang, J., F. R. Toffoletto, R. A. Wolf, S. Sazykin, R. W. Spiro, P. C. Brandt, M. G. Henderson, and H. U. Frey (2008), Rice Convection Model simulation of the 18 April 2002 sawtooth event and evidence for interchange instability, *J. Geophys. Res.*, *113*, A11214, doi:10.1029/2008JA013635.

Yoshikawa, A., O. Amm, H. Vanhamäki, and R. Fujii (2011), A self-consistent synthesis description of magnetosphere-ionosphere coupling and scale-dependent auroral process using shear Alfvén wave, *J. Geophys. Res.*, *116*, A08218, doi:10.1029/2011JA016460.

Zesta, E., L. R. Lyons, and E. Donovan (2000), The auroral signature of earthward flow bursts observed in the magnetotail, *Geophys. Res. Lett.*, *27*(20), 3241–3244.

Zhang, J.-C., R. A. Wolf, R. W. Spiro, G. M. Erickson, S. Sazykin, F. R. Toffoletto, and J. Yang (2009), Rice Convection Model simulation of the substorm-associated injection of an observed plasma bubble into the inner magnetosphere: 2. Simulation results, *J. Geophys. Res.*, *114*, A08219, doi:10.1029/2009JA014131.

Zou, S., L. R. Lyons, M. J. Nicolls, C. J. Heinselman, and S. B. Mende (2009a), Nightside ionospheric electrodynamics associated with substorms: PFISR and THEMIS ASI observations, *J. Geophys. Res.*, *114*, A12301, doi:10.1029/2009JA014259.

Zou, S., L. R. Lyons, C.-P. Wang, A. Boudouridis, J. M. Ruohoniemi, P. C. Anderson, P. L. Dyson, and J. C. Devlin (2009b), On the coupling between the Harang reversal evolution and substorm dynamics: A synthesis of SuperDARN, DMSP, and IMAGE observations, *J. Geophys. Res.*, *114*, A01205, doi:10.1029/2008JA013449.

V. Angelopoulos, Department of Earth and Space Sciences, University of California, Los Angeles, CA 90095-1567, USA.

E. Donovan, Department of Physics and Astronomy, University of Calgary, 2500 University Drive, Calgary, Alberta T2N 1N4, Canada.

M. Gkioulidou, H.-J. Kim, L. R. Lyons, Y. Nishimura, Y. Shi, C.-P. Wang, and X. Xing, Department of Atmospheric and Oceanic Sciences, University of California, Los Angeles, CA 90095-1565, USA. (larry@atmos.ucla.edu).

S. Zou, Department of Atmospheric, Oceanic and Space Sciences, University of Michigan, Ann Arbor, MI 48109, USA.

Auroral Signatures of Ionosphere-Magnetosphere Coupling at Jupiter and Saturn

L. C. Ray

Space and Atmospheres Group, Department of Physics, Imperial College London, London, UK

R. E. Ergun

Laboratory for Atmospheric and Space Physics, University of Colorado, Boulder, Colorado, USA

Department of Astrophysical and Planetary Sciences, University of Colorado, Boulder, Colorado, USA

Planetary auroral emissions are an observable signature of the coupling between the planetary magnetosphere and ionosphere. Jovian and Saturnian auroral emissions are created by a variety of processes, including those driven by the orbital motions of moons through the planetary magnetic plasma, the local creation, and subsequent pickup of plasma in the inner magnetosphere, due to the geologically active moons Io and Enceladus, respectively, and the radial transport of plasma through the middle and outer magnetosphere. Along with these internally driven phenomena, there exist auroral emissions owing to the global interaction of the planetary magnetosphere such as the release of plasma down the magnetotail. The current understanding of Jovian and Saturnian auroral processes is derived from attempting to reconcile theoretical models of ionosphere-magnetosphere coupling with physical parameters derived from auroral observations and in situ magnetospheric data. We focus on the processes limiting the strength of the ionosphere-magnetosphere coupling that predominantly occur in the ionosphere and at high latitudes along the magnetic field line.

1. INTRODUCTION

1.1. Internally Driven Magnetospheres

The Jovian and Saturnian magnetospheres have significant internal plasma sources due to their respective moons Io and Enceladus. Io orbits Jupiter at a distance of 5.9 R_J (Jovian radii) and is the most volcanically active body in the solar system outgassing ~700–3000 kg s^{-1} of neutral material, predominately sulfur dioxide, into the Jovian magnetosphere. Roughly half of this mass leaves the magnetosphere

through charge exchange and fast neutral escape [*Delamere et al.*, 2005]. The remaining ~350–1500 kg s^{-1} of plasma, composed of ionized oxygen, sulfur, and mixed compounds, populates the dense Io torus and, after a torus-residence lifetime of ~14–60 days, is transported radially outward through the magnetosphere, eventually being lost down the magnetotail of the planet (see reviews by *Thomas et al.* [2004] and *Bagenal and Delamere* [2011]). The radial profile of the plasma angular velocity stays near the planetary rotation rate ($\Omega_{Jup} = 1.76 \times 10^{-4}$ rad s^{-1}) out to ~17–20 R_J beyond which it departs significantly from this value [*McNutt et al.*, 1979]. Outside of ~15 R_J, the planetary magnetic field is distended by the current sheet created by the ion and azimuthal drift motions of the plasma. Additionally, the high-beta plasma population, $\beta > 10$ outside ~22 R_J [*Mauk et al.*, 2004], inflates the Jovian magnetosphere, expanding the system to observable subsolar magnetopause

Auroral Phenomenology and Magnetospheric Processes: Earth and Other Planets
Geophysical Monograph Series 197
10.1029/2011GM001172

distances from the planetary center of 63 R_J (during high-pressure solar wind) or 92 R_J (during low-pressure solar wind) rather than the ~43 R_J that one would predict from simple pressure balance between the planetary magnetic field and the solar wind ram pressure [*Joy et al.*, 2002].

The Saturnian moon Enceladus has a more moderate neutral outgassing rate of ~150–300 kg s^{-1} [*Hansen et al.*, 2006], primarily composed of water group molecules. The dominant ionization process is charge exchange, narrowly surpassing the combination of photoionization and electron impact ionization [*Fleshman et al.*, 2010]. Estimates of the plasma radial mass transport rate range from ~10s of kg s^{-1} to 280 kg s^{-1} (see review by *Bagenal and Delamere* [2011]). Interestingly, the radial profile of the magnetospheric plasma angular velocity shows a 20% departure from corotation outside ~4 R_S (Saturnian radii) [*Wilson et al.*, 2009]. Outside ~6 R_S, Saturn's magnetic field is radially "stretched" owing to magnetic contribution of the ring current. Past 9 R_S, the Saturnian magnetosphere becomes inflated due to internal high-beta plasma pressure, $\beta > 1$ [*Sergis et al.*, 2010], but to a lesser extent than the Jovian system. At Saturn, the observed subsolar magnetopause distances are typically 22 or 27 R_S for high- or low-pressure solar wind conditions, respectively, instead of the ~19 R_S predicted by a vacuum dipole magnetic field.

1.2. Auroral Emissions

There are three types of auroral emission observed on Jupiter, each driven by a different process (see review by *Clarke* [this volume]). These are (1) satellite aurorae, which are exemplified by the Io footprint and wake emission, Ganymede and Europa footprints, (2) the constant main aurora ("main oval"), which consists of a "ring" of emission, and (3) the variable polar emissions, which are observed poleward of the main emission. Both the internally driven satellite wake emission and the main auroral emission are signatures of steady state ionosphere-magnetosphere coupling and are the focus of this chapter.

Jupiter's satellite aurorae are created by the motion of satellites through the rapidly rotating magnetospheric plasma. The auroral emission associated with Io is the brightest and most interesting of the three satellite signatures (Io, Ganymede, and Europa). Io's auroral emission can be split into two distinct regions: the instantaneous Io spot and the steady state wake emission. The Io spot is created by an Alfvénic disturbance imposed by the Io obstacle upon the magnetic flux tube sweeping past the moon (see reviews by *Saur* [2004], *Bonfond* [this volume], and *Hess and Delamere* [this volume]). In the frame corotating with Jupiter, Io orbits with a velocity of −57 km s^{-1}, and hence, the Io-driven

auroral emissions have a retrograde relative motion. The Io wake emission is caused by the steady state current system, which is set up to transfer angular momentum from Jupiter to the local wake plasma. These currents impose a force that accelerates the plasma, which has been diverted around the obstacle, back up to corotation [*Hill and Vasyliūnas*, 2002; *Ergun et al.*, 2009]. The wake emission persists far downstream of the Io spot. Observations of the Io spot show that the mean energy of the associated precipitating electrons is ~55 keV [*Gérard et al.*, 2002]. Recent observations of the Io wake find a lower mean electron energy of ~1 keV [*Bonfond et al.*, 2009] (see Table 1).

The main auroral emission is the signature of the steady state, global current system transferring angular momentum from Jupiter to outward moving plasma. Observations of the main auroral emission show a persistent structure, which is fixed to Jupiter's longitudinal coordinate system, System III, and maps to an equatorial distance of 20–30 R_J [e.g., *Khurana*, 2001; *Clarke et al.*, 2004]. There are some variations in the main auroral emission, but it is difficult to determine whether these are due to local time effects or variations in the magnetic field structure with System III longitude as the aurora is preferentially observed on the dayside between 155° and 270° central meridian longitude (CML) [*Grodent et al.*, 2003a]. *Grodent et al.* [2003a] determined that the Jovian magnetic field controls some aspects of the main auroral emission, with the location of the auroral emission contracting toward the poles as the CML increases from 115° to 255°. *Gustin et al.* [2004] found a mean precipitating electron energy of ~30–200 keV and incident energy flux of ~2–30 mW m^{-2} for the main auroral emission. However, there are some aspects of the main auroral emission that are more clearly fixed in local time. *Radioti et al.* [2008] showed that the main auroral emission has a persistent dark region in the prenoon to noon sector, thought to be the signature of a downward current, perhaps representing closure or return currents not associated with downward precipitating electrons. Additionally, there are often bright morning storms observed in the dawn sector. The bright morning aurora is caused by much more energetic electrons with precipitating energies as large as ~460 keV and associated incident energy fluxes in the order of 100 mW m^{-2} [*Gustin et al.*, 2006]. In addition to magnetic and local time variations, the main auroral emission has been observed to shift up to 3° in latitude over time [*Grodent et al.*, 2008] and to vary with solar wind conditions [*Nichols et al.*, 2007; *Cowley et al.*, 2007; *Yates et al.*, 2012; *Delamere*, this volume].

Saturn's internally driven auroral emissions are not yet definitively identified with their corresponding magnetospheric driver and, in comparison to the Jovian system, possibly sedate. Enceladus' auroral footprint was recently

Table 1. Auroral Features With Associated Precipitating Electron Properties

Auroral Feature	Associated Process and Location[a]	Precipitating Energy Flux (mW m^{-2})	Mean Electron Energy (keV)
	Jupiter		
Io footprint	Alfvénic; 6R_J	33[b]	55[b]
Io wake	Acceleration of wake plasma; 6R_J	2–20[c]	1–2[c]
Main oval	Radial transport of iogenic plasma	2–30[d]	30–200[d]
Polar aurora	?	0–200[e]	?
	Saturn		
Enceladus footprint	Alfvénic; 4R_S	0.15–0.45[f]	?
Diffuse emission	4–11R_S	~0.3[g]	~1[g]
Main emission	Interaction with SW; ~20R_S	0.2–1.4[h]	10–18[i]

[a]Location of auroral process when mapped to the magnetosphere.
[b]*Gerard et al.* [2002].
[c]*Bonfond et al.* [2009].
[d]*Gustin et al.* [2004].
[e]Polar emission is highly variable. See discussion of *Grodent et al.* [2003b].
[f]Derived from the work of *Pryor et al.* [2011] using 1 kR = 10 mW m^{-2}.
[g]*Grodent et al.* [2010].
[h]Derived from the work of *Gustin et al.* [2009] using 1 kR = 10 mW m^{-2}.
[i]*Gustin et al.* [2009].

discovered after years of dedicated observations and is only present in a few percent of available images [*Pryor et al.*, 2011]. While there is indication of a satellite footprint, there have not yet been observations of an associated wake emission, as with the Io interaction at Jupiter. Poleward of the Enceladus footprint is a region of diffuse auroral emission, which maps to ~4–11 R_S in the magnetosphere and is consistent with electron precipitation from pitch-angle scattering of the suprathermal magnetospheric electron population [*Grodent et al.*, 2010]. However, while not yet extensively studied, this emission could also be a signature of momentum loading in the Saturnian magnetosphere. The magnetospheric plasma's angular velocity profile lags corotation outside ~3.3 R_S [*Wilson et al.*, 2009] owing to both the charge exchange between ion and neutrals and radial transport of plasma [*Saur et al.*, 2004; *Pontius and Hill*, 2009]. The brightest UV auroral signature maps to the outer magnetosphere, varying with solar wind conditions. Yet it is not clear whether the emission is the signature of field-aligned currents enforcing corotation [*Sittler et al.*, 2006] or those caused by the rotational shear flow at the open-closed boundary of the magnetopause [*Cowley et al.*, 2008; *Bunce*, this volume].

So the question is, Why does the Jovian magnetosphere have such strong auroral signatures of ionosphere-magnetosphere coupling that are clearly internally driven, while uncertainty still remains regarding the drivers for Saturn's auroral emissions? Within both magnetospheres, there exists an active moon, which populates the system with neutral particles that are subsequently ionized, picked up by

the planetary magnetic field, and transported radially outward. Both magnetospheres are inflated by their internal plasma population beyond what would be predicted by simple pressure balance between the planetary magnetic field and the solar wind ram pressure, as detailed in Table 2, and both magnetospheres are rapid rotators. However, the Saturnian magnetic field is ~20 times weaker than Jupiter's, and the upper atmosphere is more extended, due to Saturn having one third the Jovian mass and, hence, a correspondingly smaller gravitational field.

Table 2. Basic Parameters of Earth and Jupiter and Saturn

	Earth	Jupiter	Saturn
Orbital distance (AU)	1	5.2	9.5
Planetary mass (10^{24} kg)	5.9736	1,898.6	568.46
Equatorial radius (km)	6,378	71,400	60,268
Equatorial field strength (gauss)	0.3	4.2	0.2
Dipole direction (geographic)	S to N	N to S	N to S
Magnetic tilt (°)	11	9.4	<1
Rotational period (h)	24	9.9	10.7
Magnetopause distance estimated from magnetic field and solar wind pressure balance (R_P)	10.2	42.6	18.9
Observed magnetopause distance (R_P)	10	63;92[a]	22;27[b]
Auroral ionospheric Pedersen conductance (mho)	2–3	~0.1–1s	1–10s

[a]Magnetopause standoff distance has a bimodal distribution corresponding to compressed and relaxed states [*Joy et al.*, 2002].
[b]*Achilleos et al.* [2008].

2. ANGULAR MOMENTUM TRANSFER

The auroral emissions detailed above are signatures of the transfer of angular momentum from the parent planet to its surrounding magnetospheric plasma. The current system that facilitates this transfer is briefly described below, and the reader is referred to more detailed works for completeness [e.g., *Hill*, 1979; *Pontius and Hill*, 1982; *Hill*, 2001; *Hill and Vasyliūnas*, 2002; *Cowley and Bunce*, 2001, 2003; *Nichols and Cowley*, 2004; *Ergun et al.*, 2009; *Ray et al.*, 2010; *Nichols*, 2011]. In the first case of the Io wake emission, the plasma population requiring angular momentum is newly created plasma that, as a previously neutral molecule or atom, orbited Jupiter at the local Keplerian velocity. Upon ionization, the plasma is picked up by the planetary magnetic field, which is tied to the planet's rotation rate. As the background plasma has a nonzero velocity relative to the magnetic field, there exists a motional electric field ($\mathbf{E} = -\mathbf{v} \times \mathbf{B}$) in the radial direction, which maps to a latitudinal electric field in the planetary ionosphere. The ionospheric electric field corresponds to an associated latitudinal current, which diverges to flow out from the planet along the magnetic field and then radially outward in the equatorial plane, eventually returning to the planet along the magnetic field to close the circuit. The upward field-aligned currents communicate angular momentum from Jupiter to the magnetospheric plasma, with the radial current providing a $\mathbf{J} \times \mathbf{B}$ force in the rotational direction that accelerates the plasma from its initial Keplerian velocity to a combined gyration/drift motion where the guiding center angular velocity approaches coro-

tation with the planet. Current-carrying electrons precipitate into Jupiter's atmosphere, resulting in the Io wake emission. As the plasma nears corotation, the field-aligned currents transferring angular momentum subside, and the auroral emission decays. Figure 1 displays the full current circuit. The magnitude of the angular momentum transfer in the Io wake region is primarily dictated by the Pedersen conductance at the foot of the planetary magnetic field in the Jovian ionosphere. Observations show that the Io wake can extend as much as 120° in longitude downstream from the Io footprint.

This process most assuredly happens in the Saturnian system when water group molecules from Enceladus are ionized, yet there is no corresponding, detectable wake auroral signature suggesting that at Enceladus' orbit, the pickup currents are weaker than in the Jovian case. However, the ionization of neutral particles from Enceladus spans a broad radial range in the magnetosphere from ~4 to 8 R_S. Therefore, it is possible that the inner portion of the diffuse auroral emission is related to the pickup of these newly created ions.

In the next cases of the Jovian main auroral emission, and possibly both the outer region of the Saturnian diffuse aurora and the main auroral emission, these are signatures of plasma from Io or Enceladus, respectively, being transported radially outward through the parent magnetosphere. As the plasma moves outward, conservation of angular momentum dictates that it slows down. However, in the collisionless MHD approximation, the plasma is frozen into the magnetic field, and the deviation from the planetary rotation rate results in the field lines being azimuthally "bent back" near the equator.

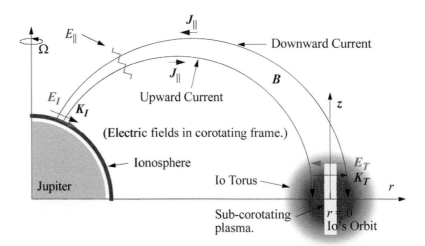

Figure 1. Current system associated with angular momentum transport in the Io wake region. $K_{I/T}$ is the height-integrated current density in the ionosphere/magnetosphere, $E_{I/T}$ is the perpendicular electric field in the ionosphere/magnetosphere, $J_\parallel$ is the field-aligned current, and $E_\parallel$ is the parallel electric field that develops at high latitudes. From the work of *Ergun et al.* [2009].

Figure 2. Upward current system associated with angular momentum transport due to radially moving plasma. $K_{I/M}$ is the height-integrated current density in the ionosphere/magnetosphere, $E_{I/M}$ is the perpendicular electric field in the ionosphere/magnetosphere, $J_\parallel$ is the field-aligned current, and $\Phi_\parallel$ is the parallel electric field that develops at high latitudes. From the work of *Ray et al.* [2010].

The curl of this field configuration is a radially outward current in the equatorial magnetosphere, closed by field-aligned currents that run between the ionosphere and magnetosphere, transferring angular momentum as shown in Figure 2. The motional electric field of the rotating magnetospheric plasma is, in a sense, "mapped" along magnetospheric field lines between these regions and, thus, partially drives ionospheric current. In the magnetosphere, the equatorial $\mathbf{J} \times \mathbf{B}$ force accelerates the plasma toward corotation, which diminishes the magnetospheric electric field. Yet unlike the case of local plasma pickup where the current system evolves in azimuth at a constant magnetospheric radius and therefore maps to a region of fairly constant magnetic field strength, here the magnetic field strength decreases as the current system evolves with radial distance from the planet. This means that at a certain radial distance, the planet will no longer be able to supply the necessary angular momentum to keep the magnetospheric plasma near corotation. When this happens, significant subcorotation of the plasma ensues. Again, the corresponding auroral emission is created by the excitation of atmospheric molecules by precipitating electrons that carry the field-aligned current between the planet and magnetospheric plasma.

There are two factors, however, which limit the current systems described above: finite ionospheric Pedersen conductance and the lack of current carriers at high latitudes along the magnetic field line. It is the interplay between these two factors that quantitatively dictates the location and intensity of the auroral emission.

2.1. Ionospheric Pedersen Conductance

The ionospheric Pedersen conductance, Σ_P, also known as the height-integrated Pedersen conductivity, controls the current density that flows through the ionosphere for the above current systems. The Pedersen conductance is associated with particle mobility in the direction parallel to the ionospheric electric field and perpendicular to the planetary magnetic field. The local Pedersen conductivity maximizes at the

altitude where the ion gyrofrequency equals the ion-neutral collision frequency. Assuming that the ionospheric electric field can be described by a simplified Ohm's law as $\mathbf{E_I} = \mathbf{K_I}/\Sigma_P$, where $\mathbf{K_I}$ is the height-integrated ionospheric current density, a small Σ_P for a fixed ionospheric electric field will decrease the magnitude of the currents that can flow through the ionosphere. This also implies that small Pedersen conductances result in relatively weak closure currents flowing between the ionosphere and magnetosphere transferring angular momentum. For low Σ_P, the magnetospheric plasma will not receive the angular momentum necessary to return toward corotation and will persistently lag the planetary rotation rate. On the other hand, for a large Σ_P, the ionosphere will readily provide the currents necessary to accelerate the magnetospheric plasma toward corotation, and we would expect the plasma to rotate closer to the planetary rate over a more extended range in radial distance.

Ideally, the ionosphere and upper neutral atmosphere would have the same rotation frequency as Jupiter's deep interior. However, at high altitudes, the ionosphere slips in its rotation due to the torque exerted on it by the currents feeding angular momentum from the ionosphere to the magnetosphere. This, in turn, reduces the rotation rate of the neutral atmosphere through ion-neutral collisions. At low altitudes, the slowing of the neutral atmosphere through ion-neutral collisions is less efficient as angular momentum is more readily transported from the deep planetary interior to the neutral atmosphere. The net result is that the low altitude ionosphere rotates near or at the planetary rotation rate, while the rotation rate of the upper ionosphere more closely reflects that of the magnetospheric plasma [*Huang and Hill*, 1989]. It is, therefore, common to discuss an "effective" conductance, which is the true Pedersen conductance reduced to account for the subcorotation of the neutral atmosphere and, hence, ionosphere in the upper atmosphere.

The ionospheric Pedersen conductances in Saturn's auroral regions are determined by combining radio occultation data with models of the neutral atmosphere and ionosphere. Calculations of the "true" Pedersen conductance range from ~1 to 10s of mho [*Moore et al.*, 2010; *Galand et al.*, 2011]. On Jupiter, estimates of the "true" Pedersen conductance range from ~0.1 to ~8 mho [*Millward et al.*, 2002] and are predominately based on modeling efforts of the atmosphere and ionosphere-magnetosphere system.

2.2. High-Latitude Plasma Density

The second restriction on the ionosphere-magnetosphere coupling current system is the centrifugal confinement of the magnetospheric plasma due to the rapid plasma rotation rate. On Jupiter, heavy ions are confined near the centrifugal equator due to large centrifugal forces [*Hill and Michel*, 1976]. While the electrons are less confined due to their smaller mass, their mobility along the magnetic field line is restricted by an ambipolar electric field, which is set up to maintain quasi-neutrality of the plasma. As the ionospheric plasma is confined to the planet due to strong gravitational forces, this restriction in the magnetospheric electron mobility leads to a lack of current-carrying plasma at high latitudes as shown in Figure 3. Consequently, field-aligned potentials develop at high magnetic latitudes to augment the electron distribution in the loss cone and, thus, increase the field-aligned current that can flow between the ionosphere and magnetosphere. The resulting current-voltage relation that describes the change in field-aligned current with field-aligned potential is nonlinear and depends on the density and temperature of the electron population at high latitudes as well as the mirror ratio at the top of the acceleration region [*Knight*, 1973; *Ray et al.*, 2009]. For the Io flux tube, the acceleration region forms at ~2.5 R_J jovicentric distance. Theoretically, it is possible for the field-aligned current density to saturate; that is, the entire electron distribution is accelerated into the loss cone such that the current density is maximized, and further increases in the field-aligned potential have no effect.

The acceleration of the magnetospheric electrons by the high-latitude field-aligned potential not only increases the field-aligned current density but also the energy of the precipitating electrons and the incident energy flux on the planetary atmosphere. This leads to enhancements in the ionospheric Pedersen conductance which, in turn, increase the angular momentum that is transferred from the planet to the magnetospheric plasma. The presence of field-aligned potentials also allows for differential rotation between the regions above and below the drop, i.e., the planetary magnetosphere and upper ionosphere, respectively.

3. NARROWNESS OF JUPITER'S MAIN AURORAL EMISSION

Understanding the interplay between the enhancement of the ionospheric Pedersen conductance and the development of field-aligned potentials is essential in describing the narrow

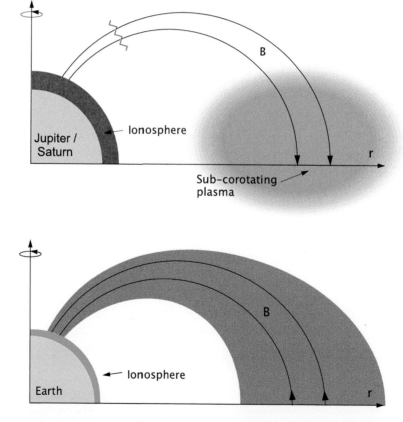

Figure 3. Cartoon of the plasma density distributions on Earth and Jupiter/Saturn. Jovian/Saturnian magnetospheric plasma is confined to the equatorial plane owing to the centrifugal confinement of the heavy ions and subsequent ambipolar electric field that develops.

main auroral emission observed on Jupiter. In the absence of field-aligned potentials, the ionosphere-magnetosphere system is perfectly coupled with the magnetosphere and upper ionosphere having the same rotation rate. However, once field-aligned potentials develop, the ionosphere and magnetosphere can differentially rotate, or slip, relative to each other. In a steady state system, the degree of differential rotation depends on the latitudinal derivative of the parallel potential, $\frac{d\Phi_\parallel}{d\theta}$, following from $\nabla \times \mathbf{E} = 0$, i.e., how the field-aligned potential changes from one flux tube to the next when moving in planetary latitude or, equivalently, radial distance in the magnetosphere. When the latitudinal derivative of the field-aligned potential is positive, the magnitude of the ionospheric electric field increases relative to that of the magnetospheric electric field, and conversely, a negative latitudinal derivative is associated with a stronger magnetospheric electric field relative to the ionospheric field.

As the field-aligned potential and field-aligned current density increase, so does the electron precipitation into the planetary atmosphere, producing a bright aurora. The enhanced precipitation also increases the Pedersen conductance, which diminishes the magnitude of the ionospheric electric field. Meanwhile, the magnetospheric electric field decreases owing to the increased angular momentum transfer allowed by the enhanced Pedersen conductance and field-aligned current density. The final limitation on the magnetospheric plasma's demand for angular momentum is the planetary magnetic field. Jupiter's north-south equatorial magnetic field strength falls off faster than a dipole from ~15 to ~50 R_J as the inner portion of the ring current field acts to reduce the north-south component of the planetary magnetic field. The radial decrease in the magnetic field strength reduces the $\mathbf{J} \times \mathbf{B}$ force such that the magnetospheric plasma again begins to lag corotation. At the location where the magnitude of the magnetospheric electric field when mapped along the field lines to the planet exceeds that of the ionospheric electric field, the latitudinal derivative of the field-aligned potential changes sign. The field-aligned potential at high latitudes declines, and with it, the field-aligned current density and ionospheric Pedersen conductance also decrease. The electron precipitation energy and electron energy flux decrease limiting the poleward extent of the auroral emission, and a bright, narrow aurora of ~1°–2° is produced [*Ray et al.*, 2010, 2012].

4. SATURN'S DIFFUSE AURORAL EMISSION

Saturn's rotationally driven auroral emission may have two components: the diffuse emission that maps to a range of magnetospheric equatorial radii spanning 4–11 R_S, the bright ring of emission mapping to ~15 R_S, or a mixture of these two observed signatures. Saturn has a lower radial mass transport rate than Jupiter and a larger Pedersen conductance. Both of these factors should indicate a strong coupling between the ionosphere and magnetosphere. However, the magnetospheric angular velocities lag corotation outside ~3.3 R_S leading to the question: What limits the transfer of angular momentum from Saturn to its magnetospheric plasma?

One possibility is that Saturn's weaker magnetic field results in the Saturnian magnetosphere being more heavily mass loaded, relative to its magnetic energy content, than the Jovian magnetosphere [*Delamere et al.*, 2007; *Vasyliunas*, 2008] and perhaps prevents the ionosphere from ever being able to "grip" its magnetosphere. A second mechanism may be collisions between the magnetospheric ions and neutrals, which could contribute to the rotational lag of the magnetospheric plasma [*Saur et al.*, 2004]. Another limitation could be the lag in the rotation rate of the neutral atmosphere in the Pedersen conducting region at the planet as any neutral lag in corotation would be communicated to the magnetospheric plasma [*Huang and Hill*, 1989]. However, the neutral atmosphere would have to subcorotate by ~20% to be consistent with the observed magnetospheric flows, which would lead to strong, possibly unsustainable, atmospheric winds.

Last, inferred energy fluxes from the diffuse auroral emission would suggest that there are no strong field-aligned potentials, ($\Phi_\parallel > $ ~3 kV), at high latitudes inside of ~11 R_S [*Grodent et al.*, 2010]. There is a large abundance of protons in the Saturnian system relative to the Jovian system owing to water group-based chemistry. The magnetospheric protons have a smaller temperature anisotropy than the water group ions [*Wilson et al.*, 2008] and, therefore, more mobility along the magnetic field. As such, current flow between the ionosphere and magnetosphere may not be as restricted as in the Jovian system, adding to the mystery of Saturn's subcorotating magnetospheric plasma.

5. CONCLUSIONS

Planetary auroral emissions are excellent diagnostic tools of ionosphere-magnetosphere coupling. However, they often lead to more questions than they solve. By trying to reconcile in situ observations of the Jovian and Saturnian magnetospheres with auroral observations, we can develop and fine tune theoretical models of the ionosphere-magnetosphere interaction.

The strong internally driven auroral emission on Jupiter is a result of the interplay between the ionospheric Pedersen conductance and the lack of current-carrying plasma at high latitudes, results of both rapid rotation and an internal plasma population. Yet at Saturn, whether the broad, diffuse emission and/or the higher latitude narrow, bright emission is the signature of the coupling between the magnetospheric plasma and planetary atmosphere remains an open question. Continued missions to the outer planets and auroral observing campaigns are the key to unlocking this mystery.

REFERENCES

Achilleos, N., C. S. Arridge, C. Bertucci, C. M. Jackman, M. K. Dougherty, K. K. Khurana, and C. T. Russell (2008), Large-scale dynamics of Saturn's magnetopause: Observations by Cassini, *J. Geophys. Res.*, *113*, A11209, doi:10.1029/2008JA013265.

Bagenal, F., and P. A. Delamere (2011), Flow of mass and energy in the magnetospheres of Jupiter and Saturn, *J. Geophys. Res.*, *116*, A05209, doi:10.1029/2010JA016294.

Bonfond, B. (2012), When moons create aurora: The satellite footprints on giant planets, in *Auroral Phenomenology and Magnetospheric Processes: Earth and Other Planets, Geophys. Monogr. Ser.*, doi:10.1029/2011GM001169, this volume.

Bonfond, B., D. Grodent, J.-C. Gérard, A. Radioti, V. Dols, P. A. Delamere, and J. T. Clarke (2009), The Io UV footprint: Location, inter-spot distances and tail vertical extent, *J. Geophys. Res.*, *114*, A07224, doi:10.1029/2009JA014312.

Bunce, E. J. (2012), Origins of Saturn's auroral emissions and their relationship to magnetosphere dynamics, in *Auroral Phenomenology and Magnetospheric Processes: Earth and Other Planets, Geophys. Monogr. Ser.*, doi:10.1029/2011GM001191, this volume.

Clarke, J. T. (2012), Auroral processes on Jupiter and Saturn, in *Auroral Phenomenology and Magnetospheric Processes: Earth and Other Planets, Geophys. Monogr. Ser.*, doi:10.1029/2011GM 001199, this volume.

Clarke, J. T., D. Grodent, S. W. H. Cowley, E. J. Bunce, P. Zarka, J. E. P. Connerney, and T. Satoh (2004), Jupiter's aurora, in *Jupiter: The Planet, Satellites and Magnetosphere*, edited by F. Bagenal, T. E. Dowling, and W. B. McKinnon, pp. 639–670, Cambridge Univ. Press, Cambridge, U. K.

Cowley, S. W. H., and E. J. Bunce (2001), Origin of the main auroral oval in Jupiter's coupled magnetosphere-ionosphere system, *Planet. Space Sci.*, *49*, 1067–1088.

Cowley, S. W. H., and E. J. Bunce (2003), Corotation-driven magnetosphere-ionosphere coupling currents in Saturn's magnetosphere and their relation to the auroras, *Ann. Geophys.*, *21*, 1691–1707, doi:10.5194/angeo-21-1691-2003.

Cowley, S. W. H., J. D. Nichols, and D. J. Andrews (2007), Modulation of Jupiter's plasma flow, polar currents, and auroral precipitation by solar wind-induced compressions and expansions of the magnetosphere: a simple theoretical model, *Ann. Geophys.*, *25*, 1433–1463, doi:10.5194/angeo-25-1433-2007.

Cowley, S. W. H., C. S. Arridge, E. J. Bunce, J. T. Clarke, A. J. Coates, M. K. Dougherty, J.-C. Gérard, D. Grodent, J. D. Nichols, and D. L. Talboys (2008), Auroral current systems in Saturn's magnetosphere: Comparison of theoretical models with Cassini and HST observations, *Ann. Geophys.*, *26*, 2613–2630, doi:10.5194/angeo-26-2613-2008.

Delamere, P. A. (2012), Auroral signatures of solar wind interaction at Jupiter, in *Auroral Phenomenology and Magnetospheric Processes: Earth and Other Planets, Geophys. Monogr. Ser.*, doi:10. 1029/2011GM001180, this volume.

Delamere, P. A., F. Bagenal, and A. Steffl (2005), Radial variations in the Io plasma torus during the Cassini era, *J. Geophys. Res.*, *110*, A12223, doi:10.1029/2005JA011251.

Delamere, P. A., F. Bagenal, V. Dols, and L. C. Ray (2007), Saturn's neutral torus versus Jupiter's plasma torus, *Geophys. Res. Lett.*, *34*, L09105, doi:10.1029/2007GL029437.

Ergun, R. E., L. Ray, P. A. Delamere, F. Bagenal, V. Dols, and Y.-J. Su (2009), Generation of parallel electric fields in the Jupiter–Io torus wake region, *J. Geophys. Res.*, *114*, A05201, doi:10.1029/ 2008JA013968.

Fleshman, B. L., P. A. Delamere, and F. Bagenal (2010), A sensitivity study of the Enceladus torus, *J. Geophys. Res.*, *115*, E04007, doi:10.1029/2009JE003372.

Galand, M., L. Moore, I. Mueller-Wodarg, M. Mendillo, and S. Miller (2011), Response of Saturn's auroral ionosphere to electron precipitation: Electron density, electron temperature, and electrical conductivity, *J. Geophys. Res.*, *116*, A09306, doi:10. 1029/2010JA016412.

Gérard, J.-C., J. Gustin, D. Grodent, P. Delamere, and J. T. Clarke (2002), Excitation of the FUV Io tail on Jupiter: Characterization of the electron precipitation, *J. Geophys. Res.*, *107*(A11), 1394, doi:10.1029/2002JA009410.

Grodent, D., J. T. Clarke, J. Kim, J. H. Waite Jr., and S. W. H. Cowley (2003a), Jupiter's main auroral oval observed with HST-STIS, *J. Geophys. Res.*, *108*(A11), 1389, doi:10.1029/2003JA 009921.

Grodent, D., J. T. Clarke, J. H. Waite Jr., S. W. H. Cowley, J.-C. Gérard, and J. Kim (2003b), Jupiter's polar auroral emissions, *J. Geophys. Res.*, *108*(A10), 1366, doi:10.1029/2003JA010017.

Grodent, D., J.-C. Gérard, A. Radioti, B. Bonfond, and A. Saglam (2008), Jupiter's changing auroral location, *J. Geophys. Res.*, *113*, A01206, doi:10.1029/2007JA012601.

Grodent, D., A. Radioti, B. Bonfond, and J.-C. Gérard (2010), On the origin of Saturn's outer auroral emission, *J. Geophys. Res.*, *115*, A08219, doi:10.1029/2009JA014901.

Gustin, J., J.-C. Gérard, D. Grodent, S. W. H. Cowley, J. T. Clarke, and A. Grard (2004), Energy-flux relationship in the FUV Jovian aurora deduced from HST-STIS spectral observations, *J. Geophys. Res.*, *109*, A10205, doi:10.1029/2003JA010365.

Gustin, J., S. W. H. Cowley, J.-C. Gérard, G. R. Gladstone, D. Grodent, and J. T. Clarke (2006), Characteristics of Jovian morning bright FUV aurora from Hubble Space Telescope/Space Telescope Imaging Spectrograph imaging and spectral observations, *J. Geophys. Res.*, *111*, A09220, doi:10.1029/2006JA 011730.

Gustin, J., J.-C. Gérard, W. Pryor, P. D. Feldman, D. Grodent, and G. Holsclaw (2009), Characteristics of Saturn's polar atmosphere and auroral electrons derived from HST/STIS, FUSE and

Cassini/UVIS spectra, *Icarus*, *200*, 176–187, doi:10.1016/j.icarus. 2008.11.013.

Hansen, C. J., L. Esposito, A. I. F. Stewart, J. Colwell, A. Hendrix, W. Pryor, D. Shemansky, and R. West (2006), Enceladus' water vapor plume, *Science*, *311*, 1422–1425, doi:10.1126/science. 1121254.

Hess, S. L. G., and P. A. Delamere (2012), Satellite-induced electron acceleration and related auroras, in *Auroral Phenomenology and Magnetospheric Processes: Earth and Other Planets*, *Geophys. Monogr. Ser.*, doi:10.1029/2011GM001175, this volume.

Hill, T. W. (1979), Inertial limit on corotation, *J. Geophys. Res.*, *84*(A11), 6554–6558.

Hill, T. W. (2001), The Jovian auroral oval, *J. Geophys. Res.*, *106*(A5), 8101–8107.

Hill, T. W., and F. C. Michel (1976), Heavy ions from the Galilean satellites and the centrifugal distortion of the Jovian magnetosphere, *J. Geophys. Res.*, *81*(25), 4561–4565.

Hill, T. W., and V. M. Vasyliunas (2002), Jovian auroral signature of Io's corotational wake, *J. Geophys. Res.*, *107*(A12), 1464, doi:10. 1029/2002JA009514.

Huang, T. S., and T. W. Hill (1989), Corotation lag of the Jovian atmosphere, ionosphere, and magnetosphere, *J. Geophys. Res.*, *94*(A4), 3761–3765.

Joy, S. P., M. G. Kivelson, R. J. Walker, K. K. Khurana, C. T. Russell, and T. Ogino (2002), Probabilistic models of the Jovian magnetopause and bow shock locations, *J. Geophys. Res.*, *107*(A10), 1309, doi:10.1029/2001JA009146.

Khurana, K. K. (2001), Influence of solar wind on Jupiter's magnetosphere deduced from currents in the equatorial plane, *J. Geophys. Res.*, *106*(A11), 25,999–26,016.

Knight, S. (1973), Parallel electric fields, *Planet. Space Sci.*, *21*, 741–750.

Mauk, B. H., D. G. Mitchell, R. W. McEntire, C. P. Paranicas, E. C. Roelof, D. J. Williams, S. M. Krimigis, and A. Lagg (2004), Energetic ion characteristics and neutral gas interactions in Jupiter's magnetosphere, *J. Geophys. Res.*, *109*, A09S12, doi:10. 1029/2003JA010270.

McNutt, R. L., Jr., J. W. Belcher, J. D. Sullivan, F. Bagenal, and H. S. Bridge (1979), Departure from rigid co-rotation of plasma in Jupiter's dayside magnetosphere, *Nature*, *280*, 803.

Millward, G., S. Miller, T. Stallard, A. D. Aylward, and N. Achilleos (2002), On the dynamics of the Jovian ionosphere and thermosphere III. The modelling of auroral conductivity, *Icarus*, *160*, 95–107, doi:10.1006/icar.2002.6951.

Moore, L., I. Mueller-Wodarg, M. Galand, A. Kliore, and M. Mendillo (2010), Latitudinal variations in Saturn's ionosphere: Cassini measurements and model comparisons, *J. Geophys. Res.*, *115*, A11317, doi:10.1029/2010JA015692.

Nichols, J. D. (2011), Magnetosphere-ionosphere coupling in Jupiter's middle magnetosphere: Computations including a self-consistent current sheet magnetic field model, *J. Geophys. Res.*, *116*, A10232, doi:10.1029/2011JA016922.

Nichols, J. D., and S. W. H. Cowley (2004), Magnetosphere-ionosphere coupling currents in Jupiter's middle magnetosphere: Effect of precipitation-induced enhancement of the ionospheric Pedersen conductivity, *Ann. Geophys.*, *22*, 1799–1827.

Nichols, J. D., E. J. Bunce, J. T. Clarke, S. W. H. Cowley, J.-C. Gérard, D. Grodent, and W. R. Pryor (2007), Response of Jupiter's UV auroras to interplanetary conditions as observed by the Hubble Space Telescope during the Cassini flyby campaign, *J. Geophys. Res.*, *112*, A02203, doi:10.1029/2006JA012005.

Pontius, D. H., Jr., and T. W. Hill (1982), Departure from corotation of the Io plasma torus: Local plasma production, *Geophys. Res. Lett.*, *9*(12), 1321–1324.

Pontius, D. H., Jr., and T. W. Hill (2009), Plasma mass loading from the extended neutral gas torus of Enceladus as inferred from the observed plasma corotation lag, *Geophys. Res. Lett.*, *36*, L23103, doi:10.1029/2009GL041030.

Pryor, W. R., et al. (2011), The auroral footprint of Enceladus on Saturn, *Nature*, *472*, 331–333, doi:10.1038/nature09928.

Radioti, A., J.-C. Gérard, D. Grodent, B. Bonfond, N. Krupp, and J. Woch (2008), Discontinuity in Jupiter's main auroral oval, *J. Geophys. Res.*, *113*, A01215, doi:10.1029/2007JA012610.

Ray, L. C., Y.-J. Su, R. E. Ergun, P. A. Delamere, and F. Bagenal (2009), Current-voltage relation of a centrifugally confined plasma, *J. Geophys. Res.*, *114*, A04214, doi:10.1029/2008JA 013969.

Ray, L. C., R. E. Ergun, P. A. Delamere, and F. Bagenal (2010), Magnetosphere-ionosphere coupling at Jupiter: Effect of field-aligned potentials on angular momentum transport, *J. Geophys. Res.*, *115*, A09211, doi:10.1029/2010JA015423.

Ray, L. C., R. E. Ergun, P. A. Delamere, and F. Bagenal (2012), Magnetosphere-ionosphere coupling at Jupiter: A parameter space study, *J. Geophys. Res.*, *117*, A01205, doi:10.1029/2011JA016899.

Saur, J. (2004), A model of Io's local electric field for a combined Alfvénic and unipolar inductor far-field coupling, *J. Geophys. Res.*, *109*, A01210, doi:10.1029/2002JA009354.

Saur, J., B. H. Mauk, A. Kaßner, and F. M. Neubauer (2004), A model for the azimuthal plasma velocity in Saturn's magnetosphere, *J. Geophys. Res.*, *109*, A05217, doi:10.1029/2003JA 010207.

Sergis, N., et al. (2010), Particle pressure, inertial force, and ring current density profiles in the magnetosphere of Saturn, based on Cassini measurements, *Geophys. Res. Lett.*, *37*, L02102, doi:10. 1029/2009GL041920.

Sittler, E. C., Jr., M. F. Blanc, and J. D. Richardson (2006), Proposed model for Saturn's auroral response to the solar wind: Centrifugal instability model, *J. Geophys. Res.*, *111*, A06208, doi:10.1029/2005JA011191.

Thomas, N., F. Bagenal, T. W. Hill, and J. K. Wilson (2004), The Io neutral clouds and plasma torus, in *Jupiter: The Planet, Satellites and Magnetosphere*, edited by F. Bagenal, T. E. Dowling, and W. B. McKinnon, pp. 561–591, Cambridge Univ. Press, Cambridge, U. K.

Vasyliunas, V. M. (2008), Comparing Jupiter and Saturn: Dimensionless input rates from plasma sources within the magnetosphere, *Ann. Geophys.*, *26*(6), 1341–1343, doi:10.5194/angeo-26-1341-2008.

214 IONOSPHERE-MAGNETOSPHERE COUPLING

Wilson, R. J., R. L. Tokar, M. G. Henderson, T. W. Hill, M. F. Thomsen, and D. H. Pontius Jr. (2008), Cassini plasma spectrometer thermal ion measurements in Saturn's inner magnetosphere, *J. Geophys. Res.*, *113*, A12218, doi:10.1029/2008JA013486.

Wilson, R. J., R. L. Tokar, and M. G. Henderson (2009), Thermal ion flow in Saturn's inner magnetosphere measured by the Cassini plasma spectrometer: A signature of the Enceladus torus?, *Geophys. Res. Lett.*, *36*, L23104, doi:10.1029/2009GL040225.

Yates, J. N., N. Achilleos, and P. Guio (2012), Influence of upstream solar wind on thermospheric flows at Jupiter, *Planet. Space Sci.*, *61*, 15–31, doi:10.1016/j.pss.2011.08.007.

R. E. Ergun, Laboratory for Atmospheric and Space Physics, Discovery Drive, Boulder, CO 80309, USA.

L. C. Ray, Space and Atmospheres Group, Department of Physics, Imperial College London, Prince Consort Road, London SW7 2AZ, UK. (l.ray@imperial.ac.uk)

Clues on Ionospheric Electrodynamics From IR Aurora at Jupiter and Saturn

Tom Stallard

Department of Physics and Astronomy, University of Leicester, Leicester, UK

Steve Miller

Atmospheric Physics Laboratory, Department of Physics and Astronomy, University College London, London, UK

Henrik Melin

Department of Physics and Astronomy, University of Leicester, Leicester, UK

Ionospheric flows within the upper atmospheres of the gas giants provide a valuable tool with which to understand the currents flowing through the ionospheres of these planets, as well as the magnetospheric origin of these currents. These flows are measured using high-resolution long-slit spectroscopy from ground-based telescopes, producing both intensity and velocity profiles that allow us to understand the flows and see how they are associated with the aurorae. Thus, it is possible to compare and contrast both morphology and flow speeds of H_3^+ in the ionospheres of Jupiter and Saturn, revealing a number of significant similarities suggesting comparable origins for ionospheric features created by currents that are associated with both internal circuits and the influence of the solar wind. There remains much controversy about the way the solar wind affects these planets, particularly at Jupiter. Only with more detailed observations will this controversy be resolved.

1. INTRODUCTION

While much of what we know about the morphology of the aurora of the gas giants (Jupiter and Saturn) comes from the detailed long-term UV observations made by the Hubble Space Telescope [*Clarke et al.*, 1996; *Grodent et al.*, 2008], IR observations have not only given us detailed information about the aurorae [*Badman et al.*, 2011; *Melin et al.*, 2011] but have also provided a depth of understanding about the physical conditions in the ionospheres of these planets [*Miller et al.*, 2006; *Stallard et al.*, 2012a].

Auroral Phenomenology and Magnetospheric Processes: Earth and Other Planets
Geophysical Monograph Series 197
10.1029/2011GM001168

Bright auroral emissions are created on both planets by energetic electrons that stream into the atmosphere from the surrounding magnetosphere along Birkeland field-aligned currents. These currents close via a Pedersen current that typically flows equatorward through the ionosphere, in turn driving a Hall drift that forces the ions to subrotate [*Hill*, 1979; *Stallard et al.*, 2001]. Auroral emissions are also produced at the magnetic footprints of some moons [*Connerney et al.*, 1993; *Pryor et al.*, 2011] and as the result of pitch angle scattering from hot plasma within the magnetosphere [*Grodent et al.*, 2010], though no ionospheric flows have been found that are associated with these aurorae.

Emission from the ionic molecule H_3^+ has been studied using ground-based telescopes, providing both images and spectra that have yielded a number of significant scientific advances. One of the most important aspects of these studies comes from high-resolution spectroscopy, allowing us to measure the line-of-sight velocity of H_3^+ in the ionosphere and, from this, to understand the current systems flowing through the upper atmospheres of these planets [*Stallard et al.*, 2001].

H_3^+ is formed through a fast chain reaction process beginning with the ionization of molecular hydrogen into H_2^+. The H_2^+ ion is then rapidly converted to H_3^+ by a strongly exothermic reaction [*Yelle and Miller*, 2004]. On both planets, the majority of ionization is in the auroral region by precipitating energetic electrons, but also occurs, globally, by solar extreme ultraviolet (EUV) radiation ionizing molecular hydrogen.

H_3^+ is, under most conditions, a highly reactive molecule, and is quickly destroyed in the presence of any species other than hydrogen or helium (since He has a lower proton affinity than H_2). However, in the upper atmosphere of Jupiter and Saturn, species concentrations are controlled by diffusion, such that they each settle out with their own scale height. In a hydrogen-rich atmosphere, this means that heavier species settle out, and at higher altitudes, H_3^+ protonating reactions cannot take place [*Yelle and Miller*, 2004]. As a result, the lifetime of the H_3^+ molecule is directly controlled by dissociative recombination, which occurs at a rate given by *McCall et al.* [2005] as 2.6×10^{-13} m^3 s^{-1}. This means that H_3^+ lifetimes are typically between a few and a few thousand seconds, for electron densities between 10^{12} and 10^9 m^{-3}, respectively. These lifetimes are a factor of 500 and 500,000 times longer than the typical radiation lifetime of a rovibrationally excited H_3^+ molecule, for which Einstein A_{if} coefficients range between ~10 and 100 s^{-1} [*Neale et al.*, 1996].

However, the IR auroral emission of H_3^+ is a thermal emission, directly affected by the conditions within the upper atmosphere [*Miller et al.*, 1990]. Simultaneous observations of the auroral region in both wavelengths have been made from Earth for Jupiter [*Clarke et al.*, 2004] and using in situ images at Saturn [*Melin et al.*, 2011]. While these show that the general morphology of the H_3^+ and UV auroral emission matches, there are still significant variations caused by the ionospheric conditions, as well as altitudinal differences, with H_3^+ brightness being enhanced with increasing altitude at Jupiter due to increases in temperature [*Lystrup et al.*, 2008]; interestingly, it appears that H_3^+ emission appears to be relatively weaker than the UV emission at higher altitudes above Saturn [*Stallard et al.*, 2012b].

Since H_3^+ has an approximate lifetime of 10 min in the upper atmosphere of the gas giants, the molecule can interact with the surrounding neutral atmosphere through collisions. H_3^+ then takes on the thermal characteristics of the surrounding neutral atmosphere, with auroral brightness strongly controlled by temperature, especially at Saturn. This interac-

tion also strongly affects ionospheric velocities, as a continual accelerating force moving ions back into corotation with the planet. As such, the measurement of ion flows within the ionosphere, moving against this background neutral atmosphere, provides a direct measure of the electric currents that cross the ionosphere.

2. VELOCITY MEASUREMENTS

While, at first appearance, there are notable brightness and morphological differences between the aurora of Jupiter and Saturn, the flow speeds of H_3^+ in the ionospheres can be directly compared and contrasted. In looking at the phenomenology of H_3^+ at each planet, a number of significant similarities have been shown to exist, so that apparently, contrasting features can be shown to have similar origins. Equally, there are also some features of intensity and velocity that may be unique to each of these planets.

Using high-resolution long-slit spectroscopy, it is possible to measure the H_3^+ velocity in the line of sight. The peak intensity and position of a particular line of H_3^+ emission can be calculated by fitting the line with a Gaussian in the wavelength direction; by repeating this process across all the spatial rows of the spectral slit, a profile of intensity and ion wind velocity can be produced (for more details of this process, including the instrumental corrections required, in the case of Jupiter, see *Stallard et al.* [2001], and of Saturn, see *Stallard et al.* [2007a]).

Figure 1 shows the intensity and velocity profiles of the auroral regions of both Jupiter and Saturn during "typical" auroral conditions, with a long-slit spectrometer aligned perpendicular to the rotational axis, cutting through the center of the main auroral oval. The ion winds that flow through the upper atmosphere of gas giants are produced by currents that connect the atmosphere with the magnetosphere, and the neutral atmosphere acts to accelerate the ionosphere into corotation with the planet where such currents do not exist. As a result, using this perpendicular cut across the planet allows the measurement of the ion flows relative to the background rotation rate of the planet.

These 1-D cuts through the auroral region can be extrapolated into 2-D maps of the ion flows across the region, as shown in Figure 2. At Jupiter, the spectral slit was scanned across the auroral region [*Stallard et al.*, 2003], while at Saturn, these can only be defined more loosely and are largely based on perpendicular cuts through the rotational pole, extrapolating from the varying positions of the slit on different nights, and the associated intensity structure measured when the slit is aligned with the rotational axis.

Using these observations, we can categorize the ion winds seen at the gas giants into six main velocity regimes:

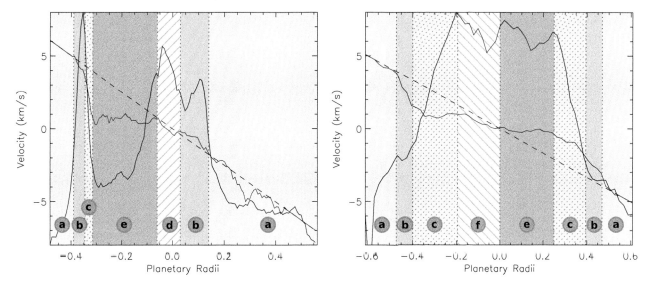

Figure 1. The typical H_3^+ line-of-sight velocity (bold line) and normalized intensity (thin line) for (left) Jupiter and (right) Saturn plotted against the rotational rate for each planet (dashed line). Each velocity profile is divided into regions demarcated by patterned backgrounds and lettering, directly relating to the lettered sections included within the text and Figures 2–4. Both intensity and velocity were calculated by fitting the spectra with a Gaussian; Jupiter data were measured on a pixel-by-pixel basis, while the Saturn data were smoothed with a 5-pixel box car function.

2.1. Region a: Equatorial Regions

This region is equatorward of the region of strong auroral emission and away from any significant currents, in which ions corotate with the planet.

2.1.1. Jupiter. EUV ionization produces enough H_3^+ emission that the ion flow can be measured in regions without any auroral component. In addition, there are also middle-latitude to low-latitude emissions that are higher than can be explained by EUV ionization alone [*Miller et al.*, 1997].

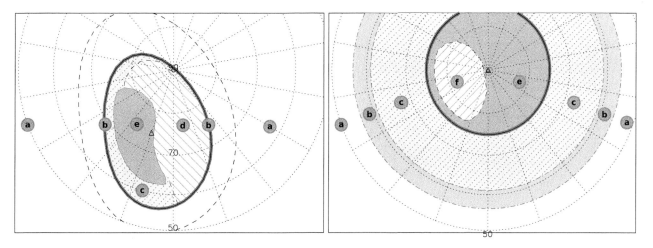

Figure 2. Maps of ion flow in the northern auroral regions of (left) Jupiter (at a central meridian longitude of 160) and (right) Saturn. The various flow regions are demarcated by patterned backgrounds and lettering, matching Figures 1, 3, and 4. In addition, the main (brightest) auroral oval for each planet is delineated (thick gray and black line), as well as the path of the magnetic mapping to Io across Jupiter (dashed line) and the mid-latitude auroral oval at Saturn (the region within the dash-dotted lines). Lines of latitude and longitude are shown (dotted lines) in steps of $10°$ and $20°$, respectively, with noon at the bottom of each map. The magnetic pole is also shown (triangle). The area covered by region f is highly variable, and so its location here is illustrative.

Measurements have shown that Jupiter does not have any significant velocity flow (>0.2 km s^{-1}) within either of these regions [*Stallard et al.*, 2001].

2.1.2. Saturn.
Equatorial H$_3^+$ emission has only been measured recently, and no velocity measurements exist in this region [*Stallard et al.*, 2012a]. However, at the edge of the auroral region, the ions can be seen to return to corotation, before the emission becomes too weak to detect.

2.2. Region b: Breakdown in Corotation

The initial breakdown in corotation poleward of equatorial corotating ions occurs. This region is directly associated with a continuous auroral oval of H$_3^+$ emission.

2.2.1. Jupiter.
The breakdown region is directly associated with the main auroral oval, which maps from at least 15 to several tens of R_J in the magnetospheric equatorial region [*Grodent et al.*, 2003a], with subrotating flows typically of the order 0.5–1.5 km s^{-1} once adjusted to take the line of sight into account. The upper limits to this velocity are measured during periods of enhanced auroral activity [*Melin et al.*, 2006], with the measured speed appearing to vary with auroral brightness [*Stallard et al.*, 2001].

2.2.2. Saturn.
The initial breakdown in corotation is associated with the mid-latitude auroral oval significantly weaker than the main auroral oval [*Stallard et al.*, 2008a], at a latitude that maps to 3–4 R_S [*Stallard et al.*, 2010], close to the location of Enceladus.

2.3. Region c: Boundary Subrotation

A region of subrotating ions that is poleward of the breakdown in corotation and equatorward of the region is associated with the solar wind but not associated with any significant auroral emission. For both planets, it is difficult to assess the extent to which this is a true velocity flow, rather than the effect of seeing blurring the boundary between two regions moving with very different velocities, with velocities scaling between those measured in the surrounding regions.

2.3.1. Jupiter.
The area covered by this region, along the dawn flank of the zero-rotation polar region, aligns this region closely to the Dark Auroral Region within the UV. However, apart from, at the most, equatorward extent, the region itself is often only a few pixels across [*Stallard et al.*, 2003].

2.3.2. Saturn.
This region, typically seen on both the dawnside and duskside of the auroral region, extends from the main auroral oval equatorward, subrotating at less than one-third corotation velocity. This region often extends to significant distances, so that the flows seen cannot be produced by a seeing-smeared boundary between the corotation breakdown and the main auroral oval and must therefore be real. It could, however, conceal significant ion flow variability that simply cannot be detected once the measurements have been affected by the Earth's atmospheric turbulence [*Stallard et al.*, 2007a].

2.4. Region d: Bright Corotation

This is a region of near corotating ions, poleward from the initial breakdown in corotation and associated with significant auroral emission.

2.4.1. Jupiter.
This velocity region is directly correlated with the auroral region known as the Bright Polar Region in H$_3^+$ emission and the Active Region in UV observations; these emission regions are associated with bright auroral arcs, on top of a moderately bright background of emission [*Grodent et al.*, 2003b]. The ion flow measured in this region appears to vary in strength, often appearing to corotate completely, while at other times subrotating at wind speeds up to 1 km s^{-1}, though the ion flows are never larger than those on the main oval [*Stallard et al.*, 2001].

2.4.2. Saturn.
Such regions of close-to-corotating ions are not measured at Saturn. However, since H$_3^+$ auroral arcs are measured at Saturn but cannot be discerned within ground-based data [*Melin et al.*, 2011], observing constraints appear to prevent the detection of any such flow regions.

2.5. Region e: Dim Subrotation

These are regions of strong subrotation or even zero rotation within the auroral polar regions.

2.5.1. Jupiter.
This region is limited to the dawnside of the polar regions, where the polar aurora is weakest, and the ionosphere appears to be stagnant within the inertial frame, producing, in the planetary frame, wind speeds of >4 km s^{-1} [*Stallard et al.*, 2001], so that the region is clearly affected by the influence of the solar wind [*Stallard et al.*, 2003; *Cowley et al.*, 2003].

2.5.2. Saturn.
This region, poleward of the main oval, cannot be directly discerned from ion flows alone, with the same subrotation of less than one-third corotation as is measured in region c [*Stallard et al.*, 2003]. However, this region sits inside the main auroral oval at Saturn, directly associating it with a region controlled by the solar wind, as described in the next section.

2.6. Region f: Polar Corotation

This is a region of corotation entirely enclosed by the surrounding subrotating ionosphere, close to the magnetic pole.

2.6.1. Jupiter. While the Jovian polar region appears to be held at zero in the inertial plane, there are some examples of subrotation within this region; however, the spatial accuracy of the data is too weak to properly detect such a region of flow [*Stallard et al.*, 2001].

2.6.2. Saturn. This region lies poleward of the main auroral oval and is not associated with any obvious auroral features. The size of this region appears to vary considerably, at times crossing the entire region inside the main auroral oval. However, it is usually concentrated on the dawnside of the polar region, with the duskside subrotating [*Stallard et al.*, 2007a]. However, during periods of major compression, associated with dawn brightening, this region cannot be observed [*Stallard et al.*, 2007a].

3. MAGNETOSPHERIC AND SOLAR WIND ORIGINS FOR ION WINDS

Since the ionospheres of the gas giants tend to be accelerated up to corotation with the planet by the surrounding neutral atmosphere, any subrotational ion flows in the ionosphere are driven by external currents. As such, the ion flows we observe, when combined with auroral emission, provide direct information about the current systems that link the ionosphere with the magnetosphere. We can also use what we know from in situ measurements of the magnetospheric configuration in order to explain the link between auroral emission and their related ion flows.

Our current understanding of the magnetospheric and solar wind origins for the different flows described in the previous section are shown schematically in Figures 3 and 4. A more detailed description of how the magnetosphere of each planet is linked to the ionosphere is given here.

3.1. Region a

The ionospheric equatorial regions of both planets are generally thought to be relatively free of currents, and so the ions in these regions are expected to corotate with the planet.

3.2. Region b

The breakdown in corotation in the ionosphere is directly associated with the breakdown in corotation in the magnetosphere, as explained by the *Hill* [1979] model. Though the

cause for this magnetospheric breakdown is different for Jupiter and Saturn, and happens at very different radial distances, the way this process drives ion flows on the planet is the same. The subrotating plasma in the magnetosphere acts as a charge moving through a magnetic field, setting up a continuous current that closes through the planet, driving particle precipitation into the atmosphere.

3.2.1. Jupiter. The magnetic field at Jupiter is strong enough that ions from Io are forced into corotation with the planet, drifting outward under centrifugal force to form a corotating plasma sheet. Angular momentum is supplied from the planet's upper atmosphere by the collision between the ionosphere and neutral thermosphere, maintaining this corotation through a "push me-pull you" effect; in the magnetosphere, as the plasma lags behind the Jovian field, it generates a current and is "instantaneously" brought back into corotation, thus switching the current off again.

As the plasma sheet moves to greater distances, its momentum increases, while the magnetic field strength driving the corotation decreases, until the mechanism breaks down catastrophically at radial distances ~20 R_J [*Hill*, 1979; *McNutt et al.*, 1981], where the magnetic field strength is too low, and the required velocity too great, for full corotation to be maintained.

This results in a very stable auroral emission, produced by the continuous particle precipitation from the current system, as well as steady ion flows in the ionosphere stable at timescales as short as a minute [*Lystrup et al.*, 2007], though they can vary over several Jovian rotations [*Stallard et al.*, 2001].

3.2.2. Saturn. The magnetosphere at Saturn is mass loaded by icy volcanic output from Enceladus. While the eruption rates of Enceladus are much lower than that of Io because the magnetic field at Saturn is weaker, the magnetosphere becomes intrinsically more heavily mass loaded than the magnetosphere of Jupiter [*Vasyliūnas*, 2008].

This means that ions produced from the torus of Enceladus never attain corotation; the neutral torus spreads inward to ~3 R_S, and any ion produced from the torus at this point, or any point further out, cannot be forced into corotation by the magnetic field. As a result, the velocity of ions in the magnetosphere break with corotation at ~3 R_S [*Wilson et al.*, 2009], the inner edge of the torus, producing Hall drift and an auroral oval within the ionosphere at the latitudes mapping between 3 and 4 R_S [*Stallard et al.*, 2010].

This current system can be compared with both the break down in corotation at Jupiter, at ~20 R_J, and with the formation of the Io spot's trailing tail, formed by mass loading from Io; Enceladus dominates Saturn's magnetic field, while ions from Io's torus are quickly forced to corotate, preventing

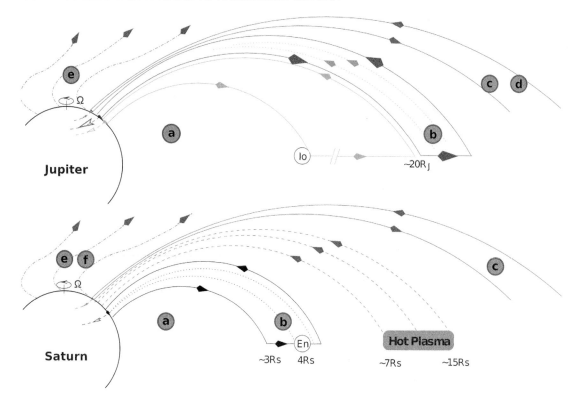

Figure 3. An illustrated representation of magnetosphere-ionosphere interconnection along magnetic field lines at Jupiter and Saturn. Currents flowing along field lines controlled by internal processes (solid lines) [*Hill*, 1979] and magnetic field lines either directly or indirectly connected with the solar wind [*Cowley et al.*, 2003, 2004; *Delamere and Bagenal*, 2010] (dash-dotted lines) are shown. In addition, for Saturn, pitch angle scattering from the hot plasma region is also represented (dashed lines) [*Grodent et al.*, 2010].

the formation of an oval of emission that completely encircles Jupiter.

3.3. Region c

This region, defined on both planets by regions of sub-rotation and relatively dark aurora, has been associated with the return flow of empty flux tubes from the *Vasyliū-nas* [1983] and, possibly, Dungey cycles, once plasmoids have escaped down the tail, as explained for region d (section 3.4). These are concentrated on the dawn flank, especially for Jupiter, due to the rotational energy of the planet pushing them to this side from midnight [*Cowley et al.*, 2003].

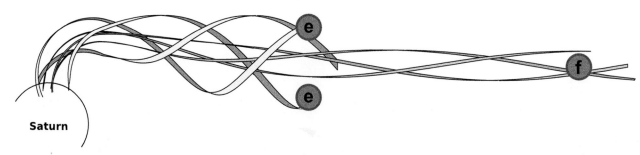

Figure 4. The *Stallard et al.* [2007b] explanation of how open field lines connect with Saturn's auroral region. Field lines connected with the solar wind subrotate and become twisted, resulting in a "core" of field lines [*Milan et al.*, 2005] that are sustained so long that they cannot effectively transfer information back to the planet, resulting in regions of corotation (labeled f) inside the open field line region (labeled e).

3.4. Region d

This region of close to corotation must map to a mid-to-far magnetospheric processes, beyond the breakdown in corotation, but remaining on closed magnetic field lines.

3.4.1. Jupiter. There remains considerable debate about the origin of the "active" region of the UV emission, with which this region of ionospheric flow is associated. *Pallier and Prangé* [2001] have suggested that this feature may be related to the cusp, an analog to what is observed in the Earth's dayside aurora, with auroral arcs mapping to dayside reconnection of the Jovian magnetic field lines with the interplanetary magnetic field. The *Cowley et al.* [2003] model produces a polar asymmetry as, in the magnetosphere, the Vasyliūnas cycle outflow of iogenic plasma flows down the dusk and midnight tails. In this case, the dusk sector of the polar ionosphere then corresponds to closed field lines, which map to the near corotating tail duskward of the outer boundary of the plasmoid.

3.4.2. Saturn. There is no observational evidence of this ion flow region at Saturn, though such flow regions could remain undetected in the region c dusk ionosphere. Indeed, *Jackman and Cowley* [2006] have theorized the existence of such flow regions, associated with the Vasyliūnas cycle, on the duskside of the aurora. Direct evidence of the Vasyliūnas cycle has now been detected within the Saturnian magnetosphere [*Masters et al.*, 2011].

3.5. Region e

On Earth, the Dungey cycle acts to produce a strong antisunward ionospheric flow, as magnetic field lines are effectively dragged across the pole by the solar wind [*Dungey*, 1961]. Since it only takes a few hours for the solar wind to pass the Earth's magnetosphere, the velocity of the flow is significantly higher than the rotation rate of the planet. However, at both Jupiter and Saturn, which have much larger magnetospheres and significantly faster rotational spin, the opposite is true. As a result, the solar wind takes many days to cross the magnetosphere. This is so slow that, to the ionosphere, the field lines affected by the solar wind appear to be stagnant, with an antisunward velocity below the observational detection limit.

3.5.1. Jupiter. Regions of zero-rotation ionospheric flow in the inertial frame of reference can only be produced through an interaction with the solar wind. The most broadly accepted theory of how the Sun controls Jupiter's upper atmosphere, first proposed by *Cowley et al.* [2003] and by the Vogt model [*Vogt et al.*, 2010; *Vogt and Kivelson*, this volume],

is that magnetic field lines in the pole are open to the solar wind, in the same way as that of the Earth's and Saturn's magnetic field lines. In this model, the polar cap region is dominated by an asymmetric Dungey cycle, similar to that of the Earth but pushed onto the dawnside of the planet by the outflow of iogenic plasma down the duskside of the magnetosphere, the effects of which are possibly seen in region d.

However, one potential problem with this theory is the significant levels of auroral emission seen in the polar regions [*Grodent et al.*, 2003b; *Stallard et al.*, 2008b], in particular the "Swirl" emission seen in the UV, short timescale apparently random bursts of emission across the region where zero-rotation ion flow is seen. This emission lies on what would be open field lines; these field lines, however, should be empty of plasma, and so this model cannot explain such emission.

A second solar wind interaction model, proposed by *Delamere and Bagenal* [2010] and discussed by *Delamere* [this volume] is that the solar wind interacts with the magnetosphere through small-scale, intermittent structures at the magnetopause boundary. This plasma-on-plasma interaction generates solar wind–imposed magnetic stresses on the magnetosphere. As a result, the solar wind interacts indirectly with the ionosphere, with "Swirl" emission produced by small clumps of closed field lines associated with Kelvin-Helmholtz instability along the boundary of the magnetosphere.

3.5.2. Saturn. Ionospheric flows match well with a Dungey cycle distorted by rotational effects as proposed by *Cowley et al.* [2004], once the effects of seeing have been accounted for [*Stallard et al.*, 2007b], but the spatial detail is not sharp enough to use ion flows to delineate this region. However, synchronous observations by the Hubble Space Telescope and Cassini have mapped the main auroral oval of Saturn to the boundary of magnetic field lines open to the solar wind, suggesting that the entire region poleward of the main auroral oval is open to the solar wind [*Bunce et al.*, 2008; *Bunce*, this volume].

3.6. Region f

When a planet's magnetic field lines are open to the solar wind, they are constrained at the solar wind end, moving steadily antisunward, while at the planetary end, they subrotate with the planet. This results in the field lines becoming twisted and producing a "core" of field lines, at the center of each of the magnetotail lobes, shielded from tail reconnection by the surrounding field lines [*Milan et al.*, 2005]. These "core" open field lines are sustained so long that they cannot effectively transfer information back to the planet, so that the ionospheric region these lines map to is dominated by the neutral atmosphere and corotates with the planet, while the newer field lines surrounding them continue to subrotate [*Stallard et al.*, 2007b].

3.6.1. Jupiter. As discussed above, there remains much debate about the mechanism of solar wind control within Jupiter's ionosphere. However, even if this region is controlled by a form of Dungey cycle, it is possible to explain the lack of any corotation in comparison with Saturn; Jupiter's solar wind–controlled region is held at a near zero rotation, with a maximum subrotation of ~10%. This could significantly inhibit the formation of a twisted tail and thus prevent an old "core" of field lines from forming.

3.6.2. Saturn. This process is able to explain the apparently contradictory presence of corotating ions inside the main auroral oval, which has been shown to mark the open-closed field line boundary. The twisting of the field lines results in an older "core" of magnetic field lines that are shielded from reconnection by smaller solar wind compressions. However, when a major solar wind compression occurs, even these shielded field lines should reconnect.

This theory has been used to explain the massive dawn brightening seen when major compressions in the solar wind impact Saturn's magnetosphere [*Cowley et al.*, 2005], as this "core" is reconnected on the nightside and rotates around, swathing the dawn polar region in emission. Here ion flow measurements are in agreement with the theory: During periods when the auroral morphology is strongly asymmetric, with a bright dawn indicating that a strong solar wind compression has occurred, the region of corotation in the pole cannot be seen; the entire auroral region subrotates [*Stallard et al.*, 2007a].

4. CONCLUSIONS

H_3^+ is an excellent diagnostic tool in helping to understand the upper atmospheres of the gas giants. In the past, it has been used as a probe of the energy distribution within the upper atmosphere, showing the temperature and cooling from the thermosphere [*Miller et al.*, 2006; *Stallard et al.*, 2012a]. Here we have described the major diagnostic tool it can be used for, measuring the ion flows within the ionosphere and how these flows are driven through interactions with the surrounding magnetosphere.

What these observations have shown is that the auroral regions of Jupiter and Saturn are far more alike than previously thought. The same magnetospheric interactions appear to occur on both planets, to a greater or lesser extent, with differences between the two planets coming from the relative

strength of these interactions. The only remaining regions that continue to produce considerable debate are the polar regions of each planet, with both emission and ion flows leading the debate about what controls these regions.

Ion wind measurements have shown that Jupiter is significantly affected by the solar wind, despite past theories predicting this would not be possible; now the debate centers on whether this region is open to the solar wind, as on Earth, or is controlled through interactions with the solar wind along the flanks of the magnetosphere. At Saturn, corotating regions poleward of the open-closed field line boundary do not fit well into current models of the magnetosphere, suggesting the twisting of the polar magnetic field lines is confusing our understanding of this region.

Measurements of the ion winds could, theoretically, provide answers to this current debate, but, as we have seen, they are limited in spatial resolution by the turbulent effects of the Earth's atmosphere. However, in the near future, ground-based observations of the auroral region using adaptive optics could make spectral measurements of Jupiter auroral region, allowing us a detailed view of the velocities within the polar regions. Beyond such observations, the NASA Juno spacecraft will move into a polar orbit around Jupiter in 2016. This will allow detailed measurements of the aurora from above, observing the emission on the planet as the spacecraft passes through the very currents that form these aurorae, as they flow along the magnetic field lines into the planet. These observations should, at long last, properly explain the origin of Jupiter's polar aurora.

Acknowledgments. This work was supported by a RCUK Fellowship for T.S. and by the UK STFC for H.M.

REFERENCES

Badman, S. V., C. Tao, A. Grocott, S. Kasahara, H. Melin, R. H. Brown, K. H. Baines, M. Fujimoto, and T. Stallard (2011), Cassini VIMS observations of latitudinal and hemispheric variations in Saturn's infrared auroral intensity, *Icarus*, *216*, 367–375, doi:10.1016/j.icarus.2011.09.031.

Bunce, E. J. (2012), Origins of Saturn's auroral emissions and their relationship to large-scale magnetosphere dynamics, in *Auroral Phenomenology and Magnetospheric Processes: Earth and Other Planets*, Geophys. Monogr. Ser., doi:10.1029/2011GM001191, this volume.

Bunce, E. J., et al. (2008), Origin of Saturn's aurora: Simultaneous observations by Cassini and the Hubble Space Telescope, *J. Geophys. Res.*, *113*, A09209, doi:10.1029/2008JA013257.

Clarke, J. T., et al. (1996), Ultraviolet imaging of Jupiter's aurora and the Io "footprint", *Science*, *274*, 404–409, doi:10.1126/science.274.5286.404.

Clarke, J. T., D. Grodent, S. W. H. Cowley, E. J. Bunce, P. Zarka, J. E. P. Connerney, and T. Satoh (2004), Jupiter's aurora, in *Jupiter: The Planet, Satellites and Magnetosphere*, edited by F. Bagenal, T. E. Dowling, and W. B. McKinnon, pp. 639–670, Cambridge Univ. Press, Cambridge, U. K.

Connerney, J. E. P., R. Baron, T. Satoh, and T. Owen (1993), Images of excited H_3^+ at the foot of the Io flux tube in Jupiter's atmosphere, *Science*, *262*, 1035–1038, doi:10.1126/science.262.5136.1035.

Cowley, S. W. H., E. J. Bunce, T. S. Stallard, and S. Miller (2003), Jupiter's polar ionospheric flows: Theoretical interpretation, *Geophys. Res. Lett.*, *30*(5), 1220, doi:10.1029/2002GL016030.

Cowley, S. W. H., E. J. Bunce, and J. M. O'Rourke (2004), A simple quantitative model of plasma flows and currents in Saturn's polar ionosphere, *J. Geophys. Res.*, *109*, A05212, doi:10.1029/2003JA010375.

Cowley, S. W. H., S. V. Badman, E. J. Bunce, J. T. Clarke, J.-C. Gérard, D. Grodent, C. M. Jackman, S. E. Milan, and T. K. Yeoman (2005), Reconnection in a rotation-dominated magnetosphere and its relation to Saturn's auroral dynamics, *J. Geophys. Res.*, *110*, A02201, doi:10.1029/2004JA010796.

Delamere, P. A. (2012), Auroral signatures of solar wind interaction at Jupiter, in *Auroral Phenomenology and Magnetospheric Processes: Earth and Other Planets, Geophys. Monogr. Ser.*, doi:10.1029/2011GM001180, this volume.

Delamere, P. A., and F. Bagenal (2010), Solar wind interaction with Jupiter's magnetosphere, *J. Geophys. Res.*, *115*, A10201, doi:10.1029/2010JA015347.

Dungey, J. W. (1961), Interplanetary field and the auroral zones, *Phys. Rev. Lett.*, *6*, 47–49, doi:10.1103/PhysRevLett.6.47.

Grodent, D., J. T. Clarke, J. Kim, J. H. Waite Jr., and S. W. H. Cowley (2003a), Jupiter's main auroral oval observed with HST-STIS, *J. Geophys. Res.*, *108*(A11), 1389, doi:10.1029/2003JA009921.

Grodent, D., J. T. Clarke, J. H. Waite Jr., S. W. H. Cowley, J.-C. Gérard, and J. Kim (2003b), Jupiter's polar auroral emissions, *J. Geophys. Res.*, *108*(A10), 1366, doi:10.1029/2003JA010017.

Grodent, D., B. Bonfond, J.-C. Gérard, A. Radioti, J. Gustin, J. T. Clarke, J. Nichols, and J. E. P. Connerney (2008), Auroral evidence of a localized magnetic anomaly in Jupiter's northern hemisphere, *J. Geophys. Res.*, *113*, A09201, doi:10.1029/2008JA013185.

Grodent, D., A. Radioti, B. Bonfond, and J.-C. Gérard (2010), On the origin of Saturn's outer auroral emission, *J. Geophys. Res.*, *115*, A08219, doi:10.1029/2009JA014901.

Hill, T. W. (1979), Inertial limit on corotation, *J. Geophys. Res.*, *84*(A11), 6554–6558.

Jackman, C. M., and S. W. H. Cowley (2006), A model of the plasma flow and current in Saturn's polar ionosphere under conditions of strong Dungey cycle driving, *Ann. Geophys.*, *24*, 1029–1055, doi:10.5194/angeo-24-1029-2006.

Lystrup, M. B., S. Miller, T. Stallard, C. G. A. Smith, and A. Aylward (2007), Variability of Jovian ion winds: An upper limit for enhanced Joule heating, *Ann. Geophys.*, *25*, 847–853, doi:10.5194/angeo-25-847-2007.

Lystrup, M. B., S. Miller, N. Dello Russo, R. J. Vervack Jr., and T. Stallard (2008), First vertical ion density profile in Jupiter's auroral atmosphere: Direct observations using the Keck II telescope, *Astrophys. J.*, *677*, 790–797, doi:10.1086/529509.

Masters, A., M. F. Thomsen, S. V. Badman, C. S. Arridge, D. T. Young, A. J. Coates, and M. K. Dougherty (2011), Supercorotating return flow from reconnection in Saturn's magnetotail, *Geophys. Res. Lett.*, *38*, L03103, doi:10.1029/2010GL046149.

McCall, B. J., et al. (2005), Storage ring measurements of the dissociative recombination rate of rotationally cold H_3^+, *J. Phys. Conf. Ser.*, *4*, 92–97, doi:10.1088/1742-6596/4/1/012.

McNutt, R. L., Jr., J. W. Belcher, and H. S. Bridge (1981), Positive ion observations in the middle magnetosphere of Jupiter, *J. Geophys. Res.*, *86*(A10), 8319–8342.

Melin, H., S. Miller, T. Stallard, C. Smith, and D. Grodent (2006), Estimated energy balance in the jovian upper atmosphere during an auroral heating event, *Icarus*, *181*, 256–265, doi:10.1016/j.icarus.2005.11.004.

Melin, H., T. Stallard, S. Miller, J. Gustin, M. Galand, S. V. Badman, W. R. Pryor, J. O'Donoghue, R. H. Brown, and K. H. Baines (2011), Simultaneous Cassini VIMS and UVIS observations of Saturn's southern aurora: Comparing emissions from H, H_2 and H_3^+ at a high spatial resolution, *Geophys. Res. Lett.*, *38*, L15203, doi:10.1029/2011GL048457.

Milan, S. E., E. J. Bunce, S. W. H. Cowley, and C. M. Jackman (2005), Implications of rapid planetary rotation for the Dungey magnetotail of Saturn, *J. Geophys. Res.*, *110*, A03209, doi:10.1029/2004JA010716.

Miller, S., R. D. Joseph, and J. Tennyson (1990), Infrared emissions of H3(+) in the atmosphere of Jupiter in the 2.1 and 4.0 micron region, *Astrophys. J.*, *360*, L55–L58, doi:10.1086/185811.

Miller, S., N. Achilleos, G. E. Ballester, H. A. Lam, J. Tennyson, T. R. Geballe, and L. M. Trafton (1997), Mid-to-low latitude H_3^+ emission from Jupiter, *Icarus*, *130*, 57–67, doi:10.1006/icar.1997.5813.

Miller, S., T. Stallard, C. Smith, G. Millward, H. Melin, M. Lystrup, and A. Aylward (2006), H_3^+: The driver of giant planet atmospheres, *Philos. Trans. R. Soc. A*, *364*, 3121–3137, doi:10.1098/rsta.2006.1877.

Neale, L., S. Miller, and J. Tennyson (1996), Spectroscopic properties of the H_3^+ molecule: A new calculated line list, *Astrophys. J.*, *464*, 516, doi:10.1086/177341.

Pallier, L., and R. Prangé (2001), More about the structure of the high latitude Jovian aurorae, *Planet. Space Sci.*, *49*, 1159–1173, doi:10.1016/S0032-0633(01)00023-X.

Pryor, W. R., et al. (2011), The auroral footprint of Enceladus on Saturn, *Nature*, *472*, 331–333, doi:10.1038/nature09928.

Stallard, T. S., S. Miller, G. Millward, and R. D. Joseph (2001), On the dynamics of the Jovian ionosphere and thermosphere: I. The measurement of ion winds, *Icarus*, *154*, 475–491, doi:10.1006/icar.2001.6681.

Stallard, T. S., S. Miller, S. W. H. Cowley, and E. J. Bunce (2003), Jupiter's polar ionospheric flows: Measured intensity and velocity variations poleward of the main auroral oval, *Geophys. Res. Lett.*, *30*(5), 1221, doi:10.1029/2002GL016031.

Stallard, T., S. Miller, H. Melin, M. Lystrup, M. Dougherty, and N. Achilleos (2007a), Saturn's auroral/polar H_3^+ infrared emission: I. General morphology and ion velocity structure, *Icarus*, *189*, 1–13, doi:10.1016/j.icarus.2006.12.027.

Stallard, T., C. Smith, S. Miller, H. Melin, M. Lystrup, A. Aylward, N. Achilleos, and M. Dougherty (2007b), Saturn's auroral/polar H_3^+ infrared emission: II. A comparison with plasma flow models, *Icarus*, *191*, 678–690, doi:10.1016/j.icarus.2007.05.016.

Stallard, T., S. Miller, H. Melin, M. Lystrup, S. W. H. Cowley, E. J. Bunce, N. Achilleos, and M. Dougherty (2008a), Jovian-like aurorae on Saturn, *Nature*, *453*, 1083–1085, doi:10.1038/nature07077.

Stallard, T., et al. (2008b), Complex structure within Saturn's infrared aurora, *Nature*, *456*, 214–217, doi:10.1038/nature07440.

Stallard, T., H. Melin, S. W. H. Cowley, S. Miller, and M. B. Lystrup (2010), Location and magnetospheric mapping of Saturn's mid-latitude infrared auroral oval, *Astrophys. J.*, *722*, L85–L89, doi:10.1088/2041-8205/722/1/L85.

Stallard, T. S., H. Melin, S. Miller, J. O'Donoghue, S. W. H. Cowley, S. V. Badman, A. Adriani, R. H. Brown, and K. H. Baines (2012a), Temperature changes and energy inputs in giant planet atmospheres: What we are learning from H_3^+, *Philos. Trans. R. Soc. A*, in press.

Stallard, T. S., H. Melin, S. Miller, S. V. Badman, R. H. Brown, and K. H. Baines (2012b), Peak emission altitude of Saturn's H_3^+ aurora, *Geophys. Res. Lett.*, *39*, L15103, doi:10.1029/2012GL052806.

Vasyliūnas, V. M. (1983), Plasma distribution and flow, in *Physics of the Jovian Magnetosphere*, edited by A. J. Dessler, p. 395, Cambridge Univ. Press, New York.

Vasyliūnas, V. M. (2008), Comparing Jupiter and Saturn: Dimensionless input rates from plasma sources within the magnetosphere, *Ann. Geophys.*, *26*, 1341–1343, doi:10.5194/angeo-26-1341-2008.

Vogt, M. F., and M. G. Kivelson (2012), Relating Jupiter's auroral features to magnetospheric sources, in *Auroral Phenomenology and Magnetospheric Processes: Earth and Other Planets*, Geophys. Monogr. Ser., doi:10.1029/2011GM001181, this volume.

Vogt, M. F., M. G. Kivelson, K. K. Khurana, S. P. Joy, and R. J. Walker (2010), Reconnection and flows in the Jovian magnetotail as inferred from magnetometer observations, *J. Geophys. Res.*, *115*, A06219, doi:10.1029/2009JA015098.

Wilson, R. J., R. L. Tokar, and M. G. Henderson (2009), Thermal ion flow in Saturn's inner magnetosphere measured by the Cassini plasma spectrometer: A signature of the Enceladus torus?, *Geophys. Res. Lett.*, *36*, L23104, doi:10.1029/2009GL040225.

Yelle, R. V., and S. Miller (2004), Jupiter's thermosphere and ionosphere, in *Jupiter: The Planet, Satellites and Magnetosphere, Cambridge Planet. Sci., vol. 1*, edited by F. Bagenal, T. E. Dowling, and W. B. McKinnon, pp. 185–218, Cambridge Univ. Press, Cambridge, U. K.

H. Melin and T. Stallard, Department of Physics and Astronomy, University of Leicester, University Road, Leicester LE1 7RH, UK. (tss@ion.le.ac.uk)

S. Miller, Atmospheric Physics Laboratory, Department of Physics and Astronomy, University College London, Gower Street, London WC1E 6BT, UK.

Section IV
Discrete Auroral Acceleration

The Acceleration Region of Stable Auroral Arcs

T. Karlsson

Space and Plasma Physics, School of Electrical Engineering, KTH, Stockholm, Sweden

The acceleration region above stable, discrete auroral arcs is reviewed. Substantial observational evidence shows that the acceleration of auroral electrons associated with these arcs is achieved by an electric potential structure above the aurora at altitudes of around 0.5–2 R_E. The morphology, internal structure, and lifetime of the predominantly U-shaped potential structure are discussed, based on observations by a number of spacecrafts. The parallel (to the geomagnetic field) electric field component of the potential structure, which accelerates the auroral electrons, is discussed in terms of relatively recent direct observations. The most important theories for how the parallel electric field is sustained are also described. The altitude distribution predicted from the various theories is compared to observations, and briefly discussed, as well as their relation to each other.

1. INTRODUCTION

The auroral display often takes the form of discrete, quiet, homogenous arcs, commonly observed during the substorm growth or late recovery phase. [*Akasofu*, 1964].The arcs are typically very elongated in an approximately east-west direction, with a length of up to many thousand kilometers, while their north-south extent is of the order of 10 km [*Knudsen et al.*, 2001]. They can be stable for remarkably long times, up to 10 h if the solar wind and magnetospheric conditions allow [*Galperin*, 2002]. The auroral emissions are created by energetic electrons of magnetospheric origin, traveling along the geomagnetic field, eventually colliding with the upper atmosphere, if the electrons are in the loss cone. At the collisions, light is emitted. The precipitating electrons carry an upward field-aligned current. The details of these processes are described in many reviews [see, e.g., *Paschmann et al.*, 2003].

One of the important open questions regarding auroral physics is how the precipitating electrons are energized, to produce a high-enough energy flux needed to produce intensities of auroral light visible by the human eye. Early optical observations of the plasma flow around auroral arcs led to the suggestion that this acceleration was achieved by electric field structures above the auroral arc [*Carlqvist and Boström*, 1970; *Hallinan and Davis*, 1970]. This chapter aims at giving a brief overview of the present observational and theoretical understanding of the auroral acceleration region associated with the stable auroral arc. Note that this stable auroral arc, as will be described below, is usually associated with monoenergetic accelerated electrons, in what is called an inverted-V structure. These electrons have a wide pitch-angle distribution, which indicates that they are of magnetospheric origin. In recent years, another type of auroral acceleration has been identified [e.g., *Chaston et al.*, 2003b], where the accelerated electrons have a broad energy range in the parallel direction, but low perpendicular energies (indicating ionospheric and/ or magnetosheath origin) [*Chaston et al.*, 2000], and are often counterstreaming. This type of electron acceleration events, produced by Alfvén waves with periods comparable to the electron transit time of the region where they are accelerated, often takes place at the poleward boundary of the auroral oval and is often associated with more dynamic auroral signatures than the stable arc [e.g., *Paschmann et al.*, 2003]. It is commonly referred to as the *Alfvénic aurora*. This type of auroral acceleration is mainly outside the scope of this chapter but will be touched upon when appropriate.

Auroral Phenomenology and Magnetospheric Processes: Earth and Other Planets

Geophysical Monograph Series 197

10.1029/2011GM001179

2. POTENTIAL STRUCTURE

Since observations of converging, perpendicular (to the geomagnetic field) electric fields above the auroral zone by the S3-3 spacecraft [*Mozer et al.*, 1977], the acceleration above the quiet aurorae has been postulated to be achieved by quasistatic, U-shaped electric potential structures (Figures 1a and 1b). The converging, perpendicular electric field at higher altitude is associated with an upward electric field at a lower altitude. The latter accelerates magnetospheric electrons downward, creating the energetic electron beam responsible for exciting the upper atmosphere constituents, which will emit the auroral light. At the same time, the parallel electric field will accelerate ions of mainly ionospheric origin upward, creating an ion beam. This basic configuration has been corroborated by many satellite observations at different altitudes. Figure 2 shows a number of examples of measurements of bipolar, converging, perpendicular electric fields from the S3-3, FAST, Polar, and Cluster spacecraft, respectively [*Mozer et al.*, 1977; *McFadden et al.*, 1999; *Mozer and Kletzing*, 1998; *Johansson et al.*, 2006]. Since it is technically difficult to measure the parallel electric field directly [*Mozer*, 1973], measurements of the perpendicular field have been used to relate the total perpendicular potential drop of the acceleration structure to signatures of acceleration in the electron and ion data.

Comparisons of electric potentials integrated along satellite passes above the region of parallel electric fields and the acceleration energy of associated ion beams have been reported on from S3-3 [*Mozer et al.*, 1980; *Redsun et al.*, 1985], Viking [*Block and Fälthammar*, 1990], Polar [*Mozer and Hull*, 2001], and FAST data [*Ergun et al.*, 2002a]. In the latter study, data are used from passes through the acceleration region, when the ion energies and the integrated electric potentials give the portion of the potential drop located below the spacecraft.

Below the U-shaped potential structure, the signature of its existence is the presence of a beam of accelerated magnetospheric electrons. Since the electrons below the center of the potential structure will have the highest energies, with progressively lower energies toward the edges of the structure, an energy-time spectrum measured by a spacecraft crossing the electron beam of the electrons will show a characteristic "inverted-V" signature [*Frank and Ackerson*, 1971].

For a spacecraft passing through the acceleration region, at any moment, the sum of the acceleration energy of the ion and electron beams will be the total field-aligned drop on that particular magnetic field line [*Redsun et al.*, 1985]. This provides a method to study the temporal stability of the total potential drop, using multipoint measurements, as will be described in section 5.

2.1. Scale Sizes, Internal Structure

On a case by case basis, good agreement has been found between the latitudinal width of the U potentials, the related inverted-V particle distributions, and optical auroral arcs [*Stenbaek-Nielsen et al.*, 1998; *Figueiredo et al.*, 2005] Few statistical investigations of the latitudinal widths of the U-shaped potentials have been reported. *Johansson et al.* [2007] give a typical latitudinal scale size of 2–5 km (all quantities in this subsection are mapped along the geomagnetic field lines to ionospheric altitude). This can be compared to the latitudinal scale sizes given for particle measurement of inverted-V events [*Newell et al.*, 1996; *Partamies et al.*, 2008] and optical observations of auroral arcs [*Knudsen et al.*, 2001]. These results are summarized in Table 1. The smaller values of *Johansson et al.* [2007] compared to the other measurements may give a clue to the typical geometry of the U potential. Since they determined the scale size of diverging electric field structures by using the full width at half maximum, it is likely that this scale size reflects large local variations in potential, rather than a smooth variation, which points to potential structures similar to Figures 1b–1f, rather than the one depicted in Figure 1a. Figure 2d [*Johansson et al.*, 2006], shows a typical converging electric field structure, where the relation between the scale size of the peak of electric field and the whole structure can be seen clearly.

On scales smaller than the overall scale, there may be internal structuring of the U potential. Figures 1c–1f show some possible morphologies. In particular, the low-altitude corrugation of the potential structure (often referred to as "fingers"), has been verified by a large number of observations [*McFadden et al.*, 1999; *Mozer and Kletzing*, 1998; *Mozer and Hull*, 2001].

2.2. Energies

The total acceleration potential of a U potential can be determined by integrating the perpendicular electric field across the structure. Several statistical studies present typical values of this acceleration potential, which then can be compared to typical energies of associated ion beams and electron inverted Vs. Some representative studies are presented in Table 2.

2.3. Auroral Density Cavity

In the region of parallel electric field, access for various populations of auroral particles is denied. Ionospheric low-energy electrons are reflected by the upward-directed electric field, whereas ionospheric ions are accelerated away from the region, becoming an upward ion beam. For the magnetospheric

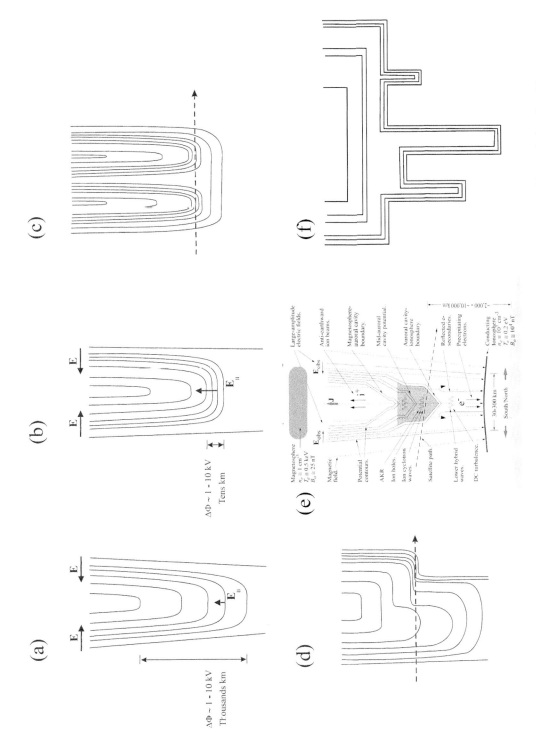

Figure 1. Variations of models of the U-shaped potential: (a)–(d) after *Hull et al.* [2003], (e) reprinte] with permission from *Ergun et al.* [2002a], copyright 2002, American Institute of Physics, and (f) after *McFadden et al.* [1999]

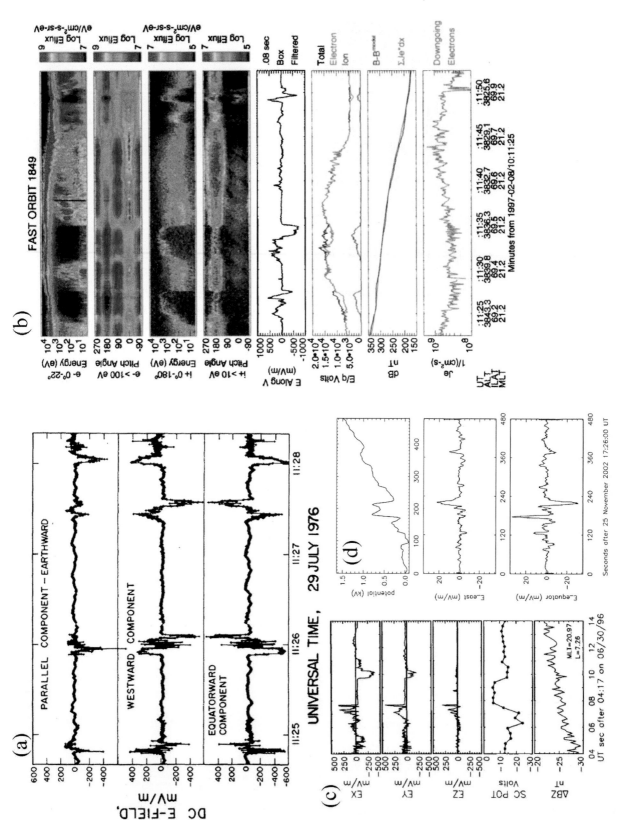

Figure 2. Examples of measurements of bipolar perpendicular electric field structures: (a) after *Mozer et al.* [1977], (b) after *McFadden et al.* [1999], (c) after *Mozer and Kletzing* [1998], and (d) after *Johansson et al.* [2006].

Table 1. Latitudinal Scale Size of Auroral Arcs and Associated Acceleration Structures

Latitudinal Scale Size (Mapped to Ionosphere) (km)	Type of Measurement	Spacecraft	Reference
2–3 (summer) 4–5 (winter)	Electric field	Cluster	*Johansson et al.* [2007]
28–35	Particle (electrons)	DMSP	*Newell et al.* [1996]
20–40	Particle (electrons)	FAST	*Partamies et al.* [2008]
45–111	Particle (ions + electrons)	Akebono	*Sakanoi et al.* [1995]
9–27	Optical	N/A	*Knudsen et al.* [2001]

particles, the situation is the opposite; magnetospheric ions are reflected, and electrons are accelerated, forming the inverted-V beam. This results in very low plasma densities in the acceleration region. The region of low density is referred to as the auroral density cavity and was first observed by *Persoon et al.* [1988], by analysis of whistler mode wave propagation. They found that auroral cavities extended at least up to an altitude of 3.6 R_E. *Janhunen et al.* [2002] report on a statistical study showing the auroral cavities to extend from an altitude of 1 to 3.25 R_E (in the nightside), with a peak of occurrence frequency at 2.25 R_E.

The auroral cavities act as a source for auroral kilometric radiation (AKR) [*Benson and Calvert*, 1979; *Gurnett and Anderson*, 1981; *Pottelette et al.*, 2001; *Strangeway et al.*, 2001], something that can be used as remote detection of the acceleration region (see section 4).

2.4. S-Shaped Potentials

Although the U shape has been the canonical shape designated to the acceleration potential, often the measurements show an asymmetry of the bipolar electric field associated with the perpendicular potential drop on each side of the structure [*Mozer et al.*, 1980; *Johansson et al.*, 2006]. In the extreme case of a monopolar electric field signature, we talk of an S-shaped potential structure. *Marklund* [1984] considers this as an effect of the superposition of the ionospheric background electric field and the high-altitude electric field, associated with the U-shaped potential, whereas *Swift* [1979] and *Mozer et al.* [1980] puts forth the idea that the S-shaped potential is an electrostatic shock associated with a particular class of solutions to the Vlasov-Poisson equations (see section 6.1.2 below). *Chiu et al.* [1981] pointed out that what looks like a U-shaped potential at high altitude may look like an S-shaped potential at lower altitudes, something that was

verified by *Marklund et al.* [2011], using Cluster multipoint measurements. The relation between the S- and U-shaped potentials is still not clear.

3. DIRECT OBSERVATIONS OF $E_\parallel$

The first direct measurements of the parallel electric field in the auroral acceleration region was made by observing the acceleration of artificially released barium ions onto auroral magnetic field lines. *Haerendel et al.* [1976] report on a parallel voltage of several kV at an altitude of 1.2 R_E.

Direct measurements using the double probe technique are problematic, due to asymmetries in the probe work functions, spacecraft shadowing of particle fluxes along the magnetic field, and uncertainties in spacecraft attitude determination, which may make an unambiguous separation of the perpendicular and parallel components of the field impossible. However, if the parallel electric field has a magnitude comparable to that of the perpendicular component, a direct measurement of $E_\parallel$ may be possible. *Mozer and Kletzing* [1998] report on four detections of parallel electric fields, using the three-axis double probe antennas on the Polar satellite. These measurements (where the authors claim that all efforts have been made to exclude spurious detections due to the difficulties mentioned above) show parallel electric fields of 200–300 mV m^{-1} at an altitude of around 1 R_E, directed away from the ionosphere, located in upward current regions, and associated with upward ion beams. The authors interpret their observations as a probing of the region where the electric field turns from a purely parallel direction to a purely perpendicular direction. Similar observations of $E_\parallel$, at lower altitude, from the FAST satellite have been reported by *Ergun et al.* [2002a].

4. ALTITUDE DISTRIBUTION OF $E_\parallel$

Several authors have reported on the altitude distribution of the field-aligned potential drop, using various methods to

Table 2. Typical Acceleration Energies Associated With Discrete Auroral Arcs

Energy (keV)	Type of Measurement	Spacecraft	Reference
1–20	Electric field	S3-3	*Redsun et al.* [1985]
1–7	Electric field	Cluster	*Johansson et al.* [2005]
1–12	Particle (electrons)	AE-D	*Lin and Hoffman* [1979]
3–30	Particle (electrons)	DMSP	*Newell et al.* [1996]
2–4	Particle (electrons)	FAST	*Partamies et al.* [2008]
1–10	Particle (ions + electrons)	DE1, DE2	*Reiff et al.* [1988]
1–10	Particle (ions + electrons)	Akebono	*Sakanoi et al.* [1995]

determine the distribution. *Lindqvist and Marklund* [1990] report an increasing perpendicular electric field (in a statistical study using double probe electric field measurements from the Viking satellite) up to an altitude of 1.7 R_E, after which the field had a constant magnitude. Since the perpendicular potential drop sampled by a satellite at a certain altitude is equal to the total parallel potential drop minus the parallel potential drop above the satellite, this indicates that there is no further potential drop above this altitude. A similar method was used by *Weimer and Gurnett* [1993], where the average perpendicular electric field measured by the DE1 satellite was increasing with altitude. Most of the gradient in the electric field strength is located at altitudes between 0.3 and 1 R_E, with evidence of an additional potential drop above 1.7 R_E. In studies as the ones above, it is important to take the geometry of the geomagnetic field lines into account. As the field lines spread out at higher altitudes, a constant potential drop between them results in an electric field strength decreasing with altitude. Expressions for electric field mapping between different altitudes in a dipole geometry are given by *Mozer* [1970] and *Weimer and Gurnett* [1993].

Using ion and electron beam energies from about 40 ion beam events, *Mozer and Hull* [2001] calculated the fraction of the total potential drop located below the satellite as a function of altitude. They concluded that the parallel potential region extended up to 2.5 R_E, with about half of the potential drop located between 1.5 and 2.0 R_E. *Reiff et al.* [1993] investigated the upper altitude limit of electron acceleration events, using DE1 data, and gives an upper limit of 2.4 R_E. It is important when interpreting these statistical results to keep in mind the difference between a statistical distribution of the parallel potential drop and the instantaneous potential drop. *Hull et al.* [2003], e.g., claim that the largest parallel electric fields identified by direct measurements are isolated within a relatively thin layer at an altitude of about 1.3 R_E.

Information on the instantaneous distribution of potential along the field line can be obtained by quasisimultaneous multipoint measurements from the Cluster satellites. *Sadeghi et al.* [2011] have used electric field, ion, and electron data to determine the fraction of acceleration potential at three different altitudes for a U potential and conclude that in one case, 52% of the total potential drop was located below 1.13 R_E, 18% between 1.13 and 1.30 R_E, and 30% above 1.30 R_E. A unique possibility to determine the instantaneous altitude extent of the acceleration potential, or at least the auroral cavity [*Benson and Calvert*, 1979], is observation of AKR. *Morioka et al.* [2007] have mapped AKR emission to altitudes along an auroral field line, by matching the AKR frequency to the local electron cyclotron frequency. Their

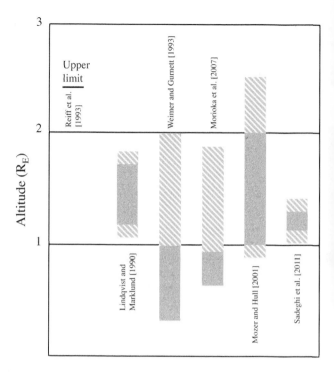

Figure 3. Altitude distribution of the auroral acceleration potential reported by various authors. Hatched regions denote uncertain or time-variable measurements.

observations show that the lower boundary of the acceleration region (if we identify the acceleration region with the auroral cavity region) has a relatively stable altitude of around 4000 km but that the upper boundary changes drastically over a short time (of the order of a minute), from about 6000 km to over 10,000 km. The change coincides in time with an onset of Pi 2 pulsations, indicative of a substorm onset or a pseudobreakup. While it is unclear if the parallel electric field can be described by static theories during these active times, it is interesting to note that the characteristic signature of the auroral cavity is exhibited. This indicates a relatively long-lived parallel electric field at these altitudes and may be an example of a temporal evolution of the plasma environment in the flux tube, as described by *McFadden et al.* [1999].

The reported observations on altitude distributions are summarized in Figure 3.

5. STATIONARITY/LIFETIMES

Regarding the temporal stability of the potential structures, this was first investigated by *Thieman and Hoffman* [1985], who used combined DE1 and DE2 observations to analyze electron data for 140 crossings of such structures. They concluded that the time scales for growth and decay of the acceleration structure were of similar order and that the lower limit for the maximum lifetime of an acceleration potential was 18 min. *Boudouridis and Spence* [2007] studied the coherence between electron and ion data precipitation data from the F6 and F8 DMSP satellites and found that the coherence is high on time scales of the order of 120 s for structures with a latitudinal scale size of 100 km or larger, whereas smaller-scale structures show a decrease in coherence on shorter time scales.

Detailed studies of the spatiotemporal behavior of selected acceleration structures in the auroral zone, with the use of Cluster data, have been performed by *Marklund et al.* [2001, 2011], *Karlsson et al.* [2004], *Figueiredo et al.* [2005], and *Sadeghi et al.* [2011]. These investigations show that the acceleration potentials can be stable on time scales from 30 s up to tens of minutes [*Figueiredo et al.*, 2005]. A new finding by *Sadeghi et al.* [2011] was that for an auroral acceleration potential increasing in time, the increase took place at the lower part of the acceleration structure (below 1.3 R_E), with the acceleration potential below that altitude increasing from 0.7 to 2.2 kV, in about 40 s, while the potential drop above that altitude stayed constant at 1 kV. During the increase of the acceleration potential, the magnetic field signatures remained virtually unchanged, with an inferred parallel current density of around 2 $\mu A\ m^{-2}$ (mapped to the ionosphere).

6. NATURE OF $E_\parallel$ AND ACCELERATION MECHANISMS

It has become clear during the last 10 or 20 years that Alfvén waves can drive certain auroral arcs [e.g., *Chaston et al.*, 2003b, 2007; *Keiling*, 2009]. For small perpendicular scale sizes, the Alfvén wave enters the dispersive domain [*Stasiewicz et al.*, 2000], which introduces a parallel wave electric field. For Alfvén waves with oscillation periods comparable to the travel time through the acceleration region, which is of the order of 1 s, the direct acceleration of electrons by the wave parallel field results in a characteristic broad energy distribution of the electrons. As described in the Introduction section, this is taken as a typical signature of the "Alfvénic aurora" [e.g., *Chaston et al.*, 2001, 2003a, 2003b].

In contrast to the Alfvénic acceleration described above, the "static" acceleration, which we may define as acceleration by a parallel electric field, which remains stationary in time on the electron transit time of 1 s, gives rise to the signature of monoenergetic acceleration of the classical inverted V. An important question in auroral physics is to determine the mechanism responsible for setting up the

parallel electric field, which accelerates the auroral electrons in the static case. One school of thought maintains that the Alfvén wave is responsible for $E_\parallel$ also in the case of "static" acceleration, only that the frequencies of the waves are considerably lower than the electron transit time. Another school considers that once the field-aligned current system is set up, other types of process may act to produce a parallel field and that the role of the Alfvén wave is mainly to govern the dynamics of the current systems. It is clear that any static field-aligned current system in the auroral zone is established by Alfvénic processes whereby a number of reflections of an Alfvén wave pulse between the ionosphere and the magnetospheric generator region produces the final state of the current system [*Goertz and Boswell*, 1979; *Lysak and Dum*, 1983]. Likewise, changes in the state of the ionospheric or magnetospheric end are communicated by Alfvén wave fronts, in a similar fashion. In light of the above, it is logical to divide the various models for generation of $E_\parallel$ into Alfvénic models and static ones. Below, we give a brief review of the more important such models.

6.1. Static Models

The basis for static models is the requirement of a field-aligned potential drop between the magnetosphere and the ionosphere to maintain field-aligned upward currents of a certain strength. Due to the mirror force, only the particles in the loss cone of an isotropic population of electrons in the magnetospheric equatorial plane will be able to travel all the way down to the ionosphere and carry a current. This is a small fraction of the original population. An upward-directed electric field will accelerate electrons downward, increasing their pitch angle, and increase the fraction of electrons in the loss cone. By calculation of adiabatic motion of the electrons (valid for static conditions), it can be shown that for a wide range of potential drops (100 V to 100 kV) the field-aligned current density j carried by the electrons is proportional to the total potential drop between the magnetosphere and the ionosphere $\Delta\Phi$:

$$\mathbf{j} = K\Delta\Phi,$$

where K is called the Knight constant or the Lyons-Evans-Lundin constant, and the relation is called the (linear) Knight relation [*Knight*, 1973; *Lundin and Sandahl*, 1978; *Fridman and Lemaire*, 1980]. The Knight relation gives information of the total potential drop required, but says nothing of where this potential drop is located along the field line.

6.1.1. Anomalous resistivity. In a collisionless plasma, it is not possible to define a classical resistivity. However, the role of binary collisions can be taken over by the interaction of electrons with various types of waves present in the auroral region. On time scales much larger than the frequency of the waves, the average effect may be similar to collision-induced resistivity. Such an *anomalous resistivity* may sustain a parallel electric field [*Papadopoulos*, 1977, and references therein]. Anomalous resistivity supporting potential drops of the order of kilovolts, as required in the acceleration region, would, however, cause a rapid heating of the local plasma. This has not been observed, and this dissipation of energy makes it difficult for a potential structure to be sustained for a long time [*Cornwall and Chiu*, 1982; *Block*, 1984; *Coroniti*, 1985]. Parallel electric fields resulting from anomalous resistivity would be distributed over a wide altitude range, where the drift velocity of the auroral electrons exceeds some critical velocity, where instabilities would trigger the anomalous instability [e.g., *Kindel and Kennel*, 1971]. *Morioka et al.* [2005] estimated that for typical conditions on auroral field lines, anomalous resistivity would be triggered at altitudes between approximately 8000 and 17,000 km.

6.1.2. Double layers and electrostatic shocks. Double layers are particular solutions to the Vlasov-Poisson equations that give two localized thin layers of opposite charge, with an electric field between them [e.g., *Block*, 1978; *Borovsky*, 1993]. They are localized structures, with a thickness in the parallel (to **B**) direction of the order of $10 \dfrac{eV}{k_B T_e} \lambda_D$ [*Fälthammar*, 2004], where V is the potential drop, T_e is the magnetospheric electron temperature, λ_D is the Debye length, and the other symbols have their usual meaning. For a potential drop of 10 kV and typical plasmasheet temperatures and densities, this is of the order of 1 km. So-called electrostatic shocks are similar to double layers, but may differ in the approximations used to obtain them as solutions to the Vlasov-Poisson equations [*Borovsky*, 1993] or the plasma normal modes used to obtain the solution [*Hudson and Mozer*, 1978]. Relatively recently, *Ergun et al.* [2000, 2002b, 2004] have obtained numerical solutions to the Vlasov-Poisson equations in a 1-D geometry with a realistic set of particle populations. The resulting solutions give two localized double layers at altitudes of 3000 and 5500 km [*Ergun et al.*, 2000], with the auroral cavity situated between the two double layers. Under certain conditions, a third double layer may be located inside the cavity [*Ergun et al.*, 2004].

6.1.3. Magnetic mirror models. The mirror point of a particle bouncing along a field line between the two hemispheres is determined by its pitch angle in the equatorial plane. If there then are different anisotropies in the ion and electron populations, such that

$$\frac{W_{i,\parallel}}{W_{e,\parallel}} \neq \frac{W_{i,\perp}}{W_{e,\perp}},$$

where W is the average thermal energy of the population, the different species will have different distributions of mirror points for the particles [*Alfvén and Fälthammar*, 1963]. This will lead to a charge separation along the field line. Since quasineutrality is a very strong requirement, electric fields are set up to move the mirror points of the particles until quasineutrality is obtained. Such charge separations will lead to parallel potential drops of the order of $k_B T_i / e$, where T_i is the magnetospheric ion temperature, and the other symbols have their usual meaning. For typical magnetospheric conditions, this is of the order of 1–5 kV [*Baumjohann et al.*, 1989]. *Jasperse and Grossbard* [2000] consider an analytical solution for $E_\parallel$ using realistic parameters and obtain a parallel potential drop of 4 kV smoothly distributed between altitudes of 1000 and 7000 km, while *Schriver et al.* [2001] obtain a potential drop of around 3 kV distributed between approximately 3000 and 6500 km. Similar results are also found by *Chiu and Schulz* [1978] and *Chiu and Cornwall* [1980].

6.2. Alfvénic Models

As mentioned above, the parallel electric field accelerating the auroral electrons might also be directly related to Alfvén waves, which, if they have a low-enough frequency, are stationary on the time scale of the electron transit time through the acceleration region. In that case, the auroral acceleration structure could be explained by exclusively invoking the physics of the Alfvén wave. A number of theories attempting this exist.

6.2.1. Kinetic Alfvén wave models. In an early model by *Goertz and Boswell* [1979], an electric field pulse with a finite perpendicular (to the magnetic field) extent is excited at the magnetospheric end of a field line. The pulse will propagate down the field line with the Alfvén speed, and due to the finite perpendicular wavelength, a parallel electric field will develop at the wave front. As the wave pulse reflects at the ionosphere, the resulting superposition of the electric fields behind the incoming and reflected pulse will produce a boundary between two regions of different amplitudes of the perpendicular electric field. While this, in principle, reproduces a U potential, the lower boundary of the potential will move with the Alfvén velocity in this model. Further reflections will move this boundary and produce electric fields that are more and more consistent with an electrostatic solution. A further prediction obtained by relating this type of Alfvénic acceleration to static auroral arcs is that regardless of the frequency of the Alfvén wave (if we imagine that the characteristic rise time of the pulse represents the wave period), the resulting parallel fields are strongly modulated, both in amplitude and altitude, at frequencies corresponding to the wave travel time between the equatorial plane and the ionosphere, which is of the order of minutes [*Goertz and Boswell*, 1979].

The term kinetic here relates to the fact that the perpendicular wave number is nonzero, which makes a kinetic treatment (as opposed to ideal MHD) of the wave essential. The nonzero perpendicular component of the wave vector also implies propagation obliquely to the magnetic field, which explains the commonly used term "oblique Alfvén waves."

Self-consistent calculations of the parallel electric field associated with field line resonances (FLR) have been carried out by *Rankin et al.* [1999], *Tikhonchuk and Rankin* [2000], and *Samson et al.* [2003]. *Samson et al.* [2003] report on parallel electric fields of the order of 1 mV m^{-1}, at altitudes of 1.4–2.2 R_E, integrating to field-aligned potential drops of the order of 6 kV, although the perpendicular electric field from the calculations are an order of magnitude smaller than those observed in the auroral region. This type of acceleration mechanism may explain the periodic restructuring and modulation of multiple auroral arcs associated with FLRs.

6.2.2. Stationary Alfvén wave models. One class of theories considers Alfvénic disturbances, which are stationary in the frame of the auroral arc. Such structures can be created when a static perturbation is threaded by a background plasma convection (similar to the Alfvén wings created in the interaction of moons and the magnetospheric plasma in, e.g., the Jupiter and Saturn magnetospheres) [*Southwood et al.*, 1980]. The perturbation travels as an Alfvén wave along the magnetic field in the frame of the convecting plasma, and the superposed plasma motion creates an oblique structure, which is reflected at the ionosphere [*Mallinckrodt and Carlson*, 1978]. Including finite electron inertia, it can be shown that a magnetic field-aligned electric field results. Note that in this model, the source of the electron energy is located on neighboring flux tubes, and not on the flux tubes associated with the actual auroral arc [*Knudsen*, 2001]. No clear prediction of the altitude distribution of the parallel electric field exists for this model.

7. DISCUSSION AND SUMMARY

It is clear that ample observational evidence exists for U- and S-shaped potential structures, stable on rather long time

scales, accelerating magnetospheric electrons to energies of several keV. In a quasistatic situation, such acceleration can be understood in a macroscopic sense as a consequence of the need to widen the loss cone in order to maintain current continuity when the thermal current is not large enough. What is less clear is what kind of microphysics is responsible for the parallel electric field, which is responsible for this acceleration. In this chapter, a number of possible mechanisms have been presented. All of these models make some testable predictions, an important example of which is the altitude distribution of $E_\parallel$. The altitude distributions of the potential drop predicted by the theories in section 6 are indicated in Figure 4. However, the situation in the auroral region is quite complex, and to discriminate between various models, several predictions need to be tested and met.

One of the most principal questions to answer is if the parallel field can be described purely by Alfvén wave physics or if other physics need to be invoked. One direct try to do this was made by *Hull et al.* [2003], who compared the ratio of two mutually perpendicular electric and magnetic field components associated with an acceleration region, with the

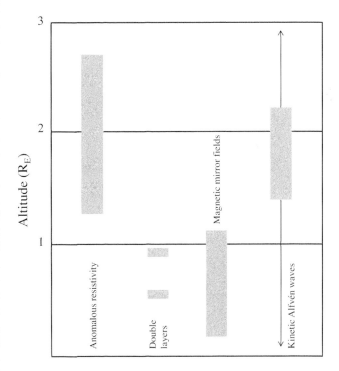

Figure 4. Altitude distribution of the auroral acceleration potential, as predicted by theories described in section 6. The gray region for the kinetic Alfvén wave entry represents the altitude extent of the parallel electric field of a field line resonance [*Samson et al.*, 2003], while the arrow represents a localized potential drop of uncertain altitude extent, which is at a location in altitude varying in time [*Goertz and Boswell*, 1979].

expected ratio of kinetic Alfvén waves, calculated by *Lysak* [1998]. Since the ratios differed by about 2 orders of magnitude, they concluded that kinetic Alfvén waves were not a plausible explanation for the parallel electric field in that case. Further comparisons of this type needs to be done, and it should be noted that the comparison of *Hull et al.* [2003] did not take into account possible effects of ionospheric reflection of the Alfvén wave. As *Lysak* [1998] points out, detailed observations of phase shifts between magnetic and electric field components may provide a way of resolving the nature of the electric field. In contrast, in an investigation of the acceleration of precipitating auroral electrons, *Chaston et al.* [2005] conclude that the acceleration takes place by ULF kinetic Alfvén waves. However, in this case, the accelerated electrons do not show typical inverted-V characteristics, but rather the broad energy range associated with the Alfvénic aurora, even if the frequencies of the waves are quite low.

Another test of the theories of the acceleration mechanism may be to check how the potential drops develop in time. The results of *Sadeghi et al.* [2011] and *Morioka et al.* [2007], which show how the altitude distribution of the potential drop develop in time, are interesting in this context. In the first instance, the potential drop at the low-altitude part of the potential drop increases in about 40 s, while the high-altitude part remains unchanged, and also, the field-aligned current remains constant. Again, this may be an example of a temporal evolution of the plasma environment in the flux tube, as invoked by *McFadden et al.* [1999] to explain the finger formation at the lower edge of the acceleration region. In the second instance, on the other hand, the high-altitude portion of the potential drop increases on a time scale of 20 s, which is interpreted by the authors as a response to increased field-aligned currents associated with substorm onset.

These observations illustrate that the distribution of the potential drop along the field line is not uniquely determined by the field-aligned current and are evolving in time. It is probably also naïve to expect that one single physical mechanism is responsible for the particle acceleration. It is more likely that the requirement of a potential drop, as determined by the Knight relation, is fulfilled in different ways during different circumstances. In particular, it seems likely, in light of Figures 2 and 3, that two or more mechanisms may dominate at different altitudes. Furthermore, the mechanism dominating at one time may not be the dominating one for the whole lifetime of the acceleration structure. When the field-aligned current is being set up, e.g., the acceleration may be Alfvénic in nature, as suggested by, e.g., *Goertz and Boswell* [1979], but after a number of Alfvén wave reflections, when the current system has been established, other physics may sustain the parallel electric field. The relative importance of the Alfvénic fields may be related to the

findings of *Keiling et al.* [2003], who estimate that one third of the Poynting flux powering the aurora is due to Alfvén waves. These kinds of ideas have been incorporated in a number of what we may call "hybrid models," where the field-aligned current dynamics, including the creation of a perpendicular fine structure, is governed by Alfvén waves, but non-Alfvénic microphysics is invoked for the creation of parallel electric fields [e.g., *Lysak and Dum*, 1983; *Goertz et al.*, 1991; *Lotko et al.*, 1998; *Streltsov*, 2007].

Most theories of auroral acceleration hardly address the 2-D structure of the acceleration region, let alone the 3-D one, where in particular, the large aspect ratio of the thin, long east-west-aligned auroral arcs remain an unsolved problem. Here much remains to be done.

REFERENCES

Akasofu, S.-I. (1964), The development of the auroral substorm, *Planet. Space Sci.*, *12*, 273–282.

Alfvén, H., and C.-G. Fälthammar (1963), *Cosmical Electrodynamics*, Clarendon Press, Oxford, U. K.

Baumjohann, W., G. Paschmann, and C. A. Cattell (1989), Average plasma properties in the central plasma sheet, *J. Geophys. Res.*, *94*(A6), 6597–6606.

Benson, R. F., and W. Calvert (1979), ISIS 1 observations at the source of auroral kilometric radiation, *Geophys. Res. Lett.*, *6*(6), 479–482.

Block, L. P. (1978), A double layer review, *Astrophys. Space Sci.*, *55*, 59–83.

Block, L. P. (1984), Three-dimensional potential structure associated with Birkeland currents, in *Magnetospheric Currents*, *Geophys. Monogr. Ser.*, vol. 28, edited by T. A. Potemra, pp. 315–324, AGU, Washington, D. C.

Block, L. P., and C.-G. Fälthammar (1990), The role of magnetic-field-aligned electric fields in auroral acceleration, *J. Geophys. Res.*, *95*(A5), 5877–5888.

Borovsky, J. E. (1993), Auroral arc thicknesses as predicted by various theories, *J. Geophys. Res.*, *98*(A4), 6101–6138.

Boudouridis, A., and H. E. Spence (2007), Separation of spatial and temporal structure of auroral particle precipitation, *J. Geophys. Res.*, *112*, A12217, doi:10.1029/2007JA012591.

Carlqvist, P., and R. Boström (1970), Space-charge regions above the aurora, *J. Geophys. Res.*, *75*(34), 7140–7146.

Chaston, C. C., C. W. Carlson, R. E. Ergun, and J. P. McFadden (2000), Alfvén waves, density cavities, and electron acceleration observed from the FAST spacecraft, *Phys. Scr. T*, *84*, 64–68.

Chaston, C. C., W. J. Peria, C. W. Carlson, R. E. Ergun, and J. P. McFadden (2001), FAST observation of inertial Alfvén waves and electron acceleration in the dayside aurora, *Phys. Chem. Earth, Part C*, *26*(1–3), 201–205.

Chaston, C. C., J. W. Bonnell, C. W. Carlson, J. P. McFadden, R. J. Strangeway, and R. E. Ergun (2003a), Kinetic effects in the acceleration of auroral electrons in small scale Alfven waves:

A FAST case study, *Geophys. Res. Lett.*, *30*(6), 1289, doi:10.1029/2002GL015777.

Chaston, C. C., L. M. Peticolas, J. W. Bonnell, C. W. Carlson, R. E. Ergun, J. P. McFadden, and R. J. Strangeway (2003b), Width and brightness of auroral arcs driven by inertial Alfven waves, *J. Geophys. Res.*, *108*(A2), 1091, doi:10.1029/2001JA007537.

Chaston, C. C., et al. (2005), Energy deposition by Alfvén waves into the dayside auroral oval: Cluster and FAST observations, *J. Geophys. Res.*, *110*, A02211, doi:10.1029/2004JA010483.

Chaston, C. C., A. J. Hull, J. W. Bonnell, C. W. Carlson, R. E. Ergun, R. J. Strangeway, and J. P. McFadden (2007), Large parallel electric fields, currents, and density cavities in dispersive Alfvén waves above the aurora, *J. Geophys. Res.*, *112*, A05215, doi:10.1029/2006JA012007.

Chiu, Y. T., and J. M. Cornwall (1980), Electrostatic model of a quiet auroral arc, *J. Geophys. Res.*, *85*(A2), 543–556.

Chiu, Y. T., and M. Schulz (1978), Self-consistent particle and parallel electrostatic field distributions in the magnetospheric-ionospheric auroral region, *J. Geophys. Res.*, *83*(A2), 629–642.

Chiu, Y. T., A. L. Newman, and J. M. Cornwall (1981), On the structures and mapping of auroral electrostatic potentials, *J. Geophys. Res.*, *86*(A12), 10,029–10,037.

Cornwall, J. M., and Y. T. Chiu (1982), Ion distribution effects of turbulence on a kinetic auroral arc model, *J. Geophys. Res.*, *87*(A3), 1517–1527.

Coroniti, F. V. (1985), Space plasma turbulence dissipation: Reality or myth?, *Space Sci. Rev.*, *42*, 399–410.

Ergun, R. E., C. W. Carlson, J. P. McFadden, F. S. Mozer, and R. J. Strangeway (2000), parallel electric fields in discrete arcs, *Geophys. Res. Lett.*, *27*(24), 4053–4056, doi:10.1029/2000GL003819.

Ergun, R. E., L. Andersson, D. S. Main, Y.-J. Su, C. W. Carlson, J. P. McFadden, F. S. Mozer, and R. J. Strangeway (2002a), Parallel electric fields in the upward current region of the aurora: Indirect and direct observations, *Phys. Plasmas*, *9*, 3685–3694.

Ergun, R. E., L. Andersson, D. Main, Y.-J. Su, D. L. Newman, M. V. Goldman, C. W. Carlson, J. P. McFadden, and F. S. Mozer (2002b), Parallel electric fields in the upward current region of the aurora: Numerical solutions, *Phys. Plasmas*, *9*(9), 3695–3704.

Ergun, R. E., L. Andersson, D. Main, Y.-J. Su, D. L. Newman, M. V. Goldman, C. W. Carlson, A. J. Hull, J. P. McFadden, and F. S. Mozer (2004), Auroral particle acceleration by strong double layers: The upward current region, *J. Geophys. Res.*, *109*, A12220, doi:10.1029/2004JA010545.

Fälthammar, C.-G. (2004), Magnetic field-aligned electric fields in collisionless space plasmas – A brief review, *Geofis. Int.*, *43*(2), 225–239.

Figueiredo, S., G. T. Marklund, T. Karlsson, T. Johansson, Y. Ebihara, M. Ejiri, N. Ivchenko, P.-A. Lindqvist, H. Nilsson, and A. Fazakerley (2005), Temporal and spatial evolution of discrete auroral arcs as seen by Cluster, *Ann. Geophys.*, *23*, 2531–2557, doi:10.5194/angeo-23-2531-2005.

Frank, L. A., and K. L. Ackerson (1971), Observations of charged particle precipitation into the auroral zone, *J. Geophys. Res.*, *76*(16), 3612–3643.

Fridman, M., and J. Lemaire (1980), Relationship between auroral electrons fluxes and field aligned electric potential difference, *J. Geophys. Res.*, *85*(A2), 664–670.

Galperin, Y. I. (2002), Multiple scales in auroral plasmas, *J. Atmos. Sol. Terr. Phys.*, *64*(2), 211–229, doi:10.1016/S1364-6826(01)00085-2.

Goertz, C. K., and R. W. Boswell (1979), Magnetosphere-ionosphere coupling, *J. Geophys. Res.*, *84*(A12), 7239–7246.

Goertz, C. K., T. Whelan, and K.-I. Nishikawa (1991), A new numerical code for simulating current-driven instabilities on auroral field lines, *J. Geophys. Res.*, *96*(A6), 9579–9593.

Gurnett, D. A., and R. R. Anderson (1981),The kilometric radio emission spectrum: Relationship to auroral acceleration processes, in *Physics of Auroral Arc Formation, Geophys. Monogr. Ser.*, vol. 25, edited by S.-I. Akasofu and J. R. Kan, pp. 341–350, AGU, Washington, D. C.

Haerendel, G., E. Rieger, A. Valanzuela, H. Föppl, H. C. Stenbaek-Nielsen, and E. M. Wescott (1976), First observation of electro-static acceleration of barium ions into the magnetosphere, in *European Programmes on Sounding-Rocket and Balloon Research in the Auroral Zone, ESA Spec. Publ., ESA SP-115*, 203–211.

Hallinan, T. J., and T. N. Davis (1970), Small-scale auroral arc distortions, *Planet. Space Sci.*, *18*(12), 1735–1744.

Hudson, M. K., and F. S. Mozer (1978), Electrostatic shocks, double layers, and anomalous resistivity in the magnetosphere, *Geophys. Res. Lett.*, *5*(2), 131–134.

Hull, A. J., J. W. Bonnell, F. S. Mozer, and J. D. Scudder (2003), A statistical study of large-amplitude parallel electric fields in the upward current region of the auroral acceleration region, *J. Geophys. Res.*, *108*(A1), 1007, doi:10.1029/2001JA007540.

Janhunen, P., A. Olsson, and H. Laakso (2002), Altitude dependence of plasma density in the auroral zone, *Ann. Geophys.*, *20*, 1743–1750.

Jasperse, J. R., and N. J. Grossbard (2000), The Alfvén-Falthammar formula for the parallel E-field and its analogue in downward auroral-current regions, *IEEE Trans. Plasma Sci.*, *28*(6), 1874–1886.

Johansson, T., T. Karlsson, G. Marklund, S. Figueiredo, and P.-A. Lindgvis (2005), A statistical study of intense electric fields at 4–7 R_E geocentric distance using Cluster, *Ann. Geophys.*, *23*, 2579–2588.

Johansson, T., G. Marklund, T. Karlsson, S. Liléo, P.-A. Lindqvist, A. Marchaudon, H. Nilsson, and A. Fazakerley (2006), On the profile of intense high-altitude auroral electric fields at magneto-spheric boundaries, *Ann. Geophys.*, *24*, 1713–1723, doi:10.5194/angeo-24-1713-2006.

Johansson, T., G. Marklund, T. Karlsson, S. Liléo, P.-A. Lindqvist, H. Nilsson, and S. Buchert (2007), Scale sizes of intense auroral electric fields observed by Cluster, *Ann. Geophys.*, *25*, 2413–2425, doi:10.5194/angeo-25-2413-2007.

Karlsson, T., G. Marklund, S. Figuieredo, T. Johansson, and S. Buchert (2004), Separating spatial and temporal variations in auroral electric and magnetic fields by using Cluster multipoint measurements, *Ann. Geophys.*, *22*, 2463–2472, doi:10.5194/angeo-22-2463-2004.

Keiling, A. (2009), Alfvén waves and their roles in the dynamics of the Earth's magnetotail: A review, *Space Sci. Rev.*, *142*, 73–156, doi:10.1007/s11214-008-9463-8.

Keiling., A., J. R. Wygant, C. A. Cattell, F. S. Mozer, and C. T. Russel (2003), The global morphology of wave Poynting flux: Powering the aurora, *Science*, *299*, 383–386, doi:10.1126/science.1080073.

Kindel, J. M., and C. F. Kennel (1971), Topside current instabilities, *J. Geophys. Res.*, *76*(13), 3055–3078.

Knight, S. (1973), Parallel electric fields, *Planet. Space Sci.*, *21*, 741–750.

Knudsen, D. J. (2001), Structure, acceleration, and energy in auroral arcs and the role of Alfvén waves, *Space Sci. Rev.*, *95*, 501–511.

Knudsen, D. J., E. F. Donovan, L. L. Cogger, B. Jackel, and W. D. Shaw (2001), Width and structure of mesoscale optical auroral arcs, *Geophys. Res. Lett.*, *28*(4), 705–708.

Lin, C. S., and R. A. Hoffman (1979), Characteristics of the inverted-V event, *J. Geophys. Res.*, *84*(A4), 1514–1524.

Lindqvist, P.-A., and G. T. Marklund (1990), A statistical study of high-altitude electric fields measured on the Viking satellite, *J. Geophys. Res.*, *95*(A5), 5867–5876.

Lotko, W., A. V. Streltsov, and C. W. Carlson (1998), Discrete auroral arc, electrostatic shock and suprathermal electrons powered by dispersive, anomalously resistive field line resonance, *Geophys. Res. Lett.*, *25*(24), 4449–4452.

Lundin, R., and I. Sandahl (1978), Some characteristics of the parallel electric field acceleration of electrons over discrete auroral arcs as observed from two rocket flights, in *European Sounding Rocket, Balloon and Related Research, with Emphasis on Experiment at High Latitudes*, ESA Spec. Publ., ESA SP-135, 125–136.

Lysak, R. L. (1998), The relationship between electrostatic shocks and kinetic Alfvén waves, *Geophys. Res. Lett.*, *25*(12), 2089–2092.

Lysak, R. L., and C. T. Dum (1983), Dynamics of magnetosphere-ionosphere coupling including turbulent transport, *J. Geophys. Res.*, *88*(A1), 365–380.

Mallinckrodt, A. J., and C. W. Carlson (1978), Relations between transverse electric fields and field-aligned currents, *J. Geophys. Res.*, *83*(A4), 1426–1432.

Marklund, G. (1984), Auroral arc classification scheme based on the observed arc-associated electric field pattern, *Planet. Space Sci.*, *32*(2), 193–211.

Marklund, G., et al. (2001), Temporal evolution of the electric field accelerating electrons away from the auroral ionosphere, *Nature*, *414*, 724–727.

Marklund, G., S. Sadeghi, T. Karlsson, P.-A. Lindqvist, H. Nilsson, C. Forsyth, A. Fazakerley, and J. Picket (2011), Altitude distribution of the auroral acceleration potential determined from Cluster satellite data at different heights, *Phys. Rev. Lett.*, *106*, 055002, doi:10.1103/PhysRevLett.106.055002.

McFadden, J. P., C. W. Carlson, and R. E. Ergun (1999), Microstructure of the auroral acceleration region as observed by FAST, *J. Geophys. Res.*, *104*(A7), 14,453–14,480.

Morioka, A., Y. S. Miyoshi, F. Tsuchiya, H. Misawa, A. Kumamoto, H. Oya, H. Matsumoto, K. Hashimoto, and T. Mukai (2005), Auroral kilometric radiation activity during magnetically quiet periods, *J. Geophys. Res.*, *110*, A11223, doi:10.1029/2005JA011204.

Morioka, A., Y. Miyoshi, F. Tsuchiya, H. Misawa, T. Sakanoi, K. Yumoto, R. R. Anderson, J. D. Menietti, and E. F. Donovan (2007), Dual structure of auroral acceleration regions at substorm onsets as derived from auroral kilometric radiation spectra, *J. Geophys. Res.*, *112*, A06245, doi:10.1029/2006JA012186.

Mozer, F. S. (1970), Electric field mapping in the ionosphere at the equatorial plane, *Planet. Space Sci.*, *18*, 259–263.

Mozer, F. S. (1973), Analyses of techniques for measuring DC and AC electric fields in the magnetosphere, *Space Sci. Rev.*, *14*, 272–313, doi:10.1007/BF02432099.

Mozer, F. S., and A. Hull (2001), Origin and geometry of upward parallel electric fields in the auroral acceleration region, *J. Geophys. Res.*, *106*, 5763–5778.

Mozer, F. S., and C. A. Kletzing (1998), Direct observation of large, quasi-static, parallel electric fields in the auroral acceleration region, *Geophys. Res. Lett.*, *25*(10), 1629–1632.

Mozer, F. S., C. W. Carlson, M. K. Hudson, R. B. Torbert, B. Parady, J. Yatteau, and M. C. Kelley (1977), Observations of paired electrostatic shocks in the polar magnetosphere, *Phys. Rev. Lett.*, *38*, 292–295, doi:10.1103/PhysRevLett.38.292.

Mozer, F., C. A. Cattell, M. K. Hudson, R. L. Lysak, M. Temerin, and R. B. Torbert (1980), Satellite measurements of low altitude auroral particle acceleration, *Space Sci. Rev.*, *27*, 155–213.

Newell, P. T., K. M. Lyons, and C.-I. Meng (1996), A large survey of electron acceleration events, *J. Geophys. Res.*, *101*(A2), 2599–2614.

Papadopoulos, K. (1977), A review of anomalous resistivity for the ionosphere, *Rev. Geophys.*, *15*(1), 113–127.

Partamies, N., E. Donovan, and D. Knudsen (2008), Statistical study of inverted-V structures in FAST data, *Ann. Geophys.*, *26*, 1439–1449, doi:10.5194/angeo-26-1439-2008.

Paschmann, G., S. Haaland, and R. Treumann (Eds.) (2003), *Auroral Plasma Physics*, Springer, New York.

Persoon, A. M., D. A. Gurnett, W. K. Peterson, J. H. Waite Jr., J. L. Burch, and J. L. Green (1988), Electron density depletions in the nightside auroral zone, *J. Geophys. Res.*, *93*(A3), 1871–1895.

Pottelette, R., R. A. Treumann, and M. Berthomier (2001), Auroral plasma turbulence and the cause of auroral kilometric radiation fine structure, *J. Geophys. Res.*, *106*, 8465–8476.

Rankin, R., J. C. Samson, and V. T. Tikhonchuk (1999), Discrete auroral arcs and nonlinear dispersive field line resonances, *Geophys. Res. Lett.*, *26*(6), 663–666.

Redsun, M. S., M. Temerin, and F. S. Mozer (1985), Classification of auroral electrostatic shocks by their ion and electron associations, *J. Geophys. Res.*, *90*(A10), 9615–9633.

Reiff, P. H., H. L. Collin, J. D. Craven, J. L. Burch, J. D. Winningham, E. G. Shelley, L. A. Frank, and M. A. Friedman (1988), Determination of auroral electrostatic potentials using high- and low-altitude particle distributions, *J. Geophys. Res.*, *93*(A7), 7441–7465.

Reiff, P. H., G. Lu, J. L. Burch, J. D. Winningham, L. A. Frank, J. D. Craven, W. K. Peterson, and R. A. Heelis (1993), On the high- and low-altitude limits of the auroral electric field region, in *Auroral Plasma Dynamics, Geophys. Monogr. Ser.*, vol. 80, edited by R. L. Lysak, pp. 143–154, AGU, Washington, D. C.

Sadeghi, S., G. T. Marklund, T. Karlsson, P.-A. Lindqvist, H. Nilsson, O. Marghitu, A. Fazakerley, and E. A. Lucek (2011), Spatiotemporal features of the auroral acceleration region as observed by Cluster, *J. Geophys. Res.*, *116*, A00K19, doi:10. 1029/2011JA016505.

Sakanoi, T., H. Fukunishi, and T. Mukai (1995), Relationship between field-aligned currents and inverted-v parallel potential drops observed at midaltitudes, *J. Geophys. Res.*, *100*(A10), 19,343–19,360.

Samson, J. C., R. Rankin, and V. T. Tikhonchuk (2003), Optical signatures of auroral arcs produced by field line resonances: Comparisons with satellite observations and modeling, *Ann. Geophys.*, *21*, 933–945.

Schriver, D., M. Ashour-Abdalla, and R. L. Richard (2001), Formation of electrostatic potential drops in the auroral zone, *Phys. Chem. Earth, Part C*, *26*(1–3), 65–70.

Southwood, D. J., M. G. Kivelson, R. J. Walker, and J. A. Slavin (1980), Io and its plasma environment, *J. Geophys. Res.*, *85*(A11), 5959–5968.

Stasiewicz, K., et al. (2000), Small scale Alfvénic structure in the aurora, *Space Sci. Rev.*, *92*, 423–533.

Stenbaek-Nielsen, H. C., T. J. Hallinan, D. L. Osborne, J. Kimball, C. Chaston, J. McFadden, G. Delory, M. Temerin, and C. W. Carlson (1998), Aircraft observations conjugate to FAST: Auroral are thicknesses, *Geophys. Res. Lett.*, *25*(12), 2073–2076.

Strangeway, R. J., R. E. Ergun, C. W. Carlson, J. P. McFadden, G. T. Delory, P. L. Pritchet (2001), Accelerated electrons as the source of auroral kilometric radiation, *Phys. Chem. Earth, Part C*, *26*(1-3), 145–149.

Streltsov, A. V. (2007), Narrowing of the discrete auroral arc by the ionosphere, *J. Geophys. Res.*, *112*, A10218, doi:10.1029/2007JA 012402.

Swift, D. W. (1979), An equipotential model for auroral arcs: The theory of two-dimensional laminar electrostatic shocks, *J. Geophys. Res.*, *84*(A11), 6427–6434.

Thieman, J. R., and R. A. Hoffman (1985), Determination of inverted-V stability from Dynamics Explorer satellite data, *J. Geophys. Res.*, *90*(A4), 3511–3516.

Tikhonchuk, V. T., and R. Rankin (2000), Electron kinetic effects in standing shear Alfvén waves in the dipolar magnetosphere, *Phys. Plasmas*, *7*(6), 2630–2645.

Weimer, D. R., and D. A. Gurnett (1993), Large-amplitude auroral electric fields measured with DE 1, *J. Geophys. Res.*, *98*(A8), 13,557–13,564.

T. Karlsson, Space and Plasma Physics, School of Electrical Engineering, KTH, SE-100 44 Stockholm, Sweden. (tomas. karlsson@ee.kth.se)

The Search for Double Layers in Space Plasmas

L. Andersson and R. E. Ergun

Laboratory for Atmospheric and Space Physics, University of Colorado, Boulder, Colorado, USA

Localized, quasi-static parallel electric fields that are created as a result of charge separation in plasmas have been studied by scientists over the last century and have become known as double layers (DLs). DLs are important because they can efficiently accelerate charge particles, dissipate energy, and cause a local break in the frozen-in condition. As a result, they are expected to be an important process in many different types of space plasmas on Earth and on many astrophysical objects. This paper presents a brief review of the history of DLs over the last century leading to the now well-established fact that they do occur naturally in space plasmas. The paper also presents some of the latest understanding of the basic properties of DLs in the aurora region and discusses some open research questions.

1. INTRODUCTION

A simple yet very effective way to accelerate charge particles is through a parallel electric field. However, charge particles have high mobility along the magnetic field, so it was believed that a parallel electric field would vanish rapidly. As a result, Hannes Alfvén formulated the frozen-in concept [*Alfvén*, 1950]. Later on, he regretted that he created this concept as he realized that localized charge separation could develop as a self-consistent plasma structure and that these structures could have a major effect on the global system with, for instance, slippage of flux tubes [*Alfvén*, 1958]. A localized charge separation with a net potential is now called a double layer (DL).

Two characteristic signatures are associated with DLs: particle acceleration and energy dissipation. These features make the DL very interesting in many different plasma environments spanning from laboratory experiments to astrophysics. There have been several reviews describing the fundamental physics associated with the DL, to which the reader is directed for a better physical understanding [*Block*, 1978; *Swift*, 1978; *Sato*, 1982; *Schamel*, 1986; *Raadu*, 1989].

With the writing of this article, it became clear that one possible area of confusion is that there are different types of DLs. In collisionless plasma, there are surface, current, and gradient types of DLs. A surface DL is created by currents to and from the surface, which results in a sheath between the surface and the plasma that may carry a net potential. Examples of surface DLs are probe/sensor interaction with plasmas [*Langmuir*, 1929], spacecraft interaction with space plasmas, and the Moon's interaction with the solar wind [*Halekas et al.*, 2003]. A gradient DL (or currentless DL) is associated with strong magnetic and/or density gradients resulting in charge separations [*Charles*, 2009; *Scime et al.*, 2010]. This type of DL is being studied actively as a potential application to ion thrusters. Finally, the current-driven DL is a result of interaction between two different plasma regions with a strong, field-aligned current. If the drift between the electrons and ions is large enough, two-stream [e.g., *Buneman*, 1959] instabilities can develop, which can lead to DLs. In this chapter, we focus on the last of these types of DL.

In current-driven plasma, there are several structures that are, on occasion, called DLs. As described by *Raadu* [1989], quasi-static theoretical descriptions for potential structures exist for solitary potential structures (e.g., electron/ion phase

Auroral Phenomenology and Magnetospheric Processes: Earth and Other Planets
Geophysical Monograph Series 197
10.1029/2011GM001170

space holes), slow ion acoustic DL, weak DL, and strong DL, etc. The focus of this chapter is on strong DL since they are very effective in accelerating particles and in dissipating energy.

Bernstein et al. [1957] derived a method to solve the Vlasov equation for self-consistent, stationary potential structures. The current-driven, quasi-static strong DL can be described as a Bernstein-Greene-Kruskal (BGK) solution. The particle populations that are required to maintain this structure include two "passing" populations and two "trapped" (reflected) populations (see Figure 1). The size of the structure is of the order of the square root of the mass ratio times the Debye length [*Block*, 1978]. Note that, since the plasma conditions can change dramatically across a DL, the Debye length can be difficult to define.

These quasi-static structures require two specific conditions: charge separation and pressure balance. The charge layers form with the correct polarity if the inflow from both sides meets the Bohm criterion [*Bohm*, 1949].

This criterion describes that to get a charge separation, additional reflected populations are required such that the time for the different populations to pass through the location of the charge separation results in a charge separation.

The pressure balance is met if the structure is in a frame in which the ion-to-electron current ratio meets the Langmuir condition [*Langmuir*, 1929; *Block*, 1972].

The difference between the weak DL and the strong DL is the relationship between the inflowing particle thermal speed (v_{th}) and the accelerated outflowing particle drift (v_d). The Langmuir condition describing the pressure balance means that a strong DL can exist in the frame where the ratio between the ion and electron current is equal to the square root of the ratio of the electron to ion mass.

For a strong DL ($v_d > v_{th}$), an unstable particle beam emerges from the DL leading to further instabilities and potential heating of the beam. As such, strong DLs are associated with waves and nonlinear features. Fundamentally, a DL is not necessarily a static structure, that is, its behavior may depend on the waves and nonlinear structures that it creates.

When describing a system associated with DLs or modeling a DL, one can start with a prescribed large-scale potential in the system or a prescribed current driving the system. Ultimately, the DL acts as the load (resistor) in the system, whereas the energy source is described as a current generator or a voltage generator. To understand the behavior of the DL (the load), one does not necessarily need a clear understanding of the energy source.

There are situations in dynamic simulations, laboratory experiments, and space observations in which the observed DL behaves as a structure that is well described by the static BGK DL solution. In these cases, we call the structure a "laminar" DL. Instabilities can still act on either side of the DL as long as the DL itself is slowly evolving in time. A "turbulent" DL, which has been identified in simulation results [*Newman et al.*, 2008], has a localized potential jump (i.e., localized parallel electric field), but the instabilities on one or both sides are so significant that they interact with the localized potential jump, and the DL is difficult to identify in the data.

The remainder of this article has the following layout. The history of DLs will be presented in chronological order with some major milestones over the last century. This is followed by a discussion of recent observations, our current understanding, and outstanding questions we have today.

2. EARLY 1900

Irving Langmuir, working for the General Electric Company in Schenectady, New York, was one of the first researchers to investigate surface-plasma interactions leading to the basics of charge separation. In work associated with surface-plasma interaction, he realized that a charge separation could develop resulting in a parallel electric field. This potential structure was called a "double sheath." *Langmuir* [1929] constructed the first self-consistent DL solution with cold particles (delta-function distributions) and experimented with current-driven discharge tubes. This early work formed the basis of the space applications today such as Langmuir probes and DL theories.

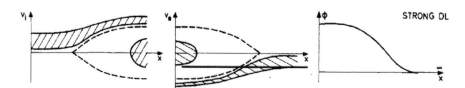

Figure 1. (left) Ion and (middle) electron-phase space diagrams for a strong double layer. Reprinted from the first part of Figure 3 of *Raadu* [1989], with permission from Elsevier. (right) The potential profile. The separatrices for the particle motions are shown as dashed lines. The diagrams illustrate that both ions and electrons have distinct passing-through and trapped populations.

3. THE ERA OF THEORY AND EXPERIMENTS (~1960)

The peak activity of laboratory research on DLs was in the 1950–1970 period. One of the early problems with DLs was their stability [*Block*, 1972; *Knorr and Goertz*, 1974]. Thus, many experiments were developed to see how DLs are created and how to control them [*Block*, 1978]. Another significant step in DL research was made when it became apparent that a DL could form completely within the plasma. A surface sheath was not needed. One way to create the DL was through the Buneman instability [*Buneman*, 1959]. This process appeared to have some type of threshold and, when triggered, was very efficient in accelerating particles. It was immediately recognized that a free-standing DL could be important in space plasmas. Its implication to space plasmas was investigated in several applications such as solar flare eruptions [*Block*, 1972; *Hasan and ter Harr*, 1986] and disruption associated with substorms [*Alfvén*, 1977; *Stenzel et al.*, 1982].

Alfvén became one of the most vocal spokesmen about the impact of DL on astrophysical objects and cosmology [*Alfvén*, 1977, 1982, 1990]. Others did more direct applications to regions such as the solar corona providing alternative methods to the BGK solution and further improvements of the theoretical descriptions [*Montgomery and Joyce*, 1969; *Block*, 1972; *Swift*, 1975; *Perkins and Sun*, 1981; *Williams*, 1986; *Sato and Miyawaki*, 1992; *Boström*, 2004].

Some of the earliest applications of numerical simulations were directed toward understanding of DLs in plasmas. Since most laboratory experiments were set up as a voltage generator, the early simulations focused on that setup [*Goertz and Joyce*, 1975; *Singh*, 1982; *Borovsky and Joyce*, 1983; *Hudson et al.*, 1983; *Lembege and Dawson*, 1989; *Borovsky*, 1992; *Singh et al.*, 2005]. Later simulations are based on current generation in the aurora region [*Newman et al.*, 2001] and in relativistic astrophysical plasmas [*Dieckmann and Bret*, 2009].

4. THE ERA OF SPACE OBSERVATIONS (~1970)

Following the success in the laboratory, space plasma physicists investigated the possibility of DLs in space. Analysis of auroral emissions indicated that the precipitating particles were accelerated. Direct observation of particles verified that the accelerated electrons were nearly monoenergetic, so the possibility that the electrons were accelerated by discrete potential structures was put forth [*Albert and Lindstrom*, 1970]. A DL is a natural candidate to carry the parallel electric field, so the search for strong DLs in the aurora was on.

Analysis of the precipitating auroral electron spectra identified a primary electron beam and secondary (scattered) electrons. The scattered electrons that were moving anti-earthward appeared to be a reflection by a parallel electric field [*Evans*, 1974]. This hypothesis ignited a debate on whether the electric field was extended or localized along the magnetic field. A theoretical description of the correlation of field-aligned currents with potential was developed by *Knight* [1973], who showed that a large-scale electric field develops naturally as a result of the combination of a magnetic mirror force and a current. The possibility of a DL, however, was not ruled out.

The next major piece of evidence came from satellites with particle observations and measurements of the perpendicular electric field. The observations suggested that the satellite crossed through a U-shaped potential [*Gurnett*, 1972] as shown in Figure 1.

While particle and field measurements suggested parallel electric fields, topside sounder experiments uncovered strong-density cavities in the topside of the *F* region [*Calvert*, 1966] and low-density cavities at high altitudes [*Hagg*, 1967; *Herzberg and Nelms*, 1969]. In situ observations identified that the density gradients were associated with perpendicular electric fields of possible potential structures [*Mozer et al.*, 1977]. However, large electric fields were also found when no density gradients were present. The observed large density gradients are easily explained by DL theory [*Block*, 1978], but the evidence was not conclusive at that time.

Active experiments using barium clouds released from sounding rockets also were used to study auroral plasmas. Some of the barium cloud/jet experiments investigated perpendicular electric fields [*Wescott et al.*, 1976] with the implication that a DL could explain the observed motion of the cloud and the existence of parallel electric fields [*Haerendel et al.*, 1976].

5. THE ERA OF WEAK DL (~1980)

The S3-3 satellite brought high-resolution observations of paired converging perpendicular DC electric fields called electrostatic shocks [*Mozer et al.*, 1977]. These structures were associated with electrostatic ion cyclotron waves and turbulence. The S3-3 satellite also made the first direct measurement of the parallel electric field associated with the paired electrostatic shocks. This measurement was possible using the 3-D electric field observations on the S3-3 satellite [*Mozer et al.*, 1977]. However, the observed parallel electric field was not convincingly in agreement with DL theory.

One of the most definitive measurements of a parallel electric field in space was an uncovering of small-amplitude electric field structures identified as weak DLs [*Temerin et al.*, 1982]. The weak DL was a bipolar structure with a small net potential, roughly a few volts. The idea that followed was that a large number (thousands) of weak DLs could produce the required net potential (kilovolts) that was inferred from the particle observations. However, a search for weak DLs leads to an estimate far below the required number. While weak DLs were an interesting phenomenon, they did not account for auroral acceleration [*Boström et al.*, 1988; *Boström*, 1992].

6. THE LOSS OF FAITH (~1990)

With no direct measurement of large localized parallel electric fields in space plasma, space plasma researchers started to come up with alternative explanations for the observations. Some publications questioned the existence of quasi-static parallel electric fields and DLs [*Bryant et al.*, 1992], but DLs were still viewed as an important candidate [*Borovsky*, 1992].

With new and exciting observations, the aurora research focused on other issues such as the effect of Alfvén waves and how ions are heated, resulting in atmospheric loss. This decade moved the research forward in many other areas but not much in understanding DLs.

7. THE ERA OF STRONG DL (~2000)

Roughly 70 years after Langmuir's work, the Polar [*Mozer and Kletzing*, 1998; *Mozer and Hull*, 2001] and FAST satellites [*Ergun et al.*, 2001; *Andersson et al.*, 2002] identified unipolar electric fields well above the instrument uncertainties. The FAST observations were accompanied by evidence of localized electron acceleration.

The first positively identified DLs through particle and electric field measurements were observed in the downward current region as a result of the DLs' antiearthward motion [*Ergun et al.*, 2001]. This motion, or the frame of the DL, was shown to be consistent with the Bohm and Langmuir conditions [*Smith and Goertz*, 1978]. The DL-accelerated electron beam was found to create waves and nonlinear plasma structures as result of electron-electron instabilities.

Two such observations can be seen in Figure 2, marked by vertical lines. The turbulent region prior to the DL almost always contains VLF waves and electron-phase space holes. The waves on the high-potential (high-altitude) side are believed to create the commonly observed VLF saucers [*Ergun et al.*, 2003]. This can be used as a remote signature of a DL. As the result of the waves and nonlinear structures, the electron beam undergoes rapid thermalization creating

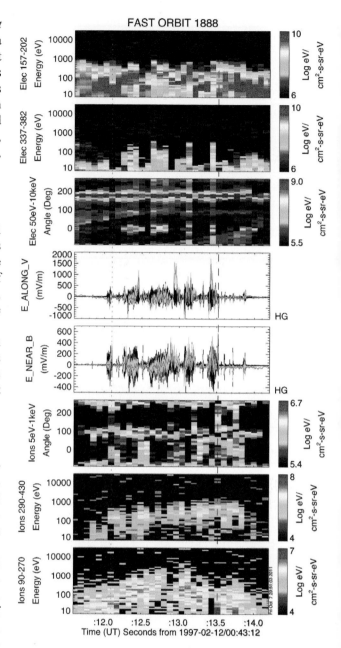

Figure 2. Multiple double layers (DLs) in the downward current region of the aurora. The top three panels are electron spectrograms and the bottom three panels are ion spectrograms. (top to bottom) Three electron spectrograms: earthward (±11°), antiearthward (±11°), and 50 eV to 10 keV pitch angle energy flux. The middle two panels represent the perpendicular (along satellite path) and the parallel electric fields. Both panels have three electric field signals (each of the colored lines was band-pass filtered differently onboard). Three ion spectrograms: 5 eV to 5 keV pitch angle, antiearthward (±70°), and earthward (±60°) ion flux. The last panel presents the perpendicular electric field. Two DLs are observed, marked by the two vertical lines.

field-aligned electrons at both 0° and 180° and heated ions at 90° (Figure 2). The DC unipolar signature at the vertical line is the DL itself where, ironically, there is no significant wave activity. The ion distribution measured at and just before (above in altitude) the DL indicates that the ions are strongly

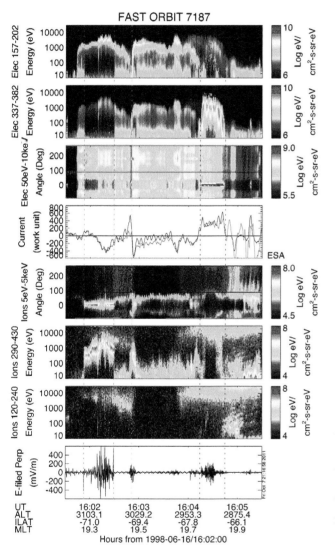

FAST ORBIT 7187

Figure 3. An example of a satellite crossing from the polar cap to the equatorward edge of the auroral oval. The electron and ion spectrograms have the same ranges as in Figure 2. Two inverted-V structures exist between ~16:01:30 and ~16:04:00 UT. The satellite moves into the aurora cavity at the vertical lines where a paired DC electric field can be seen at the boundaries with strong AC waves between the paired electric fields. The electron flux observed from ~16:04:20 to ~16:04:50 UT is during a downward current where the electron flux is modulated as a result of the narrowness of the electron beam and the instrument sector observing this flux.

heated and "plowed" in front of the DL resulting in the strong fluxes at 180° pitch angles (Figure 2). The moving DL turns out to be an efficient process for atmospheric loss [*Hwang et al.*, 2008]. On the low potential side, an earthward-traveling ion beam emerges (accelerated earthward with a significant perpendicular temperature). Upwelling thermal ions at 180° are also recorded. Ions that are heated and mirror well below the DL reach the DL and are reflected. This process results in a modified pressure cooker picture [*Gorney et al.*, 1985; *Hwang et al.*, 2008].

Strong DLs were also found in the lower boundary (Figure 3) of the upward current region's inverted-V potential structures [*Ergun et al.*, 2002; *Hull et al.*, 2003]. There are some less-definitive observations of more turbulent midcavity DLs [*Ergun et al.*, 2004] of the inverted-V potential. The lower-boundary DLs have been proven to contribute only a smaller fraction (20%–50%) of the net potential associated with upward current region [*Ergun et al.*, 2004]. It has not been demonstrated nor ruled out whether DLs are responsible for the high-altitude acceleration due to lack of high time resolution observations at these altitudes.

The DLs in the upward current region are strong for the ions but weak for the electrons, so contrary to the case in downward current region DLs, the ions control the evolution. The DLs at the lower boundary of the upward current region straddle a strong density gradient. As a result, the DLs are asymmetric [*Main et al.*, 2006]. The DLs at the lower boundary of the inverted-V potential are relatively fixed in altitude as a result of the secondary electrons and upwelling ions. The upwelling ions serve to satisfy the Bohm and Langmuir criteria.

The low-potential side of these DLs (high-altitude side) has the auroral density cavity. Strong wave activity is often associated with an ion-ion two-stream instability since both hydrogen and oxygen ions are strongly accelerated with the same energy into the cavity, which causes them to emerge with differing velocities [*Main et al.*, 2006]. The auroral cavity is also the source region for AKR radiation that can be used to remotely identify inverted-V locations and estimate the altitude location of the main parallel electric field [*Morioka et al.*, 2007].

8. DLS EVERYWHERE (>2010)

The number of observed DLs by the FAST and Polar satellites was limited since the satellites travel primarily perpendicular to the magnetic field, and the vertical scale of DLs is very narrow.

The DL observations by the Time History of Events and Macroscale Interactions during Substorms (THEMIS) mission were initially a surprise [*Ergun et al.*, 2009]. The probability of actually observing a DL increased significantly over that of the FAST mission since the THEMIS satellites dwell for long time periods on flux tubes in the magnetotail with strong currents. Furthermore, the magnetotail sound velocity is much higher than that in the low-altitude aurora (1000 km s^{-1} compared to 30 km s^{-1}). The THEMIS DLs were identified by their electric field signal alone since the temporal resolution of the particle instruments on THEMIS mission is too low. The high number of observed DLs in the magnetotail during magnetic disturbances suggests that DLs might be an important process to dissipate the energy in the Earth's magnetotail [*Ergun et al.*, 2009].

DLs also have been inferred in the outer planets. The strong DL creates electron beams. These electron beams lead to electron cyclotron emissions in the upward current regions and VLF saucers in the downward current region. With understanding from the Earth auroral radio signals, the radio signals from Jupiter indicate that DLs are an important process at the magnetosphere of Jupiter [*Hess et al.*, 2009]. Sudden changes in radio spectrograms of "millisecond" bursts at Jupiter have been interpreted as possible DLs [*Hess et al.*, 2009]. Another possible indication of DLs comes from the VLF saucer emission observed at Enceladus in Saturn's magnetosphere [*Gurnett and Pryor*, this volume].

9. WHERE WE ARE TODAY

As of this writing, several distinct types of DLs appear in the data. The DLs in the downward current region of the aurora have strong electron acceleration but appear to have weak ion acceleration, mainly because the ion temperature is much higher than the electron temperature in the vicinity of the DLs. These DLs move antiearthward at the ion-acoustic speed [*Andersson et al.*, 2002] that appears to satisfy the Bohm and Langmuir criteria. For example, the 800 V DL observed in Figure 2 at ~13.5 s is moving at ~30 km s^{-1} antiearthward.

The parallel scale length of these DLs is on the order of 10 Debye lengths, in line with theoretical predictions. The perpendicular scale length has not been directly measured. However, if DLs are the source of VLF saucers, the perpendicular scale size appears to be roughly 100 to 1000 Debye lengths. The motion of the DL dictates that their lifetime is short [*Andersson and Ergun*, 2006; *Marklund et al.*, 2001]. Numerical simulations indicate that if no warm electrons are present, the DLs are disrupted rapidly. However, a warm electron background (often seen in the downward current region of the aurora) can stabilize the DL [*Newman et al.*, 2008]. DLs in the downward current region have been observed as low as ~1500 km [*Elphic et al.*, 2000] and increase in frequency to ~4000 km, the apogee of the FAST satellite and supported by perpendicular electric field observations by

the Viking satellite [*Marklund*, 1993]. The net potential associated with the downward current region is often smaller compared to the upward current region, but as Figure 3 demonstrates, sometimes they are equal. Finally, the moving DLs and their associated wave emissions create an interesting scenario for ion heating as discussed in section 7 and, during quiet times, might be an effective process for atmospheric loss, Figure 3.

The other type of DL is associated with upward current region and partly described in section 7. The lower boundary of the inverted-V structure is fairly stationary in altitude as a result of the natural inflow of ions from the ionosphere [*Ergun et al.*, 2002; *Hull et al.*, 2003]. As a result, the DL is at the density gradient between the ionosphere and the aurora cavity. Since the lower boundary DLs in the upward current are controlled by the ions, these DLs are evolving slowly. They are also found to be oblique to the magnetic field. Interestingly, they seem to conform to the "U-shaped" potential structure (Figure 1). Multipoint measurements from the Cluster mission have provided the first glimpse of the evolution of the aurora region and the quasi-static structures in the downward [*Marklund et al.*, 2001] and the upward [*Hull et al.*, 2010] current region.

Other types of unipolar electric fields in the auroral acceleration region have been observed, but well-developed theoretical explanations are lacking. The first is observed unipolar localized parallel electric fields associated with Alfvén waves [*Ergun et al.*, 2005]. In numerical simulations of auroral Alfvén waves, solitary structures have been developed suggesting a DL type of acceleration [*Genot et al.*, 2004; *Mottez*, 2001]. This result is further supported by antiearthward field-aligned ion observations where the ion acceleration process has to be both localized in space and time suggesting that a DL type of acceleration exists in Alfvénic regions. Another puzzle is the closely spaced unipolar electric fields (seen in Figure 2 at ~12.8 s). Such closely spaced unipolar DC electric fields have not been replicated in simulations. So far, any attempt in locating two DLs close to each other has resulted in the destruction of one of them due to the instabilities created by the other (D. L. Newman, private communication, 2003).

10. SUMMARY: UNANSWERED QUESTIONS AND MOVING FORWARD

During the last 70 years, significant progress has been made on DL research, most recently characterizing the strong DLs in the aurora region. However, there are still many questions that need to be answered, both associated with the aurora region and to understand how important strong DLs are for space and astrophysical plasmas.

Some of the outstanding questions for the aurora region are (1) How do quasi-static DLs form? (2) Only a smaller part of potential drop in the upward current region can be explained today by strong DL. Can DLs also explain the rest of the potential drop? (3) Most of the downward current is associated with small perpendicular scale lengths and has short lifetimes. How can the DLs in the downward current sustain the return current? (4) What impact does the slow ion motion and ionospheric convection have on the stability of the DLs? (5) How important are DLs for the atmospheric loss? (6) Can observed parallel electric fields associated with Alfvén waves be explained by DLs? (7) What theories are needed to explain closely spaced unipolar electric fields as observed in Figure 2?

The observation of two different types of DLs in the low-altitude aurora, the observation of DLs in the magnetotail, and the implication of DLs in Jupiter's and Saturn's magnetospheres suggest that the DL is truly a universal process. With significant gaps in our understanding of DL, this area of research will continue to be important.

Acknowledgments. This work was supported by National Aeronautics and Space Administration grants NNX09AF48G and NNXIOAH46G.

REFERENCES

Albert, R. D., and P. J. Lindstrom (1970), Auroral-particle precipitation and trapping caused by electrostatic double layers in the ionosphere, *Science*, *170*, 1398–1401.

Alfvén, H. (1950), *Cosmical Electrodynamics*, Clarendon Press, Oxford, U. K.

Alfvén, H. (1958), On the theory of magnetic storms and aurorae, *Tellus*, *10*, 104–116.

Alfvén, H. (1977), Electric currents in cosmic plasmas, *Rev. Geophys.*, *15*(3), 271–284.

Alfvén, H. (1982), On hierarchical cosmology, *Astrophys. Space Sci.*, *89*, 313–324.

Alfvén, H. (1990), Cosmology in the plasma universe: An introductory exposition, *IEEE Trans. Plasma Sci.*, *18*(1), 5–10.

Andersson, L., and R. E. Ergun (2006), Acceleration of antiearthward electron fluxes in the auroral region, *J. Geophys. Res.*, *111*, A07203, doi:10.1029/2005JA011261.

Andersson, L., R. E. Ergun, D. L. Newman, J. P. McFadden, C. W. Carlson, and Y.-J. Su (2002), Characteristics of parallel electric fields in the downward current region of the aurora, *Phys. Plasma*, *9*(8), 3600–3609.

Bernstein, I. B., J. M. Greene, and M. D. Kruskal (1957), Exact nonlinear plasma oscillations, *Phys. Rev.*, *108*, 546–550.

Block, L. P. (1972), Potential double layers in the ionosphere, *Cosmic Electrodyn.*, *3*, 349.

Block, L. P. (1978), A double layer review, *Astrophys. Space Sci.*, *55*, 59–83.

Bohm, D. (1949), Minimum ionic kinetic energy for a stable sheath, in *The Characteristics of Electrical Discharges in Magnetic Fields*, edited by A. Guthrie and R. K. Wakerling, pp. 77–86, McGraw-Hill, New York.

Borovsky, J. E. (1992), Double layers do accelerate particles in the auroral zone, *Phys. Rev. Lett.*, *69*(7) 1054–1056.

Borovsky, J. E., and G. Joyce (1983), Numerically simulated two-dimensional auroral double layers, *J. Geophys. Res.*, *88*(A4), 3116–3126.

Boström, R. (1992), Observations of weak double layer on auroral field lines, *IEEE Trans. Plasma Sci.*, *20*(6), 756–763.

Boström, R. (2004), Kinetic and space charge control of current flow and voltage drops along magnetic flux tubes: 2. Space charge effects, *J. Geophys. Res.*, *109*, A01208, doi:10.1029/2003JA010078.

Boström, R., G. Gustafsson, B. Holback, G. Holmgern, H. Koskinen, and P. Kintner (1988), Characteristics of solitary waves and weak double layers in the magnetospheric plasma, *Phys. Rev. Lett.*, *61*, 82–85.

Bryant, D. A., R. Binghamn, and U. de Angelis (1992), Double layers are not particle accelerators, *Phys. Rev. Lett.*, *68*, 37–39.

Buneman, O. (1959), Dissipation of currents in ionized media, *Phys. Rev.*, *115*, 503–517.

Calvert, W. (1966), Steep horizontal electron-density gradients in the topside *F*-layer, *J. Geophys. Res.*, *71*(15), 3665–3669.

Charles, C. (2009), Λ review of recent laboratory double layer experiments, *Plasma Sources Sci. Technol.*, *16*, R1–R25.

Dieckmann, M. E., and A. Bret (2009), Particle-in-cell simulations of a strong double layer in a nonrelativistic plasma flow: Electron acceleration to ultrarelativistic speeds, *Astrophys. J.*, *694*(1), 154–164.

Elphic, R. C., J. Bonnell, R. J. Strangeway, C. W. Carlson, M. Temerin, J. P. McFadden, R. E. Ergun, and W. Peria (2000), FAST observations of upward accelerated electron beams and the downward field-aligned current region, in *Magnetospheric Current Systems*, Geophys. Monogr. Ser., vol. 118, edited by S. Ohtani et al., pp. 173–180, AGU, Washington, D. C., doi:10. 1029/GM118p0173.

Ergun, R. E., Y.-J. Su, L. Andersson, C. W. Carlson, J. P. McFadden, F. S. Mozer, D. L. Newman, M. V. Goldman, and R. J. Strangeway (2001), Direct observation of localized parallel electric fields in a space plasma, *Phys. Rev. Lett.*, *87*, 045003, doi:10. 1103/PhysRevLett.87.045003.

Ergun, R. E., L. Andersson, D. S. Main, Y.-J. Su, C. W. Carlson, J. P. McFadden, and F. S. Moser (2002), Parallel electric fields in the upward current region of the aurora: Indirect and direct observations, *Phys. Plasmas*, *9*, 3685–3694.

Ergun, R. E., C. W. Carlson, J. P. McFadden, R. J. Strangeway, M. V. Goldman, and D. L. Newman (2003), FAST observations of VLF saucers, *Phys. Plasmas*, *10*, 454.

Ergun, R. E., L. Andersson, D. Main, Y.-J. Su, D. L. Newman, M. V. Goldman, C. W. Carlson, A. J. Hull, J. P. McFadden, and F. S. Mozer (2004), Auroral particle acceleration by strong double layers: The upward current region, *J. Geophys. Res.*, *109*, A12220, doi:10.1029/2004JA010545.

Ergun, R. E., L. Andersson, Y. J. Su, D. L. Newman, M. V. Goldman, W. Lotko, C. C. Chastno, and C. W. Carlson (2005), Localized parallel electric fields associated with inertial Alfvén waves, *Phys. Plasmas*, *12*, 072901, doi:10.1063/1.1924495.

Ergun, R. E., et al. (2009), Observations of double layers in Earth's plasma sheet, *Phys. Rev. Lett.*, *102*, 155002, doi:10.1103/PhysRevLett.102.155002.

Evans, D. S. (1974), Precipitating electron fluxes formed by a magnetic field aligned potential difference, *J. Geophys. Res.*, *79*(19), 2853–2858.

Genot, V., P. Louarn, and F. Mottez (2004), Ionospheric erosion by Alfvén waves, *Ann. Geophys.*, *22*, 2081–2096.

Goertz, C. K., and G. Joyce (1975), Numerical simulations of the plasma double layer, *Astrophys. Space Sci.*, *32*, 165–173.

Gorney, D. J., Y. T. Chiu, and D. R. Croley Jr. (1985), Trapping of ion conics by downward parallel electric fields, *J. Geophys. Res.*, *90*(A5), 4205–4210.

Gurnett, D. A. (1972), Electric field and plasma observations in the magnetosphere, in *Critical Problems of Magnetospheric Physics*, edited by E. R. Dyer, pp. 123–138, Natl. Acad. of Sci., Washington, D. C.

Gurnett, D. A., and W. R. Pryor (2012), Auroral processes associated with Saturn's moon Enceladus, in *Auroral Phenomenology and Magnetospheric Processes: Earth and Other Planets*, *Geophys. Monogr. Ser.*, doi:10.1029/2011GM001174, this volume.

Haerendel, G., E. Rieger, A. Valenzuela, H. Foppl, H. C. Stenbaek-Nielsen, and E. M. Wescott (1976), First observations of electrostatic acceleration of barium ions into the magnetosphere, in *European Programmes on Sounding-Rocket and Balloon Research in the Auroral Zone*, *ESA Spec. Publ.*, ESA SP-115, 203–211.

Hagg, E. L. (1967), Electron densities of 8-100 electrons cm^{-3} deduced from Alouette II high latitude ionograms, *Can. J. Phys.*, *45*, 27–36.

Halekas, J. S., R. P. Lin, and D. L. Mitchell (2003), Inferring the scale height of the lunar nightside double layer, *Geophys. Res. Lett.*, *30*(21), 2117, doi:10.1029/2003GL018421.

Hasan, S. S., and D. ter Harr (1986), The Alfvén-Carlquist double-layer theory on solar flares, *Astrophys. Space Sci.*, *56*, 89–107.

Herzberg, L., and G. L. Nelms (1969), Ionospheric conditions following the proton flare of 7 July 1966 as deduced from topside sounding, *Ann. IQSY*, *3*, 426–436.

Hess, S., F. Mottez, and P. Zarka (2009), Effect of electric potential structures on Jovian S-burst morphology, *Geophys. Res. Lett.*, *36*, L14101, doi:10.1029/2009GL039084.

Hudson, M. K., W. Lotko, I. Roth, and E. Witt (1983), Solitary waves and double layers on auroral field lines, *J. Geophys. Res.*, *88*(A2), 916–926.

Hwang, K.-J., R. E. Ergun, L. Andersson, D. L. Newman, and C. W. Carlson (2008), Test particle simulations of the effect of moving DLs on ion outflow in the auroral downward-current region, *J. Geophys. Res.*, *113*, A01308, doi:10.1029/2007JA012640.

Hull, A. J., J. W. Bonnell, F. S. Mozer, and J. D. Scudder (2003), A statistical study of large-amplitude parallel electric fields in the upward current region of the auroral acceleration region, *J. Geophys. Res.*, *108*(A1), 1007, doi:10.1029/2001JA007540.

Hull, A. J., M. Wilber, C. C. Chaston, J. W. Bonnell, J. P. McFadden, F. S. Mozer, M. Fillingim, and M. L. Goldstein (2010), Time development of field-aligned currents, potential drops, and plasma associated with an auroral poleward boundary intensification, *J. Geophys. Res.*, *115*, A06211, doi:10.1029/2009JA014651.

Knight, S. (1973), Parallel electric fields, *Planet. Space Sci.*, *21*, 741–750.

Knorr, G., and C. K. Goertz (1974), Existence and stability of strong potential double layers, *Astrophys. Space Sci.*, *31*, 209–223.

Langmuir, I. (1929), The interaction of electron and positive ion space charges in cathode sheaths, *Phys. Rev.*, *33*, 954–989.

Lembege, B., and J. M. Dawson (1989), Formation of double layers within an oblique collisionless shock, *Phys. Rev. Lett.*, *62*, 2683–2686.

Main, D. S., D. L. Newman, and R. E. Ergun (2006), Double layers and ion phase-space holes in the auroral upward-current region, *Phys. Rev. Lett.*, *97*(18), 185001, doi:10.1103/PhysRevLett.97.185001.

Marklund, G. (1993), Viking investigations of auroral electrodynamical processes, *J. Geophys. Res.*, *98*(A2), 1691–1704.

Marklund, G., et al. (2001), Temporal evolution of the electric field accelerating electrons away from the auroral ionosphere, *Nature*, *414*, 724–727, doi:10.1038/414724a.

Montgomery, D. C., and G. Joyce (1969), Shock-like solutions of the electrostatic Vlasov equation, *J. Plasma Phys.*, *3*, 1–11.

Morioka, A., Y. Miyoshi, F. Tsuchiya, H. Misawa, T. Sakanoi, K. Yumoto, R. R. Anderson, J. D. Menietti, and E. F. Donovan (2007), Dual structure of auroral acceleration regions at substorm onsets as derived from auroral kilometric radiation spectra, *J. Geophys. Res.*, *112*, A06245, doi:10.1029/2006JA012186.

Mottez, F. (2001), Instabilities and formation of coherent structures, *Astrophys. Space Sci.*, *277*, 59–70.

Mozer, F. S., and A. Hull (2001), Origin and geometry of upward parallel electric fields in the auroral acceleration region, *J. Geophys. Res.*, *106*(A4), 5763–5778.

Mozer, F. S., and C. A. Kletzing (1998), Direct observation of large, quasi-static, parallel electric fields in the auroral acceleration region, *Geophys. Res. Lett.*, *25*(10), 1629–1632.

Mozer, F. S., C. W. Carlson, M. K. Hudson, R. B. Torbert, B. Parady, J. Yatteau, and M. C. Kelly (1977), Observations of paired electrostatic shocks in the polar magnetosphere, *Phys. Rev. Lett.*, *38*, 292–295.

Newman, D. L., M. V. Goldman, R. E. Ergun, and A. Mangeney (2001), Formation of double layers and electron holes in current-driven space plasma, *Phys. Rev. Lett.*, *87*(25), 255001, doi:10.1103/PhysRevLett.87.255001.

Newman, D. L., L. Andersson, M. V. Goldman, R. E. Ergun, and N. Sen (2008), Influence of suprathermal background electrons on strong auroral double layers: Laminar and turbulent regimes, *Phys. Plasmas*, *15*, 072903, doi:10.1063/1.2938754.

Perkins, F. W., and Y. C. Sun (1981), Double layers without current, *Phys. Rev. Lett.*, *46*(2), 115–118.

Raadu, M. A. (1989), The physics of double layers and their role in astrophysics, *Phys. Rep.*, *178*(2), 25–97.

Sato, K., and F. Miyawaki (1992), Formation of presheath and current-free double layer in a two-electron-temperature plasma, *Phys. Fluids B*, *4*(5), 1247–1254.

Sato, T. (1982), Auroral physics, in *Magnetospheric Plasma Physics*, edited by A. Nishida, pp. 197–243, D. Reidel, Dordrecht, The Netherlands.

Schamel, H. (1986), Electron holes, ion holes, and double layers, *Phys. Rep.*, *140*(3), 161–191.

Scime, E. E., et al. (2010), Time-resolved measurements of double-layer evolution in expanding plasma, *Phys. Plasmas*, *17*, 055701, doi:10.1063/1.3276773.

Singh, N. (1982), Double layer formation, *Plasma Phys.*, *24*, 639–660.

Singh, N., C. Deverapalli, A. Rajagiri, and I. Khazanov (2005), Dynamical behavior of U-shaped double layers: Cavity formation and filamentary structures, *Nonlinear Processes Geophys.*, *12*(6), 783–798.

Smith, R. A., and C. K. Goertz (1978), On the modulation of the Jovian decametric radiation by Io, 1. Acceleration of charged particles, *J. Geophys. Res.*, *83*(A6), 2617–2627.

Swift, D. W. (1975), On the formation of auroral arcs and acceleration of auroral electrons, *J. Geophys. Res.*, *80*(16), 2096–2108.

Swift, D. W. (1978), Mechanisms for the discrete aurora—A review, *Space Sci. Rev.*, *22*, 35–75.

Stenzel, R. L., W. Gekelman, and N. Wild (1982), Double layer formation during current sheet disruptions in a reconnection experiment, *Geophys. Res. Lett.*, *9*(6), 680–683.

Temerin, M., K. Cerny, W. Lotko, and F. S. Mozer (1982), Observations of double layers and solitary waves in the auroral plasma, *Phys. Rev. Lett.*, *48*(17), 1175–1179.

Wescott, E. M., H. C. Stenbaek-Nielsen, T. J. Hallinan, T. N. Davis, and H. M. Peek (1976), The Skylab barium plasma injection experiments, 2. Evidence for a double layer, *J. Geophys. Res.*, *81*(25), 4495–4502.

Williams, A. C. (1986), General Bohm and Langmuir conditions for strong double layer in plasmas, *IEEE Trans. Plasma Sci.*, *14*(6), 800–804.

L. Andersson and R. E. Ergun, Laboratory for Atmospheric and Space Physics, University of Colorado, Boulder, CO 80303, USA. (laila.andersson@lasp.colorado.edu)

Alfvén Wave Acceleration of Auroral Electrons in Warm Magnetospheric Plasma

C. E. J. Watt and R. Rankin

Department of Physics, University of Alberta, Edmonton, Alberta, Canada

Strong observational evidence exists confirming that electron acceleration in Alfvén waves is responsible for some of the auroral electron acceleration in the Earth's magnetosphere. Efficient electron acceleration favors short perpendicular wavelengths and wave propagation speeds that are comparable to the thermal speed of the plasma. These conditions are met in the plasma sheet or plasma sheet boundary layer in regions magnetically connected to the auroral oval. Motivated by high-altitude spacecraft observations, we use self-consistent numerical simulations to investigate Alfvénic acceleration in the warm plasma-dominated region above ~2 R_E altitude. We demonstrate that the nonlinear trapping of electrons in the Alfvén wave results in energy-dependent anisotropy in the electron distribution function and compare the numerical predictions to high-altitude observations. At lower altitudes, the accelerated electrons resemble the broadband energy flux enhancements observed by low-altitude spacecraft. The numerical experiments reveal that the "Alfvénic acceleration region" extends much farther from the Earth than the traditional auroral acceleration region where acceleration through potential drops is more common.

1. INTRODUCTION

Auroral displays can be dazzling in their hypnotic dance across the sky. It is clear from looking at these exciting and vivid displays that the physical processes causing these aurorae must share this dynamic nature. In addition to models of auroral plasma physics that can explain stable or slowly varying auroral arcs, it is important to investigate physical mechanisms that can explain rapid time variations. *Hasegawa* [1976] suggested that Alfvén waves with small perpendicular scale lengths could be responsible for providing a small, but finite, fluctuating parallel electric field in the magnetosphere above the aurora. This parallel electric field would accelerate electrons in the field-aligned direction, pushing them into the loss cone and on toward the ionosphere, where they would be

responsible for time-varying auroral displays with the requisite small scales perpendicular to the field.

Alfvén waves are ubiquitous throughout the Earth's magnetosphere (see review by *Keiling* [2009] and references therein) and have been observed in the magnetospheres of other magnetized planets [e.g., *Glassmeier and Espley*, 2006, and references therein]. Rocket measurements between 300 and 1000 km altitude have shown that large-amplitude Alfvén waves are observed in tandem with enhanced field-aligned electron energy flux [e.g., *Lynch et al.*, 1999]. Low-altitude spacecraft such as Dynamics Explorer 1 [*Gurnett et al.*, 1984], Freja [e.g., *Andersson et al.*, 2002a], and the Fast Auroral Snapshot (FAST) satellite [e.g., *Chaston et al.*, 1999] have provided much insight into the different forms of accelerated electrons and their association with Alfvén waves. With the wealth of data from the FAST satellite, *Chaston et al.* [2007] demonstrate that near noon and premidnight, nearly 50% of precipitating electron energy flux enhancements are accompanied by Alfvén waves.

Low-altitude observations strongly implicate Alfvén waves in the acceleration of auroral electrons, but cannot establish where this acceleration occurs in the magnetosphere above the

Auroral Phenomenology and Magnetospheric Processes: Earth and Other Planets

Geophysical Monograph Series 197

10.1029/2011GM001171

Table 1. Summary of Spacecraft Observations of Shear Alfvén Wave Activity and Electron Behavior Above the Traditional Auroral Acceleration Region

Publication	Date	Time Interval (UT)	Spacecraft (Plot Symbol)
Wygant et al. [2000]	9 May 1997	05:36–05:42	Polar (Δ)
	1 May 1997	20:25–20:35	Polar (Δ)
Keiling et al. [2002]	2 Apr 1997	12:35–12:40	Polar (Δ)
	6 Apr 1997	18:25–18:46	Polar (Δ)
Angelopoulos et al. [2002]	13 Nov 1996	11:20–11:30	Polar (Δ)
Janhunen et al. [2004]	24 Apr 1997	05:30–06:00	Polar (Δ)
Dombeck et al. [2005]	22 Oct 1999	02:05–02:20	Polar (Δ)
Wahlund et al. [2003]	28 Apr 2001	19:15–19:18	Cluster (○)
Morooka et al. [2004]	20 Jun 2001	02:00–02:30	Cluster (○)
Chaston et al. [2005a]	25 Oct 2002	14:20–15:00	Cluster (○)
Hull et al. [2010]	21 Feb 2001	06:15–06:25	Cluster (○)
Wild et al. [2011]	28 Sep 2009	00:00–01:30	Cluster (○)
Walsh et al. [2010]	1 Oct 2005	04:20–04:25	Double Star TC-2 (□)

auroral ionosphere. It is natural to first investigate the more well-known auroral acceleration region (AAR), at 1–2 R_E altitude, where parallel electrostatic fields of 1–10 kV potential drop have been shown to reside [*Reiff et al.*, 1988; *Lindqvist and Marklund*, 1990]. Some of the first test-particle modeling studies of Alfvénic acceleration focused on the AAR [e.g., *Kletzing*, 1994; *Thompson and Lysak*, 1996]. More sophisticated, self-consistent numerical simulations followed, demonstrating that the Alfvén waves in the AAR develop steepened, nonlinear structures [*Hui and Seyler*, 1992; *Clark and Seyler*, 1999; *Watt et al.*, 2004, 2005; *Watt and Rankin*, 2007a] with spiky parallel electric fields [*Watt and Rankin*, 2007b].

The Alfvén waves detected by low-altitude spacecraft most likely originate in the magnetosphere [e.g., *Keiling et al.*, 2003] or at the dayside magnetopause [*Chaston et al.*, 2005b]. It is therefore instructive to consider whether Alfvénic

acceleration occurs above the AAR. By comparing particle and wave measurements between magnetically conjugate spacecraft at >5 R_E and ~3500 km altitude, *Dombeck et al.* [2005] and *Chaston et al.* [2005a] demonstrate in two case studies that Alfvénic electron acceleration occurs between the two spacecraft. At high altitudes, an ensemble of Polar observations shows that there is consistently more energy flux in the Alfvén waves than is required to account for auroral brightenings [*Keiling et al.*, 2002]. At low altitudes, FAST typically measures more in situ electron energy flux than wave Poynting flux, strongly indicating that the energy conversion process between Alfvén waves and electrons occurs above the AAR [*Chaston*, 2006; *Janhunen et al.*, 2006].

Alfvén waves are often measured at the same time as parallel electron energization in the magnetosphere above the AAR (see list in Table 1 and locations in Figure 1). A rough

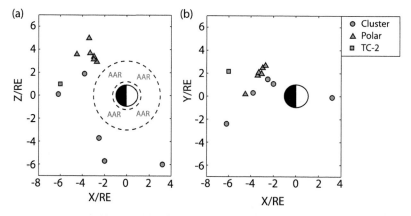

Figure 1. Locations of spacecraft in the (a) *X-Z* and (b) *X-Y* GSE planes, which have made measurements of Alfvén waves and accelerated electrons above the traditional auroral acceleration region. Observations suggest that the AAR, where quasistatic potential drops and perpendicular electric field structures reside, lies between 0.25 and 2 R_E altitude. These distances are indicated in Figure 1a by dashed lines.

estimate of the location of the AAR is indicated in Figure 1a above auroral latitudes. In this article, we will discuss the results of numerical simulations that have been motivated by these observations and show that the energy-dependent anisotropy observed in these studies at high-altitudes is a natural consequence of Alfvénic acceleration in warm plasma.

The plasma above the AAR is expected to display different characteristics from the plasma in the AAR, and these are discussed in section 2. The numerical simulation model DK-1D is described in section 3, and the results are discussed in section 4. Comparisons between the simulation output and some of the spacecraft observations indicated in Figure 1 are made in section 5. Finally, interpretation of the model results and our conclusions are presented in section 6.

2. KINETIC THEORY OF SHEAR ALFVÉN WAVES

Shear Alfvén waves are low-frequency (i.e., with frequency much less than the ion gyrofrequency) electromagnetic waves, which have a parallel electric field perturbation when the perpendicular wavelength is short. In the low-frequency limit, we assume that the plasma is essentially quasineutral. By idealizing the wave perturbations ($\propto \exp[i(\mathbf{k}.\mathbf{r} - \omega t)]$), we can approximate the continuity equation to $k_\perp J_\perp + k_\parallel J_\parallel = 0$, where $\perp$ and $\parallel$ refer to directions perpendicular and parallel to the magnetic field, respectively. Assuming the wave amplitude ($J_\perp$) and the parallel wavelength are constant, a decrease in perpendicular scale (or increase in $k_\perp$) will result in an increasing parallel current. The force that accelerates the electrons in the parallel direction to carry this current is provided by a parallel electric field. Many numerical simulations of the interaction between shear Alfvén waves and electrons have demonstrated how the wave parallel electric field can accelerate electrons to higher energies [Kletzing, 1994; Thompson and Lysak, 1996; Clark and Seyler, 1999; Chaston et al., 2000; Kletzing and Hu, 2001; Chaston et al., 2002a, 2003a, 2003b; Damiano et al., 2003; Su et al., 2004; Génot et al., 2004; Watt et al., 2004; Damiano and Wright, 2005; Watt et al., 2005, 2006; Damiano et al., 2007; Seyler and Liu, 2007; Swift, 2007; Watt and Rankin, 2008, 2009, 2010].

The magnitude of the parallel electric field and the amount of interaction between waves and electrons depend upon the plasma parameters in the magnetosphere. In our analysis, we will assume that the ion temperature is negligible ($T_e \gg T_i$) and focus on electron effects. Lysak and Lotko [1996] demonstrate that finite ion temperature effects can influence the phase velocity and, hence, the size of the parallel electric field of the shear Alfvén wave, but we leave ion temperature effects for future research.

If the local Alfvén speed v_A is larger than the thermal speed of the electrons v_{th}, then the plasma is said to be in the inertial regime [Lysak and Lotko, 1996; Nakamura, 2000]. The parallel electric field can be approximated by [Goertz, 1984]

$$\left(\frac{E_\parallel}{E_\perp}\right)_i = -\frac{\delta_e^2}{1 + k_\perp^2 \delta_e^2} k_\parallel k_\perp, \qquad (1)$$

where $\delta_e = c/\omega_{pe}$ is the electron skin depth, and $\omega_{pe} = (q_e^2 n_e/(\varepsilon_0 m_e))^{1/2}$ is the electron plasma frequency. This expression maximizes for $k_\perp \delta_e = 1$.

Alternately, if $v_A \ll v_{th}$, then the plasma is said to be in the kinetic regime, and the parallel electric field may be approximated by [Hasegawa and Mima, 1978]

$$\left(\frac{E_\parallel}{E_\perp}\right)_k = \frac{v_{th}^2}{v_A^2} k_\parallel k_\perp \delta_e^2, \qquad (2)$$

where $v_{th} = (2k_B T_e/m_e)^{1/2}$ and $v_A = B_0/(\mu_0 n_i m_i)^{1/2}$, given the background field strength B_0, the electron and ion number density $n_e = n_i$, and the ambient electron temperature T_e. Note in both cases we have assumed that T_i is effectively zero in order to highlight the electron physics.

Typically, the plasma in the AAR is in the inertial regime. Figure 2a shows an idealized distribution function constructed using parameters measured at 1700 km altitude by the FAST satellite: $B_0 = 17,000$ nT, $n_e = 10^8$ m^{-3}, and $T_e = 5$ eV. Very few of the cold electrons at low altitudes have velocities close to the phase speed of the shear Alfvén wave because the Alfvén wave speed is much higher than the thermal velocity. Interactions between the electrons and Alfvén waves can only

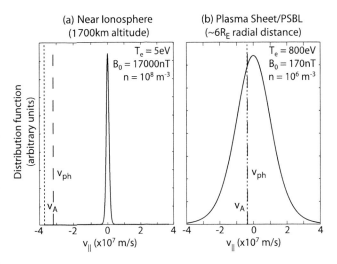

Figure 2. Electron distribution function as a function of parallel velocity. Dotted and dashed lines indicate the Alfvén speed and phase speed of the wave, respectively. Note that the negative direction indicates velocities toward the ionosphere in the simulations that will be shown in this chapter. Parameters used are indicative of (a) 1700 km altitude and (b) roughly 4 R_E altitude.

be achieved by either (1) increasing the wave amplitude or (2) increasing the perpendicular wavenumber. Nonlinear simulations have demonstrated that more electrons can be accelerated, and to higher energies, as the wave amplitude is increased [see *Watt et al.*, 2005, Figures 5 and 6; *Watt and Rankin*, 2007a, Figure 5], but that acceleration can only occur if the perpendicular wavenumber is large enough [*Watt and Rankin*, 2007a]. At higher altitudes, the plasma is dominated by warm magnetospheric plasma, and the magnetic field strength and number density have values much less than at lower altitudes. In this region, $v_A \leq v_{th}$ and the parallel electric field cannot be approximated by either of the expressions (1) or (2), or even a combination of the two [*Streltsov et al.*, 1998; *Chaston et al.*, 2003c]. Instead, estimates of the size of the parallel electric field may be obtained from linear theory [e.g., *Nakamura*, 2000], using a Lorentzian distribution function for the electrons:

$$\frac{E_\parallel}{E_\perp} = -k_\perp k_\parallel \frac{v_{th,\kappa}^2}{v_A^2} \delta_e^2 \frac{1}{2[1 - (1/(2\kappa) + \zeta Z_\kappa^*(\zeta)]}, \qquad (3)$$

where $v_{th,\kappa} = [(2\kappa - 3)/\kappa]^{1/2} (k_B T_e/m_e)^{1/2}$, and $Z_\kappa^*(\zeta)$ is the modified plasma dispersion function [*Summers and Thorne*, 1992] with argument $\zeta = \omega/(k_\parallel v_{th,\kappa})$. For interpretation of results presented later in this article, it is important to note that the magnitude of the parallel electric field in warmer plasma increases in proportion to the ratio v_{th}^2/v_A^2.

Figure 2b shows an idealized distribution function constructed using observations obtained at ~6 R_E by the Polar satellite: $B_0 = 170$ nT, $n_e = 10^6$ m^{-3}, and $T_e = 800$ eV. At higher altitudes, the wave phase speeds are much lower, and the warm plasma of the plasma sheet or plasma sheet boundary layer provides many electrons to take part in interactions with the waves. For this reason, the linear theory predicts strong damping of shear Alfvén waves in warm plasma [*Lysak and Lotko*, 1996]. We will demonstrate in the next section that this strong damping manifests itself as nonlinear trapping of electrons in the wave and show that many observed features of shear Alfvén waves and electrons in the magnetosphere can be explained by the strong interaction between the waves and the plasma.

The situation is significantly more complicated for Alfvén waves with long parallel wavelengths (a few R_E) that sample both plasma regimes above the auroral zone within a single wavelength. In these cases, nonlocal effects contribute to a larger parallel electric field than is predicted with the linear theory [*Rankin et al.*, 1999; *Tikhonchuk and Rankin*, 2000, 2002; *Lysak and Song*, 2003]. In this article, we will focus on the interpretation of waves with wavelengths less than one Earth radius, where nonlocal effects are less important.

3. NUMERICAL SIMULATIONS

The DK-1D simulation model [*Watt et al.*, 2004; *Watt and Rankin*, 2010] was designed to study the nonlinear interaction between shear Alfvén waves and electrons. It is a self-consistent 1-D Vlasov-Maxwell Eulerian code, which integrates the drift-kinetic electron equation forward in time. All perpendicular perturbations are assumed to have the form $\exp(i(k_\perp x - \omega t))$, where the perpendicular wavenumber $k_\perp$ can vary gradually with distance along the field. The governing equations, assumptions, and most up-to-date algorithms are discussed in detail in the work of *Watt and Rankin* [2010].

An example of the 1-D simulation domain, oriented along an idealized dipole geomagnetic field, is shown in Figure 3. The number density $n_0 = 10^6$ m^{-3}, temperature $T_0 = 200$ eV, and $\kappa = 5$ are initially uniform throughout the simulation domain, and the perpendicular wavenumber is scaled with the square root of the dipolar magnetic field to be equivalent to $\lambda_\perp = 4$ km at the ionosphere (110 km altitude). The large variation in v_{th}^2/v_A^2 (Figure 4a) and weak variation of $k_\perp$ (Figure 4b) over the simulation domain mimics the variation of these parameters in a more realistic magnetosphere.

A wave packet of the form $\phi = \phi_0 \exp[-(t - 3T)^2/(3T)^2] \sin(2\pi t/T)$ is added to the scalar potential at the upper boundary of the simulation domain, where the wave period $T = 2$ s [cf. *Wygant et al.*, 2002], and the wave amplitude $\phi_0 = 400$ V. The boundary conditions at the upper boundary are such that ϕ and $\partial\phi/\partial z$ are prescribed, and the moments of the incoming distribution function are constrained by the self-consistent $A_\parallel$ ($= \mu_0 J_\parallel/k_\perp^2$), and the $\partial n/\partial t$ given by the electron continuity equation. The shape of the incoming distribution function is unconstrained, and we choose a kappa distribution with $\kappa = 5$. Some small "mismatch" may occur between the incoming plasma and the potentials in cases where the interaction

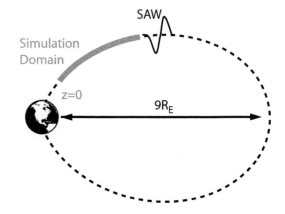

Figure 3. Diagram indicating the size and orientation of the DK-1D simulation domain.

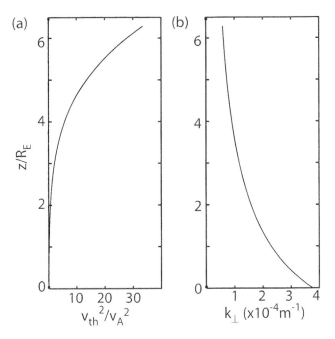

Figure 4. Variation of v_{th}^2/v_A^2 and $k_\perp$ with distance along the field z. The top of these plots corresponds to the upper simulation boundary; the bottom of these plots represents the lower boundary closest to the Earth.

between the waves and the particles is strong. Wave energy is rapidly converted into particle energy near the upper boundary to create nonlinear trapping features in phase space, as shown near the top of Figure 5a. The parallel velocity distribution function $f_e(v_\parallel, v_\perp = 0)$ (Figure 5a) and the parallel electric field $E_\parallel$ (Figure 5b) are shown as a function of distance along the field z at the same time $t = 5$ s. The wave $E_\parallel$ is strongest near the upper boundary of the simulation domain, where v_{th}/v_A maximizes, and the electrons are strongly modified [*Watt and Rankin*, 2009]. The parallel electric field diminishes quickly below $z \sim 4.5\ R_E$, following the rapid decrease in v_{th}^2/v_A^2. The evolution of the parallel electric field from high altitudes to low altitudes leads to electron acceleration as described in the diagrams in Figure 6: at high altitudes, electrons are trapped between successive crests in the parallel electric field, whereas at lower altitudes, the trapped electrons are free to escape the wave because the wave parallel electric field effectively disappears.

A typical 2-D distribution function $f(v_\parallel, v_\perp)$ from the nonlinear trapping region is shown in Figure 5c. It is significantly stretched in the parallel direction. The electrons with negative velocities are part of the population trapped by the wave, and the electrons with positive velocities carry the parallel current required by the Alfvén wave at this location. By following the earthward-accelerated electrons from Figure 5c through the simulation, we can determine the electron precipitation

expected after the wave-particle interaction is complete. Much of the trapped population at $z \sim 6\ R_E$, $t = 5$ s reaches $z \sim 1\ R_E$ at $t = 11$ s (Figure 5d). This distribution function shows at least two discrete electron beams at $v_\parallel \sim -10^7$ m s^{-1} and $v_\parallel \sim 6 \times 10^6$ m s^{-1}. The beams have spread out in perpendicular velocity due to the action of the mirror force, creating conditions that may be unstable to auroral kilometric radiation [*Roux et al.*, 1993, *Bingham and Cairns*, 2000] at this location and at lower altitudes where the mirroring is stronger.

4. ENERGY FLUX OF ACCELERATED ELECTRONS AT LOW ALTITUDE

The DK-1D simulation code self-consistently includes the energy balance between waves and electrons and can therefore be used for quantitative studies of the electron energy flux expected from shear Alfvén wave acceleration events. For example, for $E_\perp \geq 20$ mV m^{-1} with $\lambda_\perp = 4$ km (mapped to the ionosphere), and in plasma with $T_e = 200$ eV, the Alfvén wave acceleration process was able to produce an electron energy flux at the ionosphere in excess of the 1 mW m^{-2} required to observe auroral brightening [*Watt and Rankin*, 2010]. However, the efficiency of the energy transfer process depends upon the perpendicular scale size of the wave, the amplitude of the wave, the ambient plasma temperature, and the form of the electron distribution function [*Watt et al.*, 2005; *Watt and Rankin*, 2007a, 2008]. Numerical studies using idealized wave pulses and kinetic electrons have been shown to produce well-defined energy-dispersed beams [e.g., *Kletzing*, 1994; *Su et al.*, 2004; *Watt et al.*, 2005, 2006], similar to those measured by Freja [*Andersson et al.*, 2002a, 2002b] and FAST [*Chaston et al.*, 2002b; *Su et al.*, 2004]. Simulations of the interaction between sinusoidal wave structures and electrons can also result in nonlinear Landau damping and particle trapping. When this occurs, the resulting electron acceleration signatures in differential energy flux are not as clearly dispersed in energy. Figure 5e shows the earthward differential energy flux as a function of time from a location near the lower boundary of the simulation. The coherent trapping islands disintegrate as the wave propagates closer to the Earth, and the electrons from different phases of the wave can mix together. The resulting differential energy flux displays much less structure, similar to more complicated electron acceleration signatures seen in FAST data [e.g., *Chaston et al.*, 2003a].

5. HIGH ALTITUDE ACCELERATION SIGNATURES

Observations of Alfvénic wave activity in the plasma sheet and plasma sheet boundary layer indicate that they are often accompanied by electrons energized in the parallel direction

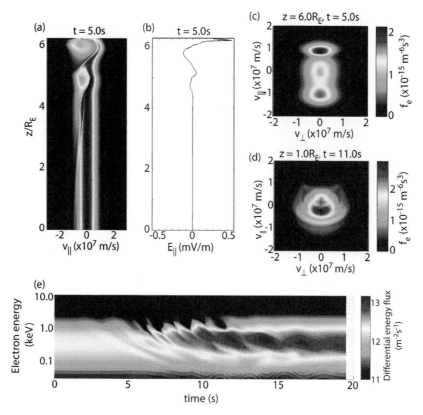

Figure 5. (a) Spatial variation of $f_e(v_\parallel)$ for $v_\perp = 0$ at $t = 5$ s, along with (b) the parallel electric field at that time. Snapshots of the 2-D distribution function $f_e(v_\parallel, v_\perp)$ at (c) $t = 5$ s near the upper boundary of the simulation and (d) the later time of $t = 11$ s near the lower boundary of the simulation. (e) Evolution of the earthward-moving differential energy flux near the lower boundary of the simulation.

(see publications in Table 1). Figures 7a and 7b show distribution functions from the Polar HYDRA instrument obtained during periods of shear Alfvén wave activity [*Wygant et al.*, 2002; *Janhunen et al.*, 2004]. The simulation distribution function shown in Figure 7c is averaged over the ~12 s integration time of the HYDRA instrument in order to make an appropriate comparison.

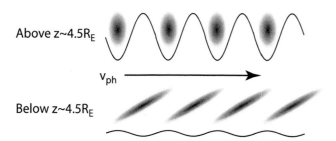

Figure 6. Diagram of nonlinear electron trapping in the parallel electric field of the Alfvén wave at high altitudes and the formation of beams of electrons at low altitudes where the parallel electric field diminishes.

All three distribution functions show similar characteristics. There is strong anisotropy favoring the parallel direction at low energies ($< 2 \times 10^7$ m s^{-1}, or < 1.1 keV), but the distribution function is essentially isotropic at higher energies. The DK-1D simulation provides an interpretation of the low-energy anisotropy. The acceleration in the direction of wave propagation is due to the electron trapping in the wave, whereas the acceleration in the opposite direction is required for the parallel current of the Alfvén wave. Interestingly, it is only the wave-particle interaction causing the electron trapping that results in net acceleration of electrons from this location. The motion of the distribution function in velocity space to carry the parallel current of the wave is temporary, and once the wave has propagated away from this location, the electrons return to their original velocities, and there is no longer anisotropy or net acceleration in the opposite direction.

6. DISCUSSION AND CONCLUSIONS

The key to Alfvén wave acceleration of auroral electrons is the perpendicular scale length of the waves. Without short

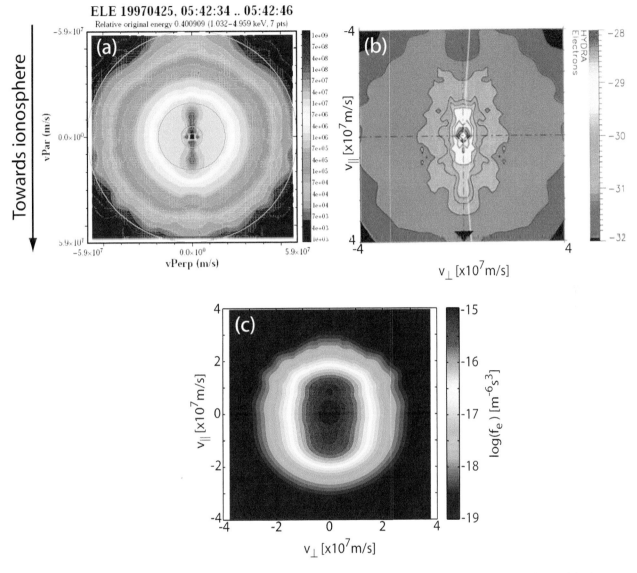

Figure 7. Observed 2-D distribution functions from (a) *Wygant et al.* [2002] and (b) *Janhunen et al.* [2004] at altitudes >4 R_E and when Alfvén waves were observed in the electric and magnetic field measurements. (c) A time-averaged 2-D distribution function obtained near the top of the simulation domain ($z = 6$ R_E). All distribution functions show strong anisotropy for energies below 1 keV ($v < 1.9 \times 10^7$ m s^{-1}) but are relatively isotropic at higher energies.

perpendicular scales, there is no parallel electric field, and no net acceleration. The key scientific target is now to identify, from a number of different hypotheses, how Alfvén waves can develop those short scales in the magnetosphere, where scale sizes are typically much larger. It has been suggested that short perpendicular scales could evolve through nonlinear turbulent wave-wave processes [e.g., *Seyler and Wu*, 2001; *Voitenko and Goossens*, 2005; *Zhao et al.*, 2010], and in fact, turbulent Alfvénic activity is a feature of the wave spectra at low altitudes, as measured by the FAST spacecraft

[*Chaston et al.*, 2008]. The nonlinear evolution of shear Alfvén waves as they interact with the ionosphere is also predicted to result in the development of short perpendicular scales [*Streltsov and Lotko*, 2004; *Streltsov*, 2007; *Lu et al.*, 2007; *Lysak and Song*, 2011], and short perpendicular wave scales are found to result as Alfvén waves propagate over strong perpendicular density gradients in the magnetosphere [*Génot et al.*, 2004; *Lysak and Song*, 2011; *Mottez and Génot*, 2011]. Finally, the short perpendicular scales required by dispersive Alfvén waves could be generated at small diffusion regions of a reconnection site [e.g., *Chaston et al.*, 2005b; *Shay et al.*, 2011]. The numerical experiments detailed in this article reduce this rich wave physics to an idealized quasisinusoidal wave packet with a fixed perpendicular wavenumber. With these caveats, it is remarkable that the results from the idealized DK-1D simulation compare so well with observations of electrons and fields from in situ spacecraft. The simulation predicts anisotropy in the low-energy distribution functions in the interaction region at high altitude, resulting from strong nonlinear interactions between the wave and the electrons, and this signature is clearly seen in more than one observation at >5 R_E radial distance (see Figure 7). When these electrons are tracked through the simulation to lower altitudes, they manifest as enhancements in electron energy flux in the 0.1 to a few keV range. These enhancements are regularly seen in low-altitude spacecraft such as Freja, FAST, and DMSP [e.g., *Newell et al.*, 2009]. Additional features such as horseshoe-shaped distribution functions [*Bingham and Cairns*, 2000] naturally arise due to the action of the mirror force on the accelerated electrons. The numerical simulations confirm that the most likely location for Alfvén wave acceleration to occur is above the traditional AAR, i.e., above 2 R_E altitude. The challenge for future numerical investigations will be to remove the assumptions in current numerical models, e.g., (1) investigate the effect of ion temperature on wave propagation and damping and (2) identify the mechanism that encourages short perpendicular scales to develop.

Acknowledgments. This work was made possible through the support of the Canadian Space Agency (CSA) and the Natural Sciences and Engineering Research Council of Canada (NSERC).

REFERENCES

Andersson, L., N. Ivchenko, J. Clemmons, A. A. Namgaladze, B. Gustavsson, J.-E. Wahlund, L. Eliasson, and R. Y. Yurik (2002a), Electron signatures and Alfvén waves, *J. Geophys. Res.*, *107* (A9), 1244, doi:10.1029/2001JA900096.

Andersson, L., J. E. Wahlund, J. Clemmons, B. Gustavsson, and L. Eliasson (2002b), Electromagnetic waves and bursty electron acceleration: Implications from Freja, *Ann. Geophys.*, *20*(2), 139–150.

Angelopoulos, V., J. A. Chapman, F. S. Mozer, J. D. Scudder, C. T. Russell, K. Tsuruda, T. Mukai, T. J. Hughes, and K. Yumoto (2002), Plasma sheet electromagnetic power generation and its dissipation along auroral field lines, *J. Geophys. Res.*, *107*(A8), 1181, doi:10.1029/2001JA900136.

Bingham, R., and R. A. Cairns (2000), Generation of auroral kilometric radiation by electron horseshoe distributions, *Phys. Plasmas*, *7*, 3089–3092.

Chaston, C. C. (2006), ULF waves and auroral electrons, in *Magnetospheric ULF Waves: Synthesis and New Directions*, Geophys. Monogr. Ser., vol. 169, edited by K. Takahashi et al., pp. 239–257, AGU, Washington, D. C., doi:10.1029/169GM16.

Chaston, C. C., C. W. Carlson, W. J. Peria, R. E. Ergun, and J. P. McFadden (1999), FAST observations of inertial Alfvén waves in the dayside aurora, *Geophys. Res. Lett.*, *26*(6), 647–650.

Chaston, C. C., C. W. Carlson, R. E. Ergun, and J. P. McFadden (2000), Alfvén waves, density cavities and electron acceleration observed from the FAST spacecraft, *Phys. Scr. T*, *84*, 64–68.

Chaston, C. C., J. W. Bonnell, L. M. Pcticolas, C. W. Carlson, J. P. McFadden, and R. E. Ergun (2002a), Driven Alfvén waves and electron acceleration: A FAST case study, *Geophys. Res. Lett.*, *29*(11), 1535, doi:10.1029/2001GL013842.

Chaston, C. C., J. W. Bonnell, C. W. Carlson, M. Berthomier, L. M. Peticolas, I. Roth, J. P. McFadden, R. E. Ergun, and R. J. Strangeway (2002b), Electron acceleration in the ionospheric Alfvén resonator, *J. Geophys. Res.*, *107*(A11), 1413, doi:10.1029/2002JA009272.

Chaston, C. C., J. W. Bonnell, C. W. Carlson, J. P. McFadden, R. E. Ergun, and R. J. Strangeway (2003a), Properties of small-scale Alfvén waves and accelerated electrons from FAST, *J. Geophys. Res.*, *108*(A4), 8003, doi:10.1029/2002JA009420.

Chaston, C. C., L. M. Peticolas, J. W. Bonnell, C. W. Carlson, R. E. Ergun, J. P. McFadden, and R. J. Strangeway (2003b), Width and brightness of auroral arcs driven by inertial Alfven waves, *J. Geophys. Res.*, *108*(A2), 1091, doi:10.1029/2001JA007537.

Chaston, C. C., J. W. Bonnell, C. W. Carlson, J. P. McFadden, R. J. Strangeway, and R. E. Ergun (2003c), Kinetic effects in the acceleration of auroral electrons in small scale Alfvén waves: A FAST case study, *Geophys. Res. Lett.*, *30*(6), 1289, doi:10.1029/2002GL015777.

Chaston, C. C., et al. (2005a), Energy deposition by Alfvén waves into the dayside auroral oval: Cluster and FAST observations, *J. Geophys. Res.*, *110*, A02211, doi:10.1029/2004JA010483.

Chaston, C. C., T. D. Phan, J. W. Bonnell, F. S. Mozer, M. Acuna, M. L. Goldstein, A. Balogh, M. Andre, H. Reme, and A. Fazakerley (2005b), Kinetic Alfvén waves observed near a reconnection X line in the Earth's magnetosphere, *Phys. Rev. Lett.*, *95*, 065002, doi:10.1103/PhysRevLett.95.065002.

Chaston, C. C., C. W. Carlson, J. P. McFadden, R. E. Ergun, and R. J. Strangeway (2007), How important are dispersive Alfvén waves for auroral particle acceleration?, *Geophys. Res. Lett.*, *34*, L07101, doi:10.1029/2006GL029144.

Chaston, C. C., C. Salem, J. W. Bonnell, C. W. Carlson, R. E. Ergun, R. J. Strangeway, and J. P. McFadden (2008), The turbulent Alfvénic aurora, *Phys. Rev. Lett.*, *100*, 175003, doi:10.1103/PhysRevLett.100.175003.

Clark, A. E., and C. E. Seyler (1999), Electron beam formation by small-scale oblique inertial Alfvén waves, *J. Geophys. Res.*, *104* (A8), 17,233–17,249.

Damiano, P. A., and A. N. Wright (2005), Two-dimensional hybrid MHD-kinetic electron simulations of an Alfvén wave pulse, *J. Geophys. Res.*, *110*, A01201, doi:10.1029/2004JA010603.

Damiano, P. A., R. D. Sydora, and J. C. Samson (2003), Hybrid magnetohydrodynamic-kinetic model of standing shear Alfvén waves, *J. Plasma Phys.*, *69*, 277–304.

Damiano, P. A., A. N. Wright, R. D. Sydora, and J. C. Samson (2007), Energy dissipation via electron energization in standing shear Alfvén waves, *Phys. Plasmas*, *14*(6), 062904, doi:10.1063/1.2744226.

Dombeck, J., C. Cattell, J. R. Wygant, A. Keiling, and J. Scudder (2005), Alfvén waves and Poynting flux observed simultaneously by Polar and FAST in the plasma sheet boundary layer, *J. Geophys. Res.*, *110*, A12S90, doi:10.1029/2005JA011269.

Genot, V., P. Louarn, and F. Mottez (2004), Alfvén wave interaction with inhomogeneous plasmas: Acceleration and energy cascade towards small-scales, *Ann. Geophys.*, *22*(6), 2081–2096.

Glassmeier, K.-H., and J. Espley (2006), ULF waves in planetary magnetospheres, in *Magnetospheric ULF Waves: Synthesis and New Directions*, Geophys. Monogr. Ser., vol. 169, edited by K. Takahashi et al., pp. 341–359, AGU, Washington, D. C., doi:10.1029/169GM22.

Goertz, C. K. (1984), Kinetic Alfvén waves on auroral field lines, *Planet. Space Sci.*, *32*, 1387–1392.

Gurnett, D. A., R. L. Huff, J. D. Menietti, J. L. Burch, J. D. Winningham, and S. D. Shawhan (1984), Correlated low-frequency electric and magnetic noise along the auroral field lines, *J. Geophys. Res.*, *89*(A10), 8971–8985.

Hasegawa, A. (1976), Particle acceleration by MHD surface wave and formation of aurora, *J. Geophys. Res.*, *81*(28), 5083–5090.

Hasegawa, A., and K. Mima (1978), Anomalous transport produced by kinetic Alfvén wave turbulence, *J. Geophys. Res.*, *83*(A3), 1117–1123.

Hui, C.-H., and C. E. Seyler (1992), Electron acceleration by Alfvén waves in the magnetosphere, *J. Geophys. Res.*, *97*(A4), 3953–3963.

Hull, A. J., M. Wilber, C. C. Chaston, J. W. Bonnell, J. P. McFadden, F. S. Mozer, M. Fillingim, and M. L. Goldstein (2010), Time development of field-aligned currents, potential drops, and plasma associated with an auroral poleward boundary intensification, *J. Geophys. Res.*, *115*, A06211, doi:10.1029/2009JA014651.

Janhunen, P., A. Olsson, J. Hanasz, C. T. Russell, H. Laakso, and J. C. Samson (2004), Different Alfvén wave acceleration processes of electrons in substorms at ~4–5 R_E and 2–3 R_E radial distance, *Ann. Geophys.*, *22*, 2213–2227.

Janhunen, P., A. Olsson, C. T. Russell, and H. Laakso (2006), Alfvénic electron acceleration in aurora occurs in global Alfvén Resonosphere region, *Space Sci. Rev.*, *122*(1–4), 89–95, doi:10.1007/s11214-006-7017-5.

Keiling, A. (2009), Alfvén waves and their roles in the dynamics of the earth's magnetotail: A review, *Space Sci. Rev.*, *142*(1–4), 73–156, doi:10.1007/s11214-008-9463-8.

Keiling, A., J. R. Wygant, C. Cattell, W. Peria, G. Parks, M. Temerin, F. S. Mozer, C. T. Russell, and C. A. Kletzing (2002), Correlation of Alfvén wave Poynting flux in the plasma sheet at 4–7 R_E with ionospheric electron energy flux, *J. Geophys. Res.*, *107*(A7), 1132, doi:10.1029/2001JA900140.

Keiling, A., J. R. Wygant, C. A. Cattell, F. S. Mozer, and C. T. Russell (2003), The global morphology of wave Poynting flux: Powering the aurora, *Science*, *299*(5605), 383–386.

Kletzing, C. A. (1994), Electron acceleration by kinetic Alfvén waves, *J. Geophys. Res.*, *99*(A6), 11,095–11,103.

Kletzing, C. A., and S. Hu (2001), Alfvén wave generated electron time dispersion, *Geophys. Res. Lett.*, *28*(4), 693–696.

Lindqvist, P.-A., and G. T. Marklund (1990), A statistical study of high-altitude electric fields measured on the Viking satellite, *J. Geophys. Res.*, *95*(A5), 5867–5876.

Lu, J. Y., R. Rankin, R. Marchand, I. J. Rae, W. Wang, S. C. Solomon, and J. Lei (2007), Electrodynamics of magnetosphere-ionosphere coupling and feedback on magnetospheric field line resonances, *J. Geophys. Res.*, *112*, A10219, doi:10.1029/2006JA012195.

Lynch, K. A., D. Pietrowski, R. B. Torbert, N. Ivchenko, G. Marklund, and F. Primdahl (1999), Multiple-point electron measurements in a nightside auroral arc: Auroral turbulence II particle observations, *Geophys. Res. Lett.*, *26*(22), 3361–3364.

Lysak, R. L., and W. Lotko (1996), On the kinetic dispersion relation for shear Alfvén waves, *J. Geophys. Res.*, *101*(A3), 5085–5094.

Lysak, R. L., and Y. Song (2003), Nonlocal kinetic theory of Alfvén waves on dipolar field lines, *J. Geophys. Res.*, *108*(A8), 1327, doi:10.1029/2003JA009859.

Lysak, R. L., and Y. Song (2011), Development of parallel electric fields at the plasma sheet boundary layer, *J. Geophys. Res.*, *116*, A00K14, doi:10.1029/2010JA016424.

Morooka, M., et al. (2004), Cluster observations of ULF waves with pulsating electron beams above the high latitude dusk-side auroral region, *Geophys. Res. Lett.*, *31*, L05804, doi:10.1029/2003GL017714.

Mottez, F., and V. Génot (2011), Electron acceleration by an Alfvénic pulse propagating in an auroral plasma cavity, *J. Geophys. Res.*, *116*, A00K15, doi:10.1029/2010JA016367.

Nakamura, T. K. (2000), Parallel electric field of a mirror kinetic Alfvén wave, *J. Geophys. Res.*, *105*(A5), 10,729–10,737.

Newell, P. T., T. Sotirelis, and S. Wing (2009), Diffuse, monoenergetic, and broadband aurora: The global precipitation budget, *J. Geophys. Res.*, *114*, A09207, doi:10.1029/2009JA014326.

Rankin, R., J. C. Samson, and V. T. Tikhonchuk (1999), Parallel electric fields in dispersive shear Alfvén waves in the dipolar magnetosphere, *Geophys. Res. Lett.*, *26*(24), 3601–3604.

Reiff, P. H., H. L. Collin, J. D. Craven, J. L. Burch, J. D. Winningham, E. G. Shelley, L. A. Frank, and M. A. Friedman (1988), Determination of auroral electrostatic potentials using high- and low-altitude particle distributions, *J. Geophys. Res.*, *93*(A7), 7441–7465.

Roux, A., A. Hilgers, H. de Feraudy, D. le Queau, P. Louarn, S. Perraut, A. Bahnsen, M. Jespersen, E. Ungstrup, and M. Andre (1993), Auroral kilometric radiation sources: In situ and remote observations from Viking, *J. Geophys. Res.*, *98*(A7), 11,657–11,670.

Seyler, C. E., and K. Wu (2001), Instability at the electron inertial scale, *J. Geophys. Res.*, *106*, 21,623–21,644.

Seyler, C. E., and K. Liu (2007), Particle energization by oblique inertial Alfvén waves in the auroral region, *J. Geophys. Res.*, *112*, A09302, doi:10.1029/2007JA012412.

Shay, M. A., J. F. Drake, J. P. Eastwood, and T. D. Phan (2011), Super-Alfvénic propagation of substorm reconnection signatures and Poynting flux, *Phys. Rev. Lett.*, *107*, 065001, doi:10.1103/PhysRevLett.107.065001.

Streltsov, A. V. (2007), Narrowing of the discrete auroral arc by the ionosphere, *J. Geophys. Res.*, *112*, A10218, doi:10.1029/2007JA012402.

Streltsov, A. V., and W. Lotko (2004), Multiscale electrodynamics of the ionosphere-magnetosphere system, *J. Geophys. Res.*, *109*, A09214, doi:10.1029/2004JA010457.

Streltsov, A. V., W. Lotko, J. R. Johnson, and C. Z. Cheng (1998), Small-scale, dispersive field line resonances in the hot magnetospheric plasma, *J. Geophys. Res.*, *103*(A11), 26,559–26,572.

Su, Y.-J., S. T. Jones, R. E. Ergun, and S. E. Parker (2004), Modeling of field-aligned electron bursts by dispersive Alfvén waves in the dayside auroral region, *J. Geophys. Res.*, *109*, A11201, doi:10.1029/2003JA010344.

Summers, D., and R. M. Thorne (1992), A new tool for analyzing microinstabilities in space plasmas modeled by a generalized Lorentzian (kappa) distribution, *J. Geophys. Res.*, *97*(A11), 16,827–16,832.

Swift, D. W. (2007), Simulation of auroral electron acceleration by inertial Alfvén waves, *J. Geophys. Res.*, *112*, A12207, doi:10.1029/2007JA012423.

Thompson, B. J., and R. L. Lysak (1996), Electron acceleration by inertial Alfvén waves, *J. Geophys. Res.*, *101*(A3), 5359–5369.

Tikhonchuk, V. T., and R. Rankin (2000), Electron kinetic effects in standing shear Alfvén waves in the dipolar magnetosphere, *Phys. Plasmas*, *7*(6), 2630–2645.

Tikhonchuk, V. T., and R. Rankin (2002), Parallel potential driven by a kinetic Alfvén wave on geomagnetic field lines, *J. Geophys. Res.*, *107*(A7), 1104, doi:10.1029/2001JA000231.

Voitenko, Y. M., and M. Goossens (2005), Nonlinear coupling of Alfvén waves with widely different cross-field wavelengths in space plasmas, *J. Geophys. Res.*, *110*, A10S01, doi:10.1029/2004JA010874.

Wahlund, J.-E., et al. (2003), Observations of auroral broadband emissions by CLUSTER, *Geophys. Res. Lett.*, *30*(11), 1563, doi:10.1029/2002GL016335.

Walsh, A. P., et al. (2010), Comprehensive ground-based and in situ observations of substorm expansion phase onset, *J. Geophys. Res.*, *115*, A00I13, doi:10.1029/2010JA015748.

Watt, C. E. J., and R. Rankin (2007a), Electron acceleration due to inertial Alfvén waves in a non-Maxwellian plasma, *J. Geophys. Res.*, *112*, A04214, doi:10.1029/2006JA011907.

Watt, C. E. J., and R. Rankin (2007b), Parallel electric fields associated with inertial Alfvén waves, *Planet. Space Sci.*, *55*, 714–721.

Watt, C. E. J., and R. Rankin (2008), Electron acceleration and parallel electric fields due to kinetic Alfvén waves in plasma with similar thermal and Alfvén speeds, *Adv. Space Res.*, *42*(5), 964–969.

Watt, C. E. J., and R. Rankin (2009), Electron trapping in shear Alfvén waves that power the aurora, *Phys. Rev. Lett.*, *102*(4), 045002, doi:10.1103/PhysRevLett.102.045002.

Watt, C. E. J., and R. Rankin (2010), Do magnetospheric shear Alfvén waves generate sufficient electron energy flux to power the aurora?, *J. Geophys. Res.*, *115*, A07224, doi:10.1029/2009JA015185.

Watt, C. E. J., R. Rankin, and R. Marchand (2004), Kinetic simulations of electron response to shear Alfvén waves in magnetospheric plasmas, *Phys. Plasmas*, *11*(4), 1277–1284.

Watt, C. E. J., R. Rankin, I. J. Rae, and D. M. Wright (2005), Self-consistent electron acceleration due to inertial Alfvén wave pulses, *J. Geophys. Res.*, *110*, A10S07, doi:10.1029/2004JA010877.

Watt, C. E. J., R. Rankin, I. J. Rae, and D. M. Wright (2006), Inertial Alfvén waves and acceleration of electrons in nonuniform magnetic fields, *Geophys. Res. Lett.*, *33*, L02106, doi:10.1029/2005GL024779.

Wild, J. A., et al. (2011), Midnight sector observations of auroral omega bands, *J. Geophys. Res.*, *116*, A00I30, doi:10.1029/2010JA015874.

Wygant, J. R., et al. (2000), Polar spacecraft based comparisons of intense electric fields and Poynting flux near and within the plasma sheet-tail lobe boundary to UVI images: An energy source for the aurora, *J. Geophys. Res.*, *105*(A8), 18,675–18,692.

Wygant, J. R., et al. (2002), Evidence for kinetic Alfvén waves and parallel electron energization at 4–6 R_E altitudes in the plasma sheet boundary layer, *J. Geophys. Res.*, *107*(A8), 1201, doi:10.1029/2001JA900113.

Zhao, J. S., D. J. Wu, and J. Y. Lu (2010), On nonlinear decay of kinetic Alfvén waves and application to some processes in space plasmas, *J. Geophys. Res.*, *115*, A12227, doi:10.1029/2010JA015630.

R. Rankin and C. E. J. Watt, Department of Physics, University of Alberta, Edmonton, AB T6G 2E1, Canada. (rrankin@ualberta.ca; watt@ualberta.ca)

Multispacecraft Observations of Auroral Acceleration by Cluster

C. Forsyth and A. N. Fazakerley

Mullard Space Science Laboratory, University College London, Dorking, UK

The electric fields in the auroral acceleration region (AAR) accelerate iono-spheric and magnetospheric particles to supra-keV energies above the aurora and are highly structured and temporally variable. While single-spacecraft missions have been a key to understanding a number of aspects of the microphysics of this region, they are fundamentally unable to deconvolve spatial and temporal varia-tions. In 2008, the Cluster spacecraft became the first multispacecraft mission to study the AAR at separations of the order of 1000 km. We review recent observa-tions by Cluster in the AAR, which show that Alfvénic acceleration regions can be either colocated with quasi-static potential drops or evolve into them, the growth of quasi-static drops occurs at low altitudes, and 20% of the total potential drop can be located inside the AAR away from the upper and lower boundaries.

1. INTRODUCTION

The excitation of atmospheric particles that creates bright, discrete auroral arcs requires an influx of supra-keV electrons from the magnetosphere that feed currents flowing through the ionosphere. Magnetospheric particles are accelerated to these energies along the magnetic field, in processes that violate basic magnetohydrodynamics as they involve field-aligned quasi-static and wave electric fields in the auroral acceleration region (AAR) [see *Karlsson*, this volume; *Watt and Rankin*, this volume]. Acceleration by quasi-static po-tential drops occurs between 1000 and 12,000 km altitude [*Reiff et al.*, 1993; *Lindqvist and Marklund*, 1990; *Lu et al.*, 1992; *Paschmann et al.*, 2003], whereas wave-driven accel-eration is predicted to occur anywhere along the magnetic field where there is a suitable particle and wave population [*Wygant et al.*, 2002; *Watt and Rankin*, 2009, 2010]. Models of the quasi-static potential in the AAR [e.g., *Ergun et al.*, 2000] suggest that this region is highly structured. Observa-tions going back to the first reports of the global nature of the optical aurora [*Akasofu*, 1964] show that they are temporally

Auroral Phenomenology and Magnetospheric Processes: Earth and Other Planets
Geophysical Monograph Series 197
© 2012. American Geophysical Union. All Rights Reserved.
10.1029/2011GM001166

variable; hence, one would expect this same variability in the AAR.

Single spacecraft observations from dedicated auroral mis-sions, such as Viking, Freja, FAST, Polar, and REIMEI provide valuable insight into the processes within the AAR. For example, many observations from the FAST mission have shown that static, stable oblique electric double layers [*Block*, 1972] are present at the lower boundary of the AAR and possibly also within the AAR [*Ergun et al.*, 2004; *Andersson and Ergun*, this volume]. Traveling intra-AAR double layers had previously been suggested as the source of fine structure in auroral kilometric radiation (AKR), detected outside the AAR, with their motion through the AAR at the ion acoustic speed resulting in the characteristic narrow-band frequency dispersed emission [*Pottelette et al.*, 2003]. How-ever, single spacecraft missions are unable to provide direct observational evidence in support of this suggestion.

Dynamics Explorer was the first multispacecraft mission to study the AAR, with two spacecraft separated by 10,000 km in altitude coming into conjunction (or close to conjunc-tion) over the auroral zone. Observations from Dynamics Explorer showed that the electron population is accelerated along the magnetic field with very little heating, consistent with a quasi-static potential drop. Ions are similarly acceler-ated by the quasi-static potential drop, but some of the kinetic energy gained by the ions is converted into thermal energy [*Reiff et al.*, 1988] (for a summary of DE results, see the work of *Burch* [1988]). These observations required two spacecraft separated along the magnetic field in order to directly compare the electron and ion distribution functions at different altitudes. A number of studies have also used fortuitous conjunctions between different missions to provide a multi-point view of auroral processes, but these, in these cases, tended to be conjunctions over large distances with only one spacecraft inside the AAR [e.g., *Janhunen et al.*, 2001; *Chaston et al.*, 2005]

While single-spacecraft missions have vastly improved our knowledge of the auroral acceleration process, their observations are limited to single points in time or space, and as such, temporal and spatial variations cannot be deconvolved from their measurements. During the early part of its mission, the four spacecraft of Cluster passed over the AAR at altitudes above 4 R_E in a string-of-pearls formation (with the spacecraft spaced out along their orbital path). This offered new insights into auroral processes, in particular, those of the downward field-aligned current region, such as showing that the ionospheric footprint of the downward current region expanded to reach a larger pool of current carriers [*Marklund et al.*, 2001; *Streltsov and Marklund*, 2006; *Cran-McGreehin et al.*, 2007], showing that the downward current region could be filamentary and that these current filaments were steady on a 1 min timescale [*Wright et al.*, 2008], showing that Alfvén waves can be ducted between downward current regions and have a perpendicular scale comparable to the separation of these regions [*Karlsson et al.*, 2004]. Since 2008, Cluster has been passing through the AAR (crossing auroral latitudes at below 2 R_E) and, for the first time, offers the opportunity to study this region with four spacecraft separated by less than a few thousand kilometers.

2. CLUSTER IN THE AURORAL ACCELERATION REGION

Cluster [*Escoubet et al.*, 2001] is the European Space Agency's first multispacecraft mission to study the magnetosphere. The four identically instrumented spacecraft were launched into a 4 × 19 R_E polar orbit with an argument of perigee that rotated 360° per year. After 2005, the spacecraft orbits evolved, with the perigee dropping and the orbital plane normal vector rotating out of the *XY* GSE plane. In the

first quarter of 2008, Cluster operations began targeting crossing of the top of the AAR, and after May 2008, the orbits were deliberately changed to optimize the multispacecraft observations in the AAR. The orbits were designed such that spacecraft formed a 1000 km tetrahedron with two spacecraft on the same orbit track and a further spacecraft at higher altitude and passing through approximately the same magnetic field lines. Figure 1 compares the Cluster 1 orbit starting at 00:00 UT on 1 December 2001 (black trace) and 1 December 2009 (red trace) in the GSE *XZ* (Figure 1a) and GSE *YZ* (Figure 1b) planes together with model magnetospheric magnetic field lines for context. Figures 1c and 1d show the magnetic foot points of Cluster 1 for the 2001 and 2009 orbits, respectively, in invariant latitude versus magnetic local time (MLT), using colored traces to indicate the altitude of the spacecraft. While the orbits in 2001 did cross onto field lines connecting to the aurora region, the crossings were at high altitude compared with the 2009 crossing, in which the foot points covered a wide range of MLTs in the auroral zone at an altitude of approximately 5000 km. Figures 1e–1h show the height distribution of those sections of the Cluster orbit that cross the nominal location of the auroral zone in different local time sectors. From 2007 onward, there was an increasing number of passes at lower altitude.

In the early part of the mission (pre-2008), the Cluster tetrahedron made high-altitude passes over the auroral zone in an elongated configuration. The spacecraft were separated by 4–5 min in total (approximately 100 s between each spacecraft), whereas each individual AAR crossing took of the order 1 min. In 2008, a concerted effort was made to make more closely spaced, optimized observations in the auroral regions. The spacecraft were maneuvered into orbits such that Cluster 3 and 4 were on approximately the same orbit but separated along it by a few hundred kilometers (equating to a temporal separation of a few minutes). Cluster 1 made higher-altitude crossings of the auroral region, but at times was magnetically conjugate to Cluster 3 or 4, separated by 1000 km or more. Cluster 2 was at approximately the same altitude as Cluster 3 and 4, but displaced transverse to the orbital plane.

Unlike missions such as FAST, the Cluster's instrument suite was not optimized for studying the microphysics of the AAR. While electric and magnetic field data can be taken at

---

Figure 1. (opposite) The evolution of the Cluster orbits and the heights at which Cluster passed over the auroral zone. Cluster orbits starting at 00:00 UT on 1 December 2001 (black) and 1 December 2009 (red) in the (a) *XZ* and (b) *YZ* GSE planes. The blue lines show a *Tsyganenko and Stern* [1996] model magnetic field. Foot point locations of Cluster 1 for (c) 2001 and (d) 2009 orbits in invariant latitude versus magnetic local time (MLT) coordinates, color coded by height. Histograms of the time that Cluster spent in the above the auroral zone versus geocentric altitude, color coded by year, and split into the (e) postnoon, (f) prenoon, (g) premidnight, and (h) postmidnight sectors. The dotted line in Figures 1e–1h shows the altitude of the original Cluster perigee, and the dashed line shows the 12,000 km upper limit of the quasi-static auroral acceleration region (AAR).

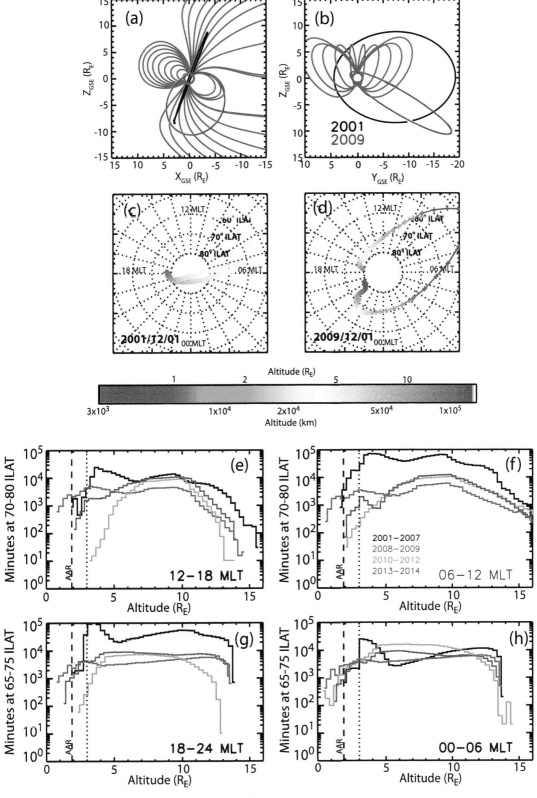

Figure 1

relatively high time resolution (>22 Hz), particle data is far more limited; for example, electron, pitch angle distributions are typically only available once every 2 s. Furthermore, Cluster can only measure the electric field in the spacecraft spin planes, and given that we expect $E.B \neq 0$ in the AAR, the electric field component along the spin axis of the spacecraft cannot be estimated by the technique assuming $E.B = 0$. However, this data is more than adequate in examining the large-scale structure of the AAR, and hence, we can compare the instantaneous distribution of the electric potential drops along the magnetic field with the statistical picture obtained from single-spacecraft data. Furthermore, because the individual spacecraft pass through the AAR at slightly different times and locations, the temporal and spatial development of these structures may be studied.

While a number of multispacecraft techniques were developed for the Cluster mission [see *Paschmann and Daly*, 1998, 2008] these cannot usually be applied in the AAR because the spacecraft separation is typically greater than the scale size of the AAR across the magnetic field, or the spacecraft tetrahedron was not optimized for these techniques. As such, the assumptions used in these techniques (e.g., the linearity of the variation of the magnetic field between spacecraft and the assumption that current sheets are planar on the scale of the spacecraft tetrahedron) are not met. However, Cluster offers the unprecedented opportunity to examine the AAR at up to four closely separated locations simultaneously or at the same location at slightly different times; thus, we can still examine the spatial and temporal differences by directly comparing the physical parameters measured by each spacecraft.

The analysis of the AAR using in situ Cluster data is a relatively new endeavor. As such, there are currently only a small number of case studies exploring the use of multipoint observations in the AAR. In the following, we will review these results from Cluster and consider how future multipoint studies of the AAR might further our understanding of the processes in this region.

3. TEMPORAL EVOLUTION OF THE AURORAL ACCELERATION POTENTIAL

Temporal variability is a well-known characteristic of the aurora, from large-scale auroral arcs that can appear invariant on the timescale of tens of minutes but that change position, to pulsating and flickering aurora, which can vary at subsecond timescales. This temporal variability is intrinsically hidden from in situ single-spacecraft observations, as spatial and temporal variations cannot be deconvolved. Many of the observations of the AAR have previously been interpreted assuming a spatial and temporal invariance, particularly noting that features such as inverted Vs in electron spectra are common and can be explained using a simple spatial electric field distribution.

The first study to deconvolve spatial and temporal measurements of auroral acceleration was by *Thieman and Hoffman* [1985]. The authors compared inverted-V signatures from high- and low-altitude passes over the auroral zone by the two Dynamics Explorer spacecraft at slightly different times. By examining the correlation between the electron spectra from the high- and low-altitude spacecraft, they inferred that inverted-V structures could be stable for 18 min (showing greater correlation than randomly selected pairs) or could vary on a timescale of less than 5 min. They also examined how the change in the peak energy of the electron distribution varied with time. However, in this analysis, they did not consider how the acceleration of electrons between the spacecraft would affect these results or how the total potential drop in the AAR varied with time. As such, it is difficult to put these results in the context of the changes in the AAR as a whole. A more interesting study would have shown how the total potential drop or its distribution along the field varied, although we note that a quantitative comparison of techniques for determining the total potential drop was not carried out until the work of *Reiff et al.* [1988].

One of the key questions in understanding the AAR, and the role it plays in the global dynamics of the magnetosphere is, "How does the AAR form?" By observing the same region of space with a number of spacecraft at slightly different times, the changes to that region can be determined. Cluster has presented the opportunity to make such observations with the AAR, either with two spacecraft separated along the same orbit track when the spacecraft formed the 1000 km tetrahedron or with the spacecraft stretched out along the orbit. These observations suggest that the damping of Alfvén waves may play a part in the formation of the quasi-static aurora.

3.1. Evolution of the Polar Cap Boundary (PCB) AAR

Following observations by single-spacecraft missions, three distinct types of AAR were identified: the upward current region, the downward current region, and the Alfvénic region [e.g., *Paschmann et al.*, 2003]. While these categories are useful for comparing similar observations from multiple AAR crossings, they do not address the underlying physics that determines how they form and interact. For example, as the auroral oval contracts, how does the quasi-static AAR form on those field lines that were previously at the PCB? Cluster has provided new insights into the interconnectedness of these regions by making multiple observations of the same AAR at small temporal separations.

A recent case study by *Hull et al.* [2010] examined how the auroral cavity associated with a poleward boundary intensification evolved over a period of nearly 3 min. Their observations were taken at high altitude (>4 R_E) above the acceleration region itself. The spacecraft formation was highly elongated along the orbital path. By examining the magnetic field and particle populations observed by Cluster, Hull et al. showed that the Alfvénic acceleration region at the PCB evolved into a quasi-static AAR. Hull et al. showed that there was a shallow density cavity at the PCB during the first spacecraft crossing, but that in subsequent crossings, hot plasma sheet ions and electrons were injected onto the PCB field lines (which had moved poleward) and that the depth of the density cavity increased. The electron spectra showed that the density cavity was formed due to the evacuation of cold electrons (which were not seen at high altitude). During the formation of the cavity, the Alfvén wave power decreased and large-scale field-aligned current structures developed as the perpendicular wavelength of the Alfvén waves approached $2\pi\lambda_e$, where λ_e is the electron inertial length.

A further PCB case study was reported by *Marklund et al.* [2011b], in which three of the Cluster spacecraft passed through the AAR at 2 R_E altitude, close to its top. The spacecraft were separated by 6 min in total and up to 600 km in altitude and showed a somewhat different picture to that of the work of *Hull et al.* [2010], particularly in relation to the changes in the PCB. As the Cluster spacecraft passed through the AAR during their event, *Marklund et al.* [2011b] were able to identify the PCB and two neighboring upward current regions associated with a bifurcated auroral oval. They reported observations of counter streaming electrons and upgoing ions at a range of energies in the PCB, which they interpreted as a signature of Alfvénic acceleration, although they provided no in situ evidence of Alfvén waves. They also observed a potential drop, calculated by integrating the electric field observations perpendicular to the magnetic field along the spacecraft path, of up to 11 kV in this region. This is comparable to the quasi-static potential drops observed in the upward current regions equatorward of the PCB in the same event. This colocation of the Alfvénic and quasi-static acceleration was observed over 6 min, although the PCB moved poleward during the interval.

The events studied by *Hull et al.* [2010] and *Marklund et al.* [2011b] occurred under different geomagnetic conditions and were observed at different altitudes. *Hull et al.* [2010] presented no observations of the electric field or potential drop from the electric field, whereas *Marklund et al.* [2011b] presented no analysis of the magnetic fields to assess the presence of Alfvén waves in the PCB, so we cannot confirm whether or not these signatures were present in both cases. However, it is interesting to speculate as to the possible reasons for the apparent differences between their conclusions. In particular, we note that the geomagnetic conditions were quite different in these two cases; the *Hull et al.* [2010] event occurring in conjunction with a short-lived poleward boundary intensification, the *Marklund et al.* [2011b] event occurring during a poleward expansion of the aurora during the apparent recovery phase of a weak substorm (as determined from auroral indices). The PBI case [*Hull et al.*, 2010] may represent a short burst of reconnection in the magnetotail, injecting plasma and intensifying the Alfvén waves on previously open field lines. In contrast, the substorm case [*Marklund et al.*, 2011b] may represent ongoing reconnection in the tail, injecting plasma and Alfvén waves onto increasingly poleward field lines, with the cavity forming in the region behind. Future multispacecraft studies should investigate whether the perpendicular scale sizes of the persistent Alfvén waves are greater than the electron inertial length, reducing the inertial damping described by *Hull et al.* [2010] and how the electron and ion inertial lengths in the AAR change during the course of substorms and nonsubstorm acceleration events.

3.2. Evolution of the Upward Current AAR

As we discussed above, the quasi-static AAR can be temporally stable on timescales of up to 18 min or temporally variable on a timescale of 5 min [*Thieman and Hoffman*, 1985]. What was unclear from this study was how the total potential drop varied over these times. *Morioka et al.* [2007, 2009] noted that AKR, typically associated with the upward current region [*Pascmann et al.*, 2003], increased at high altitude at substorm onset, whereas the low-altitude emission was present prior to the onset and during the substorm. This suggests that the potential drop in the upward current region increases at high altitude at substorm onset. A number of recent case studies using in situ data from Cluster spacecraft passing through the AAR at different times have offered insights into the variability of the upward current region.

Cluster observations of the temporal variation of the upward current AAR have been presented by *Marklund et al.* [2011b], *Sadeghi et al.* [2011], and C. Forsyth et al. (Temporal evolution and electric potential structure of the auroral acceleration region from multispacecraft measurements, submitted to *Journal of Geophysical Research*, 2012, hereinafter referred to as Forsyth et al., submitted manuscript, 2012). These case studies all present observations from neighboring pairs of upward current regions using three of the Cluster spacecraft as they pass through AAR with time differences of up to 6 min between the first and last crossing. In the studies of *Marklund et al.* [2011b] and *Sadeghi et al.* [2011], each of the three spacecraft passed through the AAR at different times and altitudes. In the Forsyth et al. (submitted manuscript, 2012) study, two of the spacecraft passed through the AAR at the same time, coming into conjunction in one of the upward current regions. The lower of these spacecraft (C3) was on the same orbit as C4, which lead C3 by 3 min. Because C3 and C1 passed through the AAR at the same time, Forsyth et al. (submitted manuscript, 2012) only had two points of comparison compared with three given by *Marklund et al.* [2011b] and *Sadeghi et al.* [2011].

In the study by *Marklund et al.* [2011b], the data from the three spacecraft presented showed similar trends, suggesting that the spacecraft observed the same regions or that the AAR was sufficiently invariant in space to directly compare the observations presented. As such, the Cluster showed that there was a long-lasting overall structure, but the details of that structure varied between the spacecraft crossings. Comparing the data from each spacecraft in the study by *Sadeghi et al.* [2011] shows strong similarities between two of the spacecraft (C1 and C4, the second two to pass through each upward current region), but that the third (C2) seems quite different, particularly the electric and magnetic field signatures. *Sadeghi et al.* [2011] give more emphasis to the idea that this discrepancy is due to the temporal variation of AAR, but given that C2 was on a different orbit track to C1 and C4 (with their footprints separated by several degrees), this may also be due to spatial differences.

Sadeghi et al. [2011] interpreted their results as showing that the growth of the potential drop tended to be at lower altitudes. This result is highly dependent on the assumption that all three spacecraft passed through the same structures, as the potential only increases between the first and second spacecraft observations, and as we noted above, these showed quite different signatures. Between the second and third spacecraft, the first upward current structure disappeared, and the second structure showed no variation in the total potential, and since the spacecraft passed through this region at different heights, any variation in the potential drop along the field could not be determined. However, the observations by *Marklund et al.* [2011b] and Forsyth et al. (submitted manuscript, 2012) support this conclusion as they also show examples of changes in the potential drop at low altitudes being more extreme than simultaneous changes at high altitude.

Statistical studies of the electric field and potential drop against altitude have previously shown that the fraction of the potential drop below a given altitude increases rapidly up to 3 R_E geocentric [*Mozer and Hull*, 2001]. They have also shown that large magnetic field–aligned electric fields (>25 mV m^{-1}) tend to be found in a narrow region around 1.28 R_E from the Earth's surface [*Hull et al.*, 2003], possibly indicative of the presence of a low-altitude double layer [*Ergun et*

al., 2000]. The observations from Cluster are consistent with this, but also seem to suggest that the growth of the potential is concentrated at low altitudes.

The processes that control the temporal variability of the AAR remain a mystery. The case studies discussed have shown that neighboring upward current regions, sometimes separated in space by less than the ion inertial length (of the order 100 km in the AAR), vary differently, with some being stable for minutes and others decaying on a subminute timescale.

4. ALTITUDINAL STRUCTURE OF THE AURORAL ACCELERATION REGION

Understanding the structure of the potential drops in the AAR is important in understanding the particle acceleration mechanisms. Models of the AAR have suggested that the potential drops are concentrated in electric double layers [*Block*, 1972] at the upper and lower boundaries of the AAR [*Ergun et al.*, 2000]. While observations have indicated the presence of double layers at low altitude [*Ergun et al.*, 2002; *Hull et al.*, 2003], other studies have suggested the presence of double layers inside the AAR [*Pottelette et al.*, 2003; *Ergun et al.*, 2004] and have also shown that average magnetic field–aligned electric fields in the AAR is of the order of 1 mV m^{-1} [*Lindqvist and Marklund*, 1990]. Given that auroral features such as AKR require a particle population that has been accelerated and mirrored and that the formation of the auroral cavity requires the evacuation of cold electrons, an important question in auroral physics is, "How is the electric potential drop in the AAR distributed along the magnetic field?"

Cluster's orbits through the AAR offer the opportunity to understand the distribution of the accelerating electrostatic potentials by determining the potential differences between multiple heights either simultaneously or near simultaneously. Previous missions, such as Dynamics Explorer, and fortuitous spacecraft conjunctions, such as Polar and FAST, have examined the potential structure over large distances (>10,000 km), typically with one or both spacecraft outside the AAR [*Reiff et al.*, 1988; *Janhunen et al.*, 2001]. The Cluster spacecraft, however, have been separated along the field by only a few 1000 km, allowing the potential gradient within the mid-AAR to be determined.

Marklund et al., [2011a], *Sadeghi et al.* [2011], and Forsyth et al. (submitted manuscript, 2012) made observations of the parallel electric potential at two heights using Cluster (typically Cluster 1 at high altitude compared to Cluster 2, 3, and 4). In the *Marklund et al.* [2011a] and *Sadeghi et al.* [2011] studies, the spacecraft passed through the AAR at different times, with a time difference of up to 5 min. As such, in order to estimate the distribution of the potential

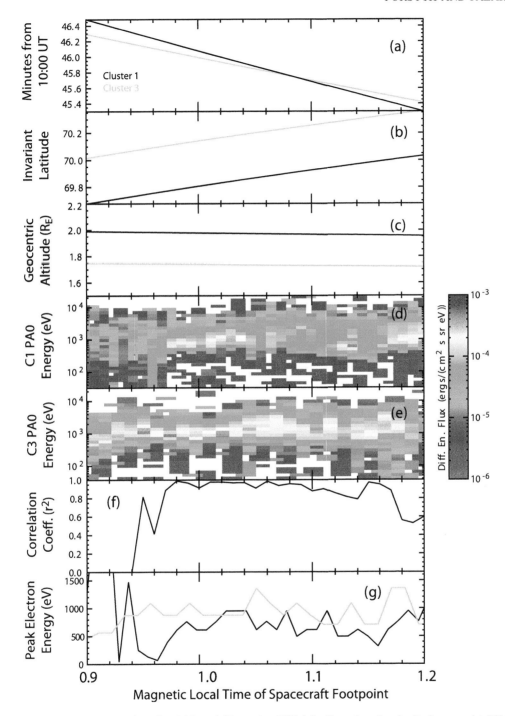

Figure 2. Data from a Cluster crossing of an AAR on 14 December 2009 (after Forsyth et al., submitted manuscript, 2012). The data are plotted against the MLT of the spacecraft foot point, calculated using the International Geomagnetic Reference Field and the *Tsyganenko and Stern* [1996] magnetic field model. (a) Times at which the spacecraft were conjugate to the given MLTs. (b) Invariant latitude of the spacecraft foot point. (c) Geocentric altitude (radial distance) of the spacecraft. In the top three plots, black indicates Cluster 1 and green indicates Cluster 3. Differential energy flux of the field-aligned (downgoing) electrons from (d) Cluster 1 and (e) Cluster 3. (f) Square of the maximum correlation coefficient from a cross-correlation analysis of the electron spectra at each local time. (g) Energy of the peak of the electron spectra. The peak energy of the electrons at C3 was generally higher than at C1. The high correlation coefficients indicate that the spectra were similar in form, indicating that the electrons were being accelerated between the two spacecraft and that the temporal and latitudinal separations were negligible.

drop along the magnetic field, it was necessary to assume that the AAR was invariant on this timescale. This argument was justified on the basis of the same total potential drop being observed by each spacecraft. In the Forsyth et al. (submitted manuscript, 2012) study, the two altitudinally separated spacecraft passed through the AAR at the same time, with the uppermost spacecraft passing over the lower spacecraft in the middle of the AAR. These three studies also assumed that the AAR was spatially invariant along its length such that differences in the invariant latitude or magnetic local time of the foot points could be ignored. For *Sadeghi et al.* [2011] and Forsyth et al. (submitted manuscript, 2012), the separation of the spacecraft foot points was small (around 0.3°), but for *Marklund et al.* [2011a], the separation was 1.7 h MLT, suggesting that the AAR was invariant over a large distance or moving at around $0.1° \text{ s}^{-1}$. Figure 2 shows data from the AAR (after Forsyth et al., submitted manuscript, 2012) showing the cross correlation between the downward going electron spectra at the two spacecraft, along with the electron spectra, showing the accelerated electron population and their peak energies.

These studies showed potential drops within the AAR of 0.6–2 kV, translating to spatially averaged electric fields of $0.15–0.8 \text{ mV m}^{-1}$. These values are much smaller than the parallel electric fields expected in double layers (of the order of tens to hundreds mV m^{-1}). We would expect double layers to have a scale size comparable to the Debye length, of the order of ones to tens of kilometers in the AAR. Cluster observations cannot be used to infer the potential distribution on scale sizes smaller than the spacecraft separation (of the order 1000 km). They can, however, be used to determine the proportion of the potential between the two spacecraft altitudes, which was typically 20% between 0.7 and 1.3 R_E altitude.

Statistical studies using single-spacecraft data have provided some insight into the distribution of the electric potential within the AAR. *Mozer and Hull* [2011] determined the statistical altitude distribution of the AAR potential and showed that the potential at low altitude increased rapidly between 2 and 3 R_E geocentric distance and more slowly above that. However, the scatter in their measurements shows that it is difficult to quantitatively compare different AAR observations, as the AAR may vary in length along the magnetic field, altitude, strength, and distribution of the potential drop. The observations from Cluster show that a substantial proportion of the potential drop can be located inside the AAR rather than at the boundaries. However, the observations presented thus far have only studied a small altitude range, and further observations need to be made at a range of altitudes to understand how much of the potential drop is within the AAR and how much is at the boundaries.

These observations will then present a more well-defined picture for theorists and simulators to consider.

5. SUMMARY

Multipoint observations by Cluster offer a new perspective on the processes and structures in the auroral acceleration region, as coordinated two-point observations by Dynamics Explorer and conjunctions between Polar and FAST have previously. For the first time, we are able to examine the temporal evolution of the AAR and the different regions of the auroral zone. These observations show that the Alfvénic acceleration in the polar cap boundary can be colocated with quasi-static acceleration signatures, the Alfvénic PCB can evolve into a quasi-static AAR, the growth of the quasi-static aurora occurs at low altitude. They also enable studies, which can probe the inner regions of the AAR, showing that 20% of the magnetic field–aligned potential drop can be located inside the AAR, although it is unclear whether this is concentrated in double layers inside the AAR or by a low-level parallel electric field permeating the whole region. This is contrary to Vlasov models of the AAR, which show that the majority of the potential drop is concentrated in double layers at the upper and lower limits of the auroral cavity.

The multipoint studies discussed above are the first to use multispacecraft observations by spacecraft separated by less than a few thousand kilometers inside the AAR. Future multispacecraft studies will be required to answer further open questions, such as, "Why do some Alfvénic PCBs evolve into quasi-static AARs, while others are superimposed on them?" and "Why is the growth of the quasi-static AAR concentrated at low altitude?" We are also hopeful that these observations will give insights into the generation of the fine-scale AKR by enabling simultaneous observations of the source regions and the emission, something that is not possible with a single spacecraft. Furthermore, to gain a full understanding of the AAR, we believe these observations need to be coupled with complementary observations of the ionosphere and magnetosphere in order to understand the drivers and impeders of the AAR and the role the AAR plays in the coupled magnetosphere-ionosphere system.

Acknowledgments. CF and ANF were supported by STFC grant ST/H00260X/1.

REFERENCES

Akasofu, S.-I. (1964), The development of the auroral substorm, *Planet. Space Sci.*, *12*, 273–282.

Andersson, L., and R. E. Ergun (2012), The search for double layers in space plasmas, in *Auroral Phenomenology and Magnetospheric*

Processes: Earth and Other Planets, Geophys. Monogr. Ser., doi:10.1029/2011GM001170, this volume.

Block, L. P. (1972), Potential double layers in the ionosphere, *Cosmic Electrodyn., 3,* 349–376.

Burch, J. L. (1988), Energetic particles and currents: Results from Dynamics Explorer, *Rev. Geophys., 26*(2), 215–228.

Chaston, C. C., et al. (2005), Energy deposition by Alfvén waves into the dayside auroral oval: Cluster and FAST observations, *J. Geophys. Res., 110,* A02211, doi:10.1029/2004JA010483.

Cran-McGreehin, A. P., A. N. Wright, and A. W. Hood (2007), Ionospheric depletion in auroral downward currents, *J. Geophys. Res., 112,* A10309, doi:10.1029/2007JA012350.

Ergun, R. E., C. W. Carlson, J. P. McFadden, F. S. Mozer, and R. J. Strangeway (2000), Parallel electric fields in discrete arcs, *Geophys. Res. Lett., 27*(24), 4053–4056, doi:10.1029/2000GL 003819.

Ergun, R. E., L. Andersson, D. S. Main, Y. Su, C. W. Carlson, J. P. McFadden, and F. S. Mozer (2002), Parallel electric fields in the upward current region of the aurora: Indirect and direct observations, *Phys. Plasmas, 9,* 3685–3694, doi:10.1063/1.1499120.

Ergun, R. E., L. Andersson, D. Main, Y.-J. Su, D. L. Newman, M. V. Goldman, C. W. Carlson, A. J. Hull, J. P. McFadden, and F. S. Mozer (2004), Auroral particle acceleration by strong double layers: The upward current region, *J. Geophys. Res., 109,* A12220, doi:10.1029/2004JA010545.

Escoubet, C. P., M. Fehringer, and M. Goldstein (2001), The Cluster mission, *Ann. Geophys., 19,* 1197–1200.

Hull, A. J., J. W. Bonnell, F. S. Mozer, and J. D. Scudder (2003), A statistical study of large-amplitude parallel electric fields in the upward current region of the auroral acceleration region, *J. Geophys. Res., 108*(A1), 1007, doi:10.1029/2001JA007540.

Hull, A. J., M. Wilber, C. C. Chaston, J. W. Bonnell, J. P. McFadden, F. S. Mozer, M. Fillingim, and M. L. Goldstein (2010), Time development of field-aligned currents, potential drops, and plasma associated with an auroral poleward boundary intensification, *J. Geophys. Res., 115,* A06211, doi:10.1029/ 2009JA014651.

Janhunen, P., A. Olsson, W. K. Peterson, H. Laakso, J. S. Pickett, T. I. Pulkkinen, and C. T. Russell (2001), A study of inverted-V auroral acceleration mechanisms using Polar/Fast Auroral Snapshot conjunctions, *J. Geophys. Res., 106,* 18,995–19,011, doi:10. 1029/2001JA900012.

Karlsson, T. (2012), The acceleration region of stable auroral arcs, in *Auroral Phenomenology and Magnetospheric Processes: Earth and Other Planets, Geophys. Monogr. Ser.,* doi:10.1029/ 2011GM001179, this volume.

Karlsson, T., G. Marklund, S. Figueiredo, T. Johansson, and S. Buchert (2004), Separating spatial and temporal variations in auroral electric and magnetic fields by Cluster multipoint measurements, *Ann. Geophys., 22,* 2463–2472, doi:10.5194/angeo-22-2463-2004.

Lindqvist, P.-A., and G. T. Marklund (1990), A statistical study of high-altitude electric fields measured on the Viking satellite, *J. Geophys. Res., 95*(A5), 5867–5876.

Lu, G., P. H. Reiff, T. E. Moore, and R. A. Heelis (1992), Upflowing ionospheric ions in the auroral region, *J. Geophys. Res., 97*(A11), 16,855–16,863.

Marklund, G. T., et al. (2001), Temporal evolution of the electric field accelerating electrons away from the auroral ionosphere, *Nature, 414,* 724–727.

Marklund, G. T., S. Sadeghi, T. Karlsson, P.-A. Lindqvist, H. Nilsson, C. Forsyth, A. Fazakerley, E. A. Lucek, and J. Pickett (2011a), Altitude distribution of the auroral acceleration potential determined from Cluster satellite data at different heights, *Phys. Rev. Lett., 106*(5), 055002, doi:10.1103/PhysRevLett.106. 055002.

Marklund, G. T., et al. (2011b), Evolution in space and time of the quasi-static acceleration potential of inverted-V aurora and its interaction with Alfvénic boundary processes, *J. Geophys. Res., 116,* A00K13, doi:10.1029/2011JA016537.

Morioka, A., Y. Miyoshi, F. Tsuchiya, H. Misawa, T. Sakanoi, K. Yumoto, R. R. Anderson, J. D. Menietti, and E. F. Donovan (2007), Dual structure of auroral acceleration regions at substorm onsets as derived from auroral kilometric radiation spectra, *J. Geophys. Res., 112,* A06245, doi:10.1029/2006JA012186.

Morioka, A., Y. Miyoshi, F. Tsuchiya, H. Misawa, K. Yumoto, G. K. Parks, R. R. Anderson, J. D. Menietti, and F. Honary (2009), Vertical evolution of auroral acceleration at substorm onset, *Ann. Geophys., 27,* 525–535, doi:10.5194/angeo-27-525-2009.

Mozer, F. S., and A. Hull (2001), Origin and geometry of upward parallel electric fields in the auroral acceleration region, *J. Geophys. Res., 106,* 5763–5778, doi:10.1029/2000JA900117.

Paschmann, G., and P. W. Daly (Eds.) (1998), Analysis methods for multi-spacecraft data, *ISSI Sci. Rep. Ser. SR-001,* vol. 1, Eur. Space Agency Publ. Div., Noordwijk, The Netherlands.

Paschmann, G., and P. W. Daly (Eds.) (2008), *Multi-Spacecraft Analysis Methods Revisited,* Eur. Space Agency Commun., Noordwijk, The Netherlands.

Paschmann, G., S. Haaland, and R. Treumann (Eds.) (2003), *Auroral Plasma Physics,* ISSI/Kluwer Acad., Dordrecht, The Netherlands.

Pottelette, R., R. A. Treumann, M. Berthomier, and J. Jasperse (2003), Electrostatic shock properties inferred from AKR fine structure, *Nonlinear Processes Geophys., 10,* 87–92.

Reiff, P. H., H. L. Collin, J. D. Craven, J. L. Burch, J. D. Winningham, E. G. Shelley, L. A. Frank, and M. A. Friedman (1988), Determination of auroral electrostatic potentials using high- and low-altitude particle distributions, *J. Geophys. Res., 93*(A7), 7441–7465.

Reiff, P. H., G. Lu, J. L. Burch, J. D. Winningham, L. A. Frank, J. D. Craven, W. K. Peterson, and R. A. Heelis (1993), On the high- and low-altitude limits of the auroral electric field region, in *Auroral Plasma Dynamics, Geophys. Monogr. Ser.,* vol. 80, edited by R. L. Lysak, pp. 143–154, AGU, Washington, D. C., doi:10.1029/GM080p0143.

Sadeghi, S., G. T. Marklund, T. Karlsson, P.-A. Lindqvist, H. Nilsson, O. Marghitu, A. Fazakerley, and E. A. Lucek (2011), Spatiotemporal features of the auroral acceleration region as observed by Cluster, *J. Geophys. Res.*, *116*, A00K19, doi:10.1029/2011JA016505.

Streltsov, A. V., and G. T. Marklund (2006), Divergent electric fields in downward current channels, *J. Geophys. Res.*, *111*, A07204, doi:10.1029/2005JA011196.

Thieman, J. R., and R. A. Hoffman (1985), Determination of inverted-V stability from Dynamics Explorer satellite data, *J. Geophys. Res.*, *90*(A4), 3511–3516.

Tsyganenko, N. A., and D. P. Stern (1996), Modeling the global magnetic field of the large-scale Birkeland current systems, *J. Geophys. Res.*, *101*(A12), 27,187–27,198.

Watt, C. E. J., and R. Rankin (2009), Electron trapping in shear Alfvén waves that power the aurora, *Phys. Rev. Lett.*, *102*(4), 045002, doi:10.1103/PhysRevLett.102.045002.

Watt, C. E. J., and R. Rankin (2010), Do magnetospheric shear Alfvén waves generate sufficient electron energy flux to power the aurora?, *J. Geophys. Res.*, *115*, A07224, doi:10.1029/2009JA015185.

Watt, C. E. J., and R. Rankin (2012), Alfvén wave acceleration of auroral electrons in warm magnetospheric plasma, in *Auroral Phenomenology and Magnetospheric Processes: Earth and Other Planets*, Geophys. Monogr. Ser., doi:10.1029/2011GM001171, this volume.

Wright, A. N., C. J. Owen, C. C. Chaston, and M. W. Dunlop (2008), Downward current electron beam observed by Cluster and FAST, *J. Geophys. Res.*, *113*, A06202, doi:10.1029/2007JA012643.

Wygant, J. R., et al. (2002), Evidence for kinetic Alfvén waves and parallel electron energization at 4–6 R_E altitudes in the plasma sheet boundary layer, *J. Geophys. Res.*, *107*(A8), 1201, doi:10.1029/2001JA900113.

A. N. Fazakerley and C. Forsyth, Mullard Space Science Laboratory, University College London, Dorking RH5 6NT, UK. (cfo@mssl.ucl.ac.uk)

Fine-Scale Characteristics of Black Aurora and Its Generation Process

T. Sakanoi,[1] Y. Obuchi,[2] Y. Ebihara,[3] Y. Miyoshi,[4] K. Asamura,[5] A. Yamazaki,[5]
Y. Kasaba,[1] M. Hirahara,[4] T. Nishiyama,[1] and S. Okano[1]

Black aurora is a small-scale (typically a few to 10 km) black structure seen in
diffuse aurora, and its generation process has been studied with immense interest.
We report the precise characteristics of black aurora based on simultaneous image
and particle measurement data and possible generation process. Thirteen black
auroral events are identified from the Reimei satellite data, and the relationship
between particle and auroral images around the satellite's magnetic footprints is
investigated in detail. We found that a number of small-scale deficiencies were
embedded in precipitating electrons from the central plasma sheet with energies
greater than 2–7 keV and that each deficiency corresponded exactly to black arcs
and black patches at the magnetic footprint. Therefore, black arcs and black patches
are not associated with a field-aligned potential (such as a divergent potential
structure) but probably originate from the suppression of pitch angle scattering. In
the black auroral region, low-energy (2–5 keV) inverted-V-type downward electrons (spanning channels that are several tens of kilometers wide) often appear to
overlap with high-energy (several keV) plasma sheet electrons.

1. INTRODUCTION

Black aurora is usually defined as the dark regions within diffuse auroras [*Davis*, 1978]. Various black aurora observations have been reported [*Trondsen and Cogger*, 1997; *Kimball and Hallinan*, 1998a, 1998b; *Peticolas et al.*, 2002; *Blixt and Kosch*, 2004], and several kinds of black auroras have been classified, including black vortex streets (black curls), black arcs, and black patches (see examples in Figure 1).

Black vortex streets show successive 0.5–5.7 km long black curls with a spatial separation of 1–4 km [*Trondsen and Cogger*, 1997; *Kimball and Hallinan*, 1998b]. In addition, black vortex streets drift eastward or westward at a speed of 0.2–5.0 km s^{-1} [*Kimball and Hallinan*, 1998b]. Each black curl has a clockwise rotational sense [*Trondsen and Cogger*, 1997; *Kimball and Hallinan*, 1998b] (while that of auroral curls is anticlockwise), which is caused by Kelvin-Helmholtz instabilities in a divergent electric field [*Marklund et al.*, 1994, 1997]. Black arcs frequently appear in the diffuse aurora around the midnight sector, showing multiple narrow channels, each a few hundred meters to 10 km wide [*Trondsen and Cogger*, 1997; *Blixt and Kosch*, 2004]. Black arcs have been reported to typically drift equatorward at a few hundred meters per second [*Trondsen and Cogger*, 1997; *Blixt and Kosch*, 2004]. Black patches (black arc segments) are also observed around the midnight sector [*Trondsen and Cogger*, 1997; *Kimball and Hallinan*, 1998a]. These authors reported that the spatial scale of black patches ranges from a few hundred meters to 14 km and that they drift eastward only at 0.2–3.4 km s^{-1}. As described above, black auroras have various forms and dynamics. However, the mechanisms for their generation and motion have not yet been clearly identified.

[1]Graduate School of Science, Tohoku University, Sendai, Japan.
[2]Genesia Corporation, Tokyo, Japan.
[3]Research Institute for Sustainable Humanosphere, Kyoto University, Kyoto, Japan.
[4]Solar-Terrestrial Environment Laboratory, Nagoya University, Nagoya, Japan.
[5]Institute of Space and Astronautical Science, Japan Aerospace Exploration Agency, Kanagawa, Japan.

Auroral Phenomenology and Magnetospheric Processes: Earth and
Other Planets
Geophysical Monograph Series 197
© 2012. American Geophysical Union. All Rights Reserved.
10.1029/2011GM001178

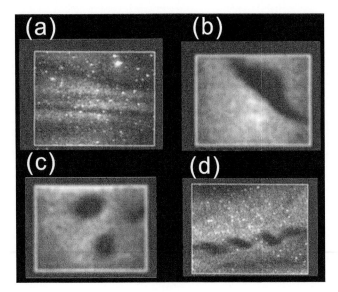

Figure 1. Examples of black auroral images taken with telephoto camera from the ground [after *Trondsen and Cogger*, 1997]. (a) Black arcs, (b) black patch, (c) black dots, and (d) black vortex street.

2. BACKGROUND ON THE GENERATION PROCESS OF BLACK AURORA

Some theories have suggested a generation mechanism that takes into account the characteristics of black auroras using data obtained by satellites. A major idea concentrates on the divergent electric field structure as shown in Figure 2a. *Marklund et al.* [1994,1997] observed a series of intense divergent electric fields (1–2 V m^{-1}) of approximately 1 km size at lower altitudes (around 800 km) and suggested the hypothesis that the downward directed electric field contributed to generating the black vortex streets. However, observational evidence has yet to be reported for such a relationship between divergent electric fields and black auroras. *Schoute-Vanneck et al.* [1990] showed that that black aurorae did not drift with the ionospheric convection, and this fact would indicate that it seems to be difficult to explain the drift with a downward directed electrical field. *Blixt et al.* [2004] reported that there was no evidence of field-aligned currents with black auroral arc from the simultaneous ground-based TV optical and incoherent scatter radar observations. *Blixt et al.* [2005] also found that the drift velocity of black aurora was proportional to the characteristic energy of the electrons in the surrounding diffuse aurora. The fact that the energy of the electron precipitation in the surrounding diffuse aurora is related to the drift velocity is directly incompatible with a downward electrical field, or at least without any physical explanation as of today. In addition, concerning the veiling of black aurora, the black aurora is more visible during the off phase of pulsating aurora [*Trondsen and Cogger*, 1997; *Kimball and Hallinan*, 1998a; *Archer et al.*, 2011]. For such a variation of the black aurora to be caused by a downward directed electrical field, there would have to be a synchronized modulation of the electrical field and the pulsating aurora. This synchronized modulation seems unlikely to occur. Further, spectral observations of black aurorae have shown that black arcs have a relatively larger intensity reduction in emission from molecular species, such as 427.8 nm (N_2^+ first negative band (1NG)), 673.0 nm (N_2 first positive band (1PG)), than from atomic oxygen (844.6 and 777.4 nm) [*Gustavsson et al.*, 2008; *Archer et al.*, 2011]. *Gustavsson et al.* [2008] showed that this was consistent with a reduction of precipitation at energies above 3–4 keV compared to the surrounding diffuse aurora, and it is inconsistent with a retarding electrical field.

Another black aurora generation mechanism was recently proposed by *Peticolas et al.* [2002], using quasisimultaneous observations between optical imaging from an aircraft and electron data from the FAST satellite as shown in Figure 2b. Black auroras were observed 40 s before and 1 min after the magnetic footprint of FAST passed the magnetic zenith, where it was measured by a narrow-field camera. The authors concluded that dropouts of downward electron energy flux were associated with black auroras and that pitch angle diffusion (causing the surrounding diffuse auroras) was suppressed at energies greater than 2 keV. The authors suggested that black auroras are generated by the small-scale suppression of pitch angle diffusion, since the surrounding diffuse electron precipitation was caused by pitch angle diffusion and not by the inverted-V-type potential structure. This idea was confirmed by simultaneous ground-based multiwavelength optical and incoherent scatter radar observations by *Gustavvson et al.* [2008].

Thus, there are possibly several types of black auroras, and their generation mechanisms have been discussed. However, the cause of black aurora, morphology, and drifting motion are still not understood well. An efficient approach to solve this problem is to perform completely simultaneous measurements of optical imaging and particles with high temporal and spatial resolutions using a single satellite.

3. SIMULTANEOUS IMAGE-PARTICLE OBSERVATION OF BLACK AURORA

Reimei is the first small (70 kg weight) scientific JAXA/ISAS satellite, which was launched in August 2005 into a sun-synchronous polar orbit in the 00:50–12:50 LT meridians at an altitude of ~640 km [*Saito et al.*, 2005]. Reimei is equipped with a three-channel monochromatic auroral camera (MAC) [*Sakanoi et al.*, 2003; *Obuchi et al.*, 2008, 2011] and

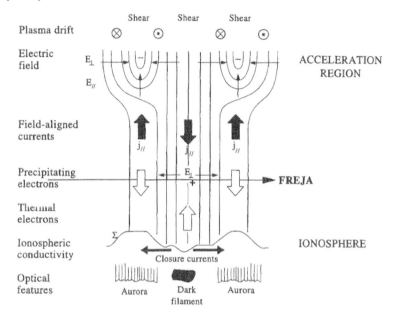

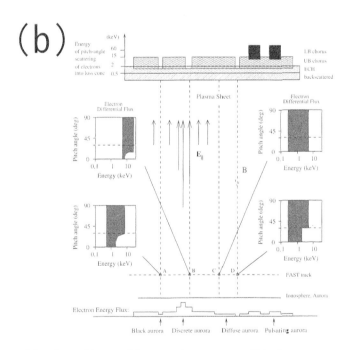

Figure 2. Conceptual drawings of the possible generation processes of black aurora proposed so far. (a) Divergent electric field structure that reduce the energy of downward going electrons [after *Marklund et al.*, 1997]. (b) The suppression of pitch angle diffusion [after *Peticolas et al.*, 2002]. See details given by *Marklund et al.* [1997] and *Peticolas et al.* [2002].

top hat electron- and ion-energy spectrum analyzers (ESA/ISA) [*Asamura et al.*, 2003, 2009]. Using the three-axis stabilized satellite attitude control system, we can point the MAC field of view (FOV) toward Reimei's magnetic footprints, thus satisfying full pitch angle coverage of ESA/ISA. These unique measurements provide us with opportunities to investigate the image/particle relationship with high temporal and kilometer spatial resolution [e.g., *Chaston et al.*, 2010].

Three channels of MAC are used to observe the auroral monochromatic images of N_2^+ 1NG (427.8 nm), O green line (557.7 nm), and N_2 1PG (670 nm) simultaneously. The exposure time is usually 60 ms, with a cycle of 120 ms. The imaging area of single frame is approximately ~70 km on a side, with a plate scale resolution of ~1.1 km, mapping at an altitude of 110 km when we point the FOV to the satellite's magnetic footprint. ESA and ISA measure auroral electrons and ions, respectively, in the energy range from 10 to 12000 eV q^{-1} with 32 logarithmic energy steps. The temporal resolution for obtaining the entire energy range is 40 ms, corresponding to 300 m mapping at an altitude of 110 km.

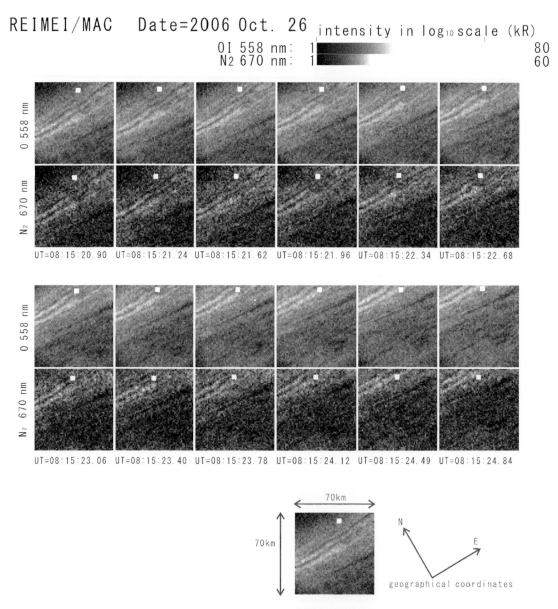

Figure 3. Example of image data of a black arc event at the wavelengths of O 557.7 nm and N_2^+ first positive band 670 nm obtained by monochromatic auroral camera on 26 October 2006. White squares in each image show the locations of the magnetic footprint of the satellite.

In this study, we inspected visually successive auroral images and movies using MAC data from November 2005 to October 2006. We found a number of black aurora events, determined based on the criteria of requiring small-scale (a few to several kilometers) rather stable black features seen in uniform, diffuse auroral emission and finally selected 13 events. Of these, nine are black arcs and four are black patches events. On these 13 events, the time and dates, locations, and types are presented in Table 1 of *Obuchi et al.* [2011].

3.1. Black Arc Event

A typical event with multiple black arcs event on 26 October 2006 is shown in Figure 3. In this figure, successive image data at the wavelengths of channels 2 (557.7 nm) and 3 (670 nm) are shown with an interval of 360 ms. The dominant motion of black structure from bottom to top is due to the satellite orbital motion from north to south in the nightside Northern Hemisphere. A white square in each image shows the location of the satellite's magnetic footprint, as derived from the International Geomagnetic Reference Field-10 model. In both the 557.7 and 670 nm data, the background diffuse aurora is seen with an intensity of ~3–5 kR. Some dark arc structures formed in the background diffuse aurora are identified slightly above the center of each image and extend from top right to bottom left. These dark structures are regarded as black arcs, which are approximately 3 to 5 km wide (the uncertainty is typically ~1 km owing to the motion of the satellite and temporal resolution of MAC). As seen in the 557.7 nm images from 08:15:22.5 and 08:15:23.7 UT, we found that the magnetic footprint of the satellite passed over five black arcs.

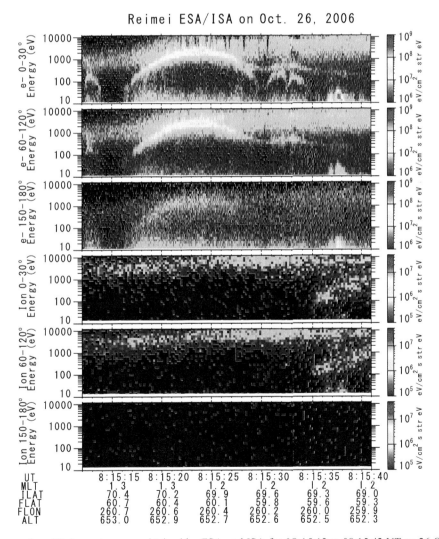

Figure 4. Energy–time (*E–t*) spectrograms obtained by ESA and ISA for 08:15:12 to 08:15:42 UT on 26 October 2006.

Figure 4 shows the energy-time (*E–t*) spectrograms of electrons and ions from ESA and ISA measurements covering the same time period, from 08:15:12 to 08:15:42 UT on 26 October 2006. In the downward electron data, we identify two types of large-scale electron precipitations: one type includes plasma sheet electrons that precipitate continuously from the beginning to the end, mostly in the keV energy range, except for some deficiencies or dropouts, and the other represents low-energy (~100–2000 eV) inverted-V electrons from 08:15:17 to 08:15:29 UT. The former plasma sheet electron precipitation is not shaped like an inverted V, indicating that it was probably precipitated by pitch angle scattering [e.g., *Villalón and Burke*, 1995]. The low-energy inverted-V electrons are less likely to contribute to the E region auroral emissions measured by MAC, since the E region aurora is generally produced by keV-range electrons. However, there low-energy inverted-V electrons are expected to contribute to the weak background auroral emission. In the ion data in the fourth and fifth panels of Figure 4, we see the Maxwellian ions in the keV range across the entire region, in addition to

~100 eV ions at ~08:15:35 UT. These characteristics support the conclusion that this region originates from the central plasma sheet [e.g., *Fukushishi et al.*, 1993].

It should be emphasized that there are small-scale deficiencies of downward electrons in energy ranges of several keV from 08:15:21 to 08:15:31 UT. The expanded *E–t* spectrogram is shown in Figure 5. In Figure 5, the period is the same as that in Figure 3, showing the black auroral images. The top panel shows the time variation of the auroral intensities of OI 557.7 nm and N_2 670 nm at the magnetic footprint. The intensities of 557.7 nm aurora are about 3–5 kR during this period, and the intensities are seen to decrease by 1.0–1.5 kR more than the surrounding diffuse aurora. On 670 nm aurora, the intensities are about 1–3 kR, and the decreases are almost similar to those of 557.7 nm aurora. Note that the ground-reflected light might be overlapped in 670 nm data since the filter bandwidth of 670 nm channel (35.95 nm) is much wider than that of 557.7 nm channel (1.68 nm) [*Sakanoi et al.*, 2003]. It is emphasized that the auroral intensities do not disappear even in the black structures. This is consistent with observations by

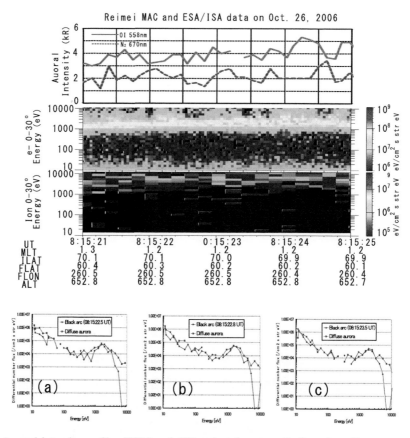

Figure 5. (top) Auroral intensity profiles (557.7 and 670 nm) at the magnetic footprints, *E–t* spectrograms (second to seventh panels), and (bottom) differential number flux profiles of downward electrons at times when black arcs appeared on 26 October 2006. We selected black arcs at (a) 08:15:22.5, (b) 08:15:22.8, (c) 08:15:23.5 UT, and the differential fluxes in the vicinity of black arc and surrounding diffuse aurora are plotted.

Gustavsson et al. [2008]. The bottom panel shows the typical three examples of local deficiencies in the downward electrons with energies greater than 3 keV, clearly corresponding to the positions of typical black arcs at 08:15:22.5, 08:15:22.8, 08:15:23.5, and 08:15:23.7 UT. However, the electron flux with energies less than 3 keV, which is associated with the inverted-V structure, does not show such deficiencies.

To investigate in detail the behavior of electron precipitation associated with black arcs, differential number fluxes of downward electrons for the selected times described above when the magnetic footprint passed over black arcs are shown in the bottom of Figure 5. The downward electron number flux is derived by averaging over pitch angles of less than 10°. In addition, the differential number flux of precipitating electrons for the surrounding diffuse aurora is also plotted in each panel, to compare it with that for black arc. In each panel, a flux peak is seen at around 2 keV, corresponding to the inverted-V structure identified in Figures 3 and 4. The differential number flux for black arcs with energies greater than 3 keV decreases quite rapidly compared with that for the surrounding diffuse aurora. On the other hand, the differential number flux for black arcs in the energy range below 3 keV shows no change compared with that for the surrounding diffuse aurora. This finding suggests that the local deficiency, or dropout, of the electron flux corresponding to black arcs is not caused by a field-aligned potential structure, such as a divergent electric field.

3.2. Statistical Characteristics of Black Aurora

Statistical characteristics, based on 13 events of black arcs and patches obtained by Reimei, are summarized as follows. The emission intensities of black arcs and patches are rather faint, on the order of 2–4 kR. Considering that inverted-V electron precipitation often overlaps with a deficiency in plasma sheet precipitation corresponding to black auroras, black arcs, and patches are not always completely black. Black arc widths are in the range of 3–10 km, while the size of black patches is 3–5 km. Black arcs and patches appear mostly in the invariant latitude range from 66° to 68°. This result is consistent with those in past studies [e.g., *Trondsen and Cogger*, 1997; *Kimball and Hallinan*, 1998b; *Peticolas et al.*, 2002]. Regarding the relationship between magnetic activity and the appearance of black arcs and patches, 11 of 19 events (including the data obtained from imaging only) were measured in magnetically quiet periods. Other events were measured for a variety of geomagnetic conditions. It is concluded that, in the Reimei observation data, no strong relationship is evident between magnetic activity and the appearance of black arcs and patches. Furthermore, past observations reported that black arcs and patches are sometimes seen associated with pulsating auroras [*Trondsen and Cogger*, 1997; *Kimball and Hallinan*,

1998b]. In Reimei observations, a pulsating aurora is sometimes observed together with black arcs or patches. Interestingly, pulsating auroras are always located equatorward of the black arcs or patches. The relationship between black aurora and pulsating aurora is beyond the scope of this study and will be investigated in the future.

4. SUMMARY AND DISCUSSION

We reported that the flux of precipitating electrons with energies greater than 2–7 keV was locally deficient at those times when the magnetic footprint of Reimei intersected black arcs or patches. Electrons with lower energies show no difference to electrons producing the diffuse aurora that surrounds black auroras. This finding suggests that the deficiency of electron flux with energies higher than 2–7 keV is not caused by deceleration due to any (divergent) electric field. In addition, the inverted-V–shaped electron precipitation with peak energy below ~3 keV often appears simultaneously and overlaps with the electron precipitation that causes diffuse auroras (6 of 13 events). It is not plausible that the small-scale divergent electric field overlaps with the convergent electric field that produces the inverted-V electrons.

Past satellite data have shown that strong electrostatic electron cyclotron harmonic waves occur in the plasma sheet and that they are able to scatter electrons in the energy range from a few hundred eV to a few keV into the loss cone [e.g., *Kennel et al.*, 1970; *Horne et al.*, 2003]. On the other hand, there are a number of studies that show whistler mode waves could scatter electrons in a range of energies from 1 to 10 keV into the loss cone by modeling and observations [e.g., *Inan et al.*, 1992; *Villalón and Burke*, 1995; *Meredith et al.*, 2001; *Miyoshi et al.*, 2010; *Nishiyama et al.*, 2011]. Our data clearly show that electron flux with energies greater than 2 keV is deficient in regions corresponding to black auroras, while electron flux corresponding to the surrounding diffuse aurora with energies below 2 keV shows no change. Therefore, pitch angle diffusion by whistler mode waves is likely suppressed in the source region with small-scale size of 3–10 km to produce black auroral features. This result is consistent with that of *Peticolas et al.* [2002] and *Gustavsson et al.* [2008]. Our results show that black arcs and patches with sizes of 3–10 km surrounded by uniform diffuse aurora are not necessarily generated by a divergent electric field. However, there should be other types of black aurora with different scales that are likely generated by different processes, including a divergent electric field.

Acknowledgments. We thank all members of the Reimei project team for their extensive efforts toward the success of the Reimei (INDEX) mission and for their support of Reimei operations.

REFERENCES

Archer, J., H. Dahlgren, N. Ivchenko, B. S. Lanchester, and G. T. Marklund (2011), Dynamics and characteristics of black aurora as observed by high-resolution ground-based imagers and radar, *Int. J. Remote Sens.*, *32*(11), 2973–2985.

Asamura, K., D. Tsujita, H. Tanaka, Y. Saito, T. Mukai, and M. Hirahara (2003), Auroral particle instrument onboard the INDEX satellite, *Adv. Space Res.*, *32*, 375–378.

Asamura, K., et al. (2009), Sheared flows and small-scale Alfvén wave generation in the auroral acceleration region, *Geophys. Res. Lett.*, *36*, L05105, doi:10.1029/2008GL036803.

Blixt, E. M., and M. J. Kosch (2004), Coordinated optical and EISCAT observations of black aurora, *Geophys. Res. Lett.*, *31*, L06813, doi:10.1029/2003GL019244.

Blixt, E. M., M. J. Kocsh, and J. Semeter (2005), Relative drift between black aurora and the ionospheric plasma, *Ann. Geophys.*, *23*, 1611–1621.

Chaston, C. C., K. Seki, T. Sakanoi, K. Asamura, and M. Hirahara (2010), Motion of aurorae, *Geophys. Res. Lett.*, *37*, L08104, doi:10.1029/2009GL042117.

Davis, T. N. (1978), Observed microstructure of auroral forms, *J. Geomagn. Geoelectr.*, *30*, 371–380.

Fukunishi, H., Y. Takahashi, T. Nagatsuma, T. Mukai, and S. Machida (1993), Latitudinal structures of nightside field-aligned currents and their relationships to the plasma sheet regions, *J. Geophys. Res.*, *98*(A7), 11,235–11,255.

Gustavsson, B., M. J. Kosch, A. Senior, A. J. Kavanagh, B. U. E. Brändström, and E. M. Blixt (2008), Combined EISCAT radar and optical multispectral and tomographic observations of black aurora, *J. Geophys. Res.*, *113*, A06308, doi:10.1029/2007JA012999.

Horne, R. B., R. M. Thorne, N. P. Meredith, and R. R. Anderson (2003), Diffuse auroral electron scattering by electron cyclotron harmonic and whistler mode waves during an isolated substorm, *J. Geophys. Res.*, *108*(A7), 1290, doi:10.1029/2002JA009736.

Inan, U. S., Y. T. Chiu, and G. T. Davidson (1992), Whistler-mode chorus and morningside aurorae, *Geophys. Res. Lett.*, *19*(7), 653–656.

Kennel, C. F., F. L. Scarf, R. W. Fredricks, J. H. McGehee, and F. V. Coroniti (1970), VLF electric field observations in the magnetosphere, *J. Geophys. Res.*, *75*(31), 6136–6152.

Kimball, J., and T. J. Hallinan (1998a), Observations of black auroral patches and of their relationship to other types of aurora, *J. Geophys. Res.*, *103*(A7), 14,671–14,682.

Kimball, J., and T. J. Hallinan (1998b), A morphological study of black vortex streets, *J. Geophys. Res.*, *103*(A7), 14,683–14,695.

Marklund, G., L. Blomberg, C. Fälthammar, and P. Lindqvist (1994), On intense diverging electric fields associated with black aurora, *Geophys. Res. Lett.*, *21*(17), 1859–1862.

Marklund, G., T. Karlsson, and J. Clemmons (1997), On low-altitude particle acceleration and intense electric fields and their relationship to black aurora, *J. Geophys. Res.*, *102*(A8), 17,509–17,522.

Meredith, N. P., R. B. Horne, and R. R. Anderson (2001), Substorm dependence of chorus amplitudes: Implications for the acceleration of electrons to relativistic energies, *J. Geophys. Res.*, *106*, 13,165–13,178, doi:10.1029/2000JA900156.

Miyoshi, Y., Y. Katoh, T. Nishiyama, T. Sakanoi, K. Asamura, and M. Hirahara (2010), Time of flight analysis of pulsating aurora electrons, considering wave-particle interactions with propagating whistler mode waves, *J. Geophys. Res.*, *115*, A10312, doi:10.1029/2009JA015127.

Nishiyama, T., T. Sakanoi, Y. Miyoshi, Y. Katoh, K. Asamura, S. Okano, and M. Hirahara (2011), The source region and its characteristic of pulsating aurora based on the Reimei observations, *J. Geophys. Res.*, *116*, A03226, doi:10.1029/2010JA015507.

Obuchi, Y., T. Sakanoi, A. Yamazaki, T. Ino, S. Okano, Y. Kasaba, M. Hirahara, Y. Kanai, and N. Takeyama (2008), Initial observations of auroras by the multi-spectral auroral camera on board the Reimei satellite, *Earth Planets Space*, *60*, 827–835.

Obuchi, Y., T. Sakanoi, K. Asamura, A. Yamazaki, Y. Kasaba, M. Hirahara, Y. Ebihara, and S. Okano (2011), Fine-scale dynamics of black auroras obtained from simultaneous imaging and particle observations with the Reimei satellite, *J. Geophys. Res.*, *116*, A00K07, doi:10.1029/2010JA016321.

Peticolas, L. M., T. J. Hallinan, H. C. Stenbaek-Nielsen, J. W. Bonnell, and C. W. Carlson (2002), A study of black aurora from aircraft-based optical observations and plasma measurements on FAST, *J. Geophys. Res.*, *107*(A8), 1217, doi:10.1029/2001JA 900157.

Saito, H., et al. (2005), An overview and initial in-orbit status of "INDEX" satellite, paper IAC-05-B5.6.B.05 presented at 56th International Astronautical Conference, Fukuoka, Japan.

Sakanoi, T., S. Okano, Y. Obuchi, T. Kobayashi, M. Ejiri, K. Asamura, and M. Hirahara (2003), Development of the multi-spectral auroral camera onboard the INDEX satellite, *Adv. Space Res.*, *32*, 379–384.

Schoute-Vanneck, H., M. W. J. Scourfield, and E. Nielsen (1990), Drifting black aurorae?, *J. Geophys. Res.*, *95*(A1), 241–246.

Trondsen, T. S., and L. L. Cogger (1997), High-resolution television observations of black aurora, *J. Geophys. Res.*, *102*(A1), 363–378.

Villalón, E., and W. J. Burke (1995), Pitch angle scattering of diffuse auroral electrons by whistler mode waves, *J. Geophys. Res.*, *100*(A10), 19,361–19,369.

K. Asamura and A. Yamazaki, Institute of Space and Astronautical Science, Japan Aerospace Exploration Agency, 3-1-1 Yoshinodai, Sagamihara, Kanagawa 229-8510, Japan.

Y. Ebihara, Research Institute for Sustainable Humanosphere, Kyoto University, Gokasho, Uji, Kyoto 611-0011, Japan.

M. Hirahara and Y. Miyoshi, Solar-Terrestrial Environment Laboratory, Nagoya University, Furocyo, Chikusa, Nagoya, Aichi 464-8601, Japan.

Y. Kasaba, T. Nishiyama, S. Okano, and T. Sakanoi, Graduate School of Science, Tohoku University, 6-3 Aramaki, Aoba, Sendai, Miyagi 980-8578, Japan. (tsakanoi@pparc.gp.tohoku.ac.jp)

Y. Obuchi, Genesia Corporation, 3-38-4-601 Shimorenjyaku, Mitaka, Tokyo 181-0013, Japan.

Two-Step Acceleration of Auroral Particles at Substorm Onset as Derived From Auroral Kilometric Radiation Spectra

Akira Morioka

Planetary Plasma and Atmospheric Research Center, Tohoku University, Sendai, Japan

Yoshizumi Miyoshi

Solar-Terrestrial Environment Laboratory, Nagoya University, Nagoya, Japan

The evolution of field-aligned auroral acceleration in the magnetosphere-ionosphere (M-I) coupling region at substorm onset is derived by means of a unique analysis of auroral kilometric radiation (AKR) frequency spectra. The derived AKR source dynamics showed two-step field-aligned acceleration at auroral substorm onset. The first step is intensification of low-altitude acceleration at an altitude of 3000–5000 km, which induces initial brightening. The second step is breakout of high-altitude acceleration (8000–12,000 km) above the preexisting low-altitude acceleration, which results in violent auroral breakup and poleward expansion. Low-altitude acceleration that is not followed by the breakout of high-altitude acceleration (one-step evolution) is a pseudosubstorm. The close relationship between development of low-altitude acceleration and gradually increasing field-aligned current (FAC) during initial brightening suggests that FAC plays an important role in linking substorm initiation in the plasma sheet with sudden auroral acceleration in the M-I coupling region.

1. INTRODUCTION

It is undoubtedly true that discrete aurora acceleration, which is the topic of this chapter, is the result of dynamic coupling of the magnetosphere to the auroral ionosphere (M-I coupling). The coupling system comprises many physical processes such as plasma transport, field-aligned current (FAC) generation, potential formation, particle acceleration, energy release and dissipation, wave generation and propagation, and their complicated interactions. The resultant auroral phenomena as well as auroral radio emissions are manifestations of the planetary M-I coupling system; they

Auroral Phenomenology and Magnetospheric Processes: Earth and Other Planets
Geophysical Monograph Series 197
10.1029/2011GM001160

are common to magnetized planets, such as Jupiter, Saturn, Uranus, and Neptune [e.g., *Zarka*, 1998]. We address, in this paper, the terrestrial auroral acceleration process in the M-I coupling region and its relationship to magnetospheric substorms, starting with two simple but fundamental questions.

The first question is related to the sudden acceleration process at substorm onset. It is well believed that inverted-V acceleration causes an auroral arc and that auroral breakup happens in or close to the inverted-V acceleration. However, it is not well understood what the difference between ordinary inverted-V acceleration and breakup acceleration is. Does the breakup acceleration develop from the inverted-V acceleration, or are they different acceleration processes? The answer to this fundamental question is not yet clear although there have been many particle and field observations in the auroral magnetosphere, which have greatly advanced our understanding of auroral particle acceleration, by using sounding rockets [e.g., *Evans*, 1974; *Arnoldy et al.*,

1974] and satellites [e.g., *Shelley et al.*, 1976; *Mizera and Fennell*, 1977] from the 1970s until recently [e.g., *McFadden et al.*, 1999; *Mozer and Hull*, 2001; *Ergun et al.*, 2004, and references therein].

The second question is related to the substorm onset process. Several models have been proposed for the major substorm onset process: the near-Earth neutral line (outside-in) model [e.g., *Baker et al.*, 1996; *Shiokawa et al.*, 1998; *Angelopoulos et al.*, 2008], in which magnetic reconnection in the midtail initiates a substorm, the current disruption (CD/inside-out) model [e.g., *Lui et al.*, 1992, 2008; *Erickson et al.*, 2000], in which cross-tail current disruption in the near-Earth tail region triggers a substorm, and the catapult model [*Machida et al.*, 2009], in which current-sheet relaxation in the plasma sheet causes a substorm. These models, anyway, implicitly assume the sudden acceleration of auroral particles in the M-I coupling region, which is not self-evident in the models. On the other hand, the sudden formation of field-aligned acceleration in the M-I coupling region is essential to completing the substorm onset process. Here we have a question. What is the relationship between substorm initiation in the plasma sheet and sudden acceleration in the M-I coupling region? This question is also basic but not yet answered convincingly.

In this paper, we first describe in detail the evolution of auroral acceleration in the M-I coupling region at substorm onset by means of a unique analysis of auroral kilometric radiation (AKR) frequency spectra. Then, we discuss the relationship between magnetospheric substorm onset and sudden auroral acceleration in the M-I coupling region.

AKR has been intensively investigated since the first comprehensive study of its phenomenological properties and of its close relationship to substorms by *Gurnett* [1974]. Subsequent detailed studies revealed that AKR is generated by auroral electron beams in the upward field-aligned electric field [*Benson et al.*, 1980; *Bahnsen et al.*, 1989; *Ungstrup et al.*, 1990]. A close association between AKR and discrete auroral arcs was clearly shown by *Gurnett* [1974] and *Kurth et al.* [1975]. The dynamic features of AKR spectra showed fairly good correlation with the development of substorms [*Kaiser and Alexander*, 1977]. *Fairfield et al.* [1999] reported that earthward flow bursts associated with substorm onsets were accompanied by AKR.

AKR shows abrupt frequency expansion into both higher and lower frequencies at substorm onset [*Kaiser and Alexander*, 1977; *Morioka et al.*, 1981; *Liou et al.*, 2000; *de Feraudy et al.*, 2001; *Hanasz et al.*, 2001]. *Hanasz et al.* [2001] and *de Feraudy et al.* [2001] considered that this frequency expansion indicates the AKR source development along the auroral magnetic field line and estimated the expansion velocity of AKR source region from the frequency

drift rate. The low-frequency component of AKR (typically tens of kHz) has been studied in terms of isolated AKR [*Steinberg et al.*, 1988], LF burst [*Kaiser et al.*, 1996; *Desch and Farrell*, 2000], and LF-AKR [*Olsson et al.*, 2004]. It has generally been agreed that the low-frequency component of AKR is accompanied by substorm onset almost without exception [e.g., *Anderson et al.*, 1997]. *Liou et al.* [2000] suggested that the expanded frequency range of AKR corresponds to a source altitude of ~2100 to 12,000 km, which is in good agreement with the distribution of the quasistatic electric field and plasma cavity along auroral field lines. *Janhunen et al.* [2004] and *Olsson et al.* [2004] have suggested that dot-shaped LF-AKR ("dot-AKR") [*de Feraudy et al.*, 2001] is produced by Alfvénic wave acceleration in the preexisting low-altitude cavity region.

These observations and accompanying theoretical works have led to the understanding that AKR emanates from accelerated auroral electron beams through wave-particle interactions at a frequency close to the local electron cyclotron frequency [e.g., *Pritchett et al.*, 2002, and references therein] at substorm. The source region of AKR was confirmed to be almost identical to the auroral particle acceleration region by direct observations using FAST [*Ergun et al.*, 1998, 2000; *Pottelette et al.*, 2001]. Thus, AKR is the only auroral phenomenon that provides information on the vertical structure of the acceleration region: AKR frequency spectra enable remote sensing of a field-aligned acceleration region and its dynamics [e.g., *Morioka et al.*, 2007, 2008] on the basis of the observed evidence that AKR is radiated (1) by accelerated electrons, (2) at the local electron cyclotron frequency, (3) along the auroral field lines, and (4) within the acceleration region [*Ergun et al.*, 1998; *Strangeway et al.*, 2001].

2. REMOTE DIAGNOSIS OF AURORAL ACCELERATION USING AKR OBSERVATIONS

Figure 1a shows an AKR source altitude versus time (*a-t*) diagram converted from the AKR frequency spectra in the frequency range from 50 to 800 kHz observed by the Geotail Plasma Wave Instrument (PWI) on 31 October 1997. The source altitude (left ordinate) was obtained by considering that the AKR emanates at the local electron cyclotron frequency, f_c, along the auroral field line of $L = 7$. The right ordinate shows the corresponding AKR frequency. Since the altitude of a constant f_c is not sensitive to invariant latitude in the polar region, the estimated AKR altitude is roughly valid even when the auroral field line of interest is not exactly $L = 7$. Thus, we can derive the structure and dynamics of the AKR source/acceleration region from the AKR *a-t* diagram. In Figure 1a, a continuous and rather stable AKR source is

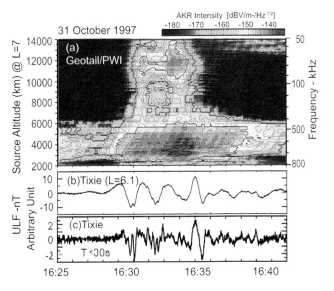

Figure 1. Auroral kilometric radiation (AKR) breakup and substorm. (a) The *a-t* diagram of AKR source along *L* = 7 field line on 31 October 1997 observed by Geotail/Plasma Wave Instrument (PWI). (b) Geomagnetic pulsation at Tixie (*L* = 6.1). (c) Short-period (<30 s) component of pulsation in Figure 1b.

evident in the altitude range from 3000 to 5000 km. We call this "low-altitude AKR" radiated in the low-altitude acceleration region [*Morioka et al.*, 2007]. When a substorm occurred as manifested by Pi2 pulsation observed at auroral station Tixie (*L* = 6.1) at 16:28 UT (Figure 1b), a new AKR source suddenly appeared at a higher altitude of 8000–14000 km, above the preexisting low-altitude AKR. We call this "high-altitude AKR" or "AKR breakup" generated in the high-altitude acceleration region. The AKR breakup showed very explosive growth, more than 1000 times within 20 s, indicating a rapid breakout of high-altitude acceleration at substorm onset [*Morioka et al.*, 2007].

3. TWO-STEP EVOLUTION OF AURORAL ACCELERATION AT SUBSTORM ONSET

Detailed observations of AKR source dynamics have revealed a two-step evolution of auroral acceleration at substorm onset. Figure 2 shows, from top to bottom, auroral UV images from the Polar ultraviolet imager (UVI), an *a-t* diagram of the AKR source from the Polar PWI for 15 min, and low-latitude pulsations at Kakioka station (raw waveform and short-period component). The AKR source dynamics (Figure 2b) showed initial enhancement of the low-altitude AKR from 15:32 UT and then breakout of high-altitude AKR (AKR breakup) at 15:36 UT. The enhancement of the low-altitude AKR accompanied initial auroral brightening

(Figure 2a, image 2) and long-period irregular pulsation (Figure 2c, blue rectangle), i.e., Pi 2 pulsation. The breakout of the high-altitude AKR accompanied auroral breakup with poleward expansion (Figure 2a, image 4) and short-period irregular pulsation (Figure 2d, yellow rectangle), i.e., Pi 1B pulsation [*Kataoka et al.*, 2009]. The time difference between the enhancements of low- and high-altitude AKR was about 4 min in this case.

This two-step evolution of auroral acceleration can also be identified from the detailed observation shown in Figure 1, where short-period pulsation (Figure 1c) appeared to be superposed on the preexisting long-period pulsation just as high-altitude AKR broke out while the enhancement of low-altitude AKR with Pi 2 pulsation had started 2 min earlier. Other examples have been shown [*Morioka et al.*, 2008, 2010].

Thus, the two-step evolution of substorm onset has been revealed from the AKR source dynamics in the M-I coupling region. The first step is the enhancement of low-altitude acceleration, which corresponds to the auroral initial brightening with Pi 2 pulsation, and the second step is the breakout of high-altitude acceleration, which corresponds to the auroral breakup and poleward expansion.

4. PSEUDOBREAKUPS AND FULL SUBSTORMS AND THEIR RELATIONSHIP TO AURORAL ACCELERATION

Figure 3b shows the keogram at Gillam (*L* = 6.4) for 2 h, showing the growth phase arc from 04:00 to 05:30 UT and the auroral breakup with poleward expansion starting at 05:30 UT. During the growth phase, at least three arc enhancements, i.e., pseudobreakups, were observed. Corresponding to each of these pseudobreakups, low-altitude AKR showed coincident enhancements (Figure 3a), as shown by the white arrows. This relationship suggests that pseudobreakup is related to activation of low-altitude acceleration. The auroral breakup and subsequent poleward expansion at 05:30 UT accompanied AKR breakup, as shown by the yellow arrow. This suggests that a substorm expansion results from the breakout of high-altitude acceleration.

Here we define the technical terms "full substorm" and "pseudosubstorm" in accordance with the work of *Ohtani et al.* [1993]. Auroral substorms that begin with initial brightening and are accompanied by auroral breakup with longitudinal and poleward expansion are designated "full substorms," and those that stay in an initially activated area followed by no global developments are called "pseudosubstorms." According to this definition, the present observation indicates that AKR breakup (intensification of both low- and high-altitude AKR) always accompanies a full substorm and that intensification of

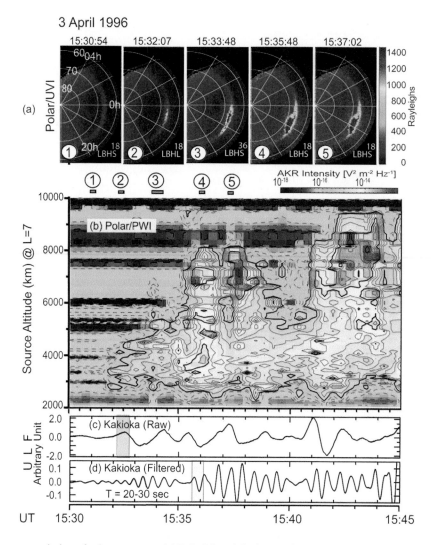

Figure 2. Two-step evolution of substorm onset. (a) Polar/ultraviolet imager images. Each frame represents northern dark hemisphere from 60° to 90° magnetic latitude. (b) The a-t diagram of AKR source observed by Polar/PWI. (c) Geomagnetic pulsation at Kakioka ($L = 1.3$). (d) Filtered component ($T = 20$–30 s) of pulsation at Kakioka. The number in each image of Figure 2a corresponds to that above thin blue lines on a-t diagram. The length of the thin blue lines indicates exposure time of corresponding UV images.

low-altitude AKR alone always accompanies a pseudosubstorm. To investigate this indication further, AKR events (Geotail/PWI observations) associated with isolated auroral intensification (Polar/UVI observations) were systematically surveyed for 3 months (December 1996 to February 1997). During this period, 13 AKR-breakup events and 7 low-altitude AKR events occurred. The results showed that 12 of the 13 AKR breakups (92%) were accompanied by full substorms and that all 7 (100%) low-altitude AKRs without high-altitude AKR were accompanied by pseudobreakups. This confirms that a full substorm can be distinguished from a pseudosubstorm by the breakout of high-altitude acceleration in the M-I

coupling region and that low-altitude acceleration is a common feature of both pseudosubstorms and full substorms.

5. FAC EVOLUTION AND TWO-STEP SUBSTORM ONSET

The evolution of FAC around auroral breakup was examined using ULF data. Figure 4 shows, from top to bottom, an a-t diagram of AKR, ordinary magnetograms at the premidnight auroral conjugate pair stations of Tjornes ($L = 6.4$, midnight magnetic local time (MMLT) = 23:45 UT) in Iceland and Syowa ($L = 6.0$, MMLT = 0:06 UT) in Antarctica,

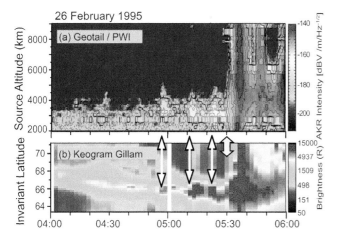

Figure 3. AKR source dynamics and pseudosubstorms and full substorms. (a) The *a-t* diagram of AKR source observed by Geotail/PWI. (b) The keogram of 5577 Å at Gillam (*L* = 6.5). White arrows indicate coincidence between low-altitude AKR sources and auroral pseudobreakups. Yellow arrow indicates breakup.

and their *dH/dt* (search coil) data. A substorm breakup can be identified from the AKR breakup at around 21:45 UT. Again, the breakup was preceded by gradually increasing low-altitude AKR; the first step of the substorm had started about 21:42 UT. In the ULF data for Tjornes and Syowa, both *dH/dt* signals showed negative excursion, as shown by the gray bars, almost simultaneously with the gradual AKR intensification (blue rectangle). This monotonic decrease in *dH/dt* before the breakup can be attributed to the exponential increase in the westward electrojet in both the northern and southern auroral ionospheres, which is connected with the exponential increase in upward FAC. Thus, we suggest that FAC is induced in the M-I coupling system and begins to increase concurrently with the enhancement of low-altitude acceleration during the first step of substorms (initial brightening). This close relationship between the development of low-altitude acceleration and the gradually increasing FAC during initial brightening suggests that FAC plays an important role in linking the substorm initiation in the plasma sheet with the sudden auroral acceleration in the M-I coupling region.

The presented monotonic decrease in *dH/dt* before breakup, which was not familiar in previous studies of substorm-related pulsations, is not unusual but rather commonly observed in search coil ULF data in the premidnight auroral latitudes [*Morioka et al.*, 2008].

After the substorm breakup at 21:45 UT in Figure 4 (the second step of the substorm), the *dH/dt* at the auroral stations showed a very irregular and large amplitude waveform, suggesting a sudden and violent increase in FAC (yellow

rectangle in Figure 4). This may be caused by a rush current into the ionosphere corresponding to the so-called current disruption in the inner plasma sheet region [*Lui et al.*, 1992].

It is interesting to note, in connection with FAC evolution, that the high-altitude acceleration at substorm breakup shown in Figure 4a seems to be induced (triggered) during the course of the development of low-altitude acceleration due to some causal relationship. In other words, high-altitude acceleration is concomitant with low-altitude acceleration. To investigate this, we performed superposed epoch analysis on the development of both high- and low-altitude acceleration. Data from

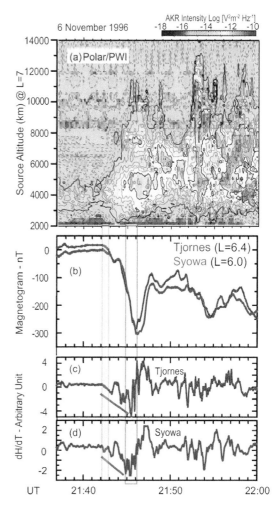

Figure 4. Two-step substorm onset and geomagnetic variation. (a) The *a-t* diagram of AKR source observed by Polar/PWI. (b) Ordinary magnetograms at Tjornes (*L* = 6.4) and geomagnetic conjugate station of Tjornes, Syowa (*L* = 6.0). (c) The *dH/dt* pulsation at Tjornes. (d) The *dH/dt* pulsation at Syowa. Gray bars indicate negative slope of *dH/dt*. Blue and yellow vertical rectangles indicate start of gradual increase in low-altitude AKR and substorm breakup, respectively.

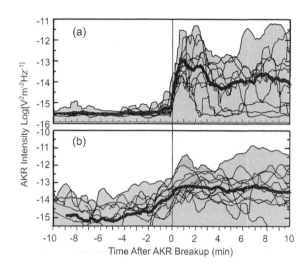

Figure 5. Superposed epoch analysis of 12 AKR evolutions. (a) Time profile of high-altitude AKR intensity. (b) Time profile of low-altitude AKR intensity. Epoch time ($t = 0$) is taken to be a sudden rise of high-altitude AKR (AKR breakup). Bold lines are logarithmic mean-value trace of power profiles.

acceleration in the altitude range of 8000–16,000 km. Pseudobreakup/initial brightening is the result of the enhanced low-altitude acceleration. A full substorm is composed of a pseudobreakup and subsequent breakout of high-altitude acceleration. That is the two-step evolution of substorm onset.

Low-altitude acceleration accompanies Pi 2 pulsation and increasing FAC as derived from dH/dt pulsation. Breakout of high-altitude acceleration accompanies large-amplitude irregular ULF disturbances, which are considered to be a manifestation of current disruption. The close relationship between gradually increasing FAC and development of low-altitude acceleration derived from ULF and AKR observations suggests that FAC plays an important role in linking substorm initiation in the plasma sheet with the sudden auroral acceleration in the M-I coupling region. *Morioka et al.* [2010] presented a scenario for the generation of low- and high-altitude acceleration regions in relation to the FAC development at substorm onset.

It is interesting to note that recent observations of bursty radio emissions from Jupiter (hectometric radiation) and Saturn

2 months of AKR observations by Polar/PWI (from 1 December 1996 to 31 January 1997) and ground Pi 2 observations were used to sample isolated substorms, and 12 substorms were obtained. The results are plotted in Figure 5, in which the upper and lower panels correspond to the superposed average-power profiles of high- and low-altitude AKR, respectively. The epoch time ($t = 0$) was taken to be the time when high-altitude AKR showed a sudden increase in intensity (AKR breakup). All the breakouts of high-altitude AKR were obviously preceded in 0.5–4 min by a gradual increase in low-altitude AKR. These statistical results support our assertion that the intensification of low-altitude acceleration plays an important role in the breakout in high-altitude acceleration.

6. SUMMARY AND CONCLUSION

AKR frequency spectra were used to derive the dynamics of field-aligned acceleration regions, on the basis of observational evidence that AKR is radiated (1) by accelerated electrons, (2) at the local electron cyclotron frequency, (3) along the auroral field line, and (4) within the acceleration region. The obtained *a-t* diagrams enable remote sensing of the vertical evolution of auroral acceleration regions at substorm, which had not been previously achieved though in situ particle and field observations. The dynamic features of auroral acceleration and their relationship to substorms are summarized in Figure 6. The auroral acceleration region consists of quasi-steady low-altitude acceleration in the altitude range of 3000–5000 km and transient high-altitude

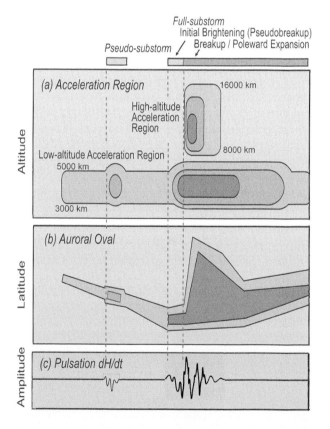

Figure 6. Schematic illustration of substorm-time evolution: (a) acceleration region, (b) auroral oval, and (c) high-latitude pulsation.

(Saturn kilometric radiation) suggest the formation of a high-altitude acceleration region and its relationship to magnetic reconnection in the magnetotail [*Louarn et al.*, 1998; *Jackman et al.*, 2009], which are analogous to the terrestrial substorms presented in this paper. Interactive studies between the Earth and other planets will improve our understanding of auroral phenomena and the plasma process in the planetary magnetosphere-ionosphere system.

Acknowledgments. We thank Y. Kasaba and the Geotail PWI team (Principal Investigator: H. Kojima) for providing the AKR data and the LEP team (Principal Investigator: Y. Saito) for providing flow burst data. We also thank E. Donovan for providing the data from the NORSTAR MSP. We are grateful to J. D. Menietti and the Polar PWI team (Principal Investigator: D. A. Gurnett) for providing the AKR data and Y. Miyashita, K. Liou, and the UVI team (Principal Investigator: G. K. Parks) for providing the UV aurora data. The authors wish to acknowledge the PI of CPMN project, K. Yumoto, SERC, Kyushu University and S. I. Solovyev, IKFIA, Russian Academy of Science, for making their Tixie data available for this study. The geomagnetic data from Syowa-Tjornes conjugate-pair stations were kindly provided by the Auroral Data Center at the National Institute of Polar Research, Japan. The 1 s resolution geomagnetic data were observed at the Kakioka Magnetic Observatory of the Japan Meteorological Agency and provided through the WDC-2 for Geomagnetism at Kyoto University, Japan. This work was supported by a Grant-in-Aid for Scientific Research (22340145) from the Ministry of Education, Culture, Sports, Science and Technology (MEXT), Japan.

REFERENCES

Anderson, R. R., et al. (1997), Observation of low frequency terrestrial type III bursts by Geotail and WIND and their association with isolated geomagnetic disturbances detected ground and space-bone instruments, in *Planetary Radio Emissions IV, Proc. Graz Conf.*, edited by H. O. Rucker, S. J. Bauer, and A. Lecacheux, pp. 241–250, Austrian Acad. of Sci. Press, Vienna.

Angelopoulos, V., et al. (2008), Tail reconnection triggering substorm onset, *Science*, *321*, 931–935.

Arnoldy, R. L., P. B. Lewis, and P. O. Isaacson (1974), Field-aligned auroral electron fluxes, *J. Geophys. Res.*, *79*(28), 4208–4221.

Bahnsen, A., B. M. Pedersen, M. Jespersen, E. Ungstrup, L. Eliasson, J. S. Murphree, R. D. Elphinstone, L. Blomberg, G. Holmgren, and L. J. Zanetti (1989), Viking observations at the source region of auroral kilometric radiation, *J. Geophys. Res.*, *94*(A6), 6643–6654.

Baker, D. N., T. I. Pulkkinen, V. Angelopoulos, W. Baumjohann, and R. L. McPherron (1996), Neutral line model of substorms: Past results and present view, *J. Geophys. Res.*, *101*(A6), 12,975–13,010.

Benson, R. F., W. Calvert, and D. M. Klumpar (1980), Simultaneous wave and particle observations in the auroral kilometric radiation source region, *Geophys. Res. Lett.*, *7*(11), 959–962.

de Feraudy, H., J. Hanasz, R. Schreiber, G. Parks, M. Brittnacher, S. Perraut, J. A. Sauvaud, F. Lefeuvre, and M. Mogilevsky (2001),

AKR bursts and substorm field line excitation, *Phys. Chem. Earth, Part C*, *26*, 151–159.

Desch, M. D., and W. M. Farrell (2000), Terrestrial LF bursts: Escape paths and wave intensification, in *Radio Astronomy at Long Wavelengths, Geophys. Monogr. Ser.*, vol. 119, edited by R. G. Stone et al., pp. 205–211, AGU, Washington, D. C., doi:10.1029/GM119p0205.

Erickson, G. M., N. C. Maynard, W. J. Burke, G. R. Wilson, and M. A. Heinemann (2000), Electromagnetics of substorm onsets in the near-geosynchronous plasma sheet, *J. Geophys. Res.*, *105* (A11), 25,265–25,290.

Ergun, R. E., et al. (1998), FAST satellite wave observations in the AKR source region, *Geophys. Res. Lett.*, *25*(12), 2061–2064.

Ergun, R. E., C. W. Carlson, J. P. McFadden, G. T. Delory, R. J. Strangeway, and P. L. Pritchett (2000), Electron-cyclotron maser driven by charged-particle acceleration from quasi-static field-aligned potentials, *Astrophys. J.*, *538*, 456–466.

Ergun, R. E., L. Andersson, D. Main, Y.-J. Su, D. L. Newman, M. V. Goldman, C. W. Carlson, A. J. Hull, J. P. McFadden, and F. S. Mozer (2004), Auroral particle acceleration by strong double layers: The upward current region, *J. Geophys. Res.*, *109*, A12220, doi:10.1029/2004JA010545.

Evans, D. S. (1974), Precipitating electron fluxes formed by a magnetic field aligned potential difference, *J. Geophys. Res.*, *79* (19), 2853–2858.

Fairfield, D. H., et al. (1999), Earthward flow bursts in the inner magnetotail and their relation to auroral brightenings, AKR intensifications, geosynchronous particle injections and magnetic activity, *J. Geophys. Res.*, *104*(A1), 355–370.

Gurnett, D. A. (1974), The Earth as a radio source: Terrestrial kilometric radiation, *J. Geophys. Res.*, *79*(28), 4227–4238.

Hanasz, J., H. de Feraudy, R. Schreiber, G. Parks, M. Brittnacher, M. M. Mogilevsky, and T. V. Romantsova (2001), Wideband bursts of auroral kilometric radiation and their association with UV auroral bulges, *J. Geophys. Res.*, *106*, 3859–3871.

Jackman, C. M., L. Lamy, M. P. Freeman, P. Zarka, B. Cecconi, W. S. Kurth, S. W. H. Cowley, and M. K. Dougherty (2009), On the character and distribution of lower-frequency radio emissions at Saturn and their relationship to substorm-like events, *J. Geophys. Res.*, *114*, A08211, doi:10.1029/2008JA013997.

Janhunen, P., A. Ollson, J. Hanasz, C. T. Russle, H. Laakso, and J. C. Samson (2004), Different Alfvén wave acceleration processes of electrons in substorms at ~4–5 R_E and 2–3 R_E radial distance, *Ann. Geo.*, *22*, 2213–2227.

Kaiser, M. L., and J. K. Alexander (1977), Relationship between auroral substorms and the occurrence of terrestrial kilometric radiation, *J. Geophys. Res.*, *82*(32), 5283–5286.

Kaiser, M. L., M. D. Desch, W. M. Farrell, J.-L. Steinberg, and M. J. Reiner (1996), LF band terrestrial radio bursts observed by WIND/WAVES, *Geophys. Res. Lett.*, *23*(10), 1283–1286.

Kataoka, R., Y. Miyoshi, and A. Morioka (2009), Hilbert-Huang Transform of geomagnetic pulsations at auroral expansion onset, *J. Geophys. Res.*, *114*, A09202, doi:10.1029/2009JA014214.

Kurth, W. S., M. M. Baumback, and D. A. Gurnett (1975), Direction-finding measurements of auroral kilometric radiation, *J. Geophys. Res.*, *80*(19), 2764–2770.

Liou, K., C.-I. Meng, A. T. Y. Lui, P. T. Newell, and R. R. Anderson (2000), Auroral kilometric radiation at substorm onset, *J. Geophys. Res.*, *105*(A11), 25,325–25,331.

Louarn, P., A. Roux, S. Perraut, W. Kurth, and D. Gurnett (1998), A study of the large-scale dynamics of the Jovian magnetosphere using the Galileo Plasma Wave Experiment, *Geophys. Res. Lett.*, *25*(15), 2905–2908.

Lui, A. T. Y., R. E. Lopez, B. J. Anderson, K. Takahashi, L. J. Zanetti, R. W. McEntire, T. A. Potemra, D. M. Klumpar, E. M. Greene, and R. Strangeway (1992), Current disruptions in the near-Earth neutral sheet region, *J. Geophys. Res.*, *97*(A2), 1461–1480.

Lui, A. T. Y., et al. (2008), Determination of the substorm initiation region from a major conjunction interval of THEMIS satellites, *J. Geophys. Res.*, *113*, A00C04, doi:10.1029/2008JA013424.

Machida, S., Y. Miyashita, A. Ieda, M. Nosé, D. Nagata, K. Liou, T. Obara, A. Nishida, Y. Saito, and T. Mukai (2009), Statistical visualization of the Earth's magnetotail based on Geotail data and the implied substorm model, *Ann. Geophys.*, *27*, 1035–1046, doi:10.5194/angeo-27-1035-2009.

McFadden, J. P., C. W. Carlson, R. E. Ergun, D. M. Klumpar, and E. Moebius (1999), Ion and electron characteristics in auroral density cavities associated with ion beams: No evidence for cold ionospheric plasma, *J. Geophys. Res.*, *104*(A7), 14,671–14,682.

Mizera, P. F., and J. F. Fennell (1977), Signatures of electric fields from high and low altitude farticles distributions, *Geophys. Res. Lett.*, *4*(8), 311–314.

Morioka, A., H. Oya, and S. Miyatake (1981), Terrestrial kilometric radiation observed by satellite Jikiken (Exos-B), *J. Geomagn. Geoelectr.*, *33*, 37–62.

Morioka, A., Y. Miyoshi, F. Tsuchiya, H. Misawa, T. Sakanoi, K. Yumoto, R. R. Anderson, J. D. Menietti, and E. F. Donovan (2007), Dual structure of auroral acceleration regions at substorm onsets as derived from auroral kilometric radiation spectra, *J. Geophys. Res.*, *112*, A06245, doi:10.1029/2006JA012186.

Morioka, A., et al. (2008), AKR breakup and auroral particle acceleration at substorm onset, *J. Geophys. Res.*, *113*, A09213, doi:10.1029/2008JA013322.

Morioka, A., et al. (2010), Two-step evolution of auroral acceleration at substorm onset, *J. Geophys. Res.*, *115*, A11213, doi:10.1029/2010JA015361.

Mozer, F. S., and A. Hull (2001), Origin and geometry of upward parallel electric fields in the auroral acceleration region, *J. Geophys. Res.*, *106*, 5763–5778.

Ohtani, S., et al. (1993), A multisatellite study of a pseudo-substorm onset in the near-Earth magnetotail, *J. Geophys. Res.*, *98*(A11), 19,355–19,367.

Olsson, A., P. Janhunen, J. Hanasz, M. Mogilevsky, S. Perraut, and J. D. Menietti (2004), Observational study of generation conditions of substorm-associated low-frequency AKR emissions, *Ann. Geophys.*, *22*, 3571–3582.

Pottelette, R., R. A. Treumann, and M. Berthomier (2001), Auroral plasma turbulence and the cause of auroral kilometric radiation fine structure, *J. Geophys. Res.*, *106*, 8465–8476.

Pritchett, P. L., R. J. Strangeway, R. E. Ergun, and C. W. Carlson (2002), Generation and propagation of cyclotron maser emissions in the finite auroral kilometric radiation source cavity, *J. Geophys. Res.*, *107*(A12), 1437, doi:10.1029/2002JA009403.

Shelley, E. G., R. D. Sharp, and R. G. Johnson (1976), Satellite observations of an ionospheric acceleration mechanism, *Geophys. Res. Lett.*, *3*(11), 654–656.

Shiokawa, K., et al. (1998), High-speed ion flow, substorm current wedge, and multiple Pi 2 pulsations, *J. Geophys. Res.*, *103*(A3), 4491–4507.

Strangeway, R. J., R. E. Ergun, C. W. Carlson, J. P. McFadden, G. T. Delory, and P. L. Pritchett (2001), Accelerated electrons as the source of auroral kilometric radiation, *Phys. Chem. Earth, Part C*, *26*, 145–149.

Steinberg, J.-L., C. Lacombe, and S. Hoang (1988), A new component of terrestrial radio emission observed from ISEE-3 and ISEE-1 in the solar wind, *Geophys. Res. Lett.*, *15*(2), 176–179.

Ungstrup, E., A. Bahnsen, H. K. Wong, M. André, and L. Matson (1990), Energy source and generation mechanism for auroral kilometric radiation, *J. Geophys. Res.*, *95*(A5), 5973–5981.

Zarka, P. (1998), Auroral radio emissions at the outer planets: Observations and theories, *J. Geophys. Res.*, *103*(E9), 20,159–20,194.

Y. Miyoshi, Solar-Terrestrial Environment Laboratory, Nagoya University, Nagoya 464-8601, Japan.

A. Morioka, Planetary Plasma and Atmospheric Research Center, Tohoku University, Sendai, Miyagi 980-8587, Japan. (morioka@pparc.geophys.tohoku.ac.jp)

Auroral Ion Precipitation and Acceleration at the Outer Planets

T. E. Cravens and N. Ozak

Department of Physics and Astronomy, University of Kansas, Lawrence, Kansas, USA

Jupiter's aurora is a powerful source of radio, IR, visible, UV, and X-ray emission. An UV aurora has also been observed on Saturn. Jovian X-ray emissions with a total power of about 1 GW were observed by the Einstein Observatory, the Roentgen satellite (ROSAT), Chandra X-ray Observatory, and XMM-Newton Observatory. Most of the X-ray power is in soft X-ray emission from the polar caps, but some harder X-ray emission from the main auroral oval was also observed and is probably due to electron bremsstrahlung emission. X-ray emission provides the main evidence that auroral energetic ion acceleration and precipitation is taking place on Jupiter. Two possible mechanisms have been suggested for the soft X-ray emission: (1) cusp entry and precipitation of solar wind heavy ions and (2) acceleration and subsequent precipitation of ions from the outer magnetosphere. Models of ion precipitation and observations of spectra containing oxygen and sulfur lines from high-charge state ions support the second mechanism. Acceleration by field-aligned potentials of about 10 MV is required for the X-ray production. Significant downward parallel currents linking the polar cap to the magnetopause region are associated with this ion precipitation.

1. INTRODUCTION

Auroral emission has been observed from many planets and satellites (e.g., Titan) [*Cravens et al.*, 2005] in the solar system and is due mainly to charged particle precipitation into the upper atmospheres of these bodies from either their magnetospheres or the solar wind [cf. *Galand and Chakrabarti*, 2002]. Auroral emission on Earth is mainly due to energetic electron precipitation, including discrete emissions associated with substorm activity, and is caused by electron acceleration by field-aligned electrical potentials and diffuse aurora in the polar cap (see the many chapters in this monograph). The topic of the current chapter is auroral ion precipitation on Jupiter and Saturn, which produces (or does not for Saturn) X-ray emission.

The most intense emission in Jupiter's aurora is in the UV part of the spectrum due to the Lyman and Werner bands of H_2 and Lyman alpha emission from atomic hydrogen with a total power of 10^{13}–10^{14} W [cf. *Clarke et al.*, 1998; *Grodent et al.*, 2001]. The main UV auroral oval is located near 65°–70° latitude, as shown in Figure 1 [*Clarke et al.*, 1998] and generates the highest intensities. Polar cap UV emissions and UV spots associated with the magnetic footprints of Io and the other Galilean satellites are also evident in this, and similar, images. Our current understanding is that the main oval emission is caused by upward field-aligned currents linking with the middle magnetosphere (radial distances averaging about $r \approx 30\ R_J$). These currents are associated with co-rotation lag of the outwardly diffusing plasma, which has been loaded down with sulfur and oxygen ions from the Io plasma torus [cf. *Hill*, 2001; *Cowley and Bunce*, 2001]. The UV spectra suggest that the electrons are accelerated to energies of 50–100 keV [e.g., *Clark et al.*, 2004]. The polar cap UV emissions are not well understood, unlike the main oval emissions, and the polar emissions can be very time variable and flare-like [*Waite et al.*, 2001].

Auroral Phenomenology and Magnetospheric Processes: Earth and Other Planets
Geophysical Monograph Series 197
10.1029/2011GM001159

Figure 1. UV image of the Jovian northern hemisphere aurora with different regions indicated. Image courtesy of J. T. Clarke.

2. BRIEF HISTORY OF EARLIER WORK ON X-RAY EMISSION ON JUPITER

X-ray emission was discovered on Jupiter in 1979 by the Einstein Observatory [*Metzger et al.*, 1983; see *Bhardwaj*, 2006], and about a decade later, many important observations were subsequently made with the Roentgen satellite (ROSAT) [e.g., *Waite et al.*, 1994]. The sulfur and oxygen fluxes measured by Voyager in the middle magnetosphere led *Gehrels and Stone* [1983] to suggest that such ions could be precipitating into the auroral atmosphere. *Horanyi et al.* [1988] modeled energetic (up to MeV) oxygen ion precipitation and the aeronomical effects (i.e., ionization rates, UV emission rates...) of this ion aurora.

The spectral resolution of the ROSAT X-ray observations [e.g., *Waite et al.*, 1994] was good enough to indicate that the X-ray emission was "soft," with photon energies between about 0.1 and 1 keV, but not good enough to clearly distinguish between a continuum-like source (as suggested by *Barbosa* [1990]) and a line-emission source, although the line model did seem to provide a somewhat better spectral fit [*Waite et al.*, 1994; *Cravens et al.*, 1995]. Recent Chandra X-ray Observatory (CXO) observations show that most of the X-ray power is due to line emission below 1 keV but that a few percent of the power, particularly above 1 keV, could be from electron bremsstrahlung [*Branduardi-Raymont et al.*, 2008].

ROSAT had poor spatial resolution but still indicated that X-ray emission came both from high latitudes (i.e., auroral,

but the latitude range was large) and lower latitudes. Both the low and high latitude emissions were attributed by *Waite et al.* [1997] to ion precipitation. *Maurellis et al.* [2000] made the alternative suggestion for the lower-latitude X-ray emission that seemed to account for the spatial morphology and lower-latitude luminosity (like the auroral X-rays, about 1 GW), scattering of solar X-ray photons from Jupiter's upper atmosphere, and also K shell fluorescence from the carbon in atmospheric methane. Later work on disk emission from Jupiter and Saturn confirmed this suggestion [e.g., *Cravens et al.*, 2005; *Bhardwaj et al.*, 2005a, 2005b].

The next step in understanding why ion precipitation could generate X-ray emission was the recognition that X-rays are produced by charge exchange collisions of highly charged heavy ions with neutrals because the product ions are left in states with high principle quantum numbers that emit in the X-ray part of the spectrum [*Cravens et al.*, 1995; *Liu and Schultz*, 1999; *Kharchenko et al.*, 1998]. *Cravens et al.* [1995] explained that the low-charge state ions (mostly O^+ and S^+ with some O^{++}, S^{++}...) [*Mauk et al.*, 2002] found in the Jovian magnetosphere undergo electron stripping (or removal) collisions with atmospheric H_2 for high energies (MeV per nucleon) resulting in higher-charge state ions. For oxygen, O^{7+} and O^{6+} ions will produce soft X-rays. The key reactions for O^{6+} can be written as

$$O^{6+} + H_2 \rightarrow O^{7+} + H_2 + e$$

$$O^{7+} + H_2 \rightarrow O^{6+*} + H_2^+.$$

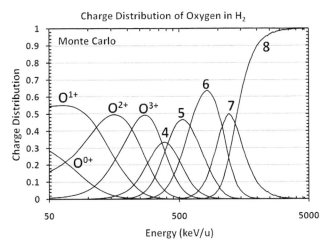

Figure 2. Equilibrium fraction for oxygen charge states in H_2 versus ion energy and very similar to results from *Cravens et al.* [1995]. From *Ozak et al.* [2010].

The O^{6+*} ion is excited and emits soft X-rays. In addition to these processes, collisional ionization and excitation of the atmospheric target gases also take place. Cravens et al. used charge exchange (CX) and stripping cross sections for O^{q+} ($q = 0$ up to 8) collisions with H_2 to calculate an equilibrium fraction versus energy (i.e., fraction of ion beam in each charge state), as shown in Figure 2. Note that the O^{6+} and O^{7+} ions are found near energies of $\approx$ 1 MeV u^{-1}. These models assumed that the ions came from the middle magnetosphere with the necessary energies.

3. CHANDRA X-RAY OBSERVATORY AND XMM-NEWTON OBSERVATIONS

Some CXO and XMM-Newton observations are briefly reviewed. See *Branduardi-Raymont et al.* [2009] for a more detailed review. Key CXO observations of Jupiter were made by *Gladstone et al.* [2002], and the high-resolution camera images clearly showed two types of soft X-ray emission: (1) evenly distributed disk emission with a total power of about 1 GW and (2) auroral emission from latitudes clearly more poleward than the main oval (also about 1 GW). The image shown in Figure 3 from CXO [*Elsner et al.*, 2005] illustrates this. The polar cap location of the emission strongly suggested that the responsible particle populations originate either in the outer magnetosphere or even on open field lines.

The measured time history of the X-ray power [*Gladstone et al.*, 2002] showed a 40 min period. That is, the X-ray emission was "pulsating." However, the periodicities in later observations were not so clear [e.g., *Elsner et al.*, 2005; *Branduardi-Raymont et al.*, 2004]. Other Jovian phenomena show ~ 40 min periodicities, including ion fluxes both inside

and outside the Jovian magnetosphere [*Anagnostopoulos et al.*, 1998] and radio emissions, called QP-40 emissions [*MacDowall et al.*, 1993].

Oxygen lines are evident in CXO spectra of Jupiter's aurora (Figure 4). The spectral region near 500–800 eV is where O^{6+} (helium-like transitions) and O^{7+} (hydrogen-like transitions) emission lines are located as predicted [e.g., *Kharchenko et al.*, 1998]. Any continuum emission (i.e., bremsstrahlung) is relatively weak. The region near 300–400 eV contains sulfur lines [see *Elsner et al.*, 2005, Table 2; *Kharchenko et al.*, 2006]. A cometary X-ray spectrum shown in the bottom panel of Figure 4 has similar oxygen lines, but unlike the Jovian spectrum, this spectrum has features near 400 eV due to solar wind carbon.

Branduardi-Raymont et al. [2008] displayed the locations on a polar map of both low-energy ($E < 2$ keV) and high-energy ($E > 2$ keV) photons using different symbols. Soft X-ray photons (line emission from charge exchange) were found mainly in the polar cap, and harder X-rays were mostly found near the main oval. The harder X-rays were interpreted as being electron bremsstrahlung photons.

Spectra measured by XMM-Newton also indicated the presence of high-charge state oxygen and sulfur lines, but with a different sulfur to oxygen ratio from CXO spectral fits [*Branduardi-Raymont et al.*, 2004, 2008]. An XMM-Newton grating spectrum was also obtained, but the auroral and disk emissions were mixed, complicating the interpretation [*Branduardi-Raymont et al.*, 2007]. Nonetheless, distinct O^{6+} and O^{7+} oxygen lines were seen, and some components had broad spectral widths consistent with fast oxygen (i.e., speeds of about 5000 km s^{-1} or energies of $\approx$ 1 MeV u^{-1}).

4. INTERPRETATION OF INITIAL CXO AND XMM AURORAL OBSERVATIONS

Cravens et al. [2003] noted that the location of the X-ray emission poleward of the main oval implied that the source particle populations were not in the middle magnetosphere as originally thought, but were on open field lines and/or in the outer magnetosphere. Two possible sources were suggested, both involving high-charge state heavy ions (e.g., oxygen): (1) a solar wind ion source and (2) a magnetospheric ion source. For the first, solar wind ions precipitate on open field lines associated with the polar cap and magnetospheric cusp. Solar wind heavy ions are already highly charged and produce X-rays following charge exchange collisions, which is the basis for the solar wind charge exchange (SWCX) mechanism for producing cometary X-rays [*Cravens*, 1997, 2002]. *Cravens et al.* [2003] estimated that to make the SWCX mechanism work for the Jovian X-ray aurora, field-aligned acceleration through a 200 kV or so electrical potential drop

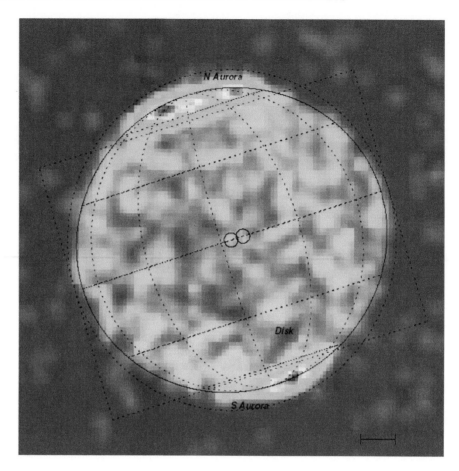

Figure 3. Chandra X-Ray Observatory (CXO) image of Jupiter. Note the disk and the auroral emission regions. From *Elsner et al.* [2005].

would be needed (via the Knight mechanism) [*Knight*, 1973; *Lyons*, 1981], not to produce the highly charged ions but to boost the flux sufficiently to provide the observed 1 GW X-ray power. It was pointed out that the solar wind protons would also be accelerated to high energies, and the associated UV emissions have not been observed (to our knowledge).

For the second mechanism, S and O ions in the outer magnetosphere are the source, but these magnetospheric ions have average energies of only 50 keV or so [*Mauk et al.*, 2002] and would be insufficient to produce the X-ray aurora. *Cravens et al.* [2003] suggested that acceleration by several MV field-aligned potentials would allow the ions to produce X-rays, as discussed earlier. The ion fluxes are also boosted by such a field-aligned acceleration process via the Knight mechanism. The same mechanism has been invoked (in different forms) for auroral electron acceleration on Earth (e.g., the current monograph) or on Jupiter [*Hill*, 2001]. *Cravens* suggested that the heavy ion precipitation was associated with the downward "return current" (into the planet) of the main cur-

rent system. Electrons that are accelerated upward to several MeV could be responsible for the observed QP-40 radio bursts [*MacDowell et al.*, 1993; *Elsner et al.*, 2005].

Bunce et al. [2004] proposed that the field-aligned potentials and associated field-aligned currents needed for the X-ray aurora are caused by pulsating reconnection at the dayside magnetopause. They estimated that, at least for fast solar wind conditions, the potentials (i.e., ion energies) and fluxes would be sufficient to generate the required X-ray aurora.

5. RECENT WORK ON THE JOVIAN X-RAY AURORA

Kharchenko et al. [2006, 2008] introduced sulfur as well as oxygen precipitation (using the relevant cross sections) in their model, given that the magnetosphere contains both species. Significant quenching effects were found for the $^3P^0$–^1S and ^3S–^1S $n = 2$ transitions of O^{6+} (i.e., the intercombination and forbidden transitions, respectively) and

Chandra ACIS-S Spectrum of Jupiter's Aurora and Comets

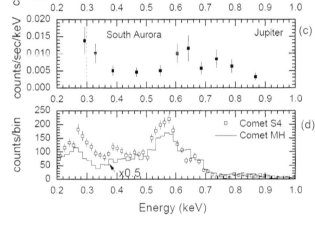

Figure 4. (middle) CXO Advanced CCD Imaging Spectrometer spectra of the north and south aurorae. (top) The instrument response function. (bottom) Cometary X-ray spectra. From *Elsner et al.* [2005].

particularly for the ^3S–^1S transition [*Kharchenko et al.*, 2008; *Ozak et al.*, 2010]. Updated cross sections for energetic sulfur and oxygen ion electron removal, charge exchange, and ionization were used for a wide range of charge states [*Kharchenko et al.*, 2008; *Hui et al.*, 2010; *Ozak et al.*, 2010]. State-specific charge exchange cross sections were also used to improve the accuracy of the predicted spectra, and Monte Carlo methods were used to determine ion charge state distributions versus altitude in the atmosphere. The Ozak et al. Monte Carlo model additionally considered the altitude dependence of the energy deposition and X-ray volume production rate, which helped with the quenching rate determination, as well as allowing opacity effects for outgoing X-ray photons to be taken into account.

Hui et al. [2010] used a detailed ion precipitation model that included not just S and O ions of various charge states, but also carbon ions, and at lower energies (200 keV or so) as well as at MeV energies in order to see if this species could contribute to the observed X-ray spectrum. They concluded that carbon lines are not making a significant contribution to

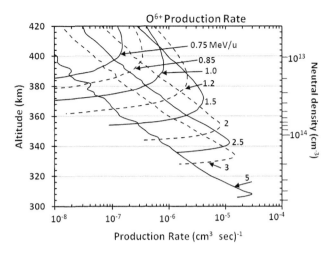

Figure 5. Production rate versus altitude for several incident oxygen energies of O^{6+} ions from charge exchange. The O^{6+} production is proportional to the X-ray emission from this species. From *Ozak et al.* [2010].

the observed Jovian X-ray spectra, thus indicating that the solar wind source is probably not important.

Ozak et al. [2010] adopted neutral upper atmosphere density profiles versus altitude based on Galileo entry probe data [*Seiff et al.*, 1996] and used the *Hui et al.* [2010] detailed X-ray transition probabilities. Figure 5 shows the production rate of O^{6+} ions created in high principle quantum number states versus altitude for a range of incident ion energies. O^{6+} accounts for a large fraction of the total X-ray volume emission rate. As expected, with higher incident ion energies, the ions penetrate deeper into the atmosphere. For example, the

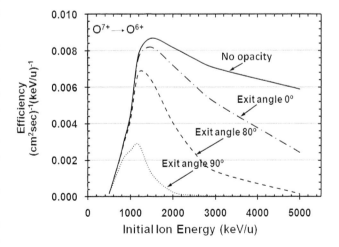

Figure 6. X-ray power efficiency for all transitions for different incident oxygen energies and for different photon exit geometries. Opacity effects were included for some curves. From *Ozak et al.* [2010].

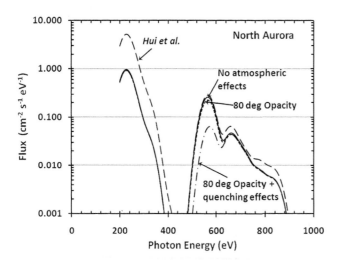

Figure 7. X-ray intensity versus photon energy from the *Ozak et al.* [2010] and *Hui et al.* [2010] models. The *Hui et al.* [2010] spectrum is a proxy for the CXO-measured spectrum.

peak production rate for O^{6+} is at an altitude $z = 390$ km for a 1 MeV u^{-1} ion. For a 2 MeV u^{-1} ion, the peak is at $z = 354$ km. The homopause is near 400 km, so that 2 MeV u^{-1} (and higher energy) ions are penetrating into the mixed atmosphere containing relatively abundant methane (a good UV and soft X-ray absorber). Production rate profiles for several S and O charge states and integrated over altitude were used to determine X-ray volume emission rates and to give predicted intensities (and spectra). The effect on outgoing radiation of the soft X-ray opacity of the atmosphere was investigated.

Ozak et al. [2010] found emission efficiencies by dividing total intensity over all directions ($4\pi I$) by the value of the initial ion energy. For example, the X-ray efficiency for oxygen charge state O^{6+} is shown in Figure 6, and the most efficient production occurs for incident ions with energies of about 1.5 MeV u^{-1}. For O^{7+} ions (not shown here), the peak efficiency is for initial energies of 2.5 MeV u^{-1}. Incident sulfur ions with energies of about 1 MeV u^{-1} are most efficient in producing S^{8+}. Opacity effects for outgoing photon angles of 0° (dash-dotted line), 80° (dashed line), and 90° (dotted line) are shown. Opacity effects become significant for incident ion energies higher than 2 MeV u^{-1} for O^{7+} ions, 1.2 MeV u^{-1} for O^{6+} ions, and 1 MeV u^{-1} for S^{8+} ions. For energies higher than 5 MeV u^{-1}, the atmosphere becomes extremely opaque at the large exit angles relevant for the auroral regions.

One of the *Ozak et al.* [2010] X-ray spectra is shown in Figure 7. The higher energy emission is due to oxygen and the lower energy emission to sulfur. The *Hui et al.* [2010] curve in Figure 7 is for a model case with the best fit (varying

incident ion energy and sulfur to oxygen ratio) to the CXO data. Clearly, the opacity affects the spectra and affects any deductions of the S to O ratios and energies. More work is still needed on this topic.

6. MAGNETOSPHERE-IONOSPHERE COUPLING ON JUPITER: IMPLICATIONS OF X-RAY AND THE ABSENCE OF A SATURNIAN X-RAY AURORA

Oxygen ions in the outer magnetosphere must be accelerated by field-aligned potentials of about 10 MV in order to produce X-ray emission. *Bunce et al.* [2004] suggested that magnetic reconnection at the dayside magnetopause of Jupiter should be able to accomplish this task for fast solar wind conditions. On the other hand, on Saturn, although disk and possibly ring X-rays were observed, an X-ray aurora was not detected. *Branduardi-Raymont et al.* [2009] set an upper limit of about 40 MW on the auroral X-ray power. *Hui et al.* [2010] modeled heavy ion precipitation on Saturn and concluded that magnetosphere-ionosphere coupling at Saturn is such that any field-aligned potential must be well below a few MV. They also concluded that no significant cusp entry and precipitation of solar wind ions is taking place on Saturn.

7. FUTURE WORK

Much work remains to be done to improve our understanding of ion precipitation and X-ray production on Jupiter, including a detailed assessment of field-aligned currents and the effects of this aurora on the atmosphere (e.g., heating, ionization UV emissions, ...). In particular, why do 10 MV field-aligned potentials develop (evidently) in the Jovian magnetosphere at high latitudes but not on Saturn?

Acknowledgments. Support from NASA Planetary Atmospheres grant NNX10AB86G is gratefully acknowledged.

REFERENCES

Anagnostopoulos, G. C., P. K. Marhavilas, E. T. Sarris, I. Karanikola, and A. Balogh (1998), Energetic ion populations and periodicities near Jupiter, *J. Geophys. Res.*, *103*(E9), 20,055–20,073.

Barbosa, D. D. (1990), Bremsstrahlung X rays from Jovian auroral electrons, *J. Geophys. Res.*, *95*(A9), 14,969–14,976.

Bhardwaj, A. (2006), X-Ray emission from Jupiter, Saturn and Earth: A short review, in *Advances in Geosciences*, vol. 3, *Planetary Science*, edited by W.-H. Ip and A. Bhardwaj, p. 215, World Sci., Hackensack, N. J.

Bhardwaj, A., G. Branduardi-Raymont, R. F. Elsner, G. R. Gladstone, G. Ramsay, P. Rodriguez, R. Soria, J. H. Waite Jr., and T. E. Cravens (2005a), Solar control on Jupiter's equatorial X-ray emissions: 26–29 November 2003 XMM-Newton observation, *Geophys. Res. Lett.*, *32*, L03S08, doi:10.1029/2004GL021497.

Bhardwaj, A., R. F. Elsner, J. H. Waite Jr., G. R. Gladstone, T. E. Cravens, and P. Ford (2005b), Chandra observation of an X-ray flare at Saturn: Evidence for direct solar control on Saturn's disk X-ray emissions, *Astrophys. J. Lett.*, *624*, L121–L124.

Branduardi-Raymont, G., R. F. Elsner, G. R. Gladstone, G. Ramsay, P. Rodriguez, R. Soria, and J. H. Waite Jr. (2004), First observation of Jupiter by XMM-Newton, *Astron. Astrophys.*, *424*, 331–337, doi:10.1051/0004-6361:20041149.

Branduardi-Raymont, G., A. Bhardwaj, R. F. Elsner, G. R. Gladstone, G. Ramsay, P. Rodriguez, R. Soria, J. H. Waite Jr., and T. E. Cravens (2007), A study of Jupiter's aurorae with XMM-Newton, *Astron. Astrophys.*, *463*, 761–774, doi:10.1051/0004-6361:20066406.

Branduardi-Raymont, G., R. F. Elsner, M. Galand, D. Grodent, T. E. Cravens, P. Ford, G. R. Gladstone, and J. H. Waite Jr. (2008), Spectral morphology of the X-ray emission from Jupiter's aurorae, *J. Geophys. Res.*, *113*, A02202, doi:10.1029/2007JA012600.

Branduardi-Raymont, G., A. Bhardwaj, R. F. Elsner, G. R. Gladstone, P. Rodriguez, J. H. Waite Jr., and T. E. Cravens (2009), X-rays from Saturn: A study with XMM-Newton and Chandra over the years 2002–05, *Astron. Astrophys.*, *510*, A73, doi:10.1051/0004-6361/200913110.

Bunce, E. J., S. W. H. Cowley, and T. K. Yeoman (2004), Jovian cusp processes: Implications for the polar aurora, *J. Geophys. Res.*, *109*, A09S13, doi:10.1029/2003JA010280.

Clarke, J. T., et al. (1998), Hubble Space Telescope imaging of Jupiter's UV aurora during the Galileo orbiter mission, *J. Geophys. Res.*, *103*(E9), 20,217–20,236.

Clarke, J. T., D. Grodent, S. W. H. Cowley, E. J. Bunce, P. Zarka, J. E. P. Connerney, and T. Satoh (2004), Jupiter's aurora, in *Jupiter: The Planet, Satellites and Magnetosphere*, edited by F. Bagenal, T. E. Dowling, and W. B. McKinnon, chap. 26, pp. 639–670, Cambridge Univ. Press, Cambridge, U. K.

Cowley, S. W. H., and E. J. Bunce (2001), Origin of the main auroral oval in Jupiter's coupled magnetosphere-ionosphere system, *Planet. Space Sci.*, *49*, 1067–1088.

Cravens, T. E. (1997), Comet Hyakutake x-ray source: Charge transfer of solar wind heavy ions, *Geophys. Res. Lett.*, *24*(1), 105–108.

Cravens, T. E. (2002), X-ray emission from comets, *Science*, *296*, 1042–1045.

Cravens, T. E., E. Howell, J. H. Waite Jr., and G. R. Gladstone (1995), Auroral oxygen precipitation at Jupiter, *J. Geophys. Res.*, *100*(A9), 17,153–17,161.

Cravens, T. E., J. H. Waite, T. I. Gombosi, N. Lugaz, G. R. Gladstone, B. H. Mauk, and R. J. MacDowall (2003), Implications of Jovian X-ray emission for magnetosphere-ionosphere coupling, *J. Geophys. Res.*, *108*(A12), 1465, doi:10.1029/2003JA010050.

Cravens, T. E., et al. (2005), Titan's ionosphere: Model comparisons with Cassini Ta data, *Geophys. Res. Lett.*, *32*, L12108, doi:10.1029/2005GL023249.

Elsner, R. F., et al. (2005), Simultaneous Chandra X-ray, Hubble Space Telescope ultraviolet, and Ulysses radio observations of Jupiter's aurora, *J. Geophys. Res.*, *110*, A01207, doi:10.1029/2004JA010717.

Galand, M., and S. Chakrabarti (2002), Auroral processes in the solar system, in *Atmospheres in the Solar System: Comparative Aeronomy, Geophys. Monogr. Ser.*, vol. 130, edited by M. Mendillo, A. Nagy, and J. H. White, pp. 55–76, AGU, Washington, D. C., doi:10.1029/130GM05.

Gehrels, N., and E. C. Stone (1983), Energetic oxygen and sulfur ions in the Jovian magnetosphere and their contribution to the auroral excitation, *J. Geophys. Res.*, *88*(A7), 5537–5550.

Gladstone, G. R., et al. (2002), A pulsating auroral x-ray hot spot on Jupiter, *Nature*, *415*, 1000–1003.

Grodent, D., J. H. Waite Jr., and J.-C. Gérard (2001), A self-consistent model of the Jovian auroral thermal structure, *J. Geophys. Res.*, *106*, 12,933–12,952.

Hill, T. W. (2001), The Jovian auroral oval, *J. Geophys. Res.*, *106*, 8101–8107.

Horanyi, M., T. E. Cravens, and J. H. Waite Jr. (1988), The precipitation of energetic heavy ions into the upper atmosphere of Jupiter, *J. Geophys. Res.*, *93*(A7), 7251–7271.

Hui, Y., T. E. Cravens, N. Ozak, and D. R. Schultz (2010), What can be learned from the absence of auroral X-ray emission from Saturn?, *J. Geophys. Res.*, *115*, A10239, doi:10.1029/2010JA015639.

Kharchenko, V., W. Liu, and A. Dalgarno (1998), X-ray and EUV emission spectra of oxygen ions precipitating into the Jovian atmosphere, *J. Geophys. Res.*, *103*(A11), 26,687–26,698.

Kharchenko, V., A. Dalgarno, D. R. Schultz, and P. C. Stancil (2006), Ion emission spectra in the Jovian X-ray aurora, *Geophys. Res. Lett.*, *33*, L11105, doi:10.1029/2006GL026039.

Kharchenko, V., A. Bhardwaj, A. Dalgarno, D. R. Schultz, and P. C. Stancil (2008), Modeling spectra of the north and south Jovian X-ray auroras, *J. Geophys. Res.*, *113*, A08229, doi:10.1029/2008JA013062.

Knight, S. (1973), Parallel electric fields, *Planet. Space Sci.*, *21*, 741–750.

Liu, W., and D. R. Schultz (1999), Jovian X-ray aurora and energetic oxygen precipitation, *Astrophys. J.*, *526*, 538–543.

Lyons, L. R. (1981), The field-aligned current versus electric potential relation and auroral electrodynamics, in *Physics of Auroral Arc Formation, Geophys. Monogr. Ser.*, vol. 25, edited by S.-I. Akasofu and J. R. Kan, pp. 252–259, AGU, Washington, D. C., doi:10.1029/GM025p0252.

MacDowall, R. J., M. J. Kaiser, M. D. Desch, W. M. Farrell, R. A. Hess, and R. G. Stone (1993), Quasiperiodic Jovian radio bursts: Observations from the Ulysses Radio and Plasma Wave Experiment, *Planet. Space Sci.*, *41*, 1059–1072.

Mauk, B. H., B. J. Anderson, and R. M. Thorne (2002), Magnetosphere-ionosphere coupling at Earth, Jupiter, and beyond, in *Atmospheres in the Solar System: Comparative Aeronomy, Geophys. Monogr. Ser.*, vol. 130, edited by M. Mendillo, A. Nagy, and J. H. White, pp. 97–114, AGU, Washington, D. C., doi:10. 1029/130GM07.

Maurellis, A. N., T. E. Cravens, G. R. Gladstone, J. H. Waite, and L. W. Acton (2000), Jovian X-ray emission from solar X-ray scattering, *Geophys. Res. Lett.*, *27*(9), 1339–1342.

Metzger, A. E., D. A. Gilman, J. L. Luthey, K. C. Hurley, H. W. Schnopper, F. D. Seward, and J. D. Sullivan (1983), The detection of X rays from Jupiter, *J. Geophys. Res.*, *88*(A10), 7731–7741.

Ozak, N., D. R. Schultz, T. E. Cravens, V. Kharchenko, and Y.-W. Hui (2010), Auroral X-ray emission at Jupiter: Depth effects, *J. Geophys. Res.*, *115*, A11306, doi:10.1029/2010JA015635.

Seiff, A., et al. (1996), Structure of the atmosphere of Jupiter: Galileo probe measurements, *Science*, *272*, 844–845.

Waite, J. H., Jr., F. Bagenal, F. Seward, C. Na, G. R. Gladstone, T. E. Cravens, K. C. Hurley, J. T. Clarke, R. Elsner, and S. A. Stern (1994), ROSAT observations of the Jupiter aurora, *J. Geophys. Res.*, *99*(A8), 14,799–14,809.

Waite, J. H., Jr., G. R. Gladstone, W. S. Lewis, P. Drossart, T. E. Cravens, A. N. Maurellis, B. H. Mauk, and S. Miller (1997), Equatorial X-ray emissions: Implications for Jupiter's high exospheric temperatures, *Science*, *276*, 104–108.

Waite, J. H., et al. (2001), An auroral flare at Jupiter, *Nature*, *410* (6830), 787–789.

T. E. Cravens and N. Ozak, Department of Physics and Astronomy, University of Kansas, Malott Hall, 1251 Wescoe Hall Drive, Lawrence, KS 66045, USA. (cravens@ku.edu)

Satellite-Induced Electron Acceleration and Related Auroras

S. L. G. Hess[1]

LATMOS, Université de Versailles St-Quentin-en-Yvelines, IPSL/CNRS, Guyancourt, France

P. A. Delamere

Laboratory for Atmospheric and Space Physics, University of Colorado, Boulder, Colorado, USA

Several cases of interactions between planetary magnetospheres and satellite atmospheres have been observed in the giant planet magnetospheres. Io, Europa, Ganymede, and Enceladus generate observable auroral emissions on their parent planet. This implies an efficient power transfer between the satellite, where the interaction occurs, and the planet, where the emission occurs. In this chapter, we discuss the power generation at the satellite, the transport of energy and momentum along the magnetic field lines via Alfvén waves, and the transfer of wave power to electrons. We relate the power generated at the satellite to the power of the observed auroral emissions.

1. INTRODUCTION

Auroral evidence of interactions between satellites and the magnetospheres of their parent planet have been gathered since the discovery of the Io-related Jovian decameter radio emissions [*Bigg*, 1964]. IR observations of Jupiter revealed aurorae at the foot of the magnetic field lines crossing Io [*Connerney et al.*, 1993]. Later observations were performed in the UV [*Clarke et al.*, 1996; *Prangé et al.*, 1996], which revealed finer structures in the Io footprint [*Gérard et al.*, 2006], as well as auroral footprints related to Europa and Ganymede [*Clarke et al.*, 2002; *Grodent et al.*, 2006, 2009]. Claims of auroral emissions triggered by Callisto have also been made, based on radio observations [*Menietti et al.*, 2001]. Recently, an Enceladus auroral footprint on Saturn was observed [*Pryor et al.*, 2011]. It thus appears that the satellite-triggered aurorae are a common phenomenon in our solar system.

Although the origin of the interaction was established as being the motion of the satellite relative to the planetary magnetic field [*Goldreich and Lynden-Bell*, 1969], its Alfvénic nature was only revealed decades later [*Neubauer*, 1980; *Goertz*, 1983], and the details of the interaction are not completely understood. Nevertheless, progress has been achieved, thanks to in situ spacecraft measurements and improved observations of the auroral emissions they trigger. Radio and UV observations of the Io aurora reveal fine structures: quasiperiodic radio bursts and bright spots inside the footprint in the UV. Both tell the same story: the interaction between Io and Jupiter is carried by Alfvén waves, which periodically accelerate electrons, generate quasiperiodic radio bursts [*Su et al.*, 2006; *Hess et al.*, 2007a], and bounce between interfaces in the system leaving auroral spots at each bounce on the Jovian ionosphere (illustrated in Figures 1a and 1b) [*Gérard et al.*, 2006; *Bonfond et al.*, 2009].

The details of the auroral structures related to the satellite-magnetosphere interactions are discussed in the chapter by *Bonfond* [this volume]. In the present chapter, we discuss the current generation and propagation between the satellites and the planets, and how they trigger the electron accelerations needed to power the observed auroral emissions.

[1]Formerly at Laboratory for Atmospheric and Space Physics, University of Colorado, Boulder, Colorado, USA.

Auroral Phenomenology and Magnetospheric Processes: Earth and Other Planets
Geophysical Monograph Series 197
10.1029/2011GM001175

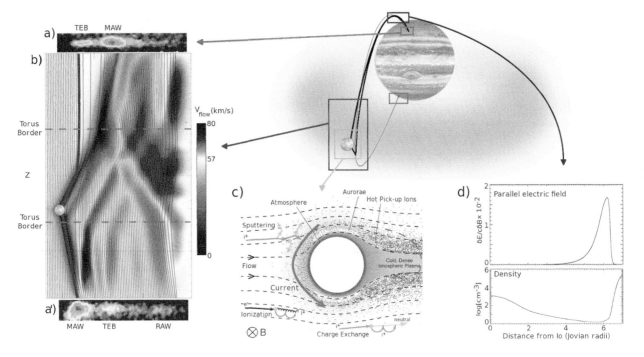

Figure 1. Sketch summarizing the interaction between a satellite (here Io) and a planetary magnetosphere and the related auroral footprints of the Io flux tube [*Gérard et al.*, 2006]. (a) The footprints are composed of subspots (main and reflected Alfvén wing spots, transhemispheric electron beam spots), which translate the Alfvénic nature of the current circuit between the satellite and the planet, simulated by *Jacobsen et al.* [2010] (b). (c) The interaction is due to the motion of the satellite with respect to the planetary magnetic field and the frozen-in plasma. Around the satellite, the plasma flow is decelerated through several processes, leading to the generation of a current, which propagates along the field lines as an Alfvén wave packet (courtesy of F. Bagenal and S. Barlett). (d) (top) The Alfvén waves develop a parallel electric field along the magnetic field lines, with (bottom) a narrow peak just above the ionosphere where the density has a minimum.

2. LOCAL INTERACTION AND CURRENT GENERATION

Satellite-magnetosphere interactions occur when a satellite orbits deep inside the magnetosphere. The satellite moves with a velocity v_{sat} relative to the ambient magnetic field B_{sat}, which leads to the generation of an electric field E_{sat} across the satellite:

$$E_{sat} = -v_{sat} \times B_{sat}. \qquad (1)$$

If the satellite is conducting, this electric field generates a current (Figure 1c). This mechanism was first proposed by *Piddington and Drake* [1968] and soon thereafter *Goldreich and Lynden-Bell* [1969] to describe the Io-Jupiter interaction, known at that time through Io-modulated radio emissions [*Bigg*, 1964]. The brief and oversimplified description of the current generation we give here is sufficient to give a correct estimate of the power generated at the satellite. However, the local interaction in the satellite vicinity may involve contributions from a magnetic field internal to the satellite [*Gurnett et al.*, 1996; *Kivelson et al.*, 1996] or to magnetospheric plasma flow perturbations caused by several processes, such as atmo-

spheric sputtering, electron impact ionization, or most importantly charge exchange [*Saur et al.*, 1999, 2002a; *Saur*, 2004; *Delamere et al.*, 2003; *Dols et al.*, 2008]. In any case, the current generator is the region that is not in corotation with the magnetic field (Figure 1c).

To compute the current running across the interaction region, one needs to know the conductivity of the whole current circuit, namely, the Pedersen conductivity of the planet's ionosphere, where the current is supposed to close, and the conductivity of the interaction region, usually limited to the Pedersen conductance of the satellite ionosphere [*Saur*, 2004]. The conductance along the magnetic field lines, which can be estimated using the Knight relation [*Knight*, 1973] or one of its derivatives [*Ergun et al.*, 2009; *Ray and Ergun*, this volume], was at first neglected, except for estimating the parallel electric field accelerating the auroral electrons. Models of electron acceleration by electric potential structures along Io flux tube have been performed [e.g., *Su et al.*, 2003].

However, these calculations were missing an important factor in that they assumed a steady state current circuit. In a magnetized plasma, the information is spread across current circuits by Alfvén waves. Hence, a steady state current system can only be established after an Alfvén wave has propagated at least once along the whole circuit. In the mean time, the current system is transient, and the current is carried by Alfvén waves rather than by global motion of the electrons [*Neubauer*, 1980; *Goertz*, 1983] (Figure 1b). This leads to a different calculation of the current delivered by the satellite-magnetosphere interaction and to a different way of accelerating electrons. In the present chapter, we ignore the case of a quasi–steady state interaction between the satellite and magnetosphere and concentrate on the transient circuits, which are responsible for all cases of satellite-related aurora observed so far, although a weaker steady state current may exist in the satellite wake.

The transient current system involving Alfvén waves leaves an imprint on the morphology of the Io-related UV aurora [*Gérard et al.*, 2006] (Figure 1a), which presents several spots with decreasing intensities. These spots are understood as the footprints of the Alfvén wing generated at Io, which bounces several times [*Neubauer*, 1980; *Goertz*, 1983; *Bonfond et al.*, 2009] (Figure 1b). An elongated continuous tail is added to the spots, which is thought to be generated by a steady state current system [*Ergun et al.*, 2009]. In the UV, the brightness of the main spots dominates the other features, which implies that the Alfvén waves carry a lot of power. The current system associated with the diffuse tail and the secondary spots involves the nonlinear dispersion of Alfvén waves with time, reflections in the plasma torus [*Jacobsen et al.*, 2007, 2010], and the nonlinear buildup of the steady state current. The Io footprint tail, which is several tens of degrees long, emits more power than the main spot, although its brightness is far weaker [*Gérard et al.*, 2006]. The power partition between the main spot and the other features of the other satellite footprints is not known because of the low tail brightness.

Satellite current circuits are large ($>10^5$ km), and the duration of the interaction between the satellite and a given flux tube, estimated as the time needed by a magnetic field line to pass the interaction region, is short (typically 1 min). Alfvén waves propagating at velocities close to the speed of light would be able to establish a steady state current system in a much smaller timescale than the interaction duration. However, the Alfvén velocity depends on the magnetic field strength B and the mass density ρ of the medium through which they travel:

$$v_a = \frac{B}{\sqrt{\mu_0 \rho + \frac{B^2}{c^2}}}. \qquad (2)$$

At the giant planets, this relativistic expression of the Alfvén velocity is necessary, because v_a is close to the speed of light

out of the equatorial plane in the inner magnetosphere of these planets. By contrast, the satellites responsible for observed aurorae all are embedded in relatively dense plasma sheet, which can be created by the satellite itself (Io, Enceladus) or by a satellite on an inner orbit (Io for Europa and Ganymede). This results in a large density in the equatorial plane and, hence, to small Alfvén velocity (as low as a few hundred of km s^{-1}). The current systems associated with the known satellite magnetosphere interactions are thus mostly transient and carried by Alfvén waves, although a weaker steady state current may exist in the satellite wake.

The Alfvén wave packet is generated at the satellite as a perturbation of the magnetic field because of the current running across the interaction region. This current, which cannot depend on remote conductances (e.g., Pedersen conductance of the planet ionosphere) is then mostly limited by the Pedersen conductance around the satellite Σ_P (from the satellite's ionosphere, magnetosphere, or a volcanic plume) and the Alfvén conductance Σ_A [*Neubauer*, 1980; *Goertz*, 1983], given by

$$\Sigma_A = \sqrt{\frac{\rho}{\mu_0 B^2}}, \qquad (3)$$

where ρ is the plasma density. The integrated current carried by the Alfvén waves can be obtained from the Ohm current-voltage relation. According to the calculations of *Saur* [2004], the current for each hemisphere is

$$J \sim 4E_{sat}R_{sat}\frac{\Sigma_A \Sigma_P}{2\Sigma_A + \Sigma_P} \sim 4E_{sat}R_{sat}\Sigma_A (\text{for } \Sigma_P \gg \Sigma_A), \qquad (4)$$

where R_{sat} is the satellite radius. In all known cases of satellite-magnetosphere interactions, the conductance around the satellite is large ($\Sigma_P \gg \Sigma_A$). The power radiated as Alfvén waves by the satellite-Jupiter interaction is estimated from the current:

$$P \sim (4E_{sat}R_{sat})^2 \Sigma_A. \qquad (5)$$

Table 1 shows the power estimates for several Jovian and Kronian satellites, computed using equation (5) and the magnetic field and density parameters given by *Neubauer* [1998] (Jovian satellites) and *Saur and Strobel* [2005] (Kronian satellites). In the Ganymede case, we used $R_{sat} = 2R_{Ganymede}$ to take into account the Ganymede magnetosphere.

Both the bounce time of the Alfvén wave and the Alfvén wave power are proportional to v_a^{-1} (the latter through Σ_A), meaning that a low Alfvén velocity ensures a long and intense transient regime and that satellite-magnetosphere interactions can generate intense auroral spots. However, the power transfer efficiency from satellite to auroral electron precipitation has to be of the order or larger than 10% according to the estimates of the power generated at the satellites and to the estimates of the power precipitated into the ionosphere (see Table 1).

Table 1. Summary of the Power Transmission Through the Torus and of the Efficiency of the Power Transferred to the Particles for Different Distributions of the Alfvén Wavelengths (for Power Laws, the Numbers Correspond to $\alpha = 5/3$ and 2, Respectively)

Satellite	Power Generated[a]	Distribution	Power Escaping the Torus[b] (%)	Power Transfer to Electrons (%)	Power Precipitated Computations[a,c]	Power Precipitated Observations
Io	~1000 GW	long scales	17	9×10^{-3}	~10^{-2} GW	a few tens of GW
		filamented	45 to 50	2 to 5	~20 GW	
Europa	~100 GW	long scales	3	2×10^{-3}	~10^{-3} GW	a few GW
		filamented	36 to 40	3 to 8	~3 GW	
Ganymede	~100 GW	long scales	2	3×10^{-4}	~10^{-4} GW	a few GW
		filamented	34 to 38	2 to 6	~2 GW	
Callisto	~10 GW	long scales	9	4×10^{-2}	~10^{-3} GW	no confirmed
		filamented	46 to 49	20 to 25	~1 GW	observation
Enceladus	~300 MW	long scales	5×10^{-3}	4×10^{-5}	~10^{-4} MW	a few tens of MW
		filamented	25 to 34	20 to 25	~70 MW	
Tethys	~1000 MW	long scales	2×10^{-3}	1×10^{-5}	~10^{-4} MW	not observed
		filamented	25 to 30	11 to 16	~100 MW	
Dione	~500 MW	long scales	6×10^{-4}	1×10^{-5}	~10^{-5} MW	not observed
		filamented	26 to 32	19 to 24	~100 MW	
Rhea	~200 MW	long scales	1×10^{-4}	4×10^{-5}	~10^{-4} MW	not observed
		filamented	25 to 30	22 to 27	~50 MW	

[a]Assuming $\Sigma_P \gg \Sigma_A$.

[b]Power reaching the acceleration region, i.e., ~1 planetary radius above the planet ionosphere.

[c]In Kronian satellite cases, it includes the power of the electrons accelerated in the opposite hemisphere, which precipitate on the main Alfvén wing spot, due to the axisymmetric magnetic field of Saturn.

Early computations [*Wright*, 1987; *Delamere et al.*, 2003] showed that ~80% of the Alfvén wave power is reflected back in the Io torus at its boundary, long before reaching the Jovian auroral region where the acceleration is thought to occur (Figure 1d and section 4). The reflection is due to the large gradient of the Alfvén velocity between the inside of the plasma torus and the outside, where the Alfvén velocity is approximately the speed of light. This means that a lower Alfvén velocity in the torus has two opposite effects regarding the power carried by the waves to the planet ionosphere (summarized in Figure 2a): (1) increasing the transition regime duration and giving a large power to the Alfvén waves and (2) inducing a reflection of the Alfvén wave power before it reaches the planet ionosphere.

3. ALFVÉN WAVE PROPAGATION

On their way to the planet ionosphere, the Alfvén waves that carry the current encounter several changes in the plasma parameters, which lead to strong variations of the Alfvén wave phase velocity (e.g., increasing magnetic field flux, plasma torus boundaries, and planetary ionosphere). The Alfvén wave phase velocity is given by [*Lysak and Song*, 2003]

$$v_{\phi,a}^2 = v_a^2 \frac{(1 + k_\perp^2 \rho_s^2)}{(1 + k_\perp^2 \lambda_e^2)}, \qquad (6)$$

where $k_\perp$ is the perpendicular component of the wave vector, ρ_s is the ion acoustic gyroradius, and $\lambda_e = c/\omega_{pe}$ is the electron inertial length. The variation of the plasma parameters causes the partial reflection of the wave packet as a function of the wavelength. Early computations used the WKB (short wavelengths) or the discontinuity (long wavelengths) approximations to compute the reflection coefficient, which led to very weak or very strong reflections, respectively. The range of wavelengths of the Alfvén waves generated by the satellite-planet interactions cover an intermediate range, not described by either of the above approximations [*Wright*, 1987]. An approximation for the reflection coefficient describing the intermediate wavelengths and in agreement with WKB and discontinuities approximations is given by *Hess et al.* [2010]:

$$R_\varepsilon(s, \mathbf{k}) = \frac{1}{\lambda_\parallel} \left(\int_{s - \frac{\lambda_\parallel}{2}}^{s + \frac{\lambda_\parallel}{2}} \frac{\nabla_s \ln(c/v_{\phi,a}(\mathbf{k}))}{2} \, ds \right)^2. \qquad (7)$$

This expression explicitly depends on the parallel wavelength ($\lambda_\parallel$) and depends on the perpendicular wavelength through $v_{\phi;a}$ (equation (6)): short wavelengths are slightly reflected, whereas long wavelengths are strongly reflected. Since the reflection coefficient depends on the parallel wavelength, the wavelength distribution of the Alfvén waves close to the satellite plays a crucial role.

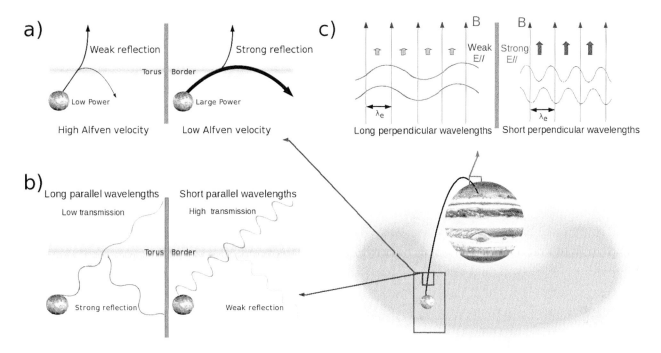

Figure 2. Sketch summarizing the power transfer during the Alfvénic phase of the satellite magnetosphere interaction. (a) In the case of a large Alfvén velocity near the satellite, the Alfvénic phase is short and does not carry much power. It mostly serves to initiate the steady state current system downstream. In the case of a low Alfvén velocity near the satellite, the Alfvénic phase is long, and the Alfvén waves carry substantial power, but most of it is trapped in the torus if there is no Alfvén wave filamentation. (b) and (c) Summary of the influence of the parallel and perpendicular wavelengths of the power transmission from the Alfvén waves generated near the satellites and the accelerated electrons. The parallel wavelength affects the Alfvén wave transmission through the density gradients, in particular, at the torus border, whereas the perpendicular wavelength affects the parallel electric field associated with the Alfvén waves.

The simplest distribution expected for a satellite-magnetosphere interaction is a Gaussian spectrum centered on waves with a satellite length-scale $k_0 = \pi/R_{sat}$:

$$f_0(k) \propto e^{-\frac{(k-k_0)^2}{2k_0^2}}. \qquad (8)$$

However, *Chust et al.* [2005] provided observations showing a possible filamentation of the Alfvén waves in the Io flux tube. Then, the wavelength distribution of the Alfvén waves should follow a power law corresponding to the turbulent filamentation of the previous Gaussian distribution [e.g., *Champeaux et al.*, 1998; *Sharma et al.*, 2008]. Power law distributions generally assume a Kolmogorov cascade, implying a spectral index of $-5/3$. This kind of cascade is, strictly speaking, only valid for nonmagnetized plasma. For highly magnetized plasma, a spectral index of -2 has been theorized by *Galtier* [2009] and has been observed by *Saur et al.* [2002b] at Jupiter. However, the threshold at which the spectral index changes is largely unknown, and the only published observation of filamented Alfvén waves [*Chust et al.*, 2005] was unable to provide an estimate. In any case,

the distribution presents a power law spectrum between the energy injection scale k_0 (satellite length scale) and the dissipation (ionic) scale $k_i = \omega_p/(m_i c)$:

$$\begin{aligned} f_{1;\alpha}(k) &\propto e^{-\frac{(k-k_0)^2}{2k_0^2}} && \text{for} \quad k < k_0 \\ f_{1;\alpha}(k) &\propto k^{-\alpha} && \text{for} \quad k_0 > k > k_i \end{aligned}, \qquad (9)$$

with $\alpha = 5/3$ or 2 depending on the spectral index assumed for the distribution. The Gaussian distribution is hereafter referred to as the long-scale distribution, in comparison to the power law distributions. The Alfvén wave power transmitted from the satellite to the acceleration region is computed by numerically integrating the Alfvén wave reflection coefficients along the magnetic field lines. The dependence of the transmitted power on the parallel wavelength is pronounced, whereas the perpendicular wavelength plays a less important role, even if smaller wavelengths are slightly less reflected than larger ones [*Hess et al.*, 2010, 2011c].

Table 1 shows the power integrated over the Alfvén wavenumber k transmitted along the magnetic field lines from the satellite to the acceleration region for the f_0, $f_{1;5/3}$, and $f_{1;2}$

distributions. To compute these values, we used the VIP4 [*Connerney et al.*, 1998] and SPV [*Davis and Smith*, 1990] internal magnetic field models for Jupiter and Saturn, respectively. The density profiles along the Jovian satellite flux tubes were approximated from the torus models of *Bagenal* [1994] and *Moncuquet et al.* [2002] and the ionospheric profile from the simulations by *Su et al.* [2003]. The density profile along the Kronian satellite flux tubes are approximated from Enceladus torus models [*Sittler et al.*, 2008; *Fleshman et al.*, 2010, and references therein] and assumes a Saturn ionosphere scale height of ~1600 km.

The power transmission along the field lines for the long-scale f_0 distribution varies more than for the filamented $f_{1;\alpha}$ distributions because the power in the f_0 distribution is more concentrated at longer wavelengths, so that a slight increase of the gradient of the Alfvén velocity will have greater impact. The power law distributions have typical transmission of 35%–50% for the Jovian satellites and 25%–30% for the Kronian satellites.

4. PARTICLE ACCELERATION

Most of the electron acceleration occurs at high latitudes [*Jones and Su*, 2008; *Hess et al.*, 2010, 2011c] and is due to the parallel electric field associated with the Alfvén waves. The parallel electric field is due to the inertial terms in the Alfvén phase velocity (equation (6)) and thus can be approximated by [*Lysak and Song*, 2003]

$$\delta E_\parallel \simeq \omega_a k_\perp \lambda_e^2 \delta B, \qquad (10)$$

where ω_a is the Alfvén frequency, and δB the magnetic field perturbation associated with the wave. The perpendicular scale is assumed proportional to the flux tube cross section ($k_\perp \propto B^{1/2}$), to be consistent with an Alfvén wave propagating inside a converging flux tube. Smaller perpendicular wavelengths result in stronger acceleration. The parallel electric field profiles for the inertial Alfvén waves peak where the density is low and the magnetic field intense. This corresponds to altitudes between ~0.5 and ~1 planetary radius from the planet's ionosphere [*Hess et al.*, 2010, 2011c, and references therein], depending on the planet and satellite involved and on the magnetic field and ionospheric density models. Figure 1d shows a typical profile of the parallel electric field associated with Alfvén waves whose wavelength is $\lambda_\parallel = \lambda_\perp = 0.1\ R_{\text{sat}}$ (Io case) and computed using the density profile along the field line shown below.

The resonance of the inertial Alfvén wave with the electrons in the auroral regions is considered in many models of particle acceleration on Earth [*Watt and Rankin*, 2008]. In the terrestrial case, the Alfvén velocity (~0.1c) is comparable

to the particle velocity (1–10 keV, ~0.1c) so that a resonant wave-particle interaction is possible. However, on Jupiter and Saturn, the Alfvén phase velocity (~c) is much larger than the characteristic particle velocities (~0.1c) due, in part, to the strong planetary magnetic field. Therefore, a resonant interaction is generally not possible, particularly for a wavelength spectrum scaled by the satellite radius.

Because most of the wave power in the acceleration region consists of parallel wavelengths ($\lambda_\parallel$) on the order of or larger than the gradient scale length of the parallel electric field ($\|\delta E_\parallel\|$), the electron acceleration is due to the limited extent of the electric field. The long parallel wavelengths ensure that electrons accelerated during one phase of the wave can escape the acceleration region before the wave phase changes. In the opposite limit, where $\lambda_\parallel \ll \|\delta E_\parallel\|$, the electrons would be accelerated and subsequently decelerated with no net particle flux out of the acceleration region. The electron distribution obtained from this acceleration process has an almost unperturbed core with extended tails parallel to B, i.e., a Kappa-like or a beam-like distribution (see discussion in the work of *Hess et al.* [2010]). The acceleration process is independent of the direction of the Alfvén wave propagation, accelerating electrons both toward and away from the planet. This implies the existence of antiplanetward electron beams, which are observed in the wakes of Io [e.g., *Williams et al.*, 1996; *Frank and Paterson*, 1999; *Mauk et al.*, 2001] and Enceladus [*Pryor et al.*, 2011]. The exact shape of the electron distribution may not be computed analytically, but is computed numerically [*Swift*, 2007; *Hess et al.*, 2007a]. We assume that (1) the acceleration occurs where the parallel electric field peaks, (2) all electrons crossing the acceleration region (assumed infinitely small along the magnetic field) gain the same energy from the parallel electric field associated to the Alfvén wave, (3) the acceleration happens only once for each electron and stands for half a period of the wave. The power transferred from the Alfvén wave to the particles is then [*Hess et al.*, 2010]

$$P_e = n v_{\text{th}} \frac{m}{2} \left(\frac{\pi}{2} \frac{(-e)\delta E_\parallel}{\omega_a m} \right)^2 A = \frac{\pi^2}{8} \frac{\omega_p^2}{\omega_a^2} \varepsilon_0 v_{\text{th}} \delta E_\parallel^2 A, \qquad (11)$$

with A the cross section of the flux tube, n the density in the acceleration region, and v_{th} the electron thermal velocity. The power of the wave on a section of the flux tube is given by the wave's Poynting flux:

$$P_w = \frac{\delta E \times \delta B}{\mu_0} A = \frac{v_{\phi,a} \delta B^2}{\mu_0} A. \qquad (12)$$

Using equations (10) and (12), the efficiency of the transfer of power from the satellite interaction to the particles, including the loss by partial reflections of the wave, is given by [*Hess et al.*, 2010, 2011c]

$$\frac{P_e}{P_w} = \int \min\left(\frac{\pi^2}{8}\frac{v_{\text{th}}}{v_{\phi,a}}k_\perp^2\lambda_e^2; 1\right)T(\mathbf{k})f(\mathbf{k})d\mathbf{k}, \quad (13)$$

where $T(\mathbf{k})$ is the Alfvén wave power transmission function computed numerically from equation (7) by integrating the wave reflection coefficient along the magnetic field line. The "min" function assures that electrons cannot gain more power than what can be provided by the wave. The power transfer efficiency depends not only on the Alfvén wave characteristics and on the magnetic field but also on the warm plasma density and temperature in the acceleration region. The temperature for the warm electrons is generally assumed to be a few hundreds of eV in both cases [*Bagenal*, 1994; *Moncuquet et al.*, 2002; *Su et al.*, 2003; *Sittler et al.*, 2008; *Fleshman et al.*, 2010]. On Jupiter, the warm component density is about 1 cm^{-3} [*Moncuquet et al.*, 2002], whereas on Saturn it varies between ~0.3 cm^{-3} at Enceladus and ~0.1 cm^{-3} at Rhea [*Saur and Strobel*, 2005, and references therein].

The efficiency of the power transfer from the satellite interaction to the electrons is shown in Table 1. For each satellite, the long-scale distribution leads to an acceleration efficiency that is orders of magnitude lower than that of the filamented cases. Filamented distributions transfer power at rates ranging from 2% up to 25%, which means that in some cases (like Rhea), almost all the power reaching the acceleration region is transferred to electrons. These differences are explained by the dependence of the acceleration efficiency on $\lambda_e^2 k_\perp^2$ (equation (13)), which depends on the interaction parameters as

$$\lambda_e^2 k_\perp^2 \propto \frac{\mu}{nR_{\text{sat}}^2}, \quad (14)$$

where μ is the mirror ratio, i.e., the ratio between the magnetic flux at the top of the planet's ionosphere and at the satellite. Thus, smaller satellites generate a more efficient electron acceleration, denser flux tubes generate less efficient acceleration, and a larger mirror ratio increases the efficiency.

Since the acceleration occurs in both directions, only half of the power is transmitted to electrons directly precipitating in the ionosphere. The other half forms transhemispheric electron beams (TEB) precipitating in the opposite hemisphere [*Bonfond et al.*, 2008]. These TEB spots are distinct from the main Alfvén wing (MAW) spot in Jovian satellite cases due to the tilt of the Jovian magnetic dipole. The Saturn magnetic dipole is aligned with the rotation axis. Hence, the Kronian satellites TEB and MAW spots form a single spot, powered half by the electron acceleration in their own hemisphere and half by the electron acceleration in the opposite hemisphere. Accordingly, electron beams are observed at the

equator close to Enceladus [*Pryor et al.*, 2011; *Gurnett and Pryor*, this volume].

5. DISCUSSION

5.1. Comparison With Observations

Table 1 shows estimates of the power precipitated into the main spots obtained from observations of the four known cases of satellite-magnetosphere interactions. Most of these estimates come from UV observations [*Bonfond et al.*, 2009; *Grodent et al.*, 2006, 2009; *Pryor et al.*, 2011]. Only the power emitted by the Io interaction can be estimated over a large spectrum of radiation, extending from UV to low-frequency radio [*Queinnec and Zarka*, 2001]. Moreover, the plasma environment surrounding Io has been explored by Galileo, unveiling the existence of keV electron beams accelerated close to Jupiter and associated with the main UV spot [e.g., *Williams et al.*, 1996; *Frank and Paterson*, 1999; *Mauk et al.*, 2001]. From these observations, *Hess et al.* [2010] estimated the power transmitted by the main Alfvén wing to the electrons to be a few 10^{10} W, both toward Jupiter and antiplanetward. Europa [*Clarke et al.*, 2002; *Grodent et al.*, 2006], Ganymede [*Clarke et al.*, 2002; *Grodent et al.*, 2009], and Enceladus [*Pryor et al.*, 2011] footprint brightness in the UV can be translated to precipitation power according to the calculation presented by *Gérard and Singh* [1982] (1 mW m^{-2} gives 10 kR) and assuming a surface for spots, which are not resolved (we used the surface of the interaction region magnetically mapped on the planet).

In any case, the filamentation of the Alfvén waves explains the power emitted in the UV by the satellite-magnetosphere interactions.

Such computations for satellite-magnetosphere interactions that have not yet been observed are also instructive. According to Table 1, Callisto emissions should only be two to three times weaker than Europa and Ganymede. This estimate supports claims of Callisto radio emissions [*Menietti et al.*, 2001]. At Saturn, Tethys and Dione interactions should lead to stronger emissions than Enceladus, but they are not observed. These satellites may have low Pedersen conductances [*Saur and Strobel*, 2005]. The calculations, performed assuming $\Sigma_P \gg \Sigma_A$, may overestimate the power generated at these satellites by orders of magnitude. In the Rhea case, the Pedersen conductivity of the satellite ionosphere should not significantly impact the result [*Saur and Strobel*, 2005]. One possible explanation for the nondetection of a Rhea footprint may then be that the faint Rhea spot is close to the bright and highly variable UV emissions from the main auroral oval of Saturn.

5.2. Further Studies

We presented above the power transfer between the satellite interaction and the electrons responsible for the MAW and TEB spots. The numerical results we present here correspond to a given longitude of the satellite. However, in the case of the Jovian satellites, the tilt of the magnetic field introduces a modulation of the interaction with longitude. The position of the satellites relative to the torus center varies with the satellite longitude, and the magnetic fields at the satellites and the mirror ratios between the satellite and the acceleration region are also modulated. The brightness modulation of the Jovian satellite footprints in UV is observed [*Serio and Clarke*, 2008; *Wannawichian et al.*, 2010]. Models of Jupiter's magnetic field [*Connerney et al.*, 1998; *Hess et al.*, 2011a] predict a variation of the mirror ratio μ by almost a factor of 2 with longitude, which translates as a modulation of almost the same amplitude of the footprint brightness (equation (14)). This modulation is close or even slightly larger than that due to the modulation of the power generation (equation (5)). Ongoing work is being performed to study in detail the origin and amplitude of the footprint brightness modulation with longitude.

The reflection coefficient along the Io flux tube peaks in a narrow region around 1 R_J of the torus center (at torus borders). This narrowness is needed to form a coherent structure (wing). However, Alfvén waves are reflected along the whole field lines. There is thus part of the Alfvén wave packet power, which does not contribute to the brightness of the MAW, TEB, or to reflected Alfvén wing spots, but rather powers the footprint tail. The Alfvén wave interference pattern in Io's wake was modeled by *Jacobsen et al.* [2007, 2010] (Figure 1b), but with a simplified model of the density and magnetic field variation between Io and Jupiter and without taking into account Alfvén wave filamentation and dissipation by electron acceleration. Hence, the actual morphology of the current system in the wake of the satellites still remains to be modeled.

The dissipation of the Alfvén wave power does not only occur by direct acceleration of the electrons. Currents are generated when the Alfvén waves reach the planet ionosphere, generating currents and power losses by Joule heating [*Codrescu et al.*, 1995], generating IR emissions [*Lystrup et al.*, 2007]. Moreover, nonlinearities in the ionospheric response to the current carried by the Alfvén waves leads to the generation of electric potential structures, which produce an additional acceleration of the electrons (up to ~ 1 keV). These potential structures are deduced from the changes in drift rate they impose to the short radio bursts generated by the Io interaction [*Hess et al.*, 2007b]. These potential structures move at the speed of sound and seem to occur quasi-periodically [*Hess et al.*, 2009] and may be related to pulsations observed in the UV spots [*Grodent et al.*, 2009], although neither the origin of the quasiperiod or the link with UV aurorae are understood.

5.3. Alfvén Wave Filamentation

For the four cases of satellite-magnetosphere interactions for which we can compare predictions and observations, the filamentation of Alfvén waves in the vicinity of the satellite appears to be necessary to explain the observed power emitted in UV from the main Alfvénic spot and in the case of Io, of radio and IR emissions as well [*Hess et al.*, 2010, 2011c]. The filamentation of current carrying Alfvén waves has also been proposed in association with the fast inward motion of empty flux tubes in the Jovian inner magnetosphere [*Hess et al.*, 2011b]. The flux of antiplanetward electrons accelerated by the Alfvén waves may be at the origin of the System III modulation of the hot electrons and of the ionization state of sulfur ions in the Io torus [*Steffl et al.*, 2006].

This filamentation explains how sufficient power can escape the plasma torus. The Alfvén velocity inside the torus controls not only both the duration of the Alfvénic transition regime and its power but also affects, in an opposite way, the efficiency of the transfer to the precipitating electrons. A low Alfvén velocity near the satellite ensures a long and powerful Alfvénic interaction, but it traps the Alfvén waves inside the torus, preventing intense auroral emissions. The filamentation of the Alfvén waves breaks this paradox by allowing the short wavelengths to pass through the torus. Moreover, these short wavelengths accelerate the electrons more efficiently, generating more intense emissions, even if a large part of the Alfvén wave power remains trapped in the equatorial plasma torus. Figure 2 summarizes the effects of shorter parallel and perpendicular wavelength on the power transmission.

REFERENCES

Bagenal, F. (1994), Empirical model of the Io plasma torus: Voyager measurements, *J. Geophys. Res.*, *99*(A6), 11,043–11,062.

Bigg, E. K. (1964), Influence of the satellite Io on Jupiter's decametric emission, *Nature*, *203*, 1008–1010.

Bonfond, B. (2012), When moons create aurora: The satellite footprints on giant planets, in *Auroral Phenomenology and Magnetospheric Processes: Earth and Other Planets*, Geophys. Monogr. Ser., doi:10.1029/2011GM001169, this volume.

Bonfond, B., D. Grodent, J.-C. Gérard, A. Radioti, J. Saur, and S. Jacobsen (2008), UV Io footprint leading spot: A key feature for understanding the UV Io footprint multiplicity?, *Geophys. Res. Lett.*, *35*, L05107, doi:10.1029/2007GL032418.

Bonfond, B., D. Grodent, J.-C. Gérard, A. Radioti, V. Dols, P. A. Delamere, and J. T. Clarke (2009), The Io UV footprint: Location, inter-spot distances and tail vertical extent, *J. Geophys. Res.*, *114*, A07224, doi:10.1029/2009JA014312.

Champeaux, S., T. Passot, and P. L. Sulem (1998), Transverse collapse of Alfvén wave-trains with small dispersion, *Phys. Plasmas*, *5*, 100–111.

Chust, T., A. Roux, W. S. Kurth, D. A. Gurnett, M. G. Kivelson, and K. K. Khurana (2005), Are Io's Alfvén wings filamented? Galileo observations, *Planet. Space Sci.*, *53*, 395–412.

Clarke, J. T., et al. (1996), Far-ultraviolet imaging of Jupiter's aurora and the Io "footprint," *Science*, *274*, 404–409.

Clarke, J. T., et al. (2002), Ultraviolet emissions from the magnetic footprints of Io, Ganymede and Europa on Jupiter, *Nature*, *415*, 997–1000.

Codrescu, M. V., T. J. Fuller-Rowell, and J. C. Foster (1995), On the importance of E-field variability for Joule heating in the high-latitude thermosphere, *Geophys. Res. Lett.*, *22*(17), 2393–2396.

Connerney, J. E. P., et al. (1993), Images of excited H_3^+ at the foot of the Io flux tube in Jupiter's atmosphere, *Science*, *262*, 1035–1038.

Connerney, J. E. P., M. H. Acuña, N. F. Ness, and T. Satoh (1998), New models of Jupiter's magnetic field constrained by the Io flux tube footprint, *J. Geophys. Res.*, *103*(A6), 11,929–11,939.

Davis, L., Jr., and E. J. Smith (1990), A model of Saturn's magnetic field based on all available data, *J. Geophys. Res.*, *95*(A9), 15,257–15,261.

Delamere, P. A., F. Bagenal, R. Ergun, and Y.-J. Su (2003), Momentum transfer between the Io plasma wake and Jupiter's ionosphere, *J. Geophys. Res.*, *108*(A6), 1241, doi:10.1029/2002JA009530.

Dols, V., P. A. Delamere, and F. Bagenal (2008), A multispecies chemistry model of Io's local interaction with the Plasma Torus, *J. Geophys. Res.*, *113*, A09208, doi:10.1029/2007JA012805.

Ergun, R. E., L. Ray, P. A. Delamere, F. Bagenal, V. Dols, and Y.-J. Su (2009), Generation of parallel electric fields in the Jupiter–Io torus wake region, *J. Geophys. Res.*, *114*, A05201, doi:10.1029/2008JA013968.

Fleshman, B. L., P. A. Delamere, and F. Bagenal (2010), A sensitivity study of the Enceladus torus, *J. Geophys. Res.*, *115*, E04007, doi:10.1029/2009JE003372.

Frank, L. A., and W. R. Paterson (1999), Intense electron beams observed at Io with the Galileo spacecraft, *J. Geophys. Res.*, *104*(A12), 28,657–28,669.

Galtier, S. (2009), Wave turbulence in magnetized plasmas, *Nonlinear Processes Geophys.*, *16*, 83–98.

Gérard, J.-C., and V. Singh (1982), A model of energy deposition of energetic electrons and EUV emission in the Jovian and Saturnian atmospheres and implications, *J. Geophys. Res.*, *87*(A6), 4525–4532.

Gérard, J.-C., A. Saglam, D. Grodent, and J. T. Clarke (2006), Morphology of the ultraviolet Io footprint emission and its control by Io's location, *J. Geophys. Res.*, *111*, A04202, doi:10.1029/2005JA011327.

Goertz, C. K. (1983), The Io-control of Jupiter's decametric radiation: The Alfvén wave model, *Adv. Space Res.*, *3*, 59–70.

Goldreich, P., and D. Lynden-Bell (1969), Io, a jovian unipolar inductor, *Astrophys. J.*, *156*, 59–78.

Grodent, D., J.-C. Gérard, J. Gustin, B. H. Mauk, J. E. P. Connerney, and J. T. Clarke (2006), Europa's FUV auroral tail on Jupiter, *Geophys. Res. Lett.*, *33*, L06201, doi:10.1029/2005GL025487.

Grodent, D., B. Bonfond, A. Radioti, J.-C. Gérard, X. Jia, J. D. Nichols, and J. T. Clarke (2009), Auroral footprint of Ganymede, *J. Geophys. Res.*, *114*, A07212, doi:10.1029/2009JA014289.

Gurnett, D. A., and W. R. Pryor (2012), Auroral processes associated with Saturn's moon Enceladus, in *Auroral Phenomenology and Magnetospheric Processes: Earth and Other Planets*, Geophys. Monogr. Ser., doi:10.1029/2011GM001174, this volume.

Gurnett, D. A., W. S. Kurth, A. Roux, S. J. Bolton, and C. F. Kennel (1996), Evidence for a magnetosphere at Ganymede from plasma-wave observations by the Galileo spacecraft, *Nature*, *384*, 535–537.

Hess, S., F. Mottez, and P. Zarka (2007a), Jovian S burst generation by Alfvén waves, *J. Geophys. Res.*, *112*, A11212, doi:10.1029/2006JA012191.

Hess, S., P. Zarka, and F. Mottez (2007b), Io-Jupiter interaction, millisecond bursts and field-aligned potentials, *Planet. Space Sci.*, *55*, 89–99.

Hess, S., P. Zarka, F. Mottez, and V. B. Ryabov (2009), Electric potential jumps in the Io-Jupiter flux tube, *Planet. Space Sci.*, *57*, 23–33.

Hess, S. L. G., P. Delamere, V. Dols, B. Bonfond, and D. Swift (2010), Power transmission and particle acceleration along the Io flux tube, *J. Geophys. Res.*, *115*, A06205, doi:10.1029/2009JA014928.

Hess, S. L. G., B. Bonfond, P. Zarka, and D. Grodent (2011a), Model of the Jovian magnetic field topology constrained by the Io auroral emissions, *J. Geophys. Res.*, *116*, A05217, doi:10.1029/2010JA016262.

Hess, S. L. G., P. A. Delamere, F. Bagenal, N. Schneider, and A. J. Steffl (2011b), Longitudinal modulation of hot electrons in the Io plasma torus, *J. Geophys. Res.*, *116*, A11215, doi:10.1029/2011JA016918.

Hess, S. L. G., P. A. Delamere, V. Dols, and L. C. Ray (2011c), Comparative study of the power transferred from satellite-magnetosphere interactions to auroral emissions, *J. Geophys. Res.*, *116*, A01202, doi:10.1029/2010JA015807.

Jacobsen, S., F. M. Neubauer, J. Saur, and N. Schilling (2007), Io's nonlinear MHD-wave field in the heterogeneous Jovian magnetosphere, *Geophys. Res. Lett.*, *34*, L10202, doi:10.1029/2006GL029187.

Jacobsen, S., J. Saur, F. M. Neubauer, B. Bonfond, J.-C. Gérard, and D. Grodent (2010), Location and spatial shape of electron beams in Io's wake, *J. Geophys. Res.*, *115*, A04205, doi:10.1029/2009JA014753.

Jones, S. T., and Y.-J. Su (2008), Role of dispersive Alfvén waves in generating parallel electric fields along the Io-Jupiter fluxtube, *J. Geophys. Res.*, *113*, A12205, doi:10.1029/2008JA 013512.

Kivelson, M. G., K. K. Khurana, C. T. Russell, R. J. Walker, J. Warnecke, F. V. Coroniti, C. Polanskey, D. J. Southwood, and G. Schubert (1996), Discovery of Ganymede's magnetic field by the Galileo spacecraft, *Nature*, *384*, 537–541.

Knight, S. (1973), Parallel electric fields, *Planet. Space Sci.*, *21*, 741–750.

Lysak, R. L., and Y. Song (2003), Kinetic theory of the Alfvén wave acceleration of auroral electrons, *J. Geophys. Res.*, *108*(A4), 8005, doi:10.1029/2002JA009406.

Lystrup, M. B., S. Miller, T. Stallard, C. G. A. Smith, and A. Aylward (2007), Variability of Jovian ion winds: An upper limit for enhanced Joule heating, *Ann. Geophys.*, *25*, 847–853.

Mauk, B. H., D. J. Williams, and A. Eviatar (2001), Understanding Io's space environment interaction: Recent energetic electron measurements from Galileo, *J. Geophys. Res.*, *106*(A11), 26,195–26,208, doi:10.1029/2000JA002508.

Menietti, J. D., D. A. Gurnett, and I. Christopher (2001), Control of Jovian radio emission by Callisto, *Geophys. Res. Lett.*, *28*(15), 3047–3050, doi:10.1029/2001GL012965.

Moncuquet, M., F. Bagenal, and N. Meyer-Vernet (2002), Latitudinal structure of outer Io plasma torus, *J. Geophys. Res.*, *107*(A9), 1260, doi:10.1029/2001JA900124.

Neubauer, F. M. (1980), Nonlinear standing Alfvén wave current system at Io: Theory, *J. Geophys. Res.*, *85*(A3), 1171–1178.

Neubauer, F. M. (1998), The sub-Alfvénic interaction of the Galilean satellites with the Jovian magnetosphere, *J. Geophys. Res.*, *103*(E9), 19,843–19,866.

Piddington, J. H., and J. F. Drake (1968), Electrodynamic effects of Jupiter's satellite Io, *Nature*, *217*, 935–937.

Prangé, R., D. Rego, D. Southwood, P. Zarka, S. Miller, and W. Ip (1996), Rapid energy dissipation and variability of the Io-Jupiter electrodynamic circuit, *Nature*, *379*, 323–325.

Pryor, W. R., et al. (2011), The auroral footprint of Enceladus on Saturn, *Nature*, *472*, 331–333.

Queinnec, J., and P. Zarka (2001), Flux, power, energy and polarization of Jovian S-bursts, *Planet. Space Sci.*, *49*, 365–376.

Ray, L. C., and R. E. Ergun (2012), Auroral signatures of ionosphere-magnetosphere coupling at Jupiter and Saturn, in *Auroral Phenomenology and Magnetospheric Processes: Earth and Other Planets*, Geophys. Monogr. Ser., doi:10.1029/2011GM001172, this volume.

Saur, J. (2004), A model of Io's local electric field for a combined Alfvénic and unipolar inductor far-field coupling, *J. Geophys. Res.*, *109*, A01210, doi:10.1029/2002JA009354.

Saur, J., and D. F. Strobel (2005), Atmospheres and plasma interactions at Saturn's largest inner icy satellites, *Astrophys. J.*, *620*, L115–L118.

Saur, J., F. M. Neubauer, D. F. Strobel, and M. E. Summers (1999), Three-dimensional plasma simulation of Io's interaction with the Io plasma torus: Asymmetric plasma flow, *J. Geophys. Res.*, *104*(A11), 25,105–25,126.

Saur, J., F. M. Neubauer, D. F. Strobel, and M. E. Summers (2002a), Interpretation of Galileo's Io plasma and field observations: I0, I24, and I27 flybys and close polar passes, *J. Geophys. Res.*, *107*(A12), 1422, doi:10.1029/2001JA005067.

Saur, J., H. Politano, A. Pouquet, and W. H. Matthaeus (2002b), Evidence for weak MHD turbulence in the middle magnetosphere of Jupiter, *Astron. Astrophys.*, *386*, 699–708.

Serio, A. W., and J. T. Clarke (2008), The variation of Io's auroral footprint brightness with the location of Io in the plasma torus, *Icarus*, *197*, 368–374.

Sharma, R. P., M. Malik, and H. D. Singh (2008), Nonlinear theory of kinetic Alfvén waves propagation and multiple filament formation, *Phys. Plasmas*, *15*(6), 062902, doi:10.1063/1.2927445.

Sittler, E. C., et al. (2008), Ion and neutral sources and sinks within Saturn's inner magnetosphere: Cassini results, *Planet. Space Sci.*, *56*, 3–18.

Steffl, A. J., P. A. Delamere, and F. Bagenal (2006), Cassini UVIS observations of the Io plasma torus: III. Observations of temporal and azimuthal variability, *Icarus*, *180*, 124–140.

Su, Y.-J., R. E. Ergun, F. Bagenal, and P. A. Delamere (2003), Io-related Jovian auroral arcs: Modeling parallel electric fields, *J. Geophys. Res.*, *108*(A2), 1094, doi:10.1029/2002JA009247.

Su, Y.-J., S. T. Jones, R. E. Ergun, F. Bagenal, S. E. Parker, P. A. Delamere, and R. L. Lysak (2006), Io-Jupiter interaction: Alfvén wave propagation and ionospheric Alfvén resonator, *J. Geophys. Res.*, *111*, A06211, doi:10.1029/2005JA011252.

Swift, D. W. (2007), Simulation of auroral electron acceleration by inertial Alfvén waves, *J. Geophys. Res.*, *112*, A12207, doi:10.1029/2007JA012423.

Wannawichian, S., J. T. Clarke, and J. D. Nichols (2010), Ten years of Hubble Space Telescope observations of the variation of the Jovian satellites' auroral footprint brightness, *J. Geophys. Res.*, *115*, A02206, doi:10.1029/2009JA014456.

Watt, C. E. J., and R. Rankin (2008), Electron acceleration and parallel electric fields due to kinetic Alfvén waves in plasma with similar thermal and Alfvén speeds, *Adv. Space Res.*, *42*, 964–969.

Williams, D. J., B. H. Mauk, R. E. McEntire, E. C. Roelof, T. P. Armstrong, B. Wilken, J. G. Roederer, S. M. Krimigis, T. A. Fritz, and L. J. Lanzerotti (1996), Electron beams and ion composition measured at Io and in its torus, *Science*, *274*, 401–403.

Wright, A. N. (1987), The interaction of Io's Alfven waves with the Jovian magnetosphere, *J. Geophys. Res.*, *92*(A9), 9963–9970.

P. A. Delamere, Laboratory for Atmospheric and Space Physics, University of Colorado, Boulder, CO 80309, USA.

S. L. G. Hess, LATMOS, Université de Versailles, IPSL/CNRS, St-Quentin en Yvelines, France. (sebastien.hess@latmos.ipsl.fr)

Auroral Processes Associated With Saturn's Moon Enceladus

D. A. Gurnett

Department of Physics and Astronomy, University of Iowa, Iowa City, Iowa, USA

W. R. Pryor

Science Department, Central Arizona College, Coolidge, Arizona, USA

Observations from the Cassini spacecraft have shown that Saturn's small moon Enceladus emits a geyser-like plume of water vapor and small icy particles from volcano-like vents in its southern polar region. It has also been shown that the interaction of this plume with the rapidly rotating magnetosphere of Saturn produces UV auroral emissions in Saturn's atmosphere near the foot of the moon's magnetic flux tube. Just how the charged particles responsible for the aurora are accelerated is a topic of considerable current interest. In this chapter, we give an overview of auroral processes associated with Enceladus. We show that the interaction of the plume with Saturn's corotating magnetospheric plasma leads to a wide variety of effects, including strong local distortions of the planetary magnetic field, the acceleration of electron beams, the generation of whistler mode radio emissions, and the excitation of a standing Alfvén wave that links Enceladus to Saturn's upper atmosphere. Many of these effects are similar to those observed near Jupiter's moon Io, which is known to produce auroral emissions near the foot of its magnetic flux tube, and to those occurring in the Earth's aurora.

1. INTRODUCTION

In early 2005, the Cassini spacecraft, which was placed in orbit around Saturn on 1 July 2004, began a series of close flybys of Saturn's moon Enceladus. This small icy moon has a radius of only 252 km and orbits Saturn near the equatorial plane at a radial distance of 3.95 R_S (radius of Saturn = 60,268 km). From previous images taken by Voyagers 1 and 2 during the 1980–1981 flybys of Saturn, it was known that the moon displayed evidence of geologic activity [*Smith et al.*, 1981]. Also, the maximum density of the E ring occurred near the orbit of Enceladus, which suggested that the moon is

the source of the small micron-sized particles in this ring [*Stone and Owen*, 1984]. Despite this evidence that something unusual was occurring at Enceladus, it came as a surprise when strong magnetic field perturbations were observed near the moon during the first Cassini close flyby of Enceladus on 17 February 2005. The initial interpretation was that the magnetic field perturbations were due to the interaction of Saturn's rapidly corotating magnetosphere with a dense cloud of gas originating from the moon [*Dougherty et al.*, 2006]. This interpretation was subsequently confirmed by Cassini imaging observations that showed a geyser-like plume of material originating from a system of volcano-like vents near the south pole of the moon [*Porco et al.*, 2006]. Two images showing the sunlight scattered from small particles in the plume are shown in Figure 1. Several distinct plumes can be seen near the moon in the first image (Figure 1a), all originating from vents in the southern polar region. Figure 1b, which has enhanced sensitivity, shows that

Auroral Phenomenology and Magnetospheric Processes: Earth and Other Planets

Geophysical Monograph Series 197

10.1029/2011GM001174

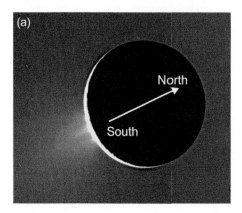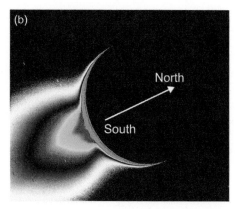

Figure 1. (a) High-phase angle clear-filter images taken by the Cassini Imaging Science Subsystem showing the plume of material ejected from the southern polar region of Enceladus. From *Porco et al.* [2006] (http://www.sciencemag.org/content/311/5766/1393.abstracta). Reprinted with permission from AAAS. (b) An enhanced color-coded version of Figure 1a.

the plume extends southward over a region that has vertical and horizontal extents comparable to the size of the moon.

Because of the remarkable processes occurring in and near Enceladus, the orbit of Cassini was subsequently adjusted to give a series of passes near and through the plume. UV absorption features observed in starlight that passed through the plume and in situ measurements of the small dust particles revealed that the plume is composed primarily of water vapor and micron and submicron water ice particles [*Hansen et al.*, 2006; *Spahn et al.*, 2006]. Mass spectrometer measurements showed that, in addition to water vapor, the plume also contains a wide range of heavier molecules, including complex hydrocarbons [*Waite et al.*, 2011]. Particle and field measurements showed that the plume has a very strong effect on the corotating magnetospheric plasma, which streams by at a nominal approach speed of 26.4 km s^{-1}. These effects include (1) charge exchange reactions between the neutral gas in the plume and the corotating magnetospheric ions, with the attendant acceleration of the newly created pickup ions to energies of several hundred eV by the corotational electric field [*Tokar et al.*, 2006], (2) local depletion of radiation belt particles due to impacts with the moon [*Jones et al.*, 2006], and (3) strong evidence that charged dust particles play an important role in the interaction [*Spahn et al.*, 2006; *Farrell et al.*, 2010; *Simon et al.*, 2011; *Kriegel et al.*, 2011; *Shafiq et al.*, 2011].

During the Voyager 1 flyby of Jupiter in 1979, it was discovered that its moon Io has active volcanism and that the resulting gas cloud around the moon interacts strongly with Jupiter's rapidly rotating magnetospheric plasma. Magnetic field measurements made during the Voyager 1 flyby of Io [*Ness et al.*, 1979; *Neubauer*, 1980] showed that a standing Alfvén wave (also called an Alfvén wing) is excited by

the moon, thereby confirming a prediction made many years earlier by *Goldreich and Lynden-Bell* [1969] to explain Io's control of Jupiter's decametric radio emissions [*Bigg*, 1964]. Several years later, aurora emissions were observed in Jupiter's atmosphere near the foot of the Io flux tube [*Connerney et al.*, 1993; *Clarke et al.*, 1996], further demonstrating the strength of the Io interaction. The recent discovery of geysers at Enceladus prompted an intense search for similar effects at Saturn. Although an initial search for auroral emissions in Saturn's atmosphere near the foot of Enceladus' magnetic field line was unsuccessful [*Wannawichian et al.*, 2008], a more extensive search by *Pryor et al.* [2011] reported clear evidence of such emissions. Pryor et al. also identified magnetic field-aligned electron beams in the downstream wake of the moon with sufficient energy flux to drive the auroral emissions. At about the same time, *Gurnett et al.* [2011] reported observations of a standing Alfvén wave excited by Enceladus and of field-aligned electron beams and radio emissions originating from the vicinity of the moon. In this chapter, we give an overview of these and other auroral processes associated with Enceladus and compare these to similar processes at Io and in the Earth's auroral regions.

2. THE AURORA FOOTPRINT OF ENCELADUS

To search for aurora near the foot of Enceladus' magnetic flux tube, *Pryor et al.* [2011] used measurements from the Cassini Ultraviolet Imaging Spectrometer (UVIS). This instrument provides spectra in both the extreme ultraviolet (EUV) and the far ultraviolet (FUV) parts of the spectrum. The EUV and FUV channels have narrow 2 × 60 and 1.5 × 60 mrad fields of view. Spatial scans of Saturn's atmosphere were obtained by slewing the spacecraft. To provide the best opportunity for detecting auroral emissions, the search was confined to times when Cassini was near periapsis at high latitudes over the mostly dark northern polar region. Of approximately 316 UV images of the region near the foot of Enceladus' flux tube, only 6 images were found with auroral spots of the right size and location to be associated with Enceladus. Two such images, obtained at subspacecraft latitudes from 74° to 65° and radial distances from 8.1 to 6.0 R_S, are shown in Figure 2. Figure 2a was taken from 02:16 to 03:28 UT on 26 August 2008, and Figure 2b was taken about an hour and a quarter later, from 03:38 to 04:50 UT. The bright circular emission at about 75° latitude is Saturn's auroral oval [*Gérard et al.*, 2004; *Grodent et al.*, 2011]. The bright spots marked by the white boxes at about 64.5°N latitude are emissions associated with Enceladus' flux tube. The boxes, which have widths of 4° in latitude and 10° in longitude, are centered on the magnetic field line through the center of Enceladus using the offset spin-aligned magnetic dipole model of *Burton et al.* [2009]. In computing these boundaries, the altitude of the emission was assumed to be 1100 km, which is the altitude that UV emissions are expected to be generated by auroral electrons impacting Saturn's mostly molecular hydrogen atmosphere [*Gérard et al.*, 2009]. As can be seen, the agreement with the predicted location of the UV emission is very good. The approximately

0.9 h shift in the local time of the auroral spot from Figure 1a to Figure 1b is caused by the orbital motion of Enceladus, which has a period of 1.37 days. Although the UV emissions are weaker and occur much less frequently than those associated with Io, these observations provide compelling evidence that Enceladus is exciting auroral emissions near the foot of its magnetic flux tube.

3. PARTICLE AND FIELD MEASUREMENTS NEAR ENCELADUS

At the time of this writing (31 October 2011), a total of 16 flybys of Enceladus have been carried out by Cassini. The flybys are designated E0 through E15. The flyby geometries are quite varied but primarily consist of two types: (1) passes obtained on equatorial orbits that are targeted to pass either upstream or downstream of the moon or, in some cases, offset to the north or south so as to pass over the poles and (2) passes on high inclination orbits that are designed to pass directly through the plume on steeply inclined north/south trajectories. Since it is not possible to describe all of the observations, we have decided to discuss two passes, E4 and E8, which provide a good overview of the observed phenomena. The spacecraft trajectories for these flybys are shown in Figure 3 using a corotational-aligned coordinate

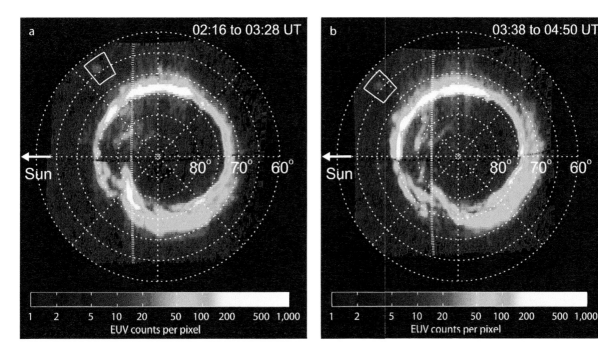

Figure 2. Two extreme UV images of the northern polar region obtained by the Ultraviolet Imaging Spectrometer on Cassini, separated by about an hour and a quarter on 26 August 2008. Reprinted by permission from Macmillan Publishers Ltd: *Nature* [*Pryor et al.*, 2011], copyright 2011.

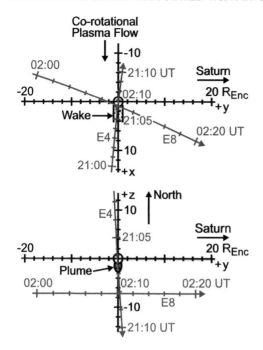

Figure 3. Spacecraft trajectories in an Enceladus-centered corotationally aligned coordinate system for the E4 and E8 flybys. The *z* axis is parallel to Saturn's rotational axis, the +*x* axis is in the direction of the nominal corotational plasma flow, and the +*y* axis, which is directed toward Saturn, completes the right-hand coordinate system.

system centered on the moon. One reason for selecting these two passes is that the field of view of the electron spectrometer (ELS), which is part of the Cassini plasma spectrometer (CAPS), was aligned nearly along the magnetic field direction. This orientation is desirable because field-aligned electron beams are expected to be the most likely source of the aurora at the foot of the magnetic field line. Very few flybys have this favorable orientation.

3.1. E4 Flyby

An overview of the particle and fields measurements obtained during the E4 flyby is shown in Figure 4. We start by discussing the electric field intensities detected by the radio and plasma wave science (RPWS) instrument. These are shown in Figure 4a, which gives a color-coded representation of the electric field intensities as a function of frequency and time. Although the RPWS instrument was designed to detect radio and plasma waves, the intense noise on the right side of the spectrogram is caused by dust particles striking the spacecraft. Because the spacecraft is moving at a very high velocity, ~17 km s^{-1}, relative to Enceladus (and to the dust particles), when a small particle strikes the spacecraft, it is instantly vaporized and ionized, thereby causing a rapidly

expanding cloud of hot electrons that produces a voltage pulse on the RPWS electric antenna. A sample voltage pulse is shown by the small plot inserted in the spectrogram. The particle size threshold for detecting such impacts is believed to be a few microns. By counting the voltage pulses, the impact rate can be determined. Around the time of peak intensity (red in the spectrogram), roughly 21:06 to 21:08 UT, the impact rate is approximately 1300 impacts per second. As can be seen from the arrow marked "north-to-south equator crossing," the highest intensities occur in the region just south of the equator, consistent with passage through the plume as determined from optical observations. It is interesting to note that there is still a significant level of impulsive dust impact noise as early as 21:00 UT, more than 20 R_S north of the moon. Although most of the dust is concentrated in the plume, some dust is observed over a very large region around the moon.

Next, we describe the plasma and magnetic field measurements obtained during the E4 flyby. An energy-time spectrogram of the electron count rate detected by the CAPS ELS is shown in Figure 4b. The corresponding pitch angles detected by the ELS are shown in Figure 4c. These are generally less than about 20° to 30°, i.e., nearly field-aligned. As can be seen, very intense electron fluxes are present during the approach to the moon, from about 20:54 to 21:05 UT. From the spacecraft trajectory in Figure 3 and noting that the magnetic field is almost directly southward (see Figure 4d), one can see that the electron fluxes are occurring along magnetic field lines that pass through the downstream wake region. The electron energy spectrum has a bimodal distribution, with an intense narrow peak at about 1 keV, and a broader peak at about 10 eV. Both have substantial temporal variations. Since the spacecraft is north of the moon and the magnetic field is directed toward the south, the pitch angles are such that the electrons are moving upward toward the moon. The magnetic field measurements in Figure 4d show that large magnetic field disturbances are present in the region where the intense electron fluxes are observed, especially in the B_x component. A negative ΔB_x disturbance relative to the magnetic field model (shown by the dashed lines) is indicative of a magnetic field line that is draped around the moon due to mass loading, as discussed by *Dougherty et al.* [2006]. There are two regions with such a draped field configuration, the first indicated by the small negative ΔB_x disturbance from about 20:56:00 to 20:58:30 UT, nearly 20 R_E downstream of the moon, and the second indicated by the larger negative ΔB_x disturbance from about 21:04:30 to 21:07:00 UT, immediately downstream of the moon. In between these two disturbances, ΔB_x is positive, indicating an antidraped magnetic field. Just what causes these relatively large downstream magnetic field fluctuations

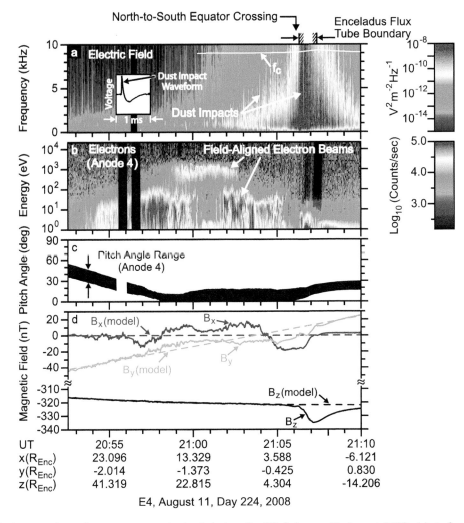

Figure 4. An overview of measurements obtained during the E4 flyby on 11 August 2008. (a) A frequency-time spectrogram of the electric field intensity detected by the radio and plasma wave science (RPWS) instrument. (b) An energy-time spectrogram of the electron count rate from the Cassini plasma spectrometer (CAP) electron spectrometer (ELS) and (c) The corresponding pitch angles. (d) A plot of the magnetic field components measured by the magnetometer (MAG).

is not known for certain. *Pryor et al.* [2011] suggested that they may be caused by Alfvén waves that are reflecting back and forth between the two hemispheres in the downstream region. However, they are also reminiscent of a downstream flow instability, such as the Kármán vortex street commonly seen downstream of a cylindrical object in a hydrodynamic flow.

3.2. E8 Flyby

An overview of the E8 particle and field measurements is shown in Figure 5, using the same basic format as in Figure 4. In sharp contrast to the E4 flyby, Figure 5a shows that

there is relatively little evidence of dust impacts in the electric field spectrogram. This is because the E8 flyby is a relatively distant flyby, passing over the south pole of the moon at a radial distance of 7.2 R_S (see Figure 3). Although there are relatively few dust impacts, a very clearly defined V-shaped radio emission can be seen that has its apex centered almost exactly on the time of closest approach. This type of radio emission is commonly observed over the terrestrial auroral zones and is called "auroral hiss" because of its close association with the aurora [see *Gurnett*, 1966; *Gurnett et al.*, 1983]. The V-shaped frequency-time characteristic is a propagation effect that arises for whistler mode waves propagating at wave normal angles near the resonance

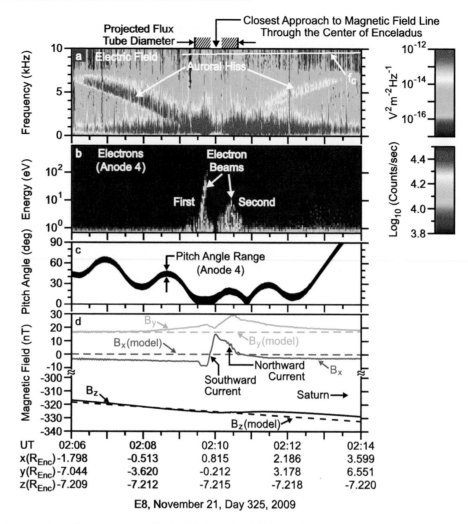

Figure 5. An overview of measurements obtained during the E8 flyby on 21 November 2009 using the format shown in Figure 4. (a) A frequency-time spectrogram of the electric field intensity detected by RPWS. (b) An energy-time spectrogram of the electron count rate from the CAP ELS and (c) the corresponding pitch angles. (d) A plot of the magnetic field components measured by the MAG. From *Gurnett et al.* [2011].

cone [*Mosier and Gurnett*, 1969; *James*, 1976]. It is well established that terrestrial auroral hiss is generated by field-aligned electron beams with energies ranging from a few tens of eV to several keV [*Ergun et al.*, 2003]. Indeed, the energy-time spectrogram in Figure 5b shows that two field-aligned electron beams are observed in the ELS data near the apex of the V-shaped emission. The magnetic field data in Figure 5d also show that two large oppositely directed ramp-like magnetic field disturbances occur in this same region. These ramp-like magnetic field variations are clearly indicative of field-aligned currents, first southward, from about 02:09:45 to 02:09:58 UT, followed almost immediately by a northward current, from 02:09:58 to 02:10:32 UT. This combination of two oppositely directed currents on a size scale

comparable to the diameter of the moon provides strong evidence that the spacecraft passed directly through a shear-mode Alfvén wave excited by the moon. For a further discussion of the Alfvén wave interpretation, see *Gurnett et al.* [2011].

Studies of auroral hiss on Earth show that the emission is produced by an electron beam-plasma interaction [*Maggs*, 1976] at the Landau resonance velocity, $v_\parallel = \omega/k_\parallel$. Terrestrial studies also show that auroral hiss can be used as a remote sensing tool to determine where the electron beam is accelerated [*Mosier and Gurnett*, 1969; *James*, 1976; *Ergun et al.*, 2003]. This technique relies on the fact that for whistler mode propagation near the resonance cone, the wave energy propagates at a known angle to the magnetic field. For the plasma parameters that exist near Enceladus, where the electron cyclotron frequency is much less than the electron plasma frequency, $f_c \ll f_p$, the angle of the ray path, ψ, relative to the magnetic field is given by the simple equation, $\sin \psi = f/f_c$, where f is the wave frequency. From this equation, one can see that the radiation from a point source is beamed along a cone-shaped surface, the opening angle of which increases with increasing frequency. It is this frequency dependence that gives the auroral hiss its characteristic V-shaped frequency-time dependence, first decreasing in frequency as the spacecraft approaches the magnetic field line through the source and then increasing as the spacecraft moves away. Since the magnetic field is known to a very good approximation in the region around the moon, it is a straightforward procedure to determine a source position that gives a good fit to the V-shaped frequency-time spectrum. Figure 6 shows the results of such a fitting procedure. Figure 6a shows the ray paths that give the best fits to the inbound and outbound branches of the V-shaped emission, and Figure 6b shows the quality of the fits, which are very good. The y' axis used in this analysis passes through the center of the moon and is

parallel to the trajectory in the x,y plane [see *Gurnett et al.*, 2011]. As can be seen, both the inbound and outbound sources are located very close to the moon, slightly south and slightly downstream, consistent with a source in or very near the plume.

4. DISCUSSION

We have described Cassini observations of aurora at the foot of the Enceladus flux tube and particle and field measurements from two close flybys of the moon that are relevant to the processes responsible for producing the aurora. The E4 flyby, which was on magnetic field lines that intersected the downstream wake, showed very intense field-aligned electron beams with energies ranging from about 10 eV to 1 keV. The beams, which are highly variable, extend over a considerable region downstream of the moon with peak energy fluxes of 1 to 2 mW m^{-2}. The field of view of the electron detector was such that it only detected electrons moving up the magnetic field line toward the moon, with no measurements in the opposite direction. Therefore, it is not clear whether the electrons are being accelerated toward the moon by some acceleration process farther down the field line near Saturn or whether they consist of magnetically reflected bidirectional beams accelerated near the moon, such as those observed by *Williams et al.* [1996] downstream of Io. In contrast, the E8 flyby, which was a distant pass over the south pole of the moon, clearly showed two field-aligned electron beams arriving from the moon. Whistler mode auroral hiss was also observed during this pass that is almost certainly generated by these beams. Ray path analyses of the auroral hiss showed that the source of these emissions, and by inference, the region where the electron beams are being accelerated, was located very close to the moon, probably in or near the plume. For a further discussion see the work of *Gurnett et al.* [2011]. Although models of the plasma interaction by *Saur et al.* [2007], *Wannawichian et al.* [2008], *Kriegel et al.* [2009], *Jia et al.* [2010], and others had suggested that an Alfvén wave might be excited by the interaction of the corotating plasma with the moon, the E8 flyby, for the first time, clearly showed that a standing Alfvén wave is, in fact, generated by Enceladus. Whether the electron beams are directly related to the Alfvén wave is not clear. The beam observed from 02:10:05 to 02:10:40 UT is in the correct location and direction to carry the northward current associated with the Alfvén wave. However, the relationship of the other beam, from 02:09:30 to 02:09:45 UT, to the Alfvén wave is not so clear, since it occurs in a region where there appears to be little or no current.

Although electron beams with significant energies and intensities have been observed in the vicinity of Enceladus,

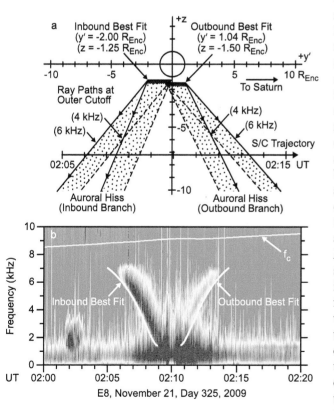

Figure 6. (a) Best-fit ray paths for the inbound and outbound branches of the V-shaped auroral hiss emission in Figure 5a and (b) corresponding best fits to the outer envelopes of the auroral hiss spectrum. From *Gurnett et al.* [2011].

it is not yet clear how they are accelerated or whether they are able to reach Saturn's atmosphere with enough energy to produce the observed aurora intensities. In fact, since the beams observed during the E4 and E8 flybys were in substantially different locations relative to the moon (downstream versus near the flux tube through the moon), it is not clear that the acceleration mechanisms are even related. For the intense beams observed moving upward on the downstream magnetic field lines during the E4 flyby, it is not known whether they can reach Saturn's atmosphere even if they have a bidirectional distribution. Although the beams were described as "field-aligned" by *Pryor et al.* [2011], the atmospheric loss cone has an angle of only 6°. Since the angular resolution of the measurements is only about 15° to 20° (see Figure 4c), one cannot be certain that a sufficient fraction of the observed energy flux, ~1 to 2 mW m^{-2}, can actually reach the foot of the field line to produce the observed aurora. For the beams observed in association with the standing Alfvén wave during the E8 flyby, there are similar uncertainties. The ray path studies of the auroral hiss observed during the E8 flyby strongly indicate that these beams are accelerated very close to Enceladus, probably in or near the plume. Since the beams observed during the E8 flyby are considerably less energetic than those observed during the E4 flyby, it is doubtful that they have enough energy to produce the aurora. We can only speculate how the electrons observed during the E8 flyby are accelerated. Possible acceleration mechanisms might include (1) electron inertial or kinetic effects associated with the Alfvén wave [*Lysak and Song*, 2003; *Watt and Rankin*, 2008], (2) parallel electric fields [*Knight*, 1973] that develop in the Alfvén wave current system, or (3) electrical effects produced by charged dust in the plume [*Kriegel et al.*, 2011; *Morooka et al.*, 2011]. In considering the origin of the aurora, one must also consider the electromagnetic energy flux carried by the Alfvén wave. For the observed amplitude of the magnetic field perturbation associated with the Alfvén wave, $\Delta B \sim 12$ nT, and the nominal plasma parameters observed during the E8 flyby ($B_0 = 325$ nT, $n = 47$ cm^{-3}, ion mass ~ 18 amu), the wave energy flux at the foot of the magnetic field line is estimated to be about 4 mW m^{-2}, which is quite significant, greater than the energy flux of the electron beams observed during the E4 flyby. This Alfvén wave energy could very well be converted to electron beam energy as the wave propagates downward and interacts with Saturn's ionosphere. For a discussion of possible processes for converting the Alfvén wave energy to field-aligned electron beams, see the work of *Hess et al.* [2011].

Acknowledgments. The research at the University of Iowa was supported by JPL contract 1415150 and the research at Central Arizona College was supported by the college, by a JPL Cassini subcontract to LASP/University of Colorado, and by a Cassini Data Analysis Program grant to Space Environment Technologies.

REFERENCES

Bigg, E. K. (1964), Influence of the satellite Io on Jupiter's decametric emission, *Nature*, *203*, 1008–1010, doi:10.1038/2031008a0.

Burton, M. E., M. K. Dougherty, and C. T. Russell (2009), Model of Saturn's internal planetary magnetic field based on Cassini observations, *Planet. Space Sci.*, *57*, 1706–1713, doi:10.1016/j.pss.2009.04.008.

Clarke, J. T., et al. (1996), Far-ultraviolet imaging of Jupiter's aurora and the Io footprint, *Science*, *274*, 404–409, doi:10.1126/science.274.5286.404.

Connerney, J. E. P., R. L. Baron, T. Satoh, and T. Owen (1993), Images of excited H$_3^+$ at the foot of the Io flux tube in Jupiter's atmosphere, *Science*, *262*, 1035–1038, doi:10.1126/science.262.5136.1035.

Dougherty, M. K., K. K. Khurana, F. M. Neubauer, C. T. Russell, J. Saur, J. S. Leisner, and M. E. Burton (2006), Identification of a dynamic atmosphere at Enceladus with the Cassini magnetometer, *Science*, *311*, 1406–1409, doi:10.1126/science.1120985.

Ergun, R. E., C. W. Carlson, J. P. McFadden, R. J. Strangeway, M. V. Goldman, and D. L. Newman (2003), Fast auroral snapshot satellite observations of very low frequency saucers, *Phys.Plasmas*, *10*, 454–462, doi:10.1063/1.1530160.

Farrell, W. M., W. S. Kurth, R. L. Tokar, J.-E. Wahlund, D. A. Gurnett, Z. Wang, R. J. MacDowall, M. W. Morooka, R. E. Johnson, and J. H. Waite Jr. (2010), Modification of the plasma in the near-vicinity of Enceladus by the enveloping dust, *Geophys. Res. Lett.*, *37*, L20202, doi:10.1029/2010GL044768.

Gérard, J.-C., D. Grodent, J. Gustin, A. Saglam, J. T. Clarke, and J. T. Trauger (2004), Characteristics of Saturn's FUV aurora observed with the Space Telescope Imaging Spectrograph, *J. Geophys. Res.*, *109*, A09207, doi:10.1029/2004JA010513.

Gérard, J.-C., B. Bonfond, J. Gustin, D. Grodent, J. T. Clarke, D. Bisikalo, and V. Shematovich (2009), Altitude of Saturn's aurora and its implications for the characteristic energy of precipitated electrons, *Geophys. Res. Lett.*, *36*, L02202, doi:10.1029/2008GL036554.

Goldreich, P., and D. Lynden-Bell (1969), Io, a Jovian unipolar inductor, *Astrophys. J.*, *156*, 59–78, doi:10.1086/149947.

Grodent, D., J. Gustin, J.-C. Gérard, A. Radioti, B. Bonfond, and W. R. Pryor (2011), Small-scale structures in Saturn's ultraviolet aurora, *J. Geophys. Res.*, *116*, A09225, doi:10.1029/2011JA016818.

Gurnett, D. A. (1966), A satellite study of VLF hiss, *J. Geophys. Res.*, *71*(23), 5599–5615, doi:10.1029/JZ071i023p05599.

Gurnett, D. A., S. D. Shawhan, and R. R. Shaw (1983), Auroral hiss, Z mode radiation, and auroral kilometric radiation in the polar magnetosphere: DE 1 observations, *J. Geophys. Res.*, *88*(A1), 329–340, doi:10.1029/JA088iA01p00329.

Gurnett, D. A., et al. (2011), Auroral hiss, electron beams and standing Alfvén wave currents near Saturn's moon Enceladus, *Geophys. Res. Lett.*, *38*, L06102, doi:10.1029/2011GL046854.

Hansen, C. J., L. Esposito, A. I. F. Stewart, J. Colwell, A. Hendrix, W. Pryor, D. Shemansky, and R. West (2006), Enceladus' water vapor plume, *Science*, *311*, 1422–1425, doi:10.1126/science.1121254.

Hess, S. L. G., P. A. Delamere, V. Dols, and L. C. Ray (2011), Comparative study of the power transferred from satellite-magnetosphere interactions to auroral emissions, *J. Geophys. Res.*, *116*, A01202, doi:10.1029/2010JA015807.

James, H. G. (1976), VLF saucers, *J. Geophys. Res.*, *81*(4), 501–514.

Jia, Y.-D., C. T. Russell, K. K. Khurana, G. Toth, J. S. Leisner, and T. I. Gombosi (2010), Interaction of Saturn's magnetosphere and its moons: 1. Interaction between corotating plasma and standard obstacles, *J. Geophys. Res.*, *115*, A04214, doi:10.1029/2009JA014630.

Jones, G. H., E. Roussos, N. Krupp, C. Paranicas, J. Woch, A. Lagg, D. G. Mitchell, S. M. Krimigis, and M. K. Dougherty (2006), Enceladus' varying imprint on the magnetosphere of Saturn, *Science*, *311*, 1412–1415, doi:10.1126/science.1121011.

Knight, S. (1973), Parallel electric fields, *Planet. Space Sci.*, *21*, 741–750, doi:10.1016/0032-0633(73)90093-7.

Kriegel, H., S. Simon, J. Müller, U. Motschmann, J. Saur, K.-H. Glassmeier, and M. K. Dougherty (2009), The plasma interaction of Enceladus: 3D hybrid simulations and comparison with Cassini MAG data, *Planet. Space Sci.*, *57*, 2113–2122, doi:10.1016/j.pss.2009.09.025.

Kriegel, H., S. Simon, U. Motschmann, J. Saur, F. M. Neubauer, A. M. Persoon, M. K. Dougherty, and D. A. Gurnett (2011), Influence of negatively charged plume grains on the structure of Enceladus' Alfvén wings: Hybrid simulations versus Cassini Magnetometer data, *J. Geophys. Res.*, *116*, A10223, doi:10.1029/2011JA016842.

Lysak, R. L., and Y. Song (2003), Kinetic theory of the Alfvén wave acceleration of auroral electrons, *J. Geophys. Res.*, *108*(A4), 8005, doi:10.1029/2002JA009406.

Maggs, J. E. (1976), Coherent generation of VLF hiss, *J. Geophys. Res.*, *81*(10), 1707–1724.

Morooka, M. W., J.-E. Wahlund, A. I. Eriksson, W. M. Farrell, D. A. Gurnett, W. S. Kurth, A. M. Persoon, M. Shafiq, M. André, and M. K. G. Holmberg (2011), Dusty plasma in the vicinity of Enceladus, *J. Geophys. Res.*, *116*, A12221, doi:10.1029/2011JA017038

Mosier, S. R., and D. A. Gurnett (1969), VLF measurements of the Poynting flux along the geomagnetic field with the Injun 5 satellite, *J. Geophys. Res.*, *74*(24), 5675–5687, doi:10.1029/JA074i024p05675.

Ness, N. F., M. H. Acuna, R. P. Lepping, L. F. Burlaga, K. W. Behannon, and F. M. Neubauer (1979), Magnetic field studies at Jupiter with Voyager 1: Preliminary results, *Science*, *204*, 982–987, doi:10.1126/science.206.4421.966.

Neubauer, F. M. (1980), Nonlinear standing Alfvén wave current system at Io: Theory, *J. Geophys. Res.*, *85*(A3), 1171–1178, doi:10.1029/JA085iA03p01171.

Porco, C. C., et al. (2006), Cassini observes the active south pole of Enceladus, *Science*, *311*, 1393–1401, doi:10.1126/science.1123013.

Pryor, W. R., et al. (2011), The Enceladus auroral footprint at Saturn, *Nature*, *472*, 331–333, doi:10.1038/nature09928.

Saur, J., F. M. Neubauer, and N. Schilling (2007), Hemisphere coupling in Enceladus' asymmetric plasma interaction, *J. Geophys. Res.*, *112*, A11209, doi:10.1029/2007JA012479.

Shafiq, M., J.-E. Wahlund, M. W. Morooka, W. S. Kurth, and W. M. Farrell (2011), Characteristics of the dust-plasma interaction near Enceladus' south pole, *Planet. Space Sci.*, *59*, 17–25, doi:10.1016/j.pss.2010.10.006.

Simon, S., J. Saur, H. Kriegel, F. M. Neubauer, U. Motschmann, and M. K. Dougherty (2011), Influence of negatively charged plume grains and hemisphere coupling currents on the structure of Enceladus' Alfvén wings: Analytical modeling of Cassini magnetometer observations, *J. Geophys. Res.*, *116*, A04221, doi:10.1029/2010JA016338.

Smith, B. A., et al. (1981), Encounter with Saturn: Voyager 1 imaging science results, *Science*, *212*, 163–191, doi:10.1126/science.212.4491.163.

Spahn, F., et al. (2006), Cassini dust measurements at Enceladus and implications for the origin of the *E*-ring, *Science*, *311*, 1416, doi:10.1126/science.1121375.

Stone, E. C., and T. C. Owen (1984), The Saturn system, in *Saturn*, edited by T. Gehrels and M. S. Matthews, p. 15, Univ. of Ariz. Press, Tucson.

Tokar, R. L., et al. (2006), The interaction of the atmosphere of Enceladus with Saturn's plasma, *Science*, *311*, 1409–1412, doi:10.1126/science.1121061.

Waite, J. H., et al. (2011), Enceladus' plume composition, paper presented at the EPSC/DPS Joint Meeting 2011, Eur. Planet. Network, Nantes, France, 2–7 Oct.

Wannawichian, S., J. T. Clarke, and D. H. Pontius Jr. (2008), Interaction evidence between Enceladus' atmosphere and Saturn's magnetosphere, *J. Geophys. Res.*, *113*, A07217, doi:10.1029/2007JA012899.

Watt, C. E. J., and R. Rankin (2008), Electron acceleration and parallel electric fields due to kinetic Alfvén waves in plasma with similar thermal and Alfvén speeds, *Adv. Space Res.*, *42*, 964–969, doi:10.1016/j.asr.2007.03.030.

Williams, D. J., B. H. Mauk, R. E. McEntire, E. C. Roelof, T. P. Armstrong, B. Wilken, J. G. Roederer, S. M. Krimigis, T. A. Fritz, and L. J. Lanzerotti (1996), Electron beams and ion composition measured at Io and in its torus, *Science*, *274*, 401–403, doi:10.1126/science.274.5286.401.

D. A. Gurnett, Department of Physics and Astronomy, University of Iowa, Iowa City, IA 52242, USA. (donald-gurnett@uiowa.edu)

W. R. Pryor, Science Department, Central Arizona College, Coolidge, AZ 85128, USA.

Section V
Aurora and Magnetospheric Dynamics

Auroral Signatures of the Dynamic Plasma Sheet

A. Keiling,[1] K. Shiokawa,[2] V. Uritsky,[3] V. Sergeev,[4] E. Zesta,[5] L. Kepko,[3] and N. Østgaard[6]

Understanding the physical connections of the coupled magnetosphere-ionosphere system will result in a more complete explanation of the aurora and will further the goal of being able to interpret the global auroral distributions as a dynamic map of the magnetosphere. Significant advances have been made in recent years toward this goal. In this chapter, we briefly review, while focusing on recent observations, selected auroral phenomena that are driven by magnetospheric processes. These include expansion of substorm aurora, plasma sheet waves and auroral modulations, ballooning instability and auroral beads, dipolarization/plasma injections and the auroral bulge, poleward boundary intensifications and auroral streamers, vortical flows and auroral spirals, and plasma flows prior to auroral onset. In addition, other auroral phenomena, such as auroral arcs, diffuse auroras, auroral asymmetry/conjugacy in the conjugate hemispheres, and auroral magnetospheric currents, are highlighted here but expanded in other chapters of this monograph for in-depth reviews.

1. INTRODUCTION

Not surprisingly, early scientific understanding of the aurora began with regions closest to the Earth. Lower-boundary altitudes and the connection between the aurora and electrical currents were established early in the last century. Spectroscopic identification of optical emission lines came later, followed by the sounding-rocket-based discovery in the early 1960s that electron beams are the principal source of excited neutral atoms and molecules. The following decades brought global-scale satellite-based images showing the entire auroral distribution from space and, more recently, detailed, high-resolution measurements of microscale plasma physics operating in the auroral acceleration zone ($<2\ R_E$ above the planet). Beyond these altitudes, there have been fewer reported magnetospheric signatures that have been linked directly to any particular auroral form. Yet it is clear that the source or generator of auroral forms and of the scattering responsible for diffuse aurora lies well beyond $2\ R_E$, relatively deep in the magnetosphere. While these processes operate in the magnetosphere, their effects are modulated by magnetosphere-ionosphere (M-I) coupling, including the auroral acceleration region, and by the ionosphere itself. Thus, a fundamental understanding of the whole system is needed in order to address even the most basic outstanding questions surrounding the aurora: (1) What determines the overall shapes, spatial scales, spacing, location, motion, and lifetimes of auroral forms? (2) How do ionospheric electrodynamics affect the aurora and the generator regions? (3) Where in the magnetotail are the generator regions of auroral structures, and what are their physical mechanisms?

While the previous parts of this monograph deal with the aurora as it relates to ionospheric electrodynamics (section 3) and the auroral acceleration region (section 4), here we review

[1]Space Sciences Laboratory, University of California, Berkeley, California, USA.

[2]Solar-Terrestrial Environment Laboratory, Nagoya University, Nagoya, Japan.

[3]NASA Goddard Space Flight Center, Greenbelt, Maryland, USA.

[4]Institute of Physics, St. Petersburg State University, St. Petersburg, Russia.

[5]Air Force Research Laboratory RVBXP, Kirtland Air Force Base, New Mexico, USA.

[6]Department of Physics and Technology, University of Bergen, Bergen, Norway.

Auroral Phenomenology and Magnetospheric Processes: Earth and Other Planets

Geophysical Monograph Series 197
10.1029/2012GM001231

recent observations of outer magnetospheric (magnetotail) signatures that have been linked directly to particular auroral forms: (1) expansion of substorm aurora, (2) plasma sheet waves and auroral modulations, (3) ballooning instability and auroral beads, (4) dipolarization/plasma injection and the auroral bulge, (5) poleward boundary intensifications and streamers, (6) vortical flow and auroral spiral, and (7) plasma flow prior to auroral onset. In addition, we briefly mention several other auroral phenomena at the end of this chapter, for which special chapters in this monograph exist (see references in each section). These phenomena are (1) auroral arcs, (2) diffuse auroras, (3) auroral asymmetry/conjugacy in the conjugate hemispheres, and (4) aurora and magnetospheric currents.

While it is unequivocal that many spacecraft missions that probed the outer magnetosphere (e.g., ISEE 1/2, CRRES, Geotail, Polar, and Cluster) in conjunction with ground-based and space-borne imagers (e.g., IMAGE) have contributed enormously to our understanding of the coupled magnetotail-ionosphere system, it can equally be said that the current Time History of Events and Macroscale Interactions during Substorms (THEMIS) project [*Angelopoulos*, 2008] has given a new boost for the ongoing auroral investigations because its uniquely designed orbits maximize coordinated investigations of space and ground observations. THEMIS is comprised of five satellites, which were launched on 17 February 2007 into equatorial orbits, and an extensive network of ground observatories, covering a 12 h local time sector over the North American continent. In addition to magnetometers, the ground observatories monitor the nightside auroral oval with fast-exposure (1 s), white-light all-sky imagers (ASIs). The satellites, on the other hand, provide the complementary measurements of fields and particles in the magnetotail.

We emphasize that we do not comprehensively review each phenomenon, but rather open a small window into the most recent results and research activities, drawing largely from research conducted with the THEMIS mission. The text and interpretations in each section of this chapter draw heavily from the key papers of each section, including excerpts. For earlier reviews, the readers are given citations in the relevant sections here. Furthermore, we focus on illustrating the observations, leaving simulations and theory to the referenced papers. Clearly, such a limited survey will only whet the appetite for the interested readers. While the focus in this chapter is on the Earth's aurora, there are many chapters in this monograph that deal with auroras on other planets, such as Mars, Jupiter, and Saturn, to which the interested reader is referred.

2. EXPANSION OF SUBSTORM AURORA

One of the most dramatic auroral displays is the auroral substorm. During its course, auroral arcs in the nightside auroral zone intensify, expand, and break up into smaller formations [*Akasofu*, 1976]. *Akasofu* [this volume] gives a historical description of the development of auroral morphology, leading to the concept of auroral substorms, and a brief review of the major features of auroral substorms. Although the details of an auroral substorm are rather complicated, a well-known characteristic is the expansion of the brightening aurora during the substorm expansion phase. Most prominent are the westward and poleward expansions, but eastward and equatorward expansions are also well documented [e.g., *Akasofu*, 1976; *Nakamura et al.*, 1993]. As inferred from ground-based all-sky auroral measurements, the expansion speeds in various directions range from less than 1 km s^{-1} to more than 10 km s^{-1} [e.g., *Solovyev et al.*, 1997; *Yago et al.*, 2005; *Shiokawa et al.*, 2005; *Liang et al.*, 2008; *Sakaguchi et al.*, 2009; *Ogasawara et al.*, 2011].

The auroral substorm has its counterpart in the magnetosphere, the magnetospheric substorm. While the goal is to understand the latter, the former allows us to indirectly monitor the latter on a global scale. If we understand the coupling between magnetosphere and ionosphere, we can use auroral observations to make inferences about the magnetosphere. For example, the bright, expanding auroras during substorm expansion phase, collectively referred to as the auroral bulge, are connected to the substorm onset region in the plasma sheet, and the speed of expansion has implications for the investigation of the onset mechanism (see section 5 of this chapter for some justification). We next give three examples, showing recent research on the westward, eastward, and equatorward auroral expansion. Although not reviewed here, we also refer the reader to the recent study by *Ogasawara et al.* [2011], who investigated 16 events of azimuthally expanding auroras.

Angelopoulos et al. [2008a] investigated the motion of a westward traveling surge (WTS) in the source plasma sheet. The WTS is the westward termination of the auroral bulge and typically shows as a large-scale fold in ASI images (see examples in the work of *Akasofu* [this volume]). Three THEMIS satellites, azimuthally aligned in the near-Earth plasma sheet, observed in sequence the substorm dipolarization and earthward flow bursts, indicating that the westward propagation speed (~250 km s^{-1}) of the substorm current wedge in space matched the propagation speed (1 h magnetic local time (MLT) min^{-1} ~ 14 km s^{-1}) of the WTS on the ground. While consistent with single-satellite observations, these observations were the first multisatellite observations to infer agreement of the expansion process in space and on the ground [*Angelopoulos et al.*, 2008a].

For a different substorm event, *Shiokawa et al.* [2009] presented simultaneous space-ground observations of the eastward extension of the initial brightening arc. The speed of the eastward arc expansion was ~2.7 km s^{-1}, as inferred from ASI images (Figure 1). The longitudinally distributed THEMIS satellites in the source region of the brightening arc observed field dipolarization, weak earthward flow, and pressure increase (Figure 1), which propagated eastward at a speed of ~50 km s^{-1} at ~12 R_E (geocentric distance). The authors thus suggested that the eastward extension of the substorm brightening arc is caused by earthward flow braking processes, which produce field dipolarization and pressure increase propagating in longitude in the near-Earth plasma sheet.

After auroral onset, the initial brightening auroral arc develops into the poleward expanding auroral bulge, which is usually bounded by a bright auroral form at its poleward edge and by the WTS at its western edge, with often less distinct auroras at its eastern edge. In contrast, its equatorward edge may terminate without any distinct arc delineating its border and, instead, shows north-south-aligned auroral forms [*Nakamura et al.*, 1993; *Sergeev et al.*, 2010]. Investigating this equatorward edge, *Sergeev et al.* [2010] concluded that the equatorward expansion of the auroral bulge corresponds to the innermost extent of earthward propagating dipolarization fronts in the magnetotail, whereas individual equatorward moving auroral enhancements correspond to the motion of individual injection fronts reaching at times distances as close to the Earth as 5.5 R_E. Using a TV camera, *Sergeev et al.* [2010] showed an initial equatorward expansion speed of 0.5–0.8 km s^{-1} (see additional data and discussions in section 5 below).

The multipoint observations of the source plasma sheet by the THEMIS satellites with conjugate ground-based auroral images provide evidence that the earthward plasma flow plays an important role in causing auroral brightening (see also section 8 below) and expansion at substorm onset. In section 5 of this chapter, we further review dipolarization and plasma injection inside the plasma sheet and their association with the expansion of the auroral bulge.

3. BALLOONING INSTABILITY AND AURORAL BEADS

The existence of wave activity in the near-Earth plasma sheet surrounding substorm onset is well established (see reviews by *Lui* [1996] and *Ohtani* [2001]). While high-frequency components have been associated with the mechanism disrupting the current sheet, low-frequency components are possibly associated with an MHD instability [e.g., *Miura et al.*, 1989; *Roux et al.*, 1991; *Cheng*, 1991; *Ohtani and Tamao*, 1993; *Pu et al.*, 1997; *Bhattacharjee et al.*, 1998; *Cheng and Lui*, 1998] that might play a role in preconditioning the current sheet leading to its subsequent disruption. The association of a ballooning instability with auroral signatures goes back to the study by *Roux et al.* [1991]. The authors interpreted magnetic field oscillations at geosynchronous distance as a westward propagating wave resulting from an Alfvén ballooning instability and suggested that this instability is the cause of the WTS. Another prominent auroral signature that has been associated with the ballooning instability is azimuthally spaced auroral forms, also called auroral beads (Figure 2). Using Viking spacecraft-based observations, *Elphinstone et al.* [1995] summarized various properties of auroral beads. The azimuthal wavelength was found to range between 132 and 583 km (corresponding to mode numbers 30 to 135) while spanning up to 8 h of local time. Auroral beads are typically observed before onset of substorms and, as such, were classified as a growth phase activity causally associated with substorm onset [*Elphinstone et al.*, 1995; *Murphree and Johnson*, 1996; *Voronkov et al.*, 1999; *Donovan et al.*, 2008; *Liang et al.*, 2008; *Liu and Liang*, 2009; *Rae et al.*, 2009].

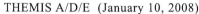

THEMIS A/D/E   (January 10, 2008)

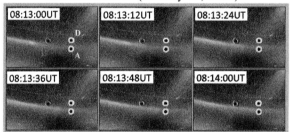

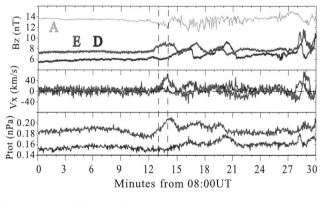

Figure 1. (top) Auroral images showing the eastward auroral extension at a substorm onset on 10 January 2008. Yellow, red, and green circles indicate the footprints of THEMIS D, E, and A, respectively. (bottom) Magnetic field, bulk flow velocity, and total pressures, as obtained by THEMIS A (green), D (blue), and E (red). The magnetic field and the flow velocities are in GSM coordinates. Only the magnetic field data were available for THEMIS A. The vertical dashed lines at 0813:00 and 0814:00 UT indicate the timings when the eastward extending arc reached the footprints of THEMIS E and D, respectively. Modified from *Shiokawa et al.* [2009].

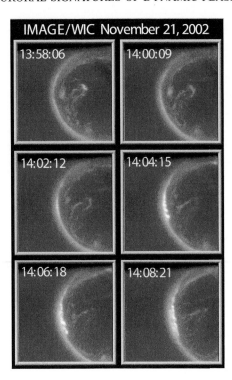

Figure 2. Sequence of IMAGE FUV/WIC images showing the evolution of the northern auroral distribution during a substorm that occurred on 21 November 2002. Auroral beads can be seen in the images taken at 14:02:12 and 14:04:15 UT. Modified from *Henderson* [2009].

Although *Elphinstone et al.* [1995] already suggested that auroral beads have their source in a ballooning instability in the near-Earth plasma sheet, direct observational evidence was not provided. In fact, the existence and role of ballooning modes has been a controversial subject in the literature. Recently, there has been new supporting evidence for ballooning modes [*Saito et al.*, 2008; *Keiling et al.*, 2008; *Liang et al.*, 2009; *Zhu et al.*, 2009]. For example, *Saito et al.* [2008] and *Keiling et al.* [2008], using single-spacecraft and multiple-spacecraft data, respectively, argued for a drift ballooning mode of measured field and particle perturbations. In spite of progress, the properties of the various types of ballooning modes are still not well characterized observationally. Moreover, their coupling to the ionosphere is poorly understood; in particular, whether or not they can cause auroral beads is not fully established. Recent THEMIS observations have shed further light on this auroral phenomenon. Here we review the event study by *Keiling et al.* [2008], for which *Raeder et al.* [this volume] conducted additional simulations. The reader is also referred to the study by *Liang et al.* [2009] who also gave evidence for the connection of ballooning mode and auroral beads.

On 23 March 2007, a substorm occurred followed by a series of spatially separated auroral intensifications. Figure 3a shows two substorm features: an expanding auroral bulge between 00 and 02 MLT (image 1) and auroral beads west of the bulge between 21 and 00 MLT at 65° latitude (images 1 and 2). The spatial separation between beads is approximately 0.4 MLT, corresponding to an azimuthal mode number of 60, which is within the range reported by *Elphinstone et al.* [1995]. Starting at ~11:12 UT, the optical intensity of the beads grew with time, until they disappeared. Shortly after disappearance (1–2 min), an intensification (image 3) occurred at the same location where the beads had occurred. This development is also reflected in the *H* component of nearby ground stations (not shown here). During the beads'

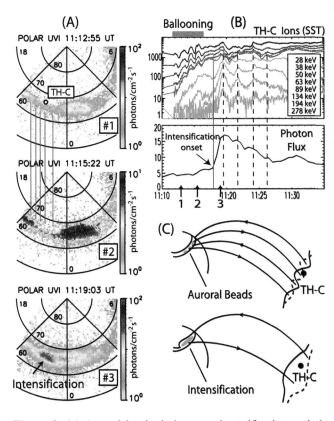

Figure 3. (a) Auroral beads during a preintensification period, observed from a global space imager (Polar UVI). Note that different color scales were used to enhance the beads. (b) Comparison of ion injection data recorded by THEMIS C and auroral luminosity. The photon flux was averaged at 65° latitude, spanning the sector 21:00 to 00:00 magnetic local time (MLT). The period of interest here is indicated by the red bar above the top panel. The arrows numbered 1 through 3 indicate the times when the UVI images were taken. (c) Cartoon illustrating the connection between the observed wave ripples in space and auroral beads and the subsequent auroral intensification. Modified from *Keiling et al.* [2008].

appearance, H showed a shallow negative slope, indicative of a moderate, increasing field-aligned current (FAC), followed by a sharp decrease during the intensification, indicative of a more intense current of the substorm current wedge.

In situ measurements from one THEMIS spacecraft (TH-C), which was near-conjugate to the auroral beads, show diamagnetic field and ion energy flux oscillations for energies less than 50 keV (Figure 3b). After 4–6 min of growing energy flux, ion injections with energies >100 keV took place. This temporal development has its counterpart in the auroral signature (e.g., onset time and duration). It was shown that the diamagnetic oscillations were westward drifting ripples on an energized plasma boundary, likely located in the transition region between dipole-like and taillike magnetic field lines. The wavelength of the ripples (ballooning mode) was consistent with the spacing of the auroral beads, allowing for magnetic mapping. Thus, a coupling was proposed, similar to that of *Roux et al.* [1991], in which upward and downward FACs connect perturbations of the ballooning mode structure with individual auroral beads (Figure 3c).

While it is clear that more observational evidence is needed to establish the coupling mechanism between auroral beads and ballooning modes, evidence also comes from numerical simulations. The MHD simulations by *Raeder et al.* [this volume] investigated the same substorm event (23 March 2007) by applying measured solar wind and interplanetary magnetic field data as input to the simulation. Ballooning mode structures appeared in the simulations several minutes before expansion phase onset. Mapped magnetically to the ionosphere, these structures resemble auroral beads, although it is pointed out that actual auroral luminosity was not derived due to the limitations of the simulation. In the simulation, the ballooning mode grows over several minutes and disintegrates approximately 2 min before the substorm expansion phase commences. Interestingly, the same temporal development was observed for the drift ballooning mode in space as well as for the auroral beads, as described above [*Keiling et al.*, 2008]. In spite of these similarities, several observations are not reproduced in the simulation. In the simulation, the ballooning mode is nonpropagating and has a mode number of 160. However, Raeder et al. alludes to the fact that their simulation does not include kinetic effects of ions, which might be the source for such differences. Moreover, we caution that the onset times of various events were not reproduced either [*Raeder et al.*, 2008].

4. PLASMA SHEET WAVES AND AURORAL MODULATIONS

The possibility of a causal connection between plasma sheet waves and optical auroral modulations has been inten-

sively debated in the literature since the early 1990s. Most studies, however, provided evidence only for one process. On the one hand, low-frequency wave activity in the central plasma sheet, which could produce detectable optical signatures, has been investigated based upon in situ observations [e.g., *Sergeev et al.*, 2004a, 2004b] as well as theoretical [e.g., *Golovchanskaya and Maltsev*, 2005; *Erkaev et al.*, 2009]. On the other hand, longitudinally propagating patches of auroral luminosity observed from the ground have been interpreted in the context of MHD waves in the magnetotail [*Safargaleev and Osipenko*, 2001; *Danielides and Kozlovsky*, 2003]. Recently, a correspondence between the two processes has been demonstrated for several events [*Uritsky et al.*, 2009] (see also section 3 above). The observed phenomenon, called longitudinally propagating arc wave (LPAW) by the authors, is characterized by azimuthally moving, periodic intensity enhancements inside preonset auroral arcs. LPAW typically travels westward in the premidnight auroral sector during the 10–20 min preceding auroral breakup with a velocity of 2–10 km s^{-1}, wave period of 40–110 s, and wavelength of 250–420 km. Magnetically conjugate in situ measurements show low-frequency plasma oscillations consistent with the parameters of the arc wave in the course of current sheet thinning. When mapped into the magnetotail, wavelength (4800–9400 km) and velocity (70–190 km s^{-1}) of the LPAW are compatible with observations and theoretical predictions for current sheet oscillations consistent with the drift wave mode (kink, sausage, ballooning, etc.) in a stretched magnetotail.

Figure 4 provides an example of LPAW in a preonset arc observed by the ground-based ASI at Fort Smith prior to the substorm onset at 08:04 UT on 3 March 2008. The wave activity started about 15 min before the onset and was found within the entire range of longitudes covered by ground-based all-sky imagers. To measure the parameters of LPAW, *Uritsky et al.* [2009] used the azimuthal (east-west) arc profiles or ewograms [*Donovan et al.*, 2006]. The detrended ewogram reveals a clear signature of westward LPAW. Figure 4d shows the wave signal extracted using the "surfing average" technique [*Uritsky et al.*, 2009] with a measured phase speed of $v_y \sim 2.3$ km s^{-1} and a period of about 100 s. During this event, the TH-C spacecraft was in the midtail plasma sheet ($X_{GSM} \sim -15\ R_E$), and it magnetically mapped within the field of view of the Fort Simpson ASI. Simultaneous observation of LPAW and flapping magnetotail oscillations in the form of magnetic and electric field oscillations, with a period of 2–3 min, is shown in Figure 5. During the growth of the oscillation, the electric field was behind the magnetic field by about a one-fourth period, which is consistent with the phase shift prediction for drift plasma modes. The MVA analysis for this wave event shows that the tail wave featured a dominantly azimuthal

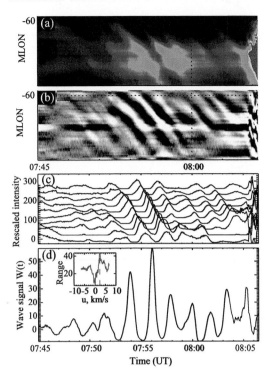

Figure 4. Longitudinal wave activity in a preonset auroral arc observed by the Fort Simpson all-sky imager (ASI) on 3 March 2008. (a) Raw ewogram representing time-evolving modulations of optical intensity along the arc. (b) Detrended ewogram with subtracted large-scale arc structure and dynamics, exhibiting stable westward propagating periodic fronts. (c) (bottom to top) Stack plot of time-varying auroral intensity at different magnetic longitudes ($-70°$ to $-60°$ MLon). (d) The wave signal extracted from the detrended ewogram using the "surfing average" technique (see text). The inset shows the dynamical range of the longitudinally propagating arc wave activity as a function of the imposed phase velocity, with a clear maximum at $v_y = 2.3$ km s^{-1}. From *Uritsky et al.* [2009].

(y) propagation front, in accordance with LPAW observations. A similar wave pattern was observed by the near-Earth TH-E spacecraft located at $X_{GSM} \sim -11 R_E$.

The plasma sheet waves were most evident in this event when both satellites underwent a transition from the central plasma sheet (plasma $\beta \sim 10$) to its boundary region (plasma $\beta < 1$) during the late growth phase. Since the auroral LPAW activity started simultaneously with the onset of the in situ wave, the observed transition is likely to reflect the global current sheet thinning process, which usually begins ~15 min prior to the onset [*Asano et al.*, 2004], the timing being roughly consistent with the emergence of LPAWs in all events shown by *Uritsky et al.* [2009]. Many wave generation mechanisms proposed for current sheet oscillations are highly contingent upon the plasma sheet geometry (e.g.,

current sheet thickness, B_z magnitude), and the observed dependence on current sheet thinning is an expected feature. More specifically, the westward traveling arc waves can be a reflection of duskward propagating magnetospheric waves, including kink, sausage, ballooning, and a variety of other drift-type (crossfield) modes in the tail. The parameters of the LPAW events reported by *Uritsky et al.* [2009] are in a good agreement with the observations and theoretical predictions for such waves, in particular, with the ballooning mode during the initial near-Earth breakup [e.g., *Park et al.*, 2010; *Zou et al.*, 2010; *Tang*, 2011].

The M-I coupling involved in the generation of LPAW is currently not fully understood. The modulation of auroral intensity by plasma sheet waves may take place in at least two ways. It is common that a drift mode (e.g., ballooning) may couple to an Alfvén wave [*Golovchanskaya and Maltsev*, 2005]; also, a drift compressional wave may modulate the loss cone and the precipitation flux of energetic particles [*Tsutomu*, 1984]. In addition, the contribution of the auroral acceleration region (1–2 R_E altitude) to the coupling should play an important part, especially for the short-wavelength LPAW modes consistent with the Kelvin-Helmholtz instability in this region [*Haerendel et al.*, 1996].

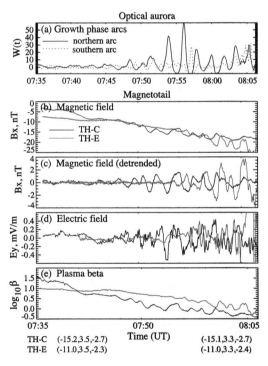

Figure 5. Simultaneous low-frequency wave oscillations (a) in the optical aurora and (b–e) in the central current sheet as measured by two THEMIS probes, prior to the substorm onset on 3 March 2008. GSM coordinates of the probes are shown below the plots. From *Uritsky et al.* [2009].

5. DIPOLARIZATION/PLASMA INJECTION AND THE AURORAL BULGE

Bright, expanding auroras during the substorm expansion phase are collectively referred to as the auroral bulge. One could argue that it is still the optical observations of the auroral bulge that provide the best tool available to monitor the development and details of the magnetospheric substorm expansion phase. However, the usefulness of this monitoring tool also depends on how well we understand the origin and mechanisms producing particular auroral forms of the auroral bulge. Although the development of the auroral bulge has been linked to the operation of current disruption and magnetic reconnection in the magnetotail [e.g., *Roux et al.*, 1991; *Yahnin et al.*, 2006], reliable evidence, which establishes this relationship, is rather limited. One piece of evidence is that auroral breakups seen by the global space-borne UVI imagers are systematically and closely associated in time with tailward reconnection outflows, when observed inside the central plasma sheet at the meridian conjugate to the breakup [*Ieda et al.*, 2008; *Miyashita et al.*, 2009].

A second piece of evidence is that the magnitude of magnetic flux threaded through the auroral bulge, at its maximal stage, is similar to the magnitude of magnetic flux accumulated in the magnetotail during the preceding growth phase, when comparing events of different intensities [*Yahnin et al.*, 2006]. Yahnin et al. also reported approximate conjugacy of the flow reversal region in the plasma sheet (mapped to the ionosphere using the T96 model [*Tsyganenko*, 1995]) with the poleward bulge latitude. This supports a scenario in which the magnetic energy storage is followed by explosive dissipation, which is primarily controlled by magnetic reconnection in the magnetotail. In that view, the bright arc at the poleward edge of the expanding bulge can be viewed as the mapping of the reconnection X line to the ionospheric. However, we also caution the reader not to ignore several objections to this simplistic view of the association with the separatrix. For example, there are doubts that the energy flux of reconnection-accelerated electrons is sufficient for providing directly the bright aurora [e.g., *Østgaard et al.*, 2009], and it is also unclear what the role of reconnection in the field-aligned electron acceleration above the ionosphere is. Moreover, a mapping of the plasma boundary in a convection-driven magnetosphere provides additional complications.

In this picture, the auroral bulge corresponds to the newly reconnected closed flux tubes in the earthward reconnection outflow, whose flux transfer contributes to dipolarization. The bulge is azimuthally confined, and observations show that the auroral brightening region expands and propagates east and west in parallel with a similar development of the dipolarized

region in the conjugate plasma sheet (see section 2 above). Particle precipitation within the bulge is enhanced considerably at energies above 30 keV, but unlike the isotropic energetic protons, the electron distributions over the loss cone are anisotropic, indicating the enhanced equatorial B_z in these plasma tubes [*Sergeev et al.*, 2010].

The poleward auroral expansion is often explained as being due to progressive magnetic reconnection and tailward retreat of the near-Earth reconnection line. However, it is also noted that because of the magnetic field line change in the dipolarized region, the ionospheric foot point of a fixed point in the equatorial plasma sheet (e.g., THEMIS spacecraft foot point in Figure 6) has to shift to more poleward latitude. To evaluate quantitatively this effect and its role in the poleward expansion, a time-dependent magnetospheric model, which closely simulates the magnetic perturbations observed in the dipolarized region, is needed. Adaptive data-based modeling has now become possible for some events due to good coverage of combined THEMIS and GOES observations. Initial results, like in Figure 6, show that the magnetic field reconfiguration during various substorm phases, rather than plasma motion in the equatorial magnetosphere, is largely responsible for the observed motion of the aurora, including poleward

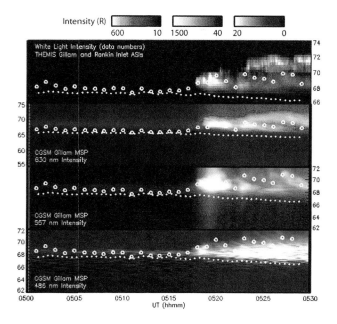

Figure 6. Auroral keograms from the Gillam observatory, prepared in three wavelengths and white light. Open circles show the THEMIS P4 probe footprints in the ionosphere, calculated from the AM03 model. Diamonds give the b2i boundary positions, obtained from optical data. From *Kubyshkina et al.* [2011].

auroral expansion during substorms [*Sergeev et al.*, 2010; *Kubyshkina et al.*, 2011].

On the other hand, the poleward auroral expansion can occur in discrete steps in the form of newly brightening regions (intensifications) at progressively higher latitudes [*Wiens and Rostoker*, 1975], and it was proposed that each intensification is connected to a different active region in the plasma sheet [e.g., *Sergeev and Yahnin*, 1979; *Lui*, 1996]. Multiple-spacecraft THEMIS observations have shown this connection for two active regions in the plasma sheet between 8 and 11 R_E [*Keiling et al.*, 2008]. Both regions underwent time-delayed major particle injections (>100 keV), which were causally connected to two progressing auroral intensifications, as observed by the global UVI imager of Polar. Although a third auroral intensification occurred, no spacecraft was probing the corresponding active plasma sheet region. The authors argued that, rather than reconnection, multiple current sheet disruption/ballooning sites developed at distances progressively farther along the magnetotail.

Much less is known about the equatorward expansion of the auroral bulge, which is also of much interest in view of its transparent association with the earthward convection in the plasma sheet and with plasma injections into the inner magnetosphere associated with flow bursts and dipolarization. Enhanced and structured auroras are observed in this region. After the breakup as well as after subsequent intensifications of poleward arcs, the structured auroras are enhanced, and activated regions expand toward the equator (Figure 7). Comparison with spacecraft observations in the conjugate near-Earth plasma sheet (using an adaptive model for the magnetic mapping) gives evidence that the equatorward edge

of the auroral bulge corresponds to the innermost extent of the earthward propagating dipolarization region in the magnetotail, whereas individual equatorward moving auroral enhancements correspond to the motion of individual injection fronts, reaching at times distances as close to the Earth as 5.5 R_E [*Sergeev et al.*, 2010]. Much more has to be learned about plasma sheet injections by studying the equatorward expanding structured auroras, a topic which was largely overlooked in past studies.

6. POLEWARD BOUNDARY INTENSIFICATIONS AND STREAMERS

Transport in the tail plasma sheet is an essential component of global magnetospheric energy, mass and magnetic flux transfer, and is often organized in azimuthally localized bursts of earthward directed fast flows (termed bursty bulk flows or BBFs) [e.g., *Angelopoulos et al.*, 1992]. Such flow channels can occur over the entire azimuthal width of the plasma sheet during all levels of geomagnetic activity, although they occur more frequently with higher AE, and can be responsible for 60%–100% [*Angelopoulos et al.*, 1992] of the measured earthward transport of mass, energy, and magnetic flux. In situ observations of tail fast flows can only offer information on a few points, at most, within the plasma sheet region, which is what makes the decoding of their auroral footprint such an important task. Several studies in the past 10 years have demonstrated general and one-to-one correlation between tail fast flows and auroral poleward boundary intensifications (PBIs) or streamers [e.g., *Fairfield et al.*, 1999; *Lyons et al.*, 1999; *Sergeev et al.*, 2000, 2004a, 2004b; *Zesta et al.*, 2000; *Nakamura et al.*, 2001; *Amm and Kauristie*, 2002; *Kauristie et al.*, 2003; *Ohtani*, 2004]. Auroral streamers are essentially PBIs that, while first appearing near the auroral poleward boundary, extend equatorward through the width of the oval. Such streamers are presumably the mapping of a fast-flow channel through the length of the magnetotail from some starting point of localized reconnection, very much in accord with the "bubble" or interchange instability scenario [*Chen and Wolf*, 1993]. While streamers are traditionally thought of as primarily north-south structures, the auroral signature of radial fast-flow channels may also appear as primarily east-west arcs due to mapping conditions from the tail and different M-I coupling processes [*Sergeev*, 2002; *Zesta et al.*, 2002].

Figure 8 gives a schematic and global view of what PBI/streamers look like in the aurora and in the magnetotail. Figure 8a is adapted from *Sergeev et al.* [2000], who postulated the existence of multiple flow channels across the width of the plasma sheet. In comparison, Figures 8b shows a nightside aurora image from the IMAGE FUV instrument

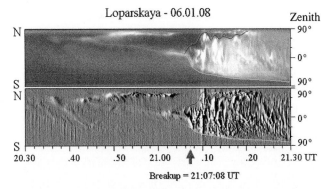

Figure 7. TV keograms from Loparskaya for 6 January 2008. Both total brightness (upper) and differential (lower) brightness keograms are presented. Breakup time is shown by an arrow. The poleward border of bright active auroras as well as the equatorward moving auroral enhancement traces, forming together the equatorward border of the auroral bulge, are marked by red and blue lines. From *Sergeev et al.* [2010].

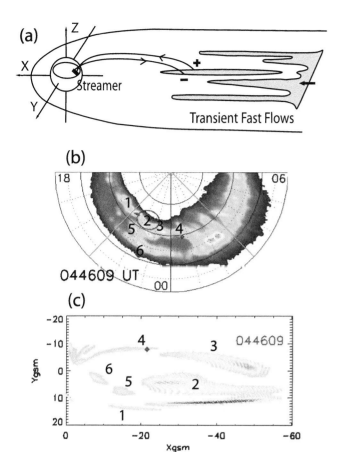

Figure 8. (a) Cartoon depiction of multiple fast-flow channels in the plasma sheet, adapted from *Sergeev et al.* [2000]. (b) Global auroral image from IMAGE FUV on 3 January 2001 when multiple poleward boundary intensification (PBI)/streamers were visible simultaneously throughout the nightside auroral oval. (c) Projections of the strongest emissions from the FUV image to the tail, reproducing a picture of multiple flow channels across the width of the plasma sheet [from *Zesta et al.*, 2006].

when multiple PBI/streamers were seen in the nightside aurora, while in situ observations by Geotail showed fast-flow channels in the plasma sheet [*Zesta et al.*, 2006]. Figure 8c shows the projection of the PBI images to the equatorial plane (region of minimum magnetic field strength B using the T96 model [*Tsyganenko*, 1995]). Only intensities above a threshold in the FUV image were projected, thus including all the individual PBI/streamers along with some equatorial auroral features, which are not PBIs. Individual features (mostly PBIs) and their projections are marked "1" through "6." If each PBI/streamer does indeed correspond to a flow channel, one can infer that many fast-flow channels exist simultaneously across the width of the plasma sheet at a given time. The similarity of the projected

PBIs/streamers in Figure 8c with the cartoon of Figure 8a is indeed striking.

Here we review a recent event study that investigated the auroral and ionospheric convection signatures of fast-flow channels, while combining them with ionospheric flow measurements from Sondrestrom incoherent scatter radar (ISR), which was in good conjunction with the foot points of the THEMIS probes [*Zesta et al.*, 2011]. The five vertical lines in Figure 9b mark key times A, B, C, D, and E. At time A, substantial tail magnetic field perturbations were observed by the THEMIS probes and a significant flow enhancement was observed by the radar (Figure 9b, bottom panel) within the polar cap, as evidenced by the low electron density levels observed by the ISR. At time B, the first fast-flow burst was observed to initiate at P1, while the ionospheric flow at magnetic latitudes $\Lambda > 71°$ (within the polar cap) significantly decreased 2 min earlier. At time C, the fast flow arrived at P4 and P3 in the inner magnetosphere at the same time as the auroral onset. Figure 9c shows the expansion shortly after with the foot points of P3 and P4 closest to the auroral onset, confirming that the fast-flow and dipolarization signatures observed in situ by them are related to this substorm onset. Time D is the onset of a series of fast flows at P1 that do not make it past P2 and are associated with a series of PBIs, shown in Figure 9d to the east of the Sondrestrom location as a series of EW PBIs.

While the first fast-flow event (times A, B, and C) was associated with a small and localized substorm onset, the second event (times D and E) was comprised of fast flows associated with a series of PBIs and having distinct characteristics. In the case of the substorm onset, ionospheric fast flows were first observed within the polar cap 19 min before onset and then were observed in situ first in midtail, at $-20\,R_E$, 4 min prior to onset. They then moved progressively closer to the Earth at $-8\,R_E$. Prolonged dipolarization only appeared in the inner magnetosphere, while the midtail probes observed only magnetic field fluctuations. The auroral onset and magnetospheric dipolarization did not happen until the tail fast flow reached the inner magnetosphere. This substorm sequence is consistent with that proposed by *Lyons et al.* [2011] [see also *Lyons et al.*, this volume], who suggest that flows well within the polar cap field lines may cross the nightside polar cap boundary into the closed field line region, as a manifestation of new reconnection flows entering the plasma sheet and moving earthward to the inner magnetosphere, where they may trigger the onset of the substorm process.

Three key differences between the tail fast flows associated with the substorm onset and the PBI events were identified by Zesta et al.: (1) the substorm tail flows penetrated to the inner magnetosphere, while the PBI fast flows attenuated before $-16\,R_E$, (2) substorm onset was at a lower latitude expanding poleward and westward, while the PBIs were in situ intensification of poleward arcs, and (3) the substorm was associated with a stronger inner magnetosphere current wedge, while the PBI was associated with a weaker, more localized, current wedge extending into the outer magnetosphere.

Finally, some recent studies [e.g., *Sergeev et al.*, 2004a, 2004b; *Kepko et al.*, 2009; *Liang et al.*, 2011] demonstrated precipitation characteristics of streamers. The accepted structure of the tail fast flow and its associated auroral signature is that upward FACs form on the westward edge of the fast-flow channel and form the auroral streamer, as illustrated in Figure 8a, which is typically the result of accelerated electron precipitation [e.g., *Sergeev et al.*, 2000, 2004a, 2004b]. *Kepko et al.* [2009] and *Liang et al.* [2011] showed that the auroral precipitation associated with streamers, although traditionally assumed to be composed of hard electrons (>1 keV), may be the result of just soft electron precipitation (see also section 8 below). *Liang et al.* [2011] were able to show from a variety of auroral observations that the source population of the precipitating aurora is only a few hundred eV. They also suggested different M-I coupling mechanisms (ELF wave particle interactions in the plasma sheet and parallel potential acceleration of soft electrons) that likely produced the precipitation signatures of the observed streamers. It is certainly interesting to determine under what conditions fast flows in the plasma sheet may produce precipitation from accelerated particles and when other processes, like wave-particle interactions, produce the precipitation signatures and which contributes more to transport and energization. Such studies will hopefully continue and provide a much needed insight into the detailed physics of the M-I coupling of tail fast flows to their ionospheric auroral signatures.

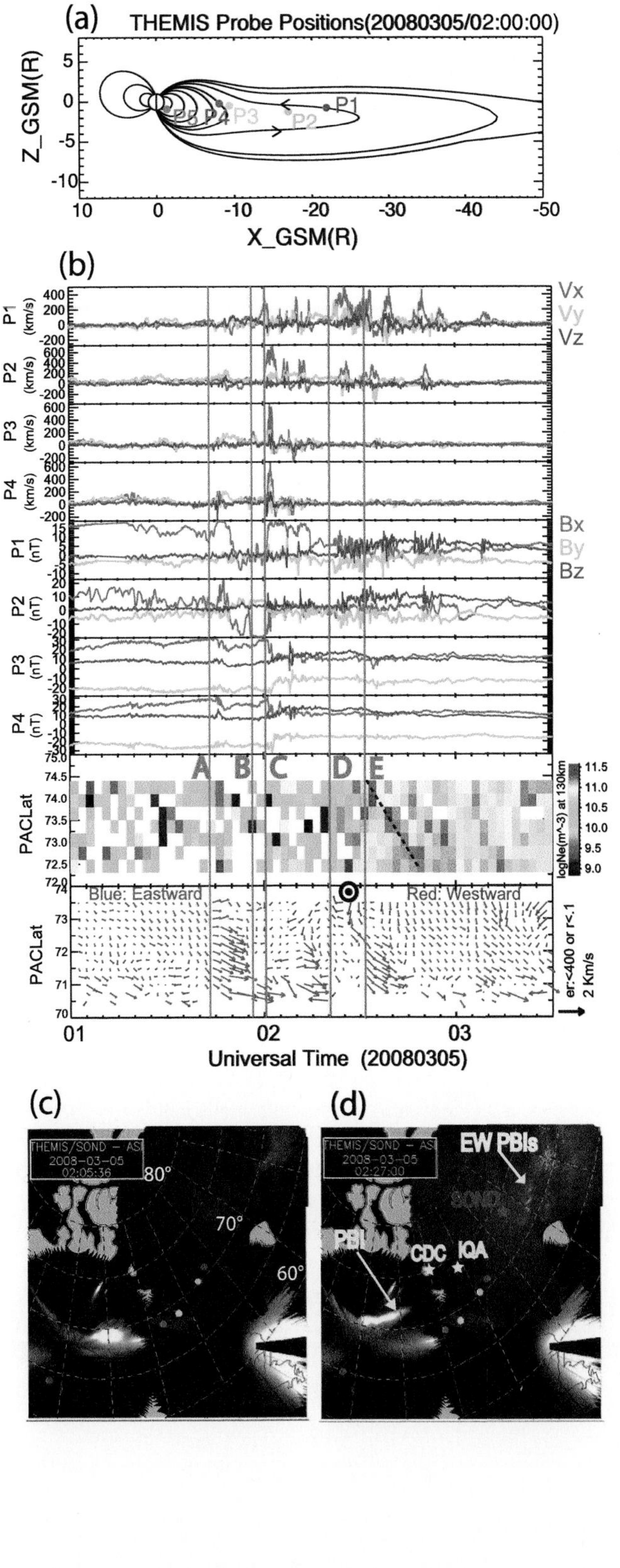

Figure 9. (a) THEMIS probe tail alignment at 02:00 UT on 5 March 2008. Probes P1–P4 were inside the tail plasma sheet from 00 UT to past 06 UT, while probe P5 was very near Earth and was not used in their study. (b) (top to bottom) The three GSM components of the ion velocity and magnetic field from P1, P2, P3, and P4. (bottom two panels) The Sondrestrom incoherent scatter radar E region electron number density at 130 km, and the F region flow vector as a function of UT and Λ, covering 70.5°–73.5°. Flow vectors with an eastward component are plotted blue, and those with a westward component are plotted red, and midnight MLT is ~02:00 UT. (c and d) Two composite auroral mosaics from the THEMIS ASIs at GILL and SNKQ and from the Sondrestrom ASI during the substorm expansion at 02:05:36 UT (first event) and during the set of PBIs at 02:27:00 UT (second event). The foot points of the THEMIS probes are shown in red, green, light blue, blue, and purple, respectively, and the geomagnetic coordinate grid is superposed as white dashed lines. Note that the THEMIS ASIs are white light cameras, while the Sondrestrom ASI observations are of 630 nm emissions and are depicted in red. From *Zesta et al.* [2011].

7. VORTICAL FLOW AND AURORAL SPIRAL

The auroral spiral is a prominent vortex structure in auroral arcs. Already in the 1970s, many of its visible properties were described by *Davis and Hallinan* [1976]. A fully developed spiral can range from 20 to 1300 km in diameter. While some auroral spirals unwind after formation, others, especially larger ones, do not unwind but decay into broken, patchy auroral forms. This temporal evolution can last from a few tens of seconds to minutes. Although ample ground-based observational descriptions of auroral spirals exist [e.g., *Davis and Hallinan*, 1976; *Untiedt et al.*, 1978; *Partamies et al.*, 2001, 2006], their magnetospheric counterparts have been investigated less often. For example, *Marklund et al.* [1998] reported Freja observations from an altitude of approximately 1700 km (i.e., outside the generator region), while the spacecraft traversed magnetic field lines adjacent to an auroral spiral and found that the electric field intensified in the direction of the auroral spiral and that the spiral was associated with intense upward FACs. An association with FACs has also been observed from ground observations [e.g., *Wescott et al.*, 1975; *Untiedt et al.*, 1978]. Some evidence suggests that a line-current-like concentration of FAC occurs within the auroral spiral [*Untiedt and Baumjohann*, 1993]. The observational evidence is supported by the model of *Hallinan* [1976] who attributes spiral formation to a FAC instability. More recent simulation work can be found in the works of *Voronkov et al.* [2000] and *Watanabe* [2010].

Recent THEMIS ground-space observations allowed for a comparison of the presumed generator region in the near-Earth plasma sheet with near-conjugate auroral spirals [*Keiling et al.*, 2009a]. Figure 10a shows a composite image from several ASI cameras spread over Northern America. In the field of view of the Gillam ASI, a large auroral spiral (200–300 km in diameter) with counterclockwise rotation had formed. Its lifespan, including formation and decay, lasted approximately 1 min. A second auroral spiral occurred about 14 min later (not shown here). The first and second appearance coincided with the substorm expansion phase and a subsequent intensification, respectively. Concurrent with both auroral spirals, a pair of equivalent ionospheric current (EIC) vortices with opposite rotations (corresponding to upward and downward currents) prevailed in the ionosphere (Figure 10b). The region of enhanced EIC currents between the centers of both EIC vortices coincided with brightened auroras and the spirals. It is noted that intense EICs can correspond to intense horizontal ionospheric plasma flows, and that an association between strong shear flow (i.e., curl of EIC) has been proposed in the past [e.g., *Steen and Collis*, 1988].

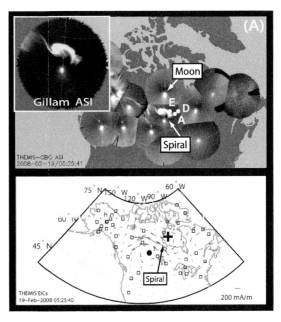

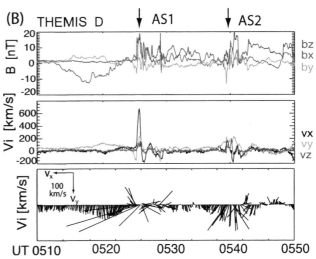

Figure 10. Ground and space data associated with an auroral spiral on 19 February 2008. (a) Mosaic of THEMIS ASI images (white light) over Northern America. Yellow dots are the footprints of the THEMIS spacecraft. The bright spot in each ASI image is the Moon. Below are equivalent ionospheric currents (EICs) in comparison to the visible aurora. The dot and cross mark the center of counterclockwise and clockwise EIC vortices, respectively. The blue vectors indicate the current direction. The opposite direction gives the plasma flow direction, assuming uniform ionospheric conductance. The open squares mark the locations of the ground stations used for the EIC calculations. (b) Spacecraft data from THEMIS D: magnetic field, ion flow velocities, vector plots of flow velocities projected onto the *X-Y* GSM plane. Modified from *Keiling et al.* [2009a].

At the time of the auroral spirals, a group of three THEMIS spacecraft (A, D, and E) were located in the nightside magnetosphere near 23 MLT and 9.5 to 12 R_E, while being close to the neutral sheet. The yellow dots in Figure 10a mark the proximate locations of the footprints. A fourth THEMIS spacecraft was located further west of the clustered spacecraft. Magnetic mapping suggests that the spacecraft foot points were near the optical spiral. Concurrent with the spiral, dipolarization, ion injection (not shown), and enhanced plasma flow occurred (see below label AS1 of Figure 10b; data from only one spacecraft are shown here). The enhanced plasma flow was identified as vortical flow, which can best be seen when plotted as projections of the velocity vectors in the X-Y (GSM) plane (last panel of Figure 10b). The other two spacecraft confirmed the existence of the flow vortex. In addition, a second, counter-rotating flow vortex, located farther west, was recorded by the fourth THEMIS spacecraft. The flow directions of the space vortices are consistent with those of the EIC vortices, suggesting a causal relationship [e.g., *Pudovkin et al.*, 1997; *Borovsky and Bonnell*, 2001]. Similar ground-space observations were recorded for the second auroral spiral (see below label AS2).

When the auroral spirals formed, the intensity of the upward FAC increased as indirectly inferred from the increasing brightness of the aurora and from associated localized H bays. In addition, *Keiling et al.* [2009a, 2009b] calculated and numerically modeled the FAC. Using the expression derived by *Hasegawa and Sato* [1979] (also in the work of *Keiling et al.* [2009b], equation (2)) for the FAC due to vorticity, it was estimated that the first vortical structure generated a current density of approximately 2.8 nA m^{-2} (14 μA m^{-2} mapped to the ionospheric altitude) or a total current of ~0.1 MA. In contrast, the numerical model accounted for not only the vortex current but also all current contributors of the substorm current wedge, as inferred from ground magnetometer data. According to the model, the upward and downward FACs, located at ~20.5 and ~23 MLT, respectively, started to develop before 05:25 UT. The meridian of the downward current coincided with those of the foot points of THEMIS A, D, and E (labels a, d, and e). The current magnitude gradually increased, reaching a peak value of ~0.7 MA. During the period of decreasing current after 05:35 UT, the current showed an upward deviation (line above the dashed line) from the general decaying trend (dashed line) at around 05:40 UT, which accounted for approximately 0.1 MA, which is similar in magnitude to the current enhancement associated with the first spiral (AS1). This time coincided with the formation of the second auroral spiral (AS2). Therefore, it is possible that the concurrent second plasma vortex in space created this current enhancement,

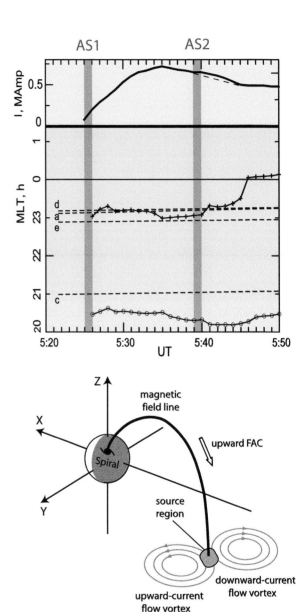

Figure 11. (top) Reconstructed current wedge current: The longitudinal locations (in MLT) of upward (circles) and downward (crosses) field-aligned currents (FACs) as well as their magnitude (in MAmp) as a function of time were modeled to fit the midlatitude magnetic variations. The vertical red bars mark 5 min intervals during which both auroral spirals occurred. The footprints of TH-A, THC, TH-D, and TH-E are indicated as dashed lines. (bottom) Illustration of the connection of an auroral spiral and its source region in the magnetosphere, located between two oppositely rotating plasma vortices. The circular shape of the vortices is idealized and is likely more stretched and irregular in reality. The source region maps into an ionospheric region of curl of EIC. From *Keiling et al.* [2009a, 2009b].

which was superimposed on the other, stronger current created by other mechanisms.

These results provide evidence that the auroral spirals were likely associated with current enhancements generated in space by the flow vortices. At the same time, other current contributors were also present that led to the large-scale substorm current. For clarification, in the scenario described above, the optical spiral was not the ionospheric footprint of the magnetospheric plasma flow vortex; rather, it was the result of some process that occurred in the transition region between the vortex centers where strong shear flows existed (i.e., the edges of the flow vortices), as illustrated in Figure 11. However, the flow vortices were essential for providing the energy and the necessary conditions for the spiral generation process. It is also cautioned that other flow patterns, as long as they provide shear flows, could potentially provide the energy for auroral spirals. To identify the mechanism, perhaps a localized current perturbation or a large-scale magnetospheric instability that causes the characteristic spiral shape, future coordinated ground-space observations will have to provide more data on the space region in the magnetotail that is fully conjugate to the auroral spiral. Obtaining such conjugacy remains a challenge given the small size of auroral spirals in comparison to the length of an auroral arc.

8. PLASMA FLOW PRIOR TO AURORAL ONSET

The timing of high-speed plasma flows in the plasma sheet with respect to auroral onset has long been used as the primary discriminator between two substorm models: the current disruption (or near-Earth initiation) model and the reconnection (or midtail initiation) model [e.g., *Ohtani*, 2004; *Sibeck and Angelopoulos*, 2008]. An earthward traveling flow burst observed in the midtail region in the minutes prior to the brightening and poleward expansion of the onset arc would be strong support in favor of the reconnection model of substorm expansion, as several prior studies have demonstrated [e.g., *Sergeev et al.*, 2005; *Angelopoulos et al.*, 2008a, 2008b; *Gabrielse et al.*, 2009; *Lin et al.*, 2009]. Despite the many events indicating high-speed plasma flows prior to auroral onset, there was always the question: Why is there no apparent auroral signature of the flow burst prior to auroral breakup? As described in section 6 earlier, PBIs, most often observed during the late expansion and recovery phases, are believed to be the auroral manifestation of flow bursts. If there are auroral signatures of flow bursts during these times, why not in the few minutes prior to auroral onset?

Recent advancements in ground-based auroral imaging have provided two separate answers to this question. The first, by *Kepko et al.* [2009], essentially argued that most prior studies examined auroral imagery in relatively energetic wavelengths and, therefore, missed a less energetic signature. Utilizing multispectral ASIs, they identified an auroral signature of a flow burst in the 630.0 nm emission prior to white light auroral onset (Figure 12). The white light images show a classic white light onset: the slow brightening of the growth phase arc followed by auroral rays, beading, and auroral expansion. At no point prior to onset is there an observable feature poleward of the onset arc location. In contrast, the 630.0 nm emissions, which are generated by low-energy electrons, showed a diffuse, equatorward moving auroral patch for several minutes prior to onset. Once this patch reached the equatorward auroral boundary, white light auroral onset followed almost immediately. THEMIS satellites were located near $X = -11 \ R_E$ and observed earthward plasma flow prior to the white light onset. The event is consistent with the traditional reconnection scenario and fills a critical missing gap in auroral observations.

In contrast, *Nishimura et al.* [2010a, 2010b] identified preonset north-south auroral structures in the white light

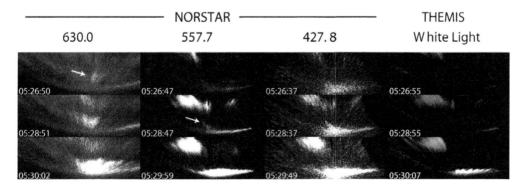

Figure 12. Images from the multispectral all-sky and THEMIS white light cameras at Gillam. North is up and west is to the left, and the perspective is above the Earth looking down. Modified from *Kepko et al.* [2009].

images. Their answer to the question is that the high temporal and spatial resolutions of the THEMIS white light ASIs compared with previous observations (e.g., global space imagers on board Polar and IMAGE) enabled identification of these precursor streamers that had been missed previously. The Nishimura et al. scenario is that auroral streamers flow down from the open/closed boundary, reaching the equatorward auroral boundary several minutes later. In the magnetosphere, these streamers correspond to earthward plasma flows. When these flows reach the inner boundary, they do not lead to auroral onset directly. Rather, if a still-undetermined set of conditions is met, the intrusion of new plasma leads to a near-Earth plasma instability and auroral expansion. This scenario is a hybrid between the two models, in effect, a triggered inside-out. We refer the reader to the work of *Lyons et al.* [this volume] for more detailed discussion.

Although both *Nishimura et al.* [2010a, 2010b] and *Kepko et al.* [2009] observed auroral signatures prior to white light auroral onset, there are a number of key observational and interpretive differences. First, the different observations reflect different energetics. The 630.0 nm emissions are created by the precipitation of low-energy electrons, while the white light emissions of the north-south streamers are likely associated with discrete acceleration. As discussed here in section 6, the north-south auroral streamer is associated with the upward FAC at the edge of the flow burst and, therefore, defines the westward (duskward) edge of the flow. In contrast, the 630.0 nm emissions likely map to the center of the flow channel. As noted by *Kepko et al.* [2009], a small, discrete arc did form on the westward boundary of the 630.0 nm emissions, consistent with the interpretation that as the flow burst reaches the near-Earth region, the FACs are increased, leading to discrete acceleration. This difference in behavior of the discrete emissions likely stems from the different characteristics of the auroral features. Note that the *Kepko et al.* [2009] event started ~1° in latitude northward of the eventual onset arc, while the *Nishimura et al.* [2010a, 2010b] events started, on average, 5° above the onset arc. Further, the 630.0 nm emissions contacted the breakup region exactly at the center of the breakup arc, while breakups associated with the north-south streamers can occur west, east, or centered on the breakup arc. Whether this latter observation is simply a reflection of statistics or whether it implies a different coupling of flows to the ionosphere remains an open question.

Despite the different interpretations of *Kepko et al.* [2009] and *Nishimura et al.* [2010a, 2010b] with respect to the details of expansion, both types of auroral observations argue for the importance of preonset, earthward plasma flows leading to auroral onset. Nevertheless, we point out that there is no consensus in the substorm community to accept one view of auroral onset mechanism. The reader is referred to, for example, *Donovan et al.* [2008], *Henderson* [2009], and *Lui* [2011] for recent studies arguing for an alternative view.

9. OTHER AURORAL PHENOMENA

9.1. Auroral Arcs

Auroral arcs are typically very elongated in an approximate east-west direction with widths of kilometers to tens of kilometers, while at times showing wavelike structures such as folds, curls, spirals, and omega bands. A more complete description of morphology and dynamics of auroral arcs is given by *Akasofu* [this volume]. While being the iconic auroral form, there is still no consensus as to the essential mechanism or mechanisms responsible for auroral arcs, albeit several models have been proposed. Although progress has been made in relating field and particle signatures in the auroral acceleration region to this auroral form [*Karlsson*, this volume], there are currently no such observations in the outer magnetosphere that can, with confidence, be associated with an individual auroral arc. Based on theoretical considerations, *Haerendel* [this volume] reviews several aspects of auroral arc formation, while *Birn et al.* [this volume] applies numerical methods to auroral arc formation.

9.2. Diffuse Auroras

In contrast to the discrete aurora, an often featureless aurora is the so-called diffuse aurora. Both electron and proton precipitations excite diffuse auroras, either caused by pitch-angle diffusion or by the breaking of the first adiabatic invariant in strongly curved magnetic fields. Dynamic features, such as the pulsating aurora, of the electron diffuse aurora are reviewed by *Lessard* [this volume] and by *Li et al.* [this volume], while the proton diffuse aurora is reviewed by *Donovan et al.* [this volume].

9.3. Auroral Asymmetry and Conjugacy in the Conjugate Hemispheres

The topics of conjugacy and asymmetry of auroras in the opposite hemispheres have obtained significant attention in recent years. Although conjugate studies date back to the 1960s, it has been a challenge to obtain data to study the dynamics of the two polar regions simultaneously. Two approaches have been pursued in the past: remote observations via spacecraft imagers and ground-based ASIs. The readers are referred to two chapters in this monograph, dealing with both approaches. *Østgaard and Laundal* [this volume] review recent studies, utilizing simultaneous global imaging data from the Polar and IMAGE spacecraft, that demonstrate the asymmetric auroral response in the conjugate hemispheres to solar wind forcing. On the other hand, *Sato et al.* [this volume] report on highly similar auroras that were simultaneously recorded with all-sky TV cameras situated at two geomagnetically conjugate points (Iceland and Antarctica). Both chapters discuss implications in the context of solar wind-magnetosphere and M-I coupling processes.

9.4. Aurora and Magnetospheric Currents

A fundamental component of auroras is the relationship between magnetospheric processes and auroral FACs. *Strangeway* [this volume] addresses this issue via the concept of force balance where the FACs provide the connection between the Lorentz force in the magnetosphere and the ionosphere. This chapter is both complementary to and a contrast with the work of *Haerendel* [this volume], since the magnetospheric processes that are responsible for FACs are essentially the same, but *Strangeway* [this volume] does not take the different types of current systems, as discussed by *Boström* [1964], as the starting point, as *Haerendel* [this volume] does.

10. CONCLUDING REMARKS

With this brief review, we have provided a limited overview of *very* recent research related to the generator regions of nightside auroras. Although significant advances have been made in the last 10 years or so, it will continue to be a key area of research for this decade and beyond since we have not reached a satisfactory level of understanding. Much more research is required to address many outstanding problems, several of which have been listed at the end of each section in this chapter. One problem that certainly applies to all reviewed phenomena is to identify the details of how the coupling between the magnetotail and specific auroral forms is achieved. Here it is important to realize that this coupling occurs in both directions. While most studies focus on the generator in the magnetotail powering the aurora in the ionosphere, we must also take into account the active role of the ionosphere when describing processes in the outer magnetosphere since they are modulated and modified by the ionosphere itself. In addition, the contribution of the auroral acceleration region, located between the ionosphere and the magnetotail, should play an important part as well.

While the various regions (ionosphere, auroral acceleration region, and generator region in the magnetotail) and their roles in controlling auroras are addressed in separate sections in this monograph, eventually, we will have to describe the entire system holistically. It is not common yet to take such a holistic approach to auroral investigations. Observationally, it is a major challenge to find suitable conjunctions with past and existing ground-based and space-borne observatories. Although a few observation-based attempts already exist, we might have to wait for future missions that routinely provide opportunities for such holistic auroral investigations. However, this holistic approach is not only an observational challenge but also a challenge for simulation work and theory, which have to overcome the difficulties of connecting the vastly different plasma regimes. Hence, important developments are also needed and expected in simulation and theoretical work in the coming decade and beyond.

Understanding the connection of the auroral generator region and the aurora will not only result in a more complete explanation of the aurora but also further the goal of being able to interpret the global auroral distributions as a dynamic map of the magnetosphere. The ultimate test of whether one really understands a complex system such as the solar wind-M-I-aurora system on Earth is to apply the underlying physical processes to other planets. However, the physical environments could not be more different. For just one example, the strong magnetic field of Jupiter is embedded in a solar wind of ~30 times weaker ram pressure and produces a vast magnetosphere with a length scale 100 times that of the Earth's. While we recognize some familiar behavior in the auroral emissions of other planets, there are also striking differences that test our basic understanding of auroral processes (see the work of *Mauk and Bagenal* [this volume] for a tutorial). In comparison to investigations of the Earth's aurora, it is an even greater challenge to obtain the same holistic data for planetary auroras, since we do not (yet) have multi-spacecraft missions (let alone ground observatories), such as Cluster and THEMIS, at other planets. While we did not touch upon auroras at other planets in this chapter, we instead refer to several other chapters in this monograph that discuss auroras on Mars, Jupiter, and Saturn.

Acknowledgments. We congratulate the entire THEMIS team (science and operation) for a successful mission, which has given auroral researchers a powerful tool for their daily work. Furthermore, Andreas Keiling thanks the members of the science program committee for their help in organizing the Aurora Chapman conference in Fairbanks, including writing the conference proposal, parts of which have been used for this chapter. This work was supported by the NASA grant NNX08AF29G and by the STEL visiting professor program at Nagoya University.

REFERENCES

Akasofu, S.-I. (1976), *Physics of Magnetospheric Substorms*, 617 pp., Reidel, Dordrecht, The Netherlands.

Akasofu, S.-I. (2012), Auroral morphology: A historical account and major auroral features during auroral substorms, in *Auroral*

Phenomenology and Magnetospheric Processes: Earth and Other Planets, *Geophys. Monogr. Ser.*, doi:10.1029/2011GM001156, this volume.

Amm, O., and K. Kauristie (2002), Ionospheric signatures of bursty bulk flows, *Surv. Geophys.*, *23*, 1–32.

Angelopoulos, V. (2008), The THEMIS mission, *Space Sci. Rev.*, *141*, 5–34, doi:10.1007/s11214-008-9336-1.

Angelopoulos, V., W. Baumjohann, C. F. Kennel, F. V. Coroniti, M. G. Kivelson, R. Pellat, R. J. Walker, H. Lühr, and G. Paschmann (1992), Bursty bulk flows in the inner central plasma sheet, *J. Geophys. Res.*, *97*(A4), 4027–4039.

Angelopoulos, V., et al. (2008a), First results from the THEMIS mission, *Space Sci. Rev.*, *141*, 453–476, doi:10.1007/s11214-008-9378-4.

Angelopoulos, V., et al. (2008b), Tail reconnection triggering substorm onset, *Science*, *321*, 931–935, doi:10.1126/science.1160495.

Asano, Y., T. Mukai, M. Hoshino, Y. Saito, H. Hayakawa, and T. Nagai (2004), Statistical study of thin current sheet evolution around substorm onset, *J. Geophys. Res.*, *109*, A05213, doi:10.1029/2004JA010413.

Bhattacharjee, A., Z. W. Ma, and X. Wang (1998), Ballooning instability of a thin current sheet in the high-Lundquist-number magnetotail, *Geophys. Res. Lett.*, *25*(6), 861–864.

Birn, J., K. Schindler, and M. Hesse (2012), Magnetotail aurora connection: The role of thin current sheets, in *Auroral Phenomenology and Magnetospheric Processes: Earth and Other Planets*, *Geophys. Monogr. Ser.*, doi:10.1029/2011GM001182, this volume.

Borovsky, J. E., and J. Bonnell (2001), The dc electrical coupling of flow vortices and flow channels in the magnetosphere to the resistive ionosphere, *J. Geophys. Res.*, *106*(A12), 28,967–28,994, doi:10.1029/1999JA000245.

Boström, R. (1964), A model of the auroral electrojets, *J. Geophys. Res.*, *69*(23), 4983–4999.

Chen, C. X., and R. A. Wolf (1993), Interpretation of high-speed flows in the plasma sheet, *J. Geophys. Res.*, *98*(A12), 21,409–21,419.

Cheng, C. Z. (1991), A kinetic-magnetohydrodynamic model for low-frequency phenomena, *J. Geophys. Res.*, *96*(A12), 21,159–21,171.

Cheng, C. Z., and A. T. Y. Lui (1998), Kinetic ballooning instability for substorm onset and current disruption observed by AMPTE/CCE, *Geophys. Res. Lett.*, *25*(21), 4091–4094.

Danielides, M. A., and A. E. Kozlovsky (2003), Rocket-born investigation of auroral patches in the evening sector during substorm recovery, *Ann. Geophys.*, *21*(6), 719–728.

Davis, T. N., and T. J. Hallinan (1976), Auroral spirals 1. Observations, *J. Geophys. Res.*, *81*(7), 3953–3958.

Donovan, E., et al. (2006), The THEMIS all-sky imaging array—System design and initial results from the prototype imager, *J. Atmos. Sol. Terr. Phys.*, *68*(13), 1472–1487.

Donovan, E., et al. (2008), Simultaneous THEMIS in situ and auroral observations of a small substorm, *Geophys. Res. Lett.*, *35*, L17S18, doi:10.1029/2008GL033794.

Donovan, E., E. Spanswick, J. Liang, J. Grant, B. Jackel, and M. Greffen (2012), Magnetospheric dynamics and the proton aurora, in *Auroral Phenomenology and Magnetospheric Processes: Earth and Other Planets*, *Geophys. Monogr. Ser.*, doi:10.1029/2011GM001241, this volume.

Elphinstone, R. D., et al. (1995), Observations in the vicinity of substorm onset: Implications for the substorm process, *J. Geophys. Res.*, *100*(A5), 7937–7969.

Erkaev, N. V., V. S. Semenov, I. V. Kubyshkin, M. V. Kubyshkina, and H. K. Biernat (2009), MHD model of the flapping motions in the magnetotail current sheet, *J. Geophys. Res.*, *114*, A03206, doi:10.1029/2008JA013728.

Fairfield, D. H., et al. (1999), Earthward flow bursts in the inner magnetotail and their relation to auroral brightenings, AKR intensifications, geosynchronous particle injections and magnetic activity, *J. Geophys. Res.*, *104*(A1), 355–370.

Gabrielse, C., et al. (2009), Timing and localization of near-Earth tail and ionospheric signatures during a substorm onset, *J. Geophys. Res.*, *114*, A00C13, doi:10.1029/2008JA013583. [Printed 115(A1), 2010].

Golovchanskaya, I. V., and Y. P. Maltsev (2005), On the identification of plasma sheet flapping waves observed by Cluster, *Geophys. Res. Lett.*, *32*, L02102, doi:10.1029/2004GL021552.

Haerendel, G. (2012), Auroral generators: A survey, in *Auroral Phenomenology and Magnetospheric Processes: Earth and Other Planets*, *Geophys. Monogr. Ser.*, doi:10.1029/2011GM001162, this volume.

Haerendel, G., B. U. Olipitz, S. Buchert, O. H. Bauer, E. Rieger, and C. LaHoz (1996), Optical and radar observations of auroral arcs with emphasis on small-scale structures, *J. Atmos. Terr. Phys.*, *58*(1–4), 71–82.

Hallinan, T. J. (1976), Auroral spirals, 2. Theory, *J. Geophys. Res.*, *81*(22), 3959–3965.

Hasegawa, A., and T. Sato (1979), Generation of field aligned current during substorm, in *Dynamics of the Magnetosphere*, edited by S.-I. Akasofu, pp. 529–542, D. Reidel, Norwell, Mass.

Henderson, M. G. (2009), Observational evidence for an inside-out substorm onset scenario, *Ann. Geophys.*, *27*, 2129–2140.

Ieda, A., et al. (2008), Longitudinal association between magnetotail reconnection and auroral breakup based on Geotail and Polar observations, *J. Geophys. Res.*, *113*, A08207, doi:10.1029/2008JA013127.

Karlsson, T. (2012), The acceleration region of stable auroral arcs, in *Auroral Phenomenology and Magnetospheric Processes: Earth and Other Planets*, *Geophys. Monogr. Ser.*, doi:10.1029/2011GM001179, this volume.

Kauristie, K., V. A. Sergeev, O. Amm, M. V. Kubyshkina, J. Jussila, E. Donovan, and K. Liou (2003), Bursty bulk flow intrusion to the inner plasma sheet as inferred from auroral observations, *J. Geophys. Res.*, *108*(A1), 1040, doi:10.1029/2002JA009371.

Keiling, A., et al. (2008), Multiple intensifications inside the auroral bulge and their association with plasma sheet activities, *J. Geophys. Res.*, *113*, A12216, doi:10.1029/2008JA013383.

Keiling, A., et al. (2009a), THEMIS ground-space observations during the development of auroral spirals, *Ann. Geophys.*, *27*, 4317–4332.

Keiling, A., et al. (2009b), Substorm current wedge driven by plasma flow vortices: THEMIS observations, *J. Geophys. Res.*, *114*, A00C22, doi:10.1029/2009JA014114. [Printed 115(A1), 2010].

Kepko, L., E. Spanswick, V. Angelopoulos, E. Donovan, J. McFadden, K.-H. Glassmeier, J. Raeder, and H. J. Singer (2009), Equatorward moving auroral signatures of a flow burst observed prior to auroral onset, *Geophys. Res. Lett.*, *36*, L24104, doi:10.1029/2009GL041476.

Kubyshkina, M., V. Sergeev, N. Tsyganenko, V. Angelopoulos, A. Runov, E. Donovan, H. Singer, U. Auster, and W. Baumjohann (2011), Time-dependent magnetospheric configuration and breakup mapping during a substorm, *J. Geophys. Res.*, *116*, A00I27, doi:10.1029/2010JA015882.

Lessard, M. R. (2012), A review of pulsating aurora, in *Auroral Phenomenology and Magnetospheric Processes: Earth and Other Planets, Geophys. Monogr. Ser.*, doi:10.1029/2011GM001187, this volume.

Li, W., J. Bortnik, Y. Nishimura, and R. M. Thorne (2012), The origin of pulsating aurora: Modulated whistler-mode chorus waves, in *Auroral Phenomenology and Magnetospheric Processes: Earth and Other Planets, Geophys. Monogr. Ser.*, doi:10.1029/2011GM001164, this volume.

Liang, J., E. F. Donovan, W. W. Liu, B. Jackel, M. Syrjäsuo, S. B. Mende, H. U. Frey, V. Angelopoulos, and M. Connors (2008), Intensification of preexisting auroral arc at substorm expansion phase onset: Wave-like disruption during the first tens of seconds, *Geophys. Res. Lett.*, *35*, L17S19, doi:10.1029/2008GL033666.

Liang, J., W. W. Liu, E. F. Donovan, and E. Spanswick (2009), In-situ observation of ULF wave activities associated with substorm expansion phase onset and current disruption, *Ann. Geophys.*, *27*, 2191–2204.

Liang, J., E. Spanswick, M. J. Nicolls, E. F. Donovan, D. Lummerzheim, and W. W. Liu (2011), Multi-instrument observations of soft electron precipitation and its association with magnetospheric flows, *J. Geophys. Res.*, *116*, A06201, doi:10.1029/2010JA015867.

Lin, N., H. U. Frey, S. B. Mende, F. S. Mozer, R. L. Lysak, Y. Song, and V. Angelopoulos (2009), Statistical study of substorm timing sequence, *J. Geophys. Res.*, *114*, A12204, doi:10.1029/2009JA014381.

Liu, W. W., and J. Liang (2009), Disruption of magnetospheric current sheet by quasi-electrostatic field, *Ann. Geophys.*, *27*(5), 1941–1950.

Lui, A. T. Y. (1996), Current disruption in the Earth's magnetosphere: Observations and models, *J. Geophys. Res.*, *101*(A6), 13,067–13,088.

Lui, A. T. Y. (2011), Reduction of the cross-tail current during near-Earth dipolarization with multisatellite observations, *J. Geophys. Res.*, *116*, A12239, doi:10.1029/2011JA017107.

Lyons, L. R., T. Nagai, G. T. Blanchard, J. C. Samson, T. Yamamoto, T. Mukai, A. Nishida, and S. Kokubun (1999), Association between Geotail plasma flows and auroral poleward boundary intensifications observed by CANOPUS photometers, *J. Geophys. Res.*, *104*(A3), 4485–4500.

Lyons, L. R., Y. Nishimura, H.-J. Kim, E. Donovan, V. Angelopoulos, G. Sofko, M. Nicolls, C. Heinselman, J. M. Ruohoniemi, and N. Nishitani (2011), Possible connection of polar cap flows to pre- and post-substorm onset PBIs and streamers, *J. Geophys. Res.*, *116*, A12225, doi:10.1029/2011JA016850.

Lyons, L. R., Y. Nishimura, X. Xing, Y. Shi, M. Gkioulidou, C.-P. Wang, H.-J. Kim, S. Zou, V. Angelopoulos, and E. Donovan (2012), Auroral disturbances as a manifestation of interplay between large-scale and mesoscale structure of magnetosphere-ionosphere electrodynamical coupling, in *Auroral Phenomenology and Magnetospheric Processes: Earth and Other Planets, Geophys. Monogr. Ser.*, doi:10.1029/2011GM001152, this volume.

Marklund, G. T., et al. (1998), Observations of the electric field fine structure associated with the westward traveling surge and large-scale auroral spirals, *J. Geophys. Res.*, *103*(A3), 4125–4144.

Mauk, B., and F. Bagenal (2012), Comparative auroral physics: Earth and other planets, in *Auroral Phenomenology and Magnetospheric Processes: Earth and Other Planets, Geophys. Monogr. Ser.*, doi:10.1029/2011GM001192, this volume.

Miura, A., S. Ohtani, and T. Tamao (1989), Ballooning instability and structure of diamagnetic hydromagnetic waves in a model magnetosphere, *J. Geophys. Res.*, *94*(A11), 15,231–15,242.

Miyashita, Y., et al. (2009), A state-of-the-art picture of substorm-associated evolution of the near-Earth magnetotail obtained from superposed epoch analysis, *J. Geophys. Res.*, *114*, A01211, doi:10.1029/2008JA013225.

Murphree, J. S., and M. L. Johnson (1996), Clues to plasma processes based on Freja UV observations, *Adv. Space Res.*, *18*, 95–105, doi:10.1016/0273-1177(95)00973-6.

Nakamura, R., T. Oguti, T. Yamamoto, and S. Kokubun (1993), Equatorward and poleward expansion of the auroras during auroral substorms, *J. Geophys. Res.*, *98*(A4), 5743–5759.

Nakamura, R., W. Baumjohann, R. Schödel, M. Brittnacher, V. A. Sergeev, M. Kubyshkina, T. Mukai, and K. Liou (2001), Earthward flow bursts, auroral streamers, and small expansions, *J. Geophys. Res.*, *106*(A6), 10,791–10,802.

Nishimura, Y., L. Lyons, S. Zou, V. Angelopoulos, and S. Mende (2010a), Substorm triggering by new plasma intrusion: THEMIS all-sky imager observations, *J. Geophys. Res.*, *115*, A07222, doi:10.1029/2009JA015166.

Nishimura, Y., et al. (2010b), Preonset time sequence of auroral substorms: Coordinated observations by all-sky imagers, satellites, and radars, *J. Geophys. Res.*, *115*, A00I08, doi:10.1029/2010JA015832.

Ogasawara, K., Y. Kasaba, Y. Nishimura, T. Hori, T. Takada, Y. Miyashita, V. Angelopoulos, S. B. Mende, and J. Bonnell (2011), Azimuthal auroral expansion associated with fast flows in the near-Earth plasma sheet: Coordinated observations of the THEMIS all-sky imagers and multiple spacecraft, *J. Geophys. Res.*, *116*, A06209, doi:10.1029/2010JA016032.

Ohtani, S.-I. (2001), Substorm trigger processes in the magnetotail: Recent observations and outstanding issues, *Space Sci. Rev.*, *95*, 347–359.

Ohtani, S.-I. (2004), Flow bursts in the plasma sheet and auroral substorm onset: Observational constraints on connection between midtail and near-Earth substorm processes, *Space Sci. Rev.*, *113*, 77–96.

Ohtani, S.-I., and T. Tamao (1993), Does the ballooning instability trigger substorms in the near-Earth magnetotail?, *J. Geophys. Res.*, *98*(A11), 19,369–19,379.

Østgaard, N., and K. M. Laundal (2012), Auroral asymmetries in the conjugate hemispheres and interhemispheric currents, in *Auroral Phenomenology and Magnetospheric Processes: Earth and Other Planets*, Geophys. Monogr. Ser., doi:10.1029/2011GM 001190, this volume.

Østgaard, N., K. Snekvik, A. L. Borg, A. Åsnes, A. Pedersen, M. Øieroset, T. Phan, and S. E. Haaland (2009), Can magnetotail reconnection produce the auroral intensities observed in the conjugate ionosphere?, *J. Geophys. Res.*, *114*, A06204, doi:10.1029/2009JA014185.

Park, M. Y., D.-Y. Lee, S. Ohtani, and K. C. Kim (2010), Statistical characteristics and significance of low-frequency instability associated with magnetic dipolarizations in the near-Earth plasma sheet, *J. Geophys. Res.*, *115*, A11203, doi:10.1029/2010JA 015566.

Partamies, N., K. Kauristie, T. I. Pulkkinen, and M. Brittnacher (2001), Statistical study of auroral spirals, *J. Geophys. Res.*, *106*(A8), 15,415–15,428, doi:10.1029/2000JA900172.

Partamies, N., K. Kauristie, E. Donovan, E. Spanswick, and K. Liou (2006), Meso-scale aurora within the expansion phase bulge, *Ann. Geophys.*, *24*, 2209–2218.

Pu, Z. Y., A. Korth, Z. X. Chen, R. H. W. Friedel, Q. G. Zong, X. M. Wang, M. H. Hong, S. Y. Fu, Z. X. Liu, and T. I. Pulkkinen (1997), MHD drift ballooning instability near the inner edge of the near-Earth plasma sheet and its application to substorm onset, *J. Geophys. Res.*, *102*, 14,397–14,406.

Pudovkin, M. I., A. Steen, and U. Brändström (1997), Vorticity in the magnetospheric plasma and its signatures in the aurora dynamics, *Space Sci. Rev.*, *80*, 411–444.

Rae, I. J., et al. (2009), Near-Earth initiation of a terrestrial substorm, *J. Geophys. Res.*, *114*, A07220, doi:10.1029/2008JA 013771.

Raeder, J., D. Larson, W. Li, L. Kepko, and T. Fuller-Rowell (2008), OpenGGCM simulations for the THEMIS mission, *Space Sci. Rev.*, *141*, 535–555, doi:10.1007/s11214-008-9421-5.

Raeder, R., P. Zhu, Y. Ge, and G. Siscoe (2012), Auroral signatures of ballooning mode near substorm onset: Open Geospace General Circulation Model simulations, in *Auroral Phenomenology and Magnetospheric Processes: Earth and Other Planets*, Geophys. Monogr. Ser., doi:10.1029/2011GM001200, this volume.

Roux, A., S. Perraut, P. Robert, A. Morane, A. Pedersen, A. Korth, G. Kremser, B. Aparicio, D. Rodgers, and R. Pellinen (1991), Plasma sheet instability related to the westward traveling surge, *J. Geophys. Res.*, *96*(A10), 17,697–17,714.

Safargaleev, V. V., and S. V. Osipenko (2001), Magnetospheric substorm precursors in pulsating and diffuse auroras, *Geomagn. Aeron.*, *41*, 756–762.

Saito, M. H., Y. Miyashita, M. Fujimoto, I. Shinohara, Y. Saito, K. Liou, and T. Mukai (2008), Ballooning mode waves prior to substorm-associated dipolarizations: Geotail observations, *Geophys. Res. Lett.*, *35*, L07103, doi:10.1029/2008GL033269.

Sakaguchi, K., K. Shiokawa, A. Ieda, R. Nomura, A. Nakajima, M. Greffen, E. Donovan, I. R. Mann, H. Kim, and M. Lessard (2009), Fine structures and dynamics in auroral initial brightening at substorm onsets, *Ann. Geophys.*, *27*, 623–630.

Sato, N., A. Kadokura, T. Motoba, K. Hosokawa, G. Bjornsson, and T. Saemundsson (2012), Ground-based aurora conjugacy and dynamic tracing of geomagnetic conjugate points, in *Auroral Phenomenology and Magnetospheric Processes: Earth and Other Planets*, Geophys. Monogr. Ser., doi:10.1029/2011GM001154, this volume.

Sergeev, V. A. (2002), Ionospheric signatures of magnetospheric particle acceleration in substorms—How to decode them?, in *Proceedings of the 6th International Conference on Substorms*, edited by R. M. Winglee, pp. 39–46, Univ. of Washington, Seattle.

Sergeev, V. A., and A. G. Yahnin (1979), The features of auroral bulge expansion, *Planet. Space Sci.*, *27*, 1429–1440, doi:10. 1016/0032-0633(79)90089-8.

Sergeev, V. A., et al. (2000), Multiple-spacecraft observation of a narrow transient plasma jet in the Earth's plasma sheet, *Geophys. Res. Lett.*, *27*(6), 851–854, doi:10.1029/1999GL010729.

Sergeev, V., A. Runov, W. Baumjohann, R. Nakamura, T. L. Zhang, A. Balogh, P. Louarnd, J.-A. Sauvaud, and H. Reme (2004a), Orientation and propagation of current sheet oscillations, *Geophys. Res. Lett.*, *31*, L05807, doi:10.1029/2003GL019346.

Sergeev, V. A., K. Liou, P. T. Newell, S.-I. Ohtani, M. R. Hairston, and F. Rich (2004b), Auroral streamers: Characteristics of associated precipitation, convection and field-aligned currents, *Ann. Geophys.*, *22*, 537–548, doi:10.5194/angeo-22-537-2004.

Sergeev, V. A., et al. (2005), Transition from substorm growth to substorm expansion phase as observed with a radial configuration of ISTP and Cluster spacecraft, *Ann. Geophys.*, *23*, 2183–2198.

Sergeev, V. A., T. A. Kornilova, I. A. Kornilov, V. Angelopoulos, M. V. Kubyshkina, M. Fillingim, R. Nakamura, J. P. McFadden, and D. Larson (2010), Auroral signatures of the plasma injection and dipolarization in the inner magnetosphere, *J. Geophys. Res.*, *115*, A02202, doi:10.1029/2009JA014522.

Shiokawa, K., K. Yago, K. Yumoto, D. G. Baishev, S. I. Solovyev, F. J. Rich, and S. B. Mende (2005), Ground and satellite observations of substorm onset arcs, *J. Geophys. Res.*, *110*, A12225, doi:10.1029/2005JA011281.

Shiokawa, K., et al. (2009), Longitudinal development of a substorm brightening arc, *Ann. Geophys.*, *27*, 1935–1940.

Sibeck, D. G., and V. Angelopoulos (2008), THEMIS science objectives and mission phases, *Space Sci. Rev.*, *141*, 35–59.

Solovyev, S. I., K. Yumoto, D. G. Baishev, and N. E. Molochushkin (1997), Generation and spectrum formation of high-latitude Pi2 geomagnetic pulsations during pseudobreakups and multiple substorm onsets, *Geomagn. Aeron.*, *37*(5), 579–586.

Steen, A., and P. N. Collis (1988), High-time resolution imaging of auroral arc deformation at substorm onset, *Planet. Space Sci.*, *36*, 715–732.

Strangeway, R. J. (2012), The relationship between magnetospheric processes and auroral field-aligned current morphology, in *Auroral Phenomenology and Magnetospheric Processes: Earth and Other Planets*, Geophys. Monogr. Ser., doi:10.1029/2012GM001211, this volume.

Tang, C.-L. (2011), The wave-like auroral structure around auroral expansion onset, *Chin. Phys. Lett.*, *28*(10), 109401, doi:10.1088/0256-307X/28/10/109401.

Tsutomu, T. (1984), Interaction of energetic particles with HM-waves in the magnetosphere, *Planet. Space Sci.*, *32*(11), 1371–1386.

Tsyganenko, N. A. (1995), Modeling the Earth's magnetospheric magnetic field confined within a realistic magnetopause, *J. Geophys. Res.*, *100*(A4), 5599–5612.

Untiedt, J., and W. Baumjohann (1993), Studies of polar current systems using the IMS Scandinavian magnetometer array, *Space Sci. Rev.*, *63*, 245–390, doi:10.1007/BF00750770.

Untiedt, J., R. Pellinen, F. Küppers, H. J. Opgenoorth, W. D. Pelster, W. Baumjohann, H. Ranta, J. Kangas, P. Czechowsky, and W. J. Heikkila (1978), Observations of the initial development of an auroral and magnetic substorm at magnetic midnight, *J. Geophys.*, *45*, 41–56.

Uritsky, V. M., J. Liang, E. Donovan, E. Spanswick, D. Knudsen, W. Liu, J. Bonnell, and K. H. Glassmeier (2009), Longitudinally propagating arc wave in the pre-onset optical aurora, *Geophys. Res. Lett.*, *36*, L21103, doi:10.1029/2009GL040777.

Voronkov, I., E. Friedrich, and J. C. Samson (1999), Dynamics of the substorm growth phase as observed using CANOPUS and SuperDARN instruments, *J. Geophys. Res.*, *104*(A12), 28,491–28,505.

Voronkov, I., E. F. Donovan, B. J. Jackel, and J. C. Samson (2000), Large-scale vortex dynamics in the evening and midnight auroral zone: Observations and simulations, *J. Geophys. Res.*, *105*(A8), 18,505–18,518, doi:10.1029/1999JA000442.

Watanabe, T.-H. (2010), Feedback instability in the magnetosphere-ionosphere coupling system: Revisited, *Phys. Plasmas*, *17*, 022904, doi:10.1063/1.3304237.

Wescott, E. M., H. C. Stenbaek-Nielsen, T. N. Davis, W. B. Murcray, H. M. Peek, and P. J. Bottoms (1975), The $L = 6.6$ Oosik barium plasma injection experiment and magnetic storm of March 7, 1972, *J. Geophys. Res.*, *80*(7), 951–967.

Wiens, R. G., and G. Rostoker (1975), Characteristics of the development of the westward electrojet during the expansive phase of magnetospheric substorms, *J. Geophys. Res.*, *80*(16), 2109–2128.

Yago, K., K. Shiokawa, K. Hayashi, and K. Yumoto (2005), Auroral particles associated with a substorm brightening arc, *Geophys. Res. Lett.*, *32*, L06104, doi:10.1029/2004GL021894.

Yahnin, A. G., I. V. Despirak, A. A. Lubchich, B. V. Kozelov, N. P. Dmitrieva, M. A. Shukhtina, and H. K. Biernat (2006), Relationship between substorm auroras and processes in the near-Earth magnetotail, *Space Sci. Rev.*, *122*, 97–106, doi:10.1007/s11214-006-5884-4.

Zesta, E., L. R. Lyons, and E. Donovan (2000), The auroral signature of earthward flow bursts observed in the magnetotail, *Geophys. Res. Lett.*, *27*(20), 3241–3244, doi:10.1029/2000GL000027.

Zesta, E., E. Donovan, L. Lyons, G. Enno, J. S. Murphree, and L. Cogger (2002), Two-dimensional structure of auroral poleward boundary intensifications, *J. Geophys. Res.*, *107*(A11), 1350, doi:10.1029/2001JA000260.

Zesta, E., L. Lyons, C.-P. Wang, E. Donovan, H. Frey, and T. Nagai (2006), Auroral poleward boundary intensifications (PBIs): Their two-dimensional structure and associated dynamics in the plasma sheet, *J. Geophys. Res.*, *111*, A05201, doi:10.1029/2004JA010640.

Zesta, E., et al. (2011), Ionospheric convection signatures of tail fast flows during substorms and Poleward Boundary Intensifications (PBI), *Geophys. Res. Lett.*, *38*, L08105, doi:10.1029/2011GL046758.

Zhu, P., J. Raeder, K. Germaschewski, and C. C. Hegna (2009), Initiation of ballooning instability in the near-Earth plasma sheet prior to the 23 March 2007 THEMIS substorm expansion onset, *Ann. Geophys.*, *27*, 1129–1138.

Zou, S., et al. (2010), Identification of substorm onset location and preonset sequence using Reimei, THEMIS GBO, PFISR, and Geotail, *J. Geophys. Res.*, *115*, A12309, doi:10.1029/2010JA015520.

A. Keiling, Space Sciences Laboratory, University of California, Berkeley, CA 94720, USA. (keiling@ssl.berkeley.edu)

L. Kepko and V. Uritsky, NASA Goddard Space Flight Center, Greenbelt, MD 20771, USA.

N. Østgaard, Department of Physics and Technology, University of Bergen, Bergen N-5020, Norway.

V. Sergeev, Institute of Physics, St. Petersburg State University, St. Petersburg 198504, Russia.

K. Shiokawa, Solar-Terrestrial Environment Laboratory, Nagoya University, Nagoya, Aichi 464-8601, Japan.

E. Zesta, Air Force Research Laboratory RVBXP, Kirtland Air Force Base, NM 87117, USA.

Magnetotail Aurora Connection: The Role of Thin Current Sheets

J. Birn

Los Alamos National Laboratory, Los Alamos, New Mexico, USA

Space Science Institute, Boulder, Colorado, USA

K. Schindler

Institut für Theoretische Physik, Ruhr-Universität Bochum, Bochum, Germany

M. Hesse

NASA Goddard Space Flight Center, Greenbelt, Maryland, USA

Connections between magnetotail structure and dynamics and auroral forms are investigated on the basis of fluid and particle simulations, as well as Vlasov theory. Our focus is on possible mechanisms to generate or enhance perpendicular electric fields associated with quasistatic electric potentials of "U" shape or "S" shape, relevant for auroral arcs. At small scales (less than ~10 km), such electric fields are associated with Hall currents and electron $\mathbf{E} \times \mathbf{B}$ drift rather than plasma flow shear, indicating the relevance of thin current sheets as possible source regions. Two-dimensional and 3-D MHD theory and simulations demonstrate that moderate deformations of the magnetotail may lead to critical states characterized by the loss of neighboring equilibrium and the formation of thin current sheets embedded in the wider plasma sheet. This provides a suitable scenario for the onset of reconnection in the tail, as well as a potential mechanism to generate or intensify auroral arcs, even prior to the onset of reconnection. Particle-in-cell simulations confirm this mechanism and further demonstrate that thin embedded electron current sheets may form from moderate compression, as well as expansion, of a thicker current sheet. This suggests that such structures might be more typical for the tail current sheet than a smooth solution and may represent a possible source region for quiescent arcs within the plasma sheet footpoint region and at the plasma sheet/polar cap boundary. One-dimensional Vlasov models of such structures indicate association with either S- or U-shaped potentials (and intermediate structures), depending on the magnitude of the magnetic field jump and the plasma beta.

Auroral Phenomenology and Magnetospheric Processes: Earth and Other Planets
Geophysical Monograph Series 197
10.1029/2011GM001182

1. INTRODUCTION

The connection between auroral forms and their presumed generator in the outer magnetosphere or magnetotail is one of the most important, yet poorly understood, links in the physical understanding of the coupled magnetosphere-ionosphere system. In situ measurements of electron precipitation support the view that there are two classes of auroral arcs [e.g., *Newell et al.*, 2009]: monoenergetic "inverted V" events that show acceleration of precipitating magnetospheric electrons in a narrow energy band [e.g., *Newell*, 2000] and events characterized by broadband spectra, presumably associated with dispersive Alfvén waves [e.g., *Chaston et al.*, 2003]. The former class is ascribed to the presence, formation, or intensification of quasistatic electric potentials of "U" shape [*Carlqvist and Boström*, 1970; *Mozer et al.*, 1980] or "S" shape [*Mizera et al.*, 1982; *Marklund et al.*, 1997; *Johansson et al.*, 2006].

These potential structures are associated not only with a parallel but also with strong perpendicular electric field components. Initially, the structures were discovered or inferred from the enhanced electron precipitation in the inverted V events, which correspond to upward-directed parallel and converging perpendicular electric field components, associated with upward field-aligned currents. However, more recently, equivalent potential structures with diverging perpendicular, associated with downward parallel, electric field components were also discovered in regions of downward current [e.g., *Marklund et al.*, 1997, 2004; *Johansson et al.*, 2007].

The downward electron acceleration by the upward electric field takes place mostly between 4000 and 8000 km altitude. However, it is unclear how far the associated perpendicular electric fields map out into the magnetosphere. Satellite observations by Polar [e.g., *Keiling et al.*, 2001] and Cluster [e.g., *Figueiredo et al.*, 2005; *Johansson et al.*, 2005, 2006, 2007] have shown that strong perpendicular electric fields can be found out to at least 5–7 R_E geocentric distance. *Keiling et al.* [2001] found that most of the events studied, largely consistent with Alfvénic aurora, were located near the plasma sheet/polar cap boundary. The events studied by *Johansson et al.* [2006] were more consistent with the quasi-static picture, with unipolar (S-shaped potential) or bipolar (U-shaped potential) electric fields; the unipolar events were located almost exclusively near the plasma sheet boundary, while the bipolar events were located mainly inside the plasma sheet.

Figure 1 illustrates the auroral acceleration scenario for a U-shaped potential (black contours). Magnetic field lines are shown in blue and electric field vectors in green. The perpendicular electric field, located in the *r,z* plane, is associated with an azimuthal velocity perturbation v_φ, which may be considered as the driver, distorting the magnetic field and thereby generating azimuthal magnetic field perturbations B_φ, which correspond to an upward field-aligned current (and reverse currents on the outside, not shown). The lowest-order Poynting vector is directed azimuthally; however, the

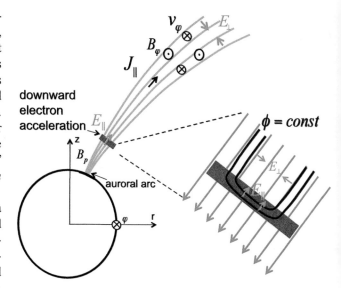

Figure 1. Schematic of the auroral acceleration region involving U-shaped potential contours (black lines). Magnetic field lines are shown in blue and electric field vectors in green.

second-order contribution from $E_\perp$ and B_φ leads to a downward Poynting flux. (It is noteworthy that in the Alfvén wave scenario, $E_\perp$ and B_φ periodically both reverse sign, such that the net Poynting flux also remains downward.)

When the transverse scale of the potential structure is sufficiently large, the velocity perturbations represent shear flows, associated with $\mathbf{E} \times \mathbf{B}$ drift of both ions and electrons [e.g., *Echim et al.*, 2007]. However, when the scale size becomes smaller than an ion inertial length or ion gyroradius, the ions decouple from the field, and only electrons exhibit the $\mathbf{E} \times \mathbf{B}$ drift. In this case, the drift corresponds to a Hall current. This regime is the focus of the present chapter. The transition between the two regimes occurs roughly at an arc width of 5–10 km. The exact value depends not only on the location of the source region in the tail but also on how stressed the configuration is in comparison to average magnetic field models.

Several small-scale regions in the tail that have a suitable location for an association with auroral arcs have been identified in chapter 7 of the special International Space Science Institute publication *Auroral Plasma Physics* [*Amm et al.*, 2002]: (1) the thin growth phase current sheet [*Pritchett and Coroniti*, 1994; *Hesse et al.*, 1996; *Birn and Schindler*, 2002; *Schindler and Birn*, 2002; *Birn et al.*, 2004c], (2) the near-Earth reconnection site [e.g., *McPherron et al.*, 1987; *Sanny et al.*, 1994], (3) the plasma sheet boundary layer (PSBL) [*Keiling et al.*, 2001; *Johansson et al.*, 2006], (4) the inner edge of the electron plasma sheet [*Elphic et al.*, 1999; *Haerendel*, 2009], and (5) current layers associated with bursty bulk flows (BBFs), which drive narrow dipolarization fronts [*Sergeev et al.*, 2009; *Runov et al.*, 2009]; (6) in addition, small-scale current sheets associated with localized electric field structures may be embedded within the tail plasma/current sheet [*McComas et al.*, 1986; *Johansson et al.*, 2006].

The statistical analysis of Cluster data by *Johansson et al.* [2007] showed that the observed perpendicular electric field structures, whether unipolar or bipolar, had thicknesses of ~4–5 km, well below the ion gyroradius or ion inertial length, which were of the order of ~20 km (all scales mapped to the ionosphere). Therefore, the ions can be expected to decouple from the magnetic field, and kinetic (Vlasov) theory or particle simulations are required to adequately investigate these structures. The potential current sheet source regions listed above are of two types, horizontal, i.e., roughly parallel to the equatorial plane (cases 1–3 and 5), or vertical, i.e., roughly parallel to the *y,z* plane (cases 4 and 5). In addition, the main magnetic field may reverse sign across the sheet (cases 1 and 2) or be essentially unidirectional with a jump in magnitude (cases 3–6).

In this chapter, we will present results on the structure of thin current sheets in the kinetic regime, based on theoretical (Vlasov) approaches and fully electrodynamic particle-in-cell (PIC) simulations. Using time-dependent boundary conditions, causing compression (or expansion) of the current sheet, we particularly present investigations of the transition from wider current sheets that still satisfy fluid approximations into the kinetic regime.

We investigate both, current sheets with a reversal of the magnetic field and current sheets without reversal. It is important to note that there is an important stability difference. According to a simple criterion [*Schindler*, 1970, 2007], 2-D field configurations are stable within a given box when a Cartesian magnetic field component exists that does not change sign within the box. According to this criterion, the current sheets without magnetic field reversal are expected to be stable (at least within 2-D). This is consistent with PIC simulations investigating the stability of bifurcated current sheets [*Camporeale and Lapenta*, 2005; *Matsui and Daughton*, 2008], which consist of pairs of sheets of such type. The simulations showed that bifurcated current sheets were more stable to tearing modes than single current sheets with antiparallel fields of similar thickness.

2. THE 2-D TAIL CURRENT SHEET

An obvious way to generate a thin current sheet is by compression of a thicker current sheet. However, applying simple 1-D scaling laws, one can conclude that the compression of the magnetotail current sheet from a quiet thickness of, say, ~6 R_E to an ion gyroscale or inertial scale of ~0.1–0.2 R_E,

requires an increase in lobe field strength by a factor of 30–60. This is totally unrealistic. One possible way out of this dilemma is discussed in sections 3 and 4. Full particle simulations of the compression (or expansion) of 1-D current sheets [*Schindler and Hesse*, 2008, 2010] show that, even though such scaling law might apply to the entire current sheet thickness, the current sheet does not maintain a smooth structure as predicted by isotropic or gyrotropic models but rather develops small-scale substructures, carrying localized electron currents associated with nongyrotropies. Such 1-D models may explain the formation of embedded thin current sheets that may be present even during quiet times, however, are not likely to apply to the substorm growth phase.

In this section, we discuss another way, which has to do with the fact that realistic magnetotail plasma sheet fields (prior to the onset of reconnection) consist of closed magnetic flux tubes that are connected to the Earth at both ends. Using 2-D equilibrium theory with isotropic pressure, *Birn and Schindler* [2002] showed analytically that a modest, nonuniform, compression of such a configuration under conservation of mass and entropy may lead to a critical state where neighboring equilibrium solutions satisfying these constraints cease to exist. The critical state is characterized by the formation of a thin embedded current sheet, which, in the MHD limit, is associated with infinite current density and locally infinitely small thickness. The analytic results were confirmed by 2-D and 3-D MHD simulations and 3-D theory [*Birn et al.*, 1998, 2004a]. Furthermore, the analytic results of *Birn and Schindler* [2002] showed that thin current sheet formation and loss of equilibrium is not restricted to tail compressions but may also be found for expansion. Recent MHD results by Otto et al. (unpublished) demonstrate that current sheet thinning may also result from azimuthal convection of plasma out of the thinning region, presumably resulting from, or associated with, dayside reconnection. This may explain why *Saito et al.* [2011] recently observed current sheet thinning in Time History of Events and Macroscale Interactions during Substorms (THEMIS) observations without significant compression.

Schindler and Birn [2002] and *Birn et al.* [2004c] used self-consistent Vlasov equilibrium theory to derive explicit models of thin embedded current sheets. A major outcome of these studies was that the self-consistent solutions are generally associated with electrostatic potentials that are constant on field lines and thus could be the source of the unipolar or bipolar perpendicular electric fields observed above arcs. These electric field structures were also found in the particle simulations [*Pritchett and Coroniti*, 1994, 1995; *Hesse et al.*, 1996].

The mechanism of thin current sheet formation was also confirmed by 2-D PIC and hybrid simulations, either from convection imposed inside the plasma sheet [*Pritchett and Coroniti*, 1994, 1995] or following from a boundary deformation [*Hesse et al.*, 1996; *Hesse and Schindler*, 2001]. In either case, the compression of the inner magnetotail caused the formation of a thin embedded current sheet and, ultimately, the onset of magnetic reconnection. A newer result of a PIC simulation of this scenario is illustrated in Figure 2. This simulation used 2×10^8 particles of each species, a mass ratio $m_i/m_e = 100$, temperature ratio $T_i/T_e = 5$, and 1600×800 grid cells in x, z. The initial current sheet width was $2L = 5d_i$ at the inner boundary. Figure 2a shows magnetic field lines and the cross-tail current density J_y (color) at $t = 64.4$ (measured in ion gyroperiods), shortly after the onset of reconnection and plasmoid formation. Figure 2b shows the electric field component E_z at the same time.

Figure 2 clearly shows the formation of a thin embedded current sheet, which extends earthward (to the left) as a bifurcated sheet. This thin current sheet is associated with a strong localized electric field component E_z of order unity (that is, ~20 mV m^{-1}, assuming a typical magnetic field of 20 nT and an Alfvén speed of 1000 km s^{-1}, at this location). The electric field also extends earthward along the magnetic field, consistent with the results from the equilibrium theory. (The outer layers of strong E_z are an artifact of the simulation, resulting from the fact that the inflowing plasma had different properties than the preexisting lobe plasma.)

3. THE 1-D PERTURBATION OF A HARRIS SHEET

In this section, we go back to the deformation of a 1-D current sheet, using kinetic theory and simulations, as investigated by

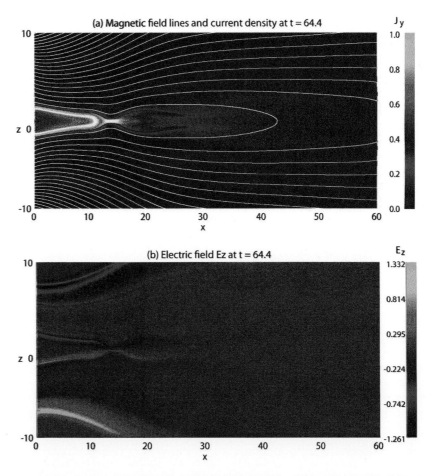

Figure 2. (a) Electric current density J_y (color) and magnetic field lines in a 2-D particle-in-cell (PIC) simulation of thin current sheet formation and reconnection, initiated by a boundary deformation, shortly after the onset of reconnection and plasmoid formation. (b) Electric field component E_z at the same time. Here and in the following figures showing PIC simulation results, quantities are normalized using the ion inertial length $d_i = c/\omega_{pi}$ and the ion gyroperiod $1/\omega_{ci}$, based on a typical density n_c and magnetic field strength B_c, such that $J_c = B_c/(\mu_0 d_i)$ and $E_c = v_A B_c$, where $v_A = B_c/\sqrt{\mu_0 m_i n_c}$.

Schindler and Hesse [2008, 2010]. The initial state was a Harris-type current sheet, with a magnetic field given by

$$B_x = \tanh(z/L). \tag{1}$$

The scale length L was chosen as $L = 5d_i$ where $d_i = c/\omega_{pi}$ is the ion inertial length and $\omega_{pi} = \sqrt{e^2 n_0 / \varepsilon_0 m_i}$ is the ion (here, proton) plasma frequency. The simulations were based on a 1-D version of a PIC code [*Hesse et al.*, 1999; *Schindler and Hesse*, 2008], employing 2.5×10^7 macroparticles of each species, 20,000 time steps, corresponding to a time interval of 400 Alfvén times, and 4000 spatial grid points. The mass ratio m_i/m_e was chosen as 25, the temperature ratio T_i/T_e as 5, the ratio v_A/c as 0.1, and the width of the simulation box in z was set to $\pm 40 d_i$. A driving electric field E_y was applied during a period of 200 Alfvén times, leading to a compression (increase of the asymptotic magnetic field) by a factor 1.3.

Figure 3, modified from *Schindler and Hesse* [2008], shows the current density J_y after 400 Alfvén times, when a new equilibrium had been reached. To reduce the noise, the output was averaged over intervals of $40/\omega_e$ centered at the times of interest. Figure 3a shows the total electric current density J_y (black solid line) with proton (red line) and electron contributions (green line). Dotted lines show the initial distributions. Figure 3b shows, again, proton (red) and electron (green) contributions to the current density together with the results for the corresponding gyrotropic model (dash-dotted lines), in which the scaling of various quantities with the magnetic field increase follows from the absence of compression along the field and the assumptions that particle numbers and magnetic moments, as well as pressure isotropy

perpendicular to the magnetic field, are conserved in each flux tube. The current density shows a bifurcated small-scale structure carried by electrons. Figure 3b also shows clearly that this structure is associated with nongyrotropy.

The development of small-scale embedded current sheets is not restricted to the compression of a current sheet. This is demonstrated by Figure 4 taken from *Schindler and Hesse* [2010]. Figure 4 shows the current density structure resulting from an expansion with a decrease of the external magnetic field by factor 0.66. The dotted black line shows the initial current density and the solid line the current density after 400 Alfvén times (the expansion subsided after ~200 Alfvén times). The dashed red and dash-dotted green lines show ion and electron contributions to the current density, respectively. Figure 4 shows some unexpected results: The expected broadening of the current sheet is shown only in the outer region, for approximately $|z| > 8$. The inner region shows a narrowing and strengthening of the ion current, while the electron current becomes multipeaked and reversed. As a consequence, the total current shows a triple peak structure. As in the case of compression, the perturbed currents are largely nongyrotropic in nature. In both compression and expansion, the current substructures involve substantial electric fields causing a significant Hall contribution to the electron currents.

4. THE 1-D CURRENT SHEETS WITHOUT FIELD REVERSAL

The small-scale embedded current structures, discussed in section 3, as well as the earthward extensions from a thin

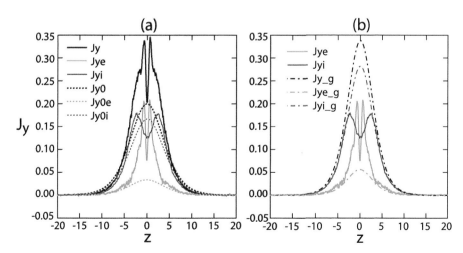

Figure 3. Compression of a 1-D current sheet: (a) total electric current density J_y (black solid line) with proton (red line) and electron contributions (green line) after 400 Alfvén times and initial current densities (dotted lines) and (b) proton (red) and electron (green) contributions to the current density. The results for the corresponding gyrotropic model are shown by dash-dotted lines. Modified after *Schindler and Hesse* [2008].

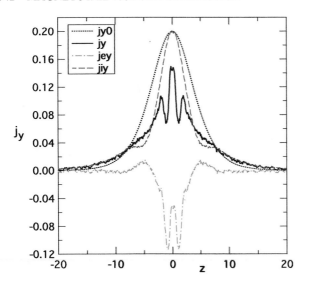

Figure 4. Expansion of a 1-D current sheet, showing the total electric current density J_y (black solid line) with proton (dashed red line) and electron contributions (dash-dotted green line) after 400 Alfvén times. The initial current density is shown by a black dotted line. After *Schindler and Hesse* [2010].

current sheet, illustrated in Figure 2a, correspond to horizontal current sheets with steps in the main magnetic field component B_x but not necessarily a reversal of its direction. Similarly, the thin layer associated with a dipolarization front (and possibly the inner boundary of the electron plasma sheet; see section 1) represent vertical current sheets with steps in B_z without necessarily reversing its direction. (Some dipolarization fronts, however, show a brief reversal of B_z.) This motivated an investigation of the kinetic structure of such current sheets by Vlasov theory. As pointed out earlier, the sheets without reversal are probably stable [*Schindler*, 1972]. Therefore, these models might be relevant for quiet arcs or for intensified arcs prior to a disruption.

Steady state solutions of such current sheets have been investigated based on 2-D Vlasov equilibrium models [*Schindler and Birn*, 2002], discussed at length by *Schindler* [2007]. In this approach, temporal variations and associated electric fields are considered as small, so that they affect the instantaneous configuration only in a parametric fashion. The distribution function for each species s then can be chosen as an arbitrary function of the constants of motion

$$f_s = F_s(H_s, P_{ys}),\qquad(2)$$

where P_y is the canonical momentum with respect to the y coordinate

$$P_{ys} = m_s v_y + q_s A(x,z)\qquad(3)$$

and H_s the Hamiltonian

$$H_s = \frac{1}{2m_s}\left(P_{xs}^2 + P_{zs}^2\right) + \frac{1}{2m_s}\left[P_{ys} - q_s A(x,z)\right]^2 + q_s \Phi(x,z)\qquad(4)$$

m_s and q_s are particle mass and charge, respectively, P_{xs} and P_{zs} denote the canonical momentum with respect to x and z, $A(x,z)$ is the magnetic flux function (or the y component of the vector potential), and $\Phi(x,z)$ the electric potential.

Under these assumptions, 2-D equilibrium configurations are governed by Ampere's law

$$-\Delta A = \mu_0 j_y(A, \Phi)\qquad(5)$$

and the condition of quasineutrality

$$\sigma(A, \Phi) = 0,\qquad(6)$$

where the electric charge density σ and current density j_y are functions of A and Φ, given by

$$\sigma(A, \Phi) = \sum_s q_s n_s = \sum_s q_s \int F_s(H_s, P_{ys})\mathrm{d}^3 v\qquad(7)$$

and

$$j_y(A, \Phi) = \sum_s q_s \int v_y F_s(H_s, P_{ys})\mathrm{d}^3 v,\qquad(8)$$

and n_s denotes the number density of species s. Note that $j_x = j_z = 0$ under the present conditions.

Consistent with the approach of *Schindler and Birn* [2002], distribution functions are chosen to be of the form

$$F_s(H, P_y) = C_s exp\left(-\frac{H_s}{kT_s}\right)g_s(P_{ys}),\qquad(9)$$

where k is Boltzmann's constant, T_s the temperature of particle species s, and $g_s(P_{ys})$ can be an arbitrary nonnegative function. In contrast to the works of *Schindler and Birn* [2002] and *Birn et al.* [2004c], however, the functions $g_s(P_{ys})$ are not chosen ad hoc to model an anticipated embedded structure, but from a condition that assumed electrons and ions to have the same number of particles with their generalized gyrocenter on any given field line. This puts a significant constraint on the relation between ion and electron distributions.

Further simplifications include the assumption that the macroscopic scale is large in the sense that the magnetic flux interval connected with the ion gyration is much smaller than a typical macroscopic flux, and an expansion for small m_e/m_i, retaining only terms of lowest order in m_e/m_i. Using a specialization to one space dimension, the potential profile then can simply be related to the magnetic field profile

$$e\Phi(A) = -\frac{kT_i}{2} \ln \frac{p(A)B(A)}{p_c B_c} + \text{const.} \qquad (10)$$

using the assumption that the potential becomes constant far away from the current sheet. The pressure p is related to the magnetic field B by pressure balance

$$p + \frac{B^2}{2\mu_0} = \hat{p} \qquad (11)$$

with constant total pressure $\hat{p}$. The magnitudes of the normalization quantities p_c and B_c in equation (10) do not affect the shape of the potential but only the additional constant.

For structures that are asymptotically homogeneous on both sides, solutions of equation (10) with equation (11) can be represented by two parameters, the plasma beta on one side of the jump, say, $\beta = 2\mu_0 p_2/B_2^2$, and the ratio between the magnetic fields on the two sides of the jump $b = B_1/B_2$. The interesting fact is that, depending on the two parameters, potentials can be of S- or U-shape (or intermediate shapes). Figure 5 illustrates where the different shapes exist in β, b

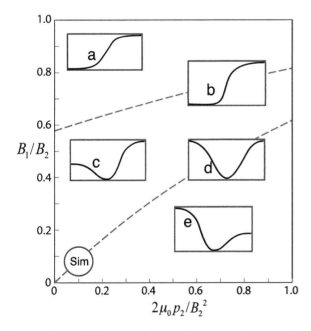

Figure 5. Potential forms associated with a magnetic step as function of the magnetic field ratio B_1/B_2 and the plasma beta, $\beta_2 = 2\mu_0 p_2/B_2^2$ on the high field side. The circle marks the parameters of a PIC simulation [*Schindler and Hesse*, 2010]. The lower dashed blue line corresponds to potentials that have the same value on either side, i.e., true U shapes (type d). The upper one represents the transition to monotonic, i.e., purely S-shaped variation (type b).

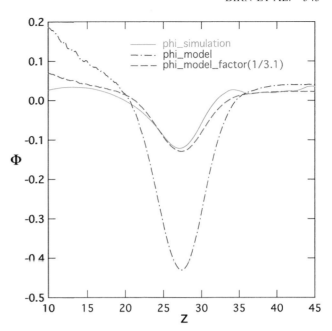

Figure 6. Potential associated with a magnetic step for $B_1/B_2 = 0.2$ and $\beta_2 = 0.2$ (red dash-dotted line) in comparison to a PIC simulation (green line). The blue dashed line represents the model result reduced by a factor $1/3.1$.

parameter space. The space is divided by two dashed lines, shown in blue. The lower one corresponds to potentials that have the same value on either side, i.e., true U shapes (type d). The upper one represents the transition to monotonic, i.e., purely S-shaped variation (type b). The magnitude of the potential differences in these solutions is of the order of the ion thermal energy, divided by e, that is, of the order of several kV.

Figure 6 shows the potential for a specific example, $B_1/B_2 = 0.08$ and $\beta_2 = 0.1$ (red dash-dotted line), using the values and the magnetic field profile from a PIC simulation (case 3 of *Schindler and Hesse* [2010]; green solid line). The two profiles agree well in shape, although the magnitudes differ by a factor of ~3, as indicated by the blue dashed line. Particularly, the location of the minimum and the fact that the profiles are close to a symmetric U shape are in good agreement. At present, we have no solid explanation for the difference in magnitude yet. One possible reason might be the difference in the form of the distribution functions between model and simulation.

5. SUMMARY

We have investigated potential connections between magnetotail structure and dynamics and auroral forms on the basis of fluid and particle simulations, as well as Vlasov theory. Our focus was on possible mechanisms to generate or enhance the perpendicular electric field associated with quasistatic electric potentials of "U" shape or "S" shape, relevant for auroral arcs. Such perpendicular electric fields are found to map out into the outer magnetosphere or magnetotail to geocentric distances of, at least, 4–7 R_E [*Keiling et al.*, 2001; *Johansson et al.*, 2005]. When the transverse scale of the potential structure is sufficiently large, the perpendicular electric field can be associated with shear flows representing $\mathbf{E} \times \mathbf{B}$ drift of both ions and electrons [e.g., *Echim et al.*, 2007]. However, when the scale size becomes smaller than an ion inertial length or ion gyroradius, the ions decouple from the field, and only electrons may exhibit the $\mathbf{E} \times \mathbf{B}$ drift. In this case, the drift corresponds to a Hall current, and the source mechanism for the perpendicular electric field involves thin current sheets and electron drifts rather than plasma flows. The transition between the two regimes occurs roughly at an arc width of 5–10 km.

In our chapter, we focused on the small-scale regime, that is, the generation and structure of thin current sheets. We demonstrated that thin current sheets may form within the thicker plasma/current sheet under modest deformations of the magnetotail. These results were found consistently from MHD theory and simulations as well as particle simulations. In the latter case, the current sheet thinning also provides a natural means to initiate collisionless reconnection.

Self-consistent Vlasov equilibrium models of thin current sheets demonstrate that the sheets generally contain electrostatic potential structures associated with strong perpendicular electric fields. PIC simulations of deformations of thick current sheets (several ion gyroradii or ion inertial lengths) showed that both compression and expansion leads to the formation of thin embedded current sheets carried by electrons and associated with nongyrotropy. This suggests that such embedded current sheet structures may be more generic than smooth current sheets [*McComas et al.*, 1986; *Johansson et al.*, 2006]. Since such current sheets are predicted to be stable (within 2-D theory) [*Schindler*, 1970, 1972], they may be a suitable mechanism to generate quiescent arcs embedded within the region connecting to the interior plasma sheet.

Vlasov equilibrium theory was also applied to 1-D sheet structures representing a jump in B without magnetic field reversal. The special solutions found could represent S- and U-shaped potentials, as well as intermediate structures, depending on the magnetic field ratio between the two sides B_1/B_2 and the plasma β. Typical magnitudes of the potential jump were of the order of a significant fraction of the ion thermal potential (kT_i/e). The form of the potential for a particular choice of B_1/B_2 and β agreed well with a result from a PIC simulation, although the simulation potential was lower by a factor of ~3. A possible reason for this difference

may be differences in the distribution functions between model and simulation. The model predicts that S-shaped potentials should be found primarily at low beta and small jumps in B, typical for the PSBL, with the electric field directed toward the interior plasma sheet, whereas U-shaped or nearly U-shaped potentials would be more prevalent under interior plasma sheet conditions, consistent with the findings of *Johansson et al.* [2006].

It should be noted that models and simulations of current sheets with a jump in B but without reversal might also apply to thin vertical current sheets at dipolarization fronts associated with BBFs [*Sergeev et al.*, 2009; *Runov et al.*, 2009]. In that case, the associated perpendicular electric field (E_x in the neutral sheet, E_z at higher z) might be the source of potentials associated with smaller-scale auroral emissions, in addition to the medium- to large-scale signatures associated with field-aligned currents driven by the flow vorticity on the outside of a BBF [*Nakamura et al.*, 2001; *Birn et al.*, 2004b; *Panov et al.*, 2010; *Birn et al.*, 2011].

A full self-consistent model of auroral arc formation should include both the source region and mechanism and the dissipation region. In this chapter, we have focused on the source mechanism for thin arcs that may be associated with thin current sheets and perpendicular electric fields in the tail. Models of the dissipation/acceleration region are mainly 1-D, disregarding the perpendicular electric field. A first attempt to construct a self-consistent 2-D model that includes both parallel and perpendicular stationary electric field components is illustrated in Figure 7. The model is

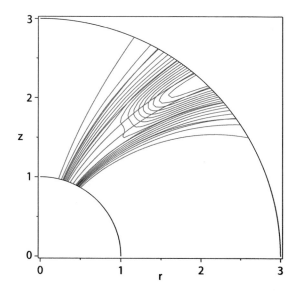

Figure 7. Magnetic field lines (blue) and equipotentials (red) in the r,z plane of a 2-D stationary potential model. The blue lines are dipole field lines; the red lines are the lines of constant electric potential.

based on the dynamics of cold electrons, treating the ions as a passive background. It uses linearization from a background state, consisting of uniform density in a dipole magnetic field. The dissipation is accomplished by an assumed localized resistivity, based on a criterion for the onset of current-driven instability. Figure 7 shows an example of potential lines obtained in this model. Blue lines are dipole field lines, and red lines are the lines of constant electric potential. In this model, the resistivity profile governs the amount of potential lines that reach the Earth and the spatial scale of the parallel electric field.

In summary, thin current sheets in the outer magnetosphere or magnetotail are a promising candidate for the source region of the perpendicular electric field associated with S- or U-shaped potentials related to auroral arcs with a transverse scale of ~10 km or less. Possible source regions include the PSBL, thin current sheets embedded in the interior of the plasma sheet, the thin growth phase current sheet, and the reconnection site.

Acknowledgments. This work was performed under the auspices of the US Department of Energy, supported by NASA's MMS/SMART Theory and Modeling and SR&T Programs.

REFERENCES

Amm, O., et al. (2002), Auroral plasma physics, *Space Sci. Rev.*, *103*(1–4), 1–475.

Birn, J., and K. Schindler (2002), Thin current sheets in the magnetotail and the loss of equilibrium, *J. Geophys. Res.*, *107*(A7), 1117, doi:10.1029/2001JA000291.

Birn, J., M. Hesse, and K. Schindler (1998), Formation of thin current sheets in space plasmas, *J. Geophys. Res.*, *103*(A4), 6843–6852.

Birn, J., J. C. Dorelli, M. Hesse, and K. Schindler (2004a), Thin current sheets and loss of equilibrium: Three-dimensional theory and simulations, *J. Geophys. Res.*, *109*, A02215, doi:10.1029/2003JA010275.

Birn, J., J. Raeder, Y. L. Wang, R. A. Wolf, and M. Hesse (2004b), On the propagation of bubbles in the geomagnetic tail, *Ann. Geophys.*, *22*, 1773–1786.

Birn, J., K. Schindler, and M. Hesse (2004c), Thin electron current sheets and their relation to auroral potentials, *J. Geophys. Res.*, *109*, A02217, doi:10.1029/2003JA010303.

Birn, J., R. Nakamura, E. V. Panov, and M. Hesse (2011), Bursty bulk flows and dipolarization in MHD simulations of magnetotail reconnection, *J. Geophys. Res.*, *116*, A01210, doi:10.1029/2010JA016083.

Camporeale, E., and G. Lapenta (2005), Model of bifurcated current sheets in the Earth's magnetotail: Equilibrium and stability, *J. Geophys. Res.*, *110*, A07206, doi:10.1029/2004JA010779.

Carlqvist, P., and R. Boström (1970), Space-charge regions above the aurora, *J. Geophys. Res.*, *75*(34), 7140–7146.

Chaston, C. C., J. W. Bonnell, C. W. Carlson, J. P. McFadden, R. E. Ergun, and R. J. Strangeway (2003), Properties of small-scale Alfvén waves and accelerated electrons from FAST, *J. Geophys. Res.*, *108*(A4), 8003, doi:10.1029/2002JA009420.

Echim, M. M., M. Roth, and J. De Keyser (2007), Sheared magnetospheric plasma flows and discrete auroral arcs: A quasi-static coupling model, *Ann. Geophys.*, *25*(1), 317–330, doi:10.5194/angeo-25-317-2007.

Elphic, R. C., M. F. Thomsen, J. E. Borovsky, and D. J. McComas (1999), Inner edge of the electron plasma sheet: Empirical models of boundary location, *J. Geophys. Res.*, *104*(A10), 22,679–22,693.

Figueiredo, S., G. T. Marklund, T. Karlsson, T. Johansson, Y. Ebihara, M. Ejiri, N. Ivchenko, P.-A. Lindqvist, H. Nilsson, and A. Fazakerley (2005), Temporal and spatial evolution of discrete auroral arcs as seen by Cluster, *Ann. Geophys.*, *23*(7), 2531–2557, doi:10.5194/angeo-23-2531-2005.

Haerendel, G. (2009), Poleward arcs of the auroral oval during substorms and the inner edge of the plasma sheet, *J. Geophys. Res.*, *114*, A06214, doi:10.1029/2009JA014138.

Hesse, M., and K. Schindler (2001), The onset of magnetic reconnection in the magnetotail, *Earth Planets Space*, *53*, 645–653.

Hesse, M., D. Winske, M. Kuznetsova, J. Birn, and K. Schindler (1996), Hybrid modeling of the formation of thin current sheets in magnetotail configurations, *J. Geomagn. Geoelectr.*, *48*, 749–763.

Hesse, M., K. Schindler, J. Birn, and M. Kuznetsova (1999), The diffusion region in collisionless magnetic reconnection, *Phys. Plasmas*, *6*, 1781–1795.

Johansson, T., T. Karlsson, G. Marklund, S. Figueiredo, P.-A. Lindqvist, and S. Buchert (2005), A statistical study of intense electric fields at 4–7 R_E geocentric distance using Cluster, *Ann. Geophys.*, *23*(7), 2579–2588, doi:10.5194/angeo-23-2579-2005.

Johansson, T., G. Marklund, T. Karlsson, S. Liléo, P.-A. Lindqvist, A. Marchaudon, H. Nilsson, and A. Fazakerley (2006), On the profile of intense high-altitude auroral electric fields at magnetospheric boundaries, *Ann. Geophys.*, *24*, 1713–1723.

Johansson, T., G. Marklund, T. Karlsson, S. Liléo, P.-A. Lindqvist, H. Nilsson, and S. Buchert (2007), Scale sizes of intense auroral electric fields observed by Cluster, *Ann. Geophys.*, *25*(11), 2413–2425, doi:10.5194/angeo-25-2413-2007.

Keiling, A., J. R. Wygant, C. Cattell, M. Johnson, M. Temerin, F. S. Mozer, C. A. Kletzing, J. Scudder, and C. T. Russell (2001), Properties of large electric fields in the plasma sheet at 4–7 R_E measured with Polar, *J. Geophys. Res.*, *106*, 5779–5798.

Marklund, G., T. Karlsson, and J. Clemmons (1997), On low-altitude particle acceleration and intense electric fields and their relationship to black aurora, *J. Geophys. Res.*, *102*(A8), 17,509–17,522.

Marklund, G. T., T. Karlsson, S. Figueiredo, T. Johansson, P.-A. Lindqvist, M. André, S. Buchert, L. M. Kistler, and A. Fazakerley (2004), Characteristics of quasi-static potential structures observed in the auroral return current region by Cluster,

Nonlinear Processes Geophys., *11*(5/6), 709–720, doi:10.5194/npg-11-709-2004.

Matsui, T., and W. Daughton (2008), Kinetic theory and simulation of collisionless tearing in bifurcated current sheets, *Phys. Plasmas*, *15*, 012901, doi:10.1063/1.2832679.

McComas, D. J., C. T. Russell, R. C. Elphic, and S. J. Bame (1986), The near-earth cross-tail current sheet: Detailed ISEE 1 and 2 case studies, *J. Geophys. Res.*, *91*(A4), 4287–4301.

McPherron, R. L., A. Nishida, and C. T. Russell (1987), Is near-Earth current sheet thinning the cause of substorm onset?, in *Quantitative Modeling of Magnetosphere-Ionosphere Coupling Processes*, edited by Y. Kamide and R. A. Wolf, p. 252, Kyoto Sangyo Univ., Kyoto, Japan.

Mizera, P. F., D. J. Gorney, and J. F. Fennell (1982), Experimental verification of an S-shaped potential structure, *J. Geophys. Res.*, *87*(A3), 1535–1539.

Mozer, F. S., C. A. Cattell, M. K. Hudson, R. L. Lysak, M. Temerin, and R. B. Torbert (1980), Satellite measurements and theories of low-altitude auroral particle acceleration, *Space Sci. Rev.*, *27*, 155–213.

Nakamura, R., W. Baumjohann, M. Brittnacher, V. A. Sergeev, M. Kubyshkina, T. Mukai, and K. Liou (2001), Flow bursts and auroral activations: Onset timing and foot point location, *J. Geophys. Res.*, *106*, 10,777–10,789.

Newell, P. T. (2000), Reconsidering the inverted-V particle signature: Relative frequency of large-scale electron acceleration events, *J. Geophys. Res.*, *105*(A7), 15,779–15,794.

Newell, P. T., T. Sotirelis, and S. Wing (2009), Diffuse, monoenergetic, and broadband aurora: The global precipitation budget, *J. Geophys. Res.*, *114*, A09207, doi:10.1029/2009JA014326.

Panov, E. V., et al. (2010), Multiple overshoot and rebound of a bursty bulk flow, *Geophys. Res. Lett.*, *37*, L08103, doi:10.1029/2009GL041971.

Pritchett, P. L., and F. V. Coroniti (1994), Convection and the formation of thin current sheets in the near-Earth plasma sheet, *Geophys. Res. Lett.*, *21*(15), 1587–1590.

Pritchett, P. L., and F. V. Coroniti (1995), Formation of thin current sheets during plasma sheet convection, *J. Geophys. Res.*, *100*(A12), 23,551–23,565.

Runov, A., V. Angelopoulos, M. I. Sitnov, V. A. Sergeev, J. Bonnell, J. P. McFadden, D. Larson, K.-H. Glassmeier, and U. Auster (2009), THEMIS observations of an earthward-propagating di-

polarization front, *Geophys. Res. Lett.*, *36*, L14106, doi:10.1029/2009GL038980.

Saito, M. H., D. Fairfield, G. Le, L.-N. Hau, V. Angelopoulos, J. P. McFadden, U. Auster, J. W. Bonnell, and D. Larson (2011), Structure, force balance, and evolution of incompressible cross-tail current sheet thinning, *J. Geophys. Res.*, *116*, A10217, doi:10.1029/2011JA016654.

Sanny, J., R. L. McPherron, C. T. Russell, D. N. Baker, T. I. Pulkkinen, and A. Nishida (1994), Growth-phase thinning of the near-Earth current sheet during the CDAW 6 substorm, *J. Geophys. Res.*, *99*(A4), 5805–5816.

Schindler, K. (1970), Stability of 2D Vlasov equilibria, in *Inter-Correlated Satellite Observations Related to Solar Events*, edited by V. Manno and D. E. Page, p. 309, D. Reidel, Dordrecht, The Netherlands.

Schindler, K. (1972), A self-consistent theory of the tail of the magnetosphere, in *Earth's Magnetospheric Processes*, edited by B. M. McCormac, p. 200, D. Reidel, Dordrecht, The Netherlands.

Schindler, K. (2007), *Physics of Space Plasma Activity*, 508 pp., Cambridge Univ. Press, Cambridge, U. K.

Schindler, K., and J. Birn (2002), Models of two-dimensional embedded thin current sheets from Vlasov theory, *J. Geophys. Res.*, *107*(A8), 1193, doi:10.1029/2001JA000304.

Schindler, K., and M. Hesse (2008), Formation of thin bifurcated current sheets by quasisteady compression, *Phys. Plasmas*, *15*, 042902, doi:10.1063/1.2907359.

Schindler, K., and M. Hesse (2010), Conditions for the formation of nongyrotropic current sheets in slowly evolving plasmas, *Phys. Plasmas*, *17*, 082103, doi:10.1063/1.3464198.

Sergeev, V., V. Angelopoulos, S. Apatenkov, J. Bonnell, R. Ergun, R. Nakamura, J. McFadden, D. Larson, and A. Runov (2009), Kinetic structure of the sharp injection/dipolarization front in the flow-braking region, *Geophys. Res. Lett.*, *36*, L21105, doi:10.1029/2009GL040658.

J. Birn, Space Science Institute, Boulder, CO 80301, USA. (jbirn@spacescience.org)

M. Hesse, NASA Goddard Space Flight Center, Greenbelt, MD 20771, USA.

K. Schindler, Institut für Theoretische Physik, Ruhr-Universität Bochum, D-44801 Bochum, Germany.

Auroral Generators: A Survey

Gerhard Haerendel

Max Planck Institute for Extraterrestrial Physics, Garching, Germany

The paper presents two classification schemes of auroral current generators. One uses the driving forces, the other uses the associated current systems. They are employed for ordering of the discussion of some of the better explored cases in the published literature. The respective scenarios are briefly described and illustrated by just one figure and a few key equations. As such, these characterizations are insufficient to convey the full meaning of the underlying concepts, which can only be appreciated by consulting the original papers. However, some of the open problems and questions are pointed out, one of the still controversial issues being the origin of the currents creating the fascinating auroral arcs in the evening auroral oval.

1. INTRODUCTION

This chapter is an attempt to classify the various generators that have been discussed in the literature as potential sources of the currents responsible for the appearance of auroral arcs and to review in brief the essential physics. Because of the limitations on the length of this contribution and the sparseness of current knowledge about the generation of auroral currents, this review will be superficial and incomplete. More details are given by *Haerendel* [2011]. All the same, this survey addresses a few intriguing problems. The classification scheme is based on driving forces and related current systems. In order to emphasize the need for proper identification of the driving forces, we will look first at a few cases of obscure generator identifications. The main part of the chapter reviews representative examples for the four classes of driving forces. The illustrating figures have been assembled in two collages under the two classes of current systems proposed by *Boström* [1964].

2. TWO KINDS OF AURORA

There are two types of aurora, diffuse aurora and structured arcs. The first type is due to particle precipitation either caused by pitch-angle diffusion or by the breaking of the first adiabatic invariant in strongly kinked magnetic fields. The second type results from the conversion of electromagnetic energy into kinetic energy of particles near the Earth. The energy is transported from a generator plasma toward the Earth in the Alfvén mode, either as propagating or as quasistationary waves. This survey deals exclusively with the second type and focuses on the source plasma, the forces acting, the shear stresses transmitting the forces toward the Earth, and the associated field-aligned currents. The energy conversion process is not a subject of this survey. In all cases, it consists of the acceleration of magnetospheric or upper ionospheric electrons and ions by electric fields, which are sustained by a wide variety of plasma wave processes powered by the magnetic energy liberated by the release of the transmitted shear stresses [*Haerendel*, 2007]. Whether an auroral arc is being generated or not is determined by the strength of the associated upward field-aligned current. Only part of the energy carried by the Alfvén waves is normally converted into the energy of auroral particles, the rest being dissipated in the ionosphere by ion-neutral collisions or ohmic losses of the closing Pedersen currents.

Auroral Phenomenology and Magnetospheric Processes: Earth and Other Planets

Geophysical Monograph Series 197

10.1029/2011GM001162

3. CLASSIFICATIONS

A plasma acting as current generator is characterized by a negative scalar product of current and electric field. This identification is not sufficient for identifying an auroral generator because the current may close entirely inside the hot plasma without touching the ionosphere. Furthermore, currents are only secondary to the forces driving them. We therefore consider an auroral generator as identified if the *force* is identified, which acts on the source plasma and cannot be fully balanced by magnetic normal forces but requires balancing shear stresses transmitting the acting forces toward the ionosphere. This transmission involves field-aligned electric currents, for which basically *two complementary configurations* exist, types I and II after the work of *Boström* [1964] (Figure 1). A logical requirement is that the generator plasma be *magnetically connected* to the region of current closure in the ionosphere, since momentum and energy are transported in the Alfvén mode essentially *along* the magnetic field lines. Temporary disconnections of the magnetospheric from the ionospheric section of a field line through the potential drop in the acceleration region (or "fracture zone") does not invalidate the magnetic connectivity of generator and load inside the current circuit. The current closure region is a load in the auroral current circuit, i.e., $j \cdot E > 0$. While the electric field in load and generator must have the same direction, the current direction is reversed. A generator definition based on the forces necessarily implies $j \cdot E < 0$ inside the generator. In the following, we use forces and current systems for classifying auroral gen-erators. There are four force classes: (1) flow drag generators, (2) flow braking (with or without vortex formation), (3) internal pressure forces, and (4) external magnetic pressure forces. These four classes will be used below for the ordering of some of the best explored generator plasmas.

3.1. Type I Current System [after Boström, 1964]

Figure 1 (top) shows the generic Type I current configuration, simply sketched as line currents. An external magnetic pressure or inertial force acts on the magnetospheric plasma and drives a current connecting to upward- and downward-directed field-aligned currents, which transport shear stresses toward the ionosphere where they act in the *normal* direction to the arc. The dominant (magnetization) current flowing at the interface is opposed to the generator current ($j \cdot E < 0$) but has zero divergence and closes inside the magnetosphere.

3.2. Type II Current System [after Boström, 1964]

Figure 1 (bottom) shows the configuration of a Type II current system. An internal pressure or inertial force acts against magnetic shear stresses. The generator current is a curvature current flowing in the same direction as the magnetization current. The shear stresses transported toward the ionosphere drive a convective flow *parallel* to the arc.

4. SOME OBSCURE IDENTIFICATIONS

The word "obscure" must not be interpreted in a derogative sense. It is here being used for cases in which the proposed auroral current configurations do not identify a generator properly, so that the true nature is obscured. One of the outstanding cases is the current disruption model of *McPherron et al.* [1973] and *Lui* [1991] because it has enjoyed enormous popularity. My criticism pertains to the aspect that the generator is obviously located outside the substorm current wedge, namely, at the tail magnetopause. The cross-tail current is simply rerouted through the ionosphere, in the first case without particular reason, in the second because of "current disruption" caused by high turbulent resistivity of the cross-tail current inside the wedge. In the latter case, the current inside the disruption region is not a generator but a load. However, the "disruption" region connects to an auroral and ionospheric energy conversion region, and forces are transmitted to the ionospheric plasma. If the generator was located outside the auroral current wedge, what could be the paths of both momentum and energy flow? They remain obscure. It is interesting that in a recent paper by *Tanaka et al.* [2010],

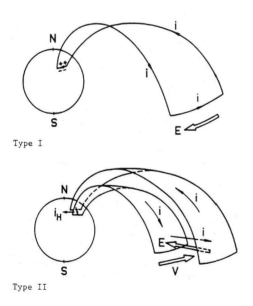

Figure 1. Two types of auroral current systems [after *Boström*, 1964].

the Region 1 field-aligned currents at substorm onset connect to currents through the dayside cusp. Again, there is no generator inside the substorm current wedge.

A similar problem exists in some cases with the reasoning behind the identification of "bursty bulk flows" (BBF) or "bubbles" with auroral streamers in the ionosphere [*Sergeev et al.*, 1999; *Kauristie et al.*, 2000; *Nakamura et al.*, 2001]. Either the electric field inside the "bubble" or BBF is simply mapped to the ionosphere or the cross-tail current is partially redirected to the ionosphere from the borders of the "bubble." Although the identification of the BBFs as sources of the auroral streamers is probably quite correct, there is no force identified driving a generator current inside these objects that would feed the attached field-aligned currents. In the next section, I return to these examples in the context of inertial currents as generators.

These examples must suffice to illustrate the key point. Without identification of the forces acting on the source plasma and bending the field thereby launching the transport of these forces earthward, a generator cannot be considered identified.

5. AURORAL GENERATORS

In this section, I will discuss a number of auroral current configurations, representative of and ordered by the four force classes as defined in section 3. These examples were not selected because the physical processes acting between generator and aurora or ionosphere can be regarded as fully understood but only because the relation between driving force and auroral dynamics appears to be sufficiently convincing. In some cases, these relations have been cast into physical equations and even quantitatively evaluated. However, it exceeds the scope of this chapter to even try summarizing these results. I will only quote a few physical relations between the generator forces and resulting field-aligned currents.

5.1. Flow Drag Generators

Following the models of *Axford and Hines* [1961] and *Dungey* [1961] for the driving of convection inside the magnetosphere by the solar wind, either by viscous interaction or as a consequence of reconnection, *Vasyliunas* [1970] presented a simple model of the resulting *convection or electric potential over the polar cap*. He circumvented elegantly the force balance problem between solar wind and ionosphere by introducing boundary conditions for the electric potential along the polar cap boundary. The field-aligned currents on the polar cap were set to zero, and the potential was derived from the divergence of the Pedersen current with

constant conductivity. Only along the polar cap boundary did the field-aligned currents not vanish.

I have chosen Vasyliunas' model because the generally observed absence of aurora on the polar cap during southward interplanetary magnetic field (IMF) B_z matches nicely the simplicity of the model. Indeed, although the solar wind is dragging the field lines extending to the polar cap (Figure 3a), the ionospheric drag forces are normally low and field-aligned currents weak, so that there is no need for postacceleration of the current carrying electrons and auroral arc formation. The forces exerted by the solar wind are largely balanced by the pressure of the plasma sheet and of the magnetic field accumulating in the tail lobes. In other words, much of the Type II current generated at the tail magnetopause closes by transverse currents inside the tail and not in the polar ionosphere.

A contrasting situation exists during northward IMF B_z when "Sun-aligned arcs" appear on the polar cap oriented more or less along the noon-midnight meridian [*Gusev and Troshichev*, 1986]. Their origin has been related to the solar wind flow acting against magnetic shear stresses inside the mantle and dragging the field lines across the polar cap in the antisolar direction. What is different from the previous situation with southward IMF? Under northward IMF, reconnection takes place with field lines from inside the polar cap. Thus, no new flux is accumulated in the tail, and the magnetic pressure of the tail lobes remains constant. Ionospheric drag is the only or dominant force balancing the drag forces against which the mantle flow is doing mechanical work. Figure 3b illustrates the interpretation of *Bonnell et al.* [1999]. The lower inset shows clearly that the associated current system is of Type II.

A third example of a flow drag generator is the *boundary layer model* of *Sonnerup* [1980]. Here the drag is exerted by viscous forces, not by connected field lines. Figure 3c shows a tailward boundary layer flow and a return flow along lower latitudes. A simple equation is used to express the force balance inside the boundary layer between the $\underline{j} \times \underline{B}$ force, balanced by the ionospheric drag force (second term) via the transfer of shear stresses, and the viscous force, represented by the third term of equation (1), without mass diffusion:

$$J_{00}B + \sigma v_x B^2 + \mu \frac{d^2 v_x}{dy^2} = 0, \qquad (1)$$

where j_{00} is a constant current representing a sunward-directed force, possibly connected with the ring current, σ is the effective integrated Pedersen conductivity for the ionospheric closure current projected up to the boundary layer, and v_x the flow velocity. The resulting field-aligned current is implicitly contained in the definition of σ. Sonnerup also considered the case of decoupling from the ionosphere by field-parallel potential drops and derived voltages of a few kV, thus capable of creating auroral arcs. Unfortunately, the role of viscous interaction in driving magnetospheric flows is not clear today. Possibly, some dayside aurora [*Sandholt et al.*, 1998] may be related to it. However, Sonnerup's theory and its subsequent elaborations constitute some of the few consistent and quantitative attempts to relate aurora physics directly with the solar wind drag.

5.2. Flow Braking

Rostoker and Boström [1976] made perhaps the earliest attempt to attribute the currents driving convection in the magnetosphere to flow braking in the tail plasma sheet. Flows toward the morning and evening flanks of the magnetosphere and their braking were seen as generators of currents of Type II, which are closing through N-S Pedersen

currents in the morning and evening auroral ovals (Figure 3d). Meanwhile, this model has been superseded by the growing understanding of substorms and related processes in the tail and distant magnetosphere. However, we will return to the driver of the oval convection in the next subsection.

The braking of earthward flows in the tail connected with dipolarization of the magnetic field has received much attention as generator of Type I current systems in the context of "bursty bulk flows" or substorm onsets. In the braking process, the magnetic field is obviously deformed by absorbing the momentum of the decelerated plasma flow. Hence, flow braking is a generator of magnetic shear stresses. The generator current is the inertial current of the decelerating plasma. *Haerendel* [1992] proposed this process as a possible generator located *inside* the substorm current wedge and calculated the resulting net current closing as westward electrojet at the poleward edge of the auroral oval (Figure 2a)

Type I Current Systems

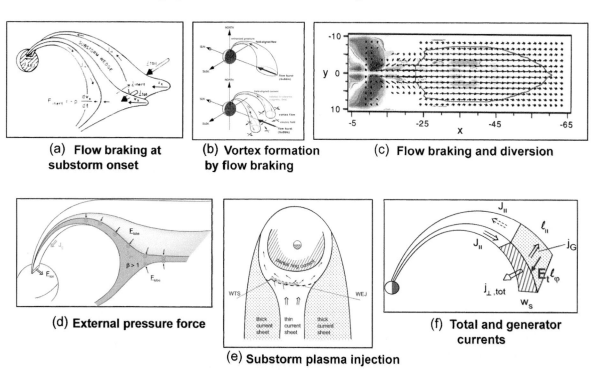

(a) **Flow braking at substorm onset**

(b) **Vortex formation by flow braking**

(c) **Flow braking and diversion**

(d) **External pressure force**

(e) **Substorm plasma injection**

(f) **Total and generator currents**

Figure 2. (a) Flow braking as substorm generator [from *Haerendel*, 1992]. (b) Vortex formation at braking of bubble flow [*Birn et al.*, 2004]. (c) Flow braking and diversion in equatorial plane and field-aligned current divergence (red, positive; blue, negative divergence) [from *Birn et al.*, 1999]. (d) Pressure force of tail lobe on hot plasma and force transmission by field-aligned currents to ionosphere [from *Haerendel*, 2009]. (e) Injection of plasma from central current sheet into outer magnetosphere and diversion into oval convection [after *Haerendel*, 2009]. (f) Currents in high-beta plasma layer as generator of auroral electrojet [after *Haerendel*, 2009].

Type II Current Systems

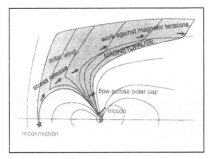

(a) **Flow drag generator**

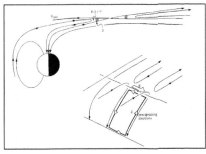

(b) **Dragging by mantle flow**

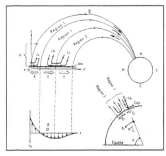

(c) **Viscous flow drag**

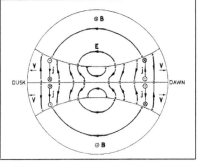

(d) **Flow braking in plasma sheet**

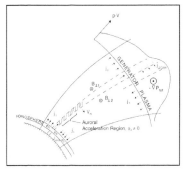

(e) **Embedded arcs**

Figure 3. (a) Cartoon of the solar wind flow drag generator during negative interplanetary magnetic field (IMF) B_z. (b) (top) Reconnection for northward IMF B_z and (bottom) transmission of shear stresses from mantle flow to ionosphere [from *Bonnell et al.*, 1999]. (c) Boundary layer and return flow generated by viscous drag of the solar wind [from *Sonnerup*, 1980]. (d) Electric fields and currents in the plasma sheet due to the braking of flows toward dusk and dawn [from *Rostoker and Boström*, 1979]. (e) Auroral current sheet embedded in a Type II current system driven by pressure forces [from *Haerendel*, 2007]. The underlying distribution of pressure times flux tube volume is shown in the upper diagram.

using an equation equivalent to the following expression [*Haerendel*, 2011]:

$$I_\phi = \delta \cdot 0.75 \text{ MA} \cdot \left(\frac{n}{1 \text{ cm}^{-3}}\right) \cdot \left(\frac{v_x}{500 \text{ km s}^{-1}}\right)^2$$

$$\times \left(\frac{20 \text{ nT}}{B}\right) \cdot \left(\frac{\ell_\parallel}{5R_E}\right). \quad (2)$$

Realistically, only the fraction, δ, of the inertial current of the order of 20% to 30% diverges into $j_\parallel$ and is involved in transferring magnetic shear stresses into the ionosphere. This means that this current would amount at best to a few hundred kA, while auroral electrojets reach 1 MA. A similar conclusion was reached by *Birn et al.* [1999].

Birn et al. [2004] modeled numerically the evolution of low-entropy flux tubes or "bubbles" after the work of *Chen and Wolf* [1993], their propagation along the tail toward the outer edge of the near-dipolar magnetosphere, and the flow braking with *vortex formation* and return flows (Figures 2b and 2b). The twisted field generated by the flow vortexes constitutes strong sources of field-aligned currents consistent with the Type I configuration. This model may well explain the origin of the auroral streamers, which have been discussed already in section 4. But, in contrast to the simple projection of the bubble electric field or re-directing of the cross-tail current, the flow braking model involves a dynamic plasma process apt to launch momentum and energy toward the ionosphere. The motion of auroral streamers therefore does not appear as an electrostatically mapped motion of the

BBFs. The related electric field, transferred by magnetic shear tresses, is inductive. Although not evident from the published paper of *Birn et al.* [2004], I suspect that the field-aligned currents calculated in their model are not strong enough to feed the auroral electrojet at substorm onset. As was stated already by *Birn et al.* [1999], the pressure forces of the slowed down and compressed plasma represent a more powerful current source. This theme will be taken up in the next subsections.

5.3. Internal Pressure Forces

Vasyliunas [1970] was probably the first who formulated the physical relations governing the connection between magnetospheric driving forces and associated currents and their closure in the ionosphere. For isotropic pressure, p, the generator current is described by

$$j_\perp = \frac{\underline{B} \times \nabla p}{B^2},\tag{3}$$

and the divergence integrated along the field yielding the field-aligned current at the ionosphere becomes

$$j_\parallel = \frac{B_{\text{ion}}}{2B_{\text{gen}}^2}(\nabla \times \underline{B}_{\text{gen}}) \cdot \nabla \int \frac{d\ell}{B}.\tag{4}$$

The meaning of the above quantities is obvious. Positive $j_\parallel$ is directed into the ionosphere. The integral is proportional to the flux tube volume. Since this volume is monotonically increasing outward, it seems that only longitudinal pressure forces can drive field-aligned currents, unless special plasma pressure gradients locally reverse the gradient of the flux tube volume. In any case, equation (4) describes well the so-called Region 2 field-aligned currents at the equatorward border of the auroral oval as driven by negative pressure gradients pointing along the oval away from midnight.

It has intrigued many researchers whether the Region 1 currents in the poleward sections of the auroral oval can also be driven by pressure gradients on closed field lines. *Rostoker and Boström* [1976] discarded this possibility as discussed above. Along similar lines, *Hasegawa and Sato* [1979] introduced field-aligned current generation by the vorticity of the plasma sheet flow as it encounters the outer magnetosphere and becomes deflected similar to what is shown in Figure 2b. In this paper, Hasegawa and Sato derived a generalization of the Vasyliunas equation (equation (4)) for field-aligned current generation by the sum of vorticity, transverse and inertial currents:

$$j_\parallel = B \int \left[\frac{\rho}{B} \cdot \frac{d}{dt}\left(\frac{\Omega}{B}\right) + \frac{2}{B^2}j_\perp \cdot \nabla B + \frac{1}{\rho B}j_{\text{in}} \cdot \nabla \rho \right] d\ell_\parallel\tag{5}$$

with the vorticity defined as $\Omega = (\nabla \times \underline{v}) \cdot \hat{b}$, where $\hat{b}$ is the unit vector along $\underline{B}$. Although the sense of the currents derived in that paper agrees with the Region 1 configuration, they are not generated by pressure gradients on closed field lines. Another attempt was made by *Yang et al.* [1994] in the context of the Rice Convection Model. In their case 3, they found pressure-driven Region 1 currents inside the plasma sheet flow in the outer magnetosphere.

The overwhelming evidence that regions 2 and 1 currents are largely (not fully) connected by N-S-oriented Pedersen currents in the ionosphere on closed field lines and the appearance of the most striking auroral arcs in the poleward section of the evening oval led *Haerendel* [2007] to simply postulate the existence of a strong inward pressure gradient in the poleward section of the oval overruling the outward gradient of the undisturbed flux tube volume (Figure 3e). Regions 1 and 2 currents are thus, to a large extent, generated by the same longitudinal pressure forces, and the related current configuration is of Type II. As shown in the cartoon of Figure 3e, the auroral arcs appear *embedded* in the upward current region where local concentrations of $j_\parallel$ are strong. However, the paper of *Haerendel* [2007] does not treat consistently the consequences of the postulated strong radially inward-directed pressure gradient nor does it present a justification of the "auroral condition" [*Haerendel*, 2007, equation 20], which amounts to the logarithmic pressure gradient outbalancing that of the flux tube volume. The fuller picture is that there are also longitudinal currents driven by the radial gradients. The associated magnetization currents close inside the magnetosphere, while the magnetic gradient and curvature currents experience a divergence because of the longitudinal pressure gradients. As shown by *Haerendel* [2008], the connected field-aligned currents are of equal strength to those driven by the radial pressure gradients because the gradient scales of driving force and divergence are just interchanged. However, it is hard to assess, in general, how much of these currents are closed in the sense of types II or I systems. This must depend on the overall pressure distribution. In the next section, we will deal with the Type I system. Summarizing this discussion, we can say that we do not yet have a full understanding of the nature of the generator of the currents that drive the oval convection and create auroral arcs embedded in the convection.

5.4. External Magnetic Pressure Forces

There is a wide variety of concepts about how the plasma ejected in the reconnection process in the near-Earth tail during substorms enters the outer, more dipolar magnetosphere and how the currents flowing in the auroral electrojet are being generated. In section 4, I have discussed the still popular concepts of current diversion or disruption, which do not properly identify the driving forces. Undoubtedly, the generator must be powered by the energy carried earthward by the reconnection exhaust. This leads to the suspicion that also the driving forces are directed earthward. Furthermore, we know since the seminal paper by *Boström* [1964] that the related current system must be of Type I. Pressure forces exerted by the hot plasma accumulating in the outer magnetosphere during night hours cannot generate the right sense of the field-aligned currents, upward in the evening and downward in the morning sectors. The underlying reason is that due to the diamagnetism, the magnetic field wants to expel the hot plasma. Therefore, it seems to be a logical conclusion that the forces that keep the hot plasma inside the magnetosphere are the same as the ones driving the generator.

This has led the author [*Haerendel*, 2009] to propose a generator model in which the driving force is exerted by the magnetic pressure of the low-density, high field tail lobes (Figure 2d). Plasma is flowing out of the central current sheet into the near-dipolar magnetosphere, whereby the formerly strongly stretched magnetic field relaxes. This would create a problem with pressure build-up and flux-tube entropy, were it not for the outflow of the plasma into the oval convection channels (Figure 2e). This is a consequence of the inward-directed $\underline{j} \times \underline{B}$ force based on the total (westward) current, $j_{\perp,\text{tot}}$, on the one hand, and the outward pressure of the partial ring current constituting an inner wall, on the other hand. There is thus a Poynting flux through the inner edge of the tail into the outer magnetosphere. However, part of the current flowing at the boundary is an *eastward-directed magnetic gradient and curvature current*. It opposes the westward electric field of the inward flow and is thus a generator current (Figure 2f). It drives a Type I system and feeds the auroral (westward) electrojet (WEJ), whereas the westward-directed and dominant magnetization current has zero divergence. The Type I current system constitutes a strong sink of energy through the auroral energy conversion process and the dissipating secondary Pedersen currents, which create an eastward flow along the WEJ. Part of the inward-directed Poynting flux is thus balanced by a downward flux into the auroral ionosphere. But the dominant contribution to the energy budget of the hot plasma is probably the expansion in the eastward and westward directions. *Haerendel* [2009] derived an equation for the magnitude of the field-aligned (sheet) currents driven by this generator. Based on the designations defined in Figure 2f, it is

$$J_{\|,\text{ion}} = \frac{\ell_{\varphi}}{\ell_{\text{div}}} \cdot \frac{\ell_{\|}}{\ell_{\text{arc}}} \cdot \frac{\beta B_{\text{gen}}}{2\mu_0}, \qquad (6)$$

where $\ell_{\varphi}/\ell_{\text{div}}$ is a measure for the fraction of the generator current closing through the ionosphere.

It may be of the order 0.2–0.3; $\ell_{\|}$ designates the effective extent of the generator along the field lines. It turns out that β of the order 2–3 can well produce current strengths consistent with observations.

At this point, the questions arises whether the plasma outflow into the two convection channels along the morning and evening ovals are solely driven by the initial push of the $\underline{j} \times \underline{B}$ force at the outer boundary (Figure 3e) or whether they are maintained by the negative pressure gradients from midnight toward the dayside. An initial push would imply that the flow is sustained by the inertia of the plasma. However, the time scales involved would be too short [*Rostoker and Boström*, 1976]. There is just not enough kinetic energy in the flow (of order 50 km s^{-1}). The pressure force is 2 orders of magnitude more powerful. We are thus thrown back to the scenario of a Type II current system maintaining the convective flow away from midnight along the oval with its embedded auroral arcs.

6. CONCLUDING REMARKS

In view of the enormous amount of knowledge existing with regard to the auroral energy conversion processes, our knowledge of the generation of the auroral currents can only be called *scanty*. Very few theories have been developed directly linking the current generation in the outer magnetosphere or tail with the auroral energy conversion. Answers from numerical modeling are still greatly limited by the insufficient resolution of the codes at the ionospheric boundaries. By presenting a more systematic overview of the types of auroral current generators illustrated by a few selected examples, this paper is meant to help whet the appetite of auroral researchers to address some of the many not answered questions.

REFERENCES

Axford, W. I., and C. O. Hines (1961), A unifying theory of high-latitude geophysical phenomena and geomagnetic storms, *Can J. Phys.*, *39*, 1433–1464.

Birn, J., M. Hesse, G. Haerendel, W. Baumjohann, and K. Shiokawa (1999), Flow braking and the substorm current wedge, *J. Geophys. Res.*, *104*, 19,895–19,904.

Birn, J., J. Raeder, Y. L. Wang, R. A. Wolf, and M. Hesse (2004), On the propagation of bubbles in the geomagnetic tail, *Ann. Geophys.*, *22*, 1773 1786.

Bonnell, J., R. C. Elphic, S. Palfery, R. J. Strangeway, W. K. Peterson, D. Klumpar, C. W. Carlson, R. E. Ergun, and J. P. McFadden (1999), Observations of polar cap arcs on FAST, *J. Geophys. Res.*, *104*(A6), 12,669–12,681.

Boström, R. (1964), A model of the auroral electrojets, *J. Geophys. Res.*, *69*(23), 4983–4999.

Chen, C. X., and R. A. Wolf (1993), Interpretation of high-speed flows in the plasma sheet, *J. Geophys. Res.*, *98*(A12), 21,409–21,419.

Dungey, J. W. (1961), Interplanetary magnetic fields and the auroral zones, *Phys. Rev. Lett.*, *6*, 47–48.

Gusev, M. G., and O. A. Troshichev (1986), Hook-shaped arcs in dayside polar cap and their relation to the IMF, *Planet. Space Sci.*, *34*, 489–496.

Haerendel, G. (1992), Disruption, ballooning or auroral avalanche —On the cause of substorms, in *Proceedings of the First International Conference on Substorms, Kiruna, Sweden, 23–27 March 1992, Eur. Space Agency Spec. Publ.*, ESA SP-337, 417–420.

Haerendel, G. (2007), Auroral arcs as sites of magnetic stress release, *J. Geophys. Res.*, *112*, A09214, doi:10.1029/2007JA012378.

Haerendel, G. (2008), Auroral arcs as current transformers, *J. Geophys. Res.*, *113*, A07205, doi:10.1029/2007JA012947.

Haerendel, G. (2009), Poleward arcs of the auroral oval during substorms and the inner edge of the plasma sheet, *J. Geophys. Res.*, *114*, A06214, doi:10.1029/2009JA014138.

Haerendel, G. (2011), Six auroral generators: A review, *J. Geophys. Res.*, *116*, A00K05, doi:10.1029/2010JA016425.

Hasegawa, A., and T. Sato (1979), Generation of field aligned current during substorm, in *Dynamics of the Magnetosphere*, edited by S.-I. Akasofu, pp. 529–542, D. Reidel, Dordrecht, Netherlands.

Kauristie, K., V. A. Sergeev, M. Kubyshkina, T. I. Pulkkinen, V. Angelopoulos, T. Phan, R. P. Lin, and J. A. Slavin (2000), Ionospheric current signatures of transient plasma sheet flows, *J. Geophys. Res.*, *105*(A5), 10,677–10,690.

Lui, A. T. Y. (1991), A synthesis of magnetospheric substorm models, *J. Geophys. Res.*, *96*(A2), 1849–1856.

McPherron, R. L., C. T. Russell, and M. P. Aubry (1973), Phenomenological model for substorms, *J. Geophys. Res.*, *78*(16), 3131–3149.

Nakamura, R., W. Baumjohann, R. Schödel, M. Brittnacher, V. A. Sergeev, M. Kubyshkina, T. Mukai, and K. Liou (2001), Earthward flow bursts, auroral streamers, and small expansions, *J. Geophys. Res.*, *106*, 10,791–10,802.

Rostoker, G., and R. Boström (1976), A mechanism for driving the gross Birkeland current configuration in the auroral oval, *J. Geophys. Res.*, *81*(1), 235–244.

Sandholt, P. E., C. J. Farrugia, J. Moen, Ø. Noraberg, B. Lybekk, T. Sten, and T. Hansen (1998), A classification of dayside auroral forms and activities as a function of interplanetary magnetic field orientation, *J. Geophys. Res.*, *103*(A10), 23,325–23,345.

Sergeev, V. A., K. Liou, C.-I. Meng, P. T. Newell, M. Brittnacher, G. Parks, and G. D. Reeves (1999), Development of auroral streamers in association with localized impulsive injections to the inner magnetotail, *Geophys. Res. Lett.*, *26*(3), 417–420.

Sonnerup, B. U. Ö. (1980), Theory of the low-latitude boundary layer, *J. Geophys. Res.*, *85*(A5), 2017–2026.

Tanaka, T., A. Nakamizo, A. Yoshikawa, S. Fujita, H. Shinagawa, H. Shimazu, T. Kikuchi, and K. K. Hashimoto (2010), Substorm convection and current system deduced from the global simulation, *J. Geophys. Res.*, *115*, A05220, doi:10.1029/2009JA014676.

Vasyliunas, V. M. (1970), Mathematical model of magnetospheric convection and its coupling to the ionosphere, in *Particles and Fields in the Magnetosphere, Astrophys. Space Sci. Libr. Ser.*, vol. 17, edited by B. M. McCormac, p. 60, D. Reidel, Dordrecht, Netherlands.

Yang, Y. S., R. W. Spiro, and R. A. Wolf (1994), Generation of Region 1 current by magnetospheric pressure gradients, *J. Geophys. Res.*, *99*(A1), 223–234.

G. Haerendel, Max Planck Institute for Extraterrestrial Physics, Garching D-85748, Germany. (hae@mpe.mpg.de)

The Relationship Between Magnetospheric Processes and Auroral Field-Aligned Current Morphology

Robert J. Strangeway

Institute of Geophysics and Planetary Physics and Earth and Space Sciences Department
University of California, Los Angeles, California, USA

The concept of force balance is applied to the coupled magnetosphere-ionosphere system, where the dynamics of the system is determined by the forces and flows rather than the electric field and currents. The electromagnetic forces are given by the Lorentz force and hence the perpendicular currents. Field-aligned currents (FACs) provide the connection between the Lorentz force in the magnetosphere and the ionosphere. From force balance, the FACs are related to pressure gradients, flow braking, or vorticity in the magnetosphere. In the ionosphere, the FACs are also related to vorticity, but conductivity gradients may play an additional role. Because the shear-mode wave carries both FAC and vorticity, it is this wave mode that provides the communication between the magnetosphere and ionosphere. In contrast to the force-balance approach, models such as the Cowling channel model for the auroral arc are usually couched in terms of electric fields and currents. Such models do not address the dynamics of the system. The concept that the Cowling channel is a source of upward Poynting flux is misleading. Furthermore, the presence of such a channel would cause modifications in the magnetospheric pressures and flows. Finally, through consideration of the relationship between FACs and the precipitating and trapped particle populations as observed at low altitudes by the FAST spacecraft, we can begin to relate auroral currents to the regions and physical processes occurring within the magnetosphere.

1. INTRODUCTION

One of the fundamental issues for auroral physics is the relationship between aurora and the underlying causal processes. It is now accepted that the aurora is a manifestation of field-aligned currents (FACs), especially in regions of accelerated electron precipitation. So the issue then becomes how to relate FACs to magnetospheric processes and regions. In this chapter, we will explore some aspects of this relationship from the perspective that the coupled magnetosphere iono-

Auroral Phenomenology and Magnetospheric Processes: Earth and Other Planets
Geophysical Monograph Series 197
10.1029/2012GM001211

sphere system is better understood in terms of forces and flows, that is, using the "B, v" paradigm [*Parker*, 1996].

The classical work of *Boström* [1964], on the other hand, uses electric fields and currents as the basis for discussing auroral arc morphology, where arc current systems are divided into "type 1," with filament-like FACs, and "type 2," with sheet-like FACs. Recently, *Haerendel* [2011] has expanded on this separation into types 1 and 2 to explore how auroral arcs are related to magnetospheric processes. This chapter is both complementary to and a contrast with the work of *Haerendel* [2011], since the magnetospheric processes that are responsible for FACs are essentially the same, but we do not take the different types of current systems as discussed by *Boström* [1964] as the starting point. Whether the FACs are sheet-like or filament-like is simply taken to be a reflection of the structure within the magnetosphere.

The outline of the chapter is as follows. In the next section, we will summarize the force-balance approach to the coupled magnetosphere-ionosphere system. This will lead to an emphasis on the Lorentz force. Since this force is given by perpendicular currents, this in turn leads to the idea that FACs are a consequence of coupling magnetospheric drivers to the ionospheric load. Because the frozen-in electron condition shows that there is a direct equivalence mechanical and electromagnetic loads, there is a connection between the work done in moving the ionospheric plasma through the collisional neutral atmosphere and the electromagnetic energy flowing into the ionosphere, as given by the Poynting flux. This is used in the third section, where we discuss whether or not auroral arcs can be a source of Poynting flux for the magnetosphere. This involves consideration of the Cowling channel model for an auroral arc, and we emphasize that such a model is incomplete as the final configuration of the current system can only be determined by including dynamics. In the fourth section, we expand on the force balance ideas in the context of particle and field observations at low altitudes and outline an approach where we map structure in the FACs to structure in the particle populations, similar to the work of *Ohtani et al.* [2010]. We present a summary in the final section.

2. MAGNETOSPHERE-IONOSPHERE COUPLING AND FORCE BALANCE

Strangeway and Raeder [2001] give a derivation of the plasma momentum equation that includes both the classical MHD terms as well as collisional terms. The magnetospheric momentum equation is given by the standard MHD equation

$$\rho \frac{D\mathbf{U}}{Dt} = \mathbf{j} \times \mathbf{B} - \nabla P, \tag{1}$$

where ρ is the plasma mass density, $\mathbf{j}$ is the current density, $\mathbf{B}$ is the magnetic field, P is the plasma pressure, assumed to be isotropic, and

$$\frac{D}{Dt} \equiv \frac{\partial}{\partial t} + \mathbf{U} \cdot \nabla \tag{2}$$

is the total time derivative following the fluid. In equation (1), we have assumed we can neglect any terms dependent on the electron mass in the inertial term on the left-hand side, as this mass is much smaller than the ion mass.

In the ionosphere, collisions dominate, while the plasma pressure is small, and

$$\rho \frac{D\mathbf{U}}{Dt} = \mathbf{j} \times \mathbf{B} - n(m_i \nu_{in} + m_e \nu_{en})(\mathbf{U}_i - \mathbf{U}_n) - m_e \nu_{en} \frac{\mathbf{j}}{e}. \tag{3}$$

In equation (3), n is the plasma number density, and we have assumed quasineutrality, i.e., $n \approx n_i \approx n_e$, where n_i and n_e are the ion and electron number densities, and m_i and m_e are the ion and electron masses, ν_{in} is the ion-neutral collision frequency, ν_{en} is the electron-neutral collision frequency, $\mathbf{U}_i$ is the ion fluid flow velocity, with $\mathbf{U}_i \approx \mathbf{U}$ because $m_e \ll m_i$, $\mathbf{U}_n$ is the neutral atmosphere velocity, and e is the magnitude of the electron charge.

If we further assume that we can neglect the electron collision terms, which is generally the case in the E and F region ionosphere, and also assume $\nu_{in}\tau \gg 1$, where τ is the time scale of interest, then

$$\mathbf{j} \times \mathbf{B} = \rho \nu_{in}(\mathbf{U} - \mathbf{U}_n). \tag{4}$$

As noted by *Parker* [1996] and *Vasyliūnas* [2001], the plasma momentum equation (1) does not contain any electric field term, and neither does the ionospheric momentum equation as given by equation (3) or (4) [*Song et al.*, 2005].

The common force in both equations (1) and (4) is the $\mathbf{j} \times \mathbf{B}$ force, often referred to as the Lorentz force. The current in the Lorentz force is, of course, perpendicular to the magnetic field. One aspect of magnetosphere-ionosphere coupling is hence how these perpendicular currents are coupled together through FACs. Such coupling is required because, in general, the ionosphere acts as a mechanical and electromagnetic load, as shown in section 3, and the ionospheric currents must be connected to generator regions in the magnetosphere.

For the magnetosphere, we can derive an equation for the FACs by first taking the curl of equation (1), assuming $\nabla \cdot \mathbf{j} = 0$:

$$(\mathbf{B} \cdot \nabla)\mathbf{j} = (\mathbf{j} \cdot \nabla)\mathbf{B} + \nabla \times \rho \frac{D\mathbf{U}}{Dt}. \tag{5}$$

The assumption that $\nabla \cdot \mathbf{j} = 0$ means that we are ignoring the displacement current. This is generally valid for time scales much longer then than the plasma-wave period, and for sufficiently dense plasmas such that the Alfvén speed is much less than the speed of light.

Taking the dot product of equation (5) with $\mathbf{B}$, and performing some additional algebra, we find

$$(\mathbf{B} \cdot \nabla)\left(\frac{\mathbf{j} \cdot \mathbf{B}}{B^2}\right) = \frac{2}{B^3}\mathbf{B} \cdot (\nabla P \times \nabla B)$$
$$+ \frac{1}{B^2}\mathbf{B} \cdot \left(\rho \frac{D\mathbf{U}}{Dt} \times \frac{\nabla V_A^2}{V_A^2}\right)$$
$$+ \frac{\rho}{B^2}\mathbf{B} \cdot \left(\nabla \times \frac{D\mathbf{U}}{Dt}\right). \tag{6}$$

The term on the left-hand side of equation (6) is the field-aligned gradient of the FAC per unit magnetic flux. The first term on the right-hand side of equation (6) is the pressure-gradient term, as derived by *Vasyliūnas* [1970] under assumptions of slow flow and steady state. This is the term used by models such as the Rice Convection Model (RCM) [*Wolf*, 1983] to derive FACs. For the second term on the right-hand side of equation (6), V_A is the Alfvén speed ($V_A^2 = B^2/\rho\mu_0$), and this terms corresponds to flow braking [*Shiokawa et al.*, 1997]. The last term is related to vorticity, $\omega = \nabla \times \mathbf{U}$.

In order to rewrite the last term in equation (6), we make use of the frozen-in approximation to the generalized Ohm's law, or equivalently, the electron fluid momentum equation. We assume that the electron pressure and electron inertia can be ignored in the electron momentum equation. Since $\mathbf{U}_i \approx \mathbf{U}$, from quasineutrality, the electron fluid flow velocity is given by $\mathbf{U}_e = \mathbf{U} - \mathbf{j}/ne$. The frozen-in condition for the electron fluid is hence

$$\mathbf{E} + \mathbf{U}_e \times \mathbf{B} = \mathbf{E} + \mathbf{U} \times \mathbf{B} - \frac{\mathbf{j} \times \mathbf{B}}{ne} = 0. \qquad (7a)$$

Assuming $\mathbf{j} \times \mathbf{B}$ is small in equation (7a) leads to the frozen-in condition for the ion fluid,

$$\mathbf{E} + \mathbf{U} \times \mathbf{B} = 0. \qquad (7b)$$

In the ionosphere, current densities can be large, and the frozen-in electron approximation, equation (7a), is more valid than equation (7b). In the magnetosphere, equation (7b) applies unless the current densities and electron pressure gradients are large. At this stage, we will assume equation (7b) applies for the magnetosphere.

From the definition of vorticity given above,

$$\nabla \times (\mathbf{U} \cdot \nabla)\mathbf{U} = -\nabla \times (\mathbf{U} \times \boldsymbol{\omega})$$
$$= \boldsymbol{\omega}(\nabla \cdot \mathbf{U}) + (\mathbf{U} \cdot \nabla)\boldsymbol{\omega} - (\boldsymbol{\omega} \cdot \nabla)\mathbf{U}, \qquad (8)$$

while the curl of the frozen-in condition, equation (7b), and Faraday's law gives

$$\frac{D\mathbf{B}}{Dt} = (\mathbf{B} \cdot \nabla)\mathbf{U} - \mathbf{B}(\nabla \cdot \mathbf{U}). \qquad (9)$$

Since $\boldsymbol{\omega} \times (\nabla \times \mathbf{U}) = 0$, it can be shown that

$$\boldsymbol{\omega} \cdot (\mathbf{B} \cdot \nabla)\mathbf{U} = \mathbf{B} \cdot (\boldsymbol{\omega} \cdot \nabla)\mathbf{U} \qquad (10)$$

and, consequently,

$$\mathbf{B} \cdot \left(\nabla \times \frac{D\mathbf{U}}{Dt} \right) = \mathbf{B} \cdot \frac{D\boldsymbol{\omega}}{Dt} - \boldsymbol{\omega} \cdot \frac{D\mathbf{B}}{Dt}. \qquad (11)$$

Substituting equation (11) into equation (6) gives

$$(\mathbf{B} \cdot \nabla)\left(\frac{\mathbf{j} \cdot \mathbf{B}}{B^2} \right) = \frac{2}{B^3} \mathbf{B} \cdot (\nabla P \times \nabla B)$$
$$+ \frac{1}{B^2} \mathbf{B} \cdot \left(\rho \frac{D\mathbf{U}}{Dt} \times \frac{\nabla V_A^2}{V_A^2} \right)$$
$$+ \frac{\rho}{B^2}\left(\mathbf{B} \cdot \frac{D\boldsymbol{\omega}}{Dt} - \boldsymbol{\omega} \cdot \frac{D\mathbf{B}}{Dt} \right), \qquad (12)$$

which more clearly shows the vorticity-dependent terms [see also *Hasegawa and Sato*, 1979]. We should also emphasize that the frozen-in condition only enters through the $D\mathbf{B}/Dt$ term.

Equation (12) can be used to derive the dispersion relation for the shear-mode Alfvén wave, under the first-order perturbation approximation. To first order, equation (12) becomes

$$\frac{\partial j_{\|}}{\partial z} = \frac{\rho_0}{B_0} \frac{\partial \omega_{\|}}{\partial t} \qquad (13)$$

since the other terms on the right-hand side of equation (12) are second order in the perturbed quantities. In equation (13), the subscript "0" denotes zero order, or unperturbed quantities, and the z axis is parallel to the ambient magnetic field $\mathbf{B}_0$. If we now take the curl of equation (9) and also assume first-order perturbations, then the parallel component of the resultant equation gives

$$\mu_0 \frac{\partial j_{\|}}{\partial t} = B_0 \frac{\partial \omega_{\|}}{\partial z}. \qquad (14)$$

Hence,

$$V_A^2 \frac{\partial^2 \omega_{\|}}{\partial z^z} = \frac{\partial^2 \omega_{\|}}{\partial t^z}, \qquad (15)$$

which is the dispersion relation for shear-mode Alfvén waves with $V_A^2 = B_0^2/\rho_0\mu_0$. It is well known that the shear mode corresponds to vorticity, but equation (15) shows why shear-mode waves are an intrinsic part of the dynamics of magnetosphere-ionosphere coupling.

For the ionosphere, on taking the curl of the ionospheric momentum equation (4),

$$(\mathbf{B} \cdot \nabla)\mathbf{j} - (\mathbf{j} \cdot \nabla)\mathbf{B} = \rho\nu_{in}(\boldsymbol{\omega} - \boldsymbol{\omega}_n) - (\mathbf{U} - \mathbf{U}_n) \times \nabla\rho\nu_{in}. \qquad (16)$$

If we consider the component of equation (16) parallel to the ambient magnetic field, then the first term gives the FAC, while the second term is usually small. The first term on the right-hand side of equation (16) is again a vorticity-related term, while the last term is associated with conductivity gradients. In the ionosphere, the relationship between the FAC and vorticity is controlled by the ion-neutral collisions.

In the magnetosphere, on the other hand, the FACs are related to vorticity through the Alfvén speed, with the added complication that pressure gradients and flow braking can also give rise to FACs. In general, then, it is by no means obvious that the two systems are always in equilibrium with each other. Understanding the dynamical linkage is one of the outstanding areas of research for magnetosphere-ionosphere coupling.

3. FIELD-ALIGNED CURRENTS AND ENERGY FLOW

In the ionosphere, equation (4) shows that $\mathbf{U} - \mathbf{U}_n$ is perpendicular to $\mathbf{j}$, which includes both Pedersen and Hall currents. As such, $\mathbf{U}$ is, in general, not perpendicular to $\mathbf{E}$, and the frozen-in electron condition (7a) is a more appropriate approximation (corresponding to vanishing electron collisions). One important consequence of equation (7a) is that, even though equation (7a) includes the $\mathbf{j} \times \mathbf{B}$ term,

$$\mathbf{j} \cdot \mathbf{E} = \mathbf{U} \cdot (\mathbf{j} \times \mathbf{B}). \qquad (17)$$

The term on the left-hand side of equation (17) is often referred to as Joule dissipation, through analogy with circuit theory, but it is more correctly the rate at which electromagnetic energy is converted to ($\mathbf{j} \cdot \mathbf{E} > 0$) or from ($\mathbf{j} \cdot \mathbf{E} < 0$) other forms of energy. The term on the right-hand side of equation (17) is the rate at which the Lorentz force does

work, and there is a direct equivalence between electrical and mechanical loads [*Strangeway and Raeder*, 2001].

A second aspect of equation (17) arises from consideration of Poynting's theorem:

$$\frac{\partial W}{\partial t} + \nabla \cdot \mathbf{S} + \mathbf{j} \cdot \mathbf{E} = 0, \qquad (18)$$

where W is the energy density of the electromagnetic fields, which is $\approx B^2 / 2\mu_0$ for an MHD fluid, and the Poynting vector $\mathbf{S} = \mathbf{E} \times \mathbf{B} / \mu_0$. If we can ignore the time variation of W, then we can relate the work done by the Lorentz force to the divergence of the electromagnetic energy flux. The Poynting vector is a useful proxy for determining electrical and mechanical generators versus for loads.

Figure 1 [after *Strangeway et al.*, 2000] presents a simplified picture of how the flows, FACs, and the electromagnetic fields are related for the ionosphere. Figure 1 is drawn in the context of the region 1 current system [*Iijima and Potemra*, 1976], with the polar cap magnetic field lines threading the high-latitude polar magnetopause, where they are convected in an antisunward direction through reconnection with the solar wind magnetic field. It should be noted that, for simplicity, in Figure 1 and throughout the rest of the chapter, the motion of the neutral atmosphere will be ignored and we assume $\mathbf{U}_n = 0$. The sketch also implicitly assumes that the current sheets are assumed to extend

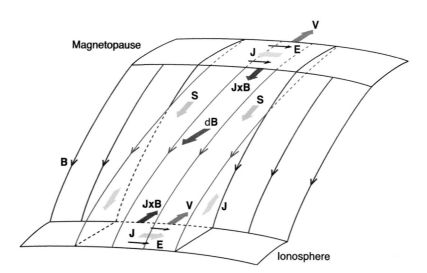

Figure 1. Simplified schematic showing the relationship between the flows, currents, and electromagnetic fields for the high-latitude ionosphere [after *Strangeway et al.*, 2000]. Figure 1 is drawn in the context of the region 1 current system, with the ionospheric closure currents resulting in antisunward flows (into the plane of the figure). It shows the background magnetic field (dark blue), the stressed magnetic field (red), the associated magnetic perturbation (dB, magenta), the flows (V, gray), the $\mathbf{J} \times \mathbf{B}$ force that drives the flow in the ionosphere and opposes the flow at the magnetopause (brown), the electric field (black arrows), and the perturbation Poynting vector (S, green). The current sheets are assumed to be infinitesimally thin, and $\mathbf{J}$ (cyan) is the current intensity.

into and out of the plane of the figure. As such, the current system is solenoidal, and the magnetic perturbation (shown in magenta) is contained entirely within the current loop. This means that the perturbation Poynting vector (green) is also inside the current loop. The current system most closely resembles the type 2 system of *Boström* [1964]. Figure 1 can be used to represent any solenoidal FAC system, and the drivers of the currents are not restricted to flows. Last, we have shown the FACs as infinitesimally thin sheets. In reality, the currents are of finite width, and the magnetic field perturbation tends to be more ramp-like, rather than a step-like discontinuity.

The ionospheric current in Figure 1 is a Pedersen current, flowing parallel to the electric field. Not shown is the corresponding Hall current, which flows in the $-\mathbf{E} \times \mathbf{B}$ direction. For current sheets that extend to infinity in the direction into or out of the page, the Hall current closes at infinity. If instead we assume finite current sheets, and further assume structure in the conductivity such that the Hall currents are unable to flow without restriction, we find a system that is often referred to as a Cowling channel. *Haerendel* [2008] has argued that the diversion of the Hall current out of the ionosphere because of conductivity gradients could be thought of as a transformer current. There are some issues with this model, which we will explore here.

Figure 2 [after *Haerendel*, 2009] emphasizes that the ionosphere is three-dimensional, and the Pedersen and Hall currents flow at different altitudes, a point also raised by *Fujii et al.* [2011]. The Cowling channel is set up as follows. We shall assume a region of enhanced conductivity associated with an auroral arc. Within this arc, the magnetosphere imposes a flow to the right (not shown in Figure 2) that is associated with a tangential electric field ($\mathbf{E}_t$). This, in turn, requires a corresponding Pedersen current, which contributes to the auroral electrojet (AEJ), and a Hall current ($\mathbf{J}_H$) that flows to the left. In the absence of any other effects, this Hall current must close within the magnetosphere through a secondary FAC system because the Hall current is blocked from flowing outside of the high-conductivity region. The alternative is to set up a secondary, polarization, electric field ($\mathbf{E}_{xi}$) that in turn has a secondary Pedersen current ($\mathbf{J}_P$) that opposes the primary Hall current. The polarization electric field has a secondary Hall current that flows in the same direction as the primary Pedersen current, resulting in the final total electrojet current.

We will now define a coordinate system with the *x* axis to the right of the figure, the *y* axis along the direction of $\mathbf{E}_t$, and the *z* axis vertically down. We will also modify Haerendel's notation to use the subscript "1" to denote the primary current system and the subscript "2" to denote the secondary

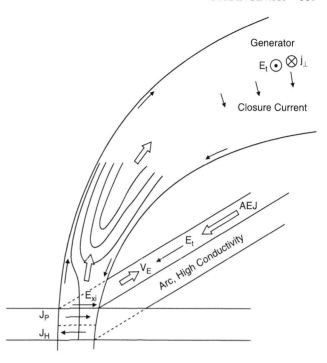

Figure 2. Three-dimensional sketch of an auroral arc [after *Haerendel*, 2009]. The arc acts as a partial Cowling channel, with an enhanced auroral electrojet set up by a polarization electric field across the arc, and a secondary current system that closes the excess Hall current in the magnetosphere.

current system. With this notation, $\mathbf{E}_t = \hat{\mathbf{y}}E_{1y}$ and the Pedersen and Hall currents for the primary current system are given by

$$\mathbf{J}_{P1} = \hat{\mathbf{y}}\Sigma_P E_{1y} \tag{19a}$$

and

$$\mathbf{J}_{H1} = -\hat{\mathbf{x}}\Sigma_H E_{1y}, \tag{19b}$$

where Σ_P and Σ_H are the height-integrated Pedersen and Hall conductivities, respectively.

We will now assume that the primary Hall current is partially canceled by a secondary Pedersen current, with an efficiency *f*. Thus, the polarization electric field associated with this Pedersen current is given by

$$\hat{\mathbf{x}}\Sigma_P E_{2x} = \mathbf{J}_{P2} = \hat{\mathbf{x}}f\Sigma_H E_{1y}, \tag{20}$$

with a resultant secondary Hall current given by

$$\mathbf{J}_{H2} = \hat{\mathbf{y}}f\frac{\Sigma_H^2}{\Sigma_P}E_{1y}. \tag{21}$$

Thus, the electrojet current is enhanced over the primary Pedersen current, and

$$\mathbf{J}_{AEJ} = \mathbf{J}_{P1} + \mathbf{J}_{H2} = \hat{\mathbf{y}}\Sigma_P\left(1 + f\frac{\Sigma_H^2}{\Sigma_P^2}\right)E_{1y}. \qquad (22)$$

The modified conductivity in equation (22) is known as the Cowling conductivity, usually with the assumption that $f = 1$. In that case, the primary Hall current and secondary Pedersen current cancel exactly, and there are no secondary FACs. Furthermore, since Σ_H is usually greater than Σ_P, the AEJ is mainly a Hall current, and this system corresponds to a *Boström* [1964] type 1 current system [*Haerendel*, 2008].

Haerendel [2008, 2009] also considers the energy dissipation associated with the Cowling channel. From equation (22), the dissipation rate for the AEJ current is given by

$$\mathbf{J}_{AEJ} \cdot \mathbf{E}_{1y} = \Sigma_P\left(1 + f\frac{\Sigma_H^2}{\Sigma_p^2}\right)E_{1y}^2, \qquad (23)$$

while the dissipation of the secondary current loop is given by

$$(\mathbf{J}_{H1} + \mathbf{J}_{P2}) \cdot \mathbf{E}_{2x} = (1 - 1/f)\Sigma_P E_{2x}^2 = f(f-1)\frac{\Sigma_H^2}{\Sigma_P}E_{1y}^2, \qquad (24)$$

where we have used equation (20) to rewrite E_{2x} in terms of E_{1y}.

In equation (24), the dissipation is negative for $0 < f < 1$, with the dissipation being zero when $f = 0$ or $f = 1$. That equation (24) can be negative is why *Haerendel* [2008, 2009] argues that the secondary current system can be a source of upward Poynting flux. But what matters is the total dissipation, and on adding equations (23) and (24),

$$\mathbf{J} \cdot \mathbf{E} = \Sigma_P\left(1 + f^2\frac{\Sigma_H^2}{\Sigma_P^2}\right)E_{1y}^2 = \Sigma_P(E_{1y}^2 + E_{2x}^2), \qquad (25)$$

as it should [see also, *Fujii et al.*, 2011]. To make this point clear, we note that for any arbitrary electric field and current that are not parallel, then $\mathbf{J} \cdot \mathbf{E} = j_x E_x + j_y E_y$ can include a negative contribution. If, for example, $\mathbf{j} \cdot \mathbf{E} > 0$ and the x axis of the coordinate system is chosen to lie between the vectors $\mathbf{j}$ and $\mathbf{E}$, then $j_x E_x > 0$ but $j_y E_y < 0$.

As shown in Figure 2, and implied by equation (4), the secondary Pedersen current must have a corresponding flow along the arc, labeled "V_E," which was not originally present in the magnetosphere. The FACs associated with the enhanced AEJ and the excess Hall current also result in additional currents within the magnetosphere. The force balance given by equation (1) therefore requires significant temporal changes in either the pressure gradients, or the flows, or both, likely mediated by waves, until a final equilibrium is achieved. That equilibrium may be considerably different than the initially assumed configuration.

4. FIELD-ALIGNED CURRENTS AND MAGNETOSPHERIC STRUCTURE

The next step in assessing the importance of the different forcing terms in equation (1) is to relate the flows and associated currents to the structure of the magnetosphere. Rather than explicitly relying on magnetic field mapping, and the associated uncertainties in such models, especially during disturbed times, we use the boundaries in the particle distributions to provide guidance as to the possible sources of the currents. This approach has been used by *Ohtani et al.* [2010] with low-altitude DMSP data. Here we will use the higher-altitude FAST spacecraft. FAST has the advantage of resolving the loss cone, while DMSP mainly measures only the precipitating particles. The information provided by measuring particle fluxes outside the loss cone may provide additional constraints on source-region mapping.

Here we will present two examples of premidnight flows as observed by FAST to demonstrate the approach being used. Figure 3 shows FAST data acquired on 16 February 2008, centered on 04:53 UT. The figure shows nearly 6 min of data, with the spacecraft passing from the polar cap, at the left of the figure, to lower latitudes, at the right of the figure. The top two panels show electron energy and angle spectra. The next two panels show ion energy and angle spectra, and the bottom panel shows the perturbation magnetic field. The green trace shows the locally eastward component, which in turn corresponds to the Lorentz force being applied to the ionosphere (see Figure 1), and the ionosphere flows westward for a positive eastward magnetic field perturbation.

The solid vertical line at 04:53 UT, marks the maximum eastward magnetic field, with the two adjacent dashed lines marking where the perturbation field changes sign. As such, then, these dashed lines mark a region of enhanced westward flow. At higher and lower latitudes, the perturbations are negative, and the inferred ionospheric flow is eastward. We do not have an explanation for the low-latitude eastward flow, but this feature warrants further investigation, beyond the scope of this chapter.

At ~04:51:30 UT, the spacecraft passes out of the polar cap into a region of upward FAC, as expected for duskside region 1 currents. The region of weaker region 1 current associated with ~1 keV electrons corresponds to eastward flows. Just prior to 04:53 UT, the flow changes from eastward to

FAST Orbit 46373

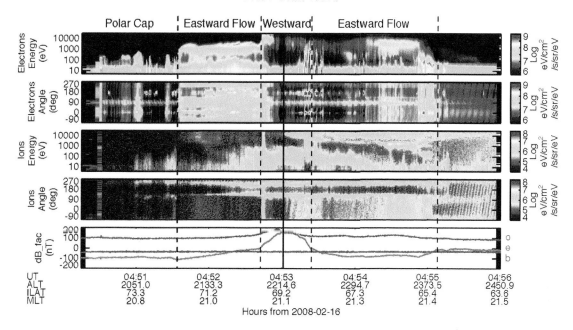

Figure 3. FAST data acquired on 16 February 2008, at ~21 magnetic local time (MLT). (top) The first and second panels show electron energy and angle spectra. The next two panels show ion energy and angle spectra, and (bottom) the perturbation magnetic field with respect to the International Geomagnetic Reference Field, cast into a local field-aligned coordinate system. The green trace shows the locally eastward (e) component of the perturbation magnetic field. The red trace shows the outward (o) component, corresponding to northward in the northern hemisphere, while the blue trace shows the model magnetic field-aligned component (b). The plots are annotated at the bottom of Figure 3 with UT, spacecraft altitude, invariant latitude (ILAT), and MLT. Both ILAT and MLT are calculated using an offset tilted dipole.

westward, and the energy of the precipitating electrons increases to ~10 keV. The region 1 current appears to correspond to boundary layer plasma sheet, with relatively weak fluxes of energetic ions.

At 04:53 UT, the FAC changes sign, having a region 2 sense. At the same time, there is an enhancement in the ~10 keV ions, which also show evidence of a filled downward loss cone. The downward current is clearly carried by upward electrons, with broadband energies up to 1 keV, and a narrow pitch angle. The downward current is associated with the presence of enhanced precipitating ion fluxes, and the peak in the eastward perturbation field occurs at the transition into a region that is more representative of the central plasma sheet. While region 2 currents are often attributed to azimuthal pressure gradients, and the FAST data give a latitudinal profile that does not provide any information concerning azimuthal gradients, the data do suggest that understanding why the flow channel appears to be colocated with the transition to the central plasma sheet requires consideration of all the terms in equation (12). In particular, a radial pressure gradient near midnight could act to slow down and deflect mainly earthward flows. This again em-

phasizes the need for a coupled model that includes plasma dynamics.

One last remark on the data in Figure 3 concerns the region of upward current near 04:55 UT that is the low-latitude boundary to the region of anomalous eastward flows. In this case, FAST measures ~100 eV precipitating electrons that carry the upward current. At the same time, the ion distribution changes from a single loss cone to a double loss cone. But trapped ion fluxes are still observed at lower latitudes, suggesting that the ion precipitation boundary is not necessarily the inner edge of the ion plasma sheet.

A second example of a duskside flow channel is given in Figure 4. While the format is similar to Figure 3, in this case, the spacecraft passes from low to high latitudes, with the polar cap to the right of Figure 3. There are many features that are similar to Figure 3. There is a region of eastward flow on closed field lines, adjacent to the polar cap. The dominant region 1 polarity current is carried by accelerated ~1 keV electrons. The current reverses at around 04:03:30 UT, where the eastward perturbation is at a maximum and is collocated with a change in the ion plasma sheet fluxes. Just after 04:02 UT, there is a brief interval of upward current that is

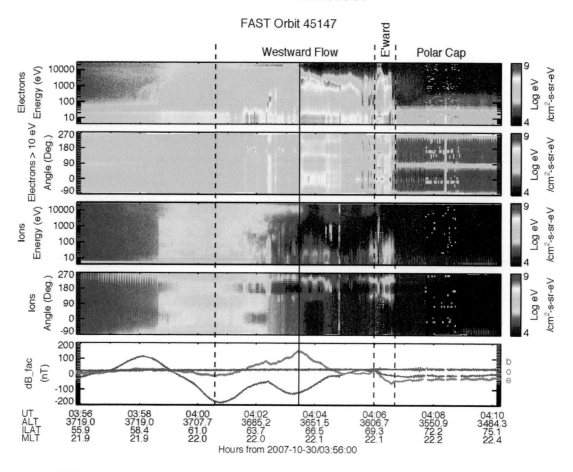

Figure 4. FAST data acquired on 30 October 2007, at ~22 MLT. Similar in format to Figure 3, although in this case the spacecraft passes from lower to higher latitudes.

associated with ~100 eV electron precipitation and the transition from a single to a double loss cone in the ions. This may correspond to the feature at 04:55 UT in Figure 3, but in this case, the upward current is embedded within a region of downward current, rather than marking the low-latitude boundary of an eastward flow channel. Although there may be some contamination from penetrating radiation, it does appear that the electron and ion plasma sheet extend to significantly lower latitudes, with the inner edge encountered at ~03:58 UT.

5. SUMMARY AND CONCLUSIONS

In this chapter, we have taken the approach that the coupled magnetosphere-ionosphere system should be considered as a mechanical system, where forces and the associated flows allow for a better understanding of the dynamical system [see also *Parker*, 1996; *Vasyliūnas*, 2001]. Magnetic field forces are given by the Lorentz

force, which in turn is given by the current density perpendicular to the magnetic field. The corresponding magnetospheric and ionospheric currents are coupled by FACs, which is why FACs are an important aspect of magnetosphere-ionosphere coupling. We further showed that within the ionosphere, the FACs are associated with either vorticity in the ionospheric flow or conductivity gradients. In the magnetosphere, the FACs are associated with plasma pressure gradients, flow braking, or vorticity. Furthermore, the shear mode, which carries FAC, also carries vorticity. Thus, we expect the shear mode to be the means by which changes are communicated between the magnetosphere and ionosphere.

Having discussed the overall approach used, we then moved to a specific issue for the coupled system: the Cowling channel. This has been explored extensively by *Haerendel* [2008, 2009]. In particular, *Haerendel* [2008] considered an auroral arc as a current transformer, with the secondary current system within the arc, which is a Cowling channel, being a source of upward Poynting flux. The Poynting flux is a useful parameter to consider since the net downward Poynting flux corresponds to the dissipation of magnetic energy within the ionosphere, at least to the extent that it is assumed that no Poynting flux flows horizontally or below the ionosphere. Furthermore, because the electron frozen-in condition is a good approximation for the topside ionosphere, the dissipation rate equals the rate at which work is done by the Lorentz force in moving the plasma through the neutrals.

It is the divergence of the Poynting flux that is physically meaningful. For quasi-static currents separating Poynting flux into upward and downward has no significance. We also noted that the presence of either a polarization electric or a secondary FAC system means that there are additional flows and FACs that must lead to a modification of the magnetosphere, with the final state indeterminate.

Finally, we considered how FACs map to boundaries within magnetosphere. This may provide clues as to which of the various processes implied by equation (12) are dominant. We presented two FAST data examples of currents and particles observed in the postdusk auroral zone. In general, the FACs conformed to the classical region 1 and 2 currents, although in one example, we found an additional low-latitude upward FAC. The transition from region 1 to 2, which is also where the maximum stress is applied to the ionosphere, was marked by a transition in the electrons, as would be expected since the region 1 currents in this local time sector are mainly carried by inverted-V electrons. The ions also showed a transition, with the >10 keV ions appearing at lower latitudes. Thus, the boundary between region 1 and region 2 currents appear to be associated with the transition between the plasma sheet boundary layer and the central plasma sheet. The transition to a double loss cone feature in the ions, while associated with a brief interval of upward current, seemed to have an arbitrary relationship to the low-latitude extent of the region 2 system. In the first example, this region of upward current was well equatorward of the major region 2 current. For the second example, the transition was in the middle of the region 2 current.

In conclusion, we have emphasized the approach that views the coupled magnetosphere-ionosphere system as a mechanically coupled system, where the forces and flows are the main factors in the system dynamics. The currents and electric fields then follow as a consequence of the mechanical coupling. This allows us to assess the appropriateness of the Cowling channel model for an auroral arc, and we are led to the conclusion that the Cowling channel can only be understood as part of the coupled system, where, again, the currents and electric field arise from the force balance. Finally, we showed two examples of low-altitude data, with

emphasis on the relationship between FACs and the particle populations. Mapping the currents to particle population boundaries in the magnetosphere may provide insight as to the physical processes responsible for the FACs and the associated auroral arcs.

Acknowledgments. We acknowledge fruitful discussions with Gerhard Haerendel, Vytenis Vasyliūnas, Eric Donovan, and Feifei Jiang. This work was supported by NASA grant NNX07AT15G and by the THEMIS mission and V. Angelopoulos through NASA contract NAS5-02099.

REFERENCES

Boström, R. (1964), A model of the auroral electrojets, *J. Geophys. Res.*, *69*(23), 4983–4999.

Fujii, R., O. Amm, A. Yoshikawa, A. Ieda, and H. Vanhamäki (2011), Reformulation and energy flow of the Cowling channel, *J. Geophys. Res.*, *116*, A02305, doi:10.1029/2010JA015989.

Haerendel, G. (2008), Auroral arcs as current transformers, *J. Geophys. Res.*, *113*, A07205, doi:10.1029/2007JA012947.

Haerendel, G. (2009), Poleward arcs of the auroral oval during substorms and the inner edge of the plasma sheet, *J. Geophys. Res.*, *114*, A06214, doi:10.1029/2009JA014138.

Haerendel, G. (2011), Six auroral generators: A review, *J. Geophys. Res.*, *116*, A00K05, doi:10.1029/2010JA016425. [Printed 117(A1), 2012].

Hasegawa, A., and T. Sato (1979), Generation of field aligned current during substorm, in *Dynamics of the Magnetosphere*, edited by S.-I. Akasofu, pp. 529–542, D. Reidel, Dordrecht, The Netherlands.

Iijima, T., and T. A. Potemra (1976), The amplitude distribution of field-aligned currents at northern high latitudes observed by Triad, *J. Geophys. Res.*, *81*(13), 2165–2174.

Ohtani, S., S. Wing, P. T. Newell, and T. Higuchi (2010), Locations of night-side precipitation boundaries relative to R2 and R1 currents, *J. Geophys. Res.*, *115*, A10233, doi:10.1029/2010JA015444.

Parker, E. N. (1996), The alternative paradigm for magnetospheric physics, *J. Geophys. Res.*, *101*(A5), 10,587–10,625.

Shiokawa, K., W. Baumjohann, and G. Haerendel (1997), Braking of high-speed flows in the near-Earth tail, *Geophys. Res. Lett.*, *24*(10), 1179–1182.

Song, P., V. M. Vasyliūnas, and L. Ma (2005), A three-fluid model of solar wind-magnetosphere-ionosphere-thermosphere coupling, in *Multiscale Coupling of Sun-Earth Processes*, edited by A. T. Y. Lui, Y. Kamide, and G. Consolini, pp. 447–456, Elsevier, Amsterdam.

Strangeway, R. J., and J. Raeder (2001), On the transition from collisionless to collisional magnetohydrodynamics, *J. Geophys. Res.*, *106*(A2), 1955–1960, doi:10.1029/2000JA900116.

Strangeway, R. J., R. C. Elphic, W. J. Peria, and C. W. Carlson (2000), FAST observations of electromagnetic stresses applied to the polar ionosphere, in *Magnetospheric Current Systems*,

Geophys. Monogr. Ser., vol. 118, edited by S. Ohtani et al., pp. 21–29, AGU, Washington, D. C., doi:10.1029/GM118p0021.

Vasyliūnas, V. M. (1970), Mathematical models of magnetospheric convection and its coupling to the ionosphere, in *Particles and Fields in the Magnetosphere*, edited by B. M. McCormack, pp. 60–71, D. Reidel, Dordrecht, The Netherlands.

Vasyliūnas, V. M. (2001), Electric field and plasma flow: What drives what?, *Geophys. Res. Lett.*, *28*(11), 2177–2180, doi:10.1029/2001GL013014.

Wolf, R. A. (1983), The quasi-static (slow-flow) region of the magnetosphere, in *Solar Terrestrial Physics*, edited by R. L. Carovillano and J. M. Forbes, p. 303, D. Reidel, Norwell, Mass.

R. J. Strangeway, Institute of Geophysics and Planetary Physics, University of California, Los Angeles, CA 90095, USA. (strangeway@igpp.ucla.edu)

Magnetospheric Dynamics and the Proton Aurora

E. Donovan, E. Spanswick, J. Liang, J. Grant, B. Jackel, and M. Greffen

Department of Physics and Astronomy, University of Calgary, Calgary, Alberta, Canada

On the nightside, the bright proton aurora forms a several degrees in latitude band of diffuse aurora near the equatorward boundary of the electron auroral oval. The precipitation is due to strong pitch angle diffusion that is thought to be the result of nonadiabatic motion of sub-keV to tens of keV central plasma sheet (CPS) protons as they traverse the tightly curved magnetic field topology in the vicinity of the neutral sheet. In this paper, we provide an overview of the relationship between the spatiotemporal evolution of the proton aurora and magnetospheric dynamics. We focus on the equatorward boundary of the proton aurora, the latitude of which has been shown to be strongly correlated with magnetic field line stretching in the inner CPS, and provide the first-ever identification of the position of the ion isotropy boundary relative to equatorial magnetospheric spacecraft.

1. INTRODUCTION

The term aurora refers to the light emitted from the upper atmosphere resulting from the precipitation of electrons and protons from outer space. Precipitating electrons collide with upper atmospheric particles, leaving them in excited states, relaxation from which produces the photons comprising the "electron aurora." Precipitating protons undergo charge-exchange collisions with atmospheric particles, creating a neutral hydrogen atom in an excited state: subsequent relaxation leads to the photons that comprise the "proton aurora."

There are fundamental differences between the electron and proton aurora. Early in the twentieth century, *Vegard* [1939] showed conclusively that hydrogen emissions in the auroral light were Doppler shifted indicating that the proton aurora was caused by particles precipitating down on the atmosphere from above. Since there is no such Doppler shift of emissions in electron aurora (the photons come from heavy atmospheric particles that have been excited by light incoming electrons), it was the proton aurora that provided the first concrete evidence of the precipitation of particles

from outer space causing the aurora. As we discuss below, this proton precipitation results from pitch angle scattering, rather than field-aligned electric fields, so the proton auroral evolution and structure is especially valuable for studying the time-evolving magnetospheric topology and dynamics. The focus of our chapter is divided between review of some well-known (though not all published) relationships between proton auroral phenomenology and magnetospheric processes, and some new results related to mapping the proton auroral distribution to the magnetotail.

Here we provide only a very cursory introduction related to how the precipitating protons interact with the atmosphere to produce the emissions and how proton aurora is observed. Precipitating protons undergo charge exchange collisions with atmospheric particles such as N_2, O_2, and O. Subsequent collisions reionize and reneutralize (taking those terms from *Eather* [1967]) the hydrogen atom/proton. The proton auroral photons are emitted from the (excited) neutral, and since this ionization/neutralization process can occur many tens of times for a given incident proton, a single proton can give rise to many photons with standard H emission wavelengths [see *Eather*, 1967, and references therein]. While the proton is in a neutral hydrogen atom, it is not magnetized. These ionization/neutralization sequences thus lead to the drifting of the proton away from the field line that it was incident on (not to be mistaken for convection), leading to the smearing of fine spatial features in the precipitating

Auroral Phenomenology and Magnetospheric Processes: Earth and Other Planets

Geophysical Monograph Series 197

10.1029/2012GM001241

proton distribution as first described by *Davidson* [1965]. There has been significant work done on how the distribution of incoming protons translates to vertical profiles of emission in the various resulting hydrogen emissions [see, e.g., *Galand and Chakrabarti*, 2006, and references therein]. This, together with spectrographic studies of the proton aurora, show broadening of the emission lines (blueward and redward) from which information can be inferred regarding the incident energy of the protons [see, e.g., *Galand and Chakrabarti*, 2006, and references therein].

The brightest emissions are the hydrogen Balmer α and β and Lyman α emissions. Lyman α is the brightest of the three, but is not useful for observations from the ground because of atmospheric absorption [see, e.g., *Galand et al.*, 2004]. It has been observed from space, notably with the Spectroscopic Imager (SI) on the NASA IMAGE satellite, which provided the only global images of the proton aurora to date [*Mende et al.*, 2000, 2001]. For ground-based observations, the Balmer α is brighter than the Balmer β (herein Hβ), but is relatively close to other (N$_2$ 1P bands) bright electron auroral emissions and so has not been widely targeted for observation. We refer the reader to excellent reviews that address interactions with the atmosphere and the spectroscopy of proton auroral emissions [see, e.g., *Galand and Chakrabarti*, 2006, and references therein].

From the ground, the proton aurora has been studied with spectrographs and photometers. Spectrographs can provide the shape of the Doppler-shifted emissions, which (as stated above) are useful for inferring information about the characteristic energies of the precipitating protons. Photometers with narrow band-pass filters are useful for tracking the total intensity of one emission band. Given the complications of observing the Balmer α, most photometer observations have been restricted to Hβ. Most Hβ observations have been produced by meridian scanning photometers (MSPs). The fact is that the proton aurora is problematic to observe from space or the ground. From the ground, spectrographs give information relevant to the precipitating energies, but are less capable of tracking the very dim emissions that often characterize the proton aurora. MSPs are highly sensitive, capable of detecting extremely dim emissions, but it is quite difficult (in many cases impossible) to separate background from the signal, and they provide no information about energy (from a single channel). From space, the Lyman α images obtained by the IMAGE SI provided us with our first view of the global proton aurora, but there are difficulties with that data. For example, the image cadence was 2 min (dictated by the satellite spin period), and the spatial resolution was comparable to the latitudinal width of the typical proton auroral band (see below). Furthermore, part of the Doppler-shifted Lyman α band was excluded to mask the

geocoronal Lyman α, so the SI intensities are not always the total due to proton aurora. In the remainder of this chapter, we focus on MSP Hβ observations, but we caution the reader that a truly comprehensive view of proton auroral dynamics will need better observations that are presently available, and with our focus, we restrict ourselves to only part of the overall story.

Our focus here is the relationship between magnetotail dynamics and topology and the proton aurora. To that end, this chapter is written as both a review of some well-known results (not all of which have been published) and as a vehicle for new material relating to the magnetospheric (equatorial) counterpart of the equatorward boundary of the proton aurora. We finish with a discussion suggesting future interesting work that should be undertaken in order to improve our ability to use the proton aurora to remote sense magnetotail dynamics.

2. THE CONNECTION BETWEEN PROTON AURORA AND MAGNETOTAIL TOPOLOGY

The keograms in Figure 1 show auroral intensities in the oxygen "greenline" or 558 nm (top panel) and proton auroral "Hβ" or 486.1 nm (bottom panel) emissions. The intensities

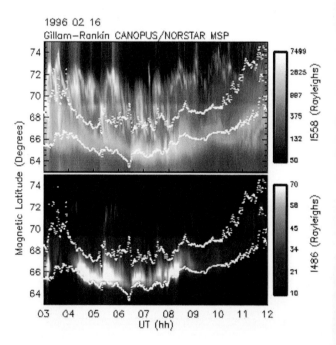

Figure 1. Merged keograms from the Rankin Inlet (MLAT = 74°) and Gillam (MLAT = 67°) meridian scanning photometers (MSPs). Nine hours of (top) 558 nm (oxygen "greenline") and (bottom) 486 nm ("H-beta" proton aurora) data, with brightness indicated in Rayleighs. Local magnetic midnight on this meridian is at roughly 06:30 UT.

are displayed "keogram"-style, with color scale indicating brightness as a function of UT and magnetic latitude (MLAT), where MLAT is according to Altitude Adjusted Corrected Geomagnetic Coordinates (AACGM) [*Baker and Wing*, 1989]. Each keogram is merged from data from two MSPs located on the same "Canadian Auroral Network for the OPEN Program Unified Study (CANOPUS) Churchill Line" magnetic meridian, one at Rankin Inlet (MLAT = 74°) and the other Gillam (MLAT = 67°). These provide slightly overlapping fields of view that together span a region extending from typically poleward of the polar cap boundary (PCB) to typically equatorward of the equatorward boundary of the auroral oval. The data shown here is from 9 h on one night starting roughly 3 h before and extending to roughly 6 h past local magnetic midnight.

We can understand much about the proton aurora and how it relates to the magnetosphere by considering observations such as these. The greenline emissions are understood to be the result of discrete and diffuse electron aurora, as well as secondary electrons produced by proton precipitation [see, e.g., *Eather*, 1967]. The greenline keogram shows a poleward boundary (sharp decrease in brightness with low intensity poleward of high intensity) that is likely at or near the PCB. The repeated (though not periodic) brightenings at this boundary are commonly referred to as poleward boundary intensifications (PBIs) [see, e.g., *Lyons et al.*, 1999]. These are thought to be causally related to reconnection at the distant neutral line and often mark the formation of either north-south structures or east-west auroral arcs that subsequently move equatorward toward the equatorward edge of the auroral oval. These equatorward moving forms, be they arcs or north-south structures, show up here as the equatorward propagating "streamers" in the greenline keogram. Most of the streamers terminate near the equatorward edge of the auroral oval and are thought to be the ionospheric signature of earthward motion of either narrow channels (north-south structures) of fast earthward flow or equatorward propagation of arcs [*Zesta et al.*, 2000, 2002].

In the bottom panel, the proton aurora is the band of luminosity with poleward and equatorward boundaries indicated by the white diamonds. These boundaries are ±1.4σ from the latitude of peak luminosity, where σ is the standard deviation of the latitude brightness profile (this specific choice of 1.4σ relates to the "optical b2i" discussed below). The same diamonds are overplotted on the greenline panel. The keograms in Figure 1 reflect commonly (though not ubiquitously) observed features of the proton aurora and its relationship to the electron aurora. The most notable of these are the following: (1) Even the "bright" proton aurora are dim in an absolute sense, with intensities shown here being typical (structured aurora of 1 kR brightness are *just*

visible to the unaided human eye, and 486 nm aurora in excess of 200 R are exceedingly rare); (2) The proton aurora is significantly less structured than the electron aurora (some of this lack of structure is attributable to the cumulative effects of the multiple charge-exchange collisions as discussed by *Davidson* [1965]); (3) The band of significant proton auroral luminosity is usually near the equatorward border of the electron auroral oval [see *Zou et al.*, in press] for a more thorough discussion of this relationship); (4) Streamers emerging from PBIs terminate either within or poleward of the band of bright proton aurora [see *Kauristie et al.*, 2003].

The population of precipitating protons is readily identifiable in data from particle detectors on low-altitude satellites transiting the auroral oval. Figure 2 includes two panels (top) showing typical FAST ESA ion data from an evening sector cut through the auroral oval. Although the ESA instrument cannot differentiate protons from other (e.g., He+, etc.) positive ions, it is usually safe to presume that most of the ions are protons (see below). The second panel shows the integrated (up to 30 keV) ion energy flux in the downgoing loss cone as a function of pitch angle and magnetic latitude. The sharp high-latitude flux change (green moderate fluxes to blue low fluxes) is the ionospheric projection of the high-latitude boundary of the ion plasma sheet (equatorward of but close to the PCB). The region of significant fluxes (indicated by green and yellow shading) is the ionospheric projection of the ion plasma sheet. The empty upgoing loss cone is centered on 180°. The downgoing loss cone spans an identical pitch angle range, but is centered on 0°. There is a boundary between an empty (equatorward of the boundary) and a full (poleward) downgoing loss cone that in this instance is at roughly 68° magnetic latitude. Poleward of this "isotropy boundary" (IB), the energy flux is isotropic except for the empty upgoing loss cone, so that fluxes inside and outside of the downgoing loss cone are the same. The empty downgoing loss cone equatorward of the IB indicates stably bounce-trapped ions, indicating conservation of the first adiabatic invariant (μ) on time scales longer than a bounce period. Poleward of the IB, the downgoing loss cone is "full," indicating stochastic process or processes that lead to violation of μ and that are responsible for strong pitch angle diffusion (meaning the loss come is filled on time scales less than a characteristic bounce period).

The top panel in Figure 2 shows differential ion energy flux as a function of energy and magnetic latitude for the same pass. One can see in this plot that the characteristic energy of the plasma sheet ions (mostly protons) generally increases with decreasing latitudes until the IB, inside of which the characteristic energy decreases sharply. As with many, or even most, such meridional passes, the trends are

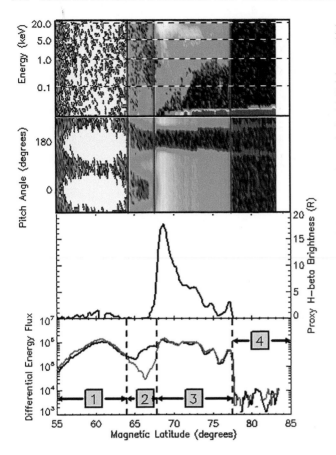

Figure 2. FAST ESA ion data from an evening sector pass through the auroral zone. (top) Integrated ion energy flux as a function of pitch angle and geomagnetic latitude. (second) Differential ion energy flux in the downgoing loss cone. The energy flux perpendicular to the magnetic field has been subtracted to remove the nearly isotropic signature of penetrating radiation in the outer radiation belt. The gray shading in the 55°–64° region masks an artifact of a slight anisotropy in the radiation belt signature. (third) Sixty-five times the downward integrated ion ESA energy flux, providing a rough proxy for H-beta proton auroral brightness (see text); (bottom) Differential energy flux (for the 8.6 keV ion ESA channel) in the direction parallel to (in red) and perpendicular to the magnetic field. The vertical dashed lines are ionospheric projections of boundaries between magnetospheric regions, with the regions being (1) equatorward of the inner edge of the ion plasma sheet; (2) stably bounce trapped plasma sheet ions; (3) strong pitch angle diffusion such that fluxes outside the upgoing loss cone are isotropic; and (4) poleward of the ion plasma sheet. The proton aurora is the ionospheric footprint of region 3, and the bright proton auroral band sits near the equatorward edge of that region.

somewhat complicated beyond these statements. Here for example, we note the emergence of a lower-energy population near the inner edge of the plasma sheet that we believe is of plasmaspheric origin. The proton aurora is the result of the

protons in the loss cone between the IB and the poleward boundary of the ion plasma sheet.

In the bottom panel of Figure 2, we show the meridian profiles of differential energy fluxes from one FAST electrostatic analyzer ion energy channel (8.6 keV). The magnetic latitude dependence of the former shows qualitatively how different magnetotail regions impress themselves on the proton precipitation. The region poleward of the ion plasma sheet shows up as a region of fluxes orders of magnitude lower than those in the plasma sheet. The differential fluxes of ions in the channel shown (8.6 keV) perpendicular to and parallel to (meaning downward in the northern hemisphere) the magnetic field are essentially identical from the poleward boundary right to the IB. Equatorward of the IB, there is a sharp drop off of downward fluxes, consistent with the IB demarking the earthward limit of whatever process or processes fill the loss cone. The region (indicated by "2" on the figure), where the perpendicular fluxes are significantly larger than the poleward fluxes, corresponds to stable bounce-trapped protons. In this case, the ion plasma sheet extends significantly earthward of the IB. The relatively large and isotropic flux values inside of the plasma sheet (region "1" in the figure) indicate the effects of penetrating radiation belt particles on the detector (not plasma sheet fluxes). The boundary between "1" and "2" is either on or poleward of the ionospheric footprint of the inner edge of the ion plasma sheet. We have shown the differential energy fluxes from one energy channel. Examination of fluxes at other energies shows essentially identical latitudinal dependence as do the 8.6 keV fluxes (the IB is slightly latitude dependent, as we discuss below). Further, the vast majority of nightside FAST passes through the auroral zone show similar morphology as this specific case.

Figure 3a is a graph of the downward integrated energy flux multiplied by 65 for the same FAST pass. As stated above, this is a rough proxy for the H-beta proton auroral brightness along the satellite track. As expected, the nonzero proton aurora along the satellite track occurs between the IB and the poleward boundary. The values are largest immediately poleward of the IB: this relatively narrow peak corresponds to the band of the "bright" proton aurora in Figure 1 (note this overflight is not the same time/location). The downward fluxes indicate there would be nonzero proton aurora all the way to the poleward boundary, but those values are at the lower limit of what can be detected.

Low-altitude satellite passes through the proton auroral region have demonstrated that the particles are in the downgoing loss cone as a consequence of strong pitch angle diffusion, meaning the differential energy flux is uniform across the downgoing loss cone [Sergeev and Tsyganenko, 1982; Sergeev et al., 1983; Donovan et al., 2003a]. Since the diffusion is due to scattering, the proton aurora is properly

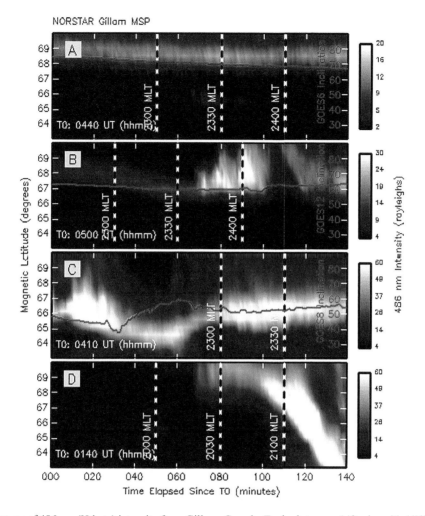

NORSTAR Gillam MSP

Figure 3. Keograms of 486 nm (H-beta) intensity from Gillam, Canada. Each plot spans 140 min, with (different) UT start times as indicated in the lower left corner of each. Note the different color-level scales, as well as the overplot of magnetic inclination as observed at GOES East (located ~2 h magnetic local time (MLT) east of the meridian scanning photometer). The inclination is shown in red (GOES data not available for event shown in Figure 3d) with the values indicated on the right-hand axis corresponding to degrees above the SM equatorial plane (plot range 20° to 90°). We also show three MLT values separated by 30 min (the station is rotating under the fixed MLT coordinate system). We show these four cases to illustrate four different manifestations of magnetospheric dynamics (or lack thereof): (a) quiet time; (b) small substorm (possibly pseudobreakup) with almost no evidence of dipolarization; (c) large substorm with major reconfiguration (dipolarization) of the inner magnetospheric magnetic field topology; and (d) solar wind pressure increase with northward interplanetary magnetic field (IMF) and subsequent (20 min later) southward turning of the IMF.

"diffuse". Such topside in situ observations also show that the bulk of the protons that cause the proton aurora have energies in the keV to several tens of keV range.

Based on such in situ and auroral observations, many authors have concluded that the precipitating protons are on closed field lines that thread the central plasma sheet (CPS) [e.g., *Eather*, 1967]. These protons are understood to enter the CPS at large distances downtail with sub-keV energies originating either from the solar wind or ionospheric outflow.

They move with quasi-static or bursty "EXB" convection, gaining energy through Fermi/Betatron acceleration as they move into ever-increasing magnitude magnetic field. The protons are scattered into the loss cone by a process or processes that violate conservation of μ. It is widely held that the dominant mechanism responsible for the pitch angle scattering are nonadiabatic effects as the protons traverse the tightly curved magnetic field lines near the vicinity of the neutral sheet. Test particle simulations have shown that this process is effective enough to fill the loss cone every bounce period for protons whose local gyroradius is larger than approximately one ninth of the local radius of curvature of the field line [*Tsyganenko*, 1982; *Sergeev et al.*, 1983]. This is often stated as the scattering is sufficient for strong pitch angle diffusion if the so-called "kappa" parameter introduced by *Büchner and Zelenyi* [1987] satisfies $\kappa < 3$, where κ^2 is the ratio of the radius of curvature to the gyroradius (and is hence energy dependent). The meridional profile of proton auroral brightness and the energy of the precipitating protons (from topside in situ observations such as those provided by FAST ESA) reflect the increasing proton energies with decreasing downtail distance. The proton aurora is dim on field lines mapping to the outer CPS due to the low energies and densities of the protons in the outer CPS. The proton aurora becomes brighter on field lines mapping to regions with larger magnetic field strength.

In a study being prepared for publication, E. Spanswick and colleagues (private communication, 2012) compare the fluxes of ions in the loss cone as observed by FAST ESA (see e.g., Figure 2) with the fluxes of ions in the loss cone observed near the magnetospheric equator by identical ESA instruments on the Time History of Events and Macroscale Interactions during Substorms (THEMIS) probes [*Angelopoulos*, 2008]. The total integrated energy flux in the downgoing loss cone is directly observed by FAST ESA whose entrance aperture angular width is smaller than the loss cone width. The entrance aperture of THEMIS ESA is much larger than the loss cone near the equator, but the downgoing flux can be inferred presuming that the fluxes are relatively constant across the loss cone, and with the knowledge that the loss cone is a small fraction of the solid angle subtended by the entrance aperture. The rationale behind their study is that FAST and THEMIS are carrying ESA instruments of the same design, and the total energy flux in the downgoing loss cone is the same just off the equator as it would be at the ionospheric end of that flux tube. They have in addition developed an entirely empirical relationship between integrated ESA downgoing ion energy flux and Hβ from numerous FAST overflights of the Gillam MSP. The relationship they established is between total energy flux (J) and 486 nm proton auroral intensity (I), and is $I = 65J$, where I and J are in Rayleighs and ergs cm^{-2} s^{-1}, respectively. Given this, one can infer the proton auroral brightness at the ionospheric end of a field line from the THEMIS ESA observation near the equator. By looking at large numbers of THEMIS ESA observations, and comparing those to large numbers of MSP observations of the proton aurora, Spanswick and colleagues are concluding that the bright band of proton aurora corresponds to CPS protons that are typically 7–12 Earth radii (R_E) in radial distance on the nightside.

3. USING THE PROTON AURORA TO REMOTE SENSE THE MAGNETOTAIL

As discussed above, the equatorward boundary of the proton aurora corresponds to the ion IB and also the b2i boundary in DMSP ion data [*Newell et al.*, 1998]. The latter is the equatorward edge of significant downward ion fluxes observed by the DMSP ion detector [*Newell et al.*, 1996]. DMSP cannot infer the IB location as it does not observe transverse fluxes, but in general, this boundary is close to if not at the IB [see, e.g., *Newell et al.*, 1998]. *Donovan et al.* [2003a] chose the name "optical b2i" for the equatorward boundary of the proton aurora as it is identified in the same way, namely, a termination of downward fluxes rather than a transition in the ratio between downward and transverse fluxes.

The fact that the scattering underlying the precipitation is thought to be caused by nonadiabatic motions in the tightly curved field lines around the neutral sheet means that the IB should correspond to the earthward limit of that scattering condition. If that limit moves earthward, then one expects the IB to move equatorward. *Sergeev and Gvozdevsky* [1995] showed this is indeed the case by demonstrating that the IB geomagnetic latitude (adjusted for the magnetic local time (MLT) dependence) is strongly correlated with the inclination of the magnetic field lines at geosynchronous orbit near midnight. Presuming that, in general, the proton aurora maps to beyond geosynchronous distance, then this increased stretching at geosynchronous distance with decreasing magnetic latitude of the IB is consistent with the idea that the precipitation is caused by pitch angle scattering related to field line curvature. It is important to note, however, that this is just a consistency check and does not prove that the scattering is related to field line curvature.

Since a more equatorward IB means more stretched field lines in the inner magnetosphere, *Sergeev and Gvozdevsky* [1995] argued that this is an ionospheric measurement that could be the basis of a proxy for the state of the inner CPS. In that paper, they presented the "MT index," which was inferred from topside ionospheric in situ observations of the ion IB. The MT index was the best estimate, based on an IB latitude made at some local time, of the IB latitude at magnetic midnight. They further demonstrated that the MT index is strongly correlated with the magnetic field inclination at geosynchronous orbit near midnight.

Referring to Figure 2, the ion IB will, in general, correspond to the equatorward boundary of the proton aurora. Motivated by this, and the MT index, *Donovan et al.* [2003a] used simultaneous observations of the proton aurora from the Gillam MSP and the magnetic field inclination at GOES 8 (then GOES "East"), to show that the equatorward boundary of the proton aurora is an excellent indicator of how stretched the magnetic field is at geosynchronous distance. They developed an empirical relationship between the latitude of the equatorward boundary of the proton aurora, which they referred to as the "optical b2i," and the magnetic inclination at GOES 8, which was located roughly 2 h MLT east of Gillam and on average ~0.8 R_E above the magnetic equator. Even with the 2-h MLT separation, they found that their relationship could "predict" the inclination at GOES 8 within 10° almost always. Significant differences between the inferred (from the proton aurora) and actual inclinations were often due to azimuthally propagating disturbances such as dipolarizations.

In developing the optical b2i, *Donovan et al.* [2003a] experimented with a number of possible boundaries that could be inferred from the proton aurora. Examples included the latitudes of the peak in brightness and variously defined equatorward and poleward boundaries. They carried out regression analysis, which allowed quantitative assessment of how well each boundary could predict the inclination at GOES 8. They concluded that of a large family of *simple* boundaries, the best predictor was obtained by a specific equatorward boundary defined to be $\Lambda_{b2i} = \Lambda_{peak} - 1.4\sigma$, where Λ_{peak} is the magnetic latitude of the peak in brightness, and σ is the standard deviation of the best fit Gaussian to the latitude profile of the proton aurora. By comparing ion data from DMSP overflights of the Gillam MSP and the proton auroral data, they showed that Λ_{b2i} is typically within 0.5° of the b2i, which is the DMSP boundary that most closely corresponds to the ion IB [see, e.g., *Newell et al.*, 1996, 1998]. It was this correspondence between the DMSP b2i and Λ_{b2i} that led *Donovan et al.* [2003a] to use the term optical b2i for their parameter. It is interesting to speculate on what it might mean that this particular boundary is the *best* indicator of the inclination at geosynchronous of any simple boundary that can be inferred from the proton aurora. One possibility is that on average the equatorward boundary of the proton aurora maps to near geosynchronous distance. This warrants further investigation.

4. MAGNETOTAIL DYNAMICS INFERRED FROM PROTON AURORAL OBSERVATIONS

Figure 3 is a stack plot of four keograms of proton auroral (486 nm) intensities from the CANOPUS (now NORSTAR) MSP at Gillam Manitoba. The time duration for each panel is 140 min, with different start times for each. The magnetic latitude range for each is the same (63°–70° AACGM). Note that magnetic midnight for Gillam is at 06:30 UT. We have chosen these four time periods to illustrate how four different magnetospheric activity scenarios are typically manifested in

the proton auroral evolution. Further, in three of the four cases, we have also shown the magnetic inclination at GOES East for the corresponding time period (in red). There is no GOES East data for the time corresponding to Figure 3d.

Figure 3a (27 December 1992 04:40–07:00 UT) corresponds to a very quiet time, quiescent auroral oval (both proton and electron, the only activity evident are some weak poleward boundary intensifications in the electron aurora well poleward of the proton aurora). This keogram shows the proton aurora during a period during which the magnetosphere was remarkably quiet based on ground magnetometer as well as GOES East and West magnetic field data. The GOES East magnetic inclination shows only diurnal variation, with values that correspond very closely to what would be expected for a dipole field (an Earth-centered dipole field would have an inclination of ~70° above the magnetic equator at the location of GOES East). The "bright" proton auroral band is, in this case, dim and shows only the diurnal variation expected from statistical studies [see, e.g., *Creutzberg et al.*, 1988]. This is the typical proton auroral behavior during times of little, if any, magnetic activity.

Figures 3b and 3c show two substorm events. Figure 3b shows the proton aurora during a relatively small substorm. As it always does during expansion phase, the proton aurora brightens and expands in latitude; however, the equatorward boundary of the proton aurora does not change in the initial 20 min following expansion phase onset. As well, the inclination of the magnetic field at GOES East undergoes only a minor change. Figure 3c is another substorm, although significantly larger. In this case, the proton aurora brightens and expands, and the equatorward boundary of the proton aurora moves significantly poleward following expansion phase onset (in that meridian). As well, the GOES East magnetic field inclination shows a significant dipolarization some 30 min before the onset in the MSP meridian.

Figure 3d (18 February 1999 01:40–04:00 UT) shows the proton aurora during a period of different magnetospheric activity. In this case, the changes in the proton aurora are due to changes in the solar wind driver, with a northward interplanetary magnetic field (IMF) pressure pulse hitting the magnetosphere at roughly 02:50 UT. The proton aurora is very dim prior to the effects of the pressure pulse. At ~02:52 UT, the proton aurora becomes significantly brighter (at the same time there is a significant increase in the electron aurora, but that response is more dynamic than is the response in the proton aurora), but the location of the equatorward boundary does not change at that time. This is confirmed by FAST ESA ion data evening sector passes that are just prior to and just after the pressure pulse arrival (they show the ion IB does not change as a consequence of the pressure pulse). Roughly 20 min later, the proton aurora begins to move rapidly equatorward. This southward motion corresponds to the arrival of a rotation from northward to southward IMF.

These four keograms demonstrate a broad range of geomagnetic activity as impressed on the proton aurora. To begin with, the proton aurora is an ever-present feature of the terrestrial environment, even during the quietest geomagnetic periods. The data shown here is consistent with the fact that the equatorward boundary of the proton aurora is an excellent proxy for the magnetic inclination in the inner CPS (see discussion around the MT index and optical b2i in the previous section). The very quiet proton aurora corresponds to an inner CPS with little cross-tail current and, hence, an essentially dipolar field at geosynchronous orbit. There must be stretching of the magnetic field leading up to the small substorm shown in Figure 3b, but the magnetotail dynamics apparently happen well beyond geosynchronous orbit and translate to no dipolarization at geosynchronous and no poleward motion of the equatorward boundary of the proton aurora. The sudden brightening that comes with the expansion phase is due to energization of CPS protons by induced electric fields as the magnetic field increases in the dipolarization.

The stretching at geosynchronous is more pronounced for the larger substorm in Figure 3c, and there is a dipolarization at geosynchronous as well as a pronounced poleward motion of the equatorward boundary of the proton aurora. The time lag between the dipolarization (occurring at a time of roughly 030 on the plot) and poleward motion (occurring at 060 on the plot) is a consequence of the time it takes the dipolarization to evolve across the 2 h of MLT separating GOES East from the MSP at Gillam.

The brightening in Figure 3d is the proton auroral signature of a sudden increase in solar wind dynamic pressure. This is a consequence of energization of the CPS proton auroral population during the compression of the entire magnetosphere. Since the loss cone in the proton auroral band is full before and after the arrival of the pressure pulse, this brightening is fundamentally different than what has been proposed to happen to the diffuse electron aurora during such events [see, e.g., *Zhou et al.*, 2003]. That is to say, the brighter proton aurora is not the result of an increase in the effectiveness of the scattering mechanism, since the CPS protons in the proton auroral region were already under strong pitch angle diffusion. Furthermore, the sudden equatorward motion of the proton aurora that begins roughly 20 min after the brightening indicates a sudden and dramatic increase in stretching of the magnetic field in the inner CPS (again, refer back to the MT index and optical b2i discussion of the previous section). As stated above, this equatorward motion starts roughly with the arrival of a southward turning

of the IMF and, hence, a rapid increase of energy input into the magnetosphere, which, in general, should lead to increased stretching of the magnetic field in the nightside magnetosphere [see, e.g., *Daglis et al.*, 2003].

Our primary intention in this section is to convey the value of the proton aurora as an indicator of dynamics in the inner CPS. Equatorward (poleward) motion of the proton auroral equatorward boundary over and above diurnal variation has a one-one correspondence with stretching (relaxation) of the magnetic field topology. Brightening indicates energization of the CPS proton population due to increasing magnetic field strength in the inner CPS. While we have not shown examples, sudden decreases in the solar wind dynamic pressure lead to sudden decreases in the proton auroral brightness. This affords us a powerful tool for studying magnetotail dynamics, if we have proton auroral observations at a number of MLTs simultaneously: we can use the proton aurora to study the 2-D (meaning "*XY*") spatio-temporal evolution of the inner CPS magnetic field topology. *Nicholson et al.* [2003] presented several event studies wherein they tracked the azimuthal evolution of the dipolarization with simultaneous proton auroral observations from MSPs in Canada and Alaska. More recently, *Gilson et al.* [2011] used IMAGE SI12 global proton auroral image sequences to explore the azimuthal (longitudinal) "splitting" of the proton aurora and its relationship to the spatiotemporal evolution of the inner CPS magnetic field topology during growth phase. The significant limitation here is the lack of multiple meridian observations of the proton aurora with the time and space resolution to adequately capture the effects of the evolution of the magnetic field topology on those meridians. The IMAGE SI12 data was obtained at 2-min cadence, with several hundred-kilometer resolution (inadequate for tracking small changes in latitude), while there are only a few MSPs providing proton auroral observations, and those located at widely separated locations. Denser deployments of MSPs and/or meridian imaging spectrographs will hopefully come in time and, with them, the opportunity to study the inner CPS dynamics in greater detail.

5. THE ION IB IN THE EQUATORIAL PLANE

As discussed above, the proton auroral precipitation is the result of strong pitch angle diffusion, which is generally held to be a consequence of violation of μ due to tight field line curvature in the vicinity of the neutral sheet. Also as stated above, this is based on test particle simulations that have shown the pitch angle scattering to be sufficient to fill the loss cone on time scales comparable to typical proton bounce periods if the "kappa" parameter satisfies the relationship $\kappa < 3$. Although there is the possibility of local minima in magnetic field strength in the neutral sheet, the magnetic field strength, in general, increases with decreasing distance from the Earth. Along with that, the local radius of curvature in the neutral sheet increases with decreasing distance. Thus, for a specific energy, the kappa parameter increases with decreasing distance. At any one location, the kappa parameter is smaller for larger energies. Thus, presuming the pitch angle scattering that fills the loss cone is due to field line curvature, one expects the IB to be closer to the Earth for larger energies. This is illustrated conceptually in Figure 4, using an empirical magnetic field model to infer equatorial radii of curvature and gyroradii.

Models such as that used to produce Figure 4 can give a heuristic picture of how gyroradius, radius of curvature, and hence kappa depend on the distance from the Earth. These models cannot, however, be expected to reproduce the

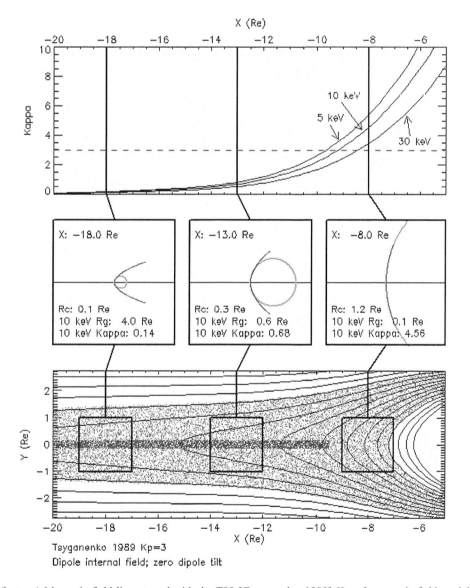

Figure 4. (bottom) Magnetic field lines traced with the T89 [*Tsyganenko*, 1989] $Kp = 3$ magnetic field model (using a dipole for the Earth's internal field). The blue and red shaded regions are meant to illustrate the ion CPS and thin current sheet regions, with the latter corresponding to curvature sufficient to violate the conservation of μ. (middle) Three panels showing field line traces in the vicinity of the neutral sheet (in red) with an inscribed circle (in gray) indicating the radius of curvature of the field line at the neutral sheet. Also shown are the gyroradius and kappa values for 10 keV protons. (top) The kappa parameter (square root of the ratio of the radius of curvature to the gyroradius) for 5, 10, and 30 keV protons.

magnetic field strength and topology around the (thin current) neutral sheet region; hence, the locations of the IB inferred from such an exercise should not be trusted. The loss cone in the vicinity of the neutral sheet is too small to allow for the determination of whether or not the loss cone is full, so that a direct identification of passage across the IB using particle observations such as those provided by THEMIS ESA is not possible. In this section, we present what we believe is the first-ever observation from which the passage of the proton $\kappa = 3$ boundary for energies in the range that cause the proton aurora across a pair of closely spaced satellites. Provided that field line curvature is the primary source of scattering that fills the loss cone, then this is the passage of the proton IB across those satellites. This observation corroborates, at least in this case, the Spanswick and colleagues statistical picture of where the proton aurora maps to in the magnetotail.

There was a small substorm with onset at roughly 02:20 UT on 28 February 2009. The expansion phase onset occurred when THEMIS A and E were in the late evening sector CPS, near the neutral sheet, ~8.1 R_E from the Earth. *Lui* [2011] has presented the THEMIS in situ and contemporaneous (electron) auroral observations in a study of the onset mechanism. Here we use observations from THEMIS A and E to explore the proton kappa parameter in the vicinity of the probes in order to infer the location of the probes relative to the ion IB.

THEMIS A and E were located at roughly the same *XY* location in GSM coordinates, but were separated by ~0.8 R_E in *Z* during the minutes around onset. Further, from the magnetic field measurements, it is clear that THEMIS E was near the neutral sheet. This situation affords us a remarkable opportunity to investigate the local curvature of the magnetic field lines while at the same time having knowledge of the magnetic field strength in the vicinity of the neutral sheet. To do so, we assume that the field line curvature can be represented by a circle. In support of this step, the magnetic field at the two satellites is close to coplanar. Further, if there is a region between the two satellites of even tighter curvature than we infer, then it would lead us to overestimating the radius of curvature. While this might be true in the late growth phase, it is almost certainly not so after the dipolarization, so our conclusions (which relate to kappa before and after the dipolarization) would be strengthened rather than weakened.

In Figure 5, we summarize our results. The top panels show the magnetic field at the two satellites at a time before and after the dipolarization. In those panels, we show with a gray thick curve the circle that has the same radius of curvature that we inferred from the magnetic field direction at the two satellites. We infer this radius of curvature from the THEMIS FGM data throughout the late growth and expansion phases. The radius of curvature is shown in the top time series panel. The two vertical red lines indicate the times corresponding to the two panels above. The sudden increase in radius of curvature around 02:23 UT is the beginning of dipolarization that marks the onset at the satellites (note we would not want to make a connection between this event and any particular onset signature in the ionosphere). At any given time, we have the radius of curvature and magnetic field strength in the neutral sheet (from THEMIS E). With these two numbers, we can determine the energy of proton for which the kappa parameter is exactly equal to 3. We call this the "critical energy," and provided that field line curvature is the cause of loss cone filling, we can infer that protons with energies larger than the critical energy are in strong diffusion, while those with lower energy are stably bounce trapped. We show this critical energy with the middle graph. On the bottom graph, we overplot this critical energy on the combined THEMIS E ion ESA/SST spectrogram.

From the results shown on the bottom graph of Figure 5, we can infer that in the late growth phase, the critical energy is below the energies of the bulk of the ion distribution observed by ESA/SST. Although the energies of the ions increase along with the dipolarization, we can see that within 2 min of the beginning of the dipolarization at this location, the critical energy has climbed above the bulk of the ion population at that location. From this, we can conclude that assuming that field line curvature is responsible for the loss cone filling, then throughout the late growth phase, the IB is earthward of the THEMIS satellite pair and that by 2 min after the start of dipolarization, the IB is tailward of the two satellites. Based on this simple observation, we can infer that in the late growth phase *in this event*, the equatorward boundary of the proton aurora maps to inside of 8 R_E radial distance in the late evening sector and that it has moved out beyond that distance after the onset. This mapping is consistent with the results of *Spanswick and colleagues*, who are finding that on average the bright proton aurora maps to outside of 7 but inside of 10 R_E under most conditions. It is unfortunate that we cannot gauge how far the IB is inside (or outside) of the satellites nor how radially thick the IB is as it crosses the satellites. Such work will need to wait for a larger constellation of satellites, but these results nevertheless offer us hope that we are starting to build up a reasonable picture of this mapping.

6. DISCUSSION AND FUTURE

The proton aurora arises as a consequence of convection, which brings protons earthward in the nightside magnetosphere, gradient, and curvature drifts, which moves the protons in the cross-tail electric field, thereby increasing their energy,

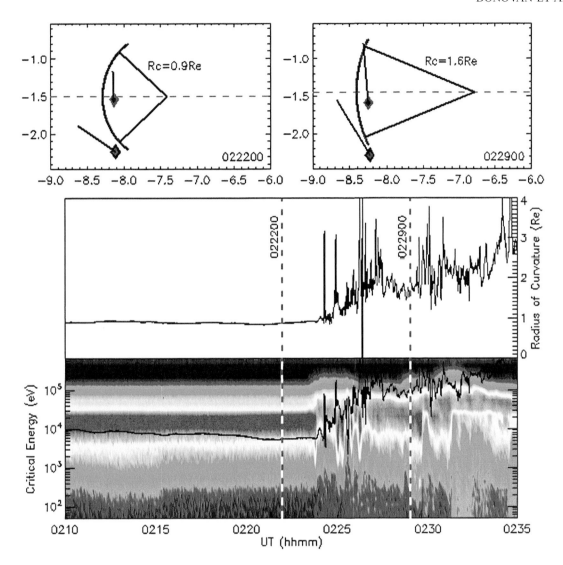

Figure 5. On 28 February 2009, THEMIS A and E were located very close in *XY*, and separated by ~0.8 R_E in *Z*. As well, THEMIS E was very near the neutral sheet. Simultaneous observations of the magnetic field at the two satellites give us the opportunity to estimate the local radius of curvature. The two panels on top illustrate this for two separate times bracketing a small substorm onset that occurred around 02:23 UT (as seen at the satellites). The three time series plots show, in order (top) the inferred radius of curvature at the satellites, (middle) the "critical energy" (see text) or energy of a proton with κ = 3 based on the local radius of curvature and magnetic field strength at THEMIS E, and (bottom) the critical energy plotted over a merged THEMIS ESA/SST spectrogram.

and pitch angle scattering that fills the loss cone on a time scale shorter than the bounce period. One can also view this process as energization as a consequence of conservation of μ as the particles move into increasing magnetic field strength in the inner magnetosphere. The larger magnetic field in the inner magnetosphere accounts for the energization and also slows the convection leading to increased proton densities. The region of strong pitch angle scattering extends from near the open-closed field line boundary inward to the energy-dependent IB. This boundary is usually tailward of the inner edge of the ion plasma sheet, but may at times be collocated with it (this separation is spanned by the region labeled "2" in Figure 2).

The loss cone is small enough in the plasma sheet that it cannot be resolved by existing in situ plasma instruments. Hence, we cannot directly observe the IB in the magnetotail. However, the inner edge of the plasma sheet can be readily observed in the magnetotail [see, *Runov et al.*, 2008]. On the other hand, the equatorward boundary of the proton aurora (the "optical b2i" as discussed above), which is readily observed in optical observations [*Donovan et al.*, 2003a] corresponds to the ionospheric projection of the IB, whereas the ionospheric projection of the plasma sheet inner edge is not discernible in the optical observations. Thus, the readily observable boundary in the ionosphere is not the same as the readily observable boundary in the magnetotail. More research needs to be carried out to explore the spatial relationship between the inner edge of ion plasma sheet and the ion IB and how that relationship depends on magnetotail activity.

The proton aurora provides us with an excellent tool for remote sensing certain aspects of magnetotail dynamics. The bright proton aurora paints the ionospheric footprint of the transition between the highly stretched tail and the less-stretched inner magnetosphere. We can expect that many fast CPS flows terminate in this "nightside transition region" (NTR), as evidenced by streamers that propagate into and terminate within the proton aurora (see Figure 1). The latitude of the equatorward boundary (the MT index or optical b2i) of the proton aurora is an excellent proxy for the magnetic topology in the inner CPS. Equatorward and poleward motion of the equatorward boundary correspond one-one with stretching and dipolarization in that MLT sector. Sudden brightening of the proton aurora corresponds to energization of the CPS proton population, often through conservation of μ during rapid increases in magnetic field strength such as during dipolarization and sudden compressions. These observations offer the possibility of tracking the 2-D (meaning "*XY*") inner CPS dynamics as they pertain to magnetic field topology, something that is not otherwise possible. That being said, the observational facilities that would enable such monitoring of the CPS dynamics are not available at present. Given the value such observations would have for understanding the relationship between convection, magnetic field topology, and plasma sheet particle populations, more ground-based instruments capable of observing the proton aurora would be a significant step forward for our field.

In the vast majority of situations where knowledge of the proton aurora is used to infer quantitative information about the magnetosphere, the presumption is made that the pitch angle scattering is due to the field line curvature criterion ($\kappa < 3$) being satisfied. There are other mechanisms that might also cause the scattering [see, e.g., *Gvozdevsky et al.*, 1997]. As discussed above, it is true that the inclination of the magnetic field at geostationary orbit is very well correlated with the ionospheric latitude of the ion IB (as inferred from the equatorward boundary of the proton aurora). The fact is, however, that this is to be expected if there is a strong correlation between the distance between the ion plasma sheet and the Earth and the latitude of the ion IB, regardless of what causes the pitch angle scattering. The ion IB latitude is observed to depend on energy. In a few cases, that latitudinal dependence has been shown to be consistent with what is expected if the dominant source of pitch angle scattering is the field line curvature. However, this has been done with ions of higher energy than those typically responsible for the proton aurora or very high-energy electrons [see, e.g., *Imhof et al.*, 1977]. *Donovan et al.* [2003b] explored this dependence of ion isotropy boundary latitude on energy using FAST ESA ion data, whose energy range is, in general, appropriate for exploring the proton aurora. They found that the vast majority of cases (not all) in the evening sector showed the dependence expected if field line curvature is causing the pitch angle scattering (namely, the IB is at lower latitude for increasing energy). Interestingly, they found that a significant fraction of the morning sector IBs showed the opposite. They concluded that it is often the case that at the equatorward boundary of the proton aurora in the morning sector, the proton precipitation is caused by a mechanism other than pitch angle scattering due to field line curvature (note that the converse is not necessarily true in the evening sector). As observations of the electron and proton aurora are seeing increasing use in remote sensing the magnetosphere, this needs to be explored more. Certainly, we have to exercise caution in interpreting the proton aurora while assuming that the pitch angle scattering is always caused by field line curvature.

When considering the "kappa" condition for scattering, it is usually assumed that this is a sharp boundary. Such interpretation is all we can do at present, but clearly we must again use caution. As pointed out by *Gilson et al.* [2011], it is not likely the case that the pitch angle diffusion is strong if kappa is slightly less than 3 and exactly zero if kappa is slightly greater than 3. The test particle simulations upon which the $\kappa = 3$ interpretation is based were carried out in the late 1970s and early 1980s and must of necessity have used relatively few test particles. Since that time, readily available computer power has increased orders of magnitude, and our understanding of the relevant magnetic field topology has increased significantly. This, too, points toward some important future research, namely, a revisiting of the test particle simulations in order to refine our understanding of, for example, how radially thick the IB ought to be in typical magnetic field topologies.

More than 30 years ago, *Lui and Burrows* [1978] used the intense proton precipitation to identify the ionospheric footprint of the transition between highly stretched and quasidipolar magnetic topologies (they referred to this NTR as the "nighttime cusp"; see their Figure 5). Since then, this has become widely accepted as a fact, being used as a cornerstone in papers where mapping of specific auroral features (e.g., the "onset arc") and their magnetospheric counterparts is useful. These are numerous, but the paper by *Samson et al.* [1992] provides an excellent example. In that paper, this mapping of the proton aurora to the NTR is used as follows. The bright band of proton aurora marks the ionospheric projection of the NTR. The onset arc is on the poleward shoulder of or just poleward of the bright proton aurora. Therefore, the onset arc maps close to the NTR. This argument is mostly sound and oft repeated (another example is by *Donovan et al.* [2008]).

Until recently, we could not say with any real confidence where the NTR is located and, hence, where the proton aurora actually maps to. Our results using two THEMIS satellites allow us to say with some certainty that in at least that one event, the IB was inside of the satellites, We finish with a final cautionary note that also points to necessary future work on the proton aurora. In general, we can study the current sheet properties tailward of the NTR using satellites such as THEMIS, Geotail, and Cluster. We have excellent observations inside of the NTR from missions such as GOES and CRRES, and RBSP promises significantly more observations in that region very soon. What we are lacking is a multisatellite picture of how the thin current sheet region blends into the inner magnetosphere. Lacking this detailed picture of the topology of the NTR means that, for now, we have important unanswered questions about the physics of this region, in general, and the proton aurora specifically.

Acknowledgments. Eric Donovan acknowledges the significant contributions of John Samson of the University of Alberta to the use of the proton aurora for remote sensing the nightside transition region (NTR). He is also grateful to William Liu, David Knudsen, Larry Lyons, and Toshi Nishimura for the many insightful discussions about auroral science and the NTR that have added enormously to this work. The authors acknowledge the CSA for its long support of the CANOPUS and now CGSM Meridian Scanning Photometer program. As well, the CSA has supported this research through three Space Science Enhancement Program grants to the University of Calgary. We are grateful to NOAA and NASA for support of the GOES, THEMIS, and FAST missions. The authors acknowledge J. McFadden, C. Carlson, D. Larson, and K.-H. Glassmeier for their work on THEMIS ESA, SST, and FGM, and FAST ESA, and Vassilis Angelopoulos for his leadership of the THEMIS mission.

REFERENCES

Angelopoulos, V. (2008), The THEMIS mission, *Space Sci. Rev.*, *141*, 5–34, doi:10.1007/s11214-008-9336-1.

Baker, K. B., and S. Wing (1989), A new magnetic coordinate system for conjugate studies at high latitudes, *J. Geophys. Res.*, *94*(A7), 9139–9143.

Büchner, J., and L. M. Zelenyi (1987), Chaotization of the electron motion as the cause of an internal magnetotail instability and substorm onset, *J. Geophys. Res.*, *92*(A12), 13,456–13,466.

Creutzberg, F., R. L. Gattinger, F. R. Harris, S. Wozniak, and A. V. Jones (1988), Auroral studies with a chain of meridian scanning photometers, 2, Mean distributions of proton and electron aurora as a function of magnetic activity, *J. Geophys. Res.*, *93*(A12), 14,591–14,601.

Daglis, I. A., E. T. Sarris, and G. Kremser (2003), Ionospheric contribution to the cross-tail current enhancement during the substorm growth phase, *J. Atmos. Sol. Terr. Phys.*, *53*, 1091–1098, doi:10.1016/0021-9169(91)90057-E.

Davidson, G. T. (1965), Expected spatial distribution of low-energy protons precipitated in the auroral zones, *J. Geophys. Res.*, *70*(5), 1061–1068.

Donovan, E. F., B. J. Jackel, I. Voronkov, T. Sotirelis, F. Creutzberg, and N. A. Nicholson (2003a), Ground-based optical determination of the b2i boundary: A basis for an optical MT-index, *J. Geophys. Res.*, *108*(A3), 1115, doi:10.1029/2001JA009198.

Donovan, E. F., B. Jackel, D. Klumpar, and R. Strangeway (2003b), Energy dependence of the isotropy boundary latitude, *Sodankylä Geophys. Obs. Publ.*, *92*, 11–14.

Donovan, E. F., et al. (2008), Simultaneous THEMIS in situ and auroral observations of a small substorm, *Geophys. Res. Lett.*, *35*, L17S18, doi:10.1029/2008GL033794.

Eather, R. H. (1967), Auroral proton precipitation and hydrogen emissions, *Rev. Geophys.*, *5*(3), 207–285.

Galand, M., and S. Chakrabarti (2006), Proton aurora observed from the ground, *J. Atmos. Sol. Terr. Phys.*, *68*, 1488–1501, doi:10.1016/j.jastp.2005.04.013.

Galand, M., J. Baumgardner, D. Pallamraju, S. Chakrabarti, U. P. Løvhaug, D. Lummerzheim, B. S. Lanchester, and M. H. Rees (2004), Spectral imaging of proton aurora and twilight at Tromsø, Norway, *J. Geophys. Res.*, *109*, A07305, doi:10.1029/2003JA010033.

Gilson, M. L., J. Raeder, E. Donovan, Y. S. Ge, and S. B. Mende (2011), Statistics of the longitudinal splitting of proton aurora during substorms, *J. Geophys. Res.*, *116*, A08226, doi:10.1029/2011JA016640.

Gvozdevsky, B. B., V. A. Sergeev, and K. Mursula (1997), Long lasting energetic proton precipitation in the inner magnetosphere after substorms, *J. Geophys. Res.*, *102*(A11), 24,333–24,338.

Imhof, W. L., J. B. Reagan, and E. E. Gaines (1977), Fine-scale spatial structure in the pitch angle distributions of energetic particles near the midnight trapping boundary, *J. Geophys. Res.*, *82*(32), 5215–5221.

Kauristie, K., V. A. Sergeev, O. Amm, M. V. Kubyshkina, J. Jussila, E. Donovan, and K. Liou (2003), Bursty bulk flow intrusion to the inner plasma sheet as inferred from auroral observations, *J. Geophys. Res.*, *108*(A1), 1040, doi:10.1029/2002JA009371.

Lui, A. T. Y. (2011), Reduction of the cross-tail current during near-Earth dipolarization with multisatellite observations, *J. Geophys. Res.*, *116*, A12239, doi:10.1029/2011JA017107.

Lui, A. T. Y., and J. R. Burrows (1978), On the location of auroral arcs near substorm onsets, *J. Geophys. Res.*, *83*(A7), 3342–3348.

Lyons, L. R., T. Nagai, G. T. Blanchard, J. C. Samson, T. Yamamoto, T. Mukai, A. Nishida, and S. Kokubun (1999), Association between Geotail plasma flows and auroral poleward boundary intensifications observed by CANOPUS photometers, *J. Geophys. Res.*, *104*(A3), 4485–4500, doi:10.1029/1998JA900140.

Mende, S. B., et al. (2000), Far ultraviolet imaging from the IMAGE spacecraft, 1, System design, *Space Sci. Rev.*, *91*, 243–270.

Mende, S. B., H. U. Frey, M. Lampton, J.-C. Gerard, B. Hubert, S. Fuselier, J. Spann, R. Gladstone, and J. L. Burch (2001), Global observations of proton and electron auroras in a substorm, *Geophys. Res. Lett.*, *28*(6), 1139–1142, doi:10.1029/2000GL012340.

Newell, P. T., Y. I. Feldstein, Y. I. Galperin, and C.-I. Meng (1996), Morphology of nightside precipitation, *J. Geophys. Res.*, *101*(A5), 10,737–10,748.

Newell, P. T., V. A. Sergeev, G. R. Bikkuzina, and S. Wing (1998), Characterizing the state of the magnetosphere: Testing the ion precipitation maxima latitude (b2i) and the ion isotropy boundary, *J. Geophys. Res.*, *103*(A3), 4739–4745.

Nicholson, N., E. Donovan, B. Jackel, L. Cogger, and D. Lummerzheim (2003), Multipoint measurements of the ion isotropy boundary, *Sodankylä Geophys. Obs. Publ.*, *92*, 37–40.

Runov, A., V. Angelopoulos, N. Ganushkina, R. Nakamura, J. McFadden, D. Larson, I. Dandouras, K.-H. Glassmeier, and C. Carr (2008), Multi-point observations of the inner boundary of the plasma sheet during geomagnetic disturbances, *Geophys. Res. Lett.*, *35*, L17S23, doi:10.1029/2008GL033982.

Samson, J. C., D. D. Wallis, T. J. Hughes, F. Creutzberg, J. M. Ruohoniemi, and R. A. Greenwald (1992), Substorm intensifications and field line resonances in the nightside magnetosphere, *J. Geophys. Res.*, *97*(A6), 8495–8518.

Sergeev, V. A., and B. B. Gvozdevsky (1995), Mt-index: A possible new index to characterize the magnetic configuration of the magnetotail, *Ann. Geophys.*, *13*, 1093–1103.

Sergeev, V. A., and N. A. Tsyganenko (1982), Energetic particle losses and trapping boundaries as deduced from calculations with a realistic magnetic field model, *Planet. Space Sci.*, *10*, 999–1006.

Sergeev, V. A., E. M. Sazhina, N. A. Tsyganenko, J. A. Lundblad, and F. Søraas (1983), Pitch-angle scattering of energetic protons in the magnetotail current sheet as the dominant source of their isotropic precipitation into the nightside ionosphere, *Planet. Space Sci.*, *31*, 1147–1155.

Tsyganenko, N. A. (1982), Pitch-angle scattering of energetic particles in the current sheet of the magnetospheric tail and stationary distribution functions, *Planet. Space Sci.*, *30*, 433–437.

Tsyganenko, N. A. (1989), A magnetospheric magnetic field model with a warped tail current sheet, *Planet. Space Sci.*, *37*, 5–20.

Vegard, L. (1939), Hydrogen lines in the auroral spectrum, *Nature*, *114*, 1089.

Zesta, E., L. R. Lyons, and E. Donovan (2000), The auroral signature of earthward flow bursts observed in the magnetotail, *Geophys. Res. Lett.*, *27*(20), 3241–3244, doi:10.1029/2000GL000027.

Zesta, E., E. Donovan, L. Lyons, G. Enno, J. S. Murphree, and L. Cogger (2002), Two-dimensional structure of auroral poleward boundary intensifications, *J. Geophys. Res.*, *107*(A11), 1350, doi:10.1029/2001JA000260.

Zhou, X.-Y., R. J. Strangeway, P. C. Anderson, D. G. Sibeck, B. T. Tsurutani, G. Haerendel, H. U. Frey, and J. K. Arballo (2003), Shock aurora: FAST and DMSP observations, *J. Geophys. Res.*, *108*(A4), 8019, doi:10.1029/2002JA009701.

Zou, Y., Y. Nishimura, L. R. Lyons, and E. F. Donovan (2012), A statistical study of the relative locations of electron and proton auroral boundaries inferred from meridian scanning photometer observations, *J. Geophys. Res.*, *117*, A06206, doi:10.1029/2011JA017357.

E. Donovan, J. Grant, M. Greffen, B. Jackel, J. Liang, and E. Spanswick, Department of Physics and Astronomy, University of Calgary, Calgary, AB T2N 1N4, Canada. (edonovan@ucalgary.ca)

The Origin of Pulsating Aurora: Modulated Whistler Mode Chorus Waves

W. Li, J. Bortnik, Y. Nishimura, and R. M. Thorne

Department of Atmospheric and Oceanic Sciences, University of California, Los Angeles, California, USA

V. Angelopoulos

Department of Earth and Space Sciences, University of California, Los Angeles, California, USA

Pulsating aurora (PA), blinking on and off on a time scale of a few to tens of seconds in the upper atmosphere, is known to be excited by modulated fluxes of precipitating electrons. The source region of the modulation is generally believed to be located in the distant equatorial magnetosphere. The most widely accepted generation mechanism of PA is modulation of precipitating electron fluxes as a result of scattering by chorus waves, the intensity of which is modulated on a similar time scale as the period of PA, in the source region near the geomagnetic equator. Recent coordinated observations of PA from the ground and chorus waves in the equatorial magnetosphere showed a remarkable correlation between the chorus wave amplitude and the auroral luminosity, thus providing strong evidence that chorus is indeed the driver of PA. Furthermore, potential mechanisms responsible for chorus intensity modulations are discussed, although this still remains an open question.

1. INTRODUCTION

Pulsating aurora (PA) is a spectacular emission, blinking on and off in the Earth's upper atmosphere. PA patches with a horizontal scale size of ~100 km occur at ~100 km altitude and switch on and off with various periodicities ranging from tenths of a second to a few tens of seconds with the strongest power spectral density of the pulsation near 2–10 s [*Parks et al.*, 1968; *Royrvik and Davis*, 1977; *Davidson*, 1990; *Sandahl et al.*, 1980; *Yamamoto*, 1988]. PA is generally weak, with a typical luminosity of a few kilorayleighs in 427.8 nm emission [*Royrvik and Davis*, 1977], different from the bright discrete aurora. It is also known that rapid variations with frequencies near 3 Hz are often superimposed on slower variations in auroral emissions and precipitating electrons

[*Oliven and Gurnett*, 1968; *Lepine et al.*, 1980; *Sandahl et al.*, 1980; *Yamamoto*, 1988]. The temporal and spatial characteristics of PA patches are highly variable over a broad and continuous spectrum [e.g., *Royrvik and Davis*, 1977; *Davidson*, 1990]. They appear as irregular shapes with typical sizes of 10–200 km, but maintain their identity from pulsation to pulsation for a few to tens of minutes [e.g., *Royrvik and Davis*, 1977]. Neighboring patches pulsate independently with respect to each other in phase and period [*Brekke*, 1971; *Oguti*, 1976]. Interestingly, PA is frequently embedded in the diffuse aurora [*Royrvik and Davis*, 1977; *Jones et al.*, 2011], which is typically structureless with low luminosity. PA typically occurs in the postmidnight sector near the equatorward boundary of the auroral oval and covers a local time range that increases with stronger geomagnetic activity, even extending into the dayside [*Royrvik and Davis*, 1977; *Oguti et al.*, 1981; *Johnstone*, 1983; *Jones et al.*, 2011].

Understanding the driver of the PA has been a long-standing problem due to its complex temporal and spatial structures. A more general review of PA has been written by *Lessard et al.* [this volume] in the context of PA phenomenology. In this

Auroral Phenomenology and Magnetospheric Processes: Earth and Other Planets

Geophysical Monograph Series 197
10.1029/2011GM001164

review paper, we provide a significant review of previous findings focused on the generation mechanism of PA, briefly present recent results regarding the potential mechanisms associated with auroral pulsations, and further discuss unsolved problems in our current understanding of the origin of PA.

2. POTENTIAL GENERATION MECHANISM OF PULSATING AURORA

Simultaneous observations of the aurora and rocket and low-altitude spacecraft measurements have revealed that auroral pulsations are caused by temporal modulation of precipitating energetic electron fluxes with energies from a few to tens of keV [*Johnstone*, 1978, 1983; *Sandahl et al.*, 1980; *McEwen et al.*, 1981; *Miyoshi et al.*, 2010; *Samara et al.*, 2010]. Energy dispersion of precipitating electrons observed in the topside ionosphere is associated with PA, since the travel time of electrons from the magnetospheric source to the ionosphere depends on their energy [e.g., *Yau et al.*, 1981; *Miyoshi et al.*, 2010; *Nishiyama et al.*, 2011].

The electron modulation source region is generally believed to be located in the equatorial magnetosphere. This conclusion is based on two primary sources of information: observations of magnetically conjugate pulsating auroral displays [*Belon et al.*, 1969; *Stenbaek-Nielsen et al.*, 1972; *Davis*, 1978] and analysis of energy dispersion in modulated precipitating electron fluxes measured by sounding rockets at low altitudes [e.g., *Yau et al.*, 1981]. Recently, *Miyoshi et al.* [2010] conducted a time-of-flight analysis of the energy dispersion of precipitating electrons observed by the low-altitude Reimei satellite and demonstrated that the modulation region of the pitch angle scattering is near the magnetic equator. *Sato et al.* [2002], however, suggested that precipitating high-energy electrons, which produce PA, are modulated by oscillation of field-aligned electric field far from the magnetic equator. *Stenbaek-Nielsen* [1980] proposed that active ionospheric processes probably play an important role in causing or modifying PA. Despite these studies [*Stenbaek-Nielsen*, 1980; *Sato et al.*, 2002], the majority of PA studies have shown the modulation source to be near the magnetic equator [e.g., *Belon et al.*, 1969; *Bryant et al.*, 1975; *Davis*, 1978; *Yau et al.*, 1981; *Sandahl et al.*, 1980; *Johnstone*, 1983; *Davidson*, 1990; *Miyoshi et al.*, 2010; *Nishimura et al.*, 2010, 2011; *Nishiyama et al.*, 2011].

Assuming that the modulation source region is near the equator, previous investigators attempted to identify the corresponding electron flux modulation in the equatorial magnetosphere [e.g., *Nemzek et al.*, 1995; *Li et al.*, 2011a]. Very few studies, however, have reported a good correlation between equatorial electron flux and corresponding PA luminosity. This could be attributed to the difficulties involved in detecting fluctuations of electron fluxes with periods close to the space-craft spin period and the limited sampling time resolution of the instrument. Furthermore, even if there are electron flux pulsations near the equator, the modulation of the electron distribution near the loss cone, which is typically less than a few degrees in the equatorial magnetosphere, is difficult to resolve. Particle detectors with extremely high resolution in pitch angle and time are required to capture electron flux modulation inside or near the loss cone in the equatorial magnetosphere.

Several mechanisms have been proposed to explain quasi-periodic precipitation of energetic electrons. A widely accepted model is based on pitch angle scattering of electrons through cyclotron resonance with plasma waves near the magnetic equator [*Coroniti and Kennel*, 1970; *Johnstone*, 1983; *Davidson*, 1990; *Miyoshi et al.*, 2010]. The two main wave modes that can resonate with plasma sheet electrons in the energy range from hundreds of eV to tens of keV are electron cyclotron harmonic (ECH) waves and whistler mode chorus waves [*Ni et al.*, 2008; *Meredith et al.*, 2009; *Li et al.*, 2010; *Liang et al.*, 2010; *Thorne*, 2010]. In particular, chorus waves, which frequently show intensity modulation on a time scale comparable to that of PA and can resonate with tens of keV electrons in the near-Earth plasma sheet, have been suggested to be specifically responsible for the pitch angle scattering of energetic electrons to illuminate PA patches in the upper atmosphere [*Gough et al.*, 1981; *Ward et al.*, 1982; *Johnstone*, 1983; *Davidson*, 1990; *Miyoshi et al.*, 2010]. Theoretical calculations indicate that a chorus wave amplitude of tens of pT, which is typically observed in the equatorial magnetosphere [e.g., *Meredith et al.*, 2003; *Li et al.*, 2009, 2011c], is sufficiently strong to cause precipitation of ~10 keV electrons into the atmosphere [e.g., *Tsurutani and Smith*, 1977; *Johnstone*, 1983; *Ni et al.*, 2008; *Thorne et al.*, 2010].

Previous studies have attempted to test wave-particle interaction theories using low-altitude rocket observations [e.g., *Yau et al.*, 1981] and high-altitude equatorial spacecraft [*Gough et al.*, 1981; *Tsuruda et al.*, 1981; *Ward et al.*, 1982; *Johnstone*, 1983]. Although a general correspondence between PA and electromagnetic waves has been observed [e.g., *Tsuruda et al.*, 1981; *Ward et al.*, 1982], a one-to-one correlation between individual bursts of chorus and auroral pulsations has not been found until very recently by *Nishimura et al.* [2010] because of the following reasons. First, it is very difficult to find the exact mapping of a distant spacecraft location to a small (~100 km) PA patch in the upper atmosphere along magnetic field lines. Second, the probability of simultaneous coordinated observation of plasma waves in the magnetosphere and high-resolution PA at low altitude is very low. Third, the presence of multiple types of plasma waves in the equatorial magnetosphere and independent PA patches makes it difficult to identify the appropriate plasma waves responsible for electron scattering that leads to PA. A one-to-one correlation between the wave intensity observed near the equator and the coordinated observation of the PA luminosity in the upper atmosphere is essential to unambiguously identify the driver of the PA, as discussed in section 3.

3. DRIVER OF THE PULSATING AURORA: WHISTLER MODE CHORUS WAVES

As discussed in section 2, chorus waves are widely accepted as a primary candidate for the driver of PA. Chorus waves are naturally occurring electromagnetic emissions, typically observed in the whistler mode. They are generated through cyclotron resonant interactions with anisotropic plasma sheet electrons [e.g., *Kennel and Petschek*, 1966; *Nunn et al.*, 1997; *Omura et al.*, 2008], which are injected into the inner magnetosphere during geomagnetically active times [*Burtis and Helliwell*, 1969; *Tsurutani and Smith*, 1977; *Meredith et al.*, 2003; *Li et al.*, 2009]. As shown in Figure 1a, they are typically observed in the low-density region outside the plasmapause from premidnight to afternoon [*Tsurutani and Smith*, 1977; *Meredith et al.*, 2003; *Li et al.*, 2009]. Spacecraft observations show that chorus frequently consists of discrete elements with rising or falling tones, each of which lasts for several tenths of seconds [e.g., *Burtis and Helliwell*, 1969; *Burton and Holzer*, 1974; *Hayakawa et al.*, 1984; *Santolik et al.*, 2003; *Li et al.*, 2011b]. The discrete elements are frequently clustered together and modulated with periodicities of a few to tens of seconds [e.g., *Li et al.*, 2011a], on a time scale comparable to that of PA. Figure 1b shows an example of rising tone chorus elements observed by the THEMIS (Time History of Events and Macroscale Interactions during Substorms) spacecraft [*Angelopoulos*, 2008], each of which is clustered together and modulated on a time scale of 10–20 s, as shown in a longer time span (Figure 1c). The source region of chorus waves is generally believed to be located near the geomagnetic equator [*LeDocq et al.*, 1998; *Lauben et al.*, 2002; *Santolik et al.*, 2003]. The frequency of chorus emissions is closely related to the equatorial electron cyclotron frequency (f_{ce}), typically occurring in a frequency range 0.1–0.8 f_{ce} [*Burtis and Helliwell*, 1969; *Tsurutani and Smith*, 1977; *Santolik et al.*, 2003]. Interestingly, chorus emissions are commonly observed in two distinct frequency bands (lower band and upper band) with a minimum wave power near 0.5 f_{ce} [*Tsurutani and Smith*, 1974; *Koons and Roeder*, 1990]. Lower-energy electrons (less than a few keV) could effectively be scattered by upper-band chorus and ECH waves, whereas lower-band chorus waves could provide efficient precipitation of high-energy electrons (greater than a few keV) through first-order cyclotron resonance [*Horne et al.*, 2003; *Ni et al.*, 2008; *Meredith et al.*, 2009; *Li et al.*, 2010].

As discussed above, although chorus has been proposed as the driver of PA for a few decades, a one-to-one correlation between chorus intensity and auroral luminosity has not been reported until very recently by *Nishimura et al.* [2010]. They utilized coordinated observations of PA by ground-based all-sky imagers (ASIs) [*Mende et al.*, 2008] and chorus waves on THEMIS A to investigate the relationship between PA luminosity at low altitudes and chorus wave intensity in the equatorial magnetosphere, as shown in Figure 2. During the period of the presence of PA, wave spectral density in both electric and magnetic field was observed by THEMIS A, which was located near the magnetic equator in the Southern Hemisphere. A typical lower-band chorus spectrum shows repetitive discrete bursts every ~10 s, as observed by THEMIS A (Figure 2b). Images from the Narsarsuaq ASI (Figure 2c) at four selected times show auroral patches with a scale size of ~100 km. A bright auroral patch (marked by the red arrow) to the west of the model footprint of the spacecraft (magenta square) is illuminated in snapshots for 01:11:00 and 1:11:18 UT in Figure 2c, when intense chorus waves are observed (vertical lines b and d in Figure 2b). Simultaneous observation of chorus wave amplitude and auroral luminosity (Figure 2d) shows an almost one-to-one correspondence with a correlation coefficient of 0.88, which provides firm evidence that intensity-modulated lower-band chorus is the driver of this PA.

More recently, *Nishimura et al.* [2011] performed a multievent study (13 events) using conjunctions between the THEMIS spacecraft and ASIs from 2007 to 2010 and further confirmed that lower-band chorus interacting with energetic electrons in the equatorial magnetosphere plays a primary role in driving PA.

4. POTENTIAL MECHANISMS RESPONSIBLE FOR CHORUS MODULATION

The discussion above provides strong support for the suggestion that modulated chorus waves are the driver of PA. Assuming that this is the case, the next question concerns the origin of the chorus wave intensity modulation that leads to the pulsation of the auroral luminosity. *Coroniti and Kennel* [1970] suggested that the quasiperiodic modulation of linear growth rates (which act as a proxy for the chorus intensity) is caused by geomagnetic pulsations. However, this theory was not supported by further satellite and ground-based observations, which showed that the hydromagnetic waves required to coincide with PA events did not exist in the equatorial magnetosphere [e.g., *Tsurutani and Smith*, 1974; *Oguti et al.*, 1986; *Li et al.*, 2011a].

Davidson [1979, 1990] developed a theory based on relaxation oscillations in the presence of energetic particles in the region of instability. In this model, strong waves result in

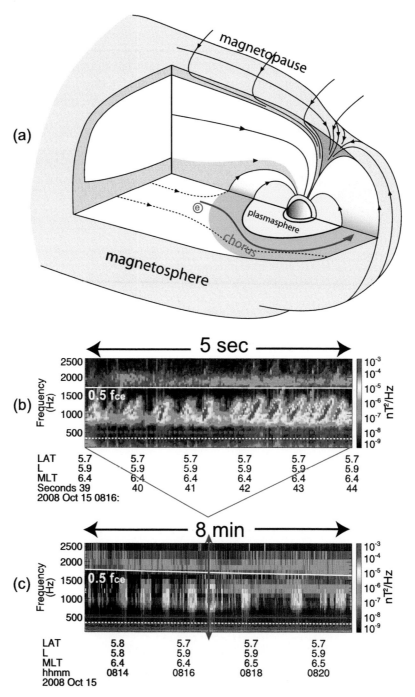

Figure 1. (a) Chorus waves (blue region) are generated outside the plasmasphere (yellow color represents the high-density plasmasphere) by the injection of electrons (red arrow) from the plasma sheet. (b) Frequency-time spectrogram of wave magnetic field spectral density observed from Time History of Events and Macroscale Interactions during Substorms (THEMIS) E in a time span of 5 s and (c) in a longer time span of 8 min. In Figures 1b and 1c, the white solid line represents $0.5 f_{ce}$, where f_{ce} is the equatorial electron cyclotron frequency.

filling the loss cone by intense pitch angle scattering, thus reducing the phase space density gradient near the loss cone, which in turn suppresses the wave growth due to the reduced anisotropy. When the loss cone electrons precipitate into the ionosphere, the condition for the wave growth is reestablished, and the cycle starts again. This process is suggested to be repeated through many cycles, thus ultimately causing modulation of precipitating electron fluxes and chorus intensity.

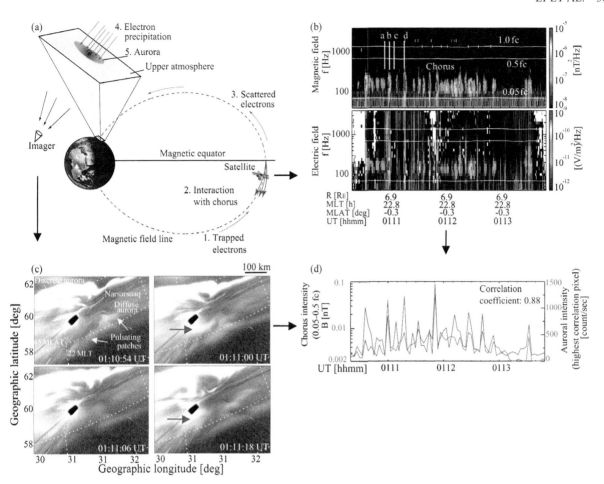

Figure 2. Coordinated observations of pulsating aurora (PA) by the Narsarsuaq all-sky imager (ASI) and THEMIS A spacecraft during 01:10:20 to 01:13:50 UT on 15 February 2009. (a) Schematic diagram showing the geometry of chorus wave propagation (red arrows), electron precipitation (blue arrows), and PA. (b) THEMIS A observation of lower-band chorus bursts showing electromagnetic field spectra. The white lines indicate 0.05, 0.5, and 1.0 f_c, where f_c is the local electron cyclotron frequency calculated from the measured magnetic field. (c) Snapshots of imager data projected onto geographic coordinates at 110 km altitude. The pulsating patch correlating with chorus is indicated by the red arrows. ASI snapshot times are also marked in Figure 2b by white vertical lines. The pink square shows the magnetic footprint of the THEMIS A spacecraft using the Tsyganenko 96 magnetic field model [*Tsyganenko and Stern*, 1996]. The spacecraft footprint was located close to the center of the imager field of view (green square on top right image). Dashed lines represent magnetic coordinates every 3° in latitude and 1 h in local time. (d) Correlation of lower-band chorus wave amplitude integrated over a frequency range of 0.05 to 0.5 f_c (red) and auroral intensity (blue) at the highest cross-correlation pixel. From *Nishimura et al.* [2010].

Observations from incoherent scatter radar [*Vickrey et al.*, 1980] and satellites [*Muldrew and Vickrey*, 1982; *Beghin et al.*, 1984] show that events characterized by whistler mode waves were presumed to occur in a region of enhanced plasma density commonly observed within the plasma trough. Based on this fact, *Demekhov and Trakhtengerts* [1994] proposed the flow-cyclotron-maser (FCM) theory to explain the wave modulation. In their model, a flux tube with enhanced cold plasma density acts as a duct "resonator" for whistler mode waves, and electrons within this duct precip-

itate into the atmosphere and cause PA. Pulsation "on" time is determined by nonlinear instability dynamics, whereas "off" time is nearly equal to the time of reaching the wave excitation threshold.

Recent THEMIS observations have revealed another potential mechanism of chorus modulation. Figure 3 shows an event observed by THEMIS D, in which quasiperiodic lower-band chorus was modulated on a time scale of ~10–20 s. Interestingly, the chorus magnetic wave amplitude (Figure 3d) was modulated between a few and ~200 pT, and strong

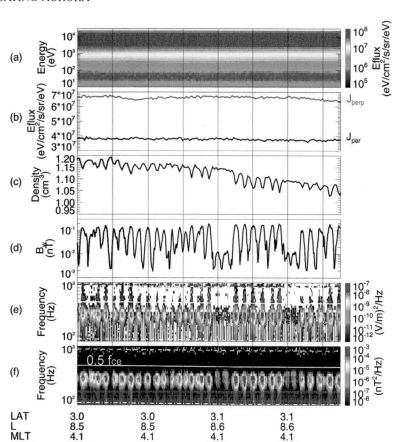

LAT   3.0   3.0   3.1   3.1
L   8.5   8.5   8.6   8.6
MLT   4.1   4.1   4.1   4.1
hhmm   0530   0532   0534   0536
2010 Jan 06

Figure 3. An event of chorus waves observed by THEMIS D on 6 January 2010. (a) Omnidirectional electron energy flux as functions of energy and time, and (b) electron energy flux over the energy range of 3–30 keV roughly perpendicular (blue) and parallel (black) to the background magnetic field. (c) Total electron density inferred from the spacecraft potential. (d) Chorus magnetic wave amplitude integrated over a frequency range of 0.05–$0.5\,f_{ce}$. Frequency-time spectrogram of (e) wave electric and (f) magnetic field spectral density. In Figures 3e and 3f, the white solid lines represent $0.5\,f_{ce}$. After *Li et al.* [2011a].

chorus amplitudes were observed over intervals of reduced total electron density (Figure 3c). The correlation coefficient between chorus wave amplitude and total electron density inferred from the spacecraft potential was -0.9, indicating that depletions in density and increases in chorus wave amplitude have an almost one-to-one correlation. The electron energy spectrogram (Figure 3a), observed simultaneously by THEMIS D, did not show the flux modulation with a good correlation with chorus wave modulation in the measurable energy range (greater than a few eV), suggesting that these density variations are probably attributed to changes in cold electron (less than a few eV) fluxes. Furthermore, resonant electron energy flux in the energy range of 3–30 keV (Figure 3b) and other key parameters related to chorus excitation (such as background magnetic field and electron anisotropy, not shown) did not show corresponding changes in

the chorus intensity modulation. The poor correlation between electron energy flux roughly parallel to the magnetic field (black line in Figure 3b) and chorus intensity may be due to the insufficient pitch angle resolution in THEMIS particle measurements. Through calculating linear growth rates of chorus waves using observed plasma parameters, *Li et al.* [2011a] also demonstrated that density depletion could lead to an intensification of chorus waves. The event shown in Figure 3 is one example of chorus modulation associated with density depletions. *Li et al.* [2011a] systematically surveyed all available chorus modulation events (with modulation periods from a few to tens of seconds) associated with density depletions from 1 June 2008 to 1 August 2010. The global distribution of these events is shown in the L-MLT (magnetic local time) domain in Figure 4a. This statistical analysis has been performed in the dominant chorus region between 5 and 12 R_E from 22 to 14 MLT. Chorus modulation events occur predominantly from premidnight to dawn, at lower L shells near midnight and extend to higher L shells at dawn. Furthermore, the occurrence rate tends to increase from premidnight to dawn and is very low after ~8 MLT. Figure 4b shows the same quantity as in Figure 4a, but in the MLT-magnetic latitude (MLAT) domain, where the footprint location (in geomagnetic latitude) of the corresponding chorus modulation event is obtained using the dipole magnetic field model for simplicity. Figure 4c shows the statistical distribution of PA observed by all-sky TVs for 34 nights from five stations [*Oguti et al.*, 1981] during the periods of $1o \leq Kp \leq 2-$, when PA events were recorded with the largest number. PA typically occurs from premidnight to dawn with a higher occurrence at later MLT. Interestingly, preferential PA regions tend to be higher in latitude from midnight toward dawn. Note that observations of PA by *Oguti et al.* [1981] were shown before 6 MLT probably due to sunlight contamination after the morning sector. Furthermore, the statistics of PA distribution by *Oguti et al.* [1981] is limited due to the short time coverage over 34 nights. Recent PA observations [*Jones et al.*, 2011] using THEMIS all-sky imager from the Gillam station within the period from September 2007 through March 2008 showed that PA occurrence increases dramatically near midnight to ~50% and continues to increase to ~60% by around 0300 MLT and remains high into the morning sector. The general trend of PA distribution in MLT by *Jones et al.* [2011] is roughly consistent with the work of *Oguti et al.* [1981], although the occurrence rate is lower than the value obtained by *Oguti et al.* [1981]. In addition, the period (from 11 January to 23 February 1980) of observations made by *Oguti et al.* [1981] was during solar maximum, whereas chorus modulation events associated with density depletions recorded in this study were during solar minimum. Nevertheless, comparison of preferential regions of chorus modulation (Figure 4b) and PA (Figure 4c) shows a remarkably similar trend, supporting the possibility that density depletions may play an important role in modulating chorus wave intensity and consequently may be responsible for driving PA. The evidence is supportive, but not yet conclusive that chorus modulation is caused solely by density depletion. In addition, the origin of the density variation itself is still not clear and needs further investigation. Nevertheless, this interesting observation may provide a potential mechanism responsible for chorus intensity modulation that leads to PA.

5. CONCLUDING REMARKS

In this review, we have summarized the mechanisms potentially responsible for driving PA in the upper atmosphere.

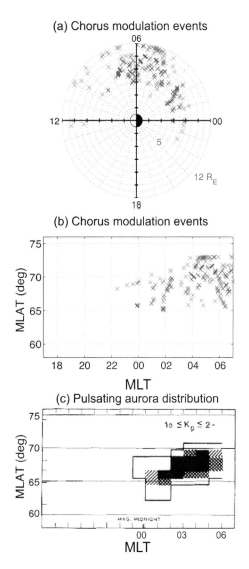

Figure 4. (a) Global distribution of chorus modulation events associated with density depletions in the dominant chorus source region between 5 and 12 R_E from 22 to 14 magnetic local time (MLT) [after *Li et al.*, 2011a]. (b) Global distribution of chorus modulation events in the MLT-MLAT domain, where the footprint locations of corresponding events in Figure 4a are calculated using the dipole magnetic field model. (c) Distribution of occurrence of PA in the *Kp* range of 1o–2– as a function of MLT and MLAT [after *Oguti et al.*, 1981]. © 1981 Canadian Science Publishing or its licensors. Reproduced with permission. The area surrounded by a thick line represents the region where the PA is observed for a 1 h interval, shaded areas represent the region of occurrences for two intervals, doubly shaded areas represent three to four intervals, and dark areas indicate five or more intervals.

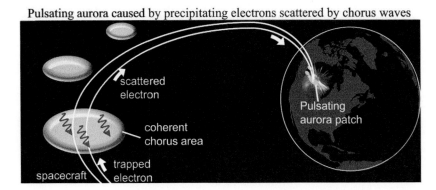

Figure 5. Schematic illustration of wave-particle interaction and resultant PA generation [after *Nishimura et al.*, 2011]. The red curved arrows represent chorus waves propagating away from the equator with a spatial coherence in a patch area (yellow elliptical region). White solid lines represent geomagnetic field lines, along which electrons can bounce back and forth. Trapped electrons moving in the direction (white arrows) opposite to the chorus propagation are scattered and subsequently precipitate into the upper atmosphere, illuminating a PA patch (green region).

Major advances in understanding the driver of the PA have been achieved in recent years due to substantial improvements in our observational capabilities. In particular, recent studies by *Nishimura et al.* [2010, 2011] using a coordinated observation of chorus waves in the equatorial magnetosphere and PA on the ground provided strong evidence that chorus is the driver of PA.

We illustrate a schematic model in Figure 5 to summarize the current understanding of PA generation process based on earlier observations and theories along with recent observations from the THEMIS mission. Generation of PA begins with the injection of anisotropic plasma sheet electrons, which provide a source of free energy for chorus excitation. In the equatorial magnetosphere, chorus intensity can be modulated on a time scale comparable to the PA modulation period. Electrons (with energy from a few to tens of keV) initially trapped in the Earth's magnetosphere are scattered toward the loss cone through resonance with those chorus waves (red curved arrows) that propagate in the direction opposite to the electron movement (white arrows). The modulation of chorus intensity in the equatorial magnetosphere results in modulated scattering rates of electrons, which ultimately precipitate into the atmosphere and illuminate an auroral patch pulsating on a time scale of a few to tens of seconds. The coherent chorus region in the equatorial magnetosphere (yellow region) probably corresponds to the PA patch in the upper atmosphere (green region).

Although a number of studies have suggested that chorus is the driver of PA, it is still an open question why chorus wave intensity is modulated with the observed periodicities and, hence, why the PA patch pulsates. Furthermore, it is still not very clear what controls the coherent size of chorus waves in the equatorial magnetosphere thus shaping PA patch in the

upper atmosphere. Through investigating the correlation of chorus wave amplitude and total electron density, we have demonstrated that density variation may play an important role in chorus intensity modulation. However, the effect of the cold plasma density and its possible influence on the persistence of the PA patch need further investigation. Future work regarding these questions will be particularly interesting in extending our understanding of the origin of PA.

Acknowledgments. The projects related to this review paper are funded by NASA grants NNX11AD75G and NNX08AI35G and NSF grants ATM 0802843 and AGS-0840178. The authors acknowledge NASA contract NAS5-02099 for the use of data from the THEMIS Mission. Specifically, we thank O. Le Contel and A. Roux for the use of SCM data, J. W. Bonnell and F. S. Mozer for the use of EFI data, S. Mende and E. Donovan for the use of the ASI data, the CSA for logistical support in fielding and data retrieval from the GBO stations, and K. H. Glassmeier, U. Auster, and W. Baumjohann for the use of FGM data provided under the lead of the Technical University of Braunschweig and with financial support through the German Ministry for Economy and Technology and the German Center for Aviation and Space (DLR) under contract 50 OC 0302.

REFERENCES

Angelopoulos, V. (2008), The THEMIS Mission, *Space Sci. Rev.*, *141*(1–4), 5–34, doi:10.1007/s11214-008-9336-1.

Beghin, C. J., J. P. Lebreton, B. N. Maehlum, P. I. J. Troim, and J. L. Michau (1984), Phenomena induced by charged particle beams, *Science*, *225*, 188–191.

Belon, A. E., J. E. Maggs, T. N. Davis, K. B. Mather, N. W. Glass, and G. F. Hughes (1969), Conjugacy of visual auroras during magnetically quiet periods, *J. Geophys. Res.*, *74*(1), 1–28.

Brekke, A. (1971), On the correlation between pulsating aurora and cosmic radio noise absorption, *Planet. Space. Sci.*, *19*, 891–896.

Bryant, D. A., M. J. Smith, and G. M. Courtier (1975), Distant modulation of electron intensity during the expansion phase of an auroral substorm, *Planet. Space Sci.*, *23*, 867–878.

Burtis, W. J., and R. A. Helliwell (1969), Banded chorus—A new type of VLF radiation observed in the magnetosphere by OGO 1 and OGO 3, *J. Geophys. Res.*, *74*(11), 3002–3010.

Burton, R. K., and R. E. Holzer (1974), The origin and propagation of chorus in the outer magnetosphere, *J. Geophys. Res.*, *79*(7), 1014–1023.

Coroniti, F. V., and C. F. Kennel (1970), Electron precipitation pulsations, *J. Geophys. Res.*, *75*(7), 1279–1289.

Davidson, G. T. (1979), Self-modulated VLF wave-electron interactions in the magnetosphere: A cause of auroral pulsations, *J. Geophys. Res.*, *84*(A11), 6517–6523.

Davidson, G. T. (1990), Pitch-angle diffusion and the origin of temporal and spatial structures in morningside aurorae, *Space Sci. Rev.*, *53*, 45–82, doi:10.1007/BF00217428.

Davis, T. N. (1978), Observed characteristics of auroral forms, *Space Sci. Rev.*, *22*, 77–113.

Demekhov, A. G., and V. Y. Trakhtengerts (1994), A mechanism of formation of pulsating aurorae, *J. Geophys. Res.*, *99*(A4), 5831–5841.

Gough, M. P., P. J. Christiansen, and R. Thomas (1981), Electrostatic emissions studied in high resolution, *Adv. Space Res.*, *1*(1), 345–351.

Hayakawa, M., Y. Yamanaka, M. Parrot, and F. Lefeuvre (1984), The wave normals of magnetospheric chorus emissions observed on board GEOS 2, *J. Geophys. Res.*, *89*(A5), 2811–2821.

Horne, R. B., R. M. Thorne, N. P. Meredith, and R. R. Anderson (2003), Diffuse auroral electron scattering by electron cyclotron harmonic and whistler mode waves during an isolated substorm, *J. Geophys. Res.*, *108*(A7), 1290, doi:10.1029/2002JA009736.

Johnstone, A. D. (1978), Pulsating aurora, *Nature*, *274*(5667), 119–126, doi:10.1038/274119a0.

Johnstone, A. D. (1983), The mechanism of pulsating aurora, *Ann. Geophys.*, *1*, 397–410.

Jones, S. L., M. R. Lessard, K. Rychert, E. Spanswick, and E. Donovan (2011), Large-scale aspects and temporal evolution of pulsating aurora, *J. Geophys. Res.*, *116*, A03214, doi:10.1029/2010JA015840.

Kennel, C. F., and H. E. Petschek (1966), Limit on stably trapped particle fluxes, *J. Geophys. Res.*, *71*(1), 1–28.

Koons, H. C., and J. L. Roeder (1990), A survey of equatorial magnetospheric wave activity between 5 and 8 R_E, *Planet. Space Sci.*, *38*(10), 1335–1341.

Lauben, D. S., U. S. Inan, T. F. Bell, and D. A. Gurnett (2002), Source characteristics of ELF/VLF chorus, *J. Geophys. Res.*, *107*(A12), 1429, doi:10.1029/2000JA003019.

LeDocq, M. J., D. A. Gurnett, and G. B. Hospodarsky (1998), Chorus source locations from VLF Poynting flux measurements with the Polar spacecraft, *Geophys. Res. Lett.*, *25*(21), 4063–4066.

Lepine, D. R., D. A. Bryant, and D. S. Hall (1980), A 2.2-Hz modulation of auroral electrons imposed at the geomagnetic equator, *Nature*, *286*, 469–471, doi:10.1038/286469a0.

Lessard, M. R. (2012), A review of pulsating aurora, in *Auroral Phenomenology and Magnetospheric Processes: Earth and Other Planets, Geophys. Monogr. Ser.*, doi:10.1029/2011GM001187, this volume.

Li, W., R. M. Thorne, V. Angelopoulos, J. Bortnik, C. M. Cully, B. Ni, O. LeContel, A. Roux, U. Auster, and W. Magnes (2009), Global distribution of whistler-mode chorus waves observed on the THEMIS spacecraft, *Geophys. Res. Lett.*, *36*, L09104, doi:10.1029/2009GL037595.

Li, W., et al. (2010), THEMIS analysis of observed equatorial electron distributions responsible for the chorus excitation, *J. Geophys. Res.*, *115*, A00F11, doi:10.1029/2009JA014845.

Li, W., J. Bortnik, R. M. Thorne, Y. Nishimura, V. Angelopoulos, and L. Chen (2011a), Modulation of whistler mode chorus waves: 2. Role of density variations, *J. Geophys. Res.*, *116*, A06206, doi:10.1029/2010JA016313.

Li, W., R. M. Thorne, J. Bortnik, Y. Shprits, Y. Nishimura, V. Angelopoulos, C. Chaston, O. Le Contel, and J. Bonnell (2011b), Typical properties of rising and falling tone chorus waves, *Geophys. Res. Lett.*, *38*, L14103, doi:10.1029/2011GL047925.

Li, W., J. Bortnik, R. M. Thorne, and V. Angelopoulos (2011c), Global distribution of wave amplitudes and wave normal angles of chorus waves using THEMIS wave observations, *J. Geophys. Res.*, *116*, A12205, doi:10.1029/2011JA017035.

Liang, J., V. Uritsky, E. Donovan, B. Ni, E. Spanswick, T. Trondsen, J. Bonnell, A. Roux, U. Auster, and D. Larson (2010), THEMIS observations of electron cyclotron harmonic emissions, ULF waves, and pulsating auroras, *J. Geophys. Res.*, *115*, A10235, doi:10.1029/2009JA015148.

McEwen, D. J., E. Yee, B. A. Whalen, and A. W. Yau (1981), Electron energy measurements in pulsating auroras, *Can. J. Phys.*, *50*, 1106–1115.

Mende, S. B., S. E. Harris, H. U. Frey, V. Angelopoulos, C. T. Russell, E. Donovan, B. Jackel, M. Greffen, and L. M. Peticolas (2008), The THEMIS array of ground-based observatories for the study of auroral substorms, *Space Sci. Rev.*, *141*, 357–387.

Meredith, N. P., R. B. Horne, R. M. Thorne, and R. R. Anderson (2003), Favored regions for chorus-driven electron acceleration to relativistic energies in the Earth's outer radiation belt, *Geophys. Res. Lett.*, *30*(16), 1871, doi:10.1029/2003GL017698.

Meredith, N. P., R. B. Horne, R. M. Thorne, and R. R. Anderson (2009), Survey of upper band chorus and ECH waves: Implications for the diffuse aurora, *J. Geophys. Res.*, *114*, A07218, doi:10.1029/2009JA014230.

Miyoshi, Y., Y. Katoh, T. Nishiyama, T. Sakanoi, K. Asamura, and M. Hirahara (2010), Time of flight analysis of pulsating aurora electrons, considering wave-particle interactions with propagating whistler mode waves, *J. Geophys. Res.*, *115*, A10312, doi:10.1029/2009JA015127.

Muldrew, D. B., and J. F. Vickrey (1982), High-latitude *F* region irregularities observed simultaneously with ISIS 1 and the Chatanika radar, *J. Geophys. Res.*, *87*(A10), 8263–8272.

Nemzek, R. J., R. Nakamura, D. N. Baker, R. D. Belian, D. J. McComas, M. F. Thomsen, and T. Yamamoto (1995),

The relationship between pulsating auroras observed from the ground and energetic electrons and plasma density measured at geosynchronous orbit, *J. Geophys. Res.*, *100*(A12), 23,935–23,944.

Nunn, D., Y. Omura, H. Matsumoto, I. Nagano, and S. Yagitani (1997), The numerical simulation of VLF chorus and discrete emissions observed on the Geotail satellite using a Vlasov code, *J. Geophys. Res.*, *102*(A12), 27,083–27,097.

Ni, B., R. M. Thorne, Y. Y. Shprits, and J. Bortnik (2008), Resonant scattering of plasma sheet electrons by whistler-mode chorus: Contribution to diffuse auroral precipitation, *Geophys. Res. Lett.*, *35*, L11106, doi:10.1029/2008GL034032.

Nishimura, Y., et al. (2010), Identifying the driver of pulsating aurora, *Science*, *330*(6000), 81–84, doi:10.1126/science.1193186.

Nishimura, Y., et al. (2011), Multievent study of the correlation between pulsating aurora and whistler mode chorus emissions, *J. Geophys. Res.*, *116*, A11221, doi:10.1029/2011JA016876.

Nishiyama, T., T. Sakanoi, Y. Miyoshi, Y. Katoh, K. Asamura, S. Okano, and M. Hirahara (2011), The source region and its characteristic of pulsating aurora based on the Reimei observations, *J. Geophys. Res.*, *116*, A03226, doi:10.1029/2010JA015507.

Oguti, T. (1976), Recurrent auroral patterns, *J. Geophys. Res.*, *81*(10), 1782–1786.

Oguti, T., S. Kokubun, K. Hayashi, K. Tsuruda, S. Machida, T. Kitamura, O. Saka, and T. Watanabe (1981), Statistics of pulsating aurora on the basis of all-sky TV data from five stations. 1. Occurrence frequency, *Can. J. Phys.*, *59*, 1150–1157.

Oguti, T., K. Hayashi, T. Yamamoto, J. Ishida, T. Higuchi, and N. Nishitani (1986), Absence of hydromagnetic waves in the magnetospheric equatorial region conjugate with pulsating auroras, *J. Geophys. Res.*, *91*(A12), 13,711–13,715.

Omura, Y., Y. Katoh, and D. Summers (2008), Theory and simulation of the generation of whistler-mode chorus, *J. Geophys. Res.*, *113*, A04223, doi:10.1029/2007JA012622.

Oliven, M. N., and D. A. Gurnett (1968), Microburst phenomena. 3. An association between microbursts and VLF chorus, *J. Geophys. Res.*, *73*(7), 2355–2362.

Parks, G. K., F. V. Coroniti, R. L. McPherron, and K. A. Anderson (1968), Studies of the magnetospheric substorm. 1. Characteristics of modulated energetic electron precipitation occurring during auroral substorms, *J. Geophys. Res.*, *73*(5), 1685–1696.

Royrvik, O., and T. N. Davis (1977), Pulsating aurora: Local and global morphology, *J. Geophys. Res.*, *82*(29), 4720–4740.

Samara, M., R. G. Michell, K. Asamura, M. Hirahara, D. L. Hampton, and H. C. Stenbaek-Nielsen (2010), Ground-based observations of diffuse auroral structures in conjunction with Reimei measurements, *Ann. Geophys.*, *28*, 873–881.

Sandahl, I., L. Eliasson, and R. Lundin (1980), Rocket observations of precipitating electrons over a pulsating aurora, *Geophys. Res. Lett.*, *7*(5), 309–312.

Santolík, O., D. A. Gurnett, J. S. Pickett, M. Parrot, and N. Cornilleau-Wehrlin (2003), Spatio-temporal structure of storm-time chorus, *J. Geophys. Res.*, *108*(A7), 1278, doi:10.1029/2002JA009791.

Sato, N., D. M. Wright, Y. Ebihara, M. Sato, Y. Murata, H. Doi, T. Saemundsson, S. E. Milan, M. Lester, and C. W. Carlson (2002), Direct comparison of pulsating aurora observed simultaneously by the FAST satellite and from the ground at Syowa, *Geophys. Res. Lett.*, *29*(21), 2041, doi:10.1029/2002GL015615.

Stenbaek-Nielsen, H. C. (1980), Pulsating aurora: The importance of the ionosphere, *Geophys. Res. Lett.*, *7*(5), 353–356.

Stenbaek-Nielsen, H. C., T. N. Davis, and N. W. Glass (1972), Relative motion of auroral conjugate points during substorms, *J. Geophys. Res.*, *77*(10), 1844–1858.

Thorne, R. M. (2010), Radiation belt dynamics: The importance of wave-particle interactions, *Geophys. Res. Lett.*, *37*, L22107, doi:10.1029/2010GL044990.

Thorne, R. M., B. Ni, X. Tao, R. B. Horne, and N. P. Meredith (2010), Scattering by chorus waves as the dominant cause of diffuse auroral precipitation, *Nature*, *467*, 943–946, doi:10.1038/nature09467.

Tsuruda, K., S. Machida, T. Oguti, S. Kokubun, K. Hayashi, T. Kitamura, O. Saka, and T. Watanabe (1981), Correlations between the very low frequency chorus and pulsating aurora observed by low-light-level television at L-4.4, *Can. J. Phys.*, *59*, 1042–1048.

Tsurutani, B. T., and E. J. Smith (1974), Postmidnight chorus: A substorm phenomenon, *J. Geophys. Res.*, *79*(1), 118–127.

Tsurutani, B. T., and E. J. Smith (1977), Two types of magnetospheric ELF chorus and their substorm dependences, *J. Geophys. Res.*, *82*(32), 5112–5128.

Tsyganenko, N. A., and D. P. Stern (1996), Modeling the global magnetic field of the large-scale Birkeland current systems, *J. Geophys. Res.*, *101*(A12), 27,187–27,198.

Vickrey, J. F., C. L. Rino, and T. A. Potemra (1980), Chatanika/Triad observations of unstable ionization enhancements in the auroral *F*-region, *Geophys. Res. Lett.*, *7*(10), 789–792.

Ward, I., M. Lester, and R. Thomas (1982), Pulsing hiss, pulsating aurora and micropulsations, *J. Atmos. Terr. Phys.*, *44*, 931–938.

Yamamoto, T. (1988), On the temporal fluctuations of pulsating auroral luminosity, *J. Geophys. Res.*, *93*(A2), 897–911.

Yau, A. W., B. A. Whalen, and D. J. McEwen (1981), Rocket-borne measurements of particle pulsation in pulsating aurora, *J. Geophys. Res.*, *86*(A7), 5673–5681.

V. Angelopoulos, Department of Earth and Space Sciences, University of California, Los Angeles, CA 90095-1567, USA.

J. Bortnik, W. Li, Y. Nishimura, and R. M. Thorne, Department of Atmospheric and Oceanic Sciences, University of California, Los Angeles, CA 90095-1565, USA. (moonli@atmos.ucla.edu).

Auroral Signatures of Ballooning Mode Near Substorm Onset: Open Geospace General Circulation Model Simulations

J. Raeder,[1] P. Zhu,[2] Y. Ge,[1] and G. Siscoe[3]

We present results from Open Geospace General Circulation Model simulations of the 23 March 2007 THEMIS "first light" substorm. We investigate the relation between ballooning modes in the tail and auroral "beads." Magnetic mapping of the ballooning structures in the tail shows an ionosphere signature that closely resembles auroral beads. The close match of the ballooning properties in the tail with observations, in particular an azimuthal wavelength of ~0.5 R_E, and the close resemblance of the mapped ballooning structures in the simulation with observed auroral beads provides evidence that beads are a signature of tail ballooning modes. Since the aurora is much easier observed globally than the tail, this result allows studying the relation of ballooning to the substorm phases using auroral observations. From the simulation, we find no direct connection between the ballooning mode and expansion phase onset. However, other instabilities that lie somewhere between ballooning and tearing may play an important role as trigger.

1. INTRODUCTION

Recent global Open Geospace General Circulation Model (OpenGGCM) simulations of substorms show that near expansion phase onset, the tail becomes ballooning mode unstable [*Raeder et al.*, 2010]. Although the ballooning mode has long been implicated as a possible trigger for substorm onset [*Lee and Wolf*, 1992; *Ohtani and Tamao*, 1993; *Vetoulis and Chen*, 1994; *Lee and Min*, 1996; *Pu et al.*, 1997, 1999; *Miura*, 2001; *Bhattacharjee et al.*, 1998; *Cheng and Zaharia*, 2004; *Zhu et al.*, 2007, 2004], it has been difficult to tie theoretical predictions to observations. The key prediction for the ballooning mode in the geomagnetic tail is a non-propagating wave structure in the azimuthal direction of fairly high wavenumber corresponding to a wavelength of ~0.5 R_E. Because the wave does not propagate, and because the orbital motion of satellites in the region of interest is rather slow, there is, under most circumstances, no unique signature to observe. *Saito et al.* [2008] presented Geotail observations that showed convected structures in the region of $X_{GSE} = -10$ to -13 R_E just before substorm onset that were consistent with ballooning mode structures. Although such observations are encouraging, they do not yet constitute proof that a ballooning mode is present.

If ballooning modes develop in the tail, they might have a visible counterpart in the ionosphere. If the azimuthal wavelength of the ballooning waves were 0.5 R_E, their distance was 10 R_E from the Earth, and if they mapped radially to the ionosphere, their azimuthal wavenumber would be $m = 2\pi 10/0.5 = 130$, and in terms of wavelength, their characteristic size at 70° magnetic latitude would be $l = \cos(70°) \times 0.5/10 = 109$ km. Such auroral structures are indeed reported, for example, by *Henderson* [2009], who shows IMAGE WIC data with auroral "beads" that have just these characteristics. Ground-based auroral "bead" observations were reported by *Liang et al.* [2008] in conjunction with substorm expansion

[1]Space Science Center and Physics Department, University of New Hampshire, Durham, New Hampshire, USA.

[2]Department of Engineering Physics and Department of Physics, University of Wisconsin-Madison, Madison, Wisconsin, USA.

[3]Center for Space Physics, Boston University, Boston, Massachusetts, USA.

Auroral Phenomenology and Magnetospheric Processes: Earth and Other Planets
Geophysical Monograph Series 197

10.1029/2011GM001200

phase onset. In subsequent work, *Liang et al.* [2009] presented a conjunction between ground-based observations of bead-like auroral structures and waves observed by Time History of Events and Macroscale Interactions during Substorms (THEMIS) in the near-Earth tail. The THEMIS waves are interpreted as the Doppler-shifted signature of ballooning modes and the corresponding auroral beads as their auroral counterpart. These observations are probably the most indicative, so far, of a link between ballooning in the tail and wavelike auroral structures. One should keep in mind, however, that such wavelike structures could have other causes such as Kelvin-Helmholtz waves. Also, the waves observed at THEMIS have frequencies in the Pi2 range. Such waves are ubiquitous at substorm onset but generally not associated with the ballooning mode [*Kepko et al.*, 2001, 2008].

Here we present OpenGGCM simulations of the 23 March 2007 THEMIS "first light" substorm [*Raeder et al.*, 2008], which show clear ballooning mode signatures near substorm onset [*Raeder et al.*, 2010]. In the latter paper, we showed that the tail signatures in the simulation are consistent with the predicted and the observed signatures, in particular, the 0.5 R_E wavelength. We now show how these signatures map to the ionosphere. In the following, we briefly describe the model, present the tail signatures, and show how they map to the ionosphere.

2. OPENGGCM MODEL

The OpenGGCM is a global coupled model of the Earth's magnetosphere, ionosphere, and thermosphere. The magnetosphere part solves the MHD equations as an initial-boundary-value problem. The MHD equations are solved to within ~3 R_E of the Earth. The region within 3 R_E is treated as a magnetosphere-ionosphere (M-I) coupling region where physical processes that couple the magnetosphere to the ionosphere-thermosphere system are parameterized using simple models and relationships. The ionosphere-thermosphere system is modeled using the NOAA Coupled Thermosphere Ionosphere Model (CTIM) [*Fuller-Rowell et al.*, 1996; *Raeder et al.*, 2001a]. The OpenGGCM has been described with some detail [see, e.g., *Raeder et al.*, 2001b; *Raeder*, 2003; *Raeder et al.*, 2008]; we thus refer the reader to these papers. In particular, *Raeder et al.* [2008] also discusses a simulation of the same substorm that is further analyzed here.

3. 23 MARCH 2007 SUBSTORM

We chose the 23 March 2007 substorm for this study because it has already been investigated by several groups [*Angelopoulos et al.*, 2008; *Runov et al.*, 2008; *Raeder et al.*,

2008; *Zhu et al.*, 2009] and because we had an OpenGGCM simulation of this substorm available [*Raeder et al.*, 2008]. In the latter paper, we showed that the simulation reproduces the salient features of a substorm that can be expected from a fluid simulation, in particular, the auroral brightening, the development of the westward traveling surge, fast flows in the tail, and the dipolarization of the magnetic field in the near-Earth tail. Of course, there are kinetic features that cannot be modeled with a fluid code, such as particle injections and auroral kilometric radiation. However, with a recently developed new model, in which the OpenGGCM is coupled with Rice Convection Model, we were also able to produce electron and proton injections for the same event, which compared quite well with observations [*Hu et al.*, 2010]. In this paper, we do therefore not address the validity of the simulation but rather investigate in detail the processes that occur just around the time of the onset of the expansion phase in the simulation.

4. TAIL DYNAMICS

The plasma and field dynamics in the tail for this substorm simulation has been shown in the paper by *Raeder et al.* [2010]. Specifically, Figures 4 and 5 in that paper show the development of the finger-like structures that are characteristic of the ballooning mode. The results presented here are from the model run with the higher resolution shown in Figure 5 of *Raeder et al.* [2010]. There, the structures were made visible by projecting quantities such as a velocity component or the imbalance between the magnetic and pressure forces in the cross-tail direction ($\tilde{F}_y = (\mathbf{j} \times \mathbf{B} - \nabla p)_y$) onto the tail current sheet. Such a projection is necessary because this is a simulation of a real event with measured solar wind and interplanetary magnetic field data as input to the simulation and with a realistic dipole tilt. Thus, the center of the current sheet is not a plane, but a surface that is tilted and undulating. However, it is still well defined by the condition $B_x = 0$. The quantity $\tilde{F}_y$ is particularly useful because it shows the ballooning mode most clearly. Figure 1 shows a different projection of the ballooning fingers, in this case, a spatial cut at $X_{GSE} = -16\ R_E$. The cuts are taken at times 10:30, 10:32, 10:34, and 10:36 UT, respectively. The figure shows clearly how the fingers develop. They start out in the center of the plasma sheet, and then spread in both azimuthal directions, but predominantly toward dawn. The spreading of the instability occurs through addition of new fingers, while their spacing, i.e., the wavelength of the mode, remains largely unchanged. Figure 5 of *Raeder et al.* [2010] also shows that at later times, the entire structure moves dawnward (compare second to third row); however, in this paper, we will only consider the early development.

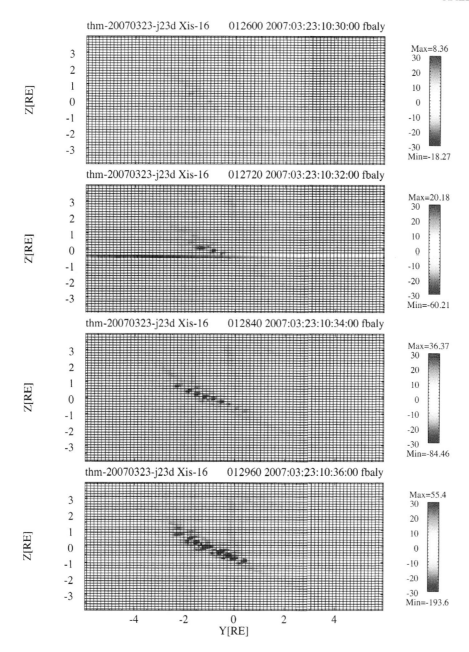

Figure 1. Cuts across the tail at $X_{GSE} = -16\ R_E$. The color coding shows the force imbalance in the y direction between the magnetic and the pressure forces in units of $f\,N\,m^{-3}$. The grid lines show the numerical grid used in the simulations. The ballooning waves are obviously sufficiently resolved.

We now wish to investigate possible auroral signatures of the ballooning growth in the tail. Auroral precipitation can be caused by several processes, in particular, by scattering of hot electrons into the loss cone and by the acceleration of electrons in auroral current sheets. The latter is associated with field-aligned currents (FACs) that originate in the magnetosphere. Other acceleration processes such as through

kinetic Alfvén waves are also possible. The OpenGGCM does include the first two processes in a self-consistent manner. For example, *Raeder et al.* [2008] show the "discrete" (i.e., FAC-accelerated) precipitation for this event. However, we do not expect the ballooning mode structures to show up in the simulated precipitation. The structures are only marginally, but sufficiently, resolved in this simulation.

Grid lines are drawn in Figure 1 to show how the ballooning fingers compare to the grid. A full wave covers approximately five to seven cells, which is well beyond the grid Nyquist frequency. However, as the field lines from these structures converge toward the ionosphere, the waves are no longer resolved. Thus, for example, the FAC generated by a ballooning finger would dissipate numerically before reaching the inner boundary of the MHD domain (at 2.7 R_E) (see *Raeder* [2003] for a discussion of M-I mapping in the OpenGGCM.) In other words, field lines that are, for example, approximately five to seven cells apart in the tail will only be approximately one cell apart at the inner boundary of the MHD domain, and thus, they would no longer be numerically resolved.

In the real M-I system, however, no such restrictions apply. It may be any of the before-mentioned processes that direct electrons toward the ionosphere. Since we are primarily interested in the precipitation pattern, we may thus make the well-founded assumption that these processes would preserve the magnetic mapping, i.e., that the electrons would move faithfully along magnetic field lines from the generator region in the tail to the topside ionosphere. We thus proceed to map the tail structures magnetically to the ionosphere.

Figure 2 shows such a mapping. Here we choose again $\tilde{F}_y$ as the quantity to map, simply for the reason that the ballooning fingers show up best in $\tilde{F}_y$. We could have used almost any other fluid/field variable, such as the pressure or a field component. The only difference would be a less pronounced signature, but the geometry would be the same. For the mapping in Figure 2, we traced field lines from the ionosphere and colored the plot according to the value of $\tilde{F}_y$ at the tip of the field line, i.e., the part of the field line farthest from Earth. This technique allows to show the structure in its entire extent since a mapping from a specific distance might have shown only part of it. Comparison with Figure 5 of *Raeder et al.* [2010] shows that depending on distance and time, only part of the unstable structure would appear in such mapping.

The times in Figure 2 are chosen differently (earlier) than in Figure 1. It turns out that the mode is already growing *somewhere* along the mapping field lines before it shows up in the specific cut of Figure 1. Specifically, the times shown here are 10:20, 10:23, 10:26, and 10:30 UT, respectively, from top to bottom in the figure. Since the mapping only shows the closed field lines, one can see the open-closed boundary as the upper edge of the colored part. Note, however, that this boundary wraps upward away from the tail center. There is also a very distinct transition further equatorward that marks the inner edge of the plasma sheet. From this time sequence, it is clear that we are still in the late growth phase of the substorm as both boundaries keep drifting

equatorward, and the polar cap still grows. The breakup occurs later. We will not analyze the breakup here any further, other than saying that it occurs *not* at the same location where the ballooning mode is present.

The morphology of the ionospheric signatures is very peculiar. First, the spacing of the "beads" (as they are often called in the literature) is approximately 2.2° in longitude, corresponding to an azimuthal wavenumber of $m \sim 160$. This compares very well to auroral images, both from the ground and from space, as shown by *Liang et al.* [2008] and by *Henderson* [2009]. That spacing is also consistent with a roughly radial mapping from the tail. Each wave crest is not simply a spot, but elongated in latitude, thus forming small streaks. These streaks all seem to be directed toward a point some 3°–4° to the north such that they seem to radiate from there. One should note, however, that these figures do not show luminosity, thus, the shape of these streaks in the real world may be different. Specifically, since the low-latitude part of each of the streaks maps closer to the Earth, and thus to hotter plasma, compared to the high-latitude parts, the visible structure may well look more like a dot. As time progresses, the streaks become more and more distorted, but also stronger, until they merge into a less well-defined structure. The similarities to the IMAGE FUV/WIC images of the 21 November 2002 substorm presented by *Henderson* [2009] are striking. In particular, the mode grows over several minutes prior to expansion phase onset and "disintegrates" ~2 min before the substorm expansion phase commences.

5. SUMMARY AND DISCUSSION

We presented results from an OpenGGCM simulation that show the development of the ballooning mode in the tail prior to expansion phase onset and how it maps into the ionosphere. We find our results to be consistent with observations in several respects, which lead to these findings:

1. The tail signatures, in particular the wavelength λ_y of the mode, are very close to observations, for example, those by *Saito et al.* [2008]. This is actually somewhat surprising because the OpenGGCM is an MHD model, while theoretical limits on λ_y are generally based on kinetic considerations.

2. Likewise, the ionosphere signatures are also in close accordance with observations. In particular, the azimuthal wavenumber matches what is typically observed. Furthermore, the timing and temporal development of the ionospheric "beads" closely matches one published observation of a substorm breakup.

3. Since the simulation produces both the tail and the ionosphere signatures and shows that they are linked, further evidence is provided that auroral "beads" are actually the counterpart of tail ballooning modes. Since the aurora is easier

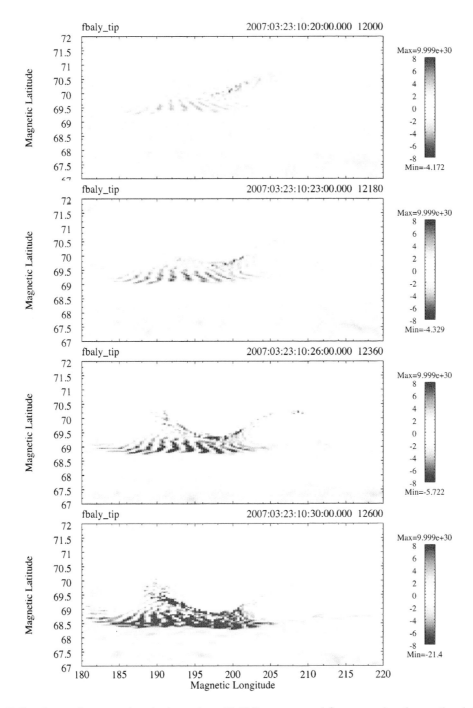

Figure 2. Ballooning modes mapped to the ionosphere. Field lines are traced from every location on the plot from the ionosphere through the magnetosphere. For field lines that are closed, the color coding is according to the same force imbalance shown in Figure 1 at the tip of the field line, in units of $fN\ m^{-3}$. The pattern seen is the expected pattern of ionospheric emissions caused by the ballooning mode. However, no statement about the luminosity can be made because we do not know the mechanism that would cause the aurora.

observed, in particular on a global level, than the tail, this will allow more extensive studies of the ballooning mode and its relation to substorm onset.

We note here that the ballooning mode occurs in the simulation not only prior to onset but also after onset. *Zhu et al.* [2009] has analyzed the stability condition using the OpenGGCM field and plasma parameters and found that the tail exceeds the marginal stability criterion at times and in regions where we find the ballooning signatures also.

In this paper, we purposely did not investigate the relation between the ballooning mode and expansion phase onset. As shown by *Raeder et al.* [2010], a second ideal-like mode (named the axial mode because it has zero azimuthal wavenumber) is present, which is more closely linked to tearing and may also be related to the so-called bubble-blob formation [*Hu et al.*, 2011]. This mode is now under investigation, and the results will be presented elsewhere. The mode is similar to the ballooning mode in the sense that it does not cause a magnetic topology change, but it also has features in common with the tearing mode, namely, that it requires some nonidealness of the plasma that allows at least for the partial decoupling of the flux tube motion from the flow. The relation between the ballooning mode and the axial mode is also still an open question. However, the simulation shows no evidence of the ballooning mode triggering tearing. If that were the case, one would expect tearing to start in the troughs of the ballooning mode where the B_z has a minimum. However, no such structuring of the tearing mode has been seen in the simulations.

Acknowledgments. This research was supported by NASA grant NAS5-02099 (THEMIS) and NSF grants ATM-0639658 and ATM-0902360. Development of the OpenGGCM has been supported by NASA grant NNG05GM57G and NSF grant ATM-0639658. G. L. S. was in part supported by NSF grant ATM-0809307. Part of the simulations was performed at the National Center for Supercomputer Applications (NCSA). Part of this work was accomplished while J. R. was on sabbatical at CESR, Toulouse, France. He would like to thank the CESR staff, and Benoit Lavraud in particular, for their hospitality and support.

REFERENCES

Angelopoulos, V., et al. (2008), First results from the THEMIS mission, *Space Sci. Rev.*, *141*, 453–476.

Bhattacharjee, A., Z. W. Ma, and X. Wang (1998), Dynamics of thin current sheets and their disruption by ballooning instabilities: A mechanism for magnetospheric substorms, *Phys. Plasmas*, *5*, 2001–2009, doi:10.1063/1.872871.

Cheng, C. Z., and S. Zaharia (2004), MHD ballooning instability in the plasma sheet, *Geophys. Res. Lett.*, *31*, L06809, doi:10.1029/2003GL018823.

Fuller-Rowell, T. J., D. Rees, S. Quegan, R. J. Moffett, M. V. Codrescu, and G. H. Millward (1996), A coupled thermosphere-ionosphere model (CTIM), in *STEP Report*, edited by R. W. Schunk, p. 217, Sci. Comm. on Sol. Terr. Phys. (SCOSTEP), NOAA/NGDC, Boulder, Colo.

Henderson, M. G. (2009), Observational evidence for an inside-out substorm onset scenario, *Ann. Geophys.*, *27*, 2129–2140.

Hu, B., F. R. Toffoletto, R. A. Wolf, S. Sazykin, J. Raeder, D. Larson, and A. Vapirev (2010), One-way coupled OpenGGCM/RCM simulation of the 23 March 2007 substorm event, *J. Geophys. Res.*, *115*, A12205, doi:10.1029/2010JA015360.

Hu, B., R. A. Wolf, F. R. Toffoletto, J. Yang, and J. Raeder (2011), Consequences of violation of frozen-in-flux: Evidence from OpenGGCM simulations, *J. Geophys. Res.*, *116*, A06223, doi:10.1029/2011JA016667.

Kepko, L., M. G. Kivelson, and K. Yumoto (2001), Flow bursts, braking, and Pi2 pulsations, *J. Geophys. Res.*, *106*, 1903–1915, doi:10.1029/2000JA000158.

Kepko, L., et al. (2008), Highly periodic stormtime activations observed by THEMIS prior to substorm onset, *Geophys. Res. Lett.*, *35*, L17S24, doi:10.1029/2008GL034235.

Lee, D.-Y., and K. W. Min (1996), On the possibility of the MHD-ballooning instability in the magnetotail-like field reversal, *J. Geophys. Res.*, *101*(A8), 17,347–17,354.

Lee, D.-Y., and R. A. Wolf (1992), Is the Earth's magnetotail balloon unstable?, *J. Geophys. Res.*, *97*(A12), 19,251–19,257.

Liang, J., E. F. Donovan, W. W. Liu, B. Jackel, M. Syrjäsuo, S. B. Mende, H. U. Frey, V. Angelopoulos, and M. Connors (2008), Intensification of preexisting auroral arc at substorm expansion phase onset: Wave-like disruption during the first tens of seconds, *Geophys. Res. Lett.*, *35*, L17S19, doi:10.1029/2008GL033666.

Liang, J., W. W. Liu, E. F. Donovan, and E. Spanswick (2009), In-situ observation of ULF wave activities associated with substorm expansion phase onset and current disruption, *Ann. Geophys.*, *27*, 2191–2204.

Miura, A. (2001), Ballooning instability as a mechanism of the near-Earth onset of substorms, *Space Sci. Rev.*, *95*, 387–398.

Ohtani, S.-I., and T. Tamao (1993), Does the ballooning instability trigger substorms in the near-Earth magnetotail?, *J. Geophys. Res.*, *98*(A11), 19,369–19,379.

Pu, Z. Y., A. Korth, Z. X. Chen, R. H. W. Friedel, Q. G. Zong, X. M. Wang, M. H. Hong, S. Y. Fu, Z. X. Liu, and T. I. Pulkkinen (1997), MHD drift ballooning instability near the inner edge of the near-Earth plasma sheet and its application to substorm onset, *J. Geophys. Res.*, *102*(A7), 14,397–14,406.

Pu, Z. Y., et al. (1999), Ballooning instability in the presence of a plasma flow: A synthesis of tail reconnection and current disruption models for the initiation of substorms, *J. Geophys. Res.*, *104*(A5), 10,235–10,248.

Raeder, J. (2003), Global magnetohydrodynamics – A tutorial, in *Space Plasma Simulation*, *Lect. Notes Phys.*, vol. 615, edited by J. Büchner, C. T. Dum, and M. Scholer, p. 212–246, Springer, Berlin.

Raeder, J., Y. Wang, and T. J. Fuller-Rowell (2001a), Geomagnetic storm simulation with a coupled magnetosphere-ionosphere-thermosphere model, in *Space Weather*, *Geophys. Monogr. Ser.*, vol. 125, edited by P. Song, H. J. Singer, and G. L. Siscoe, pp. 377–384, AGU, Washington, D. C., doi:10.1029/GM 125p0377.

Raeder, J., R. L. McPherron, L. A. Frank, S. Kokubun, G. Lu, T. Mukai, W. R. Paterson, J. B. Sigwarth, H. J. Singer, and J. A. Slavin (2001b), Global simulation of the Geospace Environment Modeling substorm challenge event, *J. Geophys. Res.*, *106*, 381–395, doi:10.1029/2000JA000605.

Raeder, J., D. Larson, W. Li, E. L. Kepko, and T. Fuller-Rowell (2008), OpenGGCM simulations for the THEMIS mission, *Space Sci. Rev.*, *141*, 535–555, doi:10.1007/s11214-008-9421-5.

Raeder, J., P. Zhu, Y. Ge, and G. Siscoe (2010), Open Geospace General Circulation Model simulation of a substorm: Axial tail instability and ballooning mode preceding substorm onset, *J. Geophys. Res.*, *115*, A00I16, doi:10.1029/2010JA015876.

Runov, A., V. Angelopoulos, N. Ganushkina, R. Nakamura, J. McFadden, D. Larson, I. Dandouras, K.-H. Glassmeier, and C. Carr (2008), Multi-point observations of the inner boundary of the plasma sheet during geomagnetic disturbances, *Geophys. Res. Lett.*, *35*, L17S23, doi:10.1029/2008GL033982.

Saito, M. H., Y. Miyashita, M. Fujimoto, I. Shinohara, Y. Saito, K. Liou, and T. Mukai (2008), Ballooning mode waves prior to substorm-associated dipolarizations: Geotail observations, *Geophys. Res. Lett.*, *35*, L07103, doi:10.1029/2008GL033269.

Vetoulis, G., and L. Chen (1994), Global structures of Alfvén-ballooning modes in magnetospheric plasmas, *Geophys. Res. Lett.*, *21*(19), 2091–2094.

Zhu, P., A. Bhattacharjee, and Z. W. Ma (2004), Finite k_y ballooning instability in the near-Earth magnetotail, *J. Geophys. Res.*, *109*, A11211, doi:10.1029/2004JA010505.

Zhu, P., C. R. Sovinec, C. C. Hegna, A. Bhattacharjee, and K. Germaschewski (2007), Nonlinear ballooning instability in the near-Earth magnetotail: Growth, structure, and possible role in substorms, *J. Geophys. Res.*, *112*, A06222, doi:10.1029/2006JA 011991.

Zhu, P., J. Raeder, K. Germaschewski, and C. C. Hegna (2009), Initiation of ballooning instability in the near-Earth plasma sheet prior to the 23 March 2007 THEMIS substorm expansion onset, *Ann. Geophys.*, *27*, 1129–1138.

Y. Ge and J. Raeder, Space Science Center and Physics Department, University of New Hampshire, 8 College Rd, Durham, NH 03824, USA. (Yasong.Ge@gmail.com, J.Raeder@unh.edu)

G. Siscoe, Center for Space Physics, Boston University, 725 Commonwealth Ave., Boston, MA 02215, USA. (siscoe@skynet.bu.edu)

P. Zhu, Department of Engineering Physics, University of Wisconsin-Madison, 1500 Engineering Dr., Madison, WI 53706, USA. (pzhu@wisc.edu)

Origins of Saturn's Auroral Emissions and Their Relationship to Large-Scale Magnetosphere Dynamics

Emma J. Bunce

Department of Physics and Astronomy, University of Leicester, Leicester, UK

In this review article, we discuss recent observations of Saturn's dynamic aurora and how these relate to large-scale current systems within the magnetosphere. We first discuss the driving mechanism of the main auroral oval in terms of a theoretical framework that takes into account both plasma subcorotation associated with mass loading in the magnetosphere and the solar wind interaction. We compare the expected field-aligned current systems from that model with images of the aurora and from in situ measurements of the high-latitude current systems measured by Cassini. We identify open questions relating to the origin and modulation of the auroral current system.

1. INTRODUCTION

The terrestrial aurora has been extensively studied from the ground and from Earth-orbiting spacecraft for many decades, and as such, our broad understanding of the driving physical mechanisms behind the phenomena is well developed. At the outer planets (e.g., Jupiter and Saturn), we have relatively much less information, both in terms of remote sensing and in situ measurements. However, the interest of the scientific community in outer planet aurora has increased in recent years due to the outstanding imaging capability of the Hubble Space Telescope (HST), the presence of in situ orbiting spacecraft such as Galileo (at Jupiter) and Cassini (at Saturn), and the subsequent ability to develop a theoretical picture of the solar wind-magnetosphere-ionosphere coupling mechanisms, which can then provide a framework for our understanding of these systems. This review chapter is focused on the development of such theoretical ideas, remote sensing observations, and in situ measurements within Saturn's magnetosphere, and thus on how we are beginning to understand the nature of the auroral emissions and their relationship

to magnetospheric dynamics. The reader is also directed toward the review of auroral dynamics at Saturn by *Kurth et al.* [2009] and references therein.

Prior to Cassini, much of what we learned about the auroral emissions at Saturn came from observations made with the HST, specifically at UV wavelengths as shown in the sequence of three images (taken approximately 2 days apart) at the bottom left of Figure 1 [see, e.g., *Cowley et al.*, 2004a; *Gérard et al.*, 2004; *Grodent et al.*, 2005], as well as ground-based telescope observations of the ion flow velocities in the auroral regions associated with IR H_3^+ ions and aurora [*Stallard et al.*, 2004]. Studies such as these have revealed an auroral zone at Saturn, which subcorotates with the planet at ~60%–70% of corotation and which is a few tens of kR in intensity in the UV. The main emission is centered at approximately 15° colatitude in both the Northern and Southern Hemispheres. The approach of Cassini to, and arrival at, Saturn has significantly advanced our understanding of the auroral images, through in situ field and particle measurements upstream of Saturn's magnetosphere in the solar wind and from within the high-latitude magnetosphere during Cassini's inclined orbit phases. During these high-latitude phases, the Cassini Ultraviolet Imaging Spectrometer (UVIS) and the visible and infrared imaging spectrometer (VIMS) have had spectacular views of the northern and southern poles of Saturn (see the top (VIMS) and lower right (UVIS) images of Figure 1).

In order to understand the aurora as a diagnostic for the large-scale magnetospheric and/or solar wind dynamics, we

Auroral Phenomenology and Magnetospheric Processes: Earth and Other Planets
Geophysical Monograph Series 197
10.1029/2011GM001191

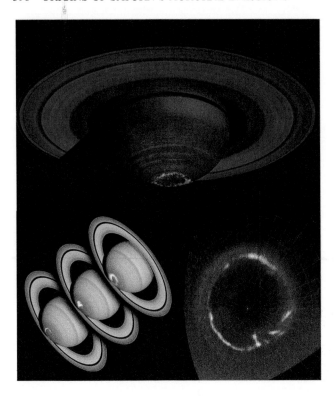

Figure 1. Saturn's dynamic auroral emissions (top) at infrared wavelengths (from NASA/Cassini/ visible and infrared imaging spectrometer team) and (bottom) at UV wavelengths (from (left) NASA/Hubble Space Telescope (HST)/Boston University (published on the cover of *Nature*, volume 433, issue 7027, 2005) and (right) NASA/Cassini/Cassini Ultraviolet Imaging Spectrometer).

must first understand the dominant processes affecting the coupled solar wind-magnetosphere-ionosphere system in the "steady state." From this point, we can address what the main source of dynamics may be and, hence, predict how the system will vary as a function of time. However, this short review does not attempt to cover small-scale magnetosphere dynamics and related structure within the auroral regions, which are evident in the imaging data [e.g., *Radioti et al.*, 2009; *Grodent et al.*, 2011]. During the Pioneer and Voyager flybys of Saturn in the 1980s, it was found that Saturn's kilometric radiation (SKR) (thought to be associated with accelerated auroral electron beams) was modulated both by the planet's rapid rotation rate of ~10.7 h and by enhancements of solar wind dynamic pressure [*Desch and Kaiser*, 1981; *Desch and Rucker*, 1983]. A logical starting point is therefore to consider the main plasma flows and currents in Saturn's magnetosphere as being produced by a combination of rotational and solar wind effects. In the next section, we will briefly introduce the theoretical framework to which the rest of the discussion in this review will refer.

2. THEORETICAL FRAMEWORK

Following the development of our understanding of how the main Jovian auroral emissions are generated, i.e., through the large-scale breakdown of corotation of plasma in the middle magnetosphere [*Cowley and Bunce*, 2001; *Hill*, 2001; *Southwood and Kivelson*, 2001], it was appropriate to question if the aurora seen at high-latitudes in Saturn's magnetosphere could be driven through a similar magnetosphere-ionosphere coupling system. The possibility was assessed using a magnetic field model based on flyby data and plasma flows from the Voyager flybys, but it was found that the field-aligned currents associated with corotation breakdown were too small (~10 nA m^{-2}) and mapped to the wrong colatitudes in the ionosphere (~20°) to account for the auroral ovals observed near 10°–17° [*Cowley and Bunce*, 2003; *Badman et al.*, 2005]. Recent Cassini UVIS observations of Saturn's aurora, however, do indicate a weaker (~2 kR) secondary oval at ~20° colatitude on the nightside [see *Grodent et al.*, 2010], which may also relate to the IR observations of Jovian-like aurora reported on by *Stallard et al.* [2008]. However, the details of this remain unclear at present and will not be pursued further in this review. *Cowley and Bunce* [2003] concluded that Saturn's "main oval" auroras are not associated with corotation-enforcement currents as they are at Jupiter, but instead are most likely to be associated with coupling to the solar wind (as on Earth). Quantitative modeling specifically suggested that Saturn's auroras are associated with a ring of upward directed field-aligned current spanning the boundary between open and closed magnetic field lines [*Cowley et al.*, 2004a, 2004b]. Three basic regimes of rotational and solar wind–driven plasma flows (solid lines) and associated upward and downward directed field-aligned currents (circled dots and crosses, respectively) are shown for the northern polar ionosphere in Figure 2, in a frame, which is fixed relative to the Sun. These flow regimes have been previously discussed in other contexts by *Hill* [1979], *Vasyliunas* [1983], and *Dungey* [1961]. The first region (at lowest colatitude) maps to the middle magnetosphere region of subcorotating plasma dominated by pickup, loss, and radial diffusion from internal sources (icy moons and rings). The associated rings of upward and downward field-aligned currents are associated with the corotation enforcement currents, the upward directed portion of which are equivalent to those that produce the main auroral emissions at Jupiter. As discussed above, these currents are found to be of insufficient strength and flowing at the incorrect colatitude to account for the main auroral emissions at Saturn. The second flow regime is a higher-latitude region of subcorotating flows where field lines are stretched out downtail and eventually pinch off, forming a plasmoid, which is subsequently released downtail

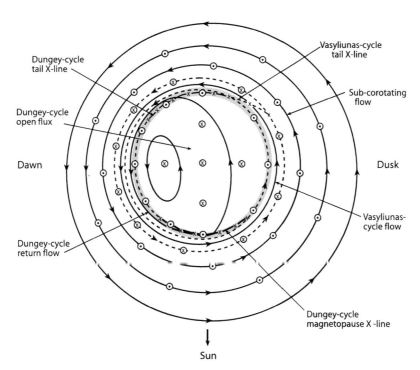

Dungey-cycle
tail X-line

Dungey-cycle
open flux

Dawn

Dungey-cycle
return flow

Vasyliunas-cycle
tail X-line

Sub-corotating
flow

Dusk

Vasyliunas-
cycle flow

Dungey-cycle
magnetopause X-line

Sun

Figure 2. Schematic diagram depicting the plasma flows and the associated field-aligned current patterns in Saturn's northern polar ionosphere. The upward directed field-aligned current pattern, which is suggested to produce the main oval at Saturn is shown shaded blue. Taken from *Cowley et al.* [2004a].

(known as the "Vasyliunas cycle"). At highest latitudes in Figure 2, we see a region of flow, which is driven by reconnection at the dayside magnetopause in which "open" field lines mapping to the tail lobes flow antisunward over the poles, and following reconnection in the tail, return to the dayside, drawn here principally via dawn, in a single-cell convection pattern (the "Dungey cycle"). The newly closed flux tubes return to the dayside via dawn due to the presence of the Vasyliunas cycle on the duskside and also due to the effect of planetary rotation on the open field lines. A slow rotation of the open field region was suggested by *Cowley et al.* [2004a] and measured by *Stallard et al.* [2004] using IR aurora data. It is this upward directed field-aligned current along the boundary between open and closed field lines (shown by the inner dashed circle shaded in blue), which is suggested to account for the main auroral oval at Saturn. These suggestions will be discussed in relation to new observations in the sections that follow.

In summary, *Cowley at al.* [2004a, 2004b] find that the (1) upward field-aligned currents are expected to occur at the open-closed field line boundary typically located at ~15° colatitude, (2) upward field-aligned current density, estimated to be ~150 nA m^{-2}, requires downward field-aligned acceleration of magnetospheric electrons (on the closed side of the boundary) through a ~10 kV potential such that the precipi-

tating electron energy flux is sufficient to produce UV intensities of a few tens of kR (where $1R = 10^{10}$ photons m^{-2} s^{-1} into 4π sr) in the upper atmosphere, and (3) the currents will be larger and the aurora brighter at dawn than at dusk when the solar wind–driven Dungey cycle is active.

Given these suggestions from the theoretical modeling work, it is thus of interest to test the hypotheses by comparing recent HST or Cassini images of the aurora with the in situ measurements of the magnetic field, particle populations, radio plasma waves, and energetic neutral atom (ENA) images measured by the Cassini spacecraft orbiting the high-latitude magnetosphere of Saturn.

3. MODULATION OF THE MAIN EMISSIONS

From experience at the Earth's magnetosphere, we might assume that the direction of the interplanetary magnetic field (IMF) and/or the solar wind dynamic pressure might influence the dynamics of Saturn's auroral oval, assuming that the upward directed current that produces it represents the boundary between magnetic field lines open to the solar wind and those which are closed. On the Earth, the brightest (substorm) auroras occur on the nightside and are produced following a prolonged interval of upstream IMF B_z southward. Of course, at Saturn, the planetary magnetic field points in the opposite direction, and hence, we might expect a similar effect following a prolonged interval of northward directed IMF. However, during an extensive campaign of HST imaging of Saturn's aurora in 2004, while the Cassini spacecraft was upstream of Saturn simultaneously measuring the properties of the solar wind, we had the best opportunity, to date, to learn that it is, in fact, the solar wind dynamic pressure, which plays a more significant role than the direction of the IMF [*Clarke et al.*, 2005; *Crary et al.*, 2005; *Kurth et al.*, 2005].

While Cassini was upstream of Saturn in 2003/2004, the solar wind magnetic field structure was consistent with that expected to be produced by corotating interaction regions (CIRs) during the declining phase of the solar cycle. In general, the data show that during this time, the IMF structure consists of two sectors during each rotation of the Sun, with crossings of the heliospheric current sheet generally embedded within few-day high field compression regions, surrounded by several day rarefaction regions [*Jackman et al.*, 2004]. During the month-long HST campaign in January 2004, a CIR passed Saturn's magnetosphere at the same time that HST imaged the Southern Hemisphere aurora. The spectacular results (seen in the bottom left sequence of HST images in Figure 1) showed that the effect of the subsequent shock-compression of the magnetosphere acted to significantly reduce the size of the main auroral oval, with the auroral emissions filling in the dawnside polar cap approximately to the pole. At the peak of the emission intensities, up to ~100 kR were measured [*Clarke et al.*, 2005]. In addition, the SKR, thought to be associated with auroral accelerated electron beams, was also significantly enhanced in response to the CIR interaction with the magnetosphere [*Kurth et al.*, 2005]. In relation to the theoretical ideas discussed above, it is supposed that the forward shock within the CIR compression region triggers an interval of rapid reconnection in Saturn's magnetic tail in which a significant fraction of the open flux in the tail lobe will be closed over an interval of several hours (less than one complete rotation of Saturn), thus significantly reducing the size of the auroral oval [*Cowley et al.*, 2005]. The rapid tail reconnection in the tail is suggested to produce a substantial amount of newly closed flux containing hot plasma from the reconnection site, forming a bulge, which extends into the polar cap. This takes place on the dawnside of midnight, due to the Vasyliunas cycle flow active near dusk. The newly closed flux tubes flow toward the planet in the tail and rotate around to dawn due to the ionospheric torque, forming a bulge on the magnetopause. In the ionosphere, the auroras will be dominated by a bright auroral patch, which extends into the midnight polar cap, with the usual aurora being present at other local times. The perturbed flow attempts to redistribute the newly closed flux around the boundary, by carrying the bulge equatorward

and dawnward. Evidence for such shock-induced auroral storms has also been presented earlier by *Prange et al.* [2004]. They report on the effect of an interplanetary shock propagating through the solar system, producing a series of planetary auroral storms on the Earth, Jupiter, and Saturn.

In addition to the solar wind dynamic pressure effects discussed above, the rate of open flux production (or reconnection voltage) has been estimated from the upstream interplanetary data during the 2004 campaign using an empirical formula based on experience on Earth [see *Badman et al.*, 2005; *Jackman et al.*, 2004]. This voltage is taken to be given by

$$\Phi = V_{SW}B_{\perp}L_0\cos^4(\theta/2), \qquad (1)$$

where V_{SW} is the radial speed of the solar wind, $B_{\perp}$ is the strength of the IMF component perpendicular to the radial flow, L_0 is a scale length taken equal to 10 Saturn radii ($R_S = 60330$ km) by analogy with Earth, and θ is the clock angle of the IMF relative to Saturn's northern magnetic axis [*Jackman et al.*, 2004]. Typical values are found to vary from 10 kV during the weak-field rarefaction region, up to 200 kV during the strong-field compression discussed above. These values have been integrated over time between individual HST image sets to estimate the total open flux produced during these intervals [see, *Badman et al.*, 2005]. Comparison with the changes in open flux obtained from the auroral images, assuming that the poleward edge of the auroral oval acts as a proxy for the open-closed field line boundary, then allows an estimate to be made of the amount of open flux closed during these intervals and, hence, the averaged tail reconnection rates. The amount of open flux was found to vary from ~13 to 49 GWb during the January 2004 interval. Intermittent intervals of tail reconnection at rates of 30–60 kV are inferred during the rarefaction regions, while compression regions are characterized by rates of 100–200 kV. While some individual episodes of pulsed dayside reconnection have been suggested from modeling and observations [*Bunce et al.*, 2005; *Gérard et al.*, 2005; *Radioti et al.*, 2011; *Badman et al.*, 2012], it seems that the brightest auroras at Saturn are seen following the passage of compression regions past Saturn's magnetosphere. The reason for this difference with the Earth's magnetosphere may be associated with the difference in the scale sizes of the two magnetospheres. The time scales for north-south fluctuations of IMF B_z are similar on Earth and on Saturn, changing rapidly over ~10 min to ~1 h (although IMF B_y turnings on time scales of few days may be a more significant factor). Similarly, dayside reconnection voltages (open flux production rates) are also similar at the two planets with values ranging between ~20 kV during rarefaction regions and ~200 kV (during compression events). However, the open flux on Saturn's tail (~50 GWb) is significantly larger than on Earth (~0.5 GWb), so the growth phase of a single "substorm-like event" could feasibly be much longer, ~1 week for Saturn compared to ~1 h on the Earth. Therefore, we might expect that Saturn's tail will typically not respond to individual intervals of northward IMF B_z but, instead, will inflate on time scales comparable to the time between the recurrent CIRs in the solar wind. Thus, it seems that compression-induced tail reconnection, while rather rare on Earth [e.g., *Boudouridis et al.*, 2003], may be the usual mode on Saturn.

4. CASSINI OBSERVATIONS OF THE HIGH-LATITUDE MAGNETOSPHERE

The observations of the aurora from the HST and the in situ solar wind measurements discussed in the previous section only provide remote sensing evidence about the physical origins of Saturn's aurora, without simultaneous in situ sampling of the magnetic field and plasma populations. In fact, it was not until the first high-latitude phase of the Cassini mission in early 2006/2007 until the large-scale field-aligned current systems connecting the magnetosphere to the ionosphere could be directly sampled for the first time (and, in fact, these measurements represented the first of their kind beyond the Earth's magnetosphere). Before reviewing the results from such observations, it is useful to first discuss the interpretation of the magnetic field signatures from single spacecraft measurements in terms of the large-scale current systems, which must be present. If we assume that the magnetic field signatures observed at high latitudes are due to quasi-steady field-aligned currents flowing between the ionosphere and the near-equatorial magnetosphere, we can then use them to estimate the direction and strength of the ionospheric Pedersen current at the feet of the field lines and, from the variations of the latter, the direction and amount of field-aligned current flowing along the field lines in the regions considered. Application of Ampère's law to a circular loop passing through the measurement point (i.e., at the spacecraft) and centered on the magnetic axis (assuming axisymmetry) allows one to estimate the total current passing through the surface enclosed by the loop:

$$\oint B \cdot dl = \iint \mu_0 J \cdot dS = \mu_0 I_{enc}. \quad (2)$$

In Figure 3, we consider the effect of a purely azimuthal magnetic field B_φ (here shown for subcorotating plasma and "lagging" magnetic field configuration, i.e., $B_\varphi < 0$ in the Northern Hemisphere and $B_\varphi > 0$ in the Southern Hemisphere) observed at some point between the ionosphere and the closure currents in the near-equatorial magnetosphere. We indicate an appropriate circular "Ampère loop" passing through the measurement point shown by dashed circles in both hemispheres. If we consider the surface formed along the field lines from the upper dashed circle to the lower dashed circle in the

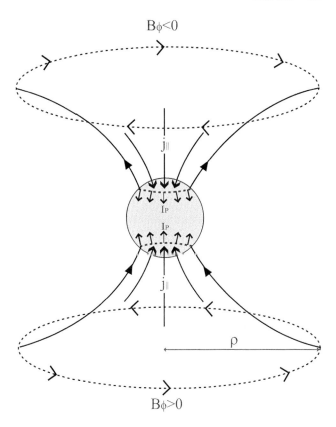

Figure 3. Schematic diagram showing the relationship between the azimuthal magnetic field and the field-aligned current according to Ampère's law. The example shown here is appropriate for subcorotating field lines, i.e., negative B_φ in the Northern Hemisphere and positive B_φ in the Southern Hemisphere. See text for details.

ionosphere, then there is no current flow across this surface as we assume the currents are entirely field-aligned, and hence, the line integral of B_φ on the upper is then equal to the net total field-aligned current, $j_\parallel$, flowing across the upper dashed loop in Figure 3 toward the planet. If the current is entirely field-aligned, then the line integral is identical on every such circle between the upper dashed loop at the measurement point and the lower dashed loop in the ionosphere. If the surface is then closed in the inner region over the top of the ionosphere, then the line integral is easily seen as the net inward field-aligned current between the pole and the latitude of the dashed circular loop in the ionosphere. From current continuity, this is also equal to the equatorward current flowing in the ionosphere at the latitude of the ring, I_P. Hence, we can write down the relationship between the magnetic field perturbation and the ionospheric Pedersen current at the feet of the field lines as

$$B_\varphi = \mp \frac{(\mu_0 I_P)}{\rho}. \quad (3)$$

Here I_P is the ionospheric Pedersen current per radian of azimuth flowing at the feet of the field lines considered, taken to be positive when directed equatorward in both hemispheres, ρ is the perpendicular distance from the axis of symmetry, and the upper sign is appropriate to the Northern Hemisphere and the lower to the Southern Hemisphere. Using this equation, we can therefore use our observations of B_φ to estimate I_P, on the basis of the two assumptions made above. The first assumption is that of approximate axisymmetry, at least locally, whose validity is indicated by the presence of highly structured azimuthal fields in thin layers in the absence of comparable structure in the other field components. The second assumption is that the observation point is located sufficiently far from the magnetospheric equator, that it lies outside the region of significant field-perpendicular magnetospheric closure currents. In addition, the value of I_P is given by

$$I_P = \rho_i i_P, \qquad (4)$$

where ρ_i is the perpendicular distance from the axis of symmetry in the ionosphere, and i_P is the height-integrated Pedersen current intensity given by

$$i_P = \Sigma_P \rho_i (\Omega_S^* - \omega) B_i, \qquad (5)$$

where Σ_P is the height-integrated Pedersen conductivity, B_i the near-vertical ionospheric field strength, Ω_S^* the angular velocity of the neutral atmosphere (possibly altered from rigid corotation by ionospheric drag), and ω is the angular velocity of the plasma, constant along a flux shell in the steady state. The Pedersen current, and hence the azimuthal field on a given flux shell, thus depends on both the Pedersen conductivity of the ionosphere and the degree to which the angular velocity of the plasma departs from the angular velocity of the neutral atmosphere. In the usual case of plasma subcorotation (the case drawn in Figure 3), such that $\omega < \Omega_S^*$, I_P is equatorward directed (positive) in both hemispheres, such that B_φ given by equation (4) is negative in the Northern Hemisphere and positive in the Southern Hemisphere, corresponding to a lagging field configuration. However, in the event that some dynamical process produces supercorotating flow, such that $\omega > \Omega_S^*$, producing a leading field configuration with B_φ positive in the Northern Hemisphere and negative in the Southern Hemisphere, I_P would then be poleward directed in both hemispheres. We further note that the integrated field-aligned current per radian of azimuth flowing along the field in a particular region can be estimated by determining the change in the value of I_P that occurs across the region. If, for example, I_P increases by amount ΔI_P on moving equatorward from one flux shell to another, then that amount of field-aligned current per radian must have flowed into the ionosphere in the region between these shells. Similarly, if I_P decreases by ΔI_P, then that

amount of field-aligned current per radian must have flowed out of the ionosphere into the magnetosphere on those flux shells. Of course, these currents are just those required to account for the changes in B_φ across the flux shells via Ampère's law from which the values of I_P were derived. The regions of upward directed field-aligned current are of particular interest since they will typically relate to the occurrence of bright ionospheric auroral emissions if the current densities are sufficiently large that magnetospheric electrons must be accelerated along the field lines by field-aligned electric fields.

During the Cassini high-latitude orbit Revolution 37 in January 2007, it was possible to make near-simultaneous measurements of the aurora from the HST and the associated field-aligned current system interpreted from the magnetic field signatures according to the above discussion [Bunce et al., 2008]. In Figure 4, we briefly summarize the results from these joint observations. We show two consecutive UV images (A and B) of Saturn's southern auroral oval observed during the HST campaign on 16 and 17 January 2007, respectively. On each image, the white track indicates the magnetically mapped footprint of the spacecraft position during Rev 37, and the red dots in each case mark the mapped position of the spacecraft at the time of the images (see Bunce et al. [2008] for details of the mapping procedure). At the time of image A, the spacecraft is located poleward of the auroral oval, and in the interval between the two images, it is clear from the mapped footprint that the spacecraft has crossed through flux tubes connecting to the auroral oval to a position slightly equatorward of the main emission shown by the red dot in image B.

At the bottom of Figure 4, we show the in situ electron energy spectrogram from Cassini Plasma Spectrometer Electron Spectrometer (CAPS-ELS) the azimuthal component of the magnetic field from the Cassini magnetometer (MAG), and the UV intensity from HST image A in red and image B in blue. The times of images A and B (shifted by the light travel time from Earth to Saturn) are marked by the solid red vertical lines. It can be clearly seen that between the two images, the B_φ component of the magnetic field (second panel) switches from positive values (representing lagging field in the Southern Hemisphere) at the time of image A to near zero (representing approximately corotating field) as the spacecraft moves equatorward. For a spacecraft moving equatorward in the Southern Hemisphere, this signature is then consistent with a decrease in the ionospheric Pedersen current as a function of time (see equation (3) above) and, hence, with an upward directed field-aligned current layer whose density in the ionosphere is estimated to be ~275 nA m^{-2} (see Bunce et al. [2008] for more details). At the same time as the transition in the magnetic field, the electron spectrogram (top panel) shows a sharp transition between a region with a distinct lack of

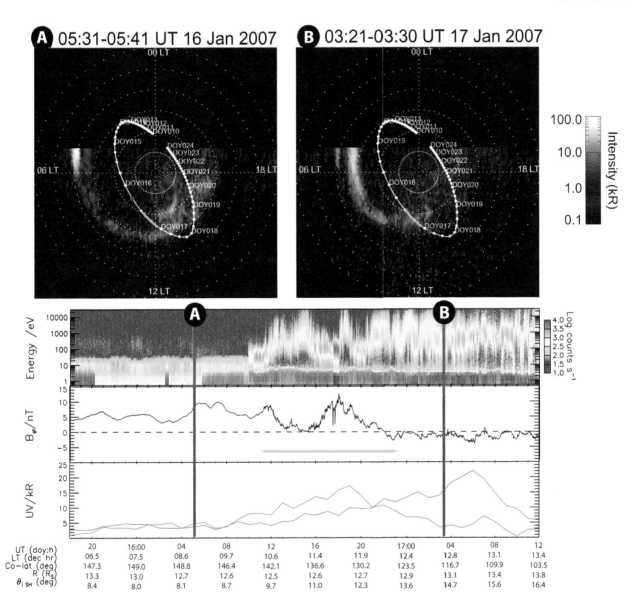

Figure 4. Joint observations from two consecutive HST (images A and B) taken in January 2007. The in situ data panels below indicate a (top) CAPS energy-time electron spectrogram, (middle) the azimuthal magnetic field (MAG) component in nT, and (bottom) the UV auroral intensity at the ionospheric footprint of the spacecraft in the Southern Hemisphere obtained from the two HST images shown, mapped magnetically as for the spacecraft footprint. Dashed circles show lines of constant colatitude at 5° intervals. Adapted from *Bunce et al.* [2008].

electrons present (shown as the mainly blue region >20 eV interpreted as the open field region up until ~10:00 UT on day 16) and a region dominated by electrons with a variety of energies, which are typical of the magnetosheath (10–100 eV) or the outer magnetosphere (100–1000 eV). We note that the electron population with energy <10 eV results from space-craft charging. These populations are then interpreted as the closed field region. This example reveals the presence of an upward directed field-aligned current system, at least on the

dayside main oval near noon, sitting at the open-closed field line boundary in good broad agreement with the theoretical considerations of *Cowley et al.* [2004a, 2004b]. The simple interpretation made by *Bunce et al.* [2008] that the lack of low-energy electrons >20 eV can be taken as the signature of the open-closed field line boundary at Saturn is, however, current-ly debated. This method of identification will be compared with other methods in the published literature, which will be discussed in more detail below.

If the peak value of the field-aligned current density in the ionosphere exceeds that provided by the total flux of electrons in magnetosphere (estimated from the ELS data), the electrons will be accelerated along field lines into the ionosphere by a field-aligned voltage [*Knight*, 1973]. This study shows that the potential required to produce the ionospheric current of ~275 nA m^{-2} is <1 kV for the magnetosheath electron population and ~10 kV for outer magnetosphere electrons. *Bunce et al.* [2008] have used these values to estimate the accelerated electron flux and, hence, the brightness of the resulting aurora at the feet of the field lines carrying the upward directed current (assuming that ~0.1 mW m^{-2} produces 1 kR of UV photons). The postaccelerated electron populations are found to produce ~1–5 kR for the magnetosheath population and ~10–50 kR for outer magnetosphere population. This estimate agrees well with the peak intensities of ~25 kR as measured by the HST in the two images A (blue) and B (red), which is estimated from the image as a function of spacecraft footprint in the bottom panel of Figure 3.

However, the data described above represents one example of the in situ field-aligned current measurements at high latitudes from Cassini with near-simultaneous HST imaging. In addition to this work, it has been the focus of a number of studies to characterize the high-latitude magnetosphere data on Saturn. These studies include investigations of the morphology, strength, and magnetically mapped location of the typical high-latitude field-aligned current patterns [*Talboys et al.*, 2009a, 2009b, 2011], the characteristics of the plasma populations and identification of electron beams and ion conics in regions of the auroral field-aligned current systems [*Mitchell et al.*, 2009a], and analyses of the auroral hiss radio emissions in relation to upgoing electron beams [see *Kopf et al.*, 2010; *Gurnett et al.*, 2010a, 2010b]. Here we focus on the advancement in our understanding of the field-aligned current systems determined from the high-latitude measurements of the magnetic field and their relationship with magnetospheric boundaries (such as the open-closed field line boundary) inferred from the low-energy electron populations. On the basis of these results, we discuss the limitations of the open-closed field line boundary identification with respect to other methods discussed by other authors [e.g., *Gurnett et al.*, 2010a, 2010b; *Masters et al.*, 2011].

There have been two phases of high-latitude orbits during which these field-aligned current systems have been extensively observed. The first encounters were during the high-latitude orbits in 2006/2007 [*Talboys et al.*, 2009a], where the magnetic field signatures associated with field-aligned currents were observed at local times predominantly on the dayside. The second set of high-latitude orbits occurred during 2008/2009 at local times on the nightside [*Talboys et al.*, 2011, 2009b].

The clearest magnetic field signatures are observed when the spacecraft traverses the high-latitude field lines at sufficiently low altitudes so that it crosses all of the field structures present from the region of open field lines at highest latitudes through to the inner part of the magnetosphere. Two general patterns of nightside azimuthal magnetic field, and hence field-aligned current signatures, were observed for the sequence of high-latitude orbits in 2008, as introduced in Figure 5 and as discussed by *Talboys et al.* [2009b].

The first pattern, referred to as type 1 (observed on ~60% of passes) exhibited generally "lagging" azimuthal field signatures where the field lines were swept back out of meridian planes. An example of type 1 field-aligned currents is shown in Figure 5a in the electron and magnetic field data (to the left) and schematically (and more generally) on the right for the Northern and Southern Hemispheres. The main current regions derived from the azimuthal magnetic field (see equation (3)) is variable but typically consists of (from high to low latitudes) downward directed currents located near or just equatorward of the open-closed field line boundary and upward directed currents that extend further into the hot plasma region on closed magnetic field lines. These "lagging" field signatures imply the transfer of angular momentum from the ionosphere to the magnetosphere, indicative of subcorotation of plasma. Statistically, throughout 2008, the type 1 upward and downward current carried ~1.5–2.5 MA rad^{-1}, respectively, along the field lines.

More complex field-aligned current signatures were observed on ~25% of the other passes, defined as type 2, and involve strong antisymmetric "leading" field signatures (swept forward out of meridian planes) in both hemispheres. Examples of the type 2 field-aligned currents are shown in Figure 5b in the electron and magnetic field measurements (to the left) and schematically to the right. The pattern of field-aligned current flow derived from the azimuthal magnetic field

Figure 5. (opposite) Examples of types 1 and 2 field-aligned currents discovered from the 2008 high-latitude orbits of Cassini. (left) The electron spectrograms (according to the same color bar as Figure 4) and magnetic field components for an example of each type of current system, followed by the (bottom) magnetically mapped ionospheric colatitude. The mapping is shifted in the northern compared to the southern ionosphere, due to the internal magnetic field of Saturn having a strong quadrupole term. The schematic picture on the right-hand side in each case shows the basic characteristics of the magnetic field perturbation and the associated field-aligned currents. Adapted from *Talboys et al.* [2009b, 2011]

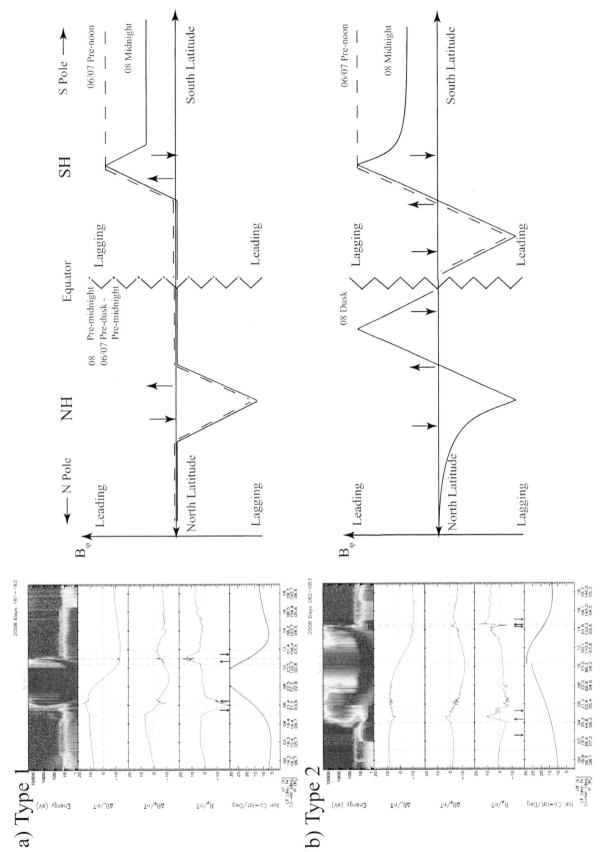

Figure 5

signatures is variable but typically consists of a distributed downward-directed field-aligned current flowing near to or just equatorward of the open field lines, followed by a major layer of upward directed field-aligned current located in the outer plasma sheet/ring current region on closed magnetic field lines, and then a further downward directed current mapping to the inner ring current. These "leading" field signatures imply the transfer of angular momentum from the magnetosphere to the ionosphere, indicative of supercorotation of plasma. Statistically, in the 2008 examples, the downward/upward/downward currents carried ~1.5/4/2 MA rad^{-1} along the field lines.

One possibility on how to interpret the existence of two distinct types of morphology in the field-aligned current pattern, suggested by *Talboys et al.* [2009b], is that the patterns may be associated with outer magnetosphere dynamics, for example, the Vasyliunas cycle (or similar). They argued that the type 2 currents were seen too frequently to be associated with the shock compression of the magnetosphere and were more likely to be associated with the (frequent) removal of plasmoids from the tail, which would lead to a region of fast, superrotating flow in the magnetosphere accounting for the "leading" magnetic field configuration. Although there are some discrepancies discussed above in the location of the upward directed field-aligned currents relative to the open-closed field line boundary compared to that predicted by the *Cowley et al.* [2004b] model, it certainly appears from the overall investigation of the 2008 data that the major field-aligned current system at high latitudes is produced in relation to outer magnetosphere dynamics, presumably driven by a combination of the solar wind interaction (the details of which are still yet to be fully understood) and the generation of flow shears in the outer magnetosphere near to the open-closed field line boundary. It is also highly probable that the high-latitude field-aligned currents are organized by the rotating magnetic field and current system (see discussion in section 5), which is not currently accounted for in the modeling work. The evidence suggests though that the main current system (and hence aurora) is certainly not associated with mass loading from inner magnetosphere as was found to be the case on Jupiter.

Overall then, *Talboys et al.* [2011] find that the upward directed field-aligned current in the observed high-latitude current system (presumably the same upward field-aligned current associated with the aurora) typically sit within the outer magnetosphere on closed field lines rather than at the open-closed field line boundary as suggested by the *Cowley et al.* [2004a, 2004b] model. The overall structure in the majority of cases (type 1), consisting of a downward field-aligned current at high latitudes followed by an upward field-aligned current at lower latitudes, is similar to that drawn in Figure 2,

although shifted with respect to the open-closed field line boundary. However, the modeling work estimated the field-aligned current profiles from variations in the angular velocity profile only and simply used a constant value of the ionospheric conductivity (1 mho). It is quite possible that there is an increased ionospheric conductivity on closed field lines resulting from particle precipitation dominating solar ionization, and hence, a constant value may not be ideal. The ionospheric Pedersen current (see equation (5)) whose variations give rise to the field-aligned current is the product of the conductivity and the angular velocity, and thus, in principle, we might expect significant field-aligned currents due to latitudinal variations of the conductivity. This means that the simple model of *Cowley et al.* [2004a, 2004b] is not necessarily incorrect in the broad sense that the interaction is likely to be driven by the solar wind interaction and relate to flow shears in the open and closed regions of the magnetosphere, but it is somewhat incomplete in details. Future development of this work might then involve the inclusion of a variable ionospheric conductivity (for example), as well as consideration of the effect of the "magnetosphere oscillations" (see next section). In addition, the "superrotation" effect discussed above that produces part of the upward field-aligned current in the outer magnetosphere at least ~25% of the time was not addressed or envisaged in the theoretical modeling. A more

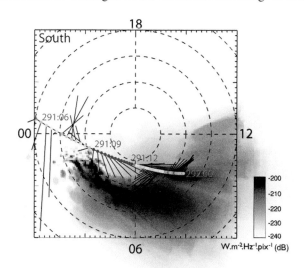

Figure 6. Magnetically mapped perturbation vectors, and the 5 keV electron flux (according to the same color bar as Figure 4), associated with the extraordinary field-aligned currents observed during Rev 89 are shown in the southern ionosphere. In the background, the radio map associated with the magnetic footprint of Saturn's kilometric radiation sources. Dashed circles indicate lines of constant colatitude at 5° intervals. Adapted from *Bunce et al.* [2010] and *Lamy et al.* [2010].

detailed understanding of the different types and temporal variations of the field-aligned current patterns is required before the theoretical model could be usefully modified.

During Revolution 89, an interesting 'extreme case' was observed in the data [*Bunce et al.*, 2010]. In Figure 6, we show a representation of the perturbation fields associated with the field-aligned currents at high latitudes observed during Rev 89, showing vectors plotted along the spacecraft track projected into the southern ionosphere (for a detailed description of how the vectors were obtained see the description by *Bunce et al.* [2010]. A vector corresponding to 10 nT in this projection has a length equivalent to the radial distance between the 5° circles. To give an indication of the plasma regime, we have also color coded the spacecraft track with the count rate of ~5 keV (CAPS) electrons using the same scale as in Figure 4. Dark blue colors therefore correspond to the open field region, while greens through reds correspond to higher-flux regions on closed field lines. The region of open field lines in this example is seen to be contracted close to the pole and surrounded by an unusual region containing hot electrons and "leading" field signatures indicative of supercorotating flow. The usual "lagging" fields indicative of subcorotation were also present at lower latitudes (seen at doy 291:06 on Figure 6), though of unusually high strength. The overall field-aligned current system thus consisted of a central region of downward current, flanked by two regions of upward current.

Interestingly, this unusual field-aligned current pattern was observed at the same time that Cassini entered into the SKR source region for the first, and only, time during the mission thus far [*Lamy et al.*, 2010; *Schippers et al.*, 2011]. In the background of Figure 6, we have placed the radio map [*Lamy et al.*, 2009] integrated along the extended time interval from 07:00 to 11:00 UT and in the frequency range 7–1000 kHz. The global distribution of the footprints of field lines supporting SKR sources follows an unusual spiral shape, starting from narrow emissions at very high latitude near 01:00 LT, then evolving to broad emission at lower latitudes toward noon. This unusual event in the field-aligned current pattern and measurement of the SKR source region is thought to have occurred following a solar wind compression of the magnetosphere. While there are, unfortunately, no auroral observations at this time, we note the similarity between the radio map "spiral pattern" and the HST image shown in Figure 1. For the event in 2004 (as reported above), we know that the dawn-filled auroral spiral followed a major compression of the magnetosphere, and it therefore seems likely that the Rev 89 event is the in situ counterpart of the auroral storm seen following a shock compression of the magnetosphere. *Mitchell et al.* [2009a, 2009b] have also shown a close relationship between solar wind compression of the magnetosphere with associated activity in the ring current and SKR brightening. This work shows an enhancement of the dawnside auroral oval suggested to be associated with a rotating partial ring current (see next section).

As discussed previously, much of the work cited above identifies the open-closed field line boundary from the low-energy ELS spectrogram data. Alternative methods for the identification of this boundary have been presented, for example, by *Gurnett et al.* [2010a, 2010b] who established the existence of a plasmapause-like density boundary at high latitudes. The evidence for this boundary originally comes from auroral hiss observations, which often show a clear upper cutoff frequency at the electron plasma frequency [e.g., *Kopf et al.*, 2010]. The electron plasma frequency is related to the electron density, and *Gurnett et al.* [2010a, 2010b] show that the upward step in the cutoff frequency corresponds to an upward step in the electron density as the spacecraft moves from high to low latitudes. Analysis of the electron anisotropy data indicates that the magnetic field lines are typically closed inside the boundary at lower latitudes and open outside the boundary at higher latitudes. Therefore, it seems that this plasma density boundary identified from the Langmuir probe (and analyzed with the Cassini LEMMS electrons for comparison) can be used to identify the open-closed field line boundary. The two examples shown in the *Gurnett et al.* [2010a, 2010b] paper identify the open-closed field line boundary at precisely the same times as *Talboys et al.* [2011] using the ELS low-energy electron data. However, more work is required to look at the identification of the boundary between open and closed field lines across various data sets, but from these few examples, it seems that the methods are in agreement. As *Gurnett et al.* [2010a, 2010b] discuss, the large-scale field-aligned currents flowing at high latitudes reside on the closed side of the boundary, again in agreement with the statistical analysis of the field-aligned currents presented by *Talboys et al.* [2011]. Importantly, *Gurnett et al.* [2010a, 2010b] have shown that the plasmapause boundary is modulated at the 10.6 and 10.8 h SKR period. The question on how the auroral current system is affected by planetary rotation will be briefly introduced in the following section.

5. ROTATIONAL MODULATION OF SATURN'S MAGNETOSPHERE

An important aspect of Saturn's magnetosphere that was not addressed in the large-scale flows and associated current systems of the *Cowley et al.* [2004a, 2004b] model, and for which Cassini has now provided clear evidence, is the system-wide oscillation that modulates the SKR emissions [*Kurth et al.*, 2007, 2008], the magnetospheric boundaries

[*Clarke et al.*, 2006, 2010], and imprints the entire magnetosphere as witnessed in the magnetic field and particle data [*Andrews et al.*, 2008; *Brandt et al.*, 2010; *Cowley et al.*, 2006; *Provan et al.*, 2009; *Southwood and Kivelson*, 2007, 2009]. *Nichols et al.* [2008] have also shown that the center of the auroral oval oscillates with a period of ~10.76 h, that is, close to the periods determined for oscillations in other magnetospheric phenomena. More recently, *Nichols et al.* [2010] have shown that both the northern and southern UV powers measured from the HST are dependent on the SKR phase, varying diurnally by factors of ~3. They also indicate that the UV variation originates from the morning half of the oval, consistent with previous observations of the SKR sources [*Galopeau et al.*, 1995; *Lamy et al.*, 2009]. *Mitchell et al.* [2009b] have demonstrated that under some magnetospheric conditions, protons and oxygen ions are accelerated once per Saturn magnetosphere rotation, at a preferred local time between midnight and dawn. They suggest that these events may result from reconnection and plasmoid formation in the magnetotail. Simultaneous auroral observations by the HST and the Cassini UVIS suggest that there is a close correlation between such magnetotail dynamics and transient auroral brightenings. The periodic ENA events also correlate strongly with SKR enhancements indicative of a connection to high-latitude auroral processes. Recent results show that there is a different SKR period at Saturn in the Northern and Southern Hemispheres [*Gurnett et al.*, 2009, 2010a, 2010b], and studies of the magnetospheric oscillations [*Andrews et al.*, 2010] show that the different rotation periods are echoed in the in situ magnetosphere data. The evidence in the magnetic field data and in other studies of ENA imaging (for example) require the existence of a rotating "partial ring current" system for the Northern and Southern Hemispheres [see *Southwood and Kivelson*, 2007; *Provan et al.*, 2009; *Mitchell et al.*, 2009b; *Andrews et al.*, 2010; *Brandt et al.*, 2010]. Given the existence of such rotating current systems with field-aligned currents thought to flow up and down the field lines in the outer magnetosphere near the boundary between the open and closed field regions, and given the fact that the auroral power correlates with the SKR phase (as discussed above), it seems highly probable that the auroral field-aligned currents will be modulated. In fact, the morphological difference between the type 1 and type 2 field-aligned current systems measured by *Talboys et al.* [2009b] may well be related to the phase of the magnetosphere oscillations (although detailed work on this topic is currently ongoing and results are not yet available). Clearly, we need to deduce the relationship between the "magnetosphere oscillations" and associated large-scale current system and the field-aligned currents measured by *Talboys et al.* [2009b] before we can fully comprehend the complexities of Saturn's auroral emissions.

6. SUMMARY

In this review, we have discussed the recent advances in our understanding of the physical drivers of Saturn's main auroral emissions, how they relate to the magnetosphere, and how they respond to large-scale solar wind and magnetospheric dynamics. We understand that the aurora is strongly driven by the solar wind interaction with the magnetosphere. The large-scale field-aligned current system, which is evident from in situ magnetic field measurements at high latitudes are located near the outer boundaries of the magnetosphere and may be associated with the plasma flow shear between open and closed magnetic field lines. However, the field-aligned current patterns are certainly more complex than initial modeling work has suggested, and details of their variations in space and time are still emerging. We do understand that the brightest auroral "storms" are driven by CIR modulation of the magnetosphere associated specifically with changes in the solar wind dynamic pressure. Open questions, which are yet to be fully addressed, include, for example, the following: (1) What is the relationship between the field-aligned currents and the magnetosphere oscillations/SKR phase? (2) What are the physical conditions under which the two main types of field-aligned current occur? (3) How does the ring current relate to the field-aligned currents at high latitudes, and how do they vary over time? (4) How does the substructure in the upward field-aligned current relate to the fine structures evident in the aurora?

Over the remaining years of the highly successful Cassini mission, we should gain significant insight into these outstanding issues, both through ongoing in situ field, plasma, and radio measurements and through the remote sensing observations of the HST, Cassini-UVIS, and Cassini-VIMS instruments.

REFERENCES

Andrews, D. J., E. J. Bunce, S. W. H. Cowley, M. K. Dougherty, G. Provan, and D. J. Southwood (2008), Planetary period oscillations in Saturn's magnetosphere: Phase relation of equatorial magnetic field oscillations and Saturn kilometric radiation modulation, *J. Geophys. Res.*, *113*, A09205, doi:10.1029/2007JA012937.

Andrews, D. J., A. J. Coates, S. W. H. Cowley, M. K. Dougherty, L. Lamy, G. Provan, and P. Zarka (2010), Magnetospheric period oscillations at Saturn: Comparison of equatorial and high-latitude magnetic field periods with north and south Saturn kilometric radiation periods, *J. Geophys. Res.*, *115*, A12252, doi:10.1029/2010JA015666.

Badman, S. V., E. J. Bunce, J. T. Clarke, S. W. H. Cowley, J.-C. Gérard, D. Grodent, and S. E. Milan (2005), Open flux estimates in Saturn's magnetosphere during the January 2004 Cassini-HST campaign, and implications for reconnection rates, *J. Geophys. Res.*, *110*, A11216, doi:10.1029/2005JA011240.

Badman, S. V., et al. (2012), Cassini observations of ion and electron beams at Saturn and their relationship to infrared auroral arcs, *J. Geophys. Res.*, *117*, A01211, doi:10.1029/2011JA017222.

Boudouridis, A., E. Zesta, R. Lyons, P. C. Anderson, and D. Lummerzheim (2003), Effect of solar wind pressure pulses on the size and strength of the auroral oval, *J. Geophys. Res.*, *108*(A4), 8012, doi:10.1029/2002JA009373.

Brandt, P. C., K. K. Khurana, D. G. Mitchell, N. Sergis, K. Dialynas, J. F. Carbary, E. C. Roelof, C. P. Paranicas, S. M. Krimigis, and B. H. Mauk (2010), Saturn's periodic magnetic field perturbations caused by a rotating partial ring current, *Geophys. Res. Lett.*, *37*, L22103, doi:10.1029/2010GL045285.

Bunce, E. J., S. W. H. Cowley, and S. E. Milan (2005), Interplanetary magnetic field control of Saturn's polar cusp aurora, *Ann. Geophys.*, *23*(4), 1405–1431.

Bunce, E. J., et al. (2008), Origin of Saturn's aurora: Simultaneous observations by Cassini and the Hubble Space Telescope, *J. Geophys. Res.*, *113*, A09209, doi:10.1029/2008JA013257.

Bunce, E. J., et al. (2010), Extraordinary field-aligned current signatures in Saturn's high-latitude magnetosphere: Analysis of Cassini data during Revolution 89, *J. Geophys. Res.*, *115*, A10238, doi:10.1029/2010JA015612.

Clarke, J. T., et al. (2005), Morphological differences between Saturn's ultraviolet aurorae and those of Earth and Jupiter, *Nature*, *433*(7027), 717–719.

Clarke, K. E., et al. (2006), Cassini observations of planetary-period oscillations of Saturn's magnetopause, *Geophys. Res. Lett.*, *33*, L23104, doi:10.1029/2006GL027821.

Clarke, K. E., D. J. Andrews, A. J. Coates, S. W. H. Cowley, and A. Masters (2010), Magnetospheric period oscillations of Saturn's bow shock, *J. Geophys. Res.*, *115*, A05202, doi:10.1029/2009JA015164.

Cowley, S. W. H., and E. J. Bunce (2001), Origin of the main auroral oval in Jupiter's coupled magnetosphere-ionosphere system, *Planet. Space Sci.*, *49*(10–11), 1067–1088.

Cowley, S. W. H., and E. J. Bunce (2003), Corotation-driven magnetosphere-ionosphere coupling currents in Saturn's magnetosphere and their relation to the auroras, *Ann. Geophys.*, *21*(8), 1691–1707.

Cowley, S. W. H., E. J. Bunce, and R. Prange (2004a), Saturn's polar ionospheric flows and their relation to the main auroral oval, *Ann. Geophys.*, *22*(4), 1379–1394.

Cowley, S. W. H., E. J. Bunce, and J. M. O'Rourke (2004b), A simple quantitative model of plasma flows and currents in Saturn's polar ionosphere, *J. Geophys. Res.*, *109*, A05212, doi:10.1029/2003JA010375.

Cowley, S. W. H., S. V. Badman, E. J. Bunce, J. T. Clarke, J.-C. Gérard, D. Grodent, C. M. Jackman, S. E. Milan, and T. K. Yeoman (2005), Reconnection in a rotation-dominated magnetosphere and its relation to Saturn's auroral dynamics, *J. Geophys. Res.*, *110*, A02201, doi:10.1029/2004JA010796.

Cowley, S. W. H., D. M. Wright, E. J. Bunce, A. C. Carter, M. K. Dougherty, G. Giampieri, J. D. Nichols, and T. R. Robinson (2006), Cassini observations of planetary-period magnetic field oscillations in Saturn's magnetosphere: Doppler shifts and phase motion, *Geophys. Res. Lett.*, *33*, L07104, doi:10.1029/2005GL025522.

Crary, F. J., et al. (2005), Solar wind dynamic pressure and electric field as the main factors controlling Saturn's aurorae, *Nature*, *433*(7027), 720–722.

Desch, M. D., and M. L. Kaiser (1981), Voyager measurement of the rotation period of Saturn's magnetic field, *Geophys. Res. Lett.*, *8*(3), 253–256.

Desch, M. D., and H. O. Rucker (1983), The relationship between Saturn kilometric radiation and the solar wind, *J. Geophys. Res.*, *88*(A11), 8999–9006.

Dungey, J. W. (1961), Interplanetary magnetic field and the auroral zones, *Phys. Rev. Lett.*, *6*, 47–48.

Galopeau, P. H. M., P. Zarka, and D. Le Quéau (1995), Source location of Saturn's kilometric radiation: The Kelvin-Helmholtz instability hypothesis, *J. Geophys. Res.*, *100*(E12), 26,397–26,410.

Gérard, J.-C., D. Grodent, J. Gustin, A. Saglam, J. T. Clarke, and J. T. Trauger (2004), Characteristics of Saturn's FUV aurora observed with the Space Telescope Imaging Spectrograph, *J. Geophys. Res.*, *109*, A09207, doi:10.1029/2004JA010513.

Gérard, J.-C., E. J. Bunce, D. Grodent, S. W. H. Cowley, J. T. Clarke, and S. V. Badman (2005), Signature of Saturn's auroral cusp: Simultaneous Hubble Space Telescope FUV observations and upstream solar wind monitoring, *J. Geophys. Res.*, *110*, A11201, doi:10.1029/2005JA011094.

Grodent, D., J.-C. Gérard, S. W. H. Cowley, E. J. Bunce, and J. T. Clarke (2005), Variable morphology of Saturn's southern ultraviolet aurora, *J. Geophys. Res.*, *110*, A07215, doi:10.1029/2004JA010983.

Grodent, D., A. Radioti, B. Bonfond, and J.-C. Gérard (2010), On the origin of Saturn's outer auroral emission, *J. Geophys. Res.*, *115*, A08219, doi:10.1029/2009JA014901.

Grodent, D., J. Gustin, J.-C. Gérard, A. Radioti, B. Bonfond, and W. R. Pryor (2011), Small-scale structures in Saturn's ultraviolet aurora, *J. Geophys. Res.*, *116*, A09225, doi:10.1029/2011JA016818.

Gurnett, D. A., A. Lecacheux, W. S. Kurth, A. M. Persoon, J. B. Groene, L. Lamy, P. Zarka, and J. F. Carbary (2009), Discovery of a north-south asymmetry in Saturn's radio rotation period, *Geophys. Res. Lett.*, *36*, L16102, doi:10.1029/2009GL039621.

Gurnett, D. A., J. B. Groene, A. M. Persoon, J. D. Menietti, S.-Y. Ye, W. S. Kurth, R. J. MacDowall, and A. Lecacheux (2010a), The reversal of the rotational modulation rates of the north and south components of Saturn kilometric radiation near equinox, *Geophys. Res. Lett.*, *37*, L24101, doi:10.1029/2010GL045796.

Gurnett, D. A., et al. (2010b), A plasmapause-like density boundary at high latitudes in Saturn's magnetosphere, *Geophys. Res. Lett.*, *37*, L16806, doi:10.1029/2010GL044466.

Hill, T. W. (1979), Inertial limit on coration, *J. Geophys. Res.*, *84*(A11), 6554–6558, doi:10.1029/JA084iA11p06554.

Hill, T. W. (2001), The Jovian auroral oval, *J. Geophys. Res.*, *106*(A5), 8101–8107, doi:10.1029/2000JA000302.

Jackman, C. M., N. Achilleos, E. J. Bunce, S. W. H. Cowley, M. K. Dougherty, G. H. Jones, S. E. Milan, and E. J. Smith (2004), Interplanetary magnetic field at ~9 AU during the declining phase of the solar cycle and its implications for Saturn's magnetospheric dynamics, *J. Geophys. Res.*, *109*, A11203, doi:10.1029/2004JA010614.

Knight, S. (1973), Parallel electric-fields, *Planet. Space Sci.*, *21*(5), 741–750.

Kopf, A. J., et al. (2010), Electron beams as the source of whistler-mode auroral hiss at Saturn, *Geophys. Res. Lett.*, *37*, L09102, doi:10.1029/2010GL042980.

Kurth, W. S., et al. (2005), An Earth-like correspondence between Saturn's auroral features and radio emission, *Nature*, *433*(7027), 722–725.

Kurth, W. S., A. Lecacheux, T. F. Averkamp, J. B. Groene, and D. A. Gurnett (2007), A Saturnian longitude system based on a variable kilometric radiation period, *Geophys. Res. Lett.*, *34*, L02201, doi:10.1029/2006GL028336.

Kurth, W. S., T. F. Averkamp, D. A. Gurnett, J. B. Groene, and A. Lecacheux (2008), An update to a Saturnian longitude system based on kilometric radio emissions, *J. Geophys. Res.*, *113*, A05222, doi:10.1029/2007JA012861.

Kurth, W. S., et al. (2009), Auroral processes, in *Saturn from Cassini-Huygens*, edited by M. K. Dougherty, L. W. Esposito, and S. M. Krimigis, pp. 333–374, Springer, Dordrecht, The Netherlands, doi:10.1007/978-1-4020-9217-6.

Lamy, L., B. Cecconi, R. Prangé, P. Zarka, J. D. Nichols, and J. T. Clarke (2009), An auroral oval at the footprint of Saturn's kilometric radio sources, colocated with the UV aurorae, *J. Geophys. Res.*, *114*, A10212, doi:10.1029/2009JA014401.

Lamy, L., et al. (2010), Properties of Saturn kilometric radiation measured within its source region, *Geophys. Res. Lett.*, *37*, L12104, doi:10.1029/2010GL043415.

Masters, A., D. G. Mitchell, A. J. Coates, and M. K. Dougherty (2011), Saturn's low-latitude boundary layer: 1. Properties and variability, *J. Geophys. Res.*, *116*, A06210, doi:10.1029/2010JA016421.

Mitchell, D. G., W. S. Kurth, G. B. Hospodarsky, N. Krupp, J. Saur, B. H. Mauk, J. F. Carbary, S. M. Krimigis, M. K. Dougherty, and D. C. Hamilton (2009a), Ion conics and electron beams associated with auroral processes on Saturn, *J. Geophys. Res.*, *114*, A02212, doi:10.1029/2008JA013621.

Mitchell, D. G., S. M. Krimigis, C. Paranicas, P. C. Brandt, J. F. Carbary, E. C. Roelof, W. S. Kurth, D. A. Gurnett, J. T. Clarke, and J. D. Nichols (2009b), Recurrent energization of plasma in the midnight-to-dawn quadrant of Saturn's magnetosphere, and its relationship to auroral UV and radio emissions, *Planet. Space Sci.*, *57*(14–15), 1732–1742.

Nichols, J. D., J. T. Clarke, S. W. H. Cowley, J. Duval, A. J. Farmer, J.-C. Gérard, D. Grodent, and S. Wannawichian (2008), Oscillation of Saturn's southern auroral oval, *J. Geophys. Res.*, *113*, A11205, doi:10.1029/2008JA013444.

Nichols, J. D., B. Cecconi, J. T. Clarke, S. W. H. Cowley, J.-C. Gérard, A. Grocott, D. Grodent, L. Lamy, and P. Zarka (2010), Variation of Saturn's UV aurora with SKR phase, *Geophys. Res. Lett.*, *37*, L15102, doi:10.1029/2010GL044057.

Prange, R., L. Pallier, K. C. Hansen, R. Howard, A. Vourlidas, G. Courtin, and C. Parkinson (2004), An interplanetary shock traced by planetary auroral storms from the Sun to Saturn, *Nature*, *432*(7013), 78–81.

Provan, G., D. J. Andrews, C. S. Arridge, A. J. Coates, S. W. H. Cowley, S. E. Milan, M. K. Dougherty, and D. M. Wright (2009), Polarization and phase of planetary-period magnetic field oscillations on high-latitude field lines in Saturn's magnetosphere, *J. Geophys. Res.*, *114*, A02225, doi:10.1029/2008JA013782.

Radioti, A., D. Grodent, J.-C. Gérard, E. Roussos, C. Paranicas, B. Bonfond, D. G. Mitchell, N. Krupp, S. Krimigis, and J. T. Clarke (2009), Transient auroral features at Saturn: Signatures of energetic particle injections in the magnetosphere, *J. Geophys. Res.*, *114*, A03210, doi:10.1029/2008JA013632.

Radioti, A., D. Grodent, J.-C. Gérard, S. E. Milan, B. Bonfond, J. Gustin, and W. Pryor (2011), Bifurcations of the main auroral ring at Saturn: Ionospheric signatures of consecutive reconnection events at the magnetopause, *J. Geophys. Res.*, *116*, A11209, doi:10.1029/2011JA016661.

Schippers, P., et al. (2011), Auroral electron distributions within and close to the Saturn kilometric radiation source region, *J. Geophys. Res.*, *116*, A05203, doi:10.1029/2011JA016461.

Southwood, D. J., and M. G. Kivelson (2001), A new perspective concerning the influence of the solar wind on the Jovian magnetosphere, *J. Geophys. Res.*, *106*(A4), 6123–6130, doi:10.1029/2000JA000236.

Southwood, D. J., and M. G. Kivelson (2007), Saturnian magnetospheric dynamics: Elucidation of a camshaft model, *J. Geophys. Res.*, *112*, A12222, doi:10.1029/2007JA012254.

Southwood, D. J., and M. G. Kivelson (2009), The source of Saturn's periodic radio emission, *J. Geophys. Res.*, *114*, A09201, doi:10.1029/2008JA013800.

Stallard, T. S., S. Miller, L. A. Trafton, T. R. Geballe, and R. D. Joseph (2004), Ion winds in Saturn's southern auroral/polar region, *Icarus*, *167*(1), 204–211.

Stallard, T., S. Miller, H. Melin, M. Lystrup, S. W. H. Cowley, E. J. Bunce, N. Achilleos, and M. Dougherty (2008), Jovian-like aurorae on Saturn, *Nature*, *453*(7198), 1083–1085.

Talboys, D. L., C. S. Arridge, E. J. Bunce, A. J. Coates, S. W. H. Cowley, and M. K. Dougherty (2009a), Characterization of auroral current systems in Saturn's magnetosphere: High-latitude Cassini observations, *J. Geophys. Res.*, *114*, A06220, doi:10.1029/2008JA013846.

Talboys, D. L., C. S. Arridge, E. J. Bunce, A. J. Coates, S. W. H. Cowley, M. K. Dougherty, and K. K. Khurana (2009b), Signatures of field-aligned currents in Saturn's nightside magnetosphere, *Geophys. Res. Lett.*, *36*, L19107, doi:10.1029/2009GL039867.

Talboys, D. L., E. J. Bunce, S. W. H. Cowley, C. S. Arridge, A. J. Coates, and M. K. Dougherty (2011), Statistical characteristics of field-aligned currents in Saturn's nightside magnetosphere, *J. Geophys. Res.*, *116*, A04213, doi:10.1029/2010JA016102.

Vasyliunas, V. M. (1983), Plasma distribution and flow, in *Physics of the Jovian Magnetosphere*, pp. 395–453, Cambridge Univ. Press, Cambridge, U. K.

E. J. Bunce, Department of Physics and Astronomy, University of Leicester, Leicester LE1 7RH, UK. (ejb10@ion.le.ac.uk)

Auroral Signatures of Solar Wind Interaction at Jupiter

P. A. Delamere

Laboratory for Atmospheric and Space Physics, University of Colorado, Boulder, Colorado, USA

Jupiter's dynamic and bright polar auroral emissions confound theoretical explanations of how the solar wind interacts with Jupiter's immense magnetosphere. Jupiter lacks a clear auroral signature of an open-closed magnetic field boundary in the polar region as one might expect for an open Earth-like magnetosphere subject to Dungey cycle convection. Instead, the polar region is filled with short-lived (~minutes) flares and swirling patches. Yet there are certain features that are persistent on time scales much longer (hours to days) than the typical variations in the orientation of the interplanetary magnetic field. This raises the question, how does the solar wind interact with Jupiter's magnetosphere? We review Jupiter's auroral observations and models for global magnetospheric dynamics and describe how each model can accommodate the various polar auroral forms.

1. INTRODUCTION

Jupiter exhibits three auroral emission regions (Figure 1) (as reviewed by *Clarke et al.* [2004] and *Clark* [this volume]). The satellite footprints manifest the complex plasma/neutral interactions of the Io plasma torus with moons and their atmospheres [*Saur et al.*, 2004; *Bonfond et al.*, 2008; *Hess et al.*, 2010]. The main auroral oval is a relatively steady auroral structure that maps to the middle magnetosphere and is related to the breakdown in corotation of the plasma torus [*Hill*, 1979; *Cowley and Bunce*, 2001; *Nichols and Cowley*, 2005; *Ray et al.*, 2010; *Ray and Ergun*, this volume]. Poleward of the main oval is an apparent battleground where the control of the polar region cannot be clearly delineated between the solar wind and the rapidly rotating magnetosphere. (We hesitate to use the terminology "polar cap" since this generally refers to a region of open flux on Earth. Due to the uncertainty in identifying open versus closed flux, we simply refer to the polar region.) Polar region emissions are bright (~30% of Jupiter's total FUV emission)

and dynamic with emission bursts on a time scale of ~100 s [*Grodent et al.*, 2003]. These emissions fill the polar region to at least some degree (see discussion below) and render the clear identification of an open/closed boundary, or polar cap, difficult. There has been considerable debate recently regarding the identification of open versus closed regions with *Vogt et al.* [2011] and *Vogt and Kivelson* [this volume] arguing for a largely open polar cap and *Delamere and Bagenal* [2010] arguing for a largely closed polar region with the solar wind interaction mediated through viscous processes at the magnetopause boundary. In this chapter, we discuss the polar auroral emissions and how different solar wind interaction models can produce the observed emissions. The paper is organized as follows: in section 2, we review auroral observations from the polar region, in section 3, we review different models for magnetospheric dynamics including both internal and external (solar wind) drivers to provide global context for the polar region, and in section 4, we provide synthesis of the observations with the competing models for magnetospheric dynamics.

2. AURORAL OBSERVATIONS

Observations of Jupiter's aurora were first made by the Voyager I UVS in 1979. Since then, many observations have been made from Earth, most notably in the UV from the IUE and from the Hubble Space Telescope (HST), and in the IR

Auroral Phenomenology and Magnetospheric Processes: Earth and Other Planets
Geophysical Monograph Series 197
10.1029/2011GM001180

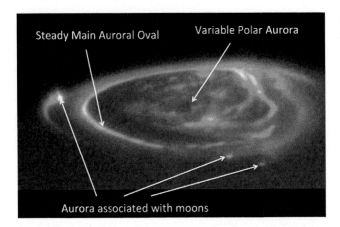

Steady Main Auroral Oval   Variable Polar Aurora

Aurora associated with moons

Figure 1. A Hubble Space Telescope UV image of Jupiter's three types of aurora: (1) The satellite-induced footprints, (2) the steady main oval, and (3) variable polar emissions.

using numerous ground-based instruments. While most observations have been made in the UV and IR, the Galileo spacecraft did provide visible images. Both UV and visible emissions are generated by direct collisional excitation from electronic transitions in H and H_2 and can be considered prompt emissions. The IR emissions, on the other hand, include collision excitation and thermal emissions from rovibrational transitions of H_3^+. As a result, the IR and UV emission features are generally similar, but significant differences have been noted that could be related to, for instance, ionospheric heating (see discussion by *Clarke et al.* [2004], *Clarke* [this volume], and *Stallard et al.* [this volume]). In the following discussion, we describe various aspects of Jupiter's polar aurora.

Initial attempts to describe the polar emissions were made by *Prangé et al.* [1998] and *Pallier and Prangé* [2001]. Using images from the HST faint object camera and reference ovals taken from the VIP4 magnetic field model [*Connerney et al.*, 1998], a preliminary identification of the northern polar cap boundary (i.e., open/closed boundary) was made. The polar cap was bounded on the duskside by "transpolar emissions" or "poleward dusk arcs" as identified by *Nichols et al.* [2009a]. In addition, bright spots observed just equatorward of the polar cap boundary and near magnetic noon led to a tentative identification of the northern polar cusp.

Following the 2000–2001 observing campaign of the HST Space Telescope Imaging Spectrograph (STIS) camera, *Grodent et al.* [2003] further refined definitions of the polar regions. In particular, they identified three distinct regions poleward of the main oval: (1) a dark region on the dawnside, (2) a swirl region at very high magnetic latitude that is characterized by short-lived (tens of seconds) patchy emission features, and (3) an active region in the noon to post-

noon sector that includes both polar flares (*Pallier and Prangé* [2001] on the cusp) and arc-like features (*Pallier and Prangé* [2001] on transpolar emissions).

A major contribution to understanding the polar region was provided by ground-based Doppler-shifted H_3^+ IR emissions measured by *Stallard et al.* [2001] and *Stallard et al.* [this volume]. The definitions of polar regions were further refined based on the observation of stagnated flow (i.e., noncorotational flow) in the dawn sector. *Stallard et al.* [2001], *Stallard et al.* [2003], and *Cowley et al.* [2003] noted that the stagnated ionospheric flows corresponded, loosely, to the swirl region defined by *Grodent et al.* [2003] and labeled this region as the "fixed Dark Polar Region," or f-DPR. The *Grodent et al.* [2003] dark region was labeled as the "rotating Dark Polar Region," or r-DPR. The dusk sector was observed to be corotating with the planet, and together with the plethora of bright auroral emission features, this region was labeled as the "Bright Polar Region." Figure 2 summarizes the UV and IR observations of *Grodent et al.* [2003] and *Stallard et al.* [2001].

Nichols et al. [2009b] noted a separate polar auroral feature based on a set of 2007 UV auroral images from HST. The feature was described as a thin, long-lived, quasi-sun-aligned polar auroral filament (PAF). The PAFs appear to bisect the polar region at very high magnetic latitude (and perhaps through the swirl region) with the section toward noon maintaining a fixed orientation and the antisunward section possibly rotating at 0%–45% of corotation. *Nichols et al.* [2009b] hesitated to associate the PAFs with anything resembling the terrestrial theta-aurora because of starkly different magnetotail dynamics and because of the relatively minor role that the interplanetary magnetic field (IMF) direction plays in Jupiter's polar ionosphere when compared with Earth. This feature will likely be critical for understanding the polar region.

While nightside observations are difficult from near-Earth platforms, *Grodent et al.* [2004] reported the observations of a bright and isolated spot appearing near the northern dusk-midnight limb and suggested that the spot may be related to tail reconnection events. Similar polar dawn spots were also suggested to be signatures of tail reconnection by *Radioti et al.* [2008]. The release of plasma down the magnetotail from such a reconnection event would be an important loss mechanism for iogenic plasma from the magnetosphere; however, *Bagenal* [2007] points out that the occurrence rate of these bright spots and estimated size of the plasmoid cannot account for the 500–1000 kg s^{-1} of plasma that must be lost down the tail.

Finally, X-ray emissions have been observed in Jupiter's polar region [*Metzger et al.*, 1983; *Waite et al.*, 1994; *Gladstone et al.*, 2002]. These X-ray emissions can be generated with

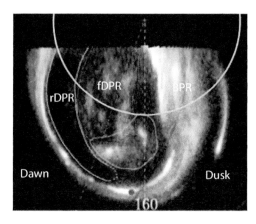

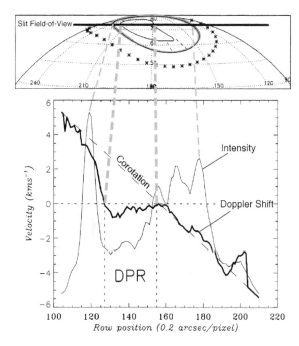

Figure 2. (left) Polar projection of the northern UV auroral region showing the shape and position of the dark region (yellow contour), the swirl region (red contour), and the active region (green contour) as they appear at central meridian longitude (CML) = 160° (marked with a vertical green dashed line). Longitude 180° is highlighted with a red dashed line. The red dot locates the magnetic footprint of Ganymede (VIP4 model) as the orbital longitude of the satellite matches the CML and therefore indicates the direction of magnetic noon at 15 R_J [*Grodent et al.*, 2003]. The purple circle is latitude 74°, the projected location of the slit field of view of the data on the right. (right) Viewing geometry (top) and Doppler-shifted H_3^+ IR emission profile (bottom) from *Stallard et al.* [2003], showing the stagnated flows in the Dark Polar Region (DPR). This DPR corresponds to the swirl region on the UV image on the left. Both images illustrate the dawn-dusk asymmetry of the polar auroral emission intensity.

a continuum source (i.e., bremmstrahlung) of precipitating electrons, or a line source expected from highly ionized ions (e.g., O^{6+} and O^{7+}). In the latter case, the high ionization state is achieved through collisional stripping and energetic ions precipitating into Jupiter's atmosphere. *Cravens et al.* [2003] and *Cravens and Ozak* [this volume] discuss both scenarios and argues in favor of line emissions from highly charged ions. The source of the precipitating ions could be (1) solar wind ions requiring a potential of ~200 keV or (2) precipitation of magnetospheric ions requiring a potential ~8 MV. In the case of magnetospheric ions, the X-ray emissions could be a signature of the return current (downward) in Jupiter's global current system and therefore could serve as an important diagnostic of magnetosphere-ionosphere coupling on closed magnetic field lines.

The interpretation of Jupiter's complex polar auroral emissions is very much model dependent. In the next section, we discuss various models for driving magnetospheric dynamics. Our goal is to provide a global context for the polar region while we attempt to relate specific auroral emissions to magnetospheric dynamics.

3. MODELS FOR DRIVING MAGNETOSPHERIC DYNAMICS

A convenient first step to quantify the solar wind interaction with Jupiter's magnetosphere is to consider the driving potentials of planetary rotation versus solar wind–driven convection. We are grateful to V. Vasyliunas (private communication, 2012) for providing an accurate history of this issue:

While the description of the corotation-convection model for Earth is often credited to *Brice and Ioannidis* [1970], the superposition of convection and corotation to obtain the complete flow pattern was discussed already in *Axford and Hines* [1961]. The use of the superposed flow to explain the formation and location of the plasmapause was proposed independently by *Nishida* [1966] and by *Brice* [1967]. In describing the combined flows *Brice* [1967] assumed the Dungey cycle flow, but *Nishida* [1966] assumed the tangential-drag ['viscous'] model of *Axford and Hines* [1961] (combined with a magnetotail

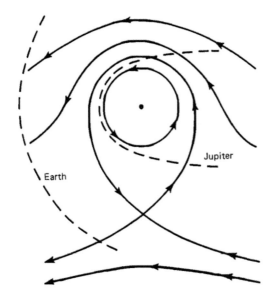

Figure 3. Corotation-convection models for Earth and Jupiter assuming sunward convection in the tail [*Cheng and Krimigis*, 1989]. The plasmapause boundary lies inside of the magnetopause boundary for Earth and outside of the magnetopause boundary for Jupiter.

opened by some unspecified process). It was the contribution of *Brice and Ioannidis* [1970] to propose that the plasmapause so determined extends at Jupiter essentially all the way to the magnetopause; *Vasyliunas* [1976] restated the argument of *Brice and Ioannidis* [1970] in a form independent of specific magnetic field and flow models.

The combined potential, in the rest frame of the planet (Jupiter), has the form

$$\phi = \frac{\Omega_J B_o R_J^2}{\rho} + \eta v_{sw} B_{sw} R_J y, \qquad (1)$$

where sw refers to solar wind, Ω_J is the planet's angular velocity, B_o is the planet's surface field strength, ρ and y are the respective radial and dawn-dusk distances in units of planetary radius, R_J, and η is an empirically determined factor (usually 0.1 to 0.2 for Earth, e.g., *Kennel and Coroniti* [1977]) that determines the reconnection potential associated with the Dungey cycle. *Cheng and Krimigis* [1989] showed that for $\eta = 0.1$, the separatrix between corotation-dominated and convection-dominated flow lies well outside of Jupiter's magnetopause (i.e., 350 R_J on the duskside and 150 R_J on the dawnside) (Figure 3). While this fact does not preclude a Dungey cycle, it does suggest that the power available from the solar wind should be dissipated in a very small polar cap [*Hill et al.*, 1983]. *Badman and Cowley* [2007] suggested that Dungey cycle voltages can be significant during solar wind compressions of the magnetosphere. Under compressed conditions, magnetic flux transport at the magnetopause boundaries could be sufficient to open the polar region and leave it open during expanded states when flux transport is minimized.

Given that the solar wind–driven flows determined via a terrestrial-based Dungey cycle are generally small compared to the planetary rotation input, the release of iogenic plasma down the tail must be dictated, in part, by the centrifugal stresses of the rapidly rotating plasma sheet. The seminal work of *Vasyliunas* [1983] sets out the basic physics of what is referred to as the "Vasyliunas cycle." The basic model (Figure 4) illustrates how plasma-laden tubes of magnetic flux are stretched out on the nightside to the point where they pinch down and reconnect, breaking off a plasmoid. This plasmoid is then ejected down the magnetotail, under the control of the solar wind. Observational evidence of plasmoid ejection down the tail has been shown in both magnetometer data [*Nishida*, 1983; *Russell et al.*, 1998, 1999, 2008; *Vogt et*

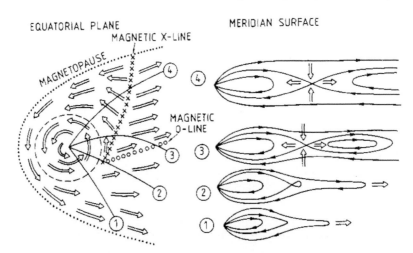

Figure 4. Schematic of the Vasyliunas cycle [*Vasyliunas*, 1983], showing the local time variation of the detachment of plasmoids.

al., 2010] and from bursts of energetic particles [*Woch et al.*, 1999, 2002; *Kronberg et al.*, 2005]. One of the great puzzles, however, is that the Vasyliunas cycle should occur independent of the solar wind input. The paucity of data at Jupiter requires a careful look at Jupiter's cousin, the rotation-driven magnetosphere of Saturn, for evidence of solar wind influences on the expected Vasyliunas cycle. Indeed, numerous studies suggest that solar wind variations can trigger tail reconnection events that should have occurred independent of external triggers (see discussions by *Mauk et al.* [2009] and *Bunce* [this volume]). The question then is, how does the solar wind interact with Jupiter's magnetosphere throughout the range of compressed and expanded states of the magnetosphere and does this trigger plasmoid ejection?

An additional consideration was presented by *Southwood and Kivelson* [2001], whereby the dynamic behavior of the magnetosphere should be considered, particularly the case of magnetospheric expansion that occurs following a solar wind compression event. Specifically, as the plasma sheet expands outward, the parallel pressure increases, causing the plasma sheet to become ballooning mode unstable. The implications are twofold for such an unstable plasma sheet: (1) the plasma sheet can pinch down, forming detached blobs and (2) the firehose condition means that Alfvén waves can no longer propagate, and communication to the ionosphere is truncated. This scenario presents a unique magnetospheric configuration where the magnetopause boundary can become amorphous and structured with regions of open flux penetrating deep into the magnetosphere. We then suggest that the solar wind interaction with the detached plasmoids would then resemble a mass pickup problem with the plasmoids being accelerated into the solar wind flow.

These considerations also led *Kivelson and Southwood* [2005] to develop a model for magnetotail flow that treats the duskside magnetosphere as ballooning unstable (due primarily to the rotational stresses on the plasma sheet generating a pressure anisotropy) with plasmoid ejection down the tail starting primarily on the dusk flank with eventual empty flux tube returning on the dawn flank (Figure 5).

We note that in addition, the stability of a magnetopause boundary that is subject to strong shear flows must be considered. *Delamere and Bagenal* [2010] discuss the implications for viscous interactions at the magnetopause boundary and argue that such viscous processes likely mediate the solar wind interaction versus large-scale Dungey-type reconnection (Figure 6). Jupiter's dawn flank is always unstable to the shear flow–driven Kelvin-Helmholtz instability (KHI) (M. Desroche, Jupiter's magnetopauase boundary, manuscript in preparation, 2012). KHI leads to driven reconnection by one of three distinct mechanisms. First, surface waves can compress a preexisting Chapman-Ferraro current

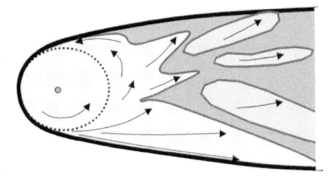

Figure 5. Magnetotail flow with a ballooning unstable plasma sheet [*Kivelson and Southwood*, 2005]. Plasmoids are released down the tail at nearly their release velocity, and empty flux tubes begin to flow sunward on the dawn flank.

layer and trigger reconnection. Second, the ensuing vortex rollup generates intense self-generated currents with substantial in-plane magnetic field perturbations that can reconnect. Finally, the self-generated currents propagating to high latitude can lead to parallel electric fields and component reconnection. All of these mechanisms can generate open flux and can also close on the magnetopause flanks leading to intermittent and small-scale reconnection (versus large-scale Dungey reconnection). The transport of plasma via KHI can also form a boundary layer akin to the low-latitude boundary layer on Earth.

The complicated entanglement of magnetic flux in the KH vortex motivated *Delamere and Bagenal* [2010] to consider global-scale KH modes on the dawn flank as a possible mechanism through which the dawn ionosphere is subject to solar wind control. The basic suggestion follows the original *Axford and Hines* [1961] viscous interaction for the terrestrial magnetosphere. Viscous processes can couple the solar wind to the magnetospheric plasma, transferring momentum from the solar wind to the magnetosphere via Maxwell stresses generated at the magnetopause boundary and via direct mass transport across the boundary. *Delamere and Bagenal* [2010] showed that in the limiting case where the magnetopause boundary is fully coupled to the solar wind (i.e., the interplanetary magnetic field cannot "slip" around the magnetosphere without generating Maxwell stresses), magnetospheric flows are governed by a net tailward force in addition to convection associated with the magnetic flux circulation governed by Dungey cycle, and/or Vasyliunas cycle, and even/or an Axford and Hines cycle. This suggestion is consistent with the perplexing tailward flow associated with the plasma torus, located deep in the magnetosphere, as well as the location of the magnetic X line in the tail. That is, the tailward Maxwell stresses work together with the

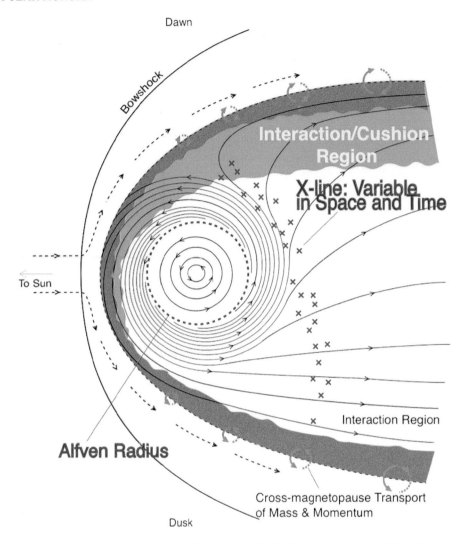

Figure 6. Composite sketch of the structure and dynamics of the Jovian magnetosphere (taken from *Delamere and Bagenal* [2010]). The spiral outflow is based on empirical models of the plasma sheet, mass conservation of the roughly 1000 kg s^{-1} of neutral gas produced by Io [*Bagenal and Delamere*, 2011; *Delamere et al.*, 2005], and by full solar wind control beyond the magnetic X line (roughly 80 R_J in the tail for the purpose of illustration). The variable X line location is based on the analysis of *Vogt et al.* [2010] and *Woch et al.* [2002]. Definitions of the "Interaction/Cushion Region" that are used loosely indicate the poorly understood nature of Jupiter's boundary layer region that is subject to the Kelvin-Helmholtz instability.

centrifugal stresses (Vasyliunas cycle) to determine the location of the X line and therefore provide an explanation for the perplexing correlation between auroral activity and solar wind conditions discussed by *Mauk et al.* [2009] in the context of Saturn.

4. SYNTHESIS OF OBSERVATIONS AND MODELS

Table 1 summarizes the polar auroral features and provides a brief identification of how these features *might* fit into models for driving magnetospheric dynamics. The solar wind (or external) drivers include "Dungey" [*Dungey*, 1961] and "Axford and Hines" [*Axford and Hines*, 1961], and the planetary (internal) driver is "Vasyliunas" [*Vasyliunas*, 1983]. These three models presumably work together in concert, and it likely requires all three models considered in unison to understand Jupiter's complex polar aurora. Nevertheless, for the sake of discussion, we keep the three models separate and provide an attempt to relate specific auroral emissions to signatures of each competing model. For example, the Dark Polar Regions on the dawnside indicate a possible single cell-convection pattern that can be related to

Table 1. Polar Auroral Features Related to Magnetospheric Drivers

Feature	Dungey (Solar Wind) [*Dungey*, 1961]	Axford and Hines (Solar Wind) [*Axford and Hines*, 1961]	Vasyliunas (Internal) [*Vasyliunas*, 1983]
Rotating-Dark Polar Region [*Stallard et al.*, 2001]	Return flow, closed flux [*Cowley et al.*, 2003]	Return flow, closed flux	Empty flux tube return flow [*Kivelson and Southwood*, 2005]
Fixed-Dark Polar Region/ swirl region [*Stallard et al.*, 2001]	Open flux [*Pallier and Prangé*, 2001; *Cowley et al.*, 2003]	Mixture of open and closed flux related to viscous interaction on dawn flank [*Delamere and Bagenal*, 2010]	
Bright Polar Region [*Stallard et al.*, 2001]	Closed flux, high-latitude edge = polar cap boundary [*Pallier and Prangé*, 2001]	Closed flux [*Pallier and Prangé*, 2001]	
Polar auroral filaments [*Nichols et al.*, 2009b]	Sliver of closed flux penetrating into another wise open polar cap?	Analogous to terrestrial transpolar arcs seen during northward interplanetary magnetic field (or closed magnetospheric configuration)?	Plasmoids drifting down the tail?
Active sector flares	Polar cusp [*Pallier and Prangé*, 2001], dayside magnetopause reconnection [*Grodent et al.*, 2003]	Return downward current [*Cravens et al.*, 2003]	
Midnight sector flares [*Grodent et al.*, 2003]	Tail reconnection		Plasmoid release

both a Dungey cycle cell and a viscous cell [*Cowley et al.*, 2003; *Delamere and Bagenal*, 2010]. In this case, these two models cannot be easily distinguished. The debate about whether the polar region is open or closed [e.g., *Vogt et al.*, 2011; *Vogt and Kivelson*, this volume] tends to separate the models more clearly where a Dungey interaction requires open field lines, and an Axford and Hines interaction can occur in a closed magnetospheric configuration. In this context, we assume a closed configuration for the viscous case and suggest, for example, that the active sector flares could be related to downward currents rather than dayside magnetopause reconnection [*Grodent et al.*, 2003; *Cravens et al.*, 2003]. The Vasyliunas cycle has been assumed to proceed independently from the solar wind interaction, but this cannot be the case if Maxwell stresses are generated on the magnetopause boundary via viscous processes, further blurring the distinction between these models. A major advance in our understanding of the nature of the solar wind interaction might lie in the PAFs. These persistent emissions are independent of the IMF orientation [*Nichols et al.*, 2009b] and likely do not have an Earth analog (e.g., transpolar theta aurora). From Table 1, it is clear that all three magnetospheric models have a stake in the PAFs.

5. SUMMARY

We summarize key topics of debate regarding Jupiter's interaction with the solar wind as follows:

1. Are the dynamic and bright polar emissions generated on open or closed field lines?

2. What is the significance of the boundary between the r-DPR and the f-DPR? For example, could this be the dawn flank magnetopause boundary (i.e., akin to the Earth's low-latitude boundary layer)?

3. Are the bright polar flares in the active sector related to magnetic reconnection at the magnetopause and/or cusp, or are these signatures of Jupiter's global return current?

4. How are PAFs formed? Is there a relation between PAFs and midnight sector flares? Are these signatures of plasmoid release in the Vasyliunas cycle?

5. Does the solar wind interaction affect the internally driven Vasyliunas cycle?

While we cannot provide answers to many of these questions at this time, it is our hope that the Juno spacecraft will provide a rich data set that will shed light on these issues.

REFERENCES

Axford, W. I., and C. O. Hines (1961), A unifying theory of high latitude geophysical phenomena and geomagnetic storms, *Can. J. Phys.*, *39*, 1433 1464.

Badman, S. V., and S. W. H. Cowley (2007), Significance of Dungey-cycle flows in Jupiter's and Saturn's magnetospheres, and their identification on closed equatorial field lines, *Ann. Geophys.*, *25*, 941–951.

Bagenal, F. (2007), The magnetosphere of Jupiter: Coupling the equator to the poles, *J. Atmos. Sol. Terr. Phys.*, *69*, 387–402, doi:10.1016/j.jastp.2006.08.012.

Bagenal, F., and P. A. Delamere (2011), Flow of mass and energy in the magnetospheres of Jupiter and Saturn, *J. Geophys. Res.*, *116*, A05209, doi:10.1029/2010JA016294.

Bonfond, B., D. Grodent, J.-C. Gérard, A. Radioti, J. Saur, and S. Jacobsen (2008), UV Io footprint leading spot: A key feature for understanding the UV Io footprint multiplicity?, *Geophys. Res. Lett.*, *35*, L05107, doi:10.1029/2007GL032418.

Brice, N. M. (1967), Bulk motion of the magnetosphere, *J. Geophys. Res.*, *72*(21), 5193–5211.

Brice, N. M., and G. A. Ioannidis (1970), The magnetospheres of Jupiter and Earth, *Icarus*, *13*, 173–183, doi:10.1016/0019-1035(70)90048-5.

Bunce, E. J. (2012), Origins of Saturn's auroral emissions and their relationship to large-scale magnetosphere dynamics, in *Auroral Phenomenology and Magnetospheric Processes: Earth and Other Planets, Geophys. Monogr. Ser.*, doi:10.1029/2011GM001191, this volume.

Cheng, A. F., and S. M. Krimigis (1989), A model of global convection in Jupiter's magnetosphere, *J. Geophys. Res.*, *94*(A9), 12,003–12,008.

Clarke, J. T. (2012), Auroral processes on Jupiter and Saturn, in *Auroral Phenomenology and Magnetospheric Processes: Earth and Other Planets, Geophys. Monogr. Ser.*, doi:10.1029/2011GM001199, this volume.

Clarke, J. T., D. Grodent, S. W. H. Cowley, E. J. Bunce, P. Zarka, J. E. P. Connerney, and T. Satoh (2004), Jupiter's aurora, in *Jupiter: The Planet, Satellites and Magnetosphere*, edited by F. Bagenal, T. E. Dowling, and W. B. McKinnon, pp. 639–670, Cambridge Univ. Press, Cambridge, U. K.

Connerney, J. E. P., M. H. Acuña, N. F. Ness, and T. Satoh (1998), New models of Jupiter's magnetic field constrained by the Io flux tube footprint, *J. Geophys. Res.*, *103*(A6), 11,929–11,939.

Cowley, S. W. H., and E. J. Bunce (2001), Origin of the main auroral oval in Jupiter's coupled magnetosphere-ionosphere system, *Planet. Space Sci.*, *49*, 1067–1088.

Cowley, S. W. H., E. J. Bunce, T. S. Stallard, and S. Miller (2003), Jupiter's polar ionospheric flows: Theoretical interpretation, *Geophys. Res. Lett.*, *30*(5), 1220, doi:10.1029/2002GL016030.

Cravens, T. E., and N. Ozak (2012), Auroral ion precipitation and acceleration at the outer planets, in *Auroral Phenomenology and Magnetospheric Processes: Earth and Other Planets, Geophys. Monogr. Ser.*, doi:10.1029/2011GM001159, this volume.

Cravens, T. E., J. H. Waite, T. I. Gombosi, N. Lugaz, G. R. Gladstone, B. H. Mauk, and R. J. MacDowall (2003), Implications of Jovian X-ray emission for magnetosphere-ionosphere coupling, *J. Geophys. Res.*, *108*(A12), 1465, doi:10.1029/2003JA010050.

Delamere, P. A., and F. Bagenal (2010), Solar wind interaction with Jupiter's magnetosphere, *J. Geophys. Res.*, *115*, A10201, doi:10.1029/2010JA015347.

Delamere, P. A., F. Bagenal, and A. Steffl (2005), Radial variations in the Io plasma torus during the Cassini era, *J. Geophys. Res.*, *110*, A12223, doi:10.1029/2005JA011251.

Dungey, J. W. (1961), Interplanetary magnetic field and the auroral zones, *Phys. Rev. Lett.*, *6*(2), 47–48, doi:10.1103/PhysRevLett.6.47.

Gladstone, G. R., et al. (2002), A pulsating auroral X-ray hot spot on Jupiter, *Nature*, *415*, 1000–1003.

Grodent, D., J. T. Clarke, J. H. Waite Jr., S. W. H. Cowley, J.-C. Gérard, and J. Kim (2003), Jupiter's polar auroral emissions, *J. Geophys. Res.*, *108*(A10), 1366, doi:10.1029/2003JA010017.

Grodent, D., J.-C. Gérard, J. T. Clarke, G. R. Gladstone, and J. H. Waite Jr. (2004), A possible auroral signature of a magnetotail reconnection process on Jupiter, *J. Geophys. Res.*, *109*, A05201, doi:10.1029/2003JA010341.

Hess, S. L. G., P. Delamere, V. Dols, B. Bonfond, and D. Swift (2010), Power transmission and particle acceleration along the Io flux tube, *J. Geophys. Res.*, *115*, A06205, doi:10.1029/2009JA014928.

Hill, T. W. (1979), Inertial limit on corotation, *J. Geophys. Res.*, *84*(A11), 6554–6558.

Hill, T. W., A. J. Dessler, and C. K. Goertz (1983), Magnetospheric models, in *Physics of the Jovian Magnetosphere*, edited by A. J. Dessler, pp. 353–394, Cambridge Univ. Press, Cambridge, U. K.

Kennel, C. F., and F. V. Coroniti (1977), Possible origins of time variability in Jupiter's outer magnetosphere, 2. Variations in solar wind magnetic field, *Geophys. Res. Lett.*, *4*(6), 215–218.

Kivelson, M. G., and D. J. Southwood (2005), Dynamical consequences of two modes of centrifugal instability in Jupiter's outer magnetosphere, *J. Geophys. Res.*, *110*, A12209, doi:10.1029/2005JA011176.

Kronberg, E. A., J. Woch, N. Krupp, A. Lagg, K. K. Khurana, and K.-H. Glassmeier (2005), Mass release at Jupiter: Substorm-like processes in the Jovian magnetotail, *J. Geophys. Res.*, *110*, A03211, doi:10.1029/2004JA010777.

Mauk, B. H., et al. (2009), Fundamental plasma processes in Saturn's magnetosphere, in *Saturn from Cassini-Huygens*, edited by M. K. Dougherty, L. W. Esposito, and S. M. Krimigis, pp. 281–331, Springer, Dordrecht, The Netherlands, doi:10.1007/978-1-4020-9217-6_11.

Metzger, A. E., D. A. Gilman, J. L. Luthey, K. C. Hurley, H. W. Schnopper, F. D. Seward, and J. D. Sullivan (1983), The detection of X rays from Jupiter, *J. Geophys. Res.*, *88*(A10), 7731–7741.

Nichols, J. D., and S. W. H. Cowley (2005), Magnetosphere-ionosphere coupling currents in Jupiter's middle magnetosphere: Effect of magnetosphere-ionosphere decoupling by field-aligned auroral voltages, *Ann. Geophys.*, *23*, 799–808, doi:10.5194/angeo-23-799-2005.

Nichols, J. D., J. T. Clarke, J. C. Gérard, D. Grodent, and K. C. Hansen (2009a), Variation of different components of Jupiter's auroral emission, *J. Geophys. Res.*, *114*, A06210, doi:10.1029/2009JA014051.

Nichols, J. D., J. T. Clarke, J. C. Gérard, and D. Grodent (2009b), Observations of Jovian polar auroral filaments, *Geophys. Res. Lett.*, *36*, L08101, doi:10.1029/2009GL037578.

Nishida, A. (1966), Formation of plasmapause, or magnetospheric plasma knee, by the combined action of magnetospheric convection and plasma escape from the tail, *J. Geophys. Res.*, *71*(23), 5669–5679.

Nishida, A. (1983), Reconnection in the Jovian magnetosphere, *Geophys. Res. Lett.*, *10*(6), 451–454.

Pallier, L., and R. Prangé (2001), More about the structure of the high latitude Jovian aurorae, *Planet. Space Sci.*, *49*, 1159–1173.

Prangé, R., D. Rego, L. Pallier, J. E. P. Connerney, P. Zarka, and J. Queinnec (1998), Detailed study of FUV Jovian auroral features with the post-COSTAR HST faint object camera, *J. Geophys. Res.*, *103*(E9), 20,195–20,215.

Radioti, A., D. Grodent, J.-C. Gérard, B. Bonfond, and J. T. Clarke (2008), Auroral polar dawn spots: Signatures of internally driven reconnection processes at Jupiter's magnetotail, *Geophys. Res. Lett.*, *35*, L03104, doi:10.1029/2007GL032460.

Ray, L. C., and R. E. Ergun (2012), Auroral signatures of ionosphere-magnetosphere coupling at Jupiter and Saturn, in *Auroral Phenomenology and Magnetospheric Processes: Earth and Other Planets*, *Geophys. Monogr. Ser.*, doi:10.1029/2011GM001172, this volume.

Ray, L. C., R. E. Ergun, P. A. Delamere, and F. Bagenal (2010), Magnetosphere-ionosphere coupling at Jupiter: Effect of field-aligned potentials on angular momentum transport, *J. Geophys. Res.*, *115*, A09211, doi:10.1029/2010JA015423.

Russell, C. T., K. K. Khurana, D. E. Huddleston, and M. G. Kivelson (1998), Localized reconnection in the near Jovian magnetotail, *Science*, *280*, 1061–1064, doi:10.1126/science.280.5366.1061.

Russell, C. T., D. E. Huddleston, K. K. Khurana, and M. G. Kivelson (1999), Structure of the Jovian magnetodisk current sheet: Initial Galileo observations, *Planet. Space Sci.*, *47*, 1101–1109.

Russell, C. T., K. K. Khurana, C. S. Arridge, and M. K. Dougherty (2008), The magnetospheres of Jupiter and Saturn and their lessons for the Earth, *Adv. Space Res.*, *41*, 1310–1318, doi:10.1016/j.asr.2007.07.037.

Saur, J., F. M. Neubauer, J. E. P. Connerney, P. Zarka, and M. G. Kivelson (2004), Plasma interactions of Io with its plasma torus, in *Jupiter: The Planet, Satellites and Magnetosphere*, edited by F. Bagenal, T. E. Dowling, and W. B. McKinnon, pp. 537–560, Cambridge Univ. Press, Cambridge, U. K.

Southwood, D. J., and M. G. Kivelson (2001), A new perspective concerning the influence of the solar wind on the Jovian magnetosphere, *J. Geophys. Res.*, *106*(A4), 6123–6130, doi:10.1029/2000JA000236.

Stallard, T., S. Miller, G. Millward, and R. D. Joseph (2001), On the dynamics of the Jovian ionosphere and thermosphere: I. The measurement of ion winds, *Icarus*, *154*, 475–491, doi:10.1006/icar.2001.6681.

Stallard, T. S., S. Miller, S. W. H. Cowley, and E. J. Bunce (2003), Jupiter's polar ionospheric flows: Measured intensity and velocity variations poleward of the main auroral oval, *Geophys. Res. Lett.*, *30*(5), 1221, doi:10.1029/2002GL016031.

Stallard, T., S. Miller, and H. Melin (2012), Clues on ionospheric electrodynamics from IR aurora at Jupiter and Saturn, in *Auroral Phenomenology and Magnetospheric Processes: Earth and Other Planets*, *Geophys. Monogr. Ser.*, doi:10.1029/2011GM001168, this volume.

Vasyliunas, V. M. (1976), Concepts of magnetospheric convection, in *The Magnetospheres of Earth and Jupiter*, edited by V. Formisano, pp. 179–188, D. Reidel, Dordrecht, The Netherlands.

Vasyliunas, V. M. (1983), Plasma distribution and flow, in *Physics of the Jovian Magnetosphere*, edited by A. J. Dessler, pp. 395–453, Cambridge Univ. Press, Cambridge, U. K.

Vogt, M. F., and M. G. Kivelson (2012), Relating Jupiter's auroral features to magnetospheric sources, in *Auroral Phenomenology and Magnetospheric Processes: Earth and Other Planets*, *Geophys. Monogr. Ser.*, doi:10.1029/2011GM001181, this volume.

Vogt, M. F., M. G. Kivelson, K. K. Khurana, S. P. Joy, and R. J. Walker (2010), Reconnection and flows in the Jovian magnetotail as inferred from magnetometer observations, *J. Geophys. Res.*, *115*, A06219, doi:10.1029/2009JA015098.

Vogt, M. F., M. G. Kivelson, K. K. Khurana, R. J. Walker, B. Bonfond, D. Grodent, and A. Radioti (2011), Improved mapping of Jupiter's auroral features to magnetospheric sources, *J. Geophys. Res.*, *116*, A03220, doi:10.1029/2010JA016148.

Waite, J. H., Jr., F. Bagenal, F. Seward, C. Na, G. R. Gladstone, T. E. Cravens, K. C. Hurley, J. T. Clarke, R. Elsner, and S. A. Stern (1994), ROSAT observations of the Jupiter aurora, *J. Geophys. Res.*, *99*(A8), 14,799–14,809.

Woch, J., et al. (1999), Plasma sheet dynamics in the Jovian magnetotail: Signatures for substorm-like processes?, *Geophys. Res. Lett.*, *26*(14), 2137–2140.

Woch, J., N. Krupp, and A. Lagg (2002), Particle bursts in the Jovian magnetosphere: Evidence for a near-Jupiter neutral line, *Geophys. Res. Lett.*, *29*(7), 1138, doi:10.1029/2001GL014080.

P. A. Delamere, Laboratory for Atmospheric and Space Physics, University of Colorado, Boulder, CO 80309, USA. (delamere@lasp.colorado.edu)

Relating Jupiter's Auroral Features to Magnetospheric Sources

Marissa F. Vogt[1]

Institute of Geophysics and Planetary Physics, University of California, Los Angeles, California, USA
Department of Earth and Space Sciences, University of California, Los Angeles, California, USA

Margaret G. Kivelson

Institute of Geophysics and Planetary Physics, University of California, Los Angeles, California, USA
Department of Earth and Space Sciences, University of California, Los Angeles, California, USA
Department of Atmospheric, Oceanic and Space Sciences, University of Michigan, Ann Arbor, Michigan, USA

The magnetospheric processes responsible for Jupiter's auroral emissions differ from those that drive the Earth's aurora. For example, Jupiter's main auroral emissions are associated with the breakdown of plasma corotation in the middle magnetosphere and not, as at the Earth, with the boundary between open and closed magnetic flux. In this chapter, we review some of the main features of Jupiter's auroral emissions and describe how they map to magnetospheric source regions or dynamic processes. Identifying the source of Jupiter's polar auroral features has been difficult because global field models are inaccurate beyond the inner magnetosphere (<30 R_J). However, a recently developed model that relates measured magnetic flux at different radial and local time locations in the magnetosphere to equivalent magnetic flux in the ionosphere (rather than tracing field lines from a global field model) provides a more accurate way to relate the polar auroral regions to magnetospheric sources and, particularly, to identify and constrain the size and location of Jupiter's polar cap. The results give insight into global dynamics and the open or closed nature of the Jovian magnetosphere.

1. INTRODUCTION

For more than a decade, ground and space-based telescope observations have produced dazzling images of Jupiter's auroral emissions in UV, IR, and visible wavelengths. These images have shown that Jupiter's aurora displays some unique features, such as satellite footprints, which are not present in the Earth's aurora. Additionally, there are some features, such as Jupiter's main oval, that may seem similar to their terrestrial counterparts, but are linked to different magnetospheric processes, than those that drive the Earth's aurora.

The purpose of this chapter is to discuss the magnetic links between Jupiter's auroral features and magnetospheric source regions. In particular, we focus on the available global field models and discuss why they are insufficient to reliably map between the ionosphere and the middle to outer magnetosphere, then we review recent work that uses flux equivalence to determine the mapping. This approach provides a more accurate way to relate the polar auroral regions to sources in the outer magnetosphere and provides insight into the sources of auroral activity.

[1]Now at Department of Physics and Astronomy, University of Leicester, Leicester, UK.

Auroral Phenomenology and Magnetospheric Processes: Earth and Other Planets
Geophysical Monograph Series 197
10.1029/2011GM001181

2. OVERVIEW OF AURORAL FEATURES

Jupiter's auroral emissions can be classified into three main types: the satellite footprints, the main oval (main emissions),

and the polar emissions [*Clarke et al.*, 1998]. All three components can be seen in Figure 1, which shows a polar projection of the UV auroral emissions in the northern hemisphere, and a comparison between the auroral observations and the

A. UV Auroral Observations

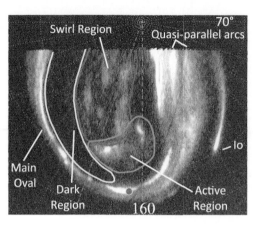

B. Flux Mapping Results, CML 160°

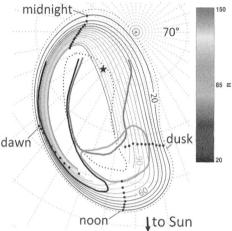

C. Composite of Mapping Results and UV Auroral Observations, CML 220°

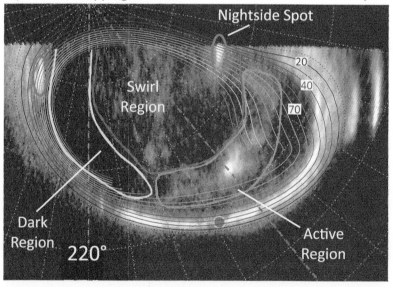

Figure 1. (a) UV observations of Jupiter's northern aurora as imaged by the Hubble Space Telescope (HST) on 16 December 2000. The three polar auroral regions, main oval, and Io footprint are labeled. From *Vogt et al.* [2011, Figure 1]. Modified from *Grodent et al.* [2003b, Figure 5]. (b) Mapping results for central meridian longitude (CML) 160° for the same viewing geometry as in Figure 1a. Contours are drawn every 10 R_J at the equator from 20 to 150 R_J and are not drawn for distances beyond the expanded magnetopause [*Joy et al.*, 2002]. The star symbol indicates the location of the magnetic pole at 9.6° colatitude and 212° SIII longitude, and black dots indicate the approximate mapping of four local times (midnight, dawn, noon, dusk). The outer black dashed line indicates the 15 R_J starting reference contour. The inner black dashed line shows the location of the fixed Dark Polar region [*Stallard et al.*, 2003]. From *Vogt et al.* [2011, Figure 14]. (c) Composite of auroral observations for CML 220° and mapping results, showing that the active region maps to just beyond the dayside magnetopause, the swirl region maps to open field lines on the nightside, and the dark region maps to open field lines near dawn. Also shown in the gray circles are a polar dawn spot and a nightside spot, which both map inside of the statistical X line, consistent with their association with inward flow from tail reconnection. From *Vogt et al.* [2011, Figure 15]. Modified from *Grodent et al.* [2003b, Figure 5].

mapping results of *Vogt et al.* [2011] (see section 5). Some of the auroral features, like the satellite footprints, have a clear and relatively well-understood link to magnetospheric source regions. Others, like the puzzling polar emissions, are more complicated and cannot be mapped reliably.

In this section, we review briefly the primary features of Jupiter's auroral emissions, focusing on the current understanding of how these features relate to magnetospheric processes. An extended discussion is given by *Clarke* [this volume].

2.1. Satellite Footprints

The most equatorward components of the Jovian auroral emissions are the satellite footprints, which, despite displacements arising from azimuthally extended interaction regions, are the features that can be most reliably linked to their magnetospheric source regions. Emissions have been observed at the ionospheric footprints of Io (5.9 R_J), Europa (9.4 R_J), and Ganymede (15 R_J) [*Connerney et al.*, 1993; *Clarke et al.*, 2002]. More details about the satellite footprints are given by *Bonfond* [this volume].

The footprints are key to our ability to map between the ionosphere and the magnetosphere because the satellites' orbital locations are known, and a footprint's ionospheric location can, therefore, be linked reliably to a source at a known radial distance in the equatorial magnetosphere. The longitudinal position can also be inferred with some small uncertainty (~1–2 h LT) as a consequence of the signal propagation time between the satellite and the Jupiter's ionosphere. Thus, satellite footprints substantially constrain global field models.

2.2. Main Oval Emissions

Poleward of the satellite footprints are the main auroral emissions. It has long been argued that the Jovian main auroral emissions do not map to the open/closed field line boundary as they do on the Earth. Instead, they are thought to be associated with the region where plasma corotation breaks down in the middle magnetosphere [e.g., *Kennel and Coroniti*, 1975; *Hill*, 1979, 2001; *Cowley and Bunce*, 2001]. In particular, the main emissions are thought to be produced in the region near and beyond ~20 R_J where corotation enforcement currents flow upward from the ionospheres, carried by downward moving accelerated electrons. Further details on the origin of the main emissions are given by *Clarke* [this volume].

2.3. Polar Aurora

At higher latitudes, the magnetospheric sources of emissions are less well established. *Grodent et al.* [2003b] have identified a number of systematic aspects of the high-latitude emissions, which they have categorized into three regions, the active, dark, and swirl regions, based on average brightness and dynamic behavior. The mapping of the polar emissions to the equator is exceptionally uncertain, partly because field models become increasingly undependable for high-latitude field lines and partly because the shapes and locations of the three regions vary with the rotation phase of Jupiter (see *Grodent et al.* [2003b], Figure 5).

The active region is very dynamic and is characterized by the presence of flares, bright spots, and arc-like features. It maps roughly to the noon LT sector and is located just poleward of the main oval. The active region flares have a brightness of a few hundred kR (by comparison, the main oval brightness is typically 50–500 kR) [*Grodent et al.*, 2003a] and a characteristic time scale of approximately minutes. Because of their location near noon LT, the spots, arcs, and flares in the active region are thought to be signatures of interaction with the solar wind or the polar cusp region [e.g., *Pallier and Prangé*, 2001; *Waite et al.*, 2001; *Grodent et al.*, 2003b].

As its name suggests, the dark region appears dim in UV images, emitting only weakly (0–10 kR) above the background level. The dark region occupies a crescent shape located just poleward of the main oval. It is fixed in LT in the dawn to prenoon local time sector, although its size contracts and expands as Jupiter rotates. This region roughly matches the area where *Pallier and Prangé* [2001] observed faint inner ovals, or arcs, which they suggested map to closed field lines in the outer magnetosphere (out to ~70 R_J).

At the center of the polar auroral emissions is the swirl region, an area of patchy, ephemeral emissions that exhibit disordered, swirling motions. Recent observations have shown the presence of "polar auroral filaments," (PAFs) thin (less than 1° width) structures, which extend in the antisunward direction from the active region through the swirl region [*Nichols et al.*, 2009]. The PAFs were observed in fewer than 10% of the observation intervals during 2007 and do not appear to occur preferentially for particular solar wind conditions, so their origin is quite puzzling, though *Nichols et al.* [2009] suggested that the PAFs could be the signature of plasmoids moving slowly down the magnetotail. *Grodent et al.* [2003b] found that the UV swirl region is nearly colocated with a feature seen in the IR emission called the fixed Dark Polar region (f-DPR), wherein ionospheric flows are nearly stagnant in a reference frame fixed to the magnetic poles as they rotate with the planet [*Stallard et al.*, 2003]. Near stagnation of flow would be expected in the region threaded by open flux tubes in an extremely extended magnetosphere, although it remains unclear what processes produce auroral emissions on open field lines.

3. JOVIAN MAGNETOSPHERIC DYNAMICS

Dynamics in the Jovian magnetosphere are predominantly rotationally driven rather than solar wind–driven as they are on the Earth. Jupiter's short rotation period (~10 h) and the vast size of the Jovian magnetosphere both contribute to the dynamical importance of rotational stresses. In this section, we briefly review the prevailing views of Jovian magnetospheric dynamics and then discuss what the aurora can teach us about global dynamics.

3.1. Dynamics in a Rotationally Driven Magnetosphere

The model of rotationally driven dynamics was first described by *Vasyliūnas* [1983]. In his model, mass-loaded flux tubes dragged from Jupiter's dayside are stretched due to centrifugal acceleration of rotating particles. The stretched flux tubes eventually pinch off, releasing mass and energy in the form of plasmoids that can escape down the tail. Observations from the energetic particle detector and the magnetometer on board Galileo have shown evidence for this internally driven mass loading and release process, referred to as the Vasyliūnas cycle [*Cowley et al.*, 2003], and have identified the location of a statistical X line separating inward and outward flow [*Woch et al.*, 2002; *Vogt et al.*, 2010]. The so-called particle "reconfiguration events" display a ~2–3 day periodicity thought to be related to the time scale of the internally driven mass loading and release process at Jupiter [*Kronberg et al.*, 2007]; the 2–3 day periodicity also appears intermittently in the magnetometer data but is not statistically significant [*Vogt et al.*, 2010]. The absence of a statistically significant internal driving period could mean that reconnection is partly influenced by external factors like the solar wind.

It is widely agreed that although the solar wind dynamic pressure controls the size and shape of the magnetosphere [*Joy et al.*, 2002] and may influence the system through viscous processes at the boundary [*Delamere and Bagenal*, 2010], the interplanetary magnetic field plays a minor role in driving dynamics at Jupiter. However, there is considerable disagreement [e.g., *Cowley et al.*, 2003; *McComas and Bagenal*, 2007; *Cowley et al.*, 2008] on just what that role may be and, in particular, the extent to which Jupiter's magnetosphere is open. However, if present, the Dungey cycle X line at Jupiter would likely be restricted to the dawnside because of the effects of corotational flow that opposes sunward flow in the dusk sectors [*Cowley et al.*, 2003; *Khurana et al.*, 2004]. However, the Vasyliūnas cycle is expected to form an X line across much of the nightside. Therefore, if signatures of in situ or auroral reconnection are observed premidnight, they are most probably internally driven. Ambiguity remains regarding reconnection signatures observed postmidnight.

3.2. How Big, and Where, Is Jupiter's Polar Cap?

A major hindrance to understanding the role of the solar wind in dynamics at Jupiter is the fact that the polar aurora does not display clear evidence of a persistent polar cap bounding the region of open flux tubes, although *Pallier and Prangé* [2001] have identified a possible auroral signature of the open-closed field line boundary. This ambiguity has led to considerable disagreement regarding the amount of open flux in Jupiter's magnetosphere.

The UV polar swirl region and the IR f-DPR are frequently interpreted as manifestations of open field lines that lie within the polar cap [*Grodent et al.*, 2003b; *Stallard et al.*, 2003]. It has been suggested that the effectively stagnant ionospheric flows of the f-DPR are associated with Dungey cycle return flows [*Cowley et al.*, 2003] whose flow speed across the ionosphere would be very slow (~km s^{-1}) because their equatorial elements in the Jovian magnetotail would require many hours or even days to convect antisunward over distances of approximately many hundreds or thousands of R_J from the nose of the magnetosphere to the distant X line.

A number of questions arise from this interpretation. For example, if the swirl region is indeed associated with open field lines, what is the mechanism that produces the UV emissions? Why is the feature identified as the boundary of the polar cap by *Pallier and Prangé* [2001] not more persistent? Are *Delamere and Bagenal* [2010] correct in their suggestion that the high variability in the size, shape, and brightness of the swirl region indicates that it is not on open field lines and that there is little steady open flux? What processes produce the *Nichols et al.* [2009] PAFs? Some of these questions become particularly significant when examined in relation to the boundaries identified by magnetic mapping and discussed in section 6.

3.3. Auroral Signatures of Jovian Tail Reconnection

Transient spots appear in auroral images in both the UV and the IR; they are thought to be associated with inward moving magnetospheric flow initiated during reconnection [*Grodent et al.*, 2004; *Radioti et al.*, 2008]. Most of these spots, referred to as polar dawn spots, are located at the equatorward edge of the dark region, near the dawn LT sector, but nightside spots, mapping roughly to the premidnight local time sector, have also been observed [*Grodent et al.*, 2004; *Radioti et al.*, 2011].

The polar dawn spots have been associated with the internally driven reconnection process because of their location, emitted power, and periodic recurrence [*Radioti et al.*, 2008, 2010]. The polar dawn spots sometimes occur with a 2–3 day periodicity similar to the reported recurrence interval of reconfiguration events seen in the particle data [*Kronberg et al.*, 2007] and persist on time scales similar to the duration of the reconnection events observed in the magnetometer data [*Vogt et al.*, 2010]. The apparent link between the magnetospheric events and the localized auroral emissions requires magnetic connectivity. Although magnetic mapping is uncertain, the magnetic model discussed in section 6 can be used to demonstrate that the proposed linkage is plausible.

4. CURRENT MAGNETIC FIELD MODELS AND THEIR LIMITATIONS

Three widely used Jovian internal magnetic field models have employed the auroral footprints of one or more Galilean moons as constraints on the properties of the field at relatively low latitudes. The VIP4 field model [*Connerney et al.*, 1998] made use of Voyager 1 and Pioneer 11 magnetic field observations and required that the model field lines traced from 5.9 R_J matched the Io footprint in the ionosphere. The VIP4 model generally does a good job of fitting the Io footprint, although in the northern hemisphere, the model deviates from the observations in the "kink" sector that gives the Io footprint its characteristic kidney bean shape.

The Grodent anomaly model (GAM) [*Grodent et al.*, 2008b] improved the agreement between the model and footprint observations in the northern hemisphere by adding a magnetic anomaly in the northern hemisphere. Inclusion of the magnetic anomaly improved the match to the Io footprint, especially in the "kink" sector, and the match to the Europa and Ganymede footprints, which shifted by approximately several degrees in the northern hemisphere, as seen in Figure 2. For all models, the inaccuracy is mostly in the direction along the curve rather than across the curve and is likely due to inaccuracies in the field models and not to the propagation time delay of the Alfvén waves [*Bonfond et al.*, 2009].

Most recently, the VIPAL internal field model [*Hess et al.*, 2011] updated the VIP4 model by placing longitudinal constraints on the mapping of the Io footprint. As a result, the VIPAL model fits the observed Io, Europa, and Ganymede footprints better than does VIP4. VIPAL does not match the satellite footprint locations as well as does the GAM in the northern hemisphere (see Figure 2), but it does predict a surface magnetic field strength that agrees better with the values deduced from observed radio emissions.

Even with recent improvements such as the inclusion of a magnetic anomaly, beyond 15 R_J where there are no satellite

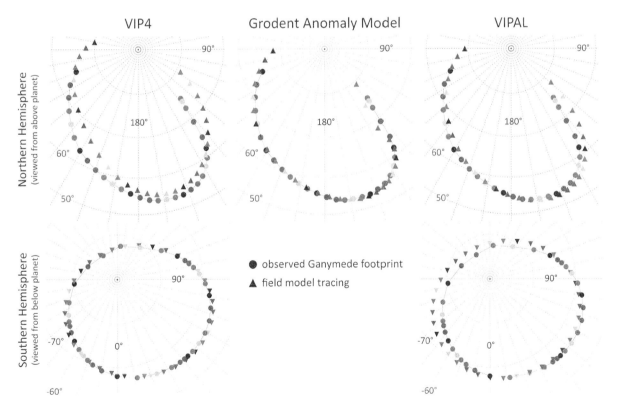

Figure 2. The locations of the observed Ganymede footprint (circles) and the ionospheric footprint of model field lines traced from 15 R_J at the equator (triangles) for three different models of Jupiter's magnetic field.

footprints to constrain the models, azimuthal currents flowing in the magnetosphere stretch field lines and compromise the mapping. Currents flowing near the magnetic equator have been incorporated into some global field models, although those models have limitations such as neglecting the effects of magnetopause currents [*Khurana*, 1997] or ignoring the 10° dipole tilt and dawn-dusk asymmetry of the equatorial magnetic field [*Alexeev and Belenkaya*, 2005]. Clearly, another approach is needed to obtain an improved field mapping in the middle and outer magnetosphere.

5. NEW MAPPING MODEL USING FLUX EQUIVALENCE

A new approach to the problem of linking polar auroral features to the equatorial magnetosphere was adopted by *Vogt et al.* [2011]. Rather than tracing equatorial magnetic field lines from the magnetosphere to the ionosphere, they quantified the distribution of magnetic flux across the nightside magnetosphere as a function of local time and radial distance in the equator from spacecraft observations and mapped the field from equator to ionosphere by means of a flux equivalence calculation. In this section, we review briefly the methods and results of the flux equivalence mapping (details are given by *Vogt et al.* [2011]).

5.1. Mapping by Flux Equivalence: Methods and Limitations

The flux equivalence calculation is based on the requirement that the magnetic flux threading a specified region at the equator must equal the magnetic flux in the area to which it maps in the ionosphere. The approach requires a reasonably accurate model for both the radial magnetic field strength in the ionosphere and the north-south component of the magnetic field at the equator.

A flux equivalence analysis had been used previously to estimate currents, flows, and magnetic mapping of Jupiter's ionosphere [*Cowley and Bunce*, 2001] making use of a simplified axisymmetric magnetic field model to estimate the magnetic flux in both the ionosphere and the magnetosphere. A key contribution of the *Vogt et al.* [2011] work is that, instead of using a model field, they fit the average observed field measurements in the equatorial plane as a function of radial distance and LT. They found that LT variations in $B_{N,equator}$, the normal component of the magnetic field across the magnetotail at the equator can be as large as a factor of ~5, so including the observed dawn-dusk asymmetries in the equatorial magnetic field is crucial for a meaningful mapping of dawn-dusk asymmetries in the auroral emissions.

The advantage of the flux equivalence calculation over using the available field models is that it allows for a more reliable mapping beyond 30 R_J into the middle and outer magnetosphere. However, this approach has limitations and is subject to inaccuracies. One source of potential inaccuracy is the internal field model used in both determining the 15 R_J reference contour at the footprint of Ganymede and in determining the magnetic field values used to calculate the flux at the ionosphere. For example, points obtained by tracing the GAM from 15 R_J fall close to the observed Ganymede footprint (see Figure 2) in latitude but have an estimated average local time inaccuracy of ~0.7 h. The GAM overestimates B_R compared to the VIPAL model, which predicts a surface magnetic field strength that better agrees with the observed radio emissions, though VIPAL does not match the Ganymede footprint as well, and such inaccuracies affect the flux mapping. Clearly, each internal field model has its relative strengths and weaknesses that affect the flux equivalence calculation, but the differences of flux mapping based on the three different field models (VIP4, GAM, VIPAL) has not yet been established.

Additional uncertainty of flux mapping arises from the data-based model B_N used to calculate the flux through the equatorial magnetosphere. That model is based on measurements acquired in differing magnetospheric states (compressed and expanded) and solar wind conditions. The assumption that B_N in the equatorial plane is unaffected, for example, by the changing size of the magnetosphere is undoubtedly an oversimplification. Consequently, one must not expect the model to apply precisely to the polar auroral observations taken at a specific moment in time, under a specific set of solar wind conditions.

5.2. Results and Comparison to Auroral Observations

The ionospheric maps obtained by the flux equivalence calculation can be compared to auroral observations [*Grodent et al.*, 2003b] to identify the magnetospheric sources of different auroral regions and features. Such a comparison is shown in Figure 1, which shows UV auroral observations for two central meridian longitudes (CMLs), i.e., the Jovian longitude facing the direction of the Earth). Figures 1b and 1c show contours corresponding to constant radial distances in the magnetosphere, every 10 R_J from 20 to 150 R_J. The white or empty area interior to the colored contours maps beyond 150 R_J on the nightside and beyond the expanded magnetopause on the dayside leading to the suggestion that it is threaded predominantly by open flux tubes.

In Figure 1, contours outlining the locations of the three polar auroral regions, following *Grodent et al.* [2003b], have been superimposed on the mapped magnetic contours. The shapes, sizes, and locations of the three polar regions change as the planet rotates or with varying solar wind conditions as is evident from a comparison of the regions of polar emissions in Figures 1a and 1c. However, although the jovigraphic location of these auroral features and their shapes change with CML, each of the three polar auroral regions maps predominantly to the same magnetospheric region(s) for the two cases.

A comparison of the mapping results with auroral observations shows that the main oval maps to different radial distances at different local times. Near dawn, the main oval maps relatively close to the planet, ~15–30 R_J; the mapped radial location then increases with increasing LT, from ~30–50 R_J at noon to ~50–60 R_J at ~15:00 LT. These results suggest that either the radial location of the corotation enforcement currents varies with local time or the outward plasma flux differs among local time sectors, or both. In studying 9 years of Hubble Space Telescope (HST) data, *Grodent et al.* [2008a] found that the main oval location observed at different times can shift by as much as 3° in latitude. They proposed that the latitudinal shift could be explained by variations in the current sheet density or thickness [e.g., *Kivelson and Southwood*, 2005] and not just by a response to changing solar wind conditions.

From the comparison between the flux equivalence mapping results and the auroral observations, *Vogt et al.* [2011] concluded that the polar auroral active region maps to Jupiter's polar cusp. Interpretations of the other mappings, still somewhat speculative, are presented in the next section.

6. APPLICATIONS OF THE FLUX EQUIVALENCE MAPPING MODEL

The flux equivalence model has many potential applications to understanding aspects of the ionospheric manifestations of the structure and dynamics of Jupiter's magnetosphere. In this section, we review briefly three of those applications.

6.1. Identifying the Size and Location of the Polar Cap

It is of interest to identify the area of the high-latitude auroral zone that maps to open field lines. Vogt et al. interpret the regions poleward of the mapped field lines from 150 R_J as only a bit larger than Jupiter's polar cap, whose locus has not yet been unambiguously identified. This leads them to propose that the swirl region links to the magnetotail, i.e., it is principally threaded with open field lines. The fixed f-DPR identified in IR measurements by *Stallard et al.* [2003] and shown as a black contour in Figure 1b also maps predominantly to the distant magnetotail, plausibly to open field lines, so it is possible that it, too, is part of the polar cap. As discussed in section 3.2, the flow is virtually stagnant in this region, a feature taken to support the interpretation. Thus, if the mag-

netic mapping is accepted, both the swirl region and the f-DPR (which largely overlaps the UV dark region) map predominantly to open tail field lines on the nightside. Although these conclusions agree with the interpretations made in several previous studies [e.g., *Cowley et al.*, 2003; *Stallard et al.*, 2003; *Grodent et al.*, 2003b], they raise an unanswered question regarding how emissions are generated on open field lines.

One way of estimating the size of the polar cap is to consider the total flux in the lobes of the magnetotail, based on the argument that their very low plasma density (<10^{-5} cm^{-3}) [*Gurnett et al.*, 1980] implies that they contain open field lines. Assuming that the polar areas mapping beyond the magnetopause or beyond 150 R_J on the nightside are predominantly on open field lines, *Vogt et al.* [2011] calculated the amount of open flux through that region. They found that the flux through the region identified as the northern hemisphere polar cap is less than ~760 GWb. If the polar cap has been accurately identified, the polar cap flux should be comparable to the tail lobe flux, which they estimate as ~740 GWb. However, accounting for the uncertainty of the flux mapping procedure, the variability associated with changing solar wind conditions, and the fact that some magnetotail flux closes beyond 150 R_J downtail, we expect that the amount of open flux is likely closer to ~600 GWb. These estimates of the open flux are somewhat larger than the upper limit values used by *Nichols et al.* [2006], who found an average 500 GWb of open flux in the tail lobes but also noted that the amount of open flux varied through the solar cycle with higher mean values near solar maximum.

Finally, Vogt et al. determined that the area of what they identified as the northern polar cap is equivalent to that of a circle around the pole with a half width of ~11°, only slightly smaller than the ~15° latitudinal half width of the polar cap on the Earth. If 20% of the flux beyond 150 R_J closes further downtail, the polar cap half width would decrease to 9.8°. By comparison, the polar cap identified by *Pallier and Prangé* [2001] in auroral images was only ~5° half width or about ~20% of the area of the polar cap identified by *Vogt et al.* [2011]. Similarly, *McComas and Bagenal* [2007], who argue that Jupiter's magnetosphere is nearly closed, proposed that the latitudinal half width of the polar cap is only ~5°. Such small areas are inconsistent with the flux mapping of *Vogt et al.* [2011]. It should be clear from the above discussion that there remains considerable uncertainty regarding the extent of the polar region that maps to open field lines.

6.2. Improved Understanding of Magnetotail Dynamics

Observations of the aurora provide an excellent tool for studying magnetospheric dynamics on a global scale because the aurora responds to activity occurring simultaneously across a large range of radial distances and local times. By comparison, spacecraft measurements provide information about only one location at an instance in time, so it is difficult to distinguish between temporally and spatially varying features. One potential use for the flux equivalence model presented here is to identify the magnetospheric activity that drives the observed polar dawn spots, introduced in section 3.3.

The auroral observations shown in Figure 1c include two spots, one in the dawn sector and one on the nightside. Both spots appear to map to equatorial regions well inside of the statistical X line [*Woch et al.*, 2002; *Vogt et al.*, 2010], consistent with their association with inward moving flow released during tail reconnection [*Radioti et al.*, 2010]. Similarly, in a recent study, *Radioti et al.* [2011] used the flux equivalence mapping model to map UV and IR auroral spots to distances inside of the statistical X line. The UV nightside spot mapped close to the position of the Galileo spacecraft, which recorded a reconnection signature in the magnetometer data at nearly the same time [*Vogt et al.*, 2010]. Additionally, the emitted power derived from the flow bubble in the

magnetic field measurements closely matches the emitted power of the nightside spot seen with HST.

It is interesting to note that in both Figure 1c and the Radioti et al. study, the nightside reconnection spots were accompanied by polar dawn spots. Could this mean that reconnection at Jupiter typically occurs simultaneously at multiple points across the tail or is the field bent back so strongly that the feet of nightside field lines are displaced in LT far more than proposed by field models? Without multipoint spacecraft measurements, and with only small statistics from which to draw inferences from the auroral data (spots at any local time appear infrequently in the auroral images), it is impossible to say. Further observations of these auroral features are needed to help answer this question and provide further insight into Jupiter's global magnetospheric dynamics.

6.3. Source Regions of Polar Flares

Quasiperiodic polar flares have recently been observed inside the main oval [*Bonfond et al.*, 2011] but have not yet

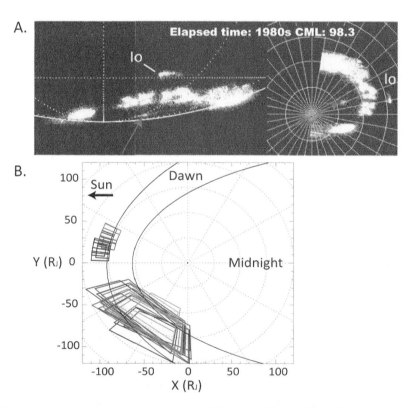

Figure 3. (a) HST observations of a flare in Jupiter's southern UV aurora. The left-hand side shows the aurora as observed from HST, with the flare identified by an arrow. The right-hand side of the image is a polar projection, and again, an arrow identifies the flare. (b) The mapped location of the flares in the equatorial plane. The thick black lines show the two preferred locations of Jupiter's magnetosphere [*Joy et al.*, 2002], and the two sets of boxes represent two different dates of observations. The size of the boxes reflects the position uncertainty of the flares because they are located close to the limb. From *Bonfond et al.* [2011].

been fully explained. These pulsating emissions were observed in the southern hemisphere during two HST observation orbits and display a typical 2–3 min recurrence period. *Bonfond et al.* [2011] used the flux equivalence mapping model to map the polar location of the flares to the dayside magnetopause, in one case near 11:00 LT and in the other case near ~16:00 LT. Figure 3 shows an example of the flares and their mapped location in the equatorial plane.

Based on the mapped location near the dayside magnetopause and the 2–3 min recurrence period, *Bonfond et al.* [2011] suggested that the periodic flares could be the signature of flux transfer events due to pulsed reconnection with the solar wind. Flux transfer events at Jupiter have been observed to have a ~1–4 min recurrence time [*Walker and Russell*, 1985], similar to the periodicity of the auroral flares. Further work on this interesting phenomenon is called for.

7. SUMMARY

This chapter has focused on the link between magnetospheric source regions and features of auroral emissions, seeking to understand the sources of the different types of emissions observed in Jupiter's aurora. We have described the links inferred from a flux equivalence model in which equatorial magnetic flux, evaluated from spacecraft measurements, is related to the equivalent magnetic flux in the ionosphere. This approach suggests, in particular, that the active region, poleward of the main oval and centered on the dayside, maps to regions that could plausibly be the dayside polar cusp and that the nearby swirl region is also on open field lines. The analysis associates the different auroral structures with sources in the middle and outer magnetospheres. Both dawn and nightside spots and quasiperiodic dayside flares have been shown to map to regions in which reconnection-associated phenomena provide plausible triggers for the emissions.

The models we have described are generic. Solar wind conditions are not known for most Jupiter observations, so the model fails to account for the roles of solar wind dynamic pressure and for possible dependence on the orientation of the interplanetary magnetic field. Despite these limitations, the methods described in this chapter have begun to elucidate the links between the magnetosphere and the ionosphere. Continued observations and new spacecraft data from JUNO are needed to provide further tests of the links proposed in these studies and to identify more unambiguously the drivers of auroral activity.

Acknowledgments. We gratefully acknowledge discussions with Krishan K. Khurana, Raymond J. Walker, Bertrand Bonfond, Denis Grodent, and Aikaterini Radioti. Each individual made several valuable contributions to the study on which this chapter is based. Additionally, much of the work in this paper was discussed during a meeting of the International Team on Planetary Magnetotails at the International Space Science Institute led by Caitriona Jackman. This work was supported in part by NASA grant NNX08AQ46G.

REFERENCES

Alexeev, I. I., and E. S. Belenkaya (2005), Modeling of the Jovian magnetosphere, *Ann. Geophys.*, *23*, 809–826.

Bonfond, B. (2012), When moons create aurora: The satellite footprints on giant planets, in *Auroral Phenomenology and Magnetospheric Processes: Earth and Other Planets, Geophys. Monogr. Ser.*, doi:10.1029/2011GM001169, this volume.

Bonfond, B., D. Grodent, J.-C. Gérard, A. Radioti, V. Dols, P. A. Delamere, and J. T. Clarke (2009), The Io UV footprint: Location, inter spot distances and tail vertical extent, *J. Geophys. Res.*, *114*, A07224, doi:10.1029/2009JA014312.

Bonfond, B., M. F. Vogt, J.-C. Gérard, D. Grodent, A. Radioti, and V. Coumans (2011), Quasi-periodic polar flares at Jupiter: A signature of pulsed dayside reconnections?, *Geophys. Res. Lett.*, *38*, L02104, doi:10.1029/2010GL045981.

Clarke, J. T. (2012), Auroral processes on Jupiter and Saturn, in *Auroral Phenomenology and Magnetospheric Processes: Earth and Other Planets, Geophys. Monogr. Ser.*, doi:10.1029/2011GM 001199, this volume.

Clarke, J. T., et al. (1998), Hubble Space Telescope imaging of Jupiter's UV aurora during the Galileo orbiter mission, *J. Geophys. Res.*, *103*(E9), 20,217–20,236.

Clarke, J. T., et al. (2002), Ultraviolet emissions from the magnetic footprints of Io, Ganymede and Europa on Jupiter, *Nature*, *415*, 997–1000.

Connerney, J. E. P., R. Baron, T. Satoh, and T. Owen (1993), Images of excited H_3^+ at the foot of the Io flux tube in Jupiter's atmosphere, *Science*, *262*, 1035–1038.

Connerney, J. E. P., M. H. Acuña, N. F. Ness, and T. Satoh (1998), New models of Jupiter's magnetic field constrained by the Io flux tube footprint, *J. Geophys. Res.*, *103*(A6), 11,929–11,939.

Cowley, S. W. H., and E. J. Bunce (2001), Origin of the main auroral oval in Jupiter's coupled magnetosphere-ionosphere system, *Planet. Space Sci.*, *49*, 1067–1088.

Cowley, S. W. H., E. J. Bunce, T. S. Stallard, and S. Miller (2003), Jupiter's polar ionospheric flows: Theoretical interpretation, *Geophys. Res. Lett.*, *30*(5), 1220, doi:10.1029/2002GL016030.

Cowley, S. W. H., S. V. Badman, S. M. Imber, and S. E. Milan (2008), Comment on "Jupiter: A fundamentally different magnetospheric interaction with the solar wind" by D. J. McComas and F. Bagenal, *Geophys. Res. Lett.*, *35*, L10101, doi:10.1029/2007GL032645.

Delamere, P. A., and F. Bagenal (2010), Solar wind interaction with Jupiter's magnetosphere, *J. Geophys. Res.*, *115*, A10201, doi:10.1029/2010JA015347.

Grodent, D., J. T. Clarke, J. Kim, J. H. Waite Jr., and S. W. H. Cowley (2003a), Jupiter's main auroral oval observed with HST-

STIS, *J. Geophys. Res.*, *108*(A11), 1389, doi:10.1029/2003JA 009921.

Grodent, D., J. T. Clarke, J. H. Waite Jr., S. W. H. Cowley, J.-C. Gérard, and J. Kim (2003b), Jupiter's polar auroral emissions, *J. Geophys. Res.*, *108*(A10), 1366, doi:10.1029/2003JA010017.

Grodent, D., J.-C. Gérard, J. T. Clarke, G. R. Gladstone, and J. H. Waite Jr. (2004), A possible auroral signature of a magnetotail reconnection process on Jupiter, *J. Geophys. Res.*, *109*, A05201, doi:10.1029/2003JA010341.

Grodent, D., J.-C. Gérard, A. Radioti, B. Bonfond, and A. Saglam (2008a), Jupiter's changing auroral location, *J. Geophys. Res.*, *113*, A01206, doi:10.1029/2007JA012601.

Grodent, D., B. Bonfond, J.-C. Gérard, A. Radioti, J. Gustin, J. T. Clarke, J. Nichols, and J. E. P. Connerney (2008b), Auroral evidence of a localized magnetic anomaly in Jupiter's northern hemisphere, *J. Geophys. Res.*, *113*, A09201, doi:10.1029/2008JA013185.

Gurnett, D. A., W. S. Kurth, and F. L. Scarf (1980), The structure of the Jovian magnetotail from plasma wave observations, *Geophys. Res. Lett.*, *7*(1), 53–56.

Hess, S. L. G., B. Bonfond, P. Zarka, and D. Grodent (2011), Model of the Jovian magnetic field topology constrained by the Io auroral emissions, *J. Geophys. Res.*, *116*, A05217, doi:10.1029/ 2010JA016262.

Hill, T. W. (1979), Inertial limit on corotation, *J. Geophys. Res.*, *84*(A11), 6554–6558.

Hill, T. W. (2001), The Jovian auroral oval, *J. Geophys. Res.*, *106*, 8101–8107, doi:10.1029/2000JA000302.

Joy, S. P., M. G. Kivelson, R. J. Walker, K. K. Khurana, C. T. Russell, and T. Ogino (2002), Probabilistic models of the Jovian magnetopause and bow shock locations, *J. Geophys. Res.*, *107*(A10), 1309, doi:10.1029/2001JA009146.

Kennel, C. F., and F. V. Coroniti (1975), Is Jupiter's magnetosphere like a pulsar's or Earth's?, *Space Sci. Rev.*, *17*, 857–883.

Khurana, K. K. (1997), Euler potential models of Jupiter's magnetospheric field, *J. Geophys. Res.*, *102*(A6), 11,295–11,306.

Khurana, K. K., M. G. Kivelson, V. M. Vasyliunas, N. Krupp, J. Woch, A. Lagg, B. H. Mauk, and W. S. Kurth (2004), The configuration of Jupiter's magnetosphere, in *Jupiter: The Planet, Satellites, and Magnetosphere*, edited by F. Bagenal, T. E. Dowling, and W. B. McKinnon, pp. 593–616, Cambridge Univ. Press, New York.

Kivelson, M. G., and D. J. Southwood (2005), Dynamical consequences of two modes of centrifugal instability in Jupiter's outer magnetosphere, *J. Geophys. Res.*, *110*, A12209, doi:10.1029/ 2005JA011176.

Kronberg, E. A., K.-H. Glassmeier, J. Woch, N. Krupp, A. Lagg, and M. K. Dougherty (2007), A possible intrinsic mechanism for the quasi-periodic dynamics of the Jovian magnetosphere, *J. Geophys. Res.*, *112*, A05203, doi:10.1029/2006JA011994.

McComas, D. J., and F. Bagenal (2007), Jupiter: A fundamentally different magnetospheric interaction with the solar wind, *Geophys. Res. Lett.*, *34*, L20106, doi:10.1029/2007GL031078.

Nichols, J. D., S. W. H. Cowley, and D. J. McComas (2006), Magnetopause reconnection rate estimates for Jupiter's magnetosphere based on interplanetary measurements at ~5AU, *Ann. Geophys.*, *24*(1), 393–406.

Nichols, J. D., J. T. Clarke, J. C. Gérard, and D. Grodent (2009), Observations of Jovian polar auroral filaments, *Geophys. Res. Lett.*, *36*, L08101, doi:10.1029/2009GL037578.

Pallier, L., and R. Prangé (2001), More about the structure of the high latitude Jovian aurorae, *Planet. Space Sci.*, *49*, 1159– 1173.

Radioti, A., D. Grodent, J.-C. Gérard, B. Bonfond, and J. T. Clarke (2008), Auroral polar dawn spots: Signatures of internally driven reconnection processes at Jupiter's magnetotail, *Geophys. Res. Lett.*, *35*, L03104, doi:10.1029/2007GL032460.

Radioti, A., D. Grodent, J.-C. Gérard, and B. Bonfond (2010), Auroral signatures of flow bursts released during magnetotail reconnection at Jupiter, *J. Geophys. Res.*, *115*, A07214, doi:10. 1029/2009JA014844.

Radioti, A., D. Grodent, J.-C. Gérard, M. F. Vogt, M. Lystrup, and B. Bonfond (2011), Nightside reconnection at Jupiter: Auroral and magnetic field observations from 26 July 1998, *J. Geophys. Res.*, *116*, A03221, doi:10.1029/2010JA016200.

Stallard, T. S., S. Miller, S. W. H. Cowley, and E. J. Bunce (2003), Jupiter's polar ionospheric flows: Measured intensity and velocity variations poleward of the main auroral oval, *Geophys. Res. Lett.*, *30*(5), 1221, doi:10.1029/2002GL016031.

Vasyliūnas, V. M. (1983), Plasma distribution and flow, in *Physics of the Jovian Magnetosphere*, edited by A. J. Dessler, p. 395, Cambridge Univ. Press, New York.

Vogt, M. F., M. G. Kivelson, K. K. Khurana, S. P. Joy, and R. J. Walker (2010), Reconnection and flows in the Jovian magnetotail as inferred from magnetometer observations, *J. Geophys. Res.*, *115*, A06219, doi:10.1029/2009JA015098.

Vogt, M. F., M. G. Kivelson, K. K. Khurana, R. J. Walker, B. Bonfond, D. Grodent, and A. Radioti (2011), Improved mapping of Jupiter's auroral features to magnetospheric sources, *J. Geophys. Res.*, *116*, A03220, doi:10.1029/2010JA016148.

Waite, J. H., Jr., et al. (2001), An auroral flare at Jupiter, *Nature*, *410*, 787–789.

Walker, R. J., and C. T. Russell (1985), Flux transfer events at the Jovian magnetopause, *J. Geophys. Res.*, *90*(A8), 7397– 7404.

Woch, J., N. Krupp, and A. Lagg (2002), Particle bursts in the Jovian magnetosphere: Evidence for a near-Jupiter neutral line, *Geophys. Res. Lett.*, *29*(7), 1138, doi:10.1029/2001GL014080.

M. G. Kivelson, Institute of Geophysics and Planetary Physics, University of California, Los Angeles, CA 90095, USA.

M. F. Vogt, Department of Physics and Astronomy, University of Leicester, Leicester LE1 7RH, UK. (mv106@le.ac.uk)

AGU Category Index

Index

Note: Page numbers with italicized *f* and *t* refer to figures and tables